Pressure Conversion Factors

	Pa	bar	atm	torr
1 Pa =	1	10^{-5}	$9.869\ 23 \times 10^{-6}$	$7.500\ 62 \times 10^{-3}$
1 bar =	10^{5}	1	0.986 923	750.062
1 atm =	$1.013\ 25 \times 10^{5}$	1.013 25	1	760
1 torr =	133.322	$1.333\ 22 \times 10^{-3}$	$1.315\ 79 \times 10^{-3}$	1

Some Commonly Used Non-SI Units

Unit	Quantity	Symbol	SI value
Angstrom	length	Å	10^{-10} m = 100 pm
Micron	length	μ	10^{-6} m
Calorie	energy	cal	4.184 J (defined)
Debye	dipole moment	D	3.3356×10^{-30} C · m
Gauss	magnetic field strength	G	10^{-4} T

Greek Alphabet

Alpha	A	α	Iota	I	ι	Rho	P	ρ
Beta	B	β	Kappa	K	κ	Sigma	Σ	σ
Gamma	Γ	γ	Lambda	Λ	λ	Tau	T	τ
Delta	Δ	δ	Mu	M	μ	Upsilon	Υ	υ
Epsilon	E	ϵ	Nu	N	ν	Phi	Φ	ϕ
Zeta	Z	ζ	Xi	Ξ	ξ	Chi	X	χ
Eta	H	η	Omicron	O	o	Psi	Ψ	ψ
Theta	Θ	θ	Pi	Π	π	Omega	Ω	ω

E_h	cm^{-1}	Hz
$2.293\ 710 \times 10^{17}$	$5.034\ 11 \times 10^{22}$	$1.509\ 189 \times 10^{33}$
$3.808\ 798 \times 10^{-4}$	83.5935	$2.506\ 069 \times 10^{12}$
$3.674\ 931 \times 10^{-2}$	8065.54	$2.417\ 988 \times 10^{14}$
1	$2.194\ 7463 \times 10^{5}$	$6.579\ 684 \times 10^{15}$
$4.556\ 335 \times 10^{-6}$	1	$2.997\ 925 \times 10^{10}$
$1.519\ 830 \times 10^{-16}$	$3.335\ 64 \times 10^{-11}$	1

Problems and Solutions

to accompany

McQuarrie and Simon

MOLECULAR THERMODYNAMICS

Problems and Solutions

to accompany

McQuarrie and Simon

MOLECULAR THERMODYNAMICS

Heather Cox
and
Carole McQuarrie

University Science Books
www.uscibooks.com

University Science Books
www.uscibooks.com

Production Manager: *Susanna Tadlock*
Designer: *Robert Ishi*
Illustrator: *John Choi*
Compositor: *Eigentype*
Printer & Binder: *Victor Graphics, Inc.*

This book is printed on acid-free paper.

ISBN 978-1-891389-07-8

Printed in the United States of America
10 9 8 7 6 5 4

Contents

Preface

This manual contains complete solutions to every one of the almost 1000 problems in *Molecular Thermodynamics* by Donald A. McQuarrie and John D. Simon.

Many of the problems in this text involve the manipulation of experimental data or empirically derived equations. In most cases, a graphing program (such as *Kaleidagraph*) is the most appropriate tool to use, and most students of physical chemistry already have experience with these programs.

For some problems, however, that are unnecessarily tedious or time-consuming using a simple graphing program, we recommend using a program like *Mathematica* or *MathCad* for related calculations and graphs. The plots of standard molar quantities against temperature using empirical equations[1], for which a function must be defined over varying temperature ranges, are simpler and pedagogically more useful with a program that allows such definitions. Also, because the program is saved, the work is easily checked and easily corrected, and the calculations of standard molar quantities using the same empirical equations[2] follow naturally. Although molecular statistical thermodynamic problems[3] can be done using paper and pencil, they are also more useful to the student when performed with a program such as *Mathematica*, thus allowing the student to spend more time working with the equations and less time writing out constants to be multiplied together.

Unfortunately, the learning curve on a program like *Mathematica* is rather steep, and instead of being a time-saving tool, it can easily become a source of frustration while one struggles to learn how to define functions or how to form a data set. We have posted several sample *Mathematica* files on the web site for this book at http://www.uscibooks.com in order to provide a practical and useful demonstration of its applications with regard to some problems in physical chemistry. This will provide the student or professor with some exposure to the available problem-solving methods, and enable those with only a basic knowledge of *Mathematica* to use it in ways that may have been difficult without a guide.

We have attempted to make this manual as accurate as possible, and would appreciate being informed of any errors that are present.

—Heather Cox
—Carole McQuarrie

---

[1] Problems 5-45, 7-15, 7-17, 7-19, 7-26, 8-25 and 8-26.

[2] Problems 7-14, 7-16, 7-18 and 7-20 through 7-25.

[3] Problems 7-30 through 7-39, 9-37 through 9-41, and 9-34 through 12-52, to name those that are the most time consuming.

Problems and Solutions

to accompany

McQuarrie and Simon

MOLECULAR THERMODYNAMICS

CHAPTER 1

The Energy Levels Of Atoms and Molecules

PROBLEMS AND SOLUTIONS

1–1. Radiation in the ultraviolet region of the electromagnetic spectrum is usually described in terms of wavelength, λ, and is given in nanometers (10^{-9} m). Calculate the values of ν, $\tilde{\nu}$, and ε for ultraviolet radiation with $\lambda = 200$ nm and compare your results with those in Figure 1.11.

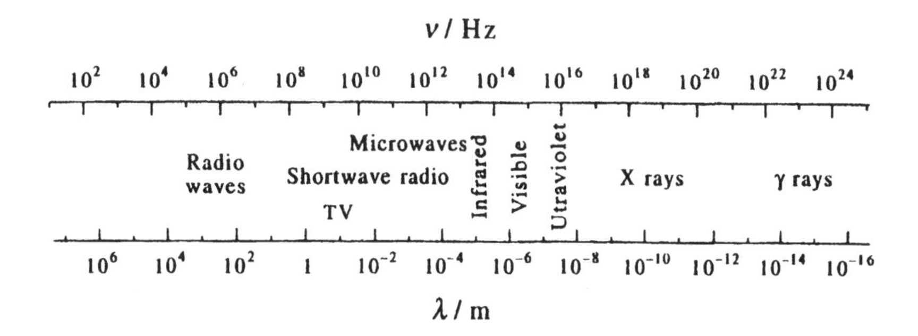

$$\nu = \frac{c}{\lambda} = \frac{2.998 \times 10^8 \text{ m}\cdot\text{s}^{-1}}{200 \times 10^{-9} \text{ m}} = 1.50 \times 10^{15} \text{ Hz}$$

$$\tilde{\nu} = \frac{1}{\lambda} = \frac{1}{200 \times 10^{-7} \text{ cm}} = 5.00 \times 10^4 \text{ cm}^{-1}$$

$$\varepsilon = \frac{hc}{\lambda} = \frac{(6.626 \times 10^{-34} \text{ J}\cdot\text{s})(2.998 \times 10^8 \text{ m}\cdot\text{s}^{-1})}{200 \times 10^{-9} \text{ m}} = 9.93 \times 10^{-19} \text{ J}$$

The frequency corresponds to the ultraviolet region in Figure 1.11.

1–2. Radiation in the infrared region is often expressed in terms of wave numbers, $\tilde{\nu} = 1/\lambda$. A typical value of $\tilde{\nu}$ in this region is 10^3 cm^{-1}. Calculate the values of ν, λ, and ε for radiation with $\tilde{\nu} = 10^3$ cm^{-1} and compare your results with those in Figure 1.11.

$$\nu = c\tilde{\nu} = (2.998 \times 10^8 \text{ m}\cdot\text{s}^{-1})(1 \times 10^5 \text{ m}^{-1}) = 3 \times 10^{13} \text{ Hz}$$

$$\lambda = \frac{1}{\tilde{\nu}} = \frac{1}{1 \times 10^5 \text{ m}^{-1}} = 1 \times 10^{-5} \text{ m}$$

$$\varepsilon = h\nu = (6.626 \times 10^{-34} \text{ J}\cdot\text{s})(3 \times 10^{13} \text{ Hz}) = 2 \times 10^{-20} \text{ J}$$

The wavelength corresponds to the infrared region in Figure 1.11.

1–3. Past the infrared region, in the direction of lower energies, is the microwave region. In this region, radiation is usually characterized by its frequency, ν, expressed in units of megahertz (MHz), where the unit, hertz (Hz), is a cycle per second. A typical microwave frequency is 2.0×10^4 MHz. Calculate the values of $\tilde{\nu}$, λ, and ε for this radiation and compare your results with those in Figure 1.11.

$$\tilde{\nu} = \frac{\nu}{c} = \frac{2.0 \times 10^{10}\ \text{s}^{-1}}{2.998 \times 10^{10}\ \text{cm}\cdot\text{s}^{-1}} = 0.67\ \text{cm}^{-1}$$

$$\lambda = \frac{c}{\nu} = \frac{2.998 \times 10^{8}\ \text{m}\cdot\text{s}^{-1}}{2.0 \times 10^{10}\ \text{Hz}} = 0.015\ \text{m}$$

$$\varepsilon = h\nu = (6.626 \times 10^{-34}\ \text{J}\cdot\text{s})(2.0 \times 10^{10}\ \text{Hz}) = 1.3 \times 10^{-23}\ \text{J}$$

The wavelength corresponds to the microwave region in Figure 1.11.

1–4. Calculate the energy of a photon for a wavelength of 100 pm (about one atomic diameter).

$$\varepsilon = \frac{hc}{\lambda} = \frac{(6.626 \times 10^{-34}\ \text{J}\cdot\text{s})(2.998 \times 10^{8}\ \text{m}\cdot\text{s}^{-1})}{1 \times 10^{-10}\ \text{m}} = 2 \times 10^{-15}\ \text{J}$$

1–5. Calculate the number of photons in a 2.00 mJ light pulse at (a) 1.06 μm, (b) 537 nm, and (c) 266 nm.

---

a.

$$\varepsilon_{\text{photon}} = h\nu = \frac{hc}{\lambda}$$

$$= \frac{(6.626 \times 10^{-34}\ \text{J}\cdot\text{s})(2.998 \times 10^{8}\ \text{m}\cdot\text{s}^{-1})}{1.06 \times 10^{-6}\ \text{m}}$$

$$= 1.87 \times 10^{-19}\ \text{J}\cdot\text{photon}^{-1}$$

Since 2.00 mJ of energy are contained in the light pulse,

$$\text{Number of photons} = \frac{2.00 \times 10^{-3}\ \text{J}}{1.87 \times 10^{-19}\ \text{J}\cdot\text{photon}^{-1}} = 1.07 \times 10^{16}\ \text{photons}$$

Parts b and c are done in the same manner to find

b. 5.41×10^{15} photons

c. 2.68×10^{15} photons

1–6. A helium-neon laser (used in supermarket scanners) emits light at 632.8 nm. Calculate the frequency of this light. What is the energy of a photon generated by this laser?

$$\nu = \frac{c}{\lambda} = \frac{2.998 \times 10^{8}\ \text{m}\cdot\text{s}^{-1}}{632.8 \times 10^{-9}\ \text{m}} = 4.738 \times 10^{14}\ \text{Hz}$$

$$\varepsilon = \frac{hc}{\lambda} = \frac{(6.626 \times 10^{-34}\ \text{J}\cdot\text{s})(2.998 \times 10^{8}\ \text{m}\cdot\text{s}^{-1})}{632.8 \times 10^{-9}\ \text{m}}$$

$$= 3.139 \times 10^{-19}\ \text{J}$$

1–7. The power output of a laser is measured in units of watts (W), where one watt is equal to one joule per second. ($1\ \text{W} = 1\ \text{J}\cdot\text{s}^{-1}$) What is the number of photons emitted per second by a 1.00 mW nitrogen laser? The wavelength emitted by a nitrogen laser is 337 nm.

$$\begin{aligned}\varepsilon_{\text{photon}} &= h\nu = \frac{hc}{\lambda} \\ &= \frac{(6.626\times10^{-34}\ \text{J}\cdot\text{s})(2.998\times10^{8}\ \text{m}\cdot\text{s}^{-1})}{337\times10^{-9}\ \text{m}} \\ &= 5.90\times10^{-19}\ \text{J}\cdot\text{photon}^{-1}\end{aligned}$$

$$\frac{1.00\times10^{-3}\ \text{J}\cdot\text{s}^{-1}}{5.90\times10^{-19}\ \text{J}\cdot\text{photon}^{-1}} = 1.70\times10^{15}\ \text{photon}\cdot\text{s}^{-1}$$

1–8. Use Equation 1.1 to calculate the value of the ionization energy of a hydrogen atom in its ground electronic state.

The ionization energy of a hydrogen atom in its ground electronic state is obtained from Equation 1.1 by letting $n = 1$, and so $\text{IE} = 2.178\ 69\times10^{-18}\ \text{J}$.

1–9. A line in the Lyman series of hydrogen has a wavelength of 1.026×10^{-7} m. Find the original energy level of the electron.

In the Lyman series, $n_1 = 1$ in Equation 1.10. Thus

$$\frac{1}{\lambda} = 109\ 680\left(1 - \frac{1}{n_2^2}\right)\ \text{cm}^{-1}$$

$$\frac{109\ 680\ \text{cm}^{-1}}{n_2^2} = -\frac{1}{1.026\times10^{-5}\ \text{cm}} + 109\ 680\ \text{cm}^{-1}$$

$$n_2 = 3$$

1–10. A ground-state hydrogen atom absorbs a photon of light that has a wavelength of 97.2 nm. It then gives off a photon that has a wavelength of 486 nm. What is the final state of the hydrogen atom?

First, we find the value of n_2, the state of the hydrogen atom that is obtained upon absorption, by using Equation 1.10 with $n_1 = 1$.

$$\frac{1}{97.2\times10^{-7}\ \text{cm}} = 109\ 680\left(1 - \frac{1}{n_2^2}\right)\ \text{cm}^{-1}$$

$$n_2 = 4$$

We can now use Equation 1.10 with $n_1 = 4$ to find the final state of the hydrogen atom:

$$\frac{1}{97.2\times10^{-7}\ \text{cm}} = 109\ 680\left(\frac{1}{4^2} - \frac{1}{n_2^2}\right)\ \text{cm}^{-1}$$

$$n_2 = 2$$

The final state of the hydrogen atom is $n = 2$.

1–11. A commonly used non-SI unit of energy is an electron volt (eV), which is the energy that an electron picks up when it passes through a potential difference of one volt. Given that a joule is a coulomb times a volt, show that $1\text{ eV} = 1.6022 \times 10^{-19}$ J.

One electron volt is equal to the charge on an electron (without the minus sign) times one volt, or

$$1\text{ eV} = (1.6022 \times 10^{-19}\text{ C})(1\text{ V}) = 1.6022 \times 10^{-19}\text{ C}\cdot\text{V} = 1.6022 \times 10^{-19}\text{ J}$$

1–12. Using the result of the previous problem, calculate the value of the ionization energy of a hydrogen atom in units of electron volts.

Using the result of Problems 1–8 and 1–11,

$$\text{IE} = (2.17869 \times 10^{-18}\text{ J})\frac{1\text{ eV}}{1.6022 \times 10^{-19}\text{ J}} = 13.598\text{ eV}$$

1–13. Using the data in Figure 1.2, show that a plot of frequency against $1/n^2$ is linear.

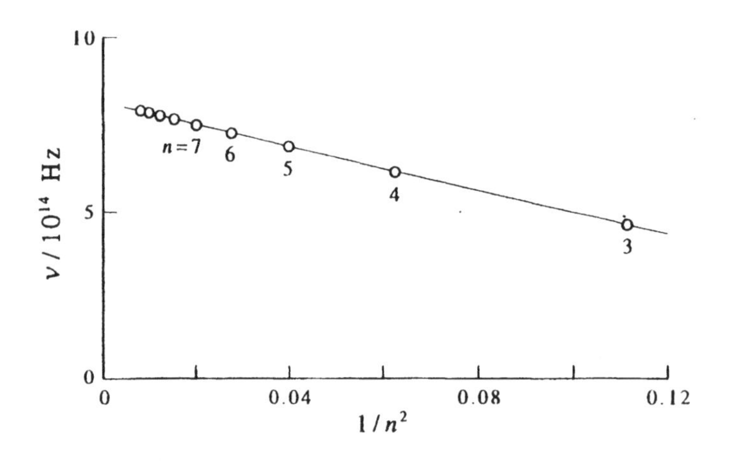

1–14. There is another series of lines in the emission spectrum of atomic hydrogen in the near infrared region called the Paschen series. This series results from $n \to 3$ transitions. Calculate the wavelength of the second line in the Paschen series, and show that this line lies in the near infrared region; that is, in the infrared region near the visible region.

The second line in the Paschen series involves a $5 \to 3$ transition. Therefore,

$$h\nu_{5\to 3} = 2.17869 \times 10^{-18}\text{ J}\left(\frac{1}{3^2} - \frac{1}{5^2}\right)$$

or

$$\nu_{5\to 3} = 2.338 \times 10^{14}\text{ s}^{-1}$$

or

$$\lambda_{5\to 3} = \frac{c}{\nu_{5\to 3}} = \frac{2.9979 \times 10^{8}\ \text{m}\cdot\text{s}^{-1}}{2.338 \times 10^{14}\ \text{s}^{-1}}$$
$$= 1.282 \times 10^{-6}\ \text{m}$$

which according to Figure 1.11 is in the infrared near the visible region.

1–15. Consider an electron in a 1*s* atomic orbital in a hydrogen atom. The average distance of this electron from the proton is given by $a_0 = 4\pi\varepsilon_0\hbar^2/m_e e^2$, where ε_0 is the permittivity of free space, $\hbar = h/2\pi$, where h is the Planck constant, m_e is the mass of an electron, and e is the protonic charge. Calculate the value of a_0, which is called the *Bohr radius*, in units of picometers.

$$a_0 = \frac{(1.112\,650 \times 10^{-10}\ \text{C}^2\cdot\text{J}^{-1}\cdot\text{m}^{-1})(1.054\,573 \times 10^{-34}\ \text{J}\cdot\text{s})^2}{(9.109\,389 \times 10^{-31}\ \text{kg})(1.602\,177 \times 10^{-19}\ \text{C})^2}$$
$$= 5.291\,772 \times 10^{-11}\ \text{m} = 52.917\,72\ \text{pm}$$

1–16. Equation 1.8 has a nice physical interpretation based upon the idea of a de Broglie wavelength. Recall that moving particles have an associated de Broglie wavelength given by $\lambda = h/mv$, where m is the mass of the particle and v is its speed. Show that if we assume that only standing de Broglie waves can fit in the interval 0 to a, then $\lambda = 2a/n$. Now use $\lambda = h/mv$ and $E = \frac{1}{2}mv^2$ to derive Equation 1.8.

The following figure shows the first three standing waves in the interval 0 to a.

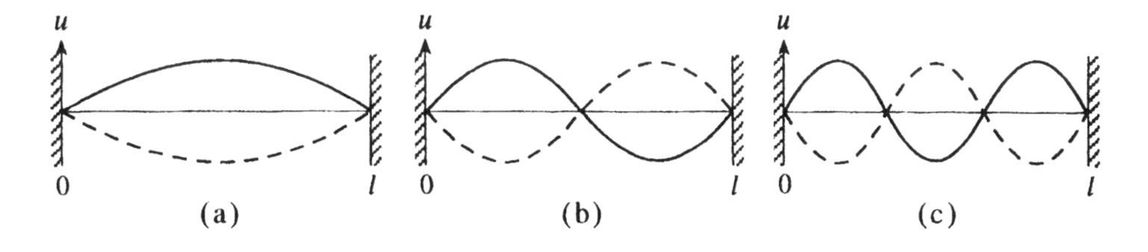

Note that the wavelengths of these standing deBroglie waves are $\lambda/2 = a$, $\lambda = a$, and $3\lambda/2 = a$. The general formula is $n\lambda/2 = a$. Substitute $\lambda = h/mv = h/m(2\varepsilon/m)^{1/2} = h/(2m\varepsilon)^{1/2}$ into $n^2\lambda^2/4 = a^2$ to get $a^2 = n^2h^2/8m\varepsilon$, or

$$\varepsilon = \frac{n^2h^2}{8ma^2}$$

1–17. The motion of macroscopic particles is governed by Newton's equation of motion, which can be written in the form

$$m\frac{d^2x}{dt^2} = f(x) \qquad (1)$$

where $x(t)$ is the position of the mass m and $f(x)$ is the force acting on the particle. Equation 1 is a differential equation whose solution gives $x(t)$, the *trajectory* of the mass. In the case of a harmonic oscillator, m is the reduced mass and $f(x) = -kx$ (Hooke's law), so that Newton's equation is

$$\mu \frac{d^2x}{dt^2} = -kx$$

Show that $x(t) = A\cos 2\pi\nu t$ satisfies this equation if $\nu = (1/2\pi)(k/\mu)^{1/2}$. This result is valid only for a macroscopic oscillator, called a *classical harmonic oscillator*.

If $x(t) = A\cos 2\pi\nu t$, then

$$\frac{dx}{dt} = -2\pi\nu A\sin 2\pi\nu t \quad \text{and} \quad \frac{d^2x}{dt^2} = -4\pi^2\nu^2 A\cos 2\pi\nu t$$

and so

$$\mu\frac{d^2x}{dt^2} = -4\pi^2\nu^2\mu A\cos 2\pi\nu t = -kA\cos 2\pi\nu t$$

Therefore, $k = 4\pi^2\nu^2\mu$, or $\nu = (1/2\pi)(k/\mu)^{1/2}$.

1–18. The kinetic energy of a classical harmonic oscillator is

$$\text{KE} = \frac{1}{2}\mu\left(\frac{dx}{dt}\right)^2$$

Using Equations 1.11 and 1.12, show that

$$\text{KE} + V(x) = \frac{kA^2}{2}$$

Interpret this result physically.

Using $x(t) = A\cos 2\pi\nu t$, we see that $dx/dt = -2\pi\nu A\sin 2\pi\nu t$, and so

$$\text{KE} = \frac{1}{2}\mu\left(\frac{dx}{dt}\right)^2 = 2\pi^2\nu^2\mu A^2\sin^2 2\pi\nu t$$

The potential energy, $V(x)$, is given by

$$V(x) = \frac{1}{2}kx^2 = \frac{kA^2}{2}\cos^2 2\pi\nu t$$

and so

$$\text{KE} + V(x) = 2\pi^2\nu^2\mu A^2\sin^2 2\pi\nu t + \frac{kA^2}{2}\cos^2 2\pi\nu t$$

But Problem 1–17 shows that $\nu^2 = k/4\pi^2\mu$, and so

$$\text{KE} + V(x) = \frac{kA^2}{2}\sin^2 2\pi\nu t + \frac{kA^2}{2}\cos^2 2\pi\nu t = \frac{kA^2}{2}$$

The total energy, which is a constant, is equal to the potential energy at its greatest value, when the kinetic energy is equal to zero.

1–19. The solution for the classical harmonic oscillator is $x(t) = A\cos 2\pi\nu t$ (Equation 1.11). Show that the displacement oscillates between $+A$ and $-A$ with a frequency ν cycle·s^{-1}. What is the period of the oscillations; that is, how long is one cycle?

Consider the solution $x(t) = A\cos 2\pi\nu t$. Because the cosine function ranges from +1 to -1, the value of x ranges from $+A$ to $-A$. To find the period of oscillation, we determine the smallest nonzero value of τ that satisfies the condition

$$\begin{aligned}\cos 2\pi\nu t &= \cos 2\pi\nu(t+\tau)\\ &= \cos 2\pi\nu t\cos 2\pi\nu\tau - \sin 2\pi\nu t\sin 2\pi\nu\tau\end{aligned}$$

This condition is met when τ simultaneously satisfies the two conditions

$$\cos 2\pi\nu\tau = 1 \qquad \sin 2\pi\nu\tau = 0$$

or, equivalently, when

$$2\pi\nu\tau = 2\pi n \qquad n = 1, 2, \ldots$$

The smallest value of τ is then $1/\nu$, which is the time it takes for the oscillator to undergo one cycle. The frequency of oscillation is

$$\nu = \frac{1}{\tau}$$

1–20. From Problem 1–19, we see that the period of a harmonic vibration is $\tau = 1/\nu$. The average of the kinetic energy over one cycle is given by

$$\langle K\rangle = \frac{1}{\tau}\int_0^\tau \frac{4\pi^2\nu^2 mA^2}{2}\sin^2 2\pi\nu t\, dt$$

Show that $\langle K\rangle = \varepsilon/2$ where ε is the total energy. Show also that $\langle V\rangle = \varepsilon/2$, where the instantaneous potential energy is given by

$$V(t) = \frac{k}{2}x^2(t) = \frac{kA^2}{2}\cos^2 2\pi\nu t$$

Interpret the result $\langle K\rangle = \langle V\rangle$.

The total energy of a harmonic oscillator is

$$\varepsilon = K(t) + V(t) = \frac{\mu}{2}\left(\frac{dx}{dt}\right)^2 + \frac{kx^2}{2}$$

Substituting $x(t) = A\cos 2\pi\nu t$ into this expression gives

$$\varepsilon = 2\mu\pi^2\nu^2A^2\sin^2 2\pi\nu t + \frac{k}{2}A^2\cos^2 2\pi\nu t$$

But Equation 1.12 says that $k = 4\pi^2\nu^2\mu$, so

$$\varepsilon = \frac{k}{2}A^2(\sin^2 2\pi\nu t + \cos^2 2\pi\nu t) = \frac{k}{2}A^2$$

Now

$$\begin{aligned}\langle K\rangle &= \frac{\mu}{2\tau}\int_0^{\tau} 4\pi^2\nu^2A^2\sin^2 2\pi\nu t\,dt \\ &= \frac{\mu\pi\nu A^2}{\tau}\int_0^{2\pi\nu\tau}\sin^2 x dx = \frac{\mu\pi\nu A^2}{\tau}\left|\frac{x}{2}+\sin 2x\right|_0^{2\pi\nu\tau} \\ &= \frac{\mu\pi\nu A^2}{\tau}(\pi\nu\tau + \sin 4\pi\nu\tau)\end{aligned}$$

But Problem 1–19 shows that $\nu\tau = 1$, so $\sin 4\pi\nu\tau = \sin 4\pi = 0$. Therefore, we have

$$\langle K\rangle = \mu\pi^2\nu^2A^2 = \frac{kA^2}{4} = \frac{\varepsilon}{2}$$

Similarly,

$$\begin{aligned}\langle V\rangle &= \frac{k}{2\tau}\int_0^{\tau} A^2\cos^2 2\pi\nu t\,dt \\ &= \frac{kA^2}{4\pi\nu\tau}\int_0^{2\pi\nu\tau}\cos^2 x dx \\ &= \frac{kA^2}{4} = \frac{\varepsilon}{2}\end{aligned}$$

The motion of a harmonic oscillator is such that $\langle K\rangle = \langle V\rangle = \varepsilon/2$ over a cycle of motion. On average, the energy is distributed evenly between the kinetic energy and the potential energy.

1–21. Calculate the value of the reduced mass of an electron in a hydrogen atom. Take the masses of the electron and proton to be 9.109390×10^{-31} kg and 1.672623×10^{-27} kg, respectively. What is the percent difference between this result and the rest mass of an electron?

The reduced mass of an electron and a proton is given by

$$\begin{aligned}\mu &= \frac{(9.109\,390 \times 10^{-31}\text{ kg})(1.672\,623 \times 10^{-27}\text{ kg})}{9.109\,390 \times 10^{-31}\text{ kg} + 1.672\,623 \times 10^{-27}\text{ kg}} \\ &= 9.104\,432 \times 10^{-31}\text{ kg}\end{aligned}$$

The mass of an electron is $9.109\,390 \times 10^{-31}$ kg, a 0.055% difference.

1–22. Quantum mechanics gives that the electronic energy of a hydrogen atom is

$$\varepsilon_n = -\frac{\mu e^4}{8\varepsilon_0^2h^2}\frac{1}{n^2} \qquad n = 1, 2, \ldots$$

where μ is the reduced mass of an electron and proton, e is the protonic charge, ε_0 is the permittivity of free space, and h is the Planck constant. Using the values given in the inside front cover, calculate the value of ε_n in terms of n^2. Compare your result with Equation 1.1.

Using the value of the reduced mass of an electron and a proton given in the previous problem

$$\begin{aligned}\varepsilon_n &= -\frac{(9.104\,432 \times 10^{-31}\text{ kg})(1.602\,177\,33 \times 10^{-19}\text{ C})^4}{8(8.854\,187\,816 \times 10^{-12}\text{ C}^2\cdot\text{J}^{-1}\cdot\text{m}^{-1})^2(6.626\,0755 \times 10^{-34}\text{ J}\cdot\text{s})^2n^2} \\ &= -\frac{2.178\,688 \times 10^{-18}\text{ J}}{n^2}\end{aligned}$$

1–23. Show that the reduced mass of two equal masses, m, is $m/2$.

If we substitute $m_1 = m_2 = m$ into Equation 1.13, we obtain

$$\mu = \frac{m^2}{2m} = \frac{m}{2}$$

1–24. In this problem, we investigate the harmonic oscillator potential as the leading term in a Taylor expansion of the actual internuclear potential, $V(R)$, about its equilibrium position, R_e. According to Problem C–13, the first few terms in this expansion are

$$V(R) = V(R_e) + \left(\frac{dV}{dR}\right)_{R=R_e}(R - R_e) + \frac{1}{2!}\left(\frac{d^2V}{dR^2}\right)_{R=R_e}(R - R_e)^2 + \frac{1}{3!}\left(\frac{d^3V}{dR^3}\right)_{R=R_e}(R - R_e)^3 + \cdots \tag{1}$$

If R is always close to R_e, then $R - R_e$ is always small. Consequently, the terms on the right side of Equation 1 get smaller and smaller. The first term in Equation 1 is a constant and depends upon where we choose the zero of energy. It is convenient to choose the zero of energy such that $V(R_e)$ equals zero and to relate $V(R)$ to this convention. Explain why there is no linear term in the displacement in Equation 1. (Note that dV/dR is essentially the force acting between the two nuclei.)
Denote $R - R_e$ by x, $(d^2V/dR^2)_{R=R_e}$ by k, and $(d^3V/dR^3)_{R=R_e}$ by γ to write Equation 1 as

$$\begin{aligned} V(x) &= \frac{1}{2}k(R - R_e)^2 + \frac{1}{6}\gamma(R - R_e)^3 + \cdots \\ &= \frac{1}{2}kx^2 + \frac{1}{6}\gamma x^3 + \cdots \end{aligned} \tag{2}$$

Argue that if we restrict ourselves to small displacements, then x will be small and we can neglect the terms beyond the quadratic term in Equation 2, showing that the general potential energy function $V(R)$ can be approximated by a harmonic-oscillator potential. We can consider corrections or extensions of the harmonic-oscillator model by the higher-order terms in Equation 2. These terms are called *anharmonic terms*.

An analytic expression that is a good approximation to an intermolecular potential energy curve is a *Morse potential*

$$V(R) = D_e(1 - e^{-\beta(R-R_e)})^2$$

where D_e and β are parameters that depend upon the molecule. The parameter D_e is the ground-state electronic energy of the molecule measured from the minimum of $V(R)$, and β is a measure of the curvature of $V(R)$ at its minimum. Derive a relation between the force constant and the parameters D_e and β. Given that $D_e = 7.31 \times 10^{-19}$ J·molecule^{-1}, $\beta = 0.0181$ pm^{-1}, and $R_e = 127.5$ pm for HCl(g), calculate the force constant of HCl(g). Plot the Morse potential for HCl(g), and plot the corresponding harmonic oscillator potential on the same graph (cf. Figure 1.5).

There is no linear term in the expansion of $V(R)$ about $R = R_e$ because R_e denotes the minimum value of $V(R)$; consequently $(dV/dR)_{R=R_e} = 0$. In addition, $-dV/dR$ is the force between the two nuclei, and this force is equal to zero at the equilibrium separation, R_e.

First let $x = R - R_e$. Then the Morse potential becomes

$$V(x) = D_e(1 - e^{-\beta x})^2$$

Now use the fact that $e^u = 1 + u + \frac{1}{2}u^2 + \cdots$ to write

$$V(x) = D_e\left[1 - \left(1 - \beta x + \frac{1}{2}\beta^2 x^2 + \cdots\right)\right]^2$$
$$= D_e\left(\beta x - \frac{1}{2}\beta^2 x^2 + \cdots\right)^2$$
$$= D_e\beta^2 x^2 - D_e\beta^3 x^3 + \cdots$$

Because $V(x) = kx^2/2$ for a harmonic oscillator,

$$k = 2D_e\beta^2 = 2(7.31 \times 10^{-19}\ \text{J}\cdot\text{molecule}^{-1})(1.81 \times 10^{10}\ \text{m}^{-1}) = 479\ \text{N}\cdot\text{m}^{-1}$$

The bond length of HCl(g) is 127.5 pm (Table 1.4). The graph below shows the Morse potential $V(x) = D_e(1 - e^{-\beta(R-R_e)})^2$ (solid line) and the harmonic oscillator potential $V(x) = \frac{1}{2}k(R - R_e)^2$ (dashed line) for HCl(g).

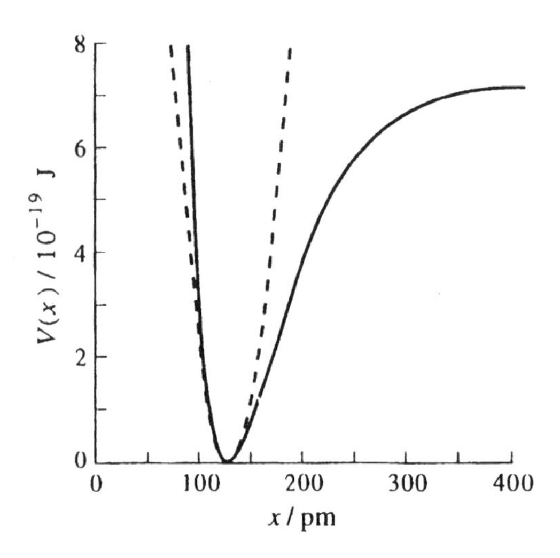

1–25. Use the result of Problem 1–24 and Equation 1.23 to show that

$$\beta = 2\pi c\tilde{\nu}\left(\frac{\mu}{2D_e}\right)^{1/2}$$

Given that $\tilde{\nu} = 2886\ \text{cm}^{-1}$ and $D_e = 440.2\ \text{kJ}\cdot\text{mol}^{-1}$ for $H^{35}Cl(g)$, calculate the value of β. Compare your result with that in Problem 1–24.

Problem 1–24 shows that $k = 2D_e\beta^2$, which upon substituting into $\tilde{\nu}_{obs} = (k/\mu)^{1/2}/2\pi c$ gives

$$\tilde{\nu}_{obs} = \frac{1}{2\pi c}\left(\frac{2D_e\beta^2}{\mu}\right)^{1/2}$$

Solving for β,

$$\beta = 2\pi c\tilde{\nu}_{obs}\left(\frac{\mu}{2D_e}\right)^{1/2}$$

For $H^{35}Cl(g)$,

$$\beta = 2\pi(2.998 \times 10^{10}\ \text{cm}\cdot\text{s}^{-1})(2886\ \text{cm}^{-1}) \left[\frac{\left(\frac{1.008 \times 34.97}{1.008 + 34.97}\right)(1.661 \times 10^{-27}\ \text{kg})}{2\left(\frac{440.2 \times 10^{3}\ \text{J}\cdot\text{mol}^{-1}}{6.022 \times 10^{23}\ \text{mol}^{-1}}\right)} \right]^{1/2}$$

$$= 1.81 \times 10^{10}\ \text{m}^{-1}$$

which is the same as the value for β given in the previous problem.

1–26. Carry out the Taylor expansion of the Morse potential in Problem 1–24 through terms in $(R - R_e)^3$. Express γ in Equation 2 of Problem 1–24 in terms of D_e and β.

First let $x = R - R_e$. Then an expansion of $V(x)$ gives

$$\begin{aligned} V(x) &= D_e\left(1 - e^{-\beta x}\right)^2 \\ &= D_e\left\{1 - \left[1 - \beta x + \frac{1}{2}\beta^2 x^2 + O(x^3)\right]\right\}^2 \\ &= D_e\left[\beta x - \frac{1}{2}\beta^2 x^2 + O(x^3)\right]^2 \\ &= D_e\beta^2 x^2\left[1 - \frac{1}{2}\beta x + O(x^2)\right]^2 \\ &= D_e\beta^2 x^2\left[1 - \beta x + O(x^2)\right] \\ &= D_e\left[\beta^2 x^2 - \beta^3 x^3 + O(x^4)\right] \end{aligned}$$

By comparing this result to Equation 2 of Problem 1–24, we see that

$$\gamma = -6D_e\beta^3$$

1–27. It turns out that the solution of the Schrödinger equation for the Morse potential (Problem 1–24) can be expressed as

$$\tilde{\varepsilon}_v = \tilde{\nu}\left(v + \frac{1}{2}\right) - \tilde{\nu}\tilde{x}\left(v + \frac{1}{2}\right)^2$$

where $v = 0, 1, 2, \ldots$ and

$$\tilde{x} = \frac{hc\tilde{\nu}}{4D_e}$$

Given that $\tilde{\nu} = 2886\ \text{cm}^{-1}$ and $D_e = 440.2\ \text{kJ}\cdot\text{mol}^{-1}$ for $H^{35}Cl(g)$, calculate the values of $\tilde{x}$ and $\tilde{\nu}\tilde{x}$.

Using the above equation for $\tilde{x}$, we have

$$\tilde{x} = \frac{(6.626 \times 10^{-34}\ \text{J}\cdot\text{s})(2.998 \times 10^{10}\ \text{cm}\cdot\text{s}^{-1})(2886\ \text{cm}^{-1})}{4\left(\frac{440.2 \times 10^{3}\ \text{J}\cdot\text{mol}^{-1}}{6.022 \times 10^{23}\ \text{mol}^{-1}}\right)}$$

$$= 0.01961$$

Therefore,

$$\tilde{\nu}\tilde{x} = (2886\ \text{cm}^{-1})(0.01961) = 56.59\ \text{cm}^{-1}$$

1–28. In the infrared spectrum of $H^{79}Br(g)$, there is an intense line at $2630\ \text{cm}^{-1}$. Use the harmonic-oscillator approximation to calculate the values of the force constant of $H^{79}Br(g)$ and the period of vibration of $H^{79}Br(g)$.

We use Equation 1.23 in the form

$$\begin{aligned} k &= (2\pi c\tilde{\nu}_{\text{obs}})^2\mu \\ &= \left[2\pi(2.9979\times10^{10}\ \text{cm}\cdot\text{s}^{-1})(2630\ \text{cm}^{-1})\right]^2 \\ &\qquad\times\left[\frac{(1.008\ \text{amu})(78.92\ \text{amu})}{79.93\ \text{amu}}\right](1.661\times10^{-27}\ \text{kg}\cdot\text{amu}^{-1}) \\ &= 406\ \text{N}\cdot\text{m}^{-1} \end{aligned}$$

According to Problem 1–19, the period of vibration is

$$\tau = \frac{1}{\nu} = \frac{1}{c\tilde{\nu}_{\text{obs}}} = \frac{1}{(2.9979\times10^{10}\ \text{cm}\cdot\text{s}^{-1})(2630\ \text{cm}^{-1})} = 1.27\times10^{-14}\ \text{s}$$

1–29. The force constant of $^{79}Br^{79}Br(g)$ is $240\ \text{N}\cdot\text{m}^{-1}$. Use the harmonic-oscillator approximation to calculate the values of the fundamental vibrational frequency and the zero-point energy (in joules per mole) of $^{79}Br^{79}Br(g)$.

Use Equation 1.19

$$\begin{aligned} \nu_{\text{obs}} &= \frac{1}{2\pi}\left(\frac{k}{\mu}\right)^{1/2} \\ &= \frac{1}{2\pi}\left\{\frac{240\ \text{N}\cdot\text{m}^{-1}}{\left[\frac{(78.92\ \text{amu})^2}{78.92\ \text{amu}+78.92\ \text{amu}}\right](1.661\times10^{-27}\ \text{kg}\cdot\text{amu}^{-1})}\right\}^{1/2} \\ &= 9.63\times10^{12}\ \text{s}^{-1} \end{aligned}$$

We use Equation 1.19 with $v = 0$ to find the zero point energy:

$$\varepsilon_0 = \tfrac{1}{2}h\nu = \tfrac{1}{2}(6.626\times10^{-34}\ \text{J}\cdot\text{s})(9.63\times10^{12}\ \text{s}^{-1}) = 3.19\times10^{-21}\ \text{J}$$

1–30. In the far-infrared spectrum of $^{39}K^{35}Cl(g)$, there is an intense line at $278.0\ \text{cm}^{-1}$. Calculate the values of the force constant and the period of vibration of $^{39}K^{35}Cl(g)$.

The reduced mass of $^{39}K^{35}Cl$ is

$$\mu = \frac{(38.964)(34.969)}{(38.964+34.969)}\ \text{amu} = 18.429\ \text{amu}$$

Use Equation 1.23:

$$\tilde{\nu} = \frac{1}{2\pi c}\left(\frac{k}{\mu}\right)^{1/2}$$

or

$$\begin{aligned} k &= (2\pi c\tilde{\nu})^2 \mu \\ &= \left[2\pi(2.998 \times 10^{10}\ \text{cm}\cdot\text{s}^{-1})(278.0\ \text{cm}^{-1})\right]^2 \\ &\qquad \times (18.492\ \text{amu})\left(1.661 \times 10^{-27}\ \text{kg}\cdot\text{amu}^{-1}\right) \\ &= 83.9\ \text{N}\cdot\text{m}^{-1} \end{aligned}$$

The period of vibration is

$$\tau = \frac{1}{\nu} = \frac{1}{c\tilde{\nu}} = 1.20 \times 10^{-13}\ \text{s}$$

1–31. Thus far, we have treated the vibrational motion of a diatomic molecule by means of a harmonic-oscillator model. We saw in Section 1–4, however, that the internuclear potential energy is not a simple parabola but is more like that illustrated in Figure 1.5. The harmonic-oscillator approximation consists of keeping only the quadratic term in the Taylor expansion of $V(R)$ (see Problem 1–24), and it predicts that there will be only one line in the vibrational spectrum of a diatomic molecule. Experimental data show there is, indeed, one dominant line (called the *fundamental*) but also lines of weaker intensity at almost integral multiples of the fundamental. These lines are called *overtones* (see Table 1.7). If the anharmonic terms in Equation 1.19 are taken into account, then a quantum mechanical calculation gives

$$\tilde{\varepsilon}_v = \tilde{\nu}(v + \tfrac{1}{2}) - \tilde{x}\tilde{\nu}(v + \tfrac{1}{2})^2 + \cdots \qquad v = 0,\ 1,\ 2,\ \ldots \tag{1}$$

where $\tilde{x}$ is called the *anharmonicity constant.* The anharmonic correction in Equation 1 is much smaller than the harmonic term because $\tilde{x} \ll 1$. Show that the levels are not equally spaced as they are for a harmonic oscillator and, in fact, that their separation decreases with increasing v. The selection rule for an anharmonic oscillator is that Δv can have any integral value, although the intensities of the $\Delta v = \pm 2,\ \pm 3,\ \ldots$ transitions are much less than for the $\Delta v = \pm 1$ transitions. Show that if we recognize that most diatomic molecules are in the ground vibrational state at room temperature, the frequencies of the observed $0 \rightarrow v$ transitions will be given by

$$\tilde{\nu}_{\text{obs}} = \tilde{\nu}v - \tilde{x}\tilde{\nu}v(v + 1) \qquad v = 1,\ 2,\ \ldots \tag{2}$$

TABLE 1.7
The vibrational spectrum of $H^{35}Cl(g)$.

Transition	$\tilde{\nu}_{\text{obs}}/\text{cm}^{-1}$	$\tilde{\nu}_{\text{obs}}/\text{cm}^{-1}$ Harmonic oscillator $\tilde{\nu} = 2885.90v$
$0 \rightarrow 1$ (fundamental)	2885.9	2885.9
$0 \rightarrow 2$ (first overtone)	5668.0	5771.8
$0 \rightarrow 3$ (second overtone)	8347.0	8657.7
$0 \rightarrow 4$ (third overtone)	10 923.1	11 543.6
$0 \rightarrow 5$ (fourth overtone)	13 396.5	14 429.5

Curve fit Equation 2 to the experimental data in Table 1.7 to find the optimum values of $\tilde{\nu}$ and $\tilde{x}\tilde{\nu}$. Use Equation 2 to calculate the values of the observed frequencies and compare your results with the experimental data.

For the $0 \rightarrow v$ transitions, we have

$$\begin{aligned}\varepsilon_v - \varepsilon_0 &= \tilde{\nu}\left(v+\frac{1}{2}\right) - \tilde{\nu}\tilde{x}\left(v+\frac{1}{2}\right)^2 - \tilde{\nu}\left(\frac{1}{2}\right) + \tilde{\nu}\tilde{x}\left(\frac{1}{4}\right)\\ &= \tilde{\nu}v - \tilde{\nu}\tilde{x}v(v+1)\end{aligned}$$

A curve fit to the data listed in the problem gives

$$\tilde{\nu}_{\text{obs}}/\text{cm}^{-1} = 2988.7v - 51.565v(v+1)$$

and so $\tilde{\nu} = 2988.7\ \text{cm}^{-1}$ and $\tilde{x} = 0.0173$.

1–32. Given that $\tilde{\nu} = 536.10\ \text{cm}^{-1}$ and $\tilde{x}\tilde{\nu} = 3.4\ \text{cm}^{-1}$ for $^{23}\text{Na}^{19}\text{F(g)}$, calculate the values of the frequencies of the first and second vibrational overtones (see Problem 1–31).

We use Equation 1 of Problem 1–31.

$$\tilde{\nu}_{\text{obs}} = \tilde{\nu}v - \tilde{\nu}\tilde{x}v(v+1) \qquad v = 1,\ 2,\ \cdots$$

Using the values of $\tilde{\nu}$ and $\tilde{x}\tilde{\nu}$ given above, we have

$$\tilde{\nu}_{\text{obs}}/\text{cm}^{-1} = 536.10v - 3.4v(v+1)$$

For the first overtone, $v = 2$, and for the second overtone, $v = 3$. Therefore,

$$\begin{aligned}\tilde{\nu}_{\text{obs}} &= 1051.8\ \text{cm}^{-1} \qquad &\text{(first overtone)}\\ &= 1567.5\ \text{cm}^{-1} \qquad &\text{(second overtone)}\end{aligned}$$

1–33. The fundamental line in the infrared spectrum of $^{12}\text{C}^{16}\text{O(g)}$ occurs at $2143.0\ \text{cm}^{-1}$, and the first overtone occurs at $4260.0\ \text{cm}^{-1}$. Calculate the values of $\tilde{\nu}$ and $\tilde{x}\tilde{\nu}$ for $^{12}\text{C}^{16}\text{O(g)}$ (see Problem 1–31).

We use Equation 1 of Problem 1–31 with $v = 1$ and $v = 2$ to write

$$\tilde{\nu}_{\text{obs}} = \tilde{\nu} - 2\tilde{x}\tilde{\nu} = 2143.0\ \text{cm}^{-1} \qquad \text{(fundamental, } v = 1\text{)}$$

$$\tilde{\nu}_{\text{obs}} = 2\tilde{\nu} - 6\tilde{x}\tilde{\nu} = 4260.0\ \text{cm}^{-1} \qquad \text{(first overtone, } v = 2\text{)}$$

Multiply the first equation by 3 and subtract the second to get

$$\tilde{\nu} = 3(2143.0\ \text{cm}^{-1}) - 4260.0\ \text{cm}^{-1} = 2169.0\ \text{cm}^{-1}$$

Multiply the first equation by 2 and subtract from the second to get

$$2\tilde{x}\tilde{\nu} = 26.0\ \text{cm}^{-1}$$

or

$$\tilde{x}\tilde{\nu} = 13.0\ \text{cm}^{-1}$$

1–34. The Morse potential is presented in Problem 1–24. Given that $D_e = 8.35 \times 10^{-19}$ J·molecule^{-1}, $\tilde{\nu} = 1556$ cm^{-1}, and $R_e = 120.7$ pm for O_2, plot a Morse potential for O_2. Plot the corresponding harmonic-oscillator potential on the same graph.

Example 1–5 shows that $k = 2D_e\beta^2$. Using Equation 1.23, we have

$$\beta^2 = \frac{k}{2D_e} = \frac{4\pi^2\nu^2\mu}{2D_e}$$

$$= \frac{4\pi^2(1556\text{ cm}^{-1})(2.9979\times10^{10}\text{ cm}\cdot\text{s}^{-1})(15.999/2\text{ amu})(1.6605\times10^{-27}\text{ kg}\cdot\text{amu}^{-1})}{2(8.35\times10^{-19}\text{ J})}$$

$$= 6.833\times10^{20}\text{ m}^{-2}$$

or $\beta = 2.614 \times 10^{10}$ m^{-1}. We now plot $V_M(R)$ according to

$$V_M(R) = (8.35\times10^{-19}\text{ J})[1 - e^{-(2.614\times10^{10}\text{ m}^{-1})(R-120.7\times10^{-12}\text{ m})}]^2$$

For the harmonic oscillator potential, $k = 4\pi^2\nu^2\mu = 1140$ N·m^{-1} and $V_{ho}(R) = k(R - R_e)^2/2$. Both $V_M(R)$ (solid line) and $V_{ho}(R)$ (dotted line) are plotted against $R/R_e = R/120.7$ pm below.

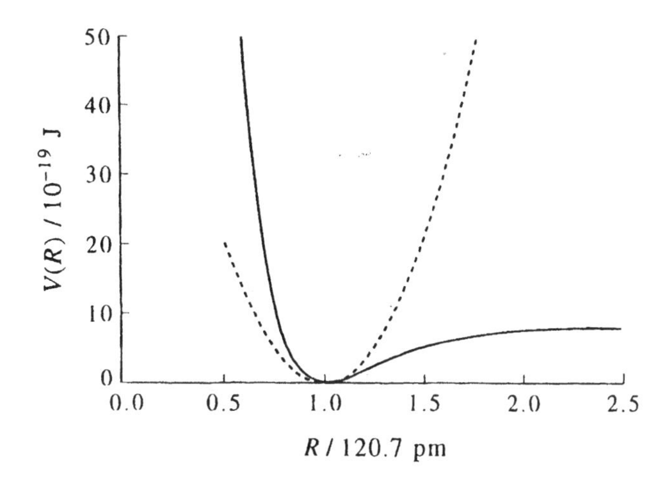

1–35. Show that the moment of inertia for a diatomic rigid rotator can be written as $I = \mu R_e^2$, where $R_e = R_1 + R_2$ (the fixed separation of the two masses), R_1 and R_2 are the distances of the two masses from the center of mass, and μ is the reduced mass.

Start with $I = m_1R_1^2 + m_2R_2^2$. Now solve $m_1R_1 = m_2R_2 = m_2(R_e - R_1)$ for R_1 to get

$$R_1 = \frac{m_2R_e}{m_1 + m_2} = \frac{m_2R_e}{M}$$

where $M = m_1 + m_2$. Use $R_2 = R_e - R_1$ to write

$$R_2 = R_e - \frac{m_2R_e}{m_1 + m_2} = \frac{m_1R_e}{M}$$

Substitute R_1 and R_2 into $I = m_1 R_1^2 + m_2 R_2^2$ to get

$$I = \frac{m_1 m_2^2 R_e^2}{M^2} + \frac{m_2 m_1^2 R_e^2}{M^2} = \frac{m_1 m_2}{M} R_e^2 = \mu R_e^2$$

1–36. In the far infrared spectrum of $H^{79}Br(g)$, there is a series of lines separated by $16.72\ cm^{-1}$. Calculate the values of the moment of inertia and the internuclear separation in $H^{79}Br(g)$.

Assuming that $H^{79}Br(g)$ can be treated as a rigid rotator,

$$\tilde{\nu} = 2\tilde{B}(J+1) \qquad J = 0, 1, 2, \ldots$$

where $\tilde{B} = h/8\pi^2 cI$. The lines in the spectrum are separated by $16.72\ cm^{-1}$, so

$$\Delta\tilde{\nu} = 2\tilde{B} = \frac{2h}{8\pi^2 cI}$$

$$16.72\ cm^{-1} = \frac{6.626 \times 10^{-34}\ J\cdot s}{4\pi^2(2.998 \times 10^{10}\ cm\cdot s^{-1})I}$$

or

$$I = 3.348 \times 10^{-47}\ kg\cdot m^2$$

We can find μ for $H^{79}Br(g)$:

$$\mu = \frac{(78.92)(1.008)}{79.93}(1.661 \times 10^{-27}\ kg) = 1.653 \times 10^{-27}\ kg$$

Now we can use the relationship $R_e = (I/\mu)^{1/2}$ to find R_e.

$$R_e = \left(\frac{3.348 \times 10^{-47}\ kg\cdot m^2}{1.653 \times 10^{-27}\ kg}\right)^{1/2} = 1.423 \times 10^{-10}\ m = 142.3\ pm$$

1–37. Given that $\tilde{\nu} = 2330\ cm^{-1}$ and that $D_0 = 78\,715\ cm^{-1}$ for $N_2(g)$, calculate the value of D_e.

Use Equation 1.24,

$$D_e = D_0 + \frac{h\nu}{2} = 78715\ cm^{-1} + \frac{2330\ cm^{-1}}{2}$$

$$= 79880\ cm^{-1} = (79880\ cm^{-1})\left(\frac{1.1963 \times 10^{-2}\ kJ\cdot mol^{-1}}{1\ cm^{-1}}\right)$$

$$= 955.6\ kJ\cdot mol^{-1}$$

1–38. The $J = 0$ to $J = 1$ transition for carbon monoxide ($^{12}C^{16}O(g)$) occurs at 1.153×10^5 MHz. Calculate the value of the bond length in carbon monoxide.

Assuming that $^{12}C^{16}O(g)$ can be treated as a rigid rotator

$$\nu = 2B(J+1) \qquad J = 0, 1, 2, \ldots$$

where $B = h/8\pi^2 I$. For the $J = 0$ to $J = 1$ transition,

$$\nu = 2B = \frac{2h}{8\pi^2 I}$$

Using $I = \mu R_e^2$, we have

$$1.153 \times 10^{11}\ \text{s}^{-1} = \frac{6.626 \times 10^{-34}\ \text{J}\cdot\text{s}}{4\pi^2 \mu R_e^2}$$

We first calculate μ and then use the above equation to find R_e.

$$\mu = \frac{192.0}{28.00}(1.661 \times 10^{-27}\ \text{kg}) = 1.139 \times 10^{-26}\ \text{kg}$$

$$R_e = \left[\frac{6.626 \times 10^{-34}\ \text{J}\cdot\text{s}}{4\pi^2(1.139 \times 10^{-26}\ \text{kg})(1.153 \times 10^{11}\ \text{s}^{-1})}\right]^{1/2}$$

$$= 1.130 \times 10^{-10}\ \text{m} = 113.0\ \text{pm}$$

1–39. The microwave spectrum of $^{39}K^{127}I(g)$ consists of a series of lines whose spacing is almost constant at 3634 MHz. Calculate the bond length of $^{39}K^{127}I(g)$.

We use the same method as in Example 1–8. From Equation 1.32,

$$\nu = 3\,634 \times 10^6\ \text{s}^{-1} = 2B = \frac{h}{4\pi^2 I}$$

Solving for I gives

$$I = \frac{6.626 \times 10^{-34}\ \text{J}\cdot\text{s}}{4\pi^2\left(3.634 \times 10^9\ \text{s}^{-1}\right)} = 4.619 \times 10^{-45}\ \text{kg}\cdot\text{m}^2$$

The reduced mass of $^{39}K^{127}I$ is

$$\mu = \frac{(38.96)(126.90)}{(38.96 + 126.90)}\ \text{amu} = 29.808\ \text{amu}$$

From the equation $I = \mu R_e^2$, we get

$$R_e = \left(\frac{I}{\mu}\right)^{1/2}$$

$$= \left(\frac{4.619 \times 10^{-45}\ \text{kg}\cdot\text{m}^2}{(29.808\ \text{amu})\left(1.661 \times 10^{-27}\ \text{kg}\cdot\text{amu}^{-1}\right)}\right)^{1/2}$$

$$= 3.055 \times 10^{-10}\ \text{m} = 305.5\ \text{pm}$$

1–40. Assuming the rotation of a diatomic molecule in the $J = 10$ state may be approximated by classical mechanics, calculate how many revolutions per second $^{23}Na^{35}Cl(g)$ makes in the $J = 10$ rotational state. The rotational constant of $^{23}Na^{35}Cl(g)$ is 6500 MHz.

The energy of a classical rotator is

$$\text{KE} = \frac{1}{2}I\omega^2$$

The quantum-mechanical energy is given by

$$\varepsilon = hBJ(J+1)$$

where $B = h/8\pi^2 I$. Equating KE with ε gives

$$\frac{I\omega^2}{2} = hBJ(J+1)$$

$$\omega^2 = \frac{2hBJ(J+1)}{I} = [2hBJ(J+1)]\left(\frac{h}{8\pi^2 B}\right)^{-1}$$

or

$$\omega = 4\pi B[J(J+1)]^{1/2} = 4\pi(6500\times 10^6\ \mathrm{s^{-1}})(110)^{1/2}$$
$$= 8.57\times 10^{11}\ \mathrm{radian\cdot s^{-1}}$$
$$= 1.36\times 10^{11}\ \mathrm{revolution\cdot s^{-1}}$$

1–41. The rigid-rotator model predicts that the lines in the rotational spectrum of a diatomic molecule should be equally spaced. The table below lists some of the observed lines in the rotational spectrum of $H^{35}Cl(g)$. The differences listed in the third column clearly show that the lines are not exactly equally spaced as the rigid-rotator approximation predicts. The discrepancy can be resolved by realizing that a chemical bond is not truly rigid. As the molecule rotates more energetically (increasing J), the centrifugal force causes the bond to stretch slightly. If this small effect is taken into account, then the energy is given by

$$\tilde{\varepsilon}_J = \frac{\varepsilon_J}{hc} = \tilde{B}J(J+1) - \tilde{D}J^2(J+1)^2 \qquad (1)$$

where $\tilde{D}$ is called the *centrifugal distortion constant*. Show that the frequencies of the absorption due to $J \rightarrow J+1$ transitions are given by

$$\tilde{\nu} = 2\tilde{B}(J+1) - 4\tilde{D}(J+1)^3 \qquad J = 0, 1, 2, \ldots \qquad (2)$$

Curve fit Equation 2 to the experimental data below and find the optimum values of $\tilde{B}$ and $\tilde{D}$. Compare the predictions of the resulting Equation 2 with the experimental data.

The rotational absorption spectrum of $H^{35}Cl(g)$.

Transition	$\tilde{\nu}_{obs}/\mathrm{cm^{-1}}$	$\Delta\tilde{\nu}_{obs}/\mathrm{cm^{-1}}$	$c\tilde{\nu}_{calc} = 2\tilde{B}(J+1)$ $\tilde{B} = 10.243\ \mathrm{cm^{-1}}$
$3\rightarrow 4$	83.03		82.72
		21.07	
$4\rightarrow 5$	104.10		103.40
		20.20	
$5\rightarrow 6$	124.30		124.08
		20.73	
$6\rightarrow 7$	145.03		144.76
		20.48	
$7\rightarrow 8$	165.51		165.44
		20.35	
$8\rightarrow 9$	185.86		186.12
		20.52	
$9\rightarrow 10$	206.38		206.80
		20.12	
$10\rightarrow 11$	226.50		227.48

Start with

$$\frac{\varepsilon_{J+1}-\varepsilon_J}{hc} = \tilde{\nu} = \tilde{B}[(J+1)(J+2) - J(J+1)] - \tilde{D}[(J+1)^2(J+2)^2 - J^2(J+1)^2]$$
$$= 2\tilde{B}(J+1) - \tilde{D}(4J^3 + 12J^2 + 12J + 4)$$
$$= 2\tilde{B}(J+1) - 4\tilde{D}(J^3 + 3J^2 + 3J + 1)$$
$$= 2\tilde{B}(J+1) - 4\tilde{D}(J+1)^3 \qquad J = 0,\ 1,\ 2,\ \cdots$$

The curve fit to the data gives

$$\tilde{\nu}/\text{cm}^{-1} = 20.806(J+1) - 0.0018(J+1)^3$$

or $\tilde{B} = 10.403\ \text{cm}^{-1}$ and $\tilde{D} = 0.00044\ \text{cm}^{-1}$.

1–42. The following data are obtained in the microwave spectrum of $^{12}C^{16}O(g)$. Use the method of Problem 1–41 to determine the values of $\tilde{B}$ and $\tilde{D}$ from these data.

Transitions	Frequency/cm^{-1}
$0 \to 1$	3.845 40
$1 \to 2$	7.690 60
$2 \to 3$	11.535 50
$3 \to 4$	15.379 90
$4 \to 5$	19.223 80
$5 \to 6$	23.066 85

Using the method of Problem 1–14, we plot $\tilde{\nu}$ vs. $J + 1$.

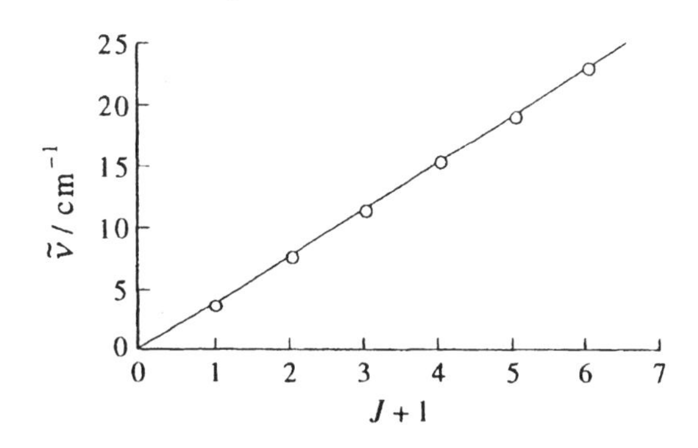

A curve fit of these data to $\tilde{\nu} = 2\tilde{B}(J+1) - 4\tilde{D}(J+1)^3$ gives $\tilde{\nu}/\text{cm}^{-1} = 3.8454(J+1) - 2.555 \times 10^{-5}(J+1)^3$ Therefore, $\tilde{B} = 1.9227\ \text{cm}^{-1}$ and $\tilde{D} = 6.39 \times 10^{-6}\ \text{cm}^{-1}$.

1–43. Given that $\tilde{B} = 8.465\ \text{cm}^{-1}$ and $\tilde{D} = 0.000346\ \text{cm}^{-1}$ for $H^{79}Br(g)$, calculate the frequency of the $J = 0 \to J = 1$, $J = 1 \to J = 2$, $J = 2 \to J = 3$, $\cdots$, $J = 6 \to J = 7$ transitions in the rotational spectrum of $H^{79}Br(g)$.

We use Equation 2 of Problem 1–41

$$\tilde{\nu} = 2\tilde{B}(J+1) - 4\tilde{D}(J+1)^3$$

with $J = 0, 1, 2, \cdots, 6$. So

$$\tilde{\nu}/\text{cm}^{-1} = 16.930(J + 1) - 0.001384(J + 1)^3$$

J	$\tilde{\nu}/\text{cm}^{-1}$
0	16.929
1	33.849
2	50.753
3	67.631
4	84.477
5	101.28
6	118.04

1–44. Determine the number of various degrees of freedom of N_2, C_2H_2, C_2H_4, C_2H_6, and C_6H_6.

		degrees of freedom		
molecule	total	translational	rotational	vibrational
N_2	6	3	2	1
C_2H_2	12	3	2	7
C_2H_4	18	3	3	12
C_2H_6	24	3	3	18
C_6H_6	36	3	3	30

1–45. Determine the total number of normal modes of vibration of HCN, CD_4, SO_3, SF_6, and $(CH_3)_2CO$.

		degrees of freedom		
molecule	total	translational	rotational	vibrational
HCN	9	3	2	4
CD_4	15	3	3	9
SO_3	12	3	3	6
SF_6	21	3	3	15
$(CH_3)_2CO$	30	3	3	24

1–46. Using the data in Table 1.6, calculate the value of the zero-point vibrational energy of a water molecule.

$$\begin{aligned}
\varepsilon_0 &= \frac{1}{2}hc(\tilde{\nu}_1 + \tilde{\nu}_2 + \tilde{\nu}_3) \\
&= \frac{(6.626 \times 10^{-34}\ \text{J}\cdot\text{s})(2.9979 \times 10^{10}\ \text{cm}\cdot\text{s}^{-1})}{2}(3725\ \text{cm}^{-1} + 3586\ \text{cm}^{-1} + 1595\ \text{cm}^{-1}) \\
&= \frac{(6.626 \times 10^{-34}\ \text{J}\cdot\text{s})(2.9979 \times 10^{10}\ \text{cm}\cdot\text{s}^{-1})}{2}(8906\ \text{cm}^{-1}) \\
&= 8.845 \times 10^{-20}\ \text{J} = 53.27\ \text{kJ}\cdot\text{mol}^{-1}
\end{aligned}$$

1–47. Using the data in Table 1.6, calculate the value of the zero-point vibrational energy of a methane molecule.

$$\begin{aligned}\varepsilon_0 &= \frac{(6.626 \times 10^{-34}\ \text{J}\cdot\text{s})(2.9979 \times 10^{10}\ \text{cm}\cdot\text{s}^{-1})}{2} \\ &\qquad \times [2898\ \text{cm}^{-1} + 2(1515\ \text{cm}^{-1}) + 3(3002\ \text{cm}^{-1}) + 3(1300\ \text{cm}^{-1})] \\ &= \frac{(6.626 \times 10^{-34}\ \text{J}\cdot\text{s})(2.9979 \times 10^{10}\ \text{cm}\cdot\text{s}^{-1})}{2}(18834\ \text{cm}^{-1}) \\ &= 1.871 \times 10^{-19}\ \text{J} = 112.6\ \text{kJ}\cdot\text{mol}^{-1}\end{aligned}$$

1–48. Given that $R_{NN} = 112.8$ pm and $R_{NO} = 118.4$ pm for $^{14}N^{14}N^{16}O$, calculate the values of the moment of inertia and $\tilde{B}$.

First we shall find the center of mass of NNO. The center of mass lies between the central nitrogen atom and the oxygen atom. Let x be the distance from the central nitrogen atom. Then, x satisfies the equation

$$\underbrace{(14.00\ \text{amu})(112.8\ \text{pm} + x)}_{\text{end nitrogen atom}} + \underbrace{(14.00\ \text{amu})(x)}_{\text{central nitrogen atom}} = \underbrace{(16.00\ \text{amu})(118.4\ \text{pm} - x)}_{\text{oxygen atom}}$$

or $x = 7.2$ pm. The moment of inertia is

$$\begin{aligned}I &= (14.00\ \text{amu})(112.8\ \text{pm} + 7.2\ \text{pm})^2 + (14.00\ \text{amu})(7.2\ \text{pm})^2 \\ &\qquad + (16.00\ \text{amu})(118.4\ \text{pm} - 7.2\ \text{pm})^2 \\ &= 4.002 \times 10^5\ \text{amu}\cdot\text{pm}^2 \\ &= (4.002 \times 10^5\ \text{amu}\cdot\text{pm}^2)(1.6605 \times 10^{-27}\ \text{kg}\cdot\text{amu}^{-1})(10^{-12}\ \text{m}\cdot\text{pm}^{-1})^2 \\ &= 6.645 \times 10^{-46}\ \text{kg}\cdot\text{m}^2\end{aligned}$$

Using Equation 1.38,

$$\begin{aligned}\tilde{B} &= \frac{h}{8\pi^2 cI} \\ &= \frac{6.626 \times 10^{-34}\ \text{J}\cdot\text{s}}{8\pi^2(2.9979 \times 10^8\ \text{m}\cdot\text{s}^{-1})(6.645 \times 10^{-46}\ \text{kg}\cdot\text{m}^2)} \\ &= 42.13\ \text{m}^{-1} = 0.4213\ \text{cm}^{-1}\end{aligned}$$

1–49. In this problem, we will see how the concept of reduced mass arises naturally when discussing the interaction of two particles. Consider two masses, m_1 and m_2, in one dimension, interacting through a potential that depends only upon their relative separation $(x_1 - x_2)$ so that $U(x_1, x_2) = U(x_1 - x_2)$. Given that the force acting upon the jth particle is $f_j = -(\partial U/\partial x_j)$, show that $f_1 = -f_2$. What law is this?

Newton's equations for m_1 and m_2 are

$$m_1 \frac{d^2x_1}{dt^2} = -\frac{\partial U}{\partial x_1} \qquad m_2 \frac{d^2x_2}{dt^2} = -\frac{\partial U}{\partial x_2}$$

Now introduce center-of-mass and relative coordinates by

$$X = \frac{m_1 x_1 + m_2 x_2}{M} \qquad x = x_1 - x_2$$

Solve for x_1 and x_2 to obtain

$$x_1 = X + \frac{m_2}{M}x \qquad x_2 = X - \frac{m_1}{M}x$$

Show that Newton's equations in these coordinates are

$$m_1 \frac{d^2X}{dt^2} + \frac{m_1 m_2}{M}\frac{d^2x}{dt^2} = -\frac{\partial U}{\partial x}$$

and

$$m_2 \frac{d^2X}{dt^2} - \frac{m_1 m_2}{M}\frac{d^2x}{dt^2} = +\frac{\partial U}{\partial x}$$

Now add these two equations to find

$$M\frac{d^2X}{dt^2} = 0$$

Interpret this result. Now divide the first equation by m_1 and the second by m_2 and subtract to obtain

$$\frac{d^2x}{dt^2} = -\left(\frac{1}{m_1} + \frac{1}{m_2}\right)\frac{\partial U}{\partial x}$$

or

$$\mu\frac{d^2x}{dt^2} = -\frac{\partial U}{\partial x}$$

where μ is the reduced mass. Interpret this result and discuss how the original two-body problem has been reduced to two one-body problems.

The forces acting upon particles 1 and 2 are

$$f_1 = -\frac{\partial U}{\partial x_1} \qquad f_2 = -\frac{\partial U}{\partial x_2} = \frac{\partial U}{\partial x_1} = -f_1$$

The center-of-mass and relative coordinates are

$$X = \frac{m_1 x_1 + m_2 x_2}{M} \qquad x = x_1 - x_2$$

Multiply the first of these equations by M and the second by m_2 and add to obtain

$$x_1 = X + \frac{m_2}{M}x$$

Multiply the first by M and the second by m_1 and subtract to obtain

$$x_2 = X - \frac{m_1}{M}x$$

Substitute these expressions for x_1 and x_2 into Newton's equations

$$m_1\frac{d^2x_1}{dt^2} = m_1\frac{d^2X}{dt^2} + \frac{m_1 m_2}{M}\frac{d^2x}{dt^2} = -\frac{\partial U}{\partial x_1} = -\frac{\partial U}{\partial x}$$

$$m_2\frac{d^2x_2}{dt^2} = m_2\frac{d^2X}{dt^2} - \frac{m_1 m_2}{M}\frac{d^2x}{dt^2} = -\frac{\partial U}{\partial x_2} = \frac{\partial U}{\partial x}$$

We add these two equaitons to obtain

$$(m_1 + m_2)\frac{d^2 X}{dt^2} = 0$$

which says that the center of mass moves uniformly. We now divide the first of Newton's equations by m_1 and the second by m_2 and subtract to obtain

$$\frac{d^2 x}{dt^2} = -\left(\frac{1}{m_1} + \frac{1}{m_2}\right)\frac{\partial U}{\partial x}$$

or

$$\mu\frac{d^2 x}{dt^2} = -\frac{\partial U}{\partial x}$$

This is the equation of motion of one body of mass μ moving under a force $-\partial U/\partial x$.

MATHCHAPTER A

Numerical Methods

PROBLEMS AND SOLUTIONS

Excel was used to create spreadsheets for the approximations in the problems below. Any spreadsheet program can be used, but programs such as Excel, where the formulas can be saved and are automatically recalculated when different values are entered, are more powerful in this case than programs which are primarily used for graphing (such as Kaleidagraph).

A–1. Solve the equation $x^5 + 2x^4 + 4x = 5$ to four significant figures for the root that lies between 0 and 1.

$$f(x) = x^5 + 2x^4 + 4x - 5$$
$$f'(x) = 5x^4 + 8x^3 + 4$$

The iterative formula for the Newton-Raphson method is

$$x_{n+1} = x_n - \frac{f(x_n)}{f'(x_n)} \tag{A.1}$$

To set up a spreadsheet for the Newton-Raphson method, let one column contain x_n, one column contain the formula for $f(x)$, and one column contain the formula for $f'(x)$. [Allow the first x_n to be input manually; thereafter, let your spreadsheet calculate the values of x_n using the equation above.] Since we wish to find the root which lies between 0 and 1, we can take x_0 to be 0.5:

n	x_n	$f(x_n)$	$f'(x_n)$
0	0.50000	−2.84375	5.3125
1	1.03529	2.62821	18.62145
2	0.894155	0.426631	12.91523
3	0.861122	0.0177315	11.85774
4	0.859627	3.4128×10^{-5}	11.81213
5	0.859624		

Thus the root is 0.8596.

A–2. Use the Newton-Raphson method to derive the iterative formula

$$x_{n+1} = \frac{1}{2}\left(x_n + \frac{A}{x_n}\right)$$

for the value of $\sqrt{A}$. This formula was discovered by a Babylonian mathematician more than 2000 years ago. Use this formula to evaluate $\sqrt{2}$ to five significant figures.

$x^2 = A$, so $x^2 - A = 0 = f(x)$ and $2x = f'(x)$. From the Newton-Raphson equation,

$$\begin{aligned} x_{n+1} &= x_n - \frac{f(x_n)}{f'(x_n)} \\ &= x_n - \frac{(x_n^2 - A)}{2x_n} = \frac{2x_n^2 - x_n^2 + A}{2x_n} \\ &= \frac{1}{2}\left(x_n + \frac{A}{x_n}\right) \end{aligned}$$

We know that $\sqrt{2}$ is between 1 and 2, so we can take x_0 to be 1.5. In three iterations, we find that $\sqrt{2} = 1.4142$ to five significant figures.

A–3. Use the Newton-Raphson method to solve the equation $e^{-x} + (x/5) = 1$ to four significant figures.

$$f(x) = e^{-x} + \frac{x}{5} - 1 = 0$$

$$f'(x) = -e^{-x} + \frac{1}{5}$$

We select 5 as x_0 because $(5/5) - 1 = 0$ and e^{-5} is a small number. Using the spreadsheet, we find

n	x_n	$f(x_n)$	$f'(x_n)$
0	5.00000	6.73795×10^{-2}	0.19326
1	4.96514	4.143×10^{-6}	0.19302
2	4.96511		

Thus the solution to the equation is 4.965 (to four significant figures).

A–4. Consider the chemical reaction described by the equation

$$CH_4(g) + H_2O(g) \rightleftharpoons CO(g) + 3\,H_2(g)$$

at 300 K. If 1.00 atm of $CH_4(g)$ and $H_2O(g)$ are introduced into a reaction vessel, the pressures at equilibrium obey the equation

$$\frac{P_{CO}P_{H_2}^3}{P_{CH_4}P_{H_2O}} = \frac{(x)(3x)^3}{(1-x)(1-x)} = 26$$

Solve this equation for x.

$$\frac{27x^4}{1 - 2x + x^2} = 26$$

$$27x^4 = 26 - 52x + 26x^2$$

The functions we will use in the spreadsheet created for Problem A–1 are then

$$f(x) = 0 = 27x^4 - 26x^2 + 52x - 26$$
$$f'(x) = 108x^3 - 52x + 52$$

We know that x must be between 0 and 1, so we can take x_0 to be 0.5. Then

n	x_n	$f(x_n)$	$f'(x_n)$
0	0.50000	−4.81250	39.50000
1	0.62184	0.31884	45.63326
2	0.61485	0.00177	45.13101
3	0.61481		

To three significant figures, $x = 0.615$ atm.

A–5. In Chapter 2, we will solve the cubic equation

$$64x^3 + 6x^2 + 12x - 1 = 0$$

Use the Newton-Raphson method to find the only real root of this equation to five significant figures.

$$f(x) = 64x^3 + 6x^2 + 12x - 1 = 0$$
$$f'(x) = 192x^2 + 12x + 12$$

The solution must be small ($-1 \leq x \leq 1$), so let us take $x_0 = -0.5$. Then

n	x_n	$f(x_n)$	$f'(x_n)$
0	−0.50000	−13.5000	54.00000
1	−0.25000	−4.62500	21.00000
2	−0.02976	−1.35352	11.81293
3	0.084817	0.10002	14.39905
4	0.0778708	1.0539×10^{-3}	14.09871
5	0.0777961	1.17×10^{-7}	14.09558
6	0.0777961		

So $x = 7.7780 \times 10^{-2}$.

A–6. Solve the equation $x^3 - 3x + 1 = 0$ for all three of its roots to four decimal places.

$$f(x) = x^3 - 3x + 1$$
$$f'(x) = 3x^2 - 3$$

Setting $f'(x)$ equal to zero, we find that the inflection points of the equation $x^3 - 3x + 1$ are 1 and -1. We can therefore set our x_0's to 0, -1.5, and 1.5.

n	x_n	$f(x_n)$	$f'(x_n)$
0	0.00000	1.00000	−3.00000
1	0.333333	0.03704	−2.66667
2	0.347222	1.956×10^{-4}	−2.63831
3	0.347296	6×10^{-9}	−2.63816
4	0.347296		

n	x_n	$f(x_n)$	$f'(x_n)$
0	1.50000	−0.12500	3.75000
1	1.53333	0.005037	4.05333
2	1.53209	7.102×10^{-6}	4.04191
3	1.53209		

n	x_n	$f(x_n)$	$f'(x_n)$
0	−1.50000	2.12500	3.75000
1	−2.06667	−1.62696	9.81333
2	−1.90088	−0.16586	7.83998
3	−1.87972	-2.5428×10^{-3}	7.60004
4	−1.87939	-6.312×10^{-7}	7.59627
5	−1.87939		

The three roots of $x^3 - 3x + 1$ are 0.3473, 1.532, and -1.879.

A–7. In Example 2–3 we will solve the cubic equation

$$\overline{V}^3 - 0.1231\overline{V}^2 + 0.02056\overline{V} - 0.001271 = 0$$

Use the Newton-Raphson method to find the root to this equation that is near $\overline{V} = 0.1$.

Let $\overline{V} = x$:

$$f(x) = x^3 - 0.1231x^2 + 0.02056x - 0.001271$$

$$f'(x) = 3x^2 - 0.2462x + 0.02056$$

We can take $x_0 = 0.120$. Then

n	x_n	$f(x_n)$	$f'(x_n)$
0	0.120	1.1516×10^{-3}	3.4216×10^{-2}
1	0.086344	2.3021×10^{-4}	2.1668×10^{-2}
2	0.075720	1.4145×10^{-5}	1.9118×10^{-2}
3	0.074980	5.6559×10^{-8}	1.8966×10^{-2}
4	0.074977	9.0564×10^{-13}	1.8965×10^{-2}
5	0.074977		

So the root to the equation near $\overline{V} = 0.120$ is $\overline{V} = 0.0750$.

A–8. In Section 2–3 we will solve the cubic equation

$$\overline{V}^3 - 0.3664\overline{V}^2 + 0.03802\overline{V} - 0.001210 = 0$$

Use the Newton-Raphson method to show that the three roots to this equation are 0.07073, 0.07897, and 0.2167.

Let $\overline{V} = x$:

$$f(x) = x^3 - 0.3664x^2 + 0.03802x - 0.001210$$

$$f'(x) = 3x^2 - 0.7328x + 0.03802$$

We can set our x_0's to 0.069, 0.080, and 0.20, conveniently close to those given in the problem text. Then

n	x_n	$f(x_n)$	$f'(x_n)$
0	0.069	-2.5414×10^{-6}	1.7398×10^{-3}
1	0.070461	-3.3701×10^{-7}	1.2805×10^{-3}
2	0.070724	-1.0719×10^{-8}	1.1991×10^{-3}
3	0.070733	-1.2323×10^{-11}	1.1964×10^{-3}
4	0.070733		

n	x_n	$f(x_n)$	$f'(x_n)$
0	0.080	-1.36×10^{-6}	-1.4040×10^{-3}
1	0.079031	-1.1951×10^{-7}	-1.1563×10^{-3}
2	0.078928	-1.3824×10^{-9}	-1.1295×10^{-3}
3	0.078927	-1.9414×10^{-13}	-1.1292×10^{-3}
4	0.078927		

n	x_n	$f(x_n)$	$f'(x_n)$
0	0.20	-2.62×10^{-4}	1.1460×10^{-2}
1	0.22286	1.3405×10^{-4}	2.3709×10^{-2}
2	0.21721	9.4788×10^{-6}	2.0388×10^{-2}
3	0.21674	6.1550×10^{-8}	2.0124×10^{-2}
4	0.21674		

A–9. Use the trapezoidal approximation and Simpson's rule to evaluate

$$I = \int_0^1 \frac{dx}{1+x^2}$$

This integral can be evaluated analytically; it is given by $\tan^{-1}(1)$, which is equal to $\pi/4$, so $I = 0.78539816$ to eight decimal places.

Set up a new spreadsheet which will use the trapezoidal approximation, and another for Simpson's rule. For Simpson's rule, your coefficients of the functions vary according to whether the variable's subscript is even or odd. I set up two columns: one for $f(x_1), f(x_3), f(x_5), \ldots$ and one for $f(x_2), f(x_4), f(x_6), \ldots$. One can then calculate $f(x_0)$ and $f(x_{2n})$ elsewhere (I used cells above my two columns) and creating an equation for the approximation to the integral then becomes trivial.

The spreadsheet for the trapezoidal approximation is much the same. This spreadsheet can be used for the remainder of the problems.

n	h	I_n(trapezoidal)	I_{2n}(Simpson's rule)
10	0.1	0.7849814972	0.7853981632
50	0.02	0.7853814967	0.7853981634
100	0.01	0.7853939967	0.7853981634

A–10. Evaluate ln 2 to six decimal places by evaluating

$$\ln 2 = \int_1^2 \frac{dx}{x}$$

What must n be to assure six-digit accuracy?

To find n to six-digit accuracy, the error must be no greater than 1×10^{-6}. Then

$$E = \frac{M(b-a)h^4}{180}$$

where **M** is the maximum value of $f^{\mathrm{IV}}(x)$. Differentiate $f(x)$ to find

$$f^{\mathrm{IV}}(x) = \frac{24}{x^5}$$

The maximum value of $f^{\mathrm{IV}}(x)$ in the interval from 1 to 2 is thus 24. Then

$$1 \times 10^{-6} = \frac{24(2-1)h^4}{180}$$

$$h = 0.0523$$

Recall $h = (b-a)/2n$. Thus $n = (b-a)/2h = 1/0.105 = 9.6$, and the smallest value of n needed is 10. We use this value of n to find $\ln 2 = 0.693147$.

A–11. Use Simpson's rule to evaluate

$$I = \int_0^\infty e^{-x^2} dx$$

and compare your result with the exact value, $\sqrt{\pi}/2$.

Use the spreadsheet created for Problem A–10 to find

$$\int_0^\infty e^{-x^2} dx = 0.886\,2269$$

We can use large values of n to find this value to greater accuracy, if need be.

A–12. Use Simpson's rule to evaluate

$$I = \int_0^{\infty} \frac{x^3 dx}{e^x - 1}$$

to six decimal places. The exact value is $\pi^4/15$.

Use the spreadsheet created for Problem A–9 to find $I = 6.493\,94$.

A–13. Use a numerical software package such as *MathCad*, *Kaleidagraph*, or *Mathematica* to evaluate the integral

$$S = 4\pi^{1/2}\left(\frac{2\alpha}{\pi}\right)^{3/4} \int_0^{\infty} r^2 e^{-r} e^{-\alpha r^2} dr$$

for values of α between 0.200 and 0.300 and show that S has a maximum value at $\alpha = 0.271$.

Here, we use values of α calculated by *Mathematica* at intervals of 0.005 and plot S against α.

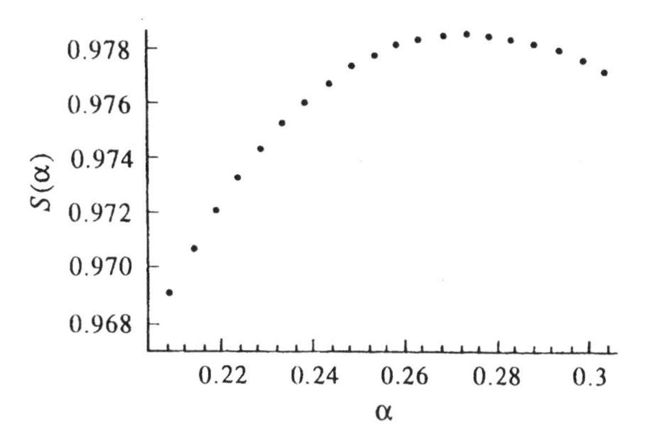

The maximum value of S is around $\alpha = 0.27$. The values of $S(\alpha)$ close to $\alpha = 0.271$ are given below.

α	0.2700	0.2705	0.2710	0.2715	0.2720
$S(\alpha)$	0.978 4029	0.978 4041	0.978 4044	0.978 4039	0.978 4024

CHAPTER 2

The Properties of Gases

PROBLEMS AND SOLUTIONS

2–1. In an issue of the journal *Science* a few years ago, a research group discussed experiments in which they determined the structure of cesium iodide crystals at a pressure of 302 gigapascals (GPa). How many atmospheres and bars is this pressure?

2.98×10^6 atm, 3.02×10^6 bar

2–2. In meteorology, pressures are expressed in units of millibars (mbar). Convert 985 mbar to torr and to atmospheres.

---

739 torr, 0.972 atm

2–3. Calculate the value of the pressure (in atm) exerted by a 33.9-foot column of water. Take the density of water to be $1.00 \text{ g}\cdot\text{mL}^{-1}$.

We first convert the height of the column to metric units: 33.9 ft = 10.33 m. Now

$$\begin{aligned} P = \rho g h &= (1.00 \text{ kg}\cdot\text{dm}^{-3})(98.067 \text{ dm}\cdot\text{s}^{-2})(103.3 \text{ dm}) \\ &= 1.013 \times 10^4 \text{ kg}\cdot\text{dm}^{-1}\cdot\text{s}^{-2} \\ &= 1.013 \times 10^5 \text{ Pa} = 1.00 \text{ atm} \end{aligned}$$

2–4. At which temperature are the Celsius and Farenheit temperature scales equal?

$-40°$

2–5. A travel guide says that to convert Celsius temperatures to Farenheit temperatures, double the Celsius temperature and add 30. Comment on this recipe.

This will provide a rough estimate of the temperature, decreasing in accuracy as temperature increases. (Of course, it is not valid for Celsius temperatures below zero degrees.) At room temperatures, it is accurate enough for ordinary purposes.

Actual T (°C)	Actual T (°F)	Travel T (°F)
0	32	30
10	50	50
20	68	70
30	86	90
40	104	110

2–6. Research in surface science is carried out using ultra-high vacuum chambers that can sustain pressures as low as 10^{-12} torr. How many molecules are there in a 1.00-cm^3 volume inside such an apparatus at 298 K? What is the corresponding molar volume $\overline{V}$ at this pressure and temperature?

We will assume ideal gas behavior, so

$$\frac{PV}{RT} = n \tag{2.1a}$$

$$\frac{(10^{-12}\text{ torr})(1.00\text{ cm}^3)}{(82.058\text{ cm}^3\cdot\text{atm}\cdot\text{mol}^{-1}\cdot\text{K}^{-1})(760\text{ torr}\cdot\text{atm}^{-1})(298\text{ K})} = n$$

$$5.38\times10^{-20}\text{ mol} = n$$

so there are 3.24×10^4 molecules in the apparatus. The molar volume is

$$\overline{V} = \frac{V}{n} = \frac{1.00\text{ cm}^3}{5.38\times10^{-20}\text{ mol}} = 1.86\times10^{19}\text{ cm}^3\cdot\text{mol}^{-1}$$

2–7. Use the following data for an unknown gas at 300 K to determine the molecular mass of the gas.

P/bar	0.1000	0.5000	1.000	1.01325	2.000
ρ/g·L^{-1}	0.1771	0.8909	1.796	1.820	3.652

The line of best fit of a plot of P/ρ versus ρ will have an intercept of RT/M. Plotting, we find that the intercept of this plot is 0.56558 bar·g^{-1}·dm^3, and so $M = 44.10$ g·mol^{-1}.

2–8. Recall from general chemistry that Dalton's law of partial pressures says that each gas in a mixture of ideal gases acts as if the other gases were not present. Use this fact to show that the partial pressure exerted by each gas is given by

$$P_j = \left(\frac{n_j}{\sum n_j}\right) P_{\text{total}} = y_j P_{\text{total}}$$

where P_j is the partial pressure of the jth gas and y_j is its mole fraction.

The ideal gas law (Equation 2.1) gives

$$P_{\text{total}}V = n_{\text{total}}RT = \sum_j n_j RT$$

and

$$P_j V = n_j RT$$

for all component gases j. Solving each expression for RT/V and equating the results gives

$$\frac{P_{\text{total}}}{\sum_j n_j} = \frac{P_j}{n_j}$$

or

$$P_j = \frac{n_j}{\sum_j n_j} P_{\text{total}} = y_j P_{\text{total}}$$

2–9. A mixture of H_2(g) and N_2(g) has a density of $0.216\ \text{g}\cdot\text{L}^{-1}$ at 300 K and 500 torr. What is the mole fraction composition of the mixture?

The density of the mixture is $0.216\ \text{g}\cdot\text{L}^{-1}$, so there are 216 g of gas present in one m^3 of gas. Take the total volume of the mixture to be $1\ \text{m}^3$. Then, using the ideal gas law (Equation 2.1), we find

$$P_{\text{tot}}V = n_{\text{tot}}RT$$

$$500\ \text{torr}\left(\frac{101\ 325\ \text{Pa}}{760\ \text{torr}}\right) 1\ \text{m}^3 = n_{\text{tot}}\left(8.3145\ \text{J}\cdot\text{mol}^{-1}\cdot\text{K}^{-1}\right)(300\ \text{K})$$

$$26.7\ \text{mol} = n_{\text{tot}}$$

There are 26.7 mol of gas per cubic meter. Let x be the number of moles of hydrogen gas. Then $n_{\text{tot}} - x$ is the number of moles of nitrogen gas. Since $M_{\text{H}_2} = 2.01588\ \text{g}\cdot\text{mol}^{-1}$ and $M_{\text{N}_2} = 28.01348\ \text{g}\cdot\text{mol}^{-1}$, we can write

$$216\ \text{g} = \left(28.01348\ \text{g}\cdot\text{mol}^{-1}\right)(26.7\ \text{mol} - x\ \text{mol}) + \left(2.01588\ \text{g}\cdot\text{mol}^{-1}\right)(x\ \text{mol})$$

$$26x = 532.6\ \text{g}$$

$$x = 20.5\ \text{g}$$

The mole fractions of each component of the mixture are therefore

$$y_{\text{H}_2} = \frac{n_{\text{H}_2}}{n_{\text{tot}}} = \frac{20.5\ \text{mol}}{26.7\ \text{mol}} = 0.77$$

and

$$y_{\text{N}_2} = \frac{n_{\text{N}_2}}{n_{\text{tot}}} = \frac{6.2\ \text{mol}}{26.7\ \text{mol}} = 0.23$$

2–10. One liter of N_2(g) at 2.1 bar and two liters of Ar(g) at 3.4 bar are mixed in a 4.0-L flask to form an ideal-gas mixture. Calculate the value of the final pressure of the mixture if the initial and final temperature of the gases are the same. Repeat this calculation if the initial temperatures of the N_2(g) and Ar(g) are 304 K and 402 K, respectively, and the final temperature of the mixture is 377 K. (Assume ideal-gas behavior.)

---

a. Initially, we have one liter of N_2 at 2.1 bar and two liters of Ar at 3.4 bar. We can use the ideal gas law (Equation 2.1) to find the number of moles of each gas:

$$n_{N_2} = \frac{P_{N_2}V_{N_2}}{RT} = \frac{(2.1 \times 10^5 \text{ Pa})(1 \times 10^{-3} \text{ m}^3)}{RT} = \frac{210 \text{ Pa}\cdot\text{m}^3}{RT}$$

$$n_{Ar} = \frac{P_{Ar}V_{Ar}}{RT} = \frac{(3.4 \times 10^5 \text{ Pa})(2 \times 10^{-3} \text{ m}^3)}{RT} = \frac{680 \text{ Pa}\cdot\text{m}^3}{RT}$$

The total moles of gas in the final mixture is the sum of the moles of each gas in the mixture, which is $(890 \text{ Pa}\cdot\text{m}^3)/RT$. So (Equation 2.1)

$$P = \frac{nRT}{V} = \frac{890 \text{ Pa}\cdot\text{m}^3}{0.0040 \text{ m}^3} = 2.2 \times 10^5 \text{ Pa} = 2.2 \text{ bar}$$

b. Here, the initial temperatures of N_2 and Ar are different from each other and from the temperature of the final mixture. From above,

$$n_{\text{total}} = n_{N_2} + n_{Ar} = \frac{210 \text{ Pa}\cdot\text{m}^3}{R(304 \text{ K})} + \frac{680 \text{ Pa}\cdot\text{m}^3}{R(402 \text{ K})}$$

Substituting into the ideal gas law (Equation 2.1),

$$P = \left[\frac{210 \text{ Pa}\cdot\text{m}^3}{R(304 \text{ K})} + \frac{680 \text{ Pa}\cdot\text{m}^3}{R(402 \text{ K})}\right]\left[\frac{R(377 \text{ K})}{0.0040 \text{ m}^3}\right] = 2.2 \times 10^5 \text{ Pa} = 2.2 \text{ bar}$$

2–11. It takes 0.3625 g of nitrogen to fill a glass container at 298.2 K and 0.0100 bar pressure. It takes 0.9175 g of an unknown homonuclear diatomic gas to fill the same bulb under the same conditions. What is this gas?

The number of moles of each gas must be the same, because P, V, and T are held constant. The number of moles of nitrogen is

$$n_{N_2} = \frac{0.3625 \text{ g}}{28.0135 \text{ g}\cdot\text{mol}^{-1}} = 1.294 \times 10^{-2} \text{ mol}$$

The molar mass of the unknown compound must be

$$M = \frac{0.9175 \text{ g}}{1.294 \times 10^{-2} \text{ mol}} = 70.903 \text{ g}\cdot\text{mol}^{-1}$$

The homonuclear diatomic gas must be chlorine (Cl_2).

2–12. Calculate the value of the molar gas constant in units of $\text{dm}^3\cdot\text{torr}\cdot\text{K}^{-1}\cdot\text{mol}^{-1}$.

$$\begin{aligned} R &= 8.31451 \text{ J}\cdot\text{mol}^{-1}\cdot\text{K}^{-1} \\ &= (8.31451 \text{ Pa}\cdot\text{m}^3\cdot\text{mol}^{-1}\cdot\text{K}^{-1})\left(\frac{10 \text{ dm}}{1 \text{ m}}\right)^3\left(\frac{760 \text{ torr}}{1.01325 \times 10^5 \text{ Pa}}\right) \\ &= 62.3639 \text{ dm}^3\cdot\text{torr}\cdot\text{K}^{-1}\cdot\text{mol}^{-1} \end{aligned}$$

2–13. Use the van der Waals equation to plot the compressibility factor, Z, against P for methane for $T = 180$ K, 189 K, 190 K, 200 K, and 250 K. *Hint*: Calculate Z as a function of $\overline{V}$ and P as a function of $\overline{V}$, and then plot Z versus P.

For methane, $a = 2.3026\ \text{dm}^6\cdot\text{bar}\cdot\text{mol}^{-2}$ and $b = 0.043067\ \text{dm}^3\cdot\text{mol}^{-1}$. By definition,

$$Z = \frac{P\overline{V}}{RT}$$

and the van der Waals equation of state is (Equation 2.5)

$$P = \frac{RT}{\overline{V} - b} - \frac{a}{\overline{V}^2}$$

We can create a parametric plot of Z versus P for the suggested temperatures, shown below. Note that the effect of molecular attraction becomes less important at higher temperatures, as observed in the legend of Figure 2.4.

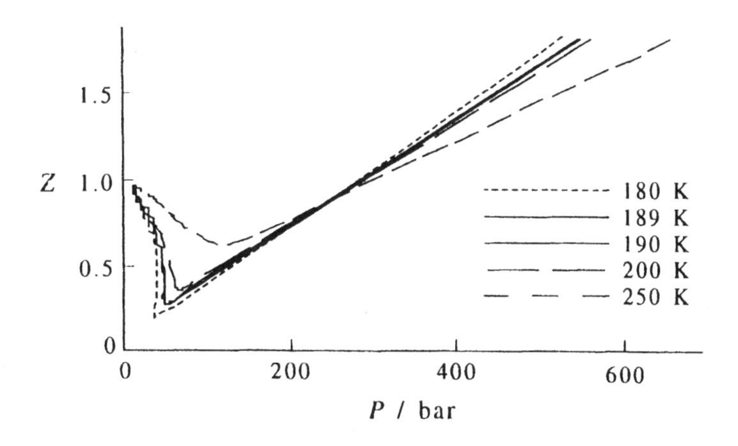

2–14. Use the Redlich-Kwong equation to plot the compressibility factor, Z, against P for methane for $T = 180$ K, 189 K, 190 K, 200 K, and 250 K. *Hint*: Calculate Z as a function of $\overline{V}$ and P as a function of $\overline{V}$, and then plot Z versus P.

For methane, $A = 32.205\ \text{dm}^6\cdot\text{bar}\cdot\text{mol}^{-2}\cdot\text{K}^{1/2}$ and $B = 0.029850\ \text{dm}^3\cdot\text{mol}^{-1}$. By definition,

$$Z = \frac{P\overline{V}}{RT}$$

and the Redlich-Kwong equation of state is (Equation 2.7)

$$P = \frac{RT}{\overline{V} - B} - \frac{A}{T^{1/2}\overline{V}(\overline{V} + B)}$$

We can create a parametric plot of Z versus P for the suggested temperatures, shown below. Note that the effect of molecular attraction becomes less important at higher temperatures, as observed in the legend of Figure 2.4.

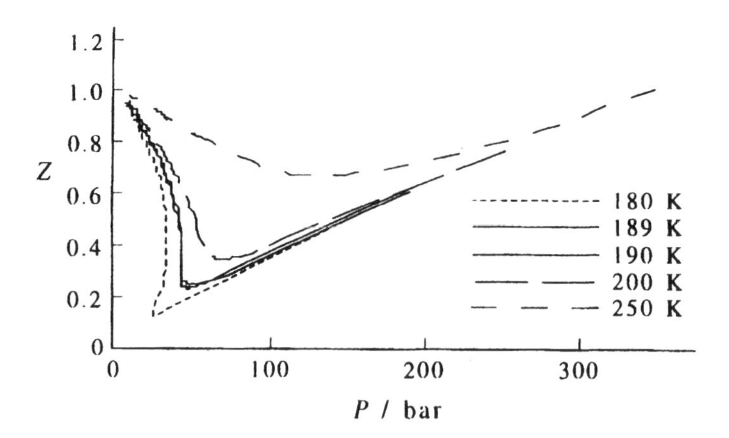

2–15. Use both the van der Waals and the Redlich-Kwong equations to calculate the molar volume of CO at 200 K and 1000 bar. Compare your result to the result you would get using the ideal-gas equation. The experimental value is $0.04009\ \text{L}\cdot\text{mol}^{-1}$.

We can use the Newton-Raphson method (MathChapter A) to solve these cubic equations of state. We can express $f(\overline{V})$ for the van der Waals equation as (Example 2–2)

$$f(\overline{V}) = \overline{V}^3 - \left(b + \frac{RT}{P}\right)\overline{V}^2 + \frac{a}{P}\overline{V} - \frac{ab}{P}$$

and $f'(\overline{V})$ as

$$f'(\overline{V}) = 3\overline{V}^2 - 2\left(b + \frac{RT}{P}\right)\overline{V} + \frac{a}{P}$$

For CO, $a = 1.4734\ \text{dm}^6\cdot\text{bar}\cdot\text{mol}^{-2}$ and $b = 0.039523\ \text{dm}^3\cdot\text{mol}^{-1}$ (Table 2.3). Then, using the Newton-Raphson method, we find that the van der Waals equation gives a result of $\overline{V} = 0.04998\ \text{dm}^3\cdot\text{mol}^{-1}$. Likewise, we can express $f(\overline{V})$ for the Redlich-Kwong equation as (Equation 2.9)

$$f(\overline{V}) = \overline{V}^3 - \frac{RT}{P}\overline{V}^2 - \left(B^2 + \frac{BRT}{P} - \frac{A}{T^{1/2}P}\right)\overline{V} - \frac{AB}{T^{1/2}P}$$

and $f'(\overline{V})$ as

$$f'(\overline{V}) = 3\overline{V}^2 - \frac{2RT}{P}\overline{V} - \left(B^2 + \frac{BRT}{P} - \frac{A}{T^{1/2}P}\right)$$

For CO, $A = 17.208\ \text{dm}^6\cdot\text{bar}\cdot\text{mol}^{-2}\cdot\text{K}^{1/2}$ and $B = 0.027394\ \text{dm}^3\cdot\text{mol}^{-1}$ (Table 2.4). Applying the Newton-Raphson method, we find that the Redlich-Kwong equation gives a result of $\overline{V} = 0.03866\ \text{dm}^3\cdot\text{mol}^{-1}$. Finally, the ideal gas equation gives (Equation 2.1)

$$\overline{V} = \frac{RT}{P} = \frac{(0.083145\ \text{dm}^3\cdot\text{bar}\cdot\text{mol}^{-1}\cdot\text{K}^{-1})(200\ \text{K})}{1000\ \text{bar}} = 0.01663\ \text{dm}^3\cdot\text{mol}^{-1}$$

The experimental value of $0.04009\ \text{dm}^3\cdot\text{mol}^{-1}$ is closest to the result given by the Redlich-Kwong equation (the two values differ by about 3%).

2–16. Compare the pressures given by (a) the ideal-gas equation, (b) the van der Waals equation, (c) the Redlich-Kwong equation, and (d) the Peng-Robinson equation for propane at 400 K and $\rho = 10.62\ \text{mol}\cdot\text{dm}^{-3}$. The experimental value is 400 bar. Take $\alpha = 9.6938\ \text{L}^2\cdot\text{mol}^{-2}$ and $\beta = 0.05632\ \text{L}\cdot\text{mol}^{-1}$ for the Peng-Robinson equation.

The molar volume corresponding to a density of 10.62 mol·dm^{-3} is 0.09416 dm^3·mol^{-1}.

a. The ideal gas equation gives a pressure of (Equation 2.1)

$$P = \frac{RT}{\overline{V}} = \frac{(0.083145\ \text{dm}^3\cdot\text{bar}\cdot\text{mol}^{-1}\cdot\text{K}^{-1})(400\ \text{K})}{0.09416\ \text{dm}^3\cdot\text{mol}^{-1}} = 353.2\ \text{bar}$$

b. The van der Waals equation gives a pressure of (Equation 2.5)

$$P = \frac{RT}{\overline{V} - b} - \frac{a}{\overline{V}^2}$$

For propane, $a = 9.3919$ dm^6·bar·mol^{-2} and $b = 0.090494$ dm^3·mol^{-1} (Table 2.3). Then

$$P = \frac{(0.083145\ \text{dm}^3\cdot\text{bar}\cdot\text{mol}^{-1}\cdot\text{K}^{-1})(400\ \text{K})}{0.09416\ \text{dm}^3\cdot\text{mol}^{-1} - 0.090494\ \text{dm}^3\cdot\text{mol}^{-1}} - \frac{9.3919\ \text{dm}^6\cdot\text{bar}\cdot\text{mol}^{-2}}{(0.09416\ \text{dm}^3\cdot\text{mol}^{-1})^2}$$
$$= 8008\ \text{bar}$$

c. The Redlich-Kwong equation gives a pressure of (Equation 2.7)

$$P = \frac{RT}{\overline{V} - B} - \frac{A}{T^{1/2}\overline{V}(\overline{V} + B)}$$

For propane, $A = 183.02$ dm^6·bar·mol^{-2}·K$^{1/2}$ and $B = 0.062723$ dm^3·mol^{-1} (Table 2.4). Then

$$P = \frac{(0.083145\ \text{dm}^3\cdot\text{bar}\cdot\text{mol}^{-1}\cdot^{-1})(400\ \text{K})}{0.09416\ \text{dm}^3\cdot\text{mol}^{-1} - 0.062723\ \text{dm}^3\cdot\text{mol}^{-1}}$$
$$- \frac{183.02\ \text{dm}^6\cdot\text{bar}\cdot\text{mol}^{-2}\cdot\text{K}^{1/2}}{(400\ \text{K})^{1/2}(0.09416\ \text{dm}^3\cdot\text{mol}^{-1})(0.09416\ \text{dm}^3\cdot\text{mol}^{-1} + 0.062723\ \text{dm}^3\cdot\text{mol}^{-1})}$$
$$= 438.4\ \text{bar}$$

d. The Peng-Robinson equation gives a pressure of (Equation 2.8)

$$P = \frac{RT}{\overline{V} - \beta} - \frac{\alpha}{\overline{V}(\overline{V} + \beta) + \beta(\overline{V} - \beta)}$$

For propane, $\alpha = 9.6938$ dm^6·bar·mol^{-2} and $\beta = 0.05632$ dm^3·mol^{-1}. Then

$$P = \frac{(0.083145\ \text{dm}^3\cdot\text{bar}\cdot\text{mol}^{-1}\cdot^{-1})(400\ \text{K})}{0.09416\ \text{dm}^3\cdot\text{mol}^{-1} - 0.05632\ \text{dm}^3\cdot\text{mol}^{-1}}$$
$$- \frac{9.6938\ \text{dm}^6\cdot\text{bar}\cdot\text{mol}^{-2}}{(0.09416)(0.09416 + 0.05632)\ \text{dm}^6\cdot\text{mol}^{-2} + (0.05632)(0.09416 - 0.05632)\ \text{dm}^6\cdot\text{mol}^{-2}}$$
$$= 284.2\ \text{bar}$$

The Redlich-Kwong equation of state gives a pressure closest to the experimentally observed pressure (the two values differ by about 10%).

2–17. Use the van der Waals equation and the Redlich-Kwong equation to calculate the value of the pressure of one mole of ethane at 400.0 K confined to a volume of 83.26 cm^3. The experimental value is 400 bar.

Here, the molar volume of ethane is 0.08326 dm^3·mol^{-1}.

a. The van der Waals equation gives a pressure of (Equation 2.5)

$$P = \frac{RT}{\overline{V} - b} - \frac{a}{\overline{V}^2}$$

For ethane, $a = 5.5818 \text{ dm}^6\cdot\text{bar}\cdot\text{mol}^{-2}$ and $b = 0.065144 \text{ dm}^3\cdot\text{mol}^{-1}$ (Table 2.3). Then

$$\begin{aligned} P &= \frac{(0.083145 \text{ dm}^3\cdot\text{bar}\cdot\text{mol}^{-1}\cdot\text{K}^{-1})(400 \text{ K})}{0.08326 \text{ dm}^3\cdot\text{mol}^{-1} - 0.065144 \text{ dm}^3\cdot\text{mol}^{-1}} - \frac{5.5818 \text{ dm}^6\cdot\text{bar}\cdot\text{mol}^{-2}}{(0.08326 \text{ dm}^3\cdot\text{mol}^{-1})^2} \\ &= 1031 \text{ bar} \end{aligned}$$

b. The Redlich-Kwong equation gives a pressure of (Equation 2.7)

$$P = \frac{RT}{\overline{V} - B} - \frac{A}{T^{1/2}\overline{V}(\overline{V} + B)}$$

For ethane, $A = 98.831 \text{ dm}^6\cdot\text{bar}\cdot\text{mol}^{-2}\cdot\text{K}^{1/2}$ and $B = 0.045153 \text{ dm}^3\cdot\text{mol}^{-1}$ (Table 2.4). Then

$$\begin{aligned} P &= \frac{(0.083145 \text{ dm}^3\cdot\text{bar}\cdot\text{mol}^{-1}\cdot\text{K}^{-1})(400 \text{ K})}{0.08326 \text{ dm}^3\cdot\text{mol}^{-1} - 0.045153 \text{ dm}^3\cdot\text{mol}^{-1}} \\ &\quad - \frac{98.831 \text{ dm}^6\cdot\text{bar}\cdot\text{mol}^{-2}\cdot\text{K}^{1/2}}{(400 \text{ K})^{1/2}(0.08326 \text{ dm}^3\cdot\text{mol}^{-1})(0.08326 + 0.045153) \text{ dm}^3\cdot\text{mol}^{-1}} \\ &= 410.6 \text{ bar} \end{aligned}$$

The value of P found using the Redlich-Kwong equation of state is the closest to the experimentally observed value (the two values differ by about 3%).

2–18. Use the van der Waals equation and the Redlich-Kwong equation to calculate the molar density of one mole of methane at 500 K and 500 bar. The experimental value is $10.06 \text{ mol}\cdot\text{L}^{-1}$.

We can use the Newton-Raphson method (MathChapter A) to solve the cubic equations of state for $\overline{V}$, and take the reciprocal to find the molar density. We use the experimentally observed molar volume of $0.09940 \text{ dm}^3\cdot\text{mol}^{-1}$ as the starting point for the iteration. We can express $f(\overline{V})$ for the van der Waals equation as (Example 2–2)

$$f(\overline{V}) = \overline{V}^3 - \left(b + \frac{RT}{P}\right)\overline{V}^2 + \frac{a}{P}\overline{V} - \frac{ab}{P}$$

and $f'(\overline{V})$ as

$$f'(\overline{V}) = 3\overline{V}^2 - 2\left(b + \frac{RT}{P}\right)\overline{V} + \frac{a}{P}$$

For methane, $a = 2.3026 \text{ dm}^6\cdot\text{bar}\cdot\text{mol}^{-2}$ and $b = 0.043067 \text{ dm}^3\cdot\text{mol}^{-1}$ (Table 2.3). Then (using the Newton-Raphson method) we find that the van der Waals equation gives a result of $\overline{V} = 0.09993 \text{ dm}^3\cdot\text{mol}^{-1}$, which corresponds to a molar density of $10.01 \text{ mol}\cdot\text{dm}^{-3}$. Likewise, we can express $f(\overline{V})$ for the Redlich-Kwong equation as (Equation 2.9)

$$f(\overline{V}) = \overline{V}^3 - \frac{RT}{P}\overline{V}^2 - \left(B^2 + \frac{BRT}{P} - \frac{A}{T^{1/2}P}\right)\overline{V} - \frac{AB}{T^{1/2}P}$$

and $f'(\overline{V})$ as

$$f'(\overline{V}) = 3\overline{V}^2 - \frac{2RT}{P}\overline{V} - \left(B^2 + \frac{BRT}{P} - \frac{A}{T^{1/2}P}\right)$$

For methane, $A = 32.205\ \text{dm}^6\cdot\text{bar}\cdot\text{mol}^{-2}\cdot\text{K}^{1/2}$ and $B = 0.029850\ \text{dm}^3\cdot\text{mol}^{-1}$ (Table 2.4). Then (using the Newton-Raphson method) we find that the Redlich-Kwong equation gives a result of $\overline{V} = 0.09729\ \text{dm}^3\cdot\text{mol}^{-1}$, which corresponds to a molar density of $10.28\ \text{mol}\cdot\text{dm}^{-3}$. The molar density of methane found using the van der Waals equation of state is within 0.5% of the experimentally observed value.

2–19. Use the Redlich-Kwong equation to calculate the pressure of methane at 200 K and a density of $27.41\ \text{mol}\cdot\text{L}^{-1}$. The experimental value is 1600 bar. What does the van der Waals equation give?

The molar volume of the methane is $0.03648\ \text{dm}^3\cdot\text{mol}^{-1}$.

a. The van der Waals equation gives a pressure of (Equation 2.5)

$$P = \frac{RT}{\overline{V} - b} - \frac{a}{\overline{V}^2}$$

For methane, $a = 2.3026\ \text{dm}^6\cdot\text{bar}\cdot\text{mol}^{-2}$ and $b = 0.043067\ \text{dm}^3\cdot\text{mol}^{-1}$ (Table 2.3). Then

$$\begin{aligned} P &= \frac{(0.083145\ \text{dm}^3\cdot\text{bar}\cdot\text{mol}^{-1}\cdot\text{K}^{-1})(200\ \text{K})}{0.03648\ \text{dm}^3\cdot\text{mol}^{-1} - 0.043067\ \text{dm}^3\cdot\text{mol}^{-1}} - \frac{2.3026\ \text{dm}^6\cdot\text{bar}\cdot\text{mol}^{-2}}{(0.03648\ \text{dm}^3\cdot\text{mol}^{-1})^2} \\ &= -4250\ \text{bar} \end{aligned}$$

b. The Redlich-Kwong equation gives a pressure of (Equation 2.7)

$$P = \frac{RT}{\overline{V} - B} - \frac{A}{T^{1/2}\overline{V}(\overline{V} + B)}$$

For ethane, $A = 32.205\ \text{dm}^6\cdot\text{bar}\cdot\text{mol}^{-2}\cdot\text{K}^{1/2}$ and $B = 0.029850\ \text{dm}^3\cdot\text{mol}^{-1}$ (Table 2.4). Then

$$\begin{aligned} P &= \frac{(0.083145\ \text{dm}^3\cdot\text{bar}\cdot\text{mol}^{-1}\cdot{}^{-1})(200\ \text{K})}{0.03648\ \text{dm}^3\cdot\text{mol}^{-1} - 0.029850\ \text{dm}^3\cdot\text{mol}^{-1}} \\ &\quad - \frac{32.205\ \text{dm}^6\cdot\text{bar}\cdot\text{mol}^{-2}\cdot\text{K}^{1/2}}{(200\ \text{K})^{1/2}(0.03648\ \text{dm}^3\cdot\text{mol}^{-1})(0.03648\ \text{dm}^3\cdot\text{mol}^{-1} + 0.029850\ \text{dm}^3\cdot\text{mol}^{-1})} \\ &= 1570\ \text{bar} \end{aligned}$$

The value of P found using the Redlich-Kwong equation of state is within 2% of the experimentally observed value. The value of P found using the van der Waals equation is obviously incorrect (as it is negative). This is a good example of the problems associated with the van der Waals equation.

2–20. The pressure of propane versus density at 400 K can be fit by the expression

$$\begin{aligned} P/\text{bar} = {} & 33.258(\rho/\text{mol}\cdot\text{L}^{-1}) - 7.5884(\rho/\text{mol}\cdot\text{L}^{-1})^2 \\ & + 1.0306(\rho/\text{mol}\cdot\text{L}^{-1})^3 - 0.058757(\rho/\text{mol}\cdot\text{L}^{-1})^4 \\ & - 0.0033566(\rho/\text{mol}\cdot\text{L}^{-1})^5 + 0.00060696(\rho/\text{mol}\cdot\text{L}^{-1})^6 \end{aligned}$$

for $0 \le \rho/\text{mol}\cdot\text{L}^{-1} \le 12.3$. Use the van der Waals equation and the Redlich-Kwong equation to calculate the pressure for $\rho = 0\ \text{mol}\cdot\text{L}^{-1}$ up to $12.3\ \text{mol}\cdot\text{L}^{-1}$. Plot your results. How do they compare to the above expression?

The van der Waals constants for propane are (Table 2.3) $a = 9.3919\ \text{dm}^6\cdot\text{bar}\cdot\text{mol}^{-2}$ and $b = 0.090494\ \text{dm}^3\cdot\text{mol}^{-1}$. From Equation 2.5, we can write the pressure calculated using the van der Waals equation of state as

$$P = \frac{RT}{\rho^{-1} - b} - \frac{a}{\rho^{-2}}$$
$$= \frac{(0.083145\ \text{dm}^3\cdot\text{bar}\cdot\text{mol}^{-1}\cdot\text{K}^{-1})(400\ \text{K})}{\rho^{-1} - 0.090494\ \text{dm}^3\cdot\text{mol}^{-1}} - \frac{9.3919\ \text{dm}^6\cdot\text{bar}\cdot\text{mol}^{-2}}{\rho^{-2}}$$

Likewise, the Redlich-Kwong constants for propane are (Table 2.4) $A = 183.02\ \text{dm}^6\cdot\text{bar}\cdot\text{mol}^{-2}\cdot\text{K}^{1/2}$ and $B = 0.062723\ \text{dm}^3\cdot\text{mol}^{-1}$. From Equation 2.7, we can write the pressure calculated using the Redlich-Kwong equation of state as

$$P = \frac{RT}{\rho^{-1} - B} - \frac{A}{T^{1/2}\rho^{-1}(\rho^{-1} + B)}$$
$$= \frac{(0.083145\ \text{dm}^3\cdot\text{bar}\cdot\text{mol}^{-1}\cdot\text{K}^{-1})(400\ \text{K})}{\rho^{-1} - 0.062723\ \text{dm}^3\cdot\text{mol}^{-1}}$$
$$- \frac{183.02\ \text{dm}^6\cdot\text{bar}\cdot\text{mol}^{-2}\cdot\text{K}^{1/2}}{(400\ \text{K})^{1/2}\rho^{-1}(\rho^{-1} + 0.062723\ \text{dm}^3\cdot\text{mol}^{-1})}$$

We plot these equations expressing pressure as a function of ρ as shown below.

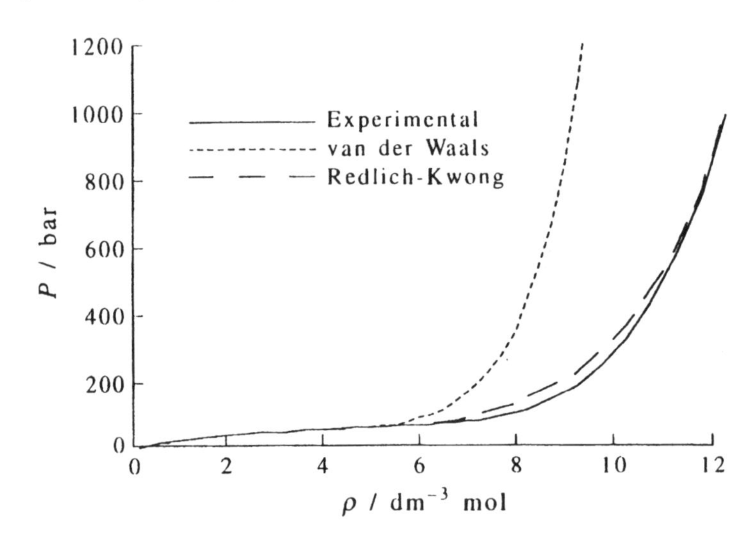

The Redlich-Kwong equation of state describes the data very well, while the van der Waals equation gives a markedly poorer approximation of the observed behavior, especially at high densities.

2–21. The Peng-Robinson equation is often superior to the Redlich-Kwong equation for temperatures near the critical temperature. Use these two equations to calculate the pressure of $CO_2(g)$ at a density of $22.0\ \text{mol}\cdot\text{L}^{-1}$ at 280 K [the critical temperature of $CO_2(g)$ is 304.2 K]. Use $\alpha = 4.192\ \text{bar}\cdot\text{L}^2\cdot\text{mol}^{-2}$ and $\beta = 0.02665\ \text{L}\cdot\text{mol}^{-1}$ for the Peng-Robinson equation.

The molar volume of CO_2 is $0.04545\ \text{dm}^3\cdot\text{mol}^{-1}$.

a. The Redlich-Kwong equation gives a pressure of (Equation 2.7)

$$P = \frac{RT}{\overline{V} - B} - \frac{A}{T^{1/2}\overline{V}(\overline{V} + B)}$$

For CO_2, $A = 64.597\ \text{dm}^6\cdot\text{bar}\cdot\text{mol}^{-2}\cdot\text{K}^{1/2}$ and $B = 0.029677\ \text{dm}^3\cdot\text{mol}^{-1}$ (Table 2.4). Then

$$P = \frac{(0.083145\ \text{dm}^3\cdot\text{bar}\cdot\text{mol}^{-1}\cdot{}^{-1})(280\ \text{K})}{0.04545\ \text{dm}^3\cdot\text{mol}^{-1} - 0.029677\ \text{dm}^3\cdot\text{mol}^{-1}} - \frac{64.597\ \text{dm}^6\cdot\text{bar}\cdot\text{mol}^{-2}\cdot\text{K}^{1/2}}{(280\ \text{K})^{1/2}(0.04545\ \text{dm}^3\cdot\text{mol}^{-1})(0.04545 + 0.029677)\ \text{dm}^3\cdot\text{mol}^{-1}}$$
$$= 345\ \text{bar}$$

b. The Peng-Robinson equation gives a pressure of (Equation 2.8)

$$P = \frac{RT}{\overline{V} - \beta} - \frac{\alpha}{\overline{V}(\overline{V} + \beta) + \beta(\overline{V} - \beta)}$$

For CO_2, $\alpha = 4.192\ \text{dm}^6\cdot\text{bar}\cdot\text{mol}^{-2}$ and $\beta = 0.02665\ \text{dm}^3\cdot\text{mol}^{-1}$. Then

$$P = \frac{(0.083145\ \text{dm}^3\cdot\text{bar}\cdot\text{mol}^{-1}\cdot{}^{-1})(280\ \text{K})}{0.04545\ \text{dm}^3\cdot\text{mol}^{-1} - 0.02665\ \text{dm}^3\cdot\text{mol}^{-1}} - \frac{4.192\ \text{dm}^6\cdot\text{bar}\cdot\text{mol}^{-2}}{[(0.04545)(0.04545 + 0.02665) + (0.02665)(0.04545 - 0.02665)]\ \text{dm}^6\cdot\text{mol}^{-2}}$$
$$= 129\ \text{bar}$$

The Peng-Robinson result is much closer to the experimental value than the value predicted by the Redlich-Kwong equation.

2–22. Show that the van der Waals equation for argon at $T = 142.69$ K and $P = 35.00$ atm can be written as

$$\overline{V}^3 - 0.3664\,\overline{V}^2 + 0.03802\,\overline{V} - 0.001210 = 0$$

where, for convenience, we have supressed the units in the coefficients. Use the Newton-Raphson method (MathChapter A) to find the three roots to this equation, and calculate the values of the density of liquid and vapor in equilibrium with each other under these conditions.

For argon, $a = 1.3307\ \text{dm}^6\cdot\text{atm}\cdot\text{mol}^{-2}$ and $b = 0.031830\ \text{dm}^3\cdot\text{mol}^{-1}$ (Table 2.3). The van der Waals equation of state can be written as (Example 2–2)

$$\overline{V}^3 - \left(b + \frac{RT}{P}\right)\overline{V}^2 + \frac{a}{P}\overline{V} - \frac{ab}{P} = 0$$
$$\overline{V}^3 - \left[0.03183 + \frac{(0.082058)(142.69)}{35.00}\right]\overline{V}^2 + \frac{1.3307}{35.00}\overline{V} - \frac{(1.3307)(0.031830)}{35.00} = 0$$
$$\overline{V}^3 - 0.3664\overline{V}^2 + 0.03802\overline{V} - 0.001210 = 0$$

where we have suppressed the units of the coefficients for convenience. (The quantity $\overline{V}$ is expressed in $\text{dm}^3\cdot\text{mol}^{-1}$.) We apply the Newton-Raphson method, using the function

$$f(\overline{V}) = \overline{V}^3 - 0.3664\overline{V}^2 + 0.03802\overline{V} - 0.001210$$

and its derivative

$$f'(\overline{V}) = 3\overline{V}^2 - 0.7328\overline{V} + 0.03802$$

to find the three roots of this equation, $0.07893\ \text{dm}^3\cdot\text{mol}^{-1}$, $0.07073\ \text{dm}^3\cdot\text{mol}^{-1}$, and $0.21674\ \text{dm}^3\cdot\text{mol}^{-1}$. The smallest root represents the molar volume of liquid argon, and the largest root represents the molar volume of the vapor. The corresponding densities are 14.14 mol·dm^{-3} and 4.614 mol·dm^{-3}, respectively.

2–23. Use the Redlich-Kwong equation and the Peng-Robinson equation to calculate the densities of the coexisting argon liquid and vapor phases at 142.69 K and 35.00 atm. Use the Redlich-Kwong constants given in Table 2.4 and take $\alpha = 1.4915$ atm·L^2·mol^{-2} and $\beta = 0.01981$ L·mol^{-1} for the Peng-Robinson equation.

---

a. For argon, $A = 16.566$ dm^6·atm·mol^{-2}·K$^{1/2}$ and $B = 0.022062$ dm^3·mol^{-1} (Table 2.4). The Redlich-Kwong equation of state can be written as (Equation 2.9)

$$0 = \overline{V}^3 - \frac{RT}{P}\overline{V}^2 - \left(B^2 + \frac{BRT}{P} - \frac{A}{T^{1/2}P}\right)\overline{V} - \frac{AB}{T^{1/2}P}$$

$$0 = \overline{V}^3 - \frac{(0.082058)(142.69)}{35.00}\overline{V}^2 - \left[(0.022062)^2 + \frac{(0.022062)(0.082058)(142.69)}{35.00} - \frac{16.566}{(142.69)^{1/2}(35.00)}\right]\overline{V} - \frac{(16.566)(0.022062)}{(142.69)^{1/2}(35.00)}$$

$$0 = \overline{V}^3 - 0.3345\overline{V}^2 + 0.03176\overline{V} - 0.0008742$$

where we have suppressed the units of the coefficients for convenience. (The quantity $\overline{V}$ is expressed in dm^3·mol^{-1}.) We apply the Newton-Raphson method, using the function

$$f(\overline{V}) = \overline{V}^3 - 0.3345\overline{V}^2 + 0.03176\overline{V} - 0.0008742$$

and its derivative

$$f'(\overline{V}) = 3\overline{V}^2 - 0.6690\overline{V} + 0.03176$$

to find the three roots of this equation to be 0.04961 dm^3·mol^{-1}, 0.09074 dm^3·mol^{-1}, and 0.19419 dm^3·mol^{-1}. The smallest root represents the molar volume of liquid argon, and the largest root represents the molar volume of the vapor. The corresponding densities are 20.13 mol·dm^{-3} and 5.148 mol·dm^{-3}, respectively.

b. The Peng-Robinson equation is given as (Equation 2.8)

$$P = \frac{RT}{\overline{V} - \beta} - \frac{\alpha}{\overline{V}(\overline{V} + \beta) + \beta(\overline{V} - \beta)}$$

This can be expressed as the cubic equation in $\overline{V}$

$$0 = \overline{V}^3 + \left(\beta - \frac{RT}{P}\right)\overline{V}^2 + \left(\frac{\alpha - 3\beta^2 P - 2\beta RT}{P}\right)\overline{V} + \frac{\beta^3 P + \beta^2 RT - \alpha\beta}{P}$$

Substituting the values given in the text of the problem, we find that the Peng-Robinson equation for argon at 142.69 K and 35.00 atm becomes

$$0 = \overline{V}^3 + \left[(0.01981) - \frac{(0.082058)(142.69)}{35.00}\right]\overline{V}^2 + \left[\frac{(1.4915) - 3(0.01981)^2(35.00) - 2(0.01981)(0.082058)(142.69)}{35.00}\right]\overline{V} + \frac{(0.01981)^3(35.00) + (0.01981)^2(0.082058)(142.69) - (1.4915)(0.01981)}{35.00}$$

$$= \overline{V}^3 - 0.3147\overline{V}^2 + 0.02818\overline{V} - 0.0007051$$

where we have suppressed the units of the coefficients for convenience. (The quantity $\overline{V}$ is expressed in $dm^3 \cdot mol^{-1}$.) We apply the Newton-Raphson method, using the function

$$f(\overline{V}) = \overline{V}^3 - 0.3147\overline{V}^2 + 0.02818\overline{V} - 0.0007051$$

and its derivative

$$f'(\overline{V}) = 3\overline{V}^2 - 0.6294\overline{V} + 0.02818$$

to find the three roots of this equation to be $0.04237\ dm^3 \cdot mol^{-1}$, $0.09257\ dm^3 \cdot mol^{-1}$, and $0.17979\ dm^3 \cdot mol^{-1}$. The smallest root represents the molar volume of liquid argon, and the largest root represents the molar volume of the vapor. The corresponding densities are $23.61\ mol \cdot dm^{-3}$ and $5.564\ mol \cdot dm^{-3}$, respectively.

2–24. Butane liquid and vapor coexist at 370.0 K and 14.35 bar. The densities of the liquid and vapor phases are $8.128\ mol \cdot L^{-1}$ and $0.6313\ mol \cdot L^{-1}$, respectively. Use the van der Waals equation, the Redlich-Kwong equation, and the Peng-Robinson equation to calculate these densities. Take $\alpha = 16.44\ bar \cdot L^2 \cdot mol^{-2}$ and $\beta = 0.07245\ L \cdot mol^{-1}$ for the Peng-Robinson equation.

---

a. For butane, $a = 13.888\ dm^6 \cdot bar \cdot mol^{-2}$ and $b = 0.11641\ dm^3 \cdot mol^{-1}$ (Table 2.3). The van der Waals equation of state can be written as (Example 2–2)

$$\overline{V}^3 - \left(b + \frac{RT}{P}\right)\overline{V}^2 + \frac{a}{P}\overline{V} - \frac{ab}{P} = 0$$

$$\overline{V}^3 - \left[0.11641 + \frac{(0.083145)(370.0)}{14.35}\right]\overline{V}^2 + \frac{13.888}{14.35}\overline{V} - \frac{(13.888)(0.11641)}{14.35} = 0$$

$$\overline{V}^3 - 2.2602\overline{V}^2 + 0.9678\overline{V} - 0.1127 = 0$$

where we have suppressed the units of the coefficients for convenience. (The quantity $\overline{V}$ is expressed in $dm^3 \cdot mol^{-1}$.) We apply the Newton-Raphson method, using the function

$$f(\overline{V}) = \overline{V}^3 - 2.2602\overline{V}^2 + 0.9678\overline{V} - 0.1127$$

and its derivative

$$f'(\overline{V}) = 3\overline{V}^2 - 4.5204\overline{V} + 0.9678$$

to find the three roots of this equation to be $0.20894\ dm^3 \cdot mol^{-1}$, $0.30959\ dm^3 \cdot mol^{-1}$, and $1.7417\ dm^3 \cdot mol^{-1}$. The smallest root represents the molar volume of liquid butane, and the largest root represents the molar volume of the vapor. The corresponding densities are $4.786\ mol \cdot dm^{-3}$ and $0.5741\ mol \cdot dm^{-3}$, respectively.

b. For butane, $A = 290.16\ dm^6 \cdot bar \cdot mol^{-2} \cdot K^{1/2}$ and $B = 0.08068\ dm^3 \cdot mol^{-1}$ (Table 2.4). The Redlich-Kwong equation of state can be written as (Equation 2.9)

$$0 = \overline{V}^3 - \frac{RT}{P}\overline{V}^2 - \left(B^2 + \frac{BRT}{P} - \frac{A}{T^{1/2}P}\right)\overline{V} - \frac{AB}{T^{1/2}P}$$

$$0 = \overline{V}^3 - \frac{(0.083145)(370.0)}{14.35}\overline{V}^2 - \left[(0.08068)^2 + \frac{(0.08068)(0.083145)(370.0)}{14.35} - \frac{290.16}{(370.0)^{1/2}(14.35)}\right]\overline{V} - \frac{(290.16)(0.08068)}{(370.0)^{1/2}(14.35)}$$

$$0 = \overline{V}^3 - 2.144\overline{V}^2 + 0.8717\overline{V} - 0.08481$$

where we have suppressed the units of the coefficients for convenience. (The quantity $\overline{V}$ is expressed in $dm^3 \cdot mol^{-1}$.) We apply the Newton-Raphson method, using the function

$$f(\overline{V}) = \overline{V}^3 - 2.144\overline{V}^2 + 0.8717\overline{V} - 0.08481$$

and its derivative

$$f'(\overline{V}) = 3\overline{V}^2 - 4.288\overline{V} + 0.8717$$

to to find the three roots of this equation to be 0.14640 $dm^3 \cdot mol^{-1}$, 0.35209 $dm^3 \cdot mol^{-1}$, and 1.6453 $dm^3 \cdot mol^{-1}$. The smallest root represents the molar volume of liquid butane, and the largest root represents the molar volume of the vapor. The corresponding densities are 6.823 $mol \cdot dm^{-3}$ and 0.6078 $mol \cdot dm^{-3}$, respectively.

c. The Peng-Robinson equation can be expressed as (Problem 2-23)

$$0 = \overline{V}^3 + \left(\beta - \frac{RT}{P}\right)\overline{V}^2 + \left(\frac{\alpha - 3\beta^2 P - 2\beta RT}{P}\right)\overline{V} + \frac{\beta^3 P + \beta^2 RT - \alpha\beta}{P}$$

Substituting the values given in the text of the problem, we find that the Peng-Robinson equation for butane at 370.0 K and 14.35 bar becomes

$$\begin{aligned} 0 = \overline{V}^3 &+ \left[(0.07245) - \frac{(0.081345)(370.0)}{14.35}\right]\overline{V}^2 \\ &+ \left[\frac{(16.44) - 3(0.07245)^2(14.35) - 2(0.07245)(0.081345)(370.0)}{14.35}\right]\overline{V} \\ &+ \frac{(0.07245)^3(14.35) + (0.07245)^2(0.081345)(370.0) - (16.44)(0.07245)}{14.35} \\ = \overline{V}^3 &- 2.071\overline{V}^2 + 0.8193\overline{V} - 0.07137 \end{aligned}$$

where we have suppressed the units of the coefficients for convenience. (The quantity $\overline{V}$ is expressed in $dm^3 \cdot mol^{-1}$.) We apply the Newton-Raphson method, using the function

$$f(\overline{V}) = \overline{V}^3 - 2.071\overline{V}^2 + 0.8193\overline{V} - 0.07137$$

and its derivative

$$f'(\overline{V}) = 3\overline{V}^2 - 4.142\overline{V} + 0.8193$$

to find the three roots of this equation to be 0.12322 $dm^3 \cdot mol^{-1}$, 0.36613 $dm^3 \cdot mol^{-1}$, and 1.5820 $dm^3 \cdot mol^{-1}$. The smallest root represents the molar volume of liquid butane, and the largest root represents the molar volume of the vapor. The corresponding densities are 8.116 $mol \cdot dm^{-3}$ and 0.6321 $mol \cdot dm^{-3}$, respectively.

Below is a table which summarizes the densities of liquid and vapor butane observed experimentally and calculated with the various equations of state above.

Equation used	Liquid $\rho/mol \cdot dm^{-3}$	Gas $\rho/mol \cdot dm^{-3}$
Experimental	8.128	0.6313
van der Waals	4.786	0.5741
Redlich-Kwong	6.831	0.6078
Peng-Robinson	8.116	0.6321

2–25. Another way to obtain expressions for the van der Waals constants in terms of critical parameters is to set $(\partial P/\partial \overline{V})_T$ and $(\partial^2 P/\partial \overline{V}^2)_T$ equal to zero at the critical point. Why are these quantities equal to zero at the critical point? Show that this procedure leads to Equations 2.12 and 2.13.

These values are equal to zero at the critical point because the critical point is an inflection point in a plot of P versus V at constant temperature.

$$P = \frac{RT}{\overline{V} - b} - \frac{a}{\overline{V}^2}$$

So

$$\left(\frac{\partial P}{\partial V}\right)_T = \frac{-RT}{\left(\overline{V} - b\right)^2} + \frac{2a}{\overline{V}^3} \tag{1}$$

$$\left(\frac{\partial^2 P}{\partial V^2}\right)_T = \frac{2RT}{\left(\overline{V} - b\right)^3} - \frac{6a}{\overline{V}^4} \tag{2}$$

If $(\partial P/\partial V)_T$ and $(\partial^2 P/\partial V^2)_T$ are zero at the critical point, then Equations 1 and 2 give

$$RT_c\overline{V}_c^{\,3} = 2a\left(\overline{V}_c - b\right)^2 \tag{3}$$

and

$$2RT_c\overline{V}_c^{\,4} = 6a\left(\overline{V}_c - b\right)^3 \tag{4}$$

Multiplying Equation 3 by $2\overline{V}_c$ gives

$$2RT_c\overline{V}_c^{\,4} = 4a\overline{V}_c\left(\overline{V}_c - b\right)^2$$

and then using Equation 4 yields

$$4a\overline{V}_c\left(\overline{V}_c - b\right)^2 = 6a\left(\overline{V}_c - b\right)^3$$

$$4\overline{V}_c = 6\overline{V}_c - 6b$$

$$3b = \overline{V}_c \tag{2.13a}$$

Substituting Equation 2.13a into Equation 3 gives

$$RT_c(3b)^3 = 2a(3b - b)^2$$

$$T_c = \frac{8ab^2}{27b^3R} = \frac{8a}{27bR} \tag{2.13c}$$

Now substitute Equations 2.13a and 2.13c into the van der Waals equation to find P_c:

$$P_c = \frac{RT_c}{\overline{V}_c - b} - \frac{a}{\overline{V}_c^{\,2}} = \frac{8aR}{27bR\,(3b - b)} - \frac{a}{(3b)^2} = \frac{a}{27b^2} \tag{2.13b}$$

Equation 2.12 follows naturally from these expressions for $\overline{V}_c$, P_c, and T_c.

2–26. Show that the Redlich-Kwong equation can be written in the form

$$\overline{V}^3 - \frac{RT}{P}\overline{V}^2 - \left(B^2 + \frac{BRT}{P} - \frac{A}{PT^{1/2}}\right)\overline{V} - \frac{AB}{PT^{1/2}} = 0$$

Now compare this equation with $(\overline{V}-\overline{V}_c)^3=0$ to get

$$3\overline{V}_c=\frac{RT_c}{P_c} \tag{1}$$

$$3\overline{V}_c^2=\frac{A}{P_cT_c^{1/2}}-\frac{BRT_c}{P_c}-B^2 \tag{2}$$

and

$$\overline{V}_c^3=\frac{AB}{P_cT_c^{1/2}} \tag{3}$$

Note that Equation 1 gives

$$\frac{P_c\overline{V}_c}{RT_c}=\frac{1}{3} \tag{4}$$

Now solve Equation 3 for A and substitute the result and Equation 4 into Equation 2 to obtain

$$B^3+3\overline{V}_cB^2+3\overline{V}_c^2B-\overline{V}_c^3=0 \tag{5}$$

Divide this equation by $\overline{V}_c^3$ and let $B/\overline{V}_c=x$ to get

$$x^3+3x^2+3x-1=0$$

Solve this cubic equation by the Newton-Raphson method (MathChapter A) to obtain $x=0.25992$, or

$$B=0.25992\overline{V}_c \tag{6}$$

Now substitute this result and Equation 4 into Equation 3 to obtain

$$A=0.42748\frac{R^2T_c^{5/2}}{P_c}$$

We start with the Redlich-Kwong equation of state,

$$P=\frac{RT}{\overline{V}-B}-\frac{A}{T^{1/2}\overline{V}(\overline{V}+B)} \tag{2.7}$$

We can rewrite the above equation as

$$P(\overline{V}-B)(\overline{V}+B)T^{1/2}\overline{V}=RT^{3/2}\overline{V}(\overline{V}+B)-A(\overline{V}-B)$$
$$PT^{1/2}\overline{V}\left(\overline{V}^2-B^2\right)=RT^{3/2}\overline{V}^2+RT^{3/2}\overline{V}B-A\overline{V}+AB$$

We express this equation as a cubic equation in $\overline{V}$:

$$\overline{V}^3-\frac{RT}{P}\overline{V}^2-\left(B^2+\frac{BRT}{P}-\frac{A}{PT^{1/2}}\right)\overline{V}-\frac{AB}{PT^{1/2}}=0 \tag{a}$$

Expanding the equation $(\overline{V}-\overline{V}_c^3)=0$ gives

$$(\overline{V}-\overline{V}_c)^3=\overline{V}^3-\overline{V}_c^{\,3}+3\overline{V}_c^{\,2}\overline{V}-3\overline{V}^2\overline{V}_c=0 \tag{b}$$

Setting the coefficients of $\overline{V}^3$, $\overline{V}^2$, $\overline{V}$, and $\overline{V}^0$ in Equations a and b equal to one another at the critical point gives

$$3\overline{V}_c = \frac{RT_c}{P_c} \tag{1}$$

$$3\overline{V}_c^2 = -B^2 - \frac{BRT_c}{P_c} + \frac{A}{P_cT_c^{1/2}} \tag{2}$$

$$\overline{V}_c^3 = \frac{AB}{P_cT_c^{1/2}} \tag{3}$$

$$\frac{P_c\overline{V}_c}{RT_c} = \frac{1}{3} \tag{4}$$

We can solve Equation 3 for A to find

$$A = \frac{\overline{V}_c^3 P_c T_c^{1/2}}{B}$$

Substituting this result into Equation 2 gives

$$3\overline{V}_c^2 = -B^2 - \frac{BRT_c\overline{V}_c}{P_c\overline{V}_c} + \frac{\overline{V}_c^3 P_c T_c^{1/2}}{BP_cT_c^{1/2}}$$

$$3\overline{V}_c^2 = -B^2 - 3B\overline{V}_c + B^{-1}\overline{V}_c^3$$

$$0 = \frac{B^3}{\overline{V}_c^3} + \frac{3B^{2^2}}{\overline{V}_c} + \frac{3B}{\overline{V}_c} - 1$$

$$0 = x^3 + 3x^2 + 3x - 1$$

where we set $x = B/\overline{V}_c$. We solved this cubic equation using the Newton-Raphson method in Example A–1 and found that $x = 0.25992$. Then $B = 0.25992\overline{V}_c$, and substituting into Equation 3 gives

$$\left(\frac{P_cV_c}{RT_c}\right) T_c^{1/2}\overline{V}_c^2 = \frac{AB}{RT_c}$$

$$T_c^{1/2}\overline{V}_c^2 = \frac{3AB}{RT_c}$$

$$A = \frac{T_c^{3/2}\overline{V}_c^2 R}{3B} = \frac{T_c^{3/2}\overline{V}R}{3\,(0.25992)}$$

$$A = \frac{P_c\overline{V}_c}{RT_c}\left(\frac{RT_c}{P_c}\right)\frac{T_c^{3/2}R}{3\,(0.25992)}$$

$$A = 0.42748\frac{R^2T_c^{5/2}}{P_c} \tag{7}$$

2–27. Use the results of the previous problem to derive Equations 2.14.

Equation 6 of Problem 2.26 gives $B = 0.25992\overline{V}_c$, and so

$$\overline{V}_c = 3.8473B \tag{2.14a}$$

Now we can use Equation 7 of Problem 2–26 to write

$$A = \frac{0.42748R^2T_c^{5/2}}{P_c} = 0.42748RT_c^{3/2}\left(\frac{RT_c}{P_c\overline{V}_c}\right)\overline{V}_c$$

Substituting Equation 4 from Problem 2–26 and Equation 2.14*a* into the above expression gives

$$A = 3(0.42748)RT_c^{3/2}(3.8473B)$$

$$T_c^{3/2} = \frac{A}{3(0.42748)(3.8473)RB}$$

$$T_c = 0.34504\left(\frac{A}{BR}\right)^{2/3}$$

which is Equation 2.14*c*. Substitute this last result into the final equation of Problem 2–26 to find

$$P_c = \frac{0.42748R^2T_c^{5/2}}{A}$$

$$= \left(\frac{0.42748}{A}\right)R^2\left[0.34504\left(\frac{A}{BR}\right)^{2/3}\right]^{5/2}$$

$$= 0.029894\frac{A^{2/3}R^{1/3}}{B^{5/3}}$$

which is Equation 2.14*b*.

2–28. Write the Peng-Robinson equation as a cubic polynomial equation in $\overline{V}$ (with the coefficient of $\overline{V}^3$ equal to one), and compare it with $(\overline{V} - \overline{V}_c)^3 = 0$ at the critical point to obtain

$$\frac{RT_c}{P_c} - \beta = 3\overline{V}_c \tag{1}$$

$$\frac{\alpha_c}{P_c} - 3\beta^2 - 2\beta\frac{RT_c}{P_c} = 3\overline{V}_c^2 \tag{2}$$

and

$$\frac{\alpha_c\beta}{P_c} - \beta^2\frac{RT_c}{P_c} - \beta^3 = \overline{V}_c^3 \tag{3}$$

(We write α_c because α depends upon the temperature.) Now eliminate α_c/P_c between Equations 2 and 3, and then use Equation 1 for $\overline{V}_c$ to obtain

$$64\beta^3 + 6\beta^2\frac{RT_c}{P_c} + 12\beta\left(\frac{RT_c}{P_c}\right)^2 - \left(\frac{RT_c}{P_c}\right)^3 = 0$$

Let $\beta/(RT_c/P_c) = x$ and get

$$64x^3 + 6x^2 + 12x - 1 = 0$$

Solve this equation using the Newton-Raphson method to obtain

$$\beta = 0.077796\frac{RT_c}{P_c}$$

Substitute this result and Equation 1 into Equation 2 to obtain

$$\alpha_c = 0.45724\frac{(RT_c)^2}{P_c}$$

Last, use Equation 1 to show that

$$\frac{P_c\overline{V}_c}{RT_c} = 0.30740$$

First, we write the Peng-Robinson equation as a cubic polynomial in $\overline{V}$, as we did in Problem 2–23.

$$0 = \overline{V}^3 + \left(\beta - \frac{RT}{P}\right)\overline{V}^2 + \left(\frac{\alpha - 3\beta^2 P - 2\beta RT}{P}\right)\overline{V} + \frac{\beta^3 P + \beta^2 RT - \alpha\beta}{P} \tag{a}$$

Expanding the equation $(\overline{V} - \overline{V}_c)^3 = 0$ gives

$$(\overline{V} - \overline{V}_c)^3 = \overline{V}^3 - \overline{V}_c^{\,3} + 3\overline{V}_c^{\,2}\overline{V} - 3\overline{V}^2\overline{V}_c = 0 \tag{b}$$

Setting the coefficients of $\overline{V}^3$, $\overline{V}^2$, $\overline{V}$, and $\overline{V}^0$ in Equations a and b equal to one another at the critical point gives

$$\beta - \frac{RT_c}{P_c} = -3\overline{V}_c \tag{1}$$

$$\frac{\alpha_c}{P_c} - 3\beta^2 - 2\beta\frac{RT_c}{P_c} = 3\overline{V}_c^{\,2} \tag{2}$$

$$\frac{\alpha_c\beta}{P_c} - \beta^3 - \frac{RT_c}{P_c}\beta^2 = \overline{V}_c^{\,3} \tag{3}$$

Solving Equation 2 for α_c/P_c gives

$$\frac{\alpha_c}{P_c} = 3\overline{V}_c^{\,2} + 3\beta^2 + 2\beta\frac{RT_c}{P_c}$$

We substitute this last result into Equation 3 to find

$$\begin{aligned}
0 &= \overline{V}_c^3 - \beta\left(3\overline{V}_c^{\,2} + 3\beta^2 + 2\beta\frac{RT_c}{P_c}\right) + \beta^2\frac{RT_c}{P_c} + \beta^3 \\
&= \overline{V}_c^3 - 2\beta^3 - \beta^2\frac{RT_c}{P_c} - 3\beta\overline{V}_c^2 \\
&= \frac{1}{27}\left(\frac{RT_c}{P_c} - \beta\right)^3 - 2\beta^3 - \beta^2\frac{RT_c}{P_c} - \frac{\beta}{3}\left(\frac{RT_c}{P_c} - \beta\right)^2 \\
&= \left(\frac{RT_c}{P_c}\right)^3 - 3\beta\left(\frac{RT_c}{P_c}\right)^2 + 3\beta^2\frac{RT_c}{P_c} - \beta^3 - 54\beta^3 - 27\beta^2\frac{RT_c}{P_c} - 9\left(\frac{RT_c}{P_c}\right)^2\beta + 18\beta\frac{RT_c}{P_c} - 9\beta^3 \\
&= \left(\frac{RT_c}{P_c}\right)^3 - 12\beta\left(\frac{RT_c}{P_c}\right)^2 - 6\beta^2\left(\frac{RT_c}{P_c}\right) - 64\beta^3
\end{aligned}$$

Set $x = \beta/\left(RT_c/P_c\right)$ and the above equation becomes

$$64x^3 + 6x^2 + 12x - 1 = 0$$

Using the Newton-Raphson method (MathChapter A), we find $x = 0.077796$ and so

$$\beta = 0.077796\frac{RT_c}{P_c}$$

We now substitute Equation 1 into Equation 2 and use the expression for β given above to write

$$\begin{aligned}
\alpha_c &= P_c\left(3\overline{V}_c^2 + 3\beta^2 + 2\beta\frac{RT_c}{P_c}\right) \\
&= P_c\left[\frac{1}{3}\left(\frac{RT_c}{P_c} - \beta\right)^2 + 3\beta^2 + 2\beta\frac{RT_c}{P_c}\right] \\
&= P_c\left(\frac{RT_c}{P_c}\right)^2\left[\frac{1}{3}(1 - 0.77796)^2 + 3(0.077796)^2 + 2(0.077796)\right] \\
&= 0.45724\frac{(RT_c)^2}{P_c}
\end{aligned}$$

Finally, substitute the expression of β given above into Equation 1 to write

$$\begin{aligned}
(1 - 0.077796)\frac{RT_c}{P_c} &= 3\overline{V}_c \\
0.92220RT_c &= 3P_c\overline{V}_c \\
0.30740 &= \frac{P_c\overline{V}_c}{RT_c}
\end{aligned}$$

2–29. Look up the boiling points of the gases listed in Table 2.5 and plot these values versus the critical temperatures T_c. Is there any correlation? Propose a reason to justify your conclusions from the plot.

A graph of boiling points versus critical temperatures of the gases listed in Table 2.5 is shown below. There appears to be a direct correlation between the boiling point of a gas and its critical temperature.

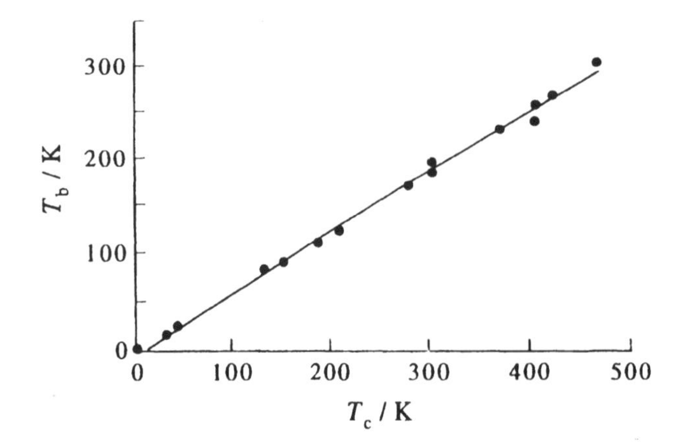

This is another illustration of the law of corresponding states: if we compare the boiling points of different gases relative to their critical temperatures, we find that all behaviors can be similarly explained (hence, the constant slope in the figure).

2–30. Show that the compressibility factor Z for the Redlich-Kwong equation can be written as in Equation 2.21.

The Redlich-Kwong equation of state is given by Equation 2.7. Thus

$$\left(\frac{\overline{V}}{RT}\right)P = \left(\frac{\overline{V}}{RT}\right)\frac{RT}{\overline{V}-B} - \frac{A}{RT^{3/2}\left(\overline{V}+B\right)}$$

or

$$Z = \frac{\overline{V}}{\overline{V}-B} - \frac{A}{RT^{3/2}\left(\overline{V}+B\right)}$$

We know from Equation 2.18 that

$$A = 0.42748\frac{R^2T_c^{5/2}}{P_c} \qquad \text{and} \qquad B = 0.086640\frac{RT_c}{P_c}$$

We can then write Z as

$$Z = \overline{V}\left(\overline{V} - 0.086640\frac{RT_c}{P_c}\right)^{-1} - \left(0.42748\frac{R^2T_c^{5/2}}{P_c}\right)\left[RT^{3/2}\left(\overline{V} + 0.086640\frac{RT_c}{P_c}\right)\right]^{-1}$$

In the solution to Problem 2.26, we showed that $3\overline{V}_cP_c = RT_c$, so

$$\begin{aligned}Z &= \frac{\overline{V}}{\overline{V} - 0.086640\left(3\overline{V}_c\right)} - \frac{0.42748T_c^{3/2}\left(3\overline{V}_c\right)}{T^{3/2}\left[\overline{V} + 0.086640\left(3\overline{V}_c\right)\right]} \\ &= \frac{\overline{V}_R}{\overline{V}_R - 0.25992} - \frac{1.28244}{T_R^{3/2}\left(\overline{V}_R + 0.25992\right)}\end{aligned}$$

2–31. Use the following data for ethane and argon at $T_R = 1.64$ to illustrate the law of corresponding states by plotting Z against $\overline{V}_R$.

Ethane (T = 500 K)		Argon (T = 247 K)	
P/bar	$\overline{V}$/L·mol^{-1}	P/atm	$\overline{V}$/L·mol^{-1}
0.500	83.076	0.500	40.506
2.00	20.723	2.00	10.106
10.00	4.105	10.00	1.999
20.00	2.028	20.00	0.9857
40.00	0.9907	40.00	0.4795
60.00	0.6461	60.00	0.3114
80.00	0.4750	80.00	0.2279
100.0	0.3734	100.0	0.1785
120.0	0.3068	120.0	0.1462
160.0	0.2265	160.0	0.1076
200.0	0.1819	200.0	0.08630
240.0	0.1548	240.0	0.07348
300.0	0.1303	300.0	0.06208
350.0	0.1175	350.0	0.05626
400.0	0.1085	400.0	0.05219
450.0	0.1019	450.0	0.04919
500.0	0.09676	500.0	0.04687
600.0	0.08937	600.0	0.04348
700.0	0.08421	700.0	0.04108

We can use Table 2.5 to find the critical molar volumes of ethane and argon (0.1480 $dm^3 \cdot mol^{-1}$ and 0.07530 $dm^3 \cdot mol^{-1}$, respectively) and use the data given in the problem text in the equations

$$Z = \frac{P\overline{V}}{RT} \qquad \text{and} \qquad \overline{V}_R = \frac{\overline{V}}{\overline{V}_c}$$

to plot Z versus $\overline{V}_R$. Note that the pressures for ethane are given in units of bar, while the pressures for argon are given in units of atm.

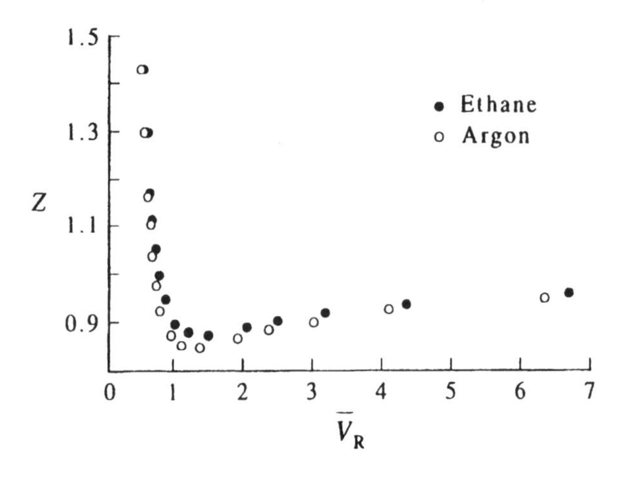

2–32. Use the data in Problem 2–31 to illustrate the law of corresponding states by plotting Z against P_R.

We can use Table 2.5 to find the critical pressures of ethane and argon (48.714 bar and 48.643 atm, respectively) and use the data given in Problem 2–31 in the equations

$$Z = \frac{P\overline{V}}{RT} \qquad \text{and} \qquad P_R = \frac{P}{P_c}$$

to plot Z versus P_R.

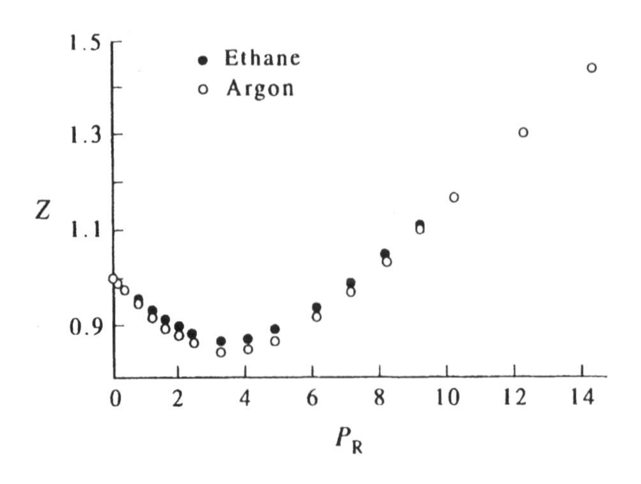

2–33. Use the data in Problem 2.31 to test the quantitative reliability of the van der Waals equation by comparing a plot of Z versus $\overline{V}_R$ from Equation 2.20 to a similar plot of the data.

We can use Equation 2.20 to express Z as a function of $\overline{V}_R$:

$$Z = \frac{\overline{V}_R}{\overline{V}_R - 1/3} - \frac{9}{8\overline{V}_R T_R}$$

We can substitute the appropriate values of $T_R = T/T_c$ (Problem 2–31) and plot Z versus $\overline{V}_R$ for both argon and ethane. The plot below shows the lines generated by applying the van der Waals equation and the actual data from Problem 2–31.

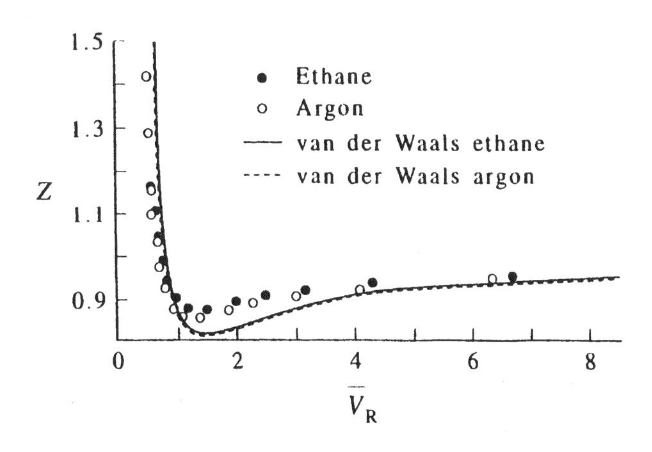

2–34. Use the data in Problem 2.31 to test the quantitative reliability of the Redlich-Kwong equation by comparing a plot of Z versus $\overline{V}_R$ from Equation 2.21 to a similar plot of the data.

We can use Equation 2.21 to express Z as a function of $\overline{V}_R$:

$$Z = \frac{\overline{V}_R}{\overline{V}_R - 0.25992} - \frac{1.2824}{T_R^{3/2}(\overline{V}_R + 0.25992)}$$

We can substitute the appropriate values of $T_R = T/T_c$ (Problem 2–31) and plot Z versus $\overline{V}_R$ for both argon and ethane. The plot below shows the lines generated by applying the Redlich-Kwong equation and the actual data from Problem 2–31.

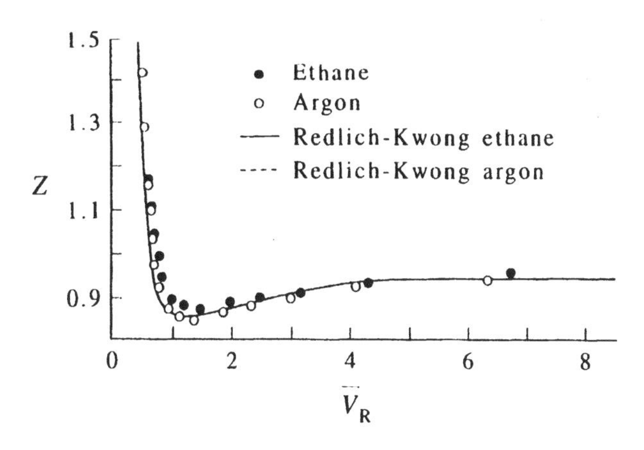

2–35. Use Figure 2.10 to estimate the molar volume of CO at 200 K and 180 bar. An accurate experimental value is 78.3 $\text{cm}^3 \cdot \text{mol}^{-1}$.

We use the critical values in Table 2.5 to write

$$T_R = \frac{T}{T_c} = \frac{200 \text{ K}}{132.85 \text{ K}} = 1.51$$

and

$$P_R = \frac{P}{P_c} = \frac{180 \text{ bar}}{34.935 \text{ bar}} = 5.15$$

From Figure 2.10,

$$Z = \frac{P\overline{V}}{RT} \approx 0.85$$

We can now find $\overline{V}$:

$$\overline{V} \approx \frac{0.85\left(0.083145 \text{ dm}^3 \cdot \text{bar} \cdot \text{mol}^{-1} \cdot \text{K}^{-1}\right)(200 \text{ K})}{180 \text{ bar}}$$

$$\overline{V} \approx 78.5 \text{ cm}^3 \cdot \text{mol}^{-1}$$

in excellent agreement with the experimental value.

2–36. Show that $B_{2V}(T) = RT B_{2P}(T)$ (see Equation 2.24).

We begin with Equations 2.22 and 2.23,

$$\frac{P\overline{V}}{RT} = 1 + \frac{B_{2V}(T)}{\overline{V}} + \frac{B_{3V}(T)}{\overline{V}^2} + O(\overline{V}^{-3})$$

and

$$\frac{P\overline{V}}{RT} = 1 + B_{2P}(T)\,P + O(P^2)$$

We now solve Equation 2.23 for P:

$$P = \frac{RT}{\overline{V}} + \frac{PRT}{\overline{V}} B_{2P}(T) + O(\overline{V}^{-2})$$

Substituting this expression for P into Equation 2.22 gives

$$1 + \frac{B_{2V}(T)}{\overline{V}} + O(\overline{V}^{-2}) = 1 + B_{2P}(T)\frac{RT}{\overline{V}} + \frac{PRT}{\overline{V}} B_{2P}(T) + O(\overline{V}^{-2})$$

and equating the coefficients of $\overline{V}^{-1}$ on both sides of the equation gives

$$B_{2V}(T) = RT\,B_{2P}(T)$$

2–37. Use the following data for NH_3(g) at 273 K to determine $B_{2P}(T)$ at 273 K.

P/bar	0.10	0.20	0.30	0.40	0.50	0.60	0.70
$(Z-1)/10^{-4}$	1.519	3.038	4.557	6.071	7.583	9.002	10.551

Ignoring terms of $O(P^2)$, we can write Equation 2.23 as

$$Z - 1 = B_{2P}(T)P$$

A plot of $(Z - 1)$ versus pressure for the data given for NH_3 at 273 K is shown below.

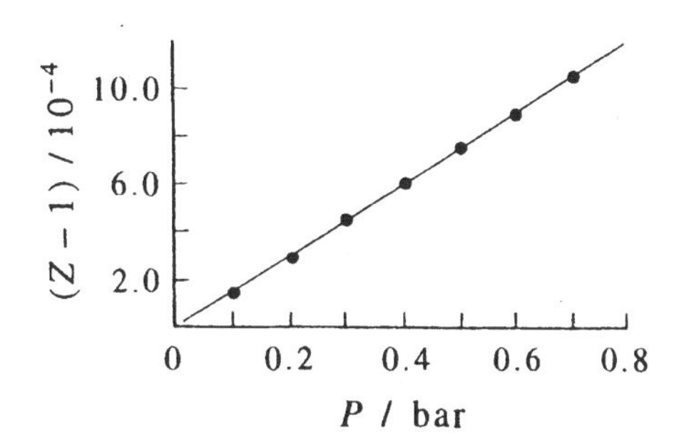

The slope of the best-fit line to the data is B_{2P} and is equal to 15.0×10^{-4} bar^{-1}.

2–38. The density of oxygen as a function of pressure at 273.15 K is listed below.

P/atm	0.2500	0.5000	0.7500	1.0000
ρ/g·dm^{-3}	0.356985	0.714154	1.071485	1.428962

Use the data to determine $B_{2V}(T)$ of oxygen. Take the atomic mass of oxygen to be 15.9994 and the value of the molar gas constant to be 8.31451 J·K^{-1}·mol^{-1} = 0.0820578 dm^3·atm·K^{-1}·mol^{-1}.

We can express the molar volume of oxygen as

$$\overline{V} = \frac{(15.9994 \text{ g·mol}^{-1})}{\rho}$$

where ρ has units of g·dm^{-3}. Using Equation 2.22 to express Z and neglecting terms of $O(\overline{V}^{-2})$, we find

$$Z - 1 = \overline{V}^{-1} B_{2V}(T)$$

or

$$\frac{P(15.994 \text{ g·mol}^{-1})}{\rho RT} - 1 = \overline{V}^{-1} B_{2V}(T)$$

A plot of $(Z - 1)$ versus $\overline{V}^{-1}$ for the data given for oxygen at 273.15 K is shown below.

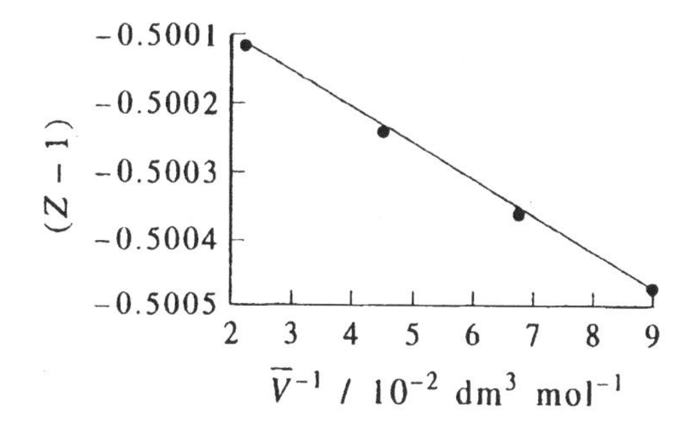

The slope of the best-fit line to the data is B_{2V} and is equal to $-5.33 \times 10^{-3}\ \text{dm}^3\cdot\text{mol}^{-1}$.

2–39. Show that the Lennard-Jones potential can be written as

$$u(r) = \varepsilon\left(\frac{r^*}{r}\right)^{12} - 2\varepsilon\left(\frac{r^*}{r}\right)^{6}$$

where r^* is the value of r at which $u(r)$ is a minimum.

The Lennard-Jones potential is (Equation 2.29)

$$u(r) = 4\varepsilon\left[\left(\frac{\sigma}{r}\right)^{12} - \left(\frac{\sigma}{r}\right)^{6}\right]$$

From Example 2.9, we know $\sigma = r^* 2^{-1/6}$, so

$$u(r) = 4\varepsilon\left[\left(\frac{r^*}{2^{1/6}r}\right)^{12} - \left(\frac{r^*}{2^{1/6}r}\right)^{6}\right]$$

$$= \frac{4\varepsilon(r^*)^{12}}{2^2 r^{12}} - \frac{4\varepsilon(r^*)^{6}}{2r^6} = \varepsilon\left(\frac{r^*}{r}\right)^{12} - 2\varepsilon\left(\frac{r^*}{r}\right)^{6}$$

2–40. Using the Lennard-Jones parameters given in Table 2.7, compare the depth of a typical Lennard-Jones potential to the strength of a covalent bond.

The parameter ε is the depth of a Lennard-Jones potential. From Table 2.7, an average value of $\varepsilon/k_B \approx 139$ K for one molecule. So, for one mole,

$$\varepsilon \approx (139\ \text{K})k_B N_A \approx \ \text{J} \approx 1\ \text{kJ}$$

In comparison, the strength of a covalent bond is on the order of 100 kJ per mole.

2–41. Compare the Lennard-Jones potentials of $H_2(g)$ and $O_2(g)$ by plotting both on the same graph.

Shown below are plots of $u(r)$ versus r for both $H_2(g)$ and $O_2(g)$. Oxygen has a deeper potential well than hydrogen and the minimum of its potential curve occurs at a higher value of r than the minimum of the potential curve of hydrogen.

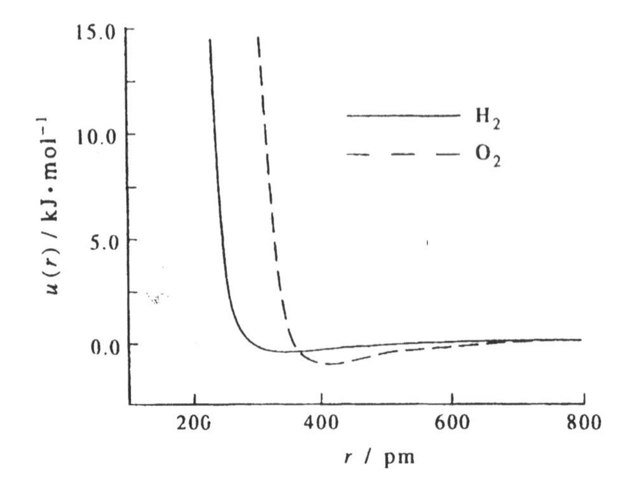

2–42. Use the data in Tables 2.5 and 2.7 to show that *roughly* $\epsilon/k = 0.75\ T_c$ and $b_0 = 0.7\ \overline{V}_c$. Thus, critical constants can be used as rough, first estimates of ϵ and b_0 $(= 2\pi N_0\sigma^3/3)$.

Let us select argon as a representative molecule. For Ar, $\varepsilon/k_B = 120$ K, $b_0 = 50.0\ \text{cm}^3\cdot\text{mol}^{-1}$, $T_c = 150.95$ K, and $\overline{V}_c = 75.3\ \text{cm}^3\cdot\text{mol}^{-1}$.

$$0.75\ T_c = 113\ \text{K} \qquad \text{compared to} \quad \varepsilon/k_B = 120\ \text{K}$$

$$0.7\ \overline{V}_c = 53\ \text{cm}^3\cdot\text{mol}^{-1} \quad \text{compared to} \quad b_0 = 50.0\ \text{cm}^3\cdot\text{mol}^{-1}$$

The equivalencies stated in the problem text hold for argon.

2–43. Prove that the second virial coefficient calculated from a general intermolecular potential of the form

$$u(r) = (\text{energy parameter}) \times f\left(\frac{r}{\text{distance parameter}}\right)$$

rigorously obeys the law of corresponding states. Does the Lennard-Jones potential satisfy this condition?

Begin with Equation 2.25,

$$B_{2V}(T) = -2\pi N_A \int_0^\infty [e^{-u(r)/k_BT} - 1]r^2 dr$$

Let $u(r) = Ef(r/r_0)$ and $T^* = k_B T$, so that we can write $B_{2V}(T)$ as

$$B_{2V}(T^*) = -2\pi N_A \int_0^\infty [e^{-Ef(r/r_0)/T^*} - 1]r^2 dr$$

Now let $r/r_0 = r_R$, where r_R is the reduced distance variable. Then $dr = r_0 dr_R$, so we can write

$$B_{2V}(T^*) = -2\pi r_0{}^3 N_A \int_0^\infty [e^{-Ef(r_R)/T^*} - 1]dr_R$$

We can divide both sides by $-2\pi r_0{}^3 N_A$ to obtain $B_{2V}{}^*$ as a function of only reduced variables:

$$B_{2V}{}^*(T^*) = \int_0^\infty [e^{-Ef(r_R)/T^*} - 1]dr_R$$

Therefore, the functional form of $u(r)$ given in the problem text rigorously obeys the law of corresponding states. The Lennard-Jones potential can be written as (Equation 2.29)

$$u(r) = 4\varepsilon\left[\left(\frac{\sigma}{r}\right)^{12} - \left(\frac{\sigma}{r}\right)^6\right] = E[x^{12} - x^6]$$

where E is an energy parameter and x is a distance parameter ($x \sim r^{-1}$). So the Lennard-Jones potential can be written as $Ef(r)$ and so satisfies the conditions of the above general intermolecular potential.

2–44. Use the following data for argon at 300.0 K to determine the value of B_{2V}. The accepted value is $-15.05\ \text{cm}^3\cdot\text{mol}^{-1}$.

P/atm	ρ/mol·L^{-1}	P/atm	ρ/mol·L^{-1}
0.01000	0.000406200	0.4000	0.0162535
0.02000	0.000812500	0.6000	0.0243833
0.04000	0.00162500	0.8000	0.0325150
0.06000	0.00243750	1.000	0.0406487
0.08000	0.00325000	1.500	0.0609916
0.1000	0.00406260	2.000	0.0813469
0.2000	0.00812580	3.000	0.122094

We can use Equation 2.22 to express $Z - 1$ (neglecting terms of $O(\overline{V}^{-2})$) as

$$Z - 1 = \overline{V}^{-1} B_{2V}(T) = \rho B_{2V}(T)$$

A plot of $(Z - 1)$ versus $\overline{V}^{-1}$ for the data given for oxygen at 273.15 K is shown below.

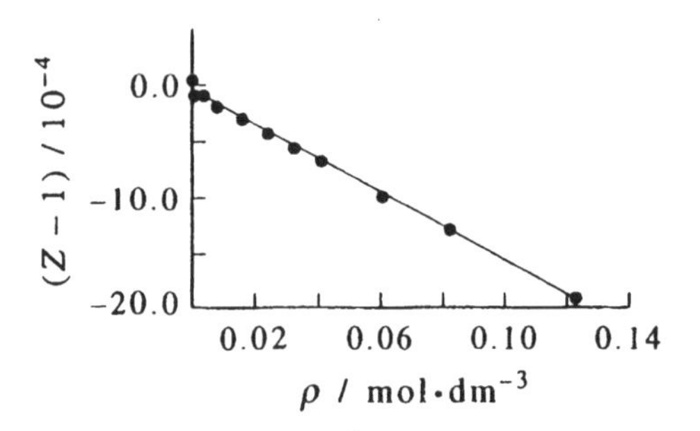

The slope of the best-fit line to the data is B_{2V} and is equal to $-15.15\ \text{cm}^3\cdot\text{mol}^{-1}$.

2–45. Using Figure 2.15 and the Lennard-Jones parameters given in Table 2.7, estimate $B_{2V}(T)$ for CH_4(g) at 0°C.

For methane, $\epsilon/k_B = 149$ K and $2\pi\sigma^3 N_A/3 = 68.1\ \text{cm}^3\cdot\text{mol}^{-1}$ (Table 2.7). Then (by definition of T^*)

$$T^* = \frac{k_B T}{\epsilon} = \frac{273.15\ \text{K}}{149\ \text{K}} = 1.83$$

From Figure 2.15, we estimate $B^*_{2V}(T^*) \approx -0.9$. Then (also by definition)

$$B_{2V}(273.15\ \text{K}) = \frac{2\pi\sigma^3 N_A B^*_{2V}(T^*)}{3} \approx -60\ \text{cm}^3\cdot\text{mol}^{-1}$$

2–46. Show that $B_{2V}(T)$ obeys the law of corresponding states for a square-well potential with a *fixed* value of λ (in other words, if all molecules had the same value of λ).

We use Equation 2.25 to express $B_{2V}(T)$:

$$B_{2V}(T) = -2\pi N_A \int_0^{\infty} \left[e^{-u(r)/k_B T} - 1\right] r^2 dr$$

where, since we have a square-well potential of fixed value λ,

$$
\begin{aligned}
u(r) &= \infty && \text{if} \quad r < \sigma \\
&= -\epsilon && \text{if} \quad \sigma < r < \lambda\sigma \\
&= 0 && \text{if} \quad \lambda\sigma < r
\end{aligned}
$$

We can now integrate $B_{2V}(T)$ over the three intervals $0 < r < \sigma, \sigma < r < \lambda\sigma$, and $\lambda\sigma < r < \infty$:

$$
\begin{aligned}
B_{2V}(T) &= -2\pi N_A \int_0^{\sigma} (0-1) r^2 dr - 2\pi N_A \int_{\sigma}^{\lambda\sigma} \left[e^{\epsilon/k_BT} - 1\right] r^2 dr + 0 \\
&= 2\pi N_A \left(\frac{\sigma^3}{3}\right) - 2\pi N_A \left(e^{\epsilon/k_BT} - 1\right) \frac{\lambda^3\sigma^3 - \sigma^3}{3} \\
&= \frac{2\pi N_A \sigma^3}{3} \left[1 - \left(\lambda^3 - 1\right)\left(e^{\epsilon/k_BT} - 1\right)\right]
\end{aligned}
$$

If we divide $B_{2V}(T)$ by σ^3 and let this quantity be a reduced value of $B_{2V}(T)$, this reduced second virial coefficient will be molecule-independent and therefore satisfy the law of corresponding states.

2–47. Using the Lennard-Jones parameters in Table 2.7, show that the following second virial cofficient data satisfy the law of corresponding states.

Argon		Nitrogen		Ethane	
T/K	$B_{2V}(T)$ $/10^{-3}$ dm^3·mol^{-1}	T/K	$B_{2V}(T)$ $/10^{-3}$ dm^3·mol^{-1}	T/K	$B_{2V}(T)$ $/10^{-3}$ dm^3·mol^{-1}
173	−64.3	143	−79.8	311	−164.9
223	−37.8	173	−51.9	344	−132.5
273	−22.1	223	−26.4	378	−110.0
323	−11.0	273	−10.3	411	−90.4
423	+1.2	323	−0.3	444	−74.2
473	4.7	373	+6.1	478	−59.9
573	11.2	423	11.5	511	−47.4
673	15.3	473	15.3		
		573	20.6		
		673	23.5		

Find the reduced parameters of each gas by dividing T by ε/k_B and B_{2V} by $2\pi\sigma^3 N_A/3$ (Table 2.7). Below, we plot $B^*_{2V}(T)$ versus T^* for each gas.

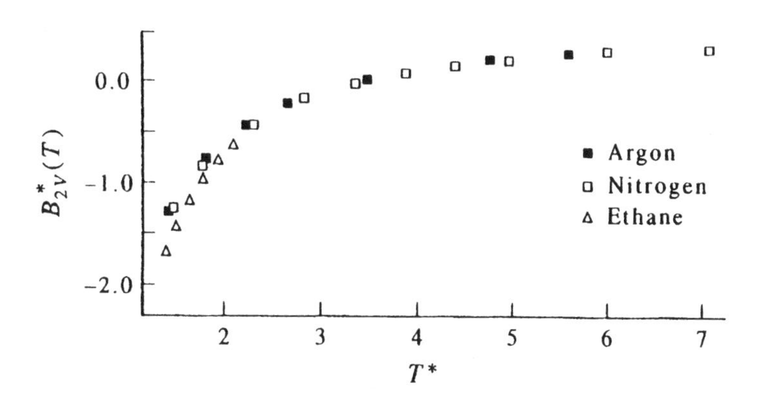

The data points for all three gases fall on the same curve, consistent with the law of corresponding states.

2–48. In Section 2–4, we expressed the van der Waals equation in reduced units by dividing P, $\overline{V}$, and T by their critical values. This suggests we can write the second virial coefficient in reduced form by dividing $B_{2V}(T)$ by $\overline{V}_c$ and T by T_c (instead of $2\pi N_A\sigma^3/3$ and ε/k as we did in Section 2–5). Reduce the second virial coefficient data given in the previous problem by using the values of $\overline{V}_c$ and T_c in Table 2.5 and show that the reduced data satisfy the law of corresponding states.

We find the reduced parameters of each gas by dividing T by T_c and B_{2V} by $\overline{V}_c$ (Table 2.5). Below, we plot $B_{2V}(T)/\overline{V}_c$ vs. T/T_c for each gas.

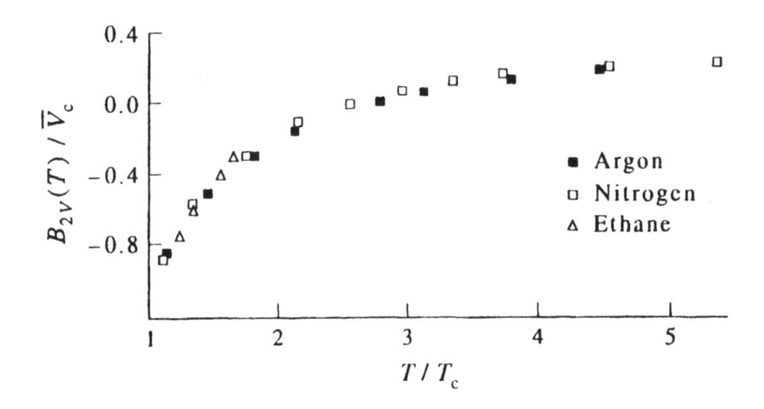

The data points for all three gases fall on the same curve, consistent with the law of corresponding states.

2–49. Listed below are experimental second virial coefficient data for argon, krypton, and xenon.

	$B_{2V}(T)/10^{-3}\text{dm}^3\cdot\text{mol}^{-1}$		
T/K	Argon	Krypton	Xenon
173.16	−63.82		
223.16	−36.79		
273.16	−22.10	−62.70	−154.75
298.16	−16.06		−130.12
323.16	−11.17	−42.78	−110.62
348.16	−7.37		−95.04
373.16	−4.14	−29.28	−82.13
398.16	−0.96		
423.16	+1.46	−18.13	−62.10
473.16	4.99	−10.75	−46.74
573.16	10.77	+0.42	−25.06
673.16	15.72	7.42	−9.56
773.16	17.76	12.70	−0.13
873.16	19.48	17.19	+7.95
973.16			14.22

Use the Lennard-Jones parameters in Table 2.7 to plot $B^*_{2V}(T^*)$, the reduced second virial coefficient, versus T^*, the reduced temperature, to illustrate the law of corresponding states.

Find the reduced parameters of each gas by dividing T by ε/k_B and B_{2V} by $2\pi\sigma^3 N_A/3$ (Table 2.7). Below, we plot $B^*_{2V}(T)$ versus T^* for each gas.

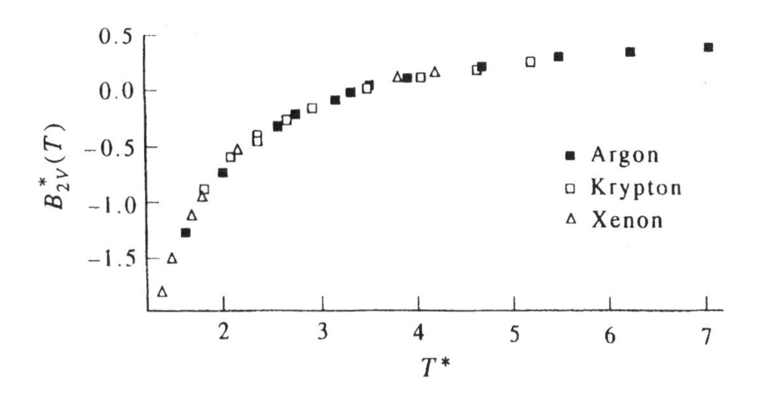

The data points for all three gases fall on the same curve, consistent with the law of corresponding states.

2–50. Use the critical temperatures and the critical molar volumes of argon, krypton, and xenon to illustrate the law of corresponding states with the data given in Problem 2–49.

We find the reduced parameters of each gas by dividing T by T_c and B_{2V} by $\overline{V}_c$ (Table 2.5). Below, we plot $B_{2V}(T)/\overline{V}_c$ vs. T/T_c for each gas.

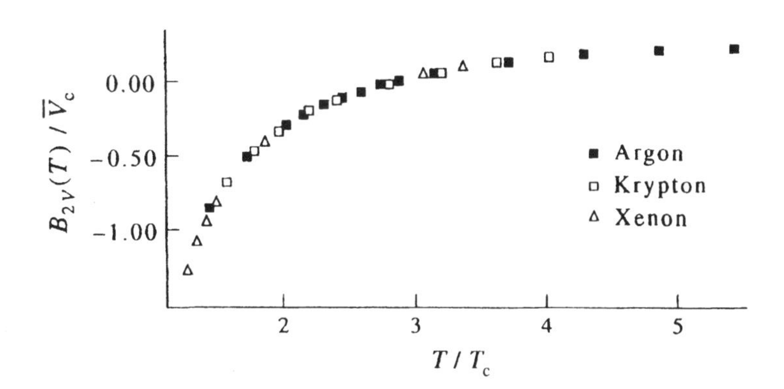

The data points for all three gases fall on the same curve, consistent with the law of corresponding states.

2–51. Evaluate $B^*_{2V}(T^*)$ in Equation 2.31 numerically from $T^* = 1.00$ to 10.0 using a packaged numerical integration program such as *MathCad* or *Mathematica*. Compare the reduced second virial coefficient data from Problem 2–49 and $B^*_{2V}(T^*)$ by plotting them all on the same graph.

Below is a table with some representative values of $B_{2V}^*(T^*)$ calculated using the numerical integration package in *Mathematica*.

T^*	$B_{2V}^*(T^*)$
1.00	−2.538081336
2.00	−0.6276252881
3.00	−0.1152339638
4.00	0.1154169217
5.00	0.2433435028
6.00	0.3229043727
7.00	0.3760884671
8.00	0.4134339539
9.00	0.4405978376
10.00	0.4608752841

The plot below shows both the curve obtained from numerically integrating $B_{2V}^*(T^*)$ and the reduced second virial coefficient data from Problem 2–49. The numerical integration is an excellent fit to the data.

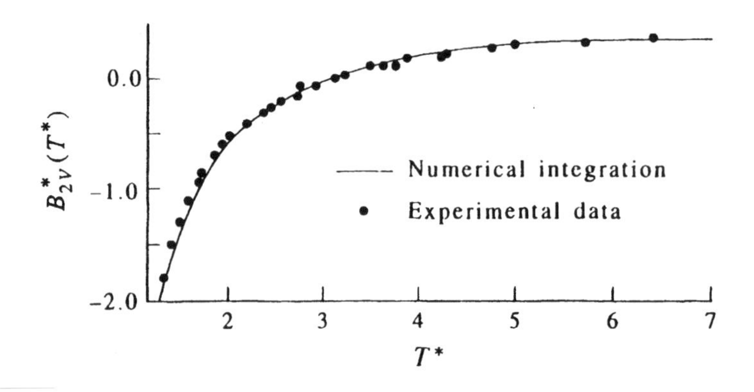

2–52. Show that the units of the right side of Equation 2.35 are energy.

$$u_{\text{induced}}(r) = -\frac{\mu_1^2\alpha_2}{(4\pi\varepsilon_0)^2r^6} - \frac{\mu_2^2\alpha_1}{(4\pi\varepsilon_0)^2r^6}$$

We know that $\alpha/4\pi\varepsilon_0$ has units of m^3, $4\pi\varepsilon_0$ has units of $\text{C}\cdot\text{V}^{-1}\cdot\text{m}^{-1}$, μ has units of $\text{C}\cdot\text{m}$, and r has units of m. Thus

$$\text{units}[u_{\text{induced}}(r)] = \frac{(\text{C}\cdot\text{m})^2\text{m}^3}{\text{C}\cdot\text{V}^{-1}\cdot\text{m}^{-1}\text{m}^6} = \text{C}\cdot\text{V} = \text{J}$$

2–53. Show that the sum of Equations 2.33, 2.35, and 2.36 gives Equation 2.37.

The sum of these three equations is

$$-\frac{\mu_1^2\alpha_2 + \mu_2^2\alpha_1}{(4\pi\varepsilon_0)^2r^6} - \frac{2\mu_1^2\mu_2^2}{(4\pi\varepsilon_0)^2(3k_\text{B}T)}\frac{1}{r^6} - \frac{3}{2}\left(\frac{I_1I_2}{I_1+I_2}\right)\frac{\alpha_1\alpha_2}{(4\pi\varepsilon_0)^2}\frac{1}{r^6}$$

For identical atoms or molecules, $I_1 = I_2 = I$, $\alpha_2 = \alpha_1 = \alpha$, and $\mu_1 = \mu_2 = \mu$, and so the sum becomes

$$\Sigma(u) = -\frac{1}{(4\pi\varepsilon_0)^2 r^6}\left(2\mu^2\alpha + \frac{2\mu^4}{3k_BT} + \frac{3I\alpha^2}{4}\right)$$

The coefficient of the r^6 term, C_6, is therefore

$$C_6 = \frac{2\mu^4}{3(4\pi\varepsilon_0)^2 k_BT} + \frac{2\alpha\mu^2}{(4\pi\varepsilon_0)^2} + \frac{3}{4}\frac{I\alpha^2}{(4\pi\varepsilon_0)^2}$$

as in Equation 2.37.

2–54. Compare the values of the coefficient of r^{-6} for $N_2(g)$ using Equation 2.37 and the Lennard-Jones parameters given in Table 2.7.

Using Equation 2.37, we find that

$$\begin{aligned}
C_6 &= \frac{2\mu^4}{3(4\pi\varepsilon_0)^2 k_BT} + \frac{2\alpha\mu^2}{(4\pi\varepsilon_0)^2} + \frac{3}{4}\frac{I\alpha^2}{(4\pi\varepsilon_0)^2}\\
&= 0 + 0 + 0.75\left(2.496\times10^{-18}\text{ J}\right)\left(1.77\times10^{-30}\text{ m}^3\right)^2\\
&= 5.86\times10^{-78}\text{ J}\cdot\text{m}^6
\end{aligned}$$

Using the Lennard-Jones parameters,

$$\begin{aligned}
C_6 &= 4\varepsilon\sigma^6 = 4\frac{\varepsilon}{k_B}k_B\sigma^6\\
&= 4\,(95.1\text{ K})\,(1.381\times10^{-23}\text{ J}\cdot\text{K}^{-1})\left(370\times10^{-12}\text{ m}\right)^6\\
&= 1.35\times10^{-77}\text{ J}\cdot\text{m}^6
\end{aligned}$$

The coefficient of r^{-6} obtained using the Lennard-Jones parameters is about twice that obtained using Equation 2.37.

2–55. Show that

$$B_{2V}(T) = B - \frac{A}{RT^{3/2}}$$

and

$$B_{3V}(T) = B^2 + \frac{AB}{RT^{3/2}}$$

for the Redlich-Kwong equation.

Begin with the Redlich-Kwong equation (Equation 2.7):

$$\begin{aligned}
P &= \frac{RT}{\overline{V} - B} - \frac{A}{T^{1/2}\overline{V}\left(\overline{V} + B\right)}\\
&= \frac{RT}{\overline{V}\left(1 - \frac{B}{\overline{V}}\right)} - \frac{A}{T^{1/2}\overline{V}^2\left(1 + \frac{B}{\overline{V}}\right)}
\end{aligned}$$

Expanding the fractions $1/(1 - B/\overline{V})$ and $1/(1 + B/\overline{V})$ (Equation C.3) gives

$$P = \frac{RT}{\overline{V}}\left[1 + \frac{B}{\overline{V}} + \frac{B^2}{\overline{V}^2} + O(\overline{V}^{-3})\right] - \frac{A}{T^{1/2}\overline{V}^2}\left[1 - \frac{B}{\overline{V}} - \frac{B^2}{\overline{V}^2} - O(\overline{V}^{-3})\right]$$

$$= \frac{RT}{\overline{V}} - \frac{A}{T^{1/2}\overline{V}^2} + \left(\frac{RT}{\overline{V}} + \frac{A}{T^{1/2}\overline{V}^2}\right)\left[\frac{B}{\overline{V}} - \frac{B^2}{\overline{V}^2} - O(\overline{V}^{-3})\right]$$

We then use the definition of Z to find that

$$Z = \frac{P\overline{V}}{RT} = 1 - \frac{A}{RT^{3/2}\overline{V}} + \left(1 + \frac{A}{RT^{3/2}\overline{V}}\right)\left[\frac{B}{\overline{V}} - \frac{B^2}{\overline{V}^2} - O(\overline{V}^{-3})\right]$$

We compare this with Equation 2.22,

$$Z = 1 + \frac{B_{2V}(T)}{\overline{V}} + \frac{B_{3V}(T)}{\overline{V}^2} + O(\overline{V}^3)$$

Setting the coefficients of $1/\overline{V}$ and $1/\overline{V}^2$ equal to one another gives

$$B_{2V} = B - \frac{A}{RT^{3/2}}$$

and

$$B_{3V} = B^2 + \frac{AB}{RT^{3/2}}$$

2–56. Show that the second and third virial coefficients of the Peng-Robinson equation are

$$B_{2V}(T) = \beta - \frac{\alpha}{RT}$$

and

$$B_{3V}(T) = \beta^2 + \frac{2\alpha\beta}{RT}$$

Begin with the Peng-Robinson equation (Equation 2.8):

$$P = \frac{RT}{\overline{V} - \beta} - \frac{\alpha}{\overline{V}(\overline{V} + \beta) + \beta(\overline{V} - \beta)}$$

$$= \frac{RT}{\overline{V}\left(1 - \frac{\beta}{\overline{V}}\right)} - \frac{\alpha}{\overline{V}^2}\left[\frac{1}{1 - \left(\frac{\beta^2}{\overline{V}^2} - \frac{2\beta}{\overline{V}}\right)}\right]$$

Expanding the fractions $1/(1 - \beta/\overline{V})$ and $1/(1 - \beta^2/\overline{V}^2 - 2\beta/\overline{V})$ (Equation C.3) gives

$$P = \frac{RT}{\overline{V}}\left[1 + \frac{\beta}{\overline{V}} + \frac{\beta^2}{\overline{V}^2} + O(\overline{V}^{-3})\right] - \frac{\alpha}{\overline{V}^2}\left[1 + \left(\frac{\beta^2}{\overline{V}^2} - \frac{2\beta}{\overline{V}}\right) + \left(\frac{\beta^2}{\overline{V}^2} - \frac{2\beta}{\overline{V}}\right)^2 + O(\overline{V}^{-3})\right]$$

$$= \frac{RT}{\overline{V}}\left[1 + \frac{\beta}{\overline{V}} + \frac{\beta^2}{\overline{V}^2} + O(\overline{V}^{-3})\right] - \frac{\alpha}{\overline{V}^2}\left[1 + \frac{2\beta}{\overline{V}} + O(\overline{V}^{-2})\right]$$

We then use the definition of Z to find that

$$Z = \frac{P\overline{V}}{RT} = 1 + \frac{\beta}{\overline{V}} + \frac{\beta^2}{\overline{V}^2} - \frac{\alpha}{\overline{V}RT} - \frac{2\alpha\beta}{RT\overline{V}^2} + O(\overline{V}^{-3})$$

We compare this with Equation 2.22,

$$Z = 1 + \frac{B_{2V}(T)}{\overline{V}} + \frac{B_{3V}(T)}{\overline{V}^2} + O(\overline{V}^3)$$

Setting the coefficients of $1/\overline{V}$ and $1/\overline{V}^2$ equal to one another gives

$$B_{2V} = \beta - \frac{\alpha}{RT}$$

and

$$B_{3V} = \beta^2 + \frac{2\alpha\beta}{RT}$$

2–57. The square-well parameters for krypton are $\varepsilon/k = 136.5$ K, $\sigma = 327.8$ pm, and $\lambda = 1.68$. Plot $B_{2V}(T)$ against T and compare your results with the data given in Problem 2–49.

From Problem 2.46, we know that, for a square-well potential,

$$\begin{aligned} B_{2V}(T) &= \frac{2\pi N_A \sigma^3}{3}\left[1 - \left(\lambda^3 - 1\right)\left(e^{\varepsilon/k_B T} - 1\right)\right] \\ &= \frac{2\pi(6.022 \times 10^{23}\ \text{mol}^{-1})(327.8\ \text{pm})^3}{3} \\ &\quad \times \left\{1 - \left[(1.68)^3 - 1\right]\left(e^{(136.5\ \text{K})/T} - 1\right]\right\} \end{aligned}$$

The plot below shows both the square-well potential curve and the experimental data from Problem 2–49 for krypton. The square-well potential is a very good fit to the data.

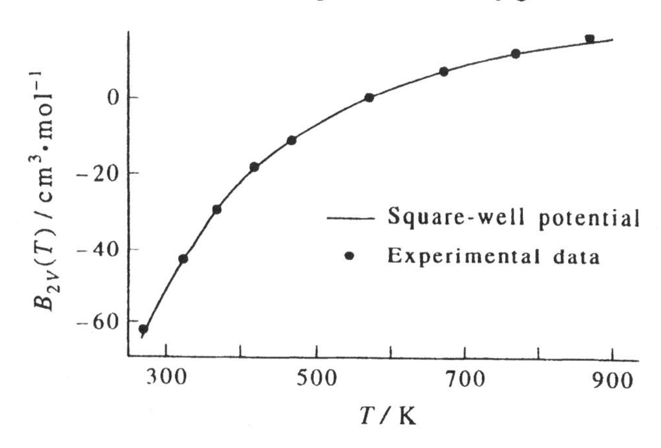

2–58. The coefficient of thermal expansion α is defined as

$$\alpha = \frac{1}{\overline{V}}\left(\frac{\partial \overline{V}}{\partial T}\right)_P$$

Show that

$$\alpha = \frac{1}{T}$$

for an ideal gas.

For an ideal gas, $P\overline{V} = RT$. Taking the partial derivative of both sides of this equation with respect to T gives

$$P\overline{V} = RT$$
$$\left(\frac{\partial \overline{V}}{\partial T}\right)_P = \frac{R}{P}$$

and so

$$\alpha = \frac{1}{\overline{V}}\left(\frac{\partial \overline{V}}{\partial T}\right)_P = \frac{R}{P\overline{V}} = \frac{1}{T}$$

2–59. The isothermal compressibility κ is defined as

$$\kappa = -\frac{1}{\overline{V}}\left(\frac{\partial \overline{V}}{\partial P}\right)_T$$

Show that

$$\kappa = \frac{1}{P}$$

for an ideal gas.

For an ideal gas, $P = RT/\overline{V}$. Taking the partial derivative of both sides of this equation with respect to P gives

$$1 = \frac{-RT}{\overline{V}^2}\left(\frac{\partial \overline{V}}{\partial P}\right)_T$$
$$-\frac{\overline{V}^2}{RT} = \left(\frac{\partial \overline{V}}{\partial P}\right)_T$$

and so

$$\kappa = -\frac{1}{\overline{V}}\left(\frac{\partial \overline{V}}{\partial P}\right)_T = \frac{\overline{V}}{RT} = \frac{1}{P}$$

Probability and Statistics

PROBLEMS AND SOLUTIONS

B–1. Consider a particle to be constrained to lie along a one-dimensional segment 0 to a. Quantum mechanics tells us that the probability that the particle is found to lie between x and $x + dx$ is given by

$$p(x)dx = \frac{2}{a}\sin^2\frac{n\pi x}{a}dx$$

where $n = 1, 2, 3, \ldots$. First show that $p(x)$ is normalized. Now calculate the average position of the particle along the line segment. The integrals that you need are (*The CRC Handbook of Chemistry and Physics* or *The CRC Standard Mathematical Tables*, CRC Press)

$$\int \sin^2 \alpha x dx = \frac{x}{2} - \frac{\sin 2\alpha x}{4\alpha}$$

and

$$\int x \sin^2 \alpha x dx = \frac{x^2}{4} - \frac{x \sin 2\alpha x}{4\alpha} - \frac{\cos 2\alpha x}{8\alpha^2}$$

If $p(x)$ is normalized, then $\int_0^a p(x)dx = 1$.

$$\begin{aligned}
\int_0^a p(x)dx &= \int_0^a \frac{2}{a}\sin^2\frac{n\pi x}{a}dx \\
&= \left[\frac{2}{a}\left(\frac{x}{2} - \frac{\sin 2n\pi a^{-1}x}{4n\pi a^{-1}}\right)\right]_0^a \\
&= \frac{2}{a}\left[\frac{a}{2} - \frac{\sin 2n\pi}{4n\pi a^{-1}} - 0 + \frac{\sin 0}{4n\pi a^{-1}}\right] \\
&= \frac{2}{a}\left(\frac{a}{2}\right) = 1
\end{aligned}$$

Thus, $p(x)$ is normalized. To find the average position of the particle along the line segment, use Equation B.12:

$$\begin{aligned}
\langle x \rangle &= \int_0^a xp(x)dx = \int_0^a x\frac{2}{a}\sin^2\frac{n\pi x}{a}dx \\
&= \frac{2}{a}\left[\frac{x^2}{4} - \frac{x \sin 2n\pi a^{-1}x}{4n\pi a^{-1}} - \frac{\cos 2n\pi a^{-1}x}{8n^2\pi^2a^{-2}}\right]_0^a \\
&= \frac{2}{a}\left[\frac{a^2}{4} - \frac{a \sin 2n\pi}{4n\pi a^{-1}} - \frac{\cos 2n\pi}{8n^2\pi^2a^{-2}} + \frac{\cos 0}{8n^2\pi^2a^{-2}}\right] \\
&= \frac{2}{a}\left[\frac{a^2}{4} - \frac{1}{8n^2\pi^2a^{-2}} + \frac{1}{8n^2\pi^2a^{-2}}\right] = \frac{2}{a}\left(\frac{a^2}{4}\right) \\
&= \frac{a}{2}
\end{aligned}$$

B–2. Calculate the variance associated with the probability distribution given in Problem B–1. The necessary integral is (*CRC tables*)

$$\int x^2 \sin^2 \alpha x dx = \frac{x^3}{6} - \left(\frac{x^2}{4\alpha} - \frac{1}{8\alpha^3}\right)\sin 2\alpha x - \frac{x\cos 2\alpha x}{4\alpha^2}$$

Use Equation B.13:

$$\begin{aligned}
\langle x^2 \rangle &= \int_0^a x^2 p(x)dx = \frac{2}{a}\int_0^a x^2 \sin^2 \frac{n\pi x}{a} dx \\
&= \frac{2}{a}\left[\frac{x^3}{6} - \left(\frac{x^2}{4n\pi a^{-1}} - \frac{1}{8n^3\pi^3 a^{-3}}\right)\sin 2n\pi a^{-1}x - \frac{x\cos 2n\pi a^{-1}x}{4n^2\pi^2 a^{-2}}\right]_0^a \\
&= \frac{2}{a}\left[\frac{a^3}{6} - \left(\frac{a^2}{4n\pi a^{-1}} - \frac{1}{8n^3\pi^3 a^{-3}}\right)\sin 2n\pi - \frac{a\cos 2n\pi}{4n^2\pi^2 a^{-2}} - 0\right] \\
&= \frac{2}{a}\left(\frac{a^3}{6} - \frac{a^3}{4n^2\pi^2}\right) \\
&= \frac{a^2}{3} - \frac{a^2}{2n^2\pi^2}
\end{aligned}$$

The variance σ^2 is given by

$$\sigma^2 = \langle x^2 \rangle - \langle x \rangle^2 \tag{B.8}$$

Using the result of Problem B–1 and the above result for $\langle x^2 \rangle$ gives

$$\begin{aligned}
\sigma^2 &= \frac{a^2}{3} - \frac{a^2}{2n^2\pi^2} - \frac{a^2}{4} \\
&= \frac{a^2}{12} - \frac{a^2}{2n^2\pi^2}
\end{aligned}$$

B–3. Using the probability distribution given in Problem B–1, calculate the probability that the particle will be found between 0 and $a/2$. The necessary integral is given in Problem B–1.

The probability that the particle will lie within the region 0 to $a/2$ is given by $\int_0^{a/2} p(x)dx$ (Equation B.10).

$$\begin{aligned}
\int_0^{a/2} p(x)dx &= \int_0^{a/2} \frac{2}{a}\sin^2 \frac{n\pi x}{a} dx \\
&= \frac{2}{a}\int_0^{a/2} \sin^2 \frac{n\pi x}{a} dx \\
&= \frac{2}{a}\left[\frac{x}{2} - \frac{\sin 2n\pi a^{-1}x}{4n\pi a^{-1}}\right]_0^{a/2} \\
&= \frac{2}{a}\left[\frac{a}{4} - \frac{\sin 2n\pi}{8n\pi a^{-1}} + \frac{\sin 0}{4n\pi a^{-1}}\right] \\
&= \frac{2}{a}\left(\frac{a}{4}\right) = \frac{1}{2}
\end{aligned}$$

The probability of the particle being found in exactly half the box is 0.5.

B–4. Prove explicitly that

$$\int_{-\infty}^{\infty} e^{-\alpha x^2} dx = 2\int_0^{\infty} e^{-\alpha x^2} dx$$

by breaking the integral from $-\infty$ to ∞ into one from $-\infty$ to 0 and another from 0 to ∞. Now let $z = -x$ in the first integral and $z = x$ in the second to prove the above relation.

$$\int_{-\infty}^{\infty} e^{-\alpha x^2} dx = \int_{-\infty}^{0} e^{-\alpha x^2} dx + \int_0^{\infty} e^{-\alpha x^2} dx$$

We can let $z = -x$ in the first integral and $z = x$ in the second and write

$$\begin{aligned}\int_{-\infty}^{\infty} e^{-\alpha x^2} dx &= -\int_{\infty}^{0} e^{-\alpha z^2} dz + \int_0^{\infty} e^{-\alpha x^2} dx \\ &= \int_0^{\infty} e^{-\alpha z^2} dz + \int_0^{\infty} e^{-\alpha z^2} dz \\ &= 2\int_0^{\infty} e^{-\alpha z^2} dz = 2\int_0^{\infty} e^{-\alpha x^2} dx\end{aligned}$$

B–5. By using the procedure in Problem B–4, show explicitly that

$$\int_{-\infty}^{\infty} x e^{-\alpha x^2} dx = 0$$

$$\int_{-\infty}^{\infty} x e^{-\alpha x^2} dx = \int_{-\infty}^{0} x e^{-\alpha x^2} dx + \int_0^{\infty} x e^{-\alpha x^2} dx$$

We can let $x = -z$ in the first integral to get

$$\begin{aligned}\int_{-\infty}^{\infty} x e^{-\alpha x^2} dx &= \int_{\infty}^{0} z e^{-\alpha z^2} dz + \int_0^{\infty} x e^{-\alpha x^2} dx \\ &= -\int_0^{\infty} z e^{-\alpha z^2} dz + \int_0^{\infty} x e^{-\alpha x^2} dx \\ &= -\int_0^{\infty} x e^{-\alpha x^2} dx + \int_0^{\infty} x e^{-\alpha x^2} dx \\ &= 0\end{aligned}$$

B–6. According to the kinetic theory of gases, the molecules in a gas travel at various speeds, and that the probability that a molecule has a speed between v and $v + dv$ is given by

$$p(v)dv = 4\pi \left(\frac{m}{2\pi k_B T}\right)^{3/2} v^2 e^{-mv^2/2k_B T} dv \qquad 0 \le v < \infty$$

where m is the mass of the particle, k_B is the Boltzmann constant (the molar gas constant R divided by the Avogadro constant), and T is the Kelvin temperature. The probability distribution of molecular speeds is called the Maxwell-Boltzmann distribution. First show that $p(v)$ is normalized, and then determine the average speed as a function of temperature. The necessary integrals are (*CRC tables*)

$$\int_0^{\infty} x^{2n} e^{-\alpha x^2} dx = \frac{1\cdot 3\cdot 5\cdots(2n-1)}{2^{n+1}\alpha^n}\left(\frac{\pi}{\alpha}\right)^{1/2} \qquad n \ge 1$$

and

$$\int_0^\infty x^{2n+1} e^{-\alpha x^2} dx = \frac{n!}{2\alpha^{n+1}}$$

where $n!$ is n factorial, or $n! = n(n-1)(n-2)\cdots(1)$.

First, we demonstrate that $p(v)$ is normalized by showing that $\int_0^\infty p(v)dv = 1$:

$$\begin{aligned}\int_0^\infty p(v)dv &= 4\pi \left(\frac{m}{2\pi k_B T}\right)^{3/2} \int_0^\infty v^2 e^{-mv^2/2k_BT} dv \\ &= 4\pi \left(\frac{m}{2\pi k_B T}\right)^{3/2} \frac{2k_BT}{4m} \left(\frac{2\pi k_B T}{m}\right)^{1/2} \\ &= \pi^{3/2} \left(\frac{m}{2\pi k_B T}\right)^{3/2} \left(\frac{2k_BT}{m}\right)^{3/2} \\ &= 1\end{aligned}$$

Using Equation B.12, we write

$$\begin{aligned}\langle v\rangle &= 4\pi \left(\frac{m}{2\pi k_B T}\right)^{3/2} \int_0^\infty v^3 e^{-mv^2/2k_BT} dv \\ &= 4\pi \left(\frac{m}{2\pi k_B T}\right)^{3/2} \left[2\left(\frac{m}{2k_BT}\right)^2\right]^{-1} \\ &= 2\pi \left(\frac{m}{2\pi k_B T}\right)^{3/2} \left(\frac{2k_BT}{m}\right)^2 \\ &= \left(\frac{8k_BT}{\pi m}\right)^{1/2}\end{aligned}$$

B–7. Use the Maxwell-Boltzmann distribution in Problem B–6 to determine the average kinetic energy of a gas-phase molecule as a function of temperature. The necessary integral is given in Problem B–6.

Kinetic energy, KE, is defined as $\text{KE} = \frac{1}{2}mv^2$, so $\langle \text{KE}\rangle = \frac{1}{2}m\langle v^2\rangle$. Using Equation B.13, we write

$$\begin{aligned}\langle v^2\rangle &= 4\pi \left(\frac{m}{2\pi k_B T}\right)^{3/2} \int_0^\infty v^4 e^{-mv^2/2k_BT} dv \\ &= 4\pi \left(\frac{m}{2\pi k_B T}\right)^{3/2} \frac{3}{8} \left(\frac{2k_BT}{m}\right)^2 \left(\frac{2\pi k_B T}{m}\right)^{1/2} \\ &= \frac{3k_BT}{m}\end{aligned}$$

And so $E = \frac{1}{2}m\langle v^2\rangle = \frac{3}{2}k_BT$.

CHAPTER 3

The Boltzmann Factor and Partition Functions

PROBLEMS AND SOLUTIONS

3–1. How would you describe an ensemble whose systems are one-liter containers of water at 25°C?

An unlimited number of one-liter containers of water in an essentially infinite heat bath at 298 K.

3–2. Show that Equation 3.8 is equivalent to $f(x+y) = f(x)f(y)$. In this problem, we will prove that $f(x) \propto e^{ax}$. First, take the logarithm of the above equation to obtain

$$\ln f(x+y) = \ln f(x) + \ln f(y)$$

Differentiate both sides with respect to x (keeping y fixed) to get

$$\left[\frac{\partial \ln f(x+y)}{\partial x}\right]_y = \frac{d \ln f(x+y)}{d(x+y)}\left[\frac{\partial(x+y)}{\partial x}\right]_y = \frac{d \ln f(x+y)}{d(x+y)}$$
$$= \frac{d \ln f(x)}{dx}$$

Now differentiate with respect to y (keeping x fixed) and show that

$$\frac{d \ln f(x)}{dx} = \frac{d \ln f(y)}{dy}$$

For this relation to be true for all x and y, each side must equal a constant, say a. Show that

$$f(x) \propto e^{ax} \qquad \text{and} \qquad f(y) \propto e^{ay}$$

Let $x = E_1 - E_2$ and $y = E_2 - E_3$. Then $x + y = E_1 - E_3$, and we can write Equation 3.8 as $f(x+y) = f(x)f(y)$. Taking the logarithm of this equation gives

$$\ln f(x+y) = \ln[f(x)f(y)]$$
$$= \ln f(x) + \ln f(y)$$

We then differentiate with respect to x and find

$$\left[\frac{\partial \ln f(x+y)}{\partial x}\right]_y = \left[\frac{\partial \ln f(x)}{\partial x}\right]_y + 0$$
$$\frac{d \ln f(x+y)}{d(x+y)}\left[\frac{\partial(x+y)}{\partial x}\right]_y = \frac{d \ln f(x)}{dx}$$
$$\frac{d \ln f(x+y)}{d(x+y)} = \frac{d \ln f(x)}{dx} \qquad (1)$$

Likewise, we differentiate with respect to y and find

$$\left[\frac{\partial \ln f(x+y)}{\partial y}\right]_x = 0 + \left[\frac{\partial \ln f(y)}{\partial y}\right]_x$$

$$\frac{d \ln f(x+y)}{d(x+y)}\left[\frac{\partial (x+y)}{\partial y}\right]_x = \frac{d \ln f(y)}{dy}$$

$$\frac{d \ln f(x+y)}{d(x+y)} = \frac{d \ln f(y)}{dy} \tag{2}$$

Equations 1 and 2 are equal to one another, and equal to a constant a (as stated in the text of the problem), so

$$\frac{d \ln f(x)}{dx} = \frac{d \ln f(y)}{dy} = a$$

We can integrate $d \ln f(x) = a dx$ to find

$$\int d \ln f(x) = \int a dx$$

$$\ln f(x) \propto ax$$

$$f(x) \propto e^{ax}$$

Similarly, integrating $d \ln f(y) = a dy$ gives $f(y) \propto e^{ay}$.

3–3. Show that $a_l/a_i = e^{\beta(E_i - E_l)}$ implies that $a_j = Ce^{-\beta E_j}$.

$$a_l/a_i = e^{\beta(E_i - E_l)} = e^{-\beta E_l} e^{\beta E_i}$$

$$a_l = \left(a_i e^{\beta E_i}\right) e^{-\beta E_l} = Ce^{-\beta E_l}$$

The subscript l is arbitrary, and so this shows the desired result.

3–4. Prove to yourself that $\sum_i e^{-\beta E_i} = \sum_j e^{-\beta E_j}$.

Expanding both sums gives

$$e^{-\beta E_1} + e^{-\beta E_2} + \ldots = e^{-\beta E_1} + e^{-\beta E_2} + \ldots$$

3–5. Show that the partition function in Example 3–1 can be written as

$$Q(\beta, B_z) = 2\cosh\left(\frac{\beta\hbar\gamma B_z}{2}\right) = 2\cosh\left(\frac{\hbar\gamma B_z}{2k_B T}\right)$$

Use the fact that $d \cosh x/dx = \sinh x$ to show that

$$\langle E\rangle = -\frac{\hbar\gamma B_z}{2}\tanh\frac{\beta\hbar\gamma B_z}{2} = -\frac{\hbar\gamma B_z}{2}\tanh\frac{\hbar\gamma B_z}{2k_B T}$$

Recall that the definition of $\cosh u$ is

$$\cosh u = \frac{e^u + e^{-u}}{2}$$

Therefore, $2\cosh u = e^u + e^{-u}$. The partition function in Example 3–1 is

$$Q(\beta, B_z) = e^{\beta\hbar\gamma B_z/2} + e^{-\beta\hbar\gamma B_z/2} = 2\cosh\frac{\beta\hbar\gamma B_z}{2}$$

We use Equation 3.20 to write

$$\begin{aligned}\langle E\rangle &= -\frac{1}{Q}\left(\frac{\partial Q}{\partial\beta}\right)_{B_z}\\ &= -\frac{1}{2\cosh\dfrac{\beta\hbar\gamma B_z}{2}}\left(\frac{2\partial\cosh\dfrac{\beta\hbar\gamma B_z}{2}}{\partial\beta}\right)_{B_z}\end{aligned}$$

Let $K = \hbar\gamma B_z/2$. Then we can write $\langle E\rangle$ as

$$\begin{aligned}\langle E\rangle &= -\frac{1}{2\cosh(K\beta)}\left[\frac{2\partial\cosh(K\beta)}{\partial\beta}\right]_{B_z}\\ &= \left[-\frac{1}{\cosh(K\beta)}\right]K\sinh(K\beta)\\ &= -K\tanh(K\beta)\\ &= -\frac{\hbar\gamma B_z}{2}\tanh\frac{\hbar\gamma B_z}{2k_BT}\end{aligned}$$

3–6. Use either the expression for $\langle E\rangle$ in Example 3–1 or the one in Problem 3–5 to show that

$$\langle E\rangle \longrightarrow -\frac{\hbar\gamma B_z}{2} \quad \text{as} \quad T\longrightarrow 0$$

and that

$$\langle E\rangle \longrightarrow 0 \quad \text{as} \quad T\longrightarrow \infty$$

A graph of $\tanh(x)$ is presented below for reference.

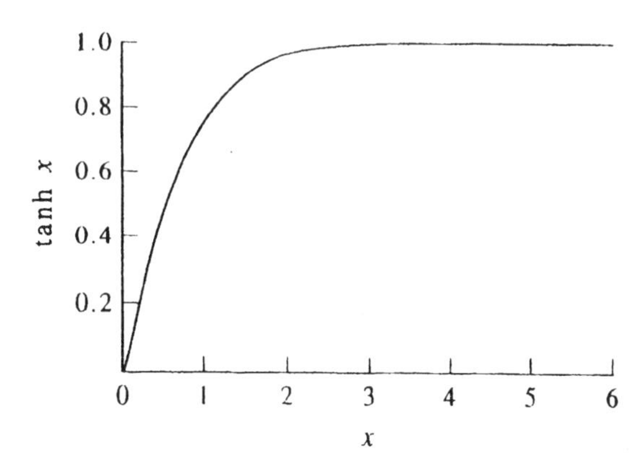

From Problem 3–5, we write

$$\langle E\rangle = -\frac{\hbar\gamma B_z}{2}\tanh\frac{\hbar\gamma B_z}{2k_BT} = -\frac{\hbar\gamma B_z}{2}\tanh\frac{\hbar\gamma B_z\beta}{2}$$

As $T \to 0, \beta \to \infty$, so

$$\lim_{T\to 0}\langle E\rangle = \lim_{\beta\to\infty} -\frac{\hbar\gamma B_z}{2}\tanh\frac{\hbar\gamma B_z\beta}{2} = -\frac{\hbar\gamma B_z}{2}$$

As $T \to \infty, \beta \to 0$, so

$$\lim_{T\to\infty}\langle E\rangle = \lim_{\beta\to 0} -\frac{\hbar\gamma B_z}{2}\tanh\frac{\hbar\gamma B_z\beta}{2} = 0$$

3–7. Generalize the results of Example 3–1 to the case of a spin-1 nucleus. Determine the low-temperature and high-temperature limits of $\langle E\rangle$.

For a spin-1 nucleus, $m_I = -1, 0,$ or 1. Therefore

$$E_0 = 0 \qquad \text{and} \qquad E_{\pm 1} = \mp\hbar\gamma B_z$$

We now apply Equation 3.14 to find

$$Q(\beta, B_z) = 1 + e^{-\beta\hbar\gamma B_z} + e^{\beta\hbar\gamma B_z} = 1 + 2\cosh(\beta\hbar\gamma B_z)$$

Using Equation 3.20 for $\langle E\rangle$ gives

$$\begin{aligned}
\langle E\rangle &= -\left(\frac{\partial \ln Q}{\partial\beta}\right)_{B_z} = -\frac{1}{Q}\left(\frac{\partial Q}{\partial\beta}\right)_{B_z} \\
&= -\frac{1}{1+2\cosh(\beta\hbar\gamma B_z)}\left[2\hbar\gamma B_z\sinh(\beta\hbar\gamma B_z)\right] \\
&= -2\hbar\gamma B_z\frac{\sinh(\beta\hbar\gamma B_z)}{1+2\cosh(\beta\hbar\gamma B_z)}
\end{aligned}$$

Below, graphs of $\sinh(x)$ and $\cosh(x)$ are presented for reference.

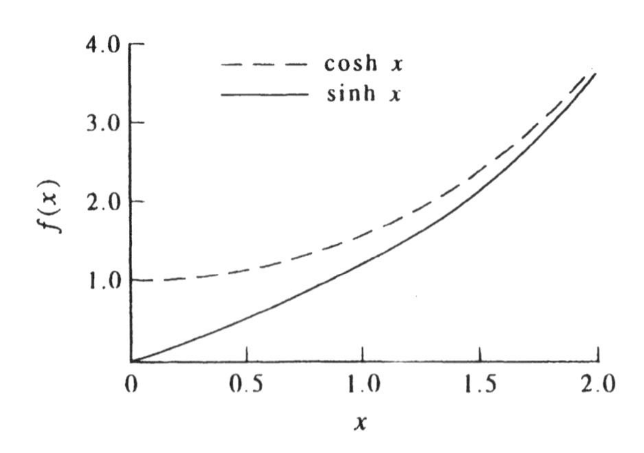

As $T \to 0, \beta \to \infty$, so

$$\lim_{T\to 0}\langle E\rangle = \lim_{\beta\to\infty} -2\hbar\gamma B_z\frac{\sinh(\beta\hbar\gamma B_z)}{1+2\cosh(\beta\hbar\gamma B_z)} = \lim_{\beta\to\infty} -\frac{\hbar\gamma B_z}{\tanh(\beta\hbar\gamma B_z)} = -\hbar\gamma B_z$$

As $T \to \infty, \beta \to 0$, so

$$\lim_{T\to\infty}\langle E\rangle = \lim_{\beta\to 0} -2\hbar\gamma B_z\frac{\sinh(\beta\hbar\gamma B_z)}{1+2\cosh(\beta\hbar\gamma B_z)} = 0$$

3–8. If N_w is the number of protons aligned with a magnetic field B_z and N_o is the number of protons opposed to the field, show that

$$\frac{N_o}{N_w} = e^{-\hbar\gamma B_z/k_B T}$$

Given that $\gamma = 26.7522 \times 10^7\ \text{rad}\cdot\text{T}^{-1}\cdot\text{s}^{-1}$ for a proton, calculate N_o/N_w as a function of temperature for a field strength of 5.0 T. At what temperature is $N_o = N_w$? Interpret this result physically.

The energy of the protons aligned with the magnetic field is $E_w = -\frac{1}{2}\hbar\gamma B_z$, and the energy of the protons aligned against the magnetic field is $E_o = \frac{1}{2}\hbar\gamma B_z$. Using Equation 3.10, we can write

$$\frac{N_o}{N_w} = \exp\left[\beta\left(-\frac{1}{2}\hbar\gamma B_z - \frac{1}{2}\hbar\gamma B_z\right)\right] = \exp\left[-\frac{\hbar\gamma B_z}{k_B T}\right]$$

For a field strength of 5.0 T,

$$\frac{N_o}{N_w} = \exp\left[\frac{-(1.05459 \times 10^{-34}\ \text{J}\cdot\text{s})(26.7522 \times 10^7\ \text{rad}\cdot\text{T}^{-1}\cdot\text{s}^{-1})(5.0\ \text{T})}{(1.38066 \times 10^{-23}\ \text{J}\cdot\text{K}^{-1})T}\right]$$

$$= \exp\left[\frac{-0.010\ \text{K}}{T}\right]$$

Let $N_o = N_w$. Then

$$\frac{N_o}{N_w} = 1 = \exp\left[\frac{-0.010\ \text{K}}{T}\right]$$

$$0 = \frac{-0.010\ \text{K}}{T}$$

and T is undefined. Therefore, we can never attain the condition $N_w = N_o$.

3–9. In Section 3–3, we derived an expression for $\langle E\rangle$ for a monatomic ideal gas by applying Equation 3.20 to $Q(N, V, T)$ given by Equation 3.22. Apply Equation 3.21 to

$$Q(N, V, T) = \frac{1}{N!}\left(\frac{2\pi m k_B T}{h^2}\right)^{3N/2} V^N$$

to derive the same result. Note that this expression for $Q(N, V, T)$ is simply Equation 3.22 with β replaced by $1/k_B T$.

Begin with Equation 3.21:

$$\langle E\rangle = k_B T^2 \left(\frac{\partial \ln Q}{\partial T}\right)_{N,V}$$

We can use the partition function given in the problem to find $(\partial \ln Q/\partial T)_{N,V}$:

$$Q(N, V, T) = \frac{V^N}{N!}\left(\frac{2\pi m k_B}{h^2}\right)^{3N/2} T^{3N/2}$$

$$\ln Q = \frac{3N}{2}\ln T + \text{terms not involving } T$$

$$\left(\frac{\partial \ln Q}{\partial T}\right)_{N,V} = \frac{3N}{2T}$$

Substituting this last result into Equation 3.21 gives

$$\langle E\rangle = k_B T^2 \frac{3N}{2T} = \frac{3}{2}(Nk_B T)$$

3–10. A gas absorbed on a surface can sometimes be modelled as a two-dimensional ideal gas. We will learn in Chapter 4 that the partition function of a two-dimensional ideal gas is

$$Q(N, A, T) = \frac{1}{N!}\left(\frac{2\pi m k_B T}{h^2}\right)^N A^N$$

where A is the area of the surface. Derive an expression for $\langle E\rangle$ and compare your result with the three-dimensional result.

We can use the partition function given in the problem to find $(\partial \ln Q/\partial T)_{N,V}$:

$$Q(N, A, T) = \frac{1}{N!}\left(\frac{2\pi m k_B T}{h^2}\right)^N A^N$$

$$\ln Q = N \ln T + \text{terms not involving } T$$

$$\left(\frac{\partial \ln Q}{\partial T}\right)_{N,V} = \frac{N}{T}$$

Substituting this last result into Equation 3.21 gives

$$\langle E\rangle = \frac{Nk_B T^2}{T} = Nk_B T$$

The value for $\langle E\rangle$ for a two-dimensional ideal gas is less than the three-dimensional ideal gas result by $Nk_B T/2$. We infer that each dimension contributes $Nk_B T/2$ to the value of $\langle E\rangle$.

3–11. Although we will not do so in this book, it is possible to derive the partition function for a monatomic van der Waals gas.

$$Q(N, V, T) = \frac{1}{N!}\left(\frac{2\pi m k_B T}{h^2}\right)^{3N/2} (V - Nb)^N e^{aN^2/Vk_B T}$$

where a and b are the van der Waals constants. Derive an expression for the energy of a monatomic van der Waals gas.

We can use the partition function given in the problem to find $(\partial \ln Q/\partial T)_{N,V}$:

$$Q(N, V, T) = \frac{(V - Nb)^N}{N!}\left(\frac{2\pi m k_B}{h^2}\right)^{3N/2} e^{aN^2/Vk_B T} T^{3N/2}$$

$$\ln Q = \frac{3N}{2}\ln T + \frac{aN^2}{Vk_B T} + \text{terms not involving } T$$

$$\left(\frac{\partial \ln Q}{\partial T}\right)_{N,V} = \frac{3N}{2T} - \frac{aN^2}{Vk_B T^2}$$

Substituting this last result into Equation 3.21 gives

$$\langle E\rangle = k_B T^2\left(\frac{3N}{2T} - \frac{aN^2}{Vk_B T^2}\right) = \frac{3}{2}Nk_B T - \frac{aN^2}{V}$$

3–12. An approximate partition function for a gas of hard spheres can be obtained from the partition function of a monatomic gas by replacing V in Equation 3.22 (and the following equation) by $V - b$, where b is related to the volume of the N hard spheres. Derive expressions for the energy and the pressure of this system.

We can use the partition function specified in the problem to find

$$Q(N, V, T) = \frac{(V-b)^N}{N!}\left(\frac{2\pi m k_B}{h^2}\right)^{3N/2} T^{3N/2}$$

$$\ln Q = \frac{3N}{2}\ln T + \text{terms not involving } T$$

Substituting into Equation 3.21, we find that the energy $\langle E\rangle$ is the same as that for a monatomic ideal gas: $3Nk_BT/2$. We can use the partition function specified in the problem to find

$$Q(N, V, T) = \frac{1}{N!}\left(\frac{2\pi m k_B T}{h^2}\right)^{3N/2}(V-b)^N$$

$$\ln Q = N\ln(V-b) + \text{terms not involving } V$$

We substitute into Equation 3.32 to find

$$\langle P\rangle = k_BT\left(\frac{\partial \ln Q}{\partial V}\right)_{N,\beta} = \frac{Nk_BT}{V-b}$$

3–13. Use the partition function in Problem 3–10 to calculate the heat capacity of a two-dimensional ideal gas.

In Problem 3–10, we found that $\langle E\rangle = Nk_BT$ for the given partition function. Since $\langle E\rangle = U$, we can substitute into Equation 3.25 to write

$$C_V = \left(\frac{\partial U}{\partial T}\right)_{N,V} = Nk_B$$

3–14. Use the partition function for a monatomic van der Waals gas given in Problem 3–11 to calculate the heat capacity of a monatomic van der Waals gas. Compare your result with that of a monatomic ideal gas.

The partition function given in Problem 3–11 is

$$Q(N, V, \beta) = \frac{1}{N!}\left(\frac{2\pi m}{h^2\beta}\right)^{3N/2}(V - Nb)^N e^{\beta a N^2/V}$$

In Problem 3.11, we found that $\langle E\rangle = 3Nk_BT/2 - aN^2/V$ for a monatomic van der Waals gas. Since $\langle E\rangle = U$, we can substitute into Equation 3.25 to write

$$C_V = \left(\frac{\partial U}{\partial T}\right)_{N,V} = \frac{3Nk_B}{2}$$

The heat capacity of a monatomic van der Waals gas is the same as that of a monatomic ideal gas.

3–15. Using the partition function given in Example 3–2, show that the pressure of an ideal diatomic gas obeys $PV = Nk_BT$, just as it does for a monatomic ideal gas.

We can use the partition function specified in Example 3–2 to find $(\partial \ln Q/\partial V)$:

$$Q(N, V, T) = \frac{1}{N!}\left(\frac{2\pi m}{h^2\beta}\right)^N V^N \left(\frac{8\pi^2 I}{h^2\beta}\right)^N \left(\frac{e^{-\beta h\nu/2}}{1-e^{-\beta h\nu}}\right)^N$$

$$\ln Q = N \ln V + \text{terms independent of } V$$

$$\left(\frac{\partial \ln Q}{\partial V}\right) = \frac{N}{V}$$

Substituting this result into Equation 3.32 gives

$$\langle P\rangle = \frac{Nk_BT}{V}$$

which is the ideal gas law.

3–16. Show that if a partition function is of the form

$$Q(N, V, T) = \frac{[q(V, T)]^N}{N!}$$

and if $q(V, T) = f(T)V$ [as it does for a monatomic ideal gas (Equation 3.22) and a diatomic ideal gas (Example 3–2)], then the ideal-gas equation of state results.

If $q(V, T) = f(T)V$, then

$$Q = \frac{f(T)^N V^N}{N!}$$

We can use this partition function to find $(\partial \ln Q/\partial V)$:

$$\ln Q = \ln\left(\frac{f(T)^N}{N!}\right) + N \ln V$$

$$= N \ln V + \text{terms independent of V}$$

$$\left(\frac{\partial \ln Q}{\partial V}\right)_{N,\beta} = \frac{N}{V}$$

Substituting this result into Equation 3.32 gives

$$\langle P\rangle = \frac{Nk_BT}{V} = \frac{nRT}{V}$$

which is the ideal gas law.

3–17. Use Equation 3.27 and the value of $\tilde{\nu}$ for O_2 given in Table 1.4 to calculate the value of the molar heat capacity of $O_2(g)$ from 300 K to 1000 K (see Figure 3.3).

From Table 1.4, $\tilde{\nu} = 1556\ \text{cm}^{-1}$. Using this value, we can write

$$\frac{hc\tilde{\nu}}{k_B} = \frac{(6.626\times10^{-34}\ \text{J}\cdot\text{s}^{-1})(2.998\times10^{10}\ \text{cm}\cdot\text{s}^{-1})(1556\ \text{cm}^{-1})}{1.3807\times10^{-23}\ \text{J}\cdot\text{K}^{-1}} = 2240\ \text{K}$$

We can substitute this value into Equation 3.27 to obtain the expression

$$\frac{\overline{C}_V}{R} = \frac{5}{2} + \left(\frac{2240\text{ K}}{T}\right)^2 \frac{e^{-2240\text{ K}/T}}{(1 - e^{-2240\text{ K}/T})^2}$$

We plot $\overline{C}_V$ versus T below.

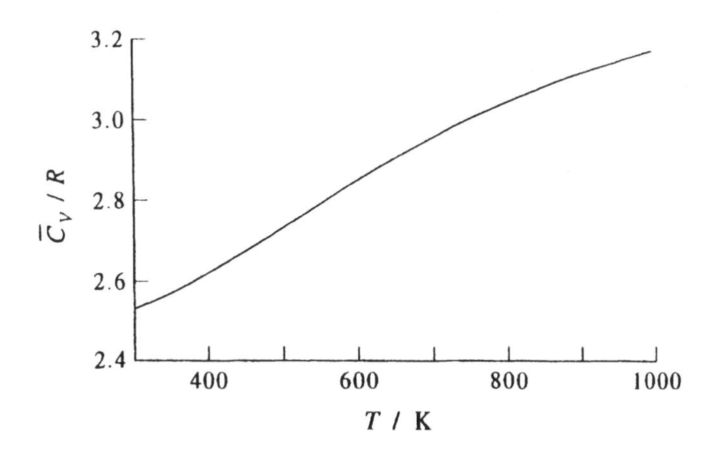

3–18. Show that the heat capacity given by Equation 3.29 in Example 3–3 obeys a law of corresponding states.

Let $h\nu/k_B = \Theta_E$. We can then write Equation 3.29 as

$$\frac{\overline{C}_V}{R} = 3\left(\frac{\Theta_E}{T}\right)^2 \frac{e^{-\Theta_E/T}}{(1 - e^{-\Theta_E/T})^2}$$

Now let $T_R = T/\Theta_E$. The above expression becomes

$$\frac{\overline{C}_V}{R} = 3\left(\frac{1}{T_R}\right)^2 \frac{e^{-1/T_R}}{(1 - e^{-1/T_R})^2}$$

This equation obeys a law of corresponding states.

3–19. Consider a system of independent, distinguishable particles that have only two quantum states with energy ε_0 (let $\varepsilon_0 = 0$) and ε_1. Show that the molar heat capacity of such a system is given by

$$\overline{C}_V = R(\beta\varepsilon)^2 \frac{e^{-\beta\varepsilon}}{(1 + e^{-\beta\varepsilon})^2}$$

and that $\overline{C}_V$ plotted against $\beta\varepsilon$ passes through a maximum value at $\beta\varepsilon$, given by the solution to $\beta\varepsilon/2 = \coth\beta\varepsilon/2$. Use a table of values of $\coth x$ (for example, the *CRC Standard Mathematical Tables*) to show that $\beta\varepsilon = 2.40$.

For the system described above, Equation 3.34 becomes

$$q = \sum_i e^{-\beta\varepsilon_i} = 1 + e^{-\beta\varepsilon_1} = 1 + e^{-\beta\varepsilon}$$

and Equation 3.43 becomes

$$\langle \varepsilon \rangle = \sum_j \varepsilon_j \frac{e^{-\beta\varepsilon_j}}{q} = \frac{\varepsilon_1 e^{-\beta\varepsilon_1}}{1+e^{-\beta\varepsilon_1}} = \frac{\varepsilon e^{-\beta\varepsilon}}{1+e^{-\beta\varepsilon}}$$

where we have dropped the subscript "1" on ε in both of the above expressions for simplicity. We can use Equation 3.42 to write $\langle E \rangle$ as

$$\langle E \rangle = N\langle \varepsilon \rangle = \frac{N\varepsilon e^{-\beta\varepsilon}}{1+e^{-\beta\varepsilon}}$$

We now substitute this last expression into Equation 3.25 to find C_V:

$$C_V = \frac{\partial \langle E \rangle}{\partial T} = -\frac{1}{k_B T^2}\left(\frac{\partial \langle E \rangle}{\partial \beta}\right)$$

$$= -\frac{N}{k_B T^2}\left[\frac{-\varepsilon^2 e^{-\beta\varepsilon}}{1+e^{-\beta\varepsilon}} + \frac{\varepsilon^2 e^{-2\beta\varepsilon}}{(1+e^{-\beta\varepsilon})^2}\right]$$

$$= -\frac{N}{k_B T^2}\left[\frac{-\varepsilon^2 e^{-\beta\varepsilon}}{(1+e^{-\beta\varepsilon})^2}\right] = Nk_B \frac{(\beta\varepsilon)^2 e^{-\beta\varepsilon}}{(1+e^{-\beta\varepsilon})^2}$$

$$\overline{C}_V = \frac{R(\beta\varepsilon)^2 e^{-\beta\varepsilon}}{(1+e^{-\beta\varepsilon})^2}$$

We now wish to find the maximum value of $\overline{C}_V$. Let $\beta\varepsilon = x$. Then

$$\frac{\overline{C}_V}{R} = \frac{x^2 e^{-x}}{(1+e^{-x})^2}$$

$$\frac{\partial(\overline{C}_V/R)}{\partial x} = \frac{2xe^{-x}}{(1+e^{-x})^2} - \frac{x^2 e^{-x}}{(1+e^{-x})^2} + \frac{2x^2 e^{-2x}}{(1+e^{-x})^3}$$

At the maximum value of $\overline{C}_V$, $(\partial \overline{C}_V/\partial x) = 0$, so

$$0 = 2 - x + 2x\frac{e^{-x}}{1+e^{-x}}$$

$$2 = x\left(1 - \frac{2e^{-x}}{1+e^{-x}}\right) = x\left(\frac{1-e^{-x}}{1+e^{-x}}\right)$$

$$\frac{1+e^{-x}}{1-e^{-x}} = \frac{x}{2}$$

$$\frac{e^{x/2}+e^{-x/2}}{e^{x/2}-e^{-x/2}} = \frac{x}{2}$$

$$\coth\frac{x}{2} = \frac{x}{2}$$

$$x = 2.40 = \beta\varepsilon$$

and $x = \beta\varepsilon = 2.40$.

3–20. Deriving the partition function for an Einstein crystal is not difficult (see Example 3–3). Each of the N atoms of the crystal is assumed to vibrate independently about its lattice position, so that the crystal is pictured as N independent harmonic oscillators, each vibrating in three directions. The partition function of a harmonic oscillator is

$$q_{ho}(T) = \sum_{v=0}^{\infty} e^{-\beta(v+\frac{1}{2})h\nu}$$

$$= e^{-\beta h\nu/2} \sum_{v=0}^{\infty} e^{-\beta v h\nu}$$

This summation is easy to evaluate if you recognize it as the so-called geometric series (MathChapter C)

$$\sum_{v=0}^{\infty} x^v = \frac{1}{1-x}$$

Show that

$$q_{\text{ho}}(T) = \frac{e^{-\beta h\nu/2}}{1 - e^{-\beta h\nu}}$$

and that

$$Q = e^{-\beta U_0} \left(\frac{e^{-\beta h\nu/2}}{1 - e^{-\beta h\nu}} \right)^{3N}$$

where U_0 simply represents the zero-of-energy, where all N atoms are infinitely separated.

For a one-dimensional harmonic oscillator,

$$q_{\text{ho}}(T) = e^{-\beta h\nu/2} \sum_{v=0}^{\infty} e^{-\beta v h\nu}$$

Let $x = e^{-\beta h\nu}$. Then

$$q_{\text{ho}}(T) = e^{-\beta h\nu/2} \sum_{v=0}^{\infty} x^v = e^{-\beta h\nu/2} \frac{1}{1-x} = \frac{e^{-\beta h\nu/2}}{1 - e^{-\beta h\nu}}$$

A three-dimensional harmonic oscillator can be viewed as three independent one-dimensional harmonic oscillators, and so in three dimensions

$$q_{\text{3D-ho}}(T) = [q_{\text{ho}}(T)]^3 = \left(\frac{e^{-\beta h\nu/2}}{1 - e^{-\beta h\nu}} \right)^3$$

Because each particle is distinguishable (due to its position in the lattice), the system partition function can be written as

$$Q = q_{\text{3D-ho},a}(V,T) q_{\text{3D-ho},b}(V,T) q_{\text{3D-ho},c}(V,T) \ldots \tag{3.33}$$

where $a, b, c, \ldots$ are independent atoms. Then

$$Q = \left(\frac{e^{-\beta h\nu/2}}{1 - e^{-\beta h\nu}} \right)^{3N} e^{-\beta U_0}$$

The $e^{-\beta U_0}$ term takes the zero-of-energy into account.

3–21. Show that

$$S = \sum_{i=1}^{2} \sum_{j=0}^{1} x^i y^j = x(1+y) + x^2(1+y) = (x+x^2)(1+y)$$

by summing over j first and then over i. Now obtain the same result by writing S as a product of two separate summations.

Summing first over j and then over i, we obtain

$$S=\sum_{i=1}^{2}\sum_{j=0}^{1}x^iy^j=\sum_{i=1}^{2}x^i\left(y^0+y^1\right)=x(1+y)+x^2(1+y)=(x+x^2)(1+y)$$

Writing S as a product of two separate summations, we find

$$S=\sum_{i=1}^{2}x^i\sum_{j=0}^{1}y^j=(x+x^2)(1+y)$$

3–22. Evaluate

$$S=\sum_{i=0}^{2}\sum_{j=0}^{1}x^{i+j}$$

by summing over j first and then over i. Now obtain the same result by writing S as a product of two separate summations.

Summing first over j and then over i, we obtain

$$S=\sum_{i=0}^{2}\sum_{j=0}^{1}x^{i+j}=\sum_{i=0}^{2}(x^{i+0}+x^{i+1})=1+x+x^2+x+x^2+x^3=1+2x+2x^2+x^3$$

Writing S as a product of two separate summations, we find

$$\sum_{i=0}^{2}x^i\sum_{j=0}^{1}x^j=(1+x+x^2)(1+x)=1+2x+2x^2+x^3$$

3–23. How many terms are there in the following summations?

a. $S=\sum_{i=1}^{3}\sum_{j=1}^{2}x^iy^j$ **b.** $S=\sum_{i=1}^{3}\sum_{j=0}^{2}x^iy^j$ **c.** $S=\sum_{i=1}^{3}\sum_{j=1}^{2}\sum_{k=1}^{2}x^iy^jz^k$

---

a. 6 terms

b. 9 terms

c. 12 terms

3–24. Consider a system of two noninteracting identical fermions, each of which has states with energies ε_1, ε_2, and ε_3. How many terms are there in the unrestricted evaluation of $Q(2, V, T)$? Enumerate the allowed total energies in the summation in Equation 3.37 (see Example 3–5). How many terms occur in $Q(2, V, T)$ when the fermion restriction is taken into account?

There are nine terms in the unrestricted evaluation. The allowed total energies given by the summation in Equation 3.37 are

$$\begin{array}{ll} \varepsilon_1+\varepsilon_3=\varepsilon_3+\varepsilon_1 & \varepsilon_1+\varepsilon_1 \\ \varepsilon_1+\varepsilon_2=\varepsilon_2+\varepsilon_1 & \varepsilon_2+\varepsilon_2 \\ \varepsilon_2+\varepsilon_3=\varepsilon_3+\varepsilon_2 & \varepsilon_2+\varepsilon_2 \end{array}$$

When the fermion restriction is taken into account (no two identical fermions can occupy the same single-particle energy state), the allowed total energies are $\varepsilon_1 + \varepsilon_3$, $\varepsilon_1 + \varepsilon_2$, and $\varepsilon_2 + \varepsilon_3$, and so there are only three terms in $Q(2, V, T)$.

3–25. Redo Problem 3–24 for the case of bosons instead of fermions.

In this case there are six allowed terms: the three alloweed in Problem 3–24 and the three in which the ε_j are the same ($\varepsilon_1 + \varepsilon_1$, $\varepsilon_2 + \varepsilon_2$, and $\varepsilon_3 + \varepsilon_3$).

3–26. Consider a system of three noninteracting identical fermions, each of which has states with energies ε_1, ε_2, and ε_3. How many terms are there in the unrestricted evaluation of $Q(3, V, T)$? Enumerate the allowed total energies in the summation of Equation 3.37 (see Example 3–5). How many terms occur in $Q(3, V, T)$ when the fermion restriction is taken into account?

There are 27 terms in the unrestricted evaluation. The allowed total energies given by the summation in Equation 3.37 are

$$
\begin{aligned}
1.\quad & \varepsilon_1 + \varepsilon_2 + \varepsilon_3 = \varepsilon_1 + \varepsilon_3 + \varepsilon_2 = \varepsilon_2 + \varepsilon_1 + \varepsilon_3 = \varepsilon_2 + \varepsilon_3 + \varepsilon_1 \\
& \qquad = \varepsilon_3 + \varepsilon_1 + \varepsilon_2 = \varepsilon_3 + \varepsilon_2 + \varepsilon_1 \\
2.\quad & \varepsilon_1 + \varepsilon_1 + \varepsilon_2 = \varepsilon_1 + \varepsilon_2 + \varepsilon_1 = \varepsilon_2 + \varepsilon_1 + \varepsilon_1 \\
3.\quad & \varepsilon_1 + \varepsilon_1 + \varepsilon_3 = \varepsilon_1 + \varepsilon_3 + \varepsilon_1 = \varepsilon_3 + \varepsilon_1 + \varepsilon_1 \\
4.\quad & \varepsilon_1 + \varepsilon_2 + \varepsilon_2 = \varepsilon_2 + \varepsilon_2 + \varepsilon_1 = \varepsilon_2 + \varepsilon_1 + \varepsilon_2 \\
5.\quad & \varepsilon_1 + \varepsilon_3 + \varepsilon_3 = \varepsilon_3 + \varepsilon_3 + \varepsilon_1 = \varepsilon_3 + \varepsilon_1 + \varepsilon_3 \\
6.\quad & \varepsilon_2 + \varepsilon_3 + \varepsilon_3 = \varepsilon_3 + \varepsilon_3 + \varepsilon_2 = \varepsilon_3 + \varepsilon_2 + \varepsilon_3 \\
7.\quad & \varepsilon_2 + \varepsilon_2 + \varepsilon_3 = \varepsilon_2 + \varepsilon_3 + \varepsilon_2 = \varepsilon_3 + \varepsilon_2 + \varepsilon_2 \\
8.\quad & \varepsilon_1 + \varepsilon_1 + \varepsilon_1 \\
9.\quad & \varepsilon_2 + \varepsilon_2 + \varepsilon_2 \\
10.\quad & \varepsilon_3 + \varepsilon_3 + \varepsilon_3
\end{aligned}
$$

When the fermion restriction is taken into account (no two identical fermions can occupy the same single-particle energy state), the only allowed total energy is $\varepsilon_1 + \varepsilon_2 + \varepsilon_3$, and so there is only one term in $Q(3, V, T)$.

3–27. Redo Problem 3–26 for the case of bosons instead of fermions.

In this case there are ten allowed terms, given by the total energies

$$
\begin{array}{ll}
\varepsilon_1 + \varepsilon_2 + \varepsilon_3 & \varepsilon_2 + \varepsilon_2 + \varepsilon_3 \\
\varepsilon_1 + \varepsilon_1 + \varepsilon_3 & \varepsilon_1 + \varepsilon_3 + \varepsilon_3 \\
\varepsilon_2 + \varepsilon_2 + \varepsilon_3 & \varepsilon_3 + \varepsilon_2 + \varepsilon_3 \\
\varepsilon_2 + \varepsilon_2 + \varepsilon_1 & \varepsilon_1 + \varepsilon_1 + \varepsilon_1 \\
\varepsilon_2 + \varepsilon_2 + \varepsilon_2 & \varepsilon_3 + \varepsilon_3 + \varepsilon_3
\end{array}
$$

3–28. Evaluate $(N/V)(h^2/8mk_BT)^{3/2}$ (see Table 3.1) for $O_2(g)$ at its normal boiling point, 90.20 K. Use the ideal-gas equation of state to calculate the density of $O_2(g)$ at 90.20 K.

Using the ideal-gas equation of state,

$$\frac{N}{V} = \frac{P}{k_B T} = \frac{1.013 \times 10^5 \text{ kg}\cdot\text{m}^{-1}\cdot\text{s}^{-2}}{(1.381 \times 10^{-23} \text{ J}\cdot\text{K}^{-1})(90.20 \text{ K})} = 8.134 \times 10^{25} \text{ kg}\cdot\text{m}^{-3}$$

Then

$$\frac{N}{V}\left(\frac{h^2}{8mk_BT}\right)^{3/2} = 8.134 \times 10^{25} \text{ kg}\cdot\text{m}^{-3}\left[\frac{(6.626 \times 10^{-34} \text{ J}\cdot\text{s})^2}{8(5.315 \times 10^{-26} \text{ kg})(1.381 \times 10^{-23} \text{ J}\cdot\text{K}^{-1})(90.20 \text{ K})}\right]^{3/2}$$
$$= 1.943 \times 10^{-6}$$

which is much less than unity.

3–29. Evaluate $(N/V)(h^2/8mk_BT)^{3/2}$ (see Table 3.1) for He(g) at its normal boiling point 4.22 K. Use the ideal-gas equation of state to calculate the density of He(g) at 4.22 K.

Using the ideal-gas equation of state,

$$\frac{N}{V} = \frac{P}{k_B T} = \frac{1.013 \times 10^5 \text{ kg}\cdot\text{m}^{-1}\cdot\text{s}^{-2}}{(1.381 \times 10^{-23} \text{ J}\cdot\text{K}^{-1})(4.22 \text{ K})} = 1.739 \times 10^{27} \text{ kg}\cdot\text{m}^{-3}$$

Then

$$\frac{N}{V}\left(\frac{h^2}{8mk_BT}\right)^{3/2} = 1.739 \times 10^{27} \text{ kg}\cdot\text{m}^{-3}\left[\frac{(6.626 \times 10^{-34} \text{ J}\cdot\text{s})^2}{8(6.646 \times 10^{-27} \text{ kg})(1.381 \times 10^{-23} \text{ J}\cdot\text{K}^{-1})(4.22 \text{ K})}\right]^{3/2}$$
$$= 0.0928$$

which is not much less than unity, because of the small mass of He and the fact that here we consider helium at 4.22 K.

3–30. Evaluate $(N/V)(h^2/8mk_BT)^{3/2}$ for the electrons in sodium metal at 298 K. Take the density of sodium to be 0.97 g·mL^{-1}. Compare your result with the value given in Table 3.1.

The mass of an electron is 9.11×10^{-31} kg, and the density of sodium is 2.54×10^{28} particles per cubic meter. Thus

$$\frac{N}{V}\left(\frac{h^2}{8mk_BT}\right)^{3/2} = (2.54 \times 10^{28} \text{ m}^3)\left[\frac{(6.626 \times 10^{-34} \text{ J}\cdot\text{s})^2}{8(9.11 \times 10^{-31} \text{ kg})(1.381 \times 10^{-23} \text{ J}\cdot\text{K}^{-1})(298 \text{ K})}\right]^{3/2} = 1420$$

which, because of the very small mass of an electron, is much greater than unity. This is essentially equivalent to the value given in Table 3.1.

3–31. Evaluate $(N/V)(h^2/8mk_BT)^{3/2}$ (see Table 3.1) for liquid hydrogen at its normal boiling point 20.3 K. The density of H_2(l) at its boiling point is 0.067 g·mL^{-1}.

The density of hydrogen gas is 2.00×10^{28} particles per cubic meter. Thus

$$\frac{N}{V}\left(\frac{h^2}{8mk_BT}\right)^{3/2} = (2.00 \times 10^{28} \text{ m}^{-3})\left[\frac{(6.626 \times 10^{-34} \text{ J}\cdot\text{s})^2}{8(3.35 \times 10^{-27} \text{ kg})(1.381 \times 10^{-23} \text{ J}\cdot\text{K}^{-1})(20.3 \text{ K})}\right]^{3/2}$$
$$= 0.286$$

3–32. Because the molecules in an ideal gas are independent, the partition function of a mixture of monatomic ideal gases is of the form

$$Q(N_1, N_2, V, T) = \frac{[q_1(V,T)]^{N_1}}{N_1!} \frac{[q_2(V,T)]^{N_2}}{N_2!}$$

where

$$q_j(V,T) = \left(\frac{2\pi m_j k_B T}{h^2}\right)^{3/2} V \qquad j = 1, 2$$

Show that

$$\langle E \rangle = \frac{3}{2}(N_1 + N_2)k_B T$$

and that

$$PV = (N_1 + N_2)k_B T$$

for a mixture of monatomic ideal gases.

We can use the given partition function to find $\ln Q$:

$$Q = \frac{q_1^{N_1} q_2^{N_2}}{N_1! \, N_2!}$$

$$\ln Q = \ln \frac{q_1^{N_1}}{N_1!} + \ln \frac{q_2^{N_2}}{N_2!} = N_1 \ln q_1 + N_2 \ln q_2 - \ln(N_2!) - \ln(N_1!)$$

Using the definition of q_j given in the problem, we can find $(\partial \ln Q/\partial \beta)_{N,V}$ and $(\partial \ln Q/\partial V)_{N,\beta}$:

$$\ln Q = -\frac{3N_1}{2} \ln \beta - \frac{3N_2}{2} \ln \beta + \text{non-}\beta\text{-related terms}$$

$$\left(\frac{\partial \ln Q}{\partial \beta}\right)_{N,V} = -\frac{3}{2}\left(\frac{N_1 + N_2}{\beta}\right)$$

$$\ln Q = N_1 \ln V + N_2 \ln V + \text{non-}V\text{-related terms}$$

$$\left(\frac{\partial \ln Q}{\partial V}\right)_{N,\beta} = \frac{N_1 + N_2}{V}$$

Now we use Equations 3.20 and 3.32 to find $\langle E \rangle$ and $\langle P \rangle$:

$$\langle E \rangle = \frac{3}{2}\left(\frac{N_1 + N_2}{\beta}\right) = \frac{3}{2}(N_1 + N_2)k_B T$$

and

$$\langle P \rangle = k_B T \left(\frac{N_1 + N_2}{V}\right)$$

$$PV = k_B T \,(N_1 + N_2)$$

3–33. We will learn in Chapter 4 that the rotational partition function of an asymmetric top molecule is given by

$$q_{\mathrm{rot}}(T) = \frac{\pi^{1/2}}{\sigma}\left(\frac{8\pi^2 I_A k_B T}{h^2}\right)^{1/2}\left(\frac{8\pi^2 I_B k_B T}{h^2}\right)^{1/2}\left(\frac{8\pi^2 I_C k_B T}{h^2}\right)^{1/2}$$

where σ is a constant and I_A, I_B, and I_C are the three (distinct) moments of inertia. Show that the rotational contribution to the molar heat capacity is $\overline{C}_{V,\mathrm{rot}} = \frac{3}{2}R$.

Using the partition function in the problem text, we can write

$$\ln q_{\mathrm{rot}} = \frac{3}{2}\ln T + \text{non-}T\text{-related terms}$$

Then we can use Equation 3.21 to write

$$\langle E\rangle_{\mathrm{rot}} = k_B T^2\left(\frac{\partial \ln q_{\mathrm{rot}}}{\partial T}\right) = \frac{3}{2}k_B T$$

and Equation 3.25 to write

$$C_{V,\mathrm{rot}} = \frac{\partial\langle E\rangle_{\mathrm{rot}}}{\partial T} = \frac{3}{2}k_B$$

For N_A particles, $\overline{C}_{V,\mathrm{rot}} = 3k_B N_A/2 = 3R/2$.

3–34. The allowed energies of a harmonic oscillator are given by $\varepsilon_v = (v + \frac{1}{2})h\nu$. The corresponding partition function is given by

$$q_{\mathrm{vib}}(T) = \sum_{v=0}^{\infty} e^{-(v+\frac{1}{2})h\nu/k_B T}$$

Let $x = e^{-h\nu/k_B T}$ and use the formula for the summation of a geometric series (Problem 3–20) to show that

$$q_{\mathrm{vib}}(T) = \frac{e^{-h\nu/2k_B T}}{1 - e^{-h\nu/k_B T}}$$

Let $x = e^{-h\nu/k_B T}$. The equation for the partition function given in the problem text then becomes

$$q_{\mathrm{vib}}(T) = \sum_{v=0}^{\infty} x^{v+1/2} = x^{1/2}\sum_{v=0}^{\infty} x^{v}$$

$$= \frac{x^{1/2}}{1-x} = \frac{e^{-h\nu/2k_B T}}{1 - e^{-h\nu/k_B T}}$$

3–35. Derive an expression for the probability that a harmonic oscillator will be found in the vth state. Calculate the probability that the first few vibrational states are occupied for HCl(g) at 300 K. (See Table 1–4 and Problem 3–34.)

$$\text{prob}\,(v = j) = \frac{e^{-\beta\varepsilon_{\mathrm{vib},j}}}{q_{\mathrm{vib}}}$$

Using the harmonic oscillator partition function, we find that

$$\text{prob}\,(v=j) = \frac{e^{-\beta h\nu v}e^{-\beta h\nu/2}(1-e^{-h\beta\nu})}{e^{-\beta h\nu/2}}$$
$$= e^{-\beta h\nu v}\left(1-e^{-\beta h\nu}\right)$$

For HCl, $h\nu/k_B = 4140$ K. The probabilities of the first three vibrational states of HCl being occupied at 300 K are

$$\text{prob}\,(v=0) = \left(1-e^{-\beta h\nu}\right) = 0.99999898$$
$$\text{prob}\,(v=1) = e^{-\beta h\nu}\left(1-e^{-\beta h\nu}\right) = 1.01\times 10^{-6}$$
$$\text{prob}\,(v=2) = e^{-2\beta h\nu}\left(1-e^{-\beta h\nu}\right) = 1.03\times 10^{-12}$$

It is clear that at room temperature, the majority of the gaseous HCl molecules are in the ground vibrational state.

3–36. Show that the fraction of harmonic oscillators in the ground vibrational state is given by

$$f_0 = 1 - e^{-h\nu/k_BT}$$

Calculate f_0 for N_2 at 300 K, 600 K, and 1000 K (see Table 1.4).

The fraction of harmonic oscillators in the ground vibrational state is the same as the probability that a harmonic oscillator will be in the ground vibrational state (from Problem 3.35):

$$f_0 = \text{prob}\,(v=0) = 1 - e^{-h\nu/k_BT}$$

For N_2, $h\nu/k_B = 3352$ K.

Temperature	f_0
300 K	1.000
600 K	0.9962
1000 K	0.9650

3–37. Use Equation 3.55 to show that the fraction of rigid rotators in the Jth rotational level is given by

$$f_J = \frac{(2J+1)e^{-\hbar^2 J(J+1)/2Ik_BT}}{q_{\text{rot}}(T)}$$

Plot the fraction in the Jth level relative to the $J=0$ level (f_J/f_0) against J for HCl(g) at 300 K. Take $\tilde{B} = 10.44\ \text{cm}^{-1}$.

$$q_{\text{rot}} = \sum_{J=0}^{\infty}(2J+1)e^{-\hbar^2 J(J+1)/2Ik_BT} \tag{3.55}$$

The number of rigid rotators in the jth rotational level is the $J=j$ term in the above sum. Thus, the fraction of rigid rotators in the Jth rotational level can be expressed as

$$f_J = \frac{(2J+1)e^{-\hbar^2 J(J+1)/2Ik_BT}}{q_{\text{rot}}}$$

and

$$\frac{f_J}{f_0} = \frac{(2J+1)e^{-\hbar^2 J(J+1)/2Ik_BT}}{1}$$

For HCl(g), $B_{HCl} = 3.13 \times 10^{11}$ Hz. So $I = h/8\pi^2 B = 2.68 \times 10^{-47}$ J. Below is a plot of (f_J/f_0) vs. J for HCl at 300 K.

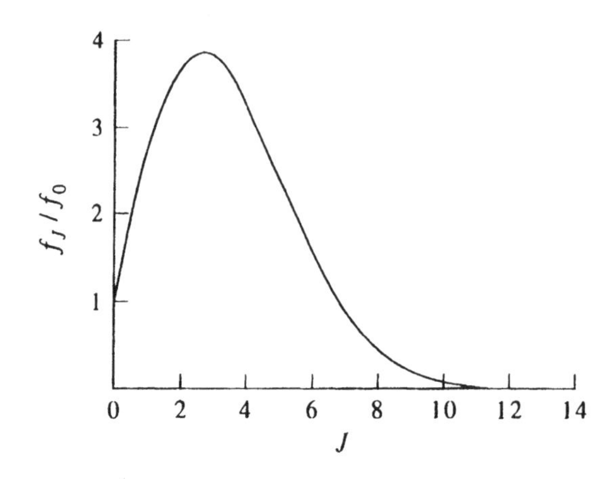

3–38. Equations 3.20 and 3.21 give the ensemble average of E, which we assert is the same as the experimentally observed value. In this problem, we will explore the standard deviation about $\langle E \rangle$ (MathChapter B). We start with either Equation 3.20 or 3.21:

$$\langle E \rangle = U = -\left(\frac{\partial \ln Q}{\partial \beta}\right)_{N,V} = k_B T^2 \left(\frac{\partial \ln Q}{\partial T}\right)_{N,V}$$

Differentiate again with respect to β or T to show that (MathChapter B)

$$\sigma_E^2 = \langle E^2 \rangle - \langle E \rangle^2 = k_B T^2 C_V$$

where C_V is the heat capacity. To explore the relative magnitude of the spread about $\langle E \rangle$, consider

$$\frac{\sigma_E}{\langle E \rangle} = \frac{(k_B T C_V)^{1/2}}{\langle E \rangle}$$

To get an idea of the size of this ratio, use the values of $\langle E \rangle$ and C_V for a (monatomic) ideal gas, namely, $\frac{3}{2}Nk_BT$ and $\frac{3}{2}Nk_B$, respectively, and show that $\sigma_E/\langle E \rangle$ goes as $N^{-1/2}$. What does this trend say about the likely observed deviations from the average macroscopic energy?

We use the following equalities (from Equation 3.18):

$$\langle E \rangle = \frac{\sum_j E_j e^{-\beta E_j}}{Q}$$

$$\langle E^2 \rangle = \frac{\sum_j E_j^2 e^{-\beta E_j}}{Q}$$

Now we can find σ_E^2:

$$\sigma_E^2 = \langle E^2 \rangle - \langle E \rangle^2 = \frac{\sum_j E_j^2 e^{-\beta E_j}}{Q} - \left(\frac{\sum_j E_j e^{-\beta E_j}}{Q}\right)^2$$

We can substitute the definition $Q = \sum_j e^{-\beta E_j}$ into Equation 3.18 and take the partial derivative with respect to β to find

$$\langle E \rangle = \frac{\sum_j E_j e^{-\beta E_j}}{\sum_j e^{-\beta E_j}}$$

$$\left(\frac{\partial \langle E \rangle}{\partial \beta}\right) = -\frac{\sum_j E_j^2 e^{-\beta E_j}}{\sum_j e^{-\beta E_j}} + \frac{(\sum_j E_j e^{-\beta E_j})(\sum_j E_j e^{-\beta E_j})}{\sum_j e^{-2\beta E_j}}$$

$$= -\frac{\sum_j E_j^2 e^{-\beta E_j}}{Q} + \left(\frac{\sum_j E_j e^{-\beta E_j}}{Q}\right)^2 = -\sigma_E^2$$

We define C_V as (Example 3–3)

$$C_V = \left(\frac{\partial \langle E \rangle}{\partial T}\right)_{N,V} = -\frac{1}{k_B T^2}\left(\frac{\partial \langle E \rangle}{\partial \beta}\right)_{N,V} = \frac{1}{k_B T^2}\sigma_E^2$$

So $\sigma_E^2 = k_B T^2 C_V$. Now consider the spread about $\langle E \rangle$ for a monatomic ideal gas:

$$\frac{\sigma_E}{\langle E \rangle} = \frac{(k_B T^2 C_V)^{1/2}}{\langle E \rangle}$$

$$= \frac{\left[k_B T^2 (\frac{3}{2} N k_B)\right]^{1/2}}{\frac{3}{2} N k_B T}$$

$$= \frac{1}{\left(\frac{3}{2}\right)^{1/2} N^{1/2}} \sim N^{-1/2}$$

Because $N^{-1/2}$ is on the order of 10^{-12} for one mole, it is very unlikely that significant deviations from the average macroscopic energy will be observed.

3–39. Following Problem 3–38, show that the variance about the average values of a *molecular* energy is given by

$$\sigma_\varepsilon^2 = \langle \varepsilon^2 \rangle - \langle \varepsilon \rangle^2 = \frac{k_B T^2 C_V}{N}$$

and that $\sigma_\varepsilon / \langle \varepsilon \rangle$ goes as order unity. What does this result say about the deviations from the average molecular energy?

We use the following equalities:

$$q = e^{-\beta \varepsilon_j}$$

$$\langle \varepsilon \rangle = \frac{\sum_j \varepsilon_j e^{-\beta \varepsilon_j}}{q}$$

$$\langle \varepsilon^2 \rangle = \frac{\sum_j \varepsilon_j^2 e^{-\beta \varepsilon_j}}{q}$$

We know that $N\langle \varepsilon \rangle = \langle E \rangle$, where N is the number of particles with molecular energy ε. Thus

$$C_V = \left(\frac{\partial \langle E \rangle}{\partial T}\right) = N\left(\frac{\partial \langle \varepsilon \rangle}{\partial T}\right)$$

We can follow the steps in Problem 3.39, subsituting ε for E and q for Q, to find

$$\sigma_\varepsilon^2 = k_B T^2 \left(\frac{\partial\langle\varepsilon\rangle}{\partial T}\right) = k_B T^2 C_V N^{-1}$$

For a monatomic ideal gas, $C_V N^{-1} = \frac{3}{2}k_B$. Now consider the spread about $\langle\varepsilon\rangle$ for a monatomic ideal gas:

$$\frac{\sigma_E}{\langle E\rangle} = \frac{\left(\frac{3}{2}k_B^2 T^2\right)^{1/2}}{\langle\varepsilon\rangle} = \frac{\left(\frac{3}{2}\right)^{1/2} k_B T}{\frac{3}{2}k_B T} = \left(\frac{3}{2}\right)^{-1/2} \approx 0.8$$

which is of order unity. The molecular energy is much more likely to deviate from the average molecular energy than the corresponding macroscopic energy is to deviate from the average macroscopic energy.

3–40. Use the result of Problem 3–38 to show that C_V is never negative.

The quantity σ^2 must be positive, because σ is a real number and the square of any real number is always positive. Thus $k_B T^2 C_V$ must be a positive number. Both k_B and T are always positive; therefore, C_V must always be positive.

3–41. The energies and degeneracies of the four lowest electronic states of Na(g) are tabulated below.

Energy/cm^{-1}	Degeneracy
0.000	2
16 956.183	2
16 973.379	4
25 739.86	2

Calculate the fraction of the atoms in each of these electronic states in a sample of Na(g) at 1000 K. Repeat this calculation for a temperature of 2500 K.

The probability of an atom being in the lth electronic state is (Equation 3.44)

$$\text{prob}_l = \frac{e^{-\varepsilon_l/k_B T}}{\sum_l e^{-\varepsilon_l/k_B T}}$$

where, for Na(g),

$$\sum_l e^{-\varepsilon_l/k_B T} = 2e^{-0/k_B T} + 2e^{-16956.183/k_B T} + 4e^{-16973.379/k_B T} + 2e^{-25739.86/k_B T}$$

To calculate the fraction of atoms in each electronic state, we multiply the state's degeneracy by $e^{-\varepsilon_i/k_B T}$ and divide by the summation above. (Note: $k_B = 0.69509\ \text{cm}^{-1}$.)

State	Fraction of atoms, 1000 K	Fraction of atoms, 2500 K
1	1.000	0.9998
2	2.545×10^{-11}	5.784×10^{-5}
3	4.966×10^{-11}	1.145×10^{-4}
4	8.273×10^{-17}	3.690×10^{-7}

3–42. The vibrational frequency of NaCl(g) is $159.23\ \text{cm}^{-1}$. Calculate the molar heat capacity of $\overline{C}_V$ at 1000 K. (See Equation 3.27.)

$$\overline{C}_V = \frac{5R}{2} + R\left(\frac{h\nu}{k_B T}\right)^2 \frac{e^{-h\nu/k_B T}}{\left(1 - e^{-h\nu/k_B T}\right)^2}$$

$\tilde{\nu}_{\text{NaCl(g)}} = 159.23\ \text{cm}^{-1}$, so $h\nu/k_B T = 0.22909$, and

$$\overline{C}_V = \left[\frac{5}{2} + (0.22909)^2 \frac{e^{0.22909}}{\left(1 - e^{0.22909}\right)^2}\right] R = 3.4956R = 29.064\ \text{J}\cdot\text{mol}^{-1}\cdot\text{K}^{-1}$$

3–43. The energies and degeneracies of the two lowest electronic states of atomic iodine are listed below.

Energy/cm^{-1}	Degeneracy
0	4
7603.2	2

What temperature is required so that 2% of the atoms are in the excited state?

As in Problem 3–41, we have

$$\text{prob}_i = \frac{e^{-\varepsilon_i/k_B T}}{\sum_i e^{-\varepsilon_i/k_B T}} \tag{3.44}$$

For 2% of the atoms to be in the excited state, we can substitute into the expression above and write

$$0.02 = 2\exp\frac{-7603.2}{0.69509T}\left(4 + 2\exp\frac{-7603.2}{0.69509T}\right)^{-1}$$

$$\ln\frac{0.04}{0.98} = \frac{-7603.2}{0.69509T}$$

$$T = 3420\ \text{K}$$

MATHCHAPTER C

Series and Limits

PROBLEMS AND SOLUTIONS

C–1. Calculate the percentage difference between e^x and $1+x$ for $x = 0.0050, 0.0100, 0.0150, \ldots, 0.1000$.

We define percentage difference as

$$\left| \frac{e^x - (1+x)}{e^x} \right|$$

Then we can create the table below:

x	e^x	$1+x$	percentage difference
0.0050	1.00501	1.0050	0.0012%
0.0100	1.01005	1.0100	0.0050%
0.0150	1.01511	1.0150	0.0111%
0.0200	1.02020	1.0200	0.0197%
0.0250	1.02532	1.0250	0.0307%
0.0300	1.03045	1.0300	0.0441%
0.0350	1.03562	1.0350	0.0598%
0.0400	1.04081	1.0400	0.0779%
0.0450	1.04603	1.0450	0.0983%
0.0500	1.05127	1.0500	0.1209%
0.0550	1.05654	1.0550	0.1458%
0.0600	1.06184	1.0600	0.1729%
0.0650	1.06716	1.0650	0.2023%
0.0700	1.07251	1.0700	0.2339%
0.0750	1.07788	1.0750	0.2676%
0.0800	1.08329	1.0800	0.3034%
0.0850	1.08872	1.0850	0.3414%
0.0900	1.09417	1.0900	0.3815%
0.0950	1.09966	1.0950	0.4237%
0.1000	1.10517	1.1000	0.4679%

C–2. Calculate the percentage difference between $\ln(1+x)$ and x for $x = 0.0050$, 0.0100, 0.0150, ..., 0.1000.

We define percentage difference as

$$\left|\frac{\ln(1+x) - x}{\ln(1+x)}\right|$$

Then we can create the table below:

x	$\ln(1+x)$	percentage difference
0.0050	0.0049875	0.2498%
0.0100	0.0099503	0.4992%
0.0150	0.0148886	0.7481%
0.0200	0.0198026	0.9967%
0.0250	0.0246926	1.2449%
0.0300	0.0295588	1.4926%
0.0350	0.0344014	1.7400%
0.0400	0.0392207	1.9869%
0.0450	0.0440169	2.2335%
0.0500	0.0487902	2.4797%
0.0550	0.0535408	2.7255%
0.0600	0.0582689	2.9709%
0.0650	0.0629748	3.2159%
0.0700	0.0676586	3.4605%
0.0750	0.0723207	3.7048%
0.0800	0.0769610	3.9487%
0.0850	0.0815800	4.1922%
0.0900	0.0861777	4.4354%
0.0950	0.0907544	4.6782%
0.1000	0.0953102	4.9206%

C–3. Write out the expansion of $(1+x)^{1/2}$ through the quadratic term.

$$\begin{aligned}(1+x)^{1/2} &= 1 + \frac{x}{2} + \frac{\frac{1}{2}\left(\frac{1}{2}-1\right)}{2}x^2 + O(x^3) \\ &= 1 + \frac{x}{2} - \frac{x^2}{8} + O(x^3)\end{aligned}$$

C–4. Evaluate the series

$$S = \sum_{v=0}^{\infty} e^{-(v+\frac{1}{2})\beta h\nu}$$

We can express this series as

$$S = \sum_{v=0}^{\infty} e^{-(v+\frac{1}{2})\beta h\nu} = e^{-\frac{1}{2}\beta h\nu} \sum_{v=0}^{\infty} e^{-v\beta h\nu}$$

Let $e^{-\beta h\nu} = x$, so

$$S = x^{\frac{1}{2}} \sum_{v=0}^{\infty} x^{v}$$

and use Equation C.3 to write

$$S = \frac{e^{-\frac{1}{2}\beta h\nu}}{1 - e^{-\beta h\nu}}$$

C–5. Show that

$$\frac{1}{(1-x)^2} = 1 + 2x + 3x^2 + 4x^3 + \cdots$$

We know (Equation C.3)

$$\frac{1}{1-x} = 1 + x + x^2 + x^3 + x^4 + \cdots$$

Differentiating both sides of this equation with respect to x gives

$$\frac{1}{(1-x)^2} = 1 + 2x + 3x^2 + 4x^3 + \cdots$$

C–6. Evaluate the series

$$S = \frac{1}{2} + \frac{1}{4} + \frac{1}{8} + \frac{1}{16} + \cdots$$

We can write S as

$$S = \sum_{n=1}^{\infty} \left(\frac{1}{2}\right)^n = \sum_{n=0}^{\infty} \left(\frac{1}{2}\right)^n - 1$$

Using the geometric series (Equation C.3), we find

$$S = \frac{1}{1 - \frac{1}{2}} - 1 = 1$$

C–7. Use Equation C.9 to derive Equations C.10 and C.11.

Let $f(x) = \sin x$. Then

$$f'(x) = \cos x \qquad f''(x) = -\sin x \quad \text{and} \quad f''' = -\cos x$$

The Maclaurin series is

$$f(x) = f(0) + \left(\frac{df}{dx}\right)_{x=0} x + \frac{1}{2!}\left(\frac{d^2 f}{dx^2}\right)_{x=0} x^2 + \frac{1}{3!}\left(\frac{d^3 f}{dx^3}\right)_{x=0} x^3 + \cdots \qquad \text{(C.9)}$$

So

$$\sin(x) = 0 + x + 0 - \frac{1}{3!}x^3 + 0 + \frac{1}{5!}x^5 + \cdots \qquad \text{(C.10)}$$

Likewise, letting $f(x) = \cos x$ and applying Equation C.9 gives

$$\cos(x) = 1 + 0 - \frac{1}{2!}x^2 + 0 + \frac{1}{4!}x^4 + 0 + \cdots \qquad \text{(C.11)}$$

C–8. Show that Equations C.2, C.10, and C.11 are consistent with the relation $e^{ix} = \cos x + i \sin x$.

We begin with the relation $e^{ix} = \cos x + i \sin x$. We use the expressions for $\sin x$ and $\cos x$ given in Equations C.10 and C.11 to write this relation as

$$\begin{aligned} e^{ix} &= 1 - \frac{x^2}{2!} + \frac{x^4}{4!} - \frac{x^6}{6!} + \cdots + i\left[x - \frac{x^3}{3!} + \frac{x^5}{5!} - \frac{x^7}{7!} + \cdots\right] \\ &= 1 + ix - \frac{x^2}{2!} - \frac{ix^3}{3!} + \frac{x^4}{4!} + \frac{ix^5}{5!} - \frac{x^6}{6!} - \frac{ix^7}{7!} + \cdots \\ &= 1 + ix + \frac{i^2x^2}{2!} + \frac{i^3x^3}{3!} + \frac{i^4x^4}{4!} + \cdots \end{aligned}$$

Let $ix = z$. Then

$$e^z = 1 + z + \frac{z^2}{2!} + \frac{z^3}{3!} + \cdots$$

in accordance with Equation C.2.

C–9. In Example 3–3, we derived a simple formula for the molar heat capacity of a solid based on a model by Einstein:

$$\overline{C}_V = 3R\left(\frac{\Theta_E}{T}\right)^2 \frac{e^{-\Theta_E/T}}{(1 - e^{-\Theta_E/T})^2}$$

where R is the molar gas constant and $\Theta_E = h\nu/k_B$ is a constant, called the Einstein temperature, that is characteristic of the solid. Show that this equation gives the Dulong and Petit limit ($\overline{C}_V \rightarrow 3R$) at high temperatures.

Let $x = \Theta_E/T$. Then

$$\overline{C}_V = 3Rx^2 \frac{e^{-x}}{(1 - e^{-x})^2}$$

At high temperatures, $x \rightarrow 0$. We can use a series expansion of e^x (Equation C.2) and write

$$\begin{aligned} \overline{C}_V &= \lim_{x\rightarrow 0} 3Rx^2 \frac{e^{-x}}{(1 - e^{-x})^2} \\ &= 3R \lim_{x\rightarrow 0} \frac{1 - x + \cdots}{(1 - 1 + x + \cdots)^2} = 3R \end{aligned}$$

C–10. Evaluate the limit of

$$f(x) = \frac{e^{-x}\sin^2 x}{x^2}$$

as $x \to 0$.

We can use Equations C.2 and C.10 to write this function as a series expansion:

$$f(x) = \frac{[1 - x + O(x^2)][x + O(x^2)]^2}{x^2} = \frac{x^2 + O(x^3)}{x^2}$$

As $x \to 0$, $f(x) \to 1$.

C–11. Evaluate the integral

$$I = \int_0^a x^2 e^{-x}\cos^2 x\,dx$$

for small values of a by expanding I in powers of a through quadratic terms.

$$\begin{aligned}
I &= \int_0^a x^2 e^{-x}\cos^2 x\,dx \\
&= \int_0^a x^2\left[1 - x + \frac{x^2}{2!} - \frac{x^3}{3!} + O(x^4)\right]\left[1 - \frac{x^2}{2!} + O(x^4)\right]^2 dx \\
&= \int_0^a \left[x^2 - x^3 + O(x^4)\right]\left[1 - \frac{x^2}{2!} + O(x^4)\right]^2 dx \\
&= \int_0^a \left[x^2 - x^3 + O(x^4)\right]\left[1 - x^2 + O(x^4)\right] dx \\
&= \int_0^a \left[x^2 - x^3 + O(x^4)\right] dx \\
&= \frac{a^3}{3} + O(a^4)
\end{aligned}$$

C–12. Prove that the series for $\sin x$ converges for all values of x.

We can use the ratio test (Equation C.5) and the Maclaurin series for $\sin x$ (Equation C.10):

$$\begin{aligned}
r &= \lim_{n\to\infty}\left|\frac{u_{n+1}}{u_n}\right| \\
&= \lim_{n\to\infty}\left|\frac{x^{2n+1}}{(2n+1)!}\,\frac{(2n-1)!}{x^{2n-1}}\right| \\
&= \lim_{n\to\infty}\left|\frac{x^2}{(2n+1)(2n)}\right| = 0
\end{aligned}$$

Because $r < 1$, the series converges for all values of x.

C–13. A Maclaurin series is an expansion about the point $x = 0$. A series of the form

$$f(x) = c_0 + c_1(x - x_0) + c_2(x - x_0)^2 + \cdots$$

is an expansion about the point x_0 and is called a Taylor series. First show that $c_0 = f(x_0)$. Now differentiate both sides of the above expansion with respect to x and then let $x = x_0$ to show that $c_1 = (df/dx)_{x=x_0}$. Now show that

$$c_n = \frac{1}{n!}\left(\frac{d^n f}{dx^n}\right)_{x=x_0}$$

and so

$$f(x) = f(x_0) + \left(\frac{df}{dx}\right)_{x_0}(x - x_0) + \frac{1}{2}\left(\frac{d^2 f}{dx^2}\right)_{x_0}(x - x_0)^2 + \cdots$$

At x_0, we find

$$f(x) = c_0 + c_1(x - x_0) + c_2(x - x_0)^2 + \cdots$$
$$f(x_0) = c_0$$

Differentiating $f(x)$ with respect to x gives

$$f'(x) = c_1 + 2c_2(x - x_0) + \cdots$$
$$f'(x_0) = c_1$$

Likewise, the second derivative of $f(x)$ at x_0 is $2c_2$, the third derivative of $f(x)$ at x_0 is $3!c_3$, and, generally,

$$\left(\frac{d^n f}{dx^n}\right)_{x=x_0} = n!c_n$$
$$c_n = \frac{1}{n!}\left(\frac{d^n f}{dx^n}\right)_{x=x_0}$$

Substituting into the Taylor series for c_n, we find

$$f(x) = f(x_0) + \left(\frac{df}{dx}\right)_{x=x_0}(x - x_0) + \frac{1}{2}\left(\frac{d^2 f}{dx^2}\right)_{x=x_0}(x - x_0)^2 + \cdots$$

C–14. Later on, we will need to sum the series

$$s_1 = \sum_{v=0}^{\infty} v x^v$$

and

$$s_2 = \sum_{v=0}^{\infty} v^2 x^v$$

To sum the first one, start with (Equation C.3)

$$s_0 = \sum_{v=0}^{\infty} x^v = \frac{1}{1 - x}$$

Differentiate with respect to x and then multiply by x to obtain

$$s_1 = \sum_{v=0}^{\infty} v x^v = x\frac{ds_0}{dx} = x\frac{d}{dx}\left(\frac{1}{1 - x}\right) = \frac{x}{(1 - x)^2}$$

Using the same approach, show that

$$s_2 = \sum_{\upsilon=0}^{\infty} \upsilon^2 x^{\upsilon} = \frac{x + x^2}{(1-x)^3}$$

Following the procedure outlined in the problem, we find

$$s_0 = \sum_{n=0}^{\infty} x^n = \frac{1}{1-x}$$

$$\frac{ds_0}{dx} = \sum_{n=0}^{\infty} n x^{n-1} = \frac{1}{(1-x)^2}$$

$$x\frac{ds_0}{dx} = \sum_{n=0}^{\infty} n x^n = \frac{x}{(1-x)^2} = s_1$$

Likewise,

$$s_1 = \sum_{n=0}^{\infty} n x^n = \frac{x}{(1-x)^2}$$

$$\frac{ds_1}{dx} = \sum_{n=0}^{\infty} n^2 x^{n-1} = \frac{1}{(1-x)^2} + \frac{2x}{(1-x)^3}$$

$$x\frac{ds_1}{dx} = \sum_{n=0}^{\infty} n^2 x^n = x\left[\frac{1-x+2x}{(1-x)^3}\right]$$

$$= \frac{x+x^2}{(1-x)^3}$$

CHAPTER 4

Partition Functions and Ideal Gases

PROBLEMS AND SOLUTIONS

4–1. Equation 4.7 shows that $\langle\varepsilon_{\text{trans}}\rangle = \frac{3}{2}k_{\text{B}}T$ in three dimensions, and Problem 4–3 shows that $\langle\varepsilon_{\text{trans}}\rangle = \frac{1}{2}k_{\text{B}}T$ in one dimension and $\frac{2}{2}k_{\text{B}}T$ in two dimensions. Show that typical values of translational quantum numbers at room temperature are $O(10^9)$ for $m = 10^{-26}$ kg, $a = 1$ dm, and $T = 300$ K.

The average translational energy at 300 K is on the order of $k_{\text{B}}T = 4.142 \times 10^{-21}$ J. Recall that

$$\langle\varepsilon\rangle = \frac{h^2n^2}{8ma^2}$$

So

$$n^2 \approx \frac{8ma^2}{h^2}k_{\text{B}}T = \frac{8(10^{-26}\text{ kg})(0.1\text{ m})^2}{(6.626 \times 10^{-34}\text{ J}\cdot\text{s})^2}(4.142 \times 10^{-21}\text{ J})$$

$$n \approx 2.75 \times 10^9$$

In two and three dimensions, $\langle\varepsilon_{\text{trans}}\rangle$ depends on the sum of the squares of the respective quantum numbers. So for comparable values of a for each dimension, n will be of $O(10^9)$.

4–2. Show that the difference between the successive terms in the summation in Equation 4.4 is very small for $m = 10^{-26}$ kg, $a = 1$ dm, and $T = 300$ K. Recall from Problem 4–1 that typical values of n are $O(10^9)$.

Equation 4.4 gives

$$q_{\text{trans}}(V, T) = \left[\sum_{n=1}^{\infty}\exp\left(\frac{-\beta h^2n^2}{8ma^2}\right)\right]^3$$

The difference between terms in $q_{\text{trans}}(V, T)$ is then

$$e^{-A(n+1)^2} - e^{-An^2} = e^{-An^2}\left[e^{-A(2n+1)} - 1\right]$$

where $A = \beta h^2/8ma^2 \sim 10^{-19}$. Therefore,

$$e^{-A(2n+1)} - 1 \approx 10^{-10}$$

4–3. Show that

$$q_{\text{trans}}(a, T) = \left(\frac{2\pi m k_B T}{h^2}\right)^{1/2} a$$

in one dimension and that

$$q_{\text{trans}}(a, T) = \left(\frac{2\pi m k_B T}{h^2}\right) a^2$$

in two dimensions. Use these results to show that $\langle \varepsilon_{\text{trans}} \rangle$ has a contribution of $k_B T/2$ to its total value for each dimension.

Remember that $\int_0^\infty e^{-\alpha n^2} dn = \left(\frac{\pi}{4\alpha}\right)^{1/2}$. Then, for one dimension,

$$q_{\text{trans}}(a, T) = \int_0^\infty e^{-\beta h^2 n^2/8ma^2} dn = \left(\frac{2\pi m k_B T}{h^2}\right)^{1/2} a$$

And for two dimensions,

$$q_{\text{trans}}(a, T) = \left(\int_0^\infty e^{-\beta h^2 n^2/8ma^2} dn\right)^2 = \left(\frac{2\pi m k_B T}{h^2}\right) a^2$$

Now

$$\langle \epsilon_{\text{trans}} \rangle = k_B T^2 \left(\frac{\partial \ln q_{\text{trans}}}{\partial T}\right)_V$$

The partition function is proportional to $T^{n/2}$, where n is the dimension. So

$$\left(\frac{\partial \ln q_{\text{trans}}}{\partial T}\right)_V = \frac{n}{2T}$$

and

$$\langle \epsilon_{\text{trans}} \rangle = k_B T^2 \frac{n}{2T} = \frac{n k_B T}{2}$$

4–4. Using the data in Table 1.2, calculate the fraction of sodium atoms in the first excited state at 300 K, 1000 K, and 2000 K.

We can use the second line of Equation 4.10 to calculate the fraction of sodium atoms in the first excited state, with $g_{e1} = 2$, $g_{e2} = 2$, $g_{e3} = 4$, and $g_{e4} = 2$:

$$f_2 = \frac{2e^{-\beta\epsilon_{e2}}}{2 + 2e^{-\beta\epsilon_{e2}} + 4e^{-\beta\epsilon_{e3}} + 2e^{-\beta\epsilon_{e4}} + \ldots}$$

Using the data in Table 1.2, we find that the numerator of this fraction is

$$2\exp\left[-\frac{16\,956.183\ \text{cm}^{-1}}{(0.6950\ \text{cm}^{-1}\cdot\text{K}^{-1})T}\right]$$

and the denominator is

$$2+2\exp\left[-\frac{16\,956.183\text{ cm}^{-1}}{(0.6950\text{ cm}^{-1}\cdot\text{K}^{-1})T}\right]+4\exp\left[-\frac{16\,973.379\text{ cm}^{-1}}{(0.6950\text{ cm}^{-1}\cdot\text{K}^{-1})T}\right]$$
$$+2\exp\left[-\frac{25\,739.86\text{ cm}^{-1}}{(0.6950\text{ cm}^{-1}\cdot\text{K}^{-1})T}\right]+\cdots$$

Using these values, we find that the values of f_2 for the various temperatures are

$$f_2(T=300\text{ K})=4.8\times10^{-36}$$
$$f_2(T=1000\text{ K})=2.5\times10^{-11}$$
$$f_2(T=2000\text{ K})=5.0\times10^{-6}$$

4–5. Using the data in Table 4.1, evaluate the fraction of lithium atoms in the first excited state at 300 K, 1000 K, and 2000 K.

We can use the second line of Equation 4.10 to calculate the fraction of lithium atoms in the first excited state, with $g_{e1}=2$, $g_{e2}=2$, $g_{e3}=4$, and $g_{e4}=2$:

$$f_2=\frac{2e^{-\beta\epsilon_{e2}}}{2+2e^{-\beta\epsilon_{e2}}+4e^{-\beta\epsilon_{e3}}+2e^{-\beta\epsilon_{e4}}+\dots}$$

Using the data in Table 1.4, we find that the numerator of this fraction is

$$2\exp\left[-\frac{14\,903.66\text{ cm}^{-1}}{(0.6950\text{ cm}^{-1}\cdot\text{K}^{-1})T}\right]$$

and the denominator is

$$2+2\exp\left[-\frac{14\,903.66\text{ cm}^{-1}}{(0.6950\text{ cm}^{-1}\cdot\text{K}^{-1})T}\right]+4\exp\left[-\frac{14\,904.00\text{ cm}^{-1}}{(0.6950\text{ cm}^{-1}\cdot\text{K}^{-1})T}\right]$$
$$+2\exp\left[-\frac{27\,206.12\text{ cm}^{-1}}{(0.6950\text{ cm}^{-1}\cdot\text{K}^{-1})T}\right]+\cdots$$

Using these values, we find that the values of f_2 for the various temperatures are

$$f_2(T=300\text{ K})=9.0\times10^{-32}$$
$$f_2(T=1000\text{ K})=4.9\times10^{-10}$$
$$f_2(T=2000\text{ K})=2.2\times10^{-5}$$

4–6. Show that each dimension contributes $R/2$ to the molar translational heat capacity.

In Problem 4.3, we showed that $\langle\epsilon_{\text{trans}}\rangle$ has a contribution of $k_BT/2$ from each dimension. From Chapter 3,

$$C_V=\left(\frac{\partial\langle E\rangle}{\partial T}\right)_{N,V}=N\left(\frac{\partial\langle\epsilon\rangle}{\partial T}\right)_{N,V}$$

Because

$$\left(\frac{\partial \langle \epsilon \rangle}{\partial T}\right)_{N,V} = \left[\frac{\partial (k_B T/2)}{\partial T}\right]_{N,V} = \frac{k_B}{2}$$

each dimension contributes $Nk_B/2 = R/2$ to the molar translational heat capacity.

4–7. Using the values of Θ_{vib} and D_0 in Table 4.2, calculate the vaues of D_e for CO, NO, and K_2.

We can use the definitions $D_e = D_0 + h\nu/2$ and $\Theta_{vib} = h\nu/k_B$ to write

$$D_e = D_0 + \frac{k_B \Theta_{vib}}{2} = D_0 + \frac{R\Theta_{vib}}{2}$$

$$D_e(\text{CO}) = 1070\ \text{kJ}\cdot\text{mol}^{-1} + \frac{(8.314 \times 10^{-3}\ \text{kJ}\cdot\text{mol}^{-1}\cdot\text{K}^{-1})(3103\ \text{K})}{2}$$

$$= 1083\ \text{kJ}\cdot\text{mol}^{-1}$$

$$D_e(\text{NO}) = 626.8\ \text{kJ}\cdot\text{mol}^{-1} + \frac{(8.314 \times 10^{-3}\ \text{kJ}\cdot\text{mol}^{-1}\cdot\text{K}^{-1})(2719\ \text{K})}{2}$$

$$= 638.1\ \text{kJ}\cdot\text{mol}^{-1}$$

$$D_e(\text{K}_2) = 53.5\ \text{kJ}\cdot\text{mol}^{-1} + \frac{(8.314 \times 10^{-3}\ \text{kJ}\cdot\text{mol}^{-1}\cdot\text{K}^{-1})(133\ \text{K})}{2}$$

$$= 54.1\ \text{kJ}\cdot\text{mol}^{-1}$$

4–8. Calculate the characteristic vibrational temperature Θ_{vib} for $H_2(g)$ and $D_2(g)$ ($\tilde{\nu}_{H_2} = 4401\ \text{cm}^{-1}$ and $\tilde{\nu}_{D_2} = 3112\ \text{cm}^{-1}$).

From the definition of Θ_{vib}, we can write $\Theta_{vib} = hc\tilde{\nu}/k_B$. Then

$$\Theta_{vib}(\text{H}_2) = \frac{(6.626 \times 10^{-34}\ \text{J}\cdot\text{s})(2.9979 \times 10^{10}\ \text{cm}\cdot\text{s}^{-1})(4401\ \text{cm}^{-1})}{1.381 \times 10^{-23}\ \text{J}\cdot\text{K}^{-1}} = 6332\ \text{K}$$

$$\Theta_{vib}(\text{D}_2) = \frac{(6.626 \times 10^{-34}\ \text{J}\cdot\text{s})(2.9979 \times 10^{10}\ \text{cm}\cdot\text{s}^{-1})(3112\ \text{cm}^{-1})}{1.381 \times 10^{-23}\ \text{J}\cdot\text{K}^{-1}} = 4476\ \text{K}$$

4–9. Plot the vibrational contribution to the molar heat capacity of $Cl_2(g)$ from 250 K to 1000 K.

Use Equation 4.26 to write $\overline{C}_{V,vib}$ as a function of T:

$$\overline{C}_{V,vib} = R\left(\frac{\Theta_{vib}}{T}\right)^2 \frac{e^{-\Theta_{vib}/T}}{(1 - e^{-\Theta_{vib}/T})^2}$$

For Cl_2, we use $\Theta_{vib} = 805$ K (Table 4.2) in the above equation and plot $\overline{C}_{V,vib}$ versus T.

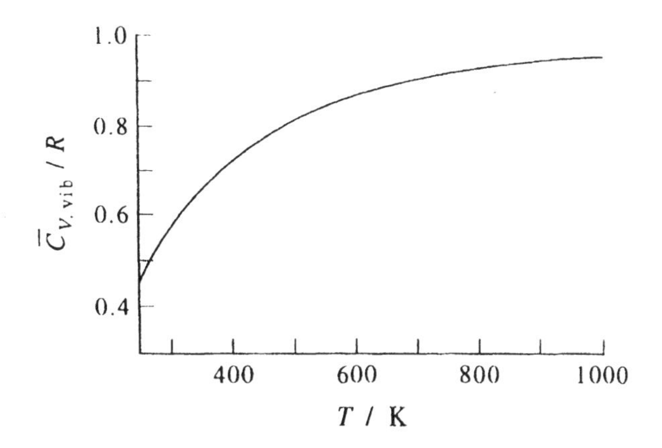

4–10. Plot the fraction of HCl(g) molecules in the first few vibrational states at 300 K and 1000 K.

Use Equation 4.28, substituting $\Theta_{\text{vib}} = 4227$ K (Table 4.2), to write f_v as a function of v, and plot. At 300 K $f_{v>0} = 7.6 \times 10^{-7}$ and at 1000 K $f_{v>0} = 1.46 \times 10^{-2}$.

$$f_v = (1 - e^{-\Theta_{\text{vib}}/T})e^{-v\Theta_{\text{vib}}/T}$$

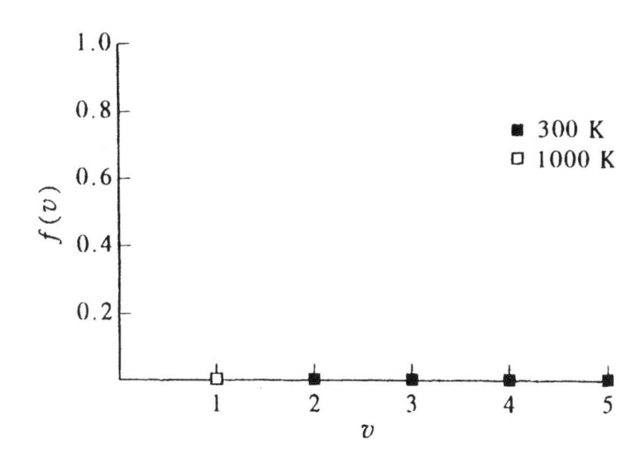

4–11. Calculate the fraction of molecules in the ground vibrational state and in all the excited states at 300 K for each of the molecules in Table 4.2.

The fraction of molecules in the ground vibrational state is given by (Equation 4.27)

$$f_0 = 1 - e^{-\Theta_{\text{vib}}/T}$$

and the fraction in all the excited states is $1 - f_0$, or $e^{-\Theta_{vib}/T}$. We can use the Θ_{vib} for the molecules given in Table 4.2 to find f_0 and $f_{v>0}$.

molecule	f_0	$f_{v>0}$
H_2	1.0000	1.007×10^{-9}
D_2	1.0000	4.356×10^{-7}
Cl_2	0.93167	6.834×10^{-2}
Br_2	0.78633	0.2137
I_2	0.064180	0.3582
O_2	0.99946	5.421×10^{-4}
N_2	0.99999	1.3051×10^{-5}
CO	0.99997	3.221×10^{-5}
NO	0.99988	1.158×10^{-4}
HCl	1.0000	7.600×10^{-7}
HBr	1.0000	3.294×10^{-6}
HI	0.99998	1.871×10^{-5}
Na_2	0.053389	0.4661
K_2	0.35811	0.6419

4–12. Calculate the value of the characteristic rotational temperature Θ_{rot} for $H_2(g)$ and $D_2(g)$. (The bond lengths of H_2 and D_2 are 74.16 pm.) The atomic mass of deuterium is 2.014.

The reduced masses of hydrogen and deuterium, respectively, are

$$\mu(H_2) = \frac{(1.674 \times 10^{-27}\ \text{kg})^2}{2(1.674 \times 10^{-27}\ \text{kg})} = 8.370 \times 10^{-28}\ \text{kg}$$

and

$$\mu(D_2) = \frac{(3.344 \times 10^{-27}\ \text{kg})^2}{2(3.344 \times 10^{-27}\ \text{kg})} = 1.672 \times 10^{-27}\ \text{kg}$$

Now use the formula $\Theta_{rot} = \hbar^2/2\mu R^2 k_B$ (Equation 4.32) to find the value of Θ_{rot} for hydrogen and deuterium:

$$\Theta_{rot}(H_2) = \frac{(1.055 \times 10^{-34}\ \text{J}\cdot\text{s})^2}{2(8.370 \times 10^{-28}\ \text{kg})(74.16 \times 10^{-12}\ \text{m})^2(1.38066 \times 10^{-23}\ \text{J}\cdot\text{K}^{-1})}$$
$$= 87.56\ \text{K}$$

$$\Theta_{rot}(D_2) = \frac{(1.055 \times 10^{-34}\ \text{J}\cdot\text{s})^2}{2(1.674 \times 10^{-27}\ \text{kg})(74.16 \times 10^{-12}\ \text{m})^2(1.38066 \times 10^{-23}\ \text{J}\cdot\text{K}^{-1})}$$
$$= 43.78\ \text{K}$$

4–13. The average molar rotational energy of a diatomic molecule is RT. Show that typical values of J are given by $J(J+1) = T/\Theta_{rot}$. What are typical values of J for $N_2(g)$ at 300 K?

If $\langle E\rangle = RT$, then $\epsilon = k_B T$. From Equation 4.30a,

$$J(J+1) = \frac{2Ik_B T}{\hbar^2} = \frac{T}{\Theta_{rot}}$$

For N_2 at 300 K,

$$J(J+1) \approx 104$$

and $J \approx 9$ or 10.

4–14. There is a mathematical procedure to calculate the error in replacing a summation by an integral as we do for the translational and rotational partition functions. The formula is called the Euler-Maclaurin summation formula and goes as follows:

$$\sum_{n=a}^{b} f(n) = \int_a^b f(n)dn + \frac{1}{2}\{f(b) + f(a)\} - \frac{1}{12}\left\{\left.\frac{df}{dn}\right|_{n=a} - \left.\frac{df}{dn}\right|_{n=b}\right\}$$
$$+ \frac{1}{720}\left\{\left.\frac{d^3f}{dn^3}\right|_{n=a} - \left.\frac{d^3f}{dn^3}\right|_{n=b}\right\} + \cdots$$

Apply this formula to Equation 4.33 to obtain

$$q_{rot}(T) = \frac{T}{\Theta_{rot}}\left\{1 + \frac{1}{3}\left(\frac{\Theta_{rot}}{T}\right) + \frac{1}{15}\left(\frac{\Theta_{rot}}{T}\right)^2 + O\left[\left(\frac{\Theta_{rot}}{T}\right)^3\right]\right\}$$

Calculate the correction to replacing Equation 4.33 by an integral for $N_2(g)$ at 300 K; $H_2(g)$ at 300 K (being so light, H_2 is an extreme example).

$$q_{rot}(T) = \sum_{J=0}^{\infty}(2J+1)e^{-\Theta_{rot}J(J+1)/T}$$
$$= \int_0^{\infty} dJ(2J+1)e^{-\Theta_{rot}J(J+1)/T} + \frac{1}{2}(f(\infty) + f(0))$$
$$- \frac{1}{12}\left[\left.\frac{df}{dJ}\right|_{J=0} - \left.\frac{df}{dJ}\right|_{J=\infty}\right] + \frac{1}{720}\left[\left.\frac{d^3f}{dJ^3}\right|_{J=0} - \left.\frac{d^3f}{dJ^3}\right|_{J=\infty}\right] + \cdots$$

Let $u = J(J+1)$ and $du = (2J+1)dJ$. Then at $J = 0$, $u = 0$, and at $J = \infty$, $u = \infty$. Also, find the first and third derivatives of $f(J)$:

$$f(J) = (2J+1)e^{-\Theta_{rot}J(J+1)/T}$$
$$\frac{df}{dJ} = \frac{-\Theta_{rot}}{T}(2J+1)^2 e^{-\Theta_{rot}J(J+1)/T} + 2e^{-\Theta_{rot}J(J+1)/T}$$
$$\frac{d^2f}{dJ^2} = 4\left(\frac{-\Theta_{rot}}{T}\right)(2J+1)e^{-\Theta_{rot}J(J+1)/T} + \left(\frac{\Theta_{rot}}{T}\right)^2(2J+1)^3 e^{-\Theta_{rot}J(J+1)/T}$$
$$+ 2\left(\frac{-\Theta_{rot}}{T}\right)(2J+1)e^{-\Theta_{rot}J(J+1)/T}$$
$$\frac{d^3f}{dJ^3} = 8\left(\frac{-\Theta_{rot}}{T}\right)e^{-\Theta_{rot}J(J+1)/T} + 4\left(\frac{\Theta_{rot}}{T}\right)e^{-\Theta_{rot}J(J+1)/T} + O\left[\left(\frac{\Theta_{rot}}{T}\right)^2\right]$$

$$q_{rot} = \int_0^{\infty}(2J+1)e^{-\Theta_{rot}J(J+1)/T} + \frac{1}{2}(1) - \frac{1}{12}\left(\frac{-\Theta_{rot}}{T} + 2\right)$$

$$+\frac{1}{720}\left(-8\frac{\Theta_{\rm rot}}{T}-4\frac{\Theta_{\rm rot}}{T}\right)+O\left\{\left(\frac{\Theta_{\rm rot}}{T}\right)^2\right\}$$

$$=\frac{T}{\Theta_{\rm rot}}\left\{1+\frac{1}{3}\left(\frac{\Theta_{\rm rot}}{T}\right)+\frac{1}{15}\left(\frac{\Theta_{\rm rot}}{T}\right)^2+O\left[\left(\frac{\Theta_{\rm rot}}{T}\right)^2\right]\right\}$$

For N_2 the correction factor to $q_{\rm rot}$ at 300 K is 0.32%; for H_2, the correction factor is 9.45%.

4–15. Apply the Euler-Maclaurin summation formula (Problem 4–14) to the one-dimensional version of Equation 4.4 to obtain

$$q_{\rm trans}(a,T)=\left(\frac{2\pi mk_{\rm B}T}{h^2}\right)^{1/2}a+\left[\frac{1}{2}+\frac{h^2}{48ma^2k_{\rm B}T}\right]e^{-h^2/8ma^2k_{\rm B}T}$$

Show that the correction amounts to about 10^{-8}% for $m = 10^{-26}$ kg, $a = 1$ dm, and $T = 300$ K.

The one-dimensional version of Equation 4.4 is

$$q=\sum_{n=1}^{\infty}\exp\left(-\frac{\beta h^2n^2}{8ma^2}\right)=\sum_{n=1}^{\infty}e^{-bn^2}$$

where we let $b = \beta h^2/8ma^2$. The pertinent derivatives of q are

$$\frac{dq}{dn}=-2bne^{-bn^2}$$

$$\frac{d^2q}{dn^2}=-2be^{-bn^2}+4b^2n^2e^{-bn^2}$$

$$\frac{d^3q}{dn^3}=4b^2ne^{-bn^2}-8b^3n^3e^{-bn^2}$$

We can approximate

$$\int_1^{\infty}e^{-bn^2}dn\approx\int_0^{\infty}e^{-bn^2}dn$$

and use the Euler-Maclaurin summation formula from Problem 4–14 to find

$$q=\int_0^{\infty}e^{-bn^2}dn+\frac{1}{2}(e^{-b}-0)-\frac{1}{12}(-2be^{-b})+\frac{1}{720}(4b^2e^{-b}-8b^3e^{-b})$$

$$=\left(\frac{2\pi ma^2k_{\rm B}T}{h^2}\right)^{1/2}+\left(\frac{1}{2}+\frac{1}{6}b+\frac{1}{180}b^2-\frac{1}{90}b^3\right)e^{-b}$$

$$=\left(\frac{2\pi mk_{\rm B}T}{h^2}\right)^{1/2}a+\left[\frac{1}{2}+\frac{h^2}{48ma^2k_{\rm B}T}+O(b^2)\right]e^{-h^2/8ma^2k_{\rm B}T}$$

For $m = 10^{-26}$ kg, $a = 1$ dm, and $T = 300$ K, simply replacing the sum by an integral gives a value of 2.43×10^9. The correction term is 0.5, which is 2×10^{-8}% of the value of the sum.

4–16. We were able to evaluate the vibrational partition function for a harmonic oscillator exactly by recognizing the summation as a geometric series. Apply the Euler-Maclaurin summation formula (Problem 4–14) to this case and show that

$$\sum_{v=0}^{\infty}e^{-\beta(v+\frac{1}{2})h\nu}=e^{-\Theta_{\rm vib}/2T}\sum_{v=0}^{\infty}e^{-v\Theta_{\rm vib}/T}$$

$$=e^{-\Theta_{\rm vib}/2T}\left[\frac{T}{\Theta_{\rm vib}}+\frac{1}{2}+\frac{\Theta_{\rm vib}}{12T}+\cdots\right]$$

Show that the corrections to replacing the summation by an integration are very large for $O_2(g)$ at 300 K. Fortunately, we don't need to replace the summation by an integration in this case.

Recall that $\Theta_{vib} = h\nu/k_B$, so

$$\sum_{v=0}^{\infty} e^{-\beta(v+\frac{1}{2})h\nu} = e^{-\Theta_{vib}/2T} \sum_{v=0}^{\infty} e^{-v\Theta_{vib}/T}$$

Then

$$f(v) = e^{-v\Theta_{vib}/T} \qquad f'(v) = -\frac{\Theta_{vib}}{T} e^{-v\Theta_{vib}/T} \qquad f'''(v) = -\left(\frac{\Theta_{vib}}{T}\right)^3 e^{-v\Theta_{vib}/T}$$

and applying the Euler-Maclaurin summation formula yields

$$\begin{aligned}\sum_{v=0}^{\infty} e^{-v\Theta_{vib}/T} &= \int_0^{\infty} e^{-v\Theta_{vib}/T} dv + \frac{1}{2}(1) - \frac{1}{12}\left(-\frac{\Theta_{vib}}{T}\right) + \frac{1}{720}\left(-\frac{\Theta_{vib}}{T}\right) + O(T^{-3}) \\ &= \frac{T}{\Theta_{vib}} + \frac{1}{2} + \frac{\Theta_{vib}}{12T} + O(T^{-3})\end{aligned}$$

Then

$$\sum_{v=0}^{\infty} e^{-\beta(v+\frac{1}{2})h\nu} = e^{-\Theta_{vib}/2T}\left[\frac{T}{\Theta_{vib}} + \frac{1}{2} + \frac{\Theta_{vib}}{12T} + O(T^{-3})\right]$$

For O_2, $\Theta_{vib} = 2256$ K. Using Equation 4.23, we find that

$$q_{vib}(T) = \frac{e^{-2256\text{ K}/600\text{ K}}}{1 - e^{-2256\text{ K}/600\text{ K}}} = 0.0238$$

and using the correction, we have

$$q_{vib}(T) = e^{-2256\text{ K}/600\text{ K}}\left(\frac{300\text{ K}}{2256\text{ K}} + \frac{1}{2} + \frac{2256\text{ K}}{3600\text{ K}}\right) = 0.0293$$

which is about a 20% difference.

4–17. Plot the fraction of NO(g) molecules in the various rotational levels at 300 K and at 1000 K.

Use Equation 4.35, substituting $\Theta_{rot} = 2.39$ K (Table 4.2), to write f_J as a function of J, and plot.

$$f_J = (2J+1)(\Theta_{rot}/T)e^{-\Theta_{rot}J(J+1)/T}$$

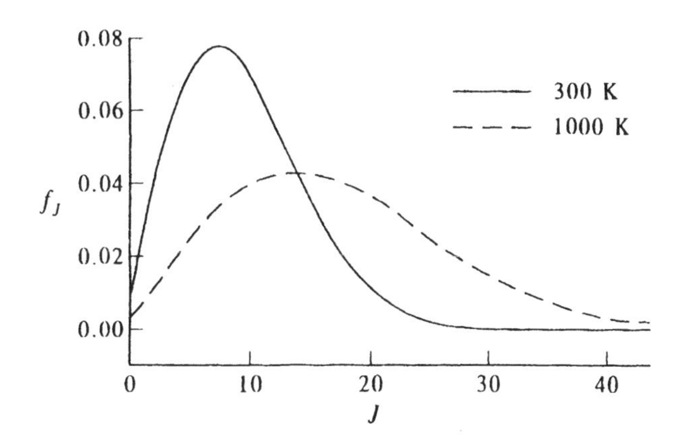

4–18. Show that the values of J at the maximum of a plot of f_J versus J (Equation 4.35) is given by

$$J_{\text{max}} \approx \left(\frac{T}{2\Theta_{\text{rot}}}\right)^{1/2} - \frac{1}{2}$$

Hint: Treat J as a continuous variable. Use this result to verify the values of J at the maxima in the plots in Problem 4–17.

$$f_J = (2J+1)\frac{\Theta_{\text{rot}}}{T}e^{-\Theta_{\text{rot}}J(J+1)/T}$$

At the maximum of a plot of f_J versus J, the slope is zero, so

$$\frac{df}{dJ} = \frac{2\Theta_{\text{rot}}}{T}e^{-\Theta_{\text{rot}}J_{\text{max}}(J_{\text{max}}+1)/T} - (2J_{\text{max}}+1)^2\left(\frac{\Theta_{\text{rot}}}{T}\right)^2 e^{-\Theta_{\text{rot}}J_{\text{max}}(J_{\text{max}}+1)/T} = 0$$

We can solve this equation for J_{max}:

$$(2J_{\text{max}}+1)^2\left(\frac{\Theta_{\text{rot}}}{T}\right)^2 e^{-\Theta_{\text{rot}}J_{\text{max}}(J_{\text{max}}+1)/T} = \frac{2\Theta_{\text{rot}}}{T}e^{-\Theta_{\text{rot}}J_{\text{max}}(J_{\text{max}}+1)/T}$$

$$(2J_{\text{max}}+1)^2 = \frac{2T}{\Theta_{\text{rot}}}$$

$$J_{\text{max}} = \left(\frac{T}{2\Theta_{\text{rot}}}\right)^{1/2} - \frac{1}{2}$$

For NO(g) at 300 K and 1000 K, $\Theta_{\text{rot}} = 2.39$ K, so the values of J_{max} given by the above equation are $J_{\text{max}} \approx 7$ and $J_{\text{max}} \approx 14$, respectively, in agreement with the plot in Problem 4–17.

4–19. The experimental heat capacity of $N_2(g)$ can be fit to the empirical formula

$$\overline{C}_V(T)/R = 2.283 + (6.291 \times 10^{-4}\ \text{K}^{-1})T - (5.0 \times 10^{-10}\ \text{K}^{-2})T^2$$

over the temperature range 300 K < T < 1500 K. Plot $\overline{C}_V(T)/R$ versus T over this range using Equation 4.41, and compare your results with the experimental curve.

For N_2, $\Theta_{\text{vib}} = 3374$ K, so we plot the experimental equation given in the problem text and the theoretical equation

$$\frac{\overline{C}_V(T)}{R} = \frac{5}{2} + \left(\frac{\Theta_{\text{vib}}}{T}\right)^2 \frac{e^{-\Theta_{\text{vib}}T}}{(1-e^{-\Theta_{\text{vib}}T})^2}$$

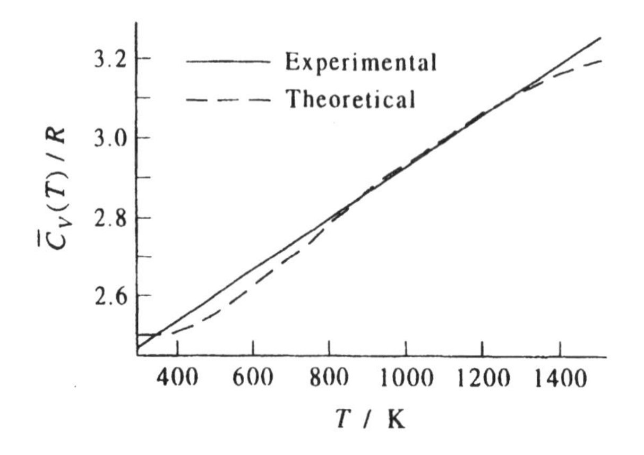

4–20. The experimental heat capacity of CO(g) can be fit to the empirical formula

$$\overline{C}_V(T)/R = 2.192 + (9.240 \times 10^{-4}\ \mathrm{K}^{-1})T - (1.41 \times 10^{-7}\ \mathrm{K}^{-2})T^2$$

over the temperature range 300 K < T < 1500 K. Plot $\overline{C}_V(T)/R$ versus T over this range using Equation 4.41, and compare your results with the experimental curve.

For CO, $\Theta_{\text{vib}} = 3103$ K, so we plot the experimental equation given in the problem text and the theoretical equation

$$\frac{\overline{C}_V(T)}{R} = \frac{5}{2} + \left(\frac{\Theta_{\text{vib}}}{T}\right)^2 \frac{e^{-\Theta_{\text{vib}}T}}{(1 - e^{-\Theta_{\text{vib}}T})^2}$$

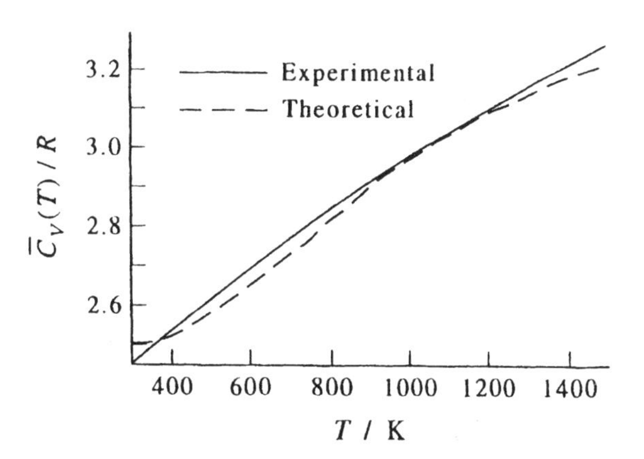

4–21. Calculate the contribution of each normal mode to the molar vibrational heat capacity of H_2O(g) at 600 K.

For H_2O, the values of Θ_{vib} for the three normal modes are 2290 K, 5160 K, and 5360 K. For $\Theta_{\text{vib},j} = 5360$ K,

$$\frac{\overline{C}_{V,j}}{R} = \left(\frac{5360}{600}\right)^2 \frac{e^{-5360/600}}{(1 - e^{-5360/600})^2} = 1.05 \times 10^{-2}$$

For $\Theta_{\text{vib},j} = 5160$ K,

$$\frac{\overline{C}_{V,j}}{R} = \left(\frac{5160}{600}\right)^2 \frac{e^{-5160/600}}{(1 - e^{-5160/600})^2} = 1.36 \times 10^{-2}$$

For $\Theta_{\text{vib},j} = 2290$ K,

$$\frac{\overline{C}_{V,j}}{R} = \left(\frac{2290}{600}\right)^2 \frac{e^{-2290/600}}{(1 - e^{-2290/600})^2} = 0.335$$

4–22. In analogy to the characteristic vibrational temperature, we can define a characteristic electronic temperature by

$$\Theta_{\text{elec},j} = \frac{\varepsilon_{ej}}{k_{\text{B}}}$$

where ε_{ej} is the energy of the jth excited electronic state relative to the ground state. Show that if we define the ground state to be the zero of energy, then

$$q_{\text{elec}} = g_0 + g_1 e^{-\Theta_{\text{elec},1}/T} + g_2 e^{-\Theta_{\text{elec},2}/T} + \cdots$$

The first and second excited electronic states of O(g) lie 158.2 cm^{-1} and 226.5 cm^{-1} above the ground electronic state. Given $g_0 = 5$, $g_1 = 3$, and $g_2 = 1$, calculate the values of $\Theta_{\text{elec},1}$, $\Theta_{\text{elec},2}$, and q_{elec} (ignoring any higher states) for O(g) at 5000 K.

Substituting the values given in the problem into the definition of $\Theta_{\text{elec},j}$ gives

$$\Theta_{\text{elec},1} = \frac{158.2\text{ cm}^{-1}}{0.69509\text{ cm}^{-1}\cdot\text{K}^{-1}} = 227.6\text{ K}$$

$$\Theta_{\text{elec},2} = \frac{226.5\text{ cm}^{-1}}{0.69509\text{ cm}^{-1}\cdot\text{K}^{-1}} = 325.8\text{ K}$$

We can write q_{elec} as (Equation 4.8)

$$q_{\text{elec}} = \sum g_{ej} e^{\varepsilon_{ej}/k_\text{B}T}$$

$$= g_0 + g_1 e^{-\Theta_{\text{elec},1}/T} + g_2 e^{-\Theta_{\text{elec},2}/T} + \cdots$$

$$q_{\text{elec}} = 5 + 3e^{-227.6/5000} + 1e^{-325.8/5000} = 8.803$$

4–23. Determine the symmetry numbers for H_2O, HOD, CH_4, SF_6, C_2H_2, and C_2H_4.

Symmetry numbers of selected molecules

Molecule	Symmetry Number
H_2O	2
HOD	1
CH_4	12
SF_6	24
C_2H_2	2
C_2H_4	4

4–24. The HCN(g) molecule is a linear molecule, and the following constants determined spectroscopically are $I = 18.816 \times 10^{-47}$ kg·m^2, $\tilde{\nu}_1 = 2096.7$ cm^{-1} (HC–N stretch), $\tilde{\nu}_2 = 713.46$ cm^{-1} (H–C–N bend, two-fold degeneracy), and $\tilde{\nu}_3 = 3311.47$ cm^{-1} (H–C stretch). Calculate the values of Θ_{rot} and Θ_{vib} and $\overline{C}_V$ at 3000 K.

We can use the definitions of Θ_{rot} and Θ_{vib} to write

$$\Theta_{\text{rot}} = \frac{\hbar^2}{2Ik_\text{B}} = 2.1405\text{ K}$$

$$\Theta_{\text{vib}} = \frac{hc\tilde{\nu}}{k_\text{B}}$$

$$\Theta_{\text{vib},1} = 3016\text{ K (HC-N stretch)}$$

$$\Theta_{\text{vib},2} = 1026\text{ K (H-C-N bend)}$$

$$\Theta_{\text{vib},3} = 4765\text{ K (H-C stretch)}$$

For linear polyatomic molecules, Equation 4.59 holds, and so

$$\begin{aligned}\frac{\overline{C}_V}{R} &= \frac{5}{2} + \sum_{j=1}^{4} \overline{C}_{\text{vib},j} \\ &= \frac{5}{2} + \left(\frac{3017}{3000}\right)^2 \frac{e^{-3017/3000}}{(1-e^{-3017/3000})^2} + 2\left(\frac{1026}{3000}\right)^2 \frac{e^{-1026/3000}}{(1-e^{-1026/3000})^2} \\ &\quad + \left(\frac{4764}{3000}\right)^2 \frac{e^{-4764/3000}}{(1-e^{-4764/3000})^2} \\ &= 2.5 + 0.92 + 1.98 + 0.81 = 6.21\end{aligned}$$

4–25. The acetylene molecule is linear, the C≡C bond length is 120.3 pm, and the C–H bond length is 106.0 pm. What is the symmetry number of acetylene? Determine the moment of inertia (Section 13–8) of acetylene and calculate the value of Θ_{rot}. The fundamental frequencies of the normal modes are $\tilde{\nu}_1 = 1975\ \text{cm}^{-1}$, $\tilde{\nu}_2 = 3370\ \text{cm}^{-1}$, $\tilde{\nu}_3 = 3277\ \text{cm}^{-1}$, $\tilde{\nu}_4 = 729\ \text{cm}^{-1}$, and $\tilde{\nu}_5 = 600\ \text{cm}^{-1}$. The normal modes $\tilde{\nu}_4$ and $\tilde{\nu}_5$ are doubly degenerate. All the other modes are nondegenerate. Calculate $\Theta_{\text{vib},j}$ and $\overline{C}_V$ at 300 K.

The symmetry number of acetylene is 2 (Problem 4.24). Choose the coordinate axis to bisect the center of the triple bond. Then

$$\begin{aligned}I = \sum_i m_i z_i^2 &= 2\left(\frac{120.3 \times 10^{-12}\ \text{m}}{2}\right)^2 (1.995 \times 10^{-26}\ \text{kg}) \\ &\quad +2\left(\frac{120.3 \times 10^{-12}\ \text{m}}{2} + 106.0 \times 10^{-12}\ \text{m}\right)^2 (1.673 \times 10^{-27}\ \text{kg}) \\ &= 2.368 \times 10^{-46}\ \text{kg}\cdot\text{m}^2\end{aligned}$$

We can use the definitions of Θ_{rot} and Θ_{vib} to write

$$\begin{aligned}\Theta_{\text{rot}} &= \frac{\hbar^2}{2Ik_B} = 1.702\ \text{K} \\ \Theta_{\text{vib}} &= \frac{hc\tilde{\nu}}{k_B} \\ \Theta_{\text{vib},2} &= 2842\ \text{K} \\ \Theta_{\text{vib},2} &= 4849\ \text{K} \\ \Theta_{\text{vib},3} &= 4715\ \text{K} \\ \Theta_{\text{vib},4} &= 1049\ \text{K} \\ \Theta_{\text{vib},5} &= 863.3\ \text{K}\end{aligned}$$

The vibrational molar heat capacities are given by (Equation 4.26)

$$\overline{C}_{V,\text{vib}} = R\left(\frac{\Theta_{\text{vib}}}{T}\right)^2 \frac{e^{-\Theta_{\text{vib}}/T}}{(1-e^{-\Theta_{\text{vib}}/T})^2}$$

Because acetylene is a linear polyatomic molecule, we can use Equation 4.59 to find the molar heat capacity at 300 K:

$$\begin{aligned}\frac{\overline{C}_V}{R} &= \frac{5}{2} + \sum_{j=1}^{7} \overline{C}_{\text{vib},j} \\ &= \frac{5}{2} + \overline{C}_{\text{vib},1} + \overline{C}_{\text{vib},2} + \overline{C}_{\text{vib},3} + 2\overline{C}_{\text{vib},4} + 2\overline{C}_{\text{vib},5} \\ &= \frac{5}{2} + 6.92 \times 10^{-3} + 2.50 \times 10^{-5} + 3.69 \times 10^{-5} + 2(0.394) + 2(0.523) \\ &= 4.34\end{aligned}$$

4–26. Plot the summand in Equation 4.53 versus J, and show that the most important values of J are large for $T \gg \Theta_{\text{rot}}$. We use this fact in going from Equation 4.53 to Equation 4.54.

The summand in Equation 4.53 is

$$(2J+1)^2 e^{-\hbar^2 J(J+1)/2Ik_BT} = (2J+1)^2 e^{-J(J+1)\Theta_{\text{rot}}/T}$$

Let $x = \Theta_{\text{rot}}/T$, so that for $\Theta_{\text{rot}} \ll T$ x is small, and then plot the summand versus J for different values of x.

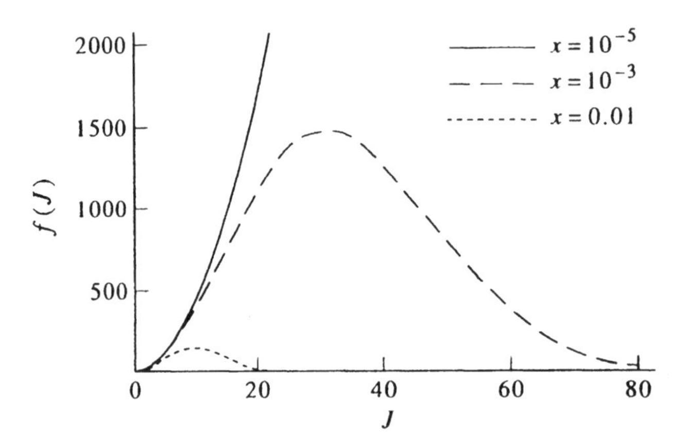

We can see that for $\Theta_{\text{rot}} \ll T$, the most important values of J (as far as contributions to the summand are concerned) are large, so Equation 4.54 holds quite well when this condition is met.

4–27. Use the Euler-Maclaurin summation formula (Problem 4–14) to show that

$$q_{\text{rot}}(T) = \frac{\pi^{1/2}}{\sigma}\left(\frac{T}{\Theta_{\text{rot}}}\right)^{3/2} + \frac{1}{6} + O\left(\frac{\Theta_{\text{rot}}}{T}\right)$$

for a spherical top molecule. Show that the correction to replacing Equation 4.53 by an integral is about 1% for CH_4 and 0.001% for CCl_4 at 300 K.

For a spherical top molecule,

$$q_{\text{rot}}(T) = \sum_{J=0}^{\infty}(2J+1)^2 e^{-\hbar^2 J(J+1)/2Ik_BT} \tag{4.53}$$

The pertinent derivatives are

$$f(J) = (2J+1)^2 e^{-\Theta_{rot}J(J+1)/T}$$
$$\frac{df}{dJ} = 4(2J+1)e^{-\Theta_{rot}(J^2+J)/T} - \frac{\Theta_{rot}}{T}(2J+1)^3 e^{-\Theta_{rot}(J^2+J)/T}$$
$$\frac{d^2f}{dJ^2} = 8e^{-\Theta_{rot}(J^2+J)/T} - \frac{10\Theta_{rot}}{T}(2J+1)^2 e^{-\Theta_{rot}(J^2+J)/T}$$
$$+\left(\frac{\Theta_{rot}}{T}\right)^2 (2J+1)^4 e^{-\Theta_{rot}(J^2+J)/T}$$

Applying the Euler-Maclaurin summation formula gives

$$q_{rot}(T) = \frac{1}{\sigma}\int_0^\infty e^{-\Theta_{rot}(J^2+J)/T} + \frac{1}{2}(1) - \frac{1}{12}\left[4 + O\left(\frac{\Theta_{rot}}{T}\right)\right] + O\left(\frac{\Theta_{rot}}{T}\right)$$
$$= \frac{\pi^{1/2}}{\sigma}\left(\frac{T}{\Theta_{rot}}\right)^{3/2} + \frac{1}{6} + O\left(\frac{\Theta_{rot}}{T}\right)$$

where we have included the symmetry number σ in the integral, as was done in the text. For CH_4 at 300 K the integral has a value of 37.07, and the correction term is about 1% of that; for CCl_4 at 300 K the integral has a value of about 32 500, and the correction term is about 0.001% of that.

4–28. The N–N and N–O bond lengths in the (linear) molecule N_2O are 109.8 pm and 121.8 pm, respectively. Calculate the center of mass and the moment of inertia of $^{14}N^{14}N^{16}O$. Compare your answer with the value obtained from Θ_{rot} in Table 4.4.

Choose the coordinate axis to bisect the central nitrogen atom. Then the moment of inertia (using the isotopic masses from the *CRC Handbook*) is

$$I = \sum_i m_i z_i^2 = (109.8 \times 10^{-12}\ \text{m})^2(2.325 \times 10^{-26}\ \text{kg})$$
$$+(121.8 \times 10^{-12}\ \text{m})^2(2.656 \times 10^{-26}\ \text{kg})$$
$$= 6.746 \times 10^{-46}\ \text{kg}\cdot\text{m}^2$$

and the center of mass of the molecule (relative to the central nitrogen atom) is

$$X = 0 + \frac{14.003}{44.001}(109.8 \times 10^{-12}\ \text{m}) + \frac{15.995}{44.013}(121.8 \times 10^{-12}\ \text{m})$$
$$= 7.922 \times 10^{-11}\ \text{m}$$

The center of mass of the molecule is 79.22 pm away from the central nitrogen atom, along the N–O bond. The value of Θ_{rot} calculated from the above values is

$$\Theta_{rot} = \frac{\hbar^2}{2Ik_B} = 0.597\ \text{K}$$

This value is within 1% of that in Table 4.4.

4–29. NO_2(g) is a bent triatomic molecule. The following data determined from spectroscopic measurements are $\tilde{\nu}_1 = 1319.7\ \text{cm}^{-1}$, $\tilde{\nu}_2 = 749.8\ \text{cm}^{-1}$, $\tilde{\nu}_3 = 1617.75\ \text{cm}^{-1}$, $\tilde{A}_0 = 8.0012\ \text{cm}^{-1}$, $\tilde{B}_0 = 0.43304\ \text{cm}^{-1}$, and $\tilde{C}_0 = 0.41040\ \text{cm}^{-1}$. Determine the three characteristic vibrational temperatures and the characteristic rotational temperatures for each of the principle axes of NO_2(g) at 1000 K. Calculate the value of $\overline{C}_V$ at 1000 K.

We can use Equation 4.49 to find $\Theta_{\text{vib},j}$:

$$\Theta_{\text{vib},j} = \frac{hc\nu_j}{k_B} = \frac{hc\tilde{\nu}_j}{k_B}$$

$$\Theta_{\text{vib},1} = \frac{(6.626 \times 10^{-34}\ \text{J}\cdot\text{s})(2.998 \times 10^{10}\ \text{cm}\cdot\text{s}^{-1})(1319.7\ \text{cm}^{-1})}{1.381 \times 10^{-23}\ \text{J}\cdot\text{K}^{-1}} = 1899\ \text{K}$$

$$\Theta_{\text{vib},2} = \frac{(6.626 \times 10^{-34}\ \text{J}\cdot\text{s})(2.998 \times 10^{10}\ \text{cm}\cdot\text{s}^{-1})(749.8\ \text{cm}^{-1})}{1.381 \times 10^{-23}\ \text{J}\cdot\text{K}^{-1}} = 1079\ \text{K}$$

$$\Theta_{\text{vib},3} = \frac{(6.626 \times 10^{-34}\ \text{J}\cdot\text{s})(2.998 \times 10^{10}\ \text{cm}\cdot\text{s}^{-1})(1617.75\ \text{cm}^{-1})}{1.381 \times 10^{-23}\ \text{J}\cdot\text{K}^{-1}} = 2328\ \text{K}$$

and Equation 4.32 for Θ_{rot}:

$$\Theta_{\text{rot},A} = \frac{(6.626 \times 10^{-34}\ \text{J}\cdot\text{s})(2.998 \times 10^{10}\ \text{cm}\cdot\text{s}^{-1})(8.0012\ \text{cm}^{-1})}{1.381 \times 10^{-23}\ \text{J}\cdot\text{K}^{-1}} = 11.51\ \text{K}$$

$$\Theta_{\text{rot},B} = \frac{(6.626 \times 10^{-34}\ \text{J}\cdot\text{s})(2.998 \times 10^{10}\ \text{cm}\cdot\text{s}^{-1})(0.43304\ \text{cm}^{-1})}{1.381 \times 10^{-23}\ \text{J}\cdot\text{K}^{-1}} = 0.6230\ \text{K}$$

$$\Theta_{\text{rot},C} = \frac{(6.626 \times 10^{-34}\ \text{J}\cdot\text{s})(2.998 \times 10^{10}\ \text{cm}\cdot\text{s}^{-1})(0.41040\ \text{cm}^{-1})}{1.381 \times 10^{-23}\ \text{J}\cdot\text{K}^{-1}} = 0.5905\ \text{K}$$

Finally, we use Equation 4.59 to determine the value of $\overline{C}_V$ at 1000 K.

$$\begin{aligned}\frac{\overline{C}_V}{R} &= 2\left(\frac{3}{2}\right) + (1.899)^2 \frac{e^{-1.899}}{(1 - e^{-1.899})^2} + (1.079)^2 \frac{e^{-1.079}}{(1 - e^{-1.079})^2} \\ &\qquad + (2.328)^2 \frac{e^{-2.328}}{(1 - e^{-2.328})^2} \\ &= 5.304\end{aligned}$$

4–30. The experimental heat capacity of $NH_3(g)$ can be fit to the empirical formula

$$\overline{C}_V(T)/R = 2.115 + (3.919 \times 10^{-3}\ \text{K}^{-1})T - (3.66 \times 10^{-7}\ \text{K}^{-2})T^2$$

over the temperature range $300\ \text{K} < T < 1500\ \text{K}$. Plot $\overline{C}_V(T)/R$ versus T over this range using Equation 4.62 and the molecular parameters in Table 4.4, and compare your results with the experimental curve.

For NH_3, we plot the experimental equation given in the problem text and the theoretical equation

$$\frac{\overline{C}_V(T)}{R} = \frac{3}{2} + \frac{3}{2} + \sum_{j=1}^{6} \left(\frac{\Theta_{\text{vib},j}}{T}\right)^2 \frac{e^{-\Theta_{\text{vib},j}/T}}{(1 - e^{-\Theta_{\text{vib},j}/T})^2} \tag{4.62}$$

where $\Theta_{vib,1} = 4800$ K, $\Theta_{vib,2} = 1360$ K, $\Theta_{vib,3} = \Theta_{vib,4} = 4880$ K, and $\Theta_{vib,5} = \Theta_{vib,6} = 2330$ K (Table 4.4).

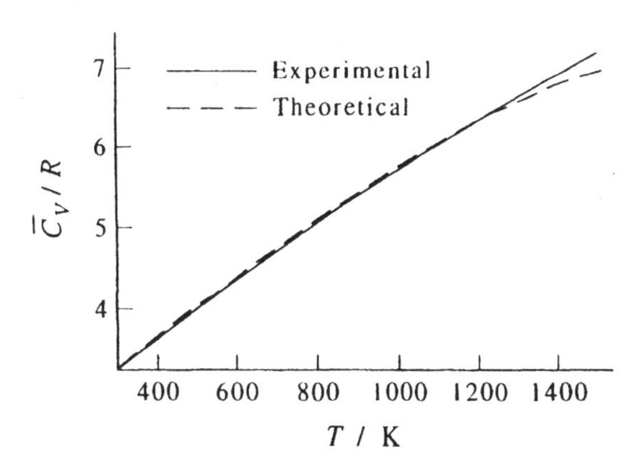

4–31. The experimental heat capacity of $SO_2(g)$ can be fit to the empirical formula

$$\overline{C}_V(T)/R = 6.8711 - \frac{1454.62\ \text{K}}{T} + \frac{160\,351\ \text{K}^2}{T^2}$$

over the temperature range 300 K $< T <$ 1500 K. Plot $\overline{C}_V(T)/R$ versus T over this range using Equation 4.62 and the molecular parameters in Table 4.4, and compare your results with the experimental curve.

For SO_2, we plot the experimental equation given in the problem text and the theoretical equation

$$\frac{\overline{C}_V(T)}{R} = \frac{3}{2} + \frac{3}{2} + \sum_{j=1}^{3}\left(\frac{\Theta_{vib,j}}{T}\right)^2 \frac{e^{-\Theta_{vib,j}/T}}{(1 - e^{-\Theta_{vib,j}/T})^2} \tag{4.62}$$

where $\Theta_{vib,1} = 1660$ K, $\Theta_{vib,2} = 750$ K, and $\Theta_{vib,3} = 1960$ K (Table 4.4).

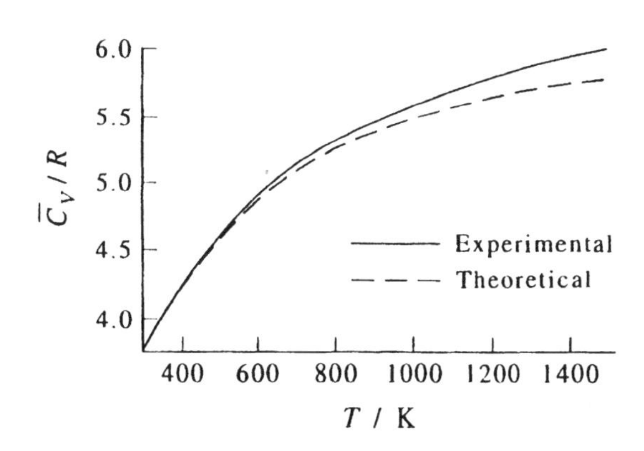

4–32. The experimental heat capacity of $CH_4(g)$ can be fit to the empirical formula

$$\overline{C}_V(T)/R = 1.099 + (7.27 \times 10^{-3}\ \text{K}^{-1})T + (1.34 \times 10^{-7}\ \text{K}^{-2})T^2 - (8.67 \times 10^{-10}\ \text{K}^{-3})T^3$$

over the temperature range 300 K $< T <$ 1500 K. Plot $\overline{C}_V(T)/R$ versus T over this range using Equation 4.62 and the molecular parameters in Table 4.4, and compare your results with the experimental curve.

For CH_4, we plot the experimental equation given in the problem text and the theoretical equation

$$\frac{\overline{C}_V(T)}{R} = \frac{3}{2} + \frac{3}{2} + \sum_{j=1}^{9} \left(\frac{\Theta_{\text{vib},j}}{T}\right)^2 \frac{e^{-\Theta_{\text{vib},j}/T}}{(1 - e^{-\Theta_{\text{vib},j}/T})^2} \tag{4.62}$$

where $\Theta_{\text{vib},1} = 4170$ K, $\Theta_{\text{vib},2} = \Theta_{\text{vib},3} = 2180$ K, $\Theta_{\text{vib},4} = \Theta_{\text{vib},5} = \Theta_{\text{vib},6} = 4320$ K, and $\Theta_{\text{vib},7} = \Theta_{\text{vib},8} = \Theta_{\text{vib},9} = 1870$ K (Table 4.4).

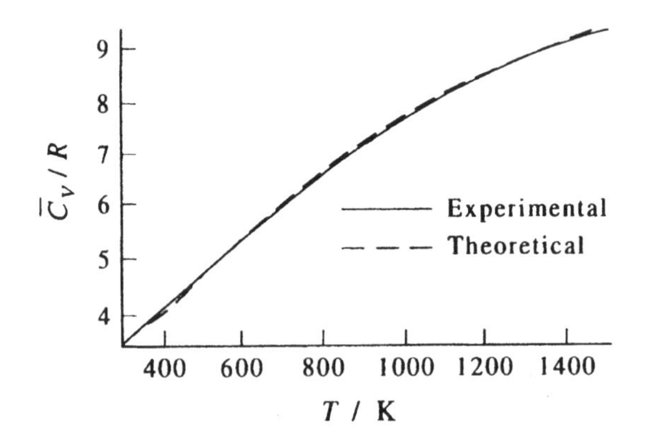

4–33. Show that the moment of inertia of a diatomic molecule is μR_e^2, where μ is the reduced mass, and R_e is the equilibrium bond length.

Let the point labelled in the figure be the center of mass of the molecule. r_1 and r_2 are the distances from the center of mass to masses 1 and 2 (with masses m_1 and m_2), respectively; R_e is the bond length and M is the total mass of the molecule.

We can then write (by the definition of the center of mass)

$$r_1 = \frac{m_2 r_2}{m_1}$$

Now, because $R_e = r_1 + r_2$ and $M = m_1 + m_2$,

$$R_e = \left(\frac{m_2}{m_1} + 1\right) r_2 = \frac{M}{m_1} r_2$$

Finally, because $I = \sum_j m_j r_j^2$ and $\mu = m_1 m_2 / M$,

$$\begin{aligned} I = m_1 r_1^2 + m_2 r_2^2 &= \left[m_1 \left(\frac{m_2}{m_1}\right)^2 + m_2 \right] r_2^2 \\ &= \left(\frac{m_2^2}{m_1} + m_2\right) \frac{m_1^2 R_e^2}{M^2} = \left(\frac{m_2 M}{m_1}\right) \frac{m_1^2 R_e^2}{M^2} \\ &= \frac{m_1 m_2}{M} R_e^2 = \mu R_e^2 \end{aligned}$$

4–34. Given that the values of Θ_{rot} and Θ_{vib} for H_2 are 85.3 K and 6332 K, respectively, calculate these quantities for HD and D_2. *Hint*: Use the Born-Oppenheimer approximation.

In the Born-Oppenheimer approximation, the potential curve of a diatomic molecule is independent of the isotopes of the constituent atoms. Then, in the formula $I = \mu R_0^2$, R_0 is the same for D_2, H_2, and DH. Therefore, in the harmonic oscillator-rigid rotator approximation,

$$\Theta_{\text{vib}} = \frac{h\nu}{k_B} \sim \mu^{-1/2} \qquad \text{and} \qquad \Theta_{\text{rot}} = \frac{\hbar^2}{2Ik_B} \sim \mu^{-1}$$

We can calculate the reduced masses of H_2, D_2, and DH in atomic masses:

$$\mu_{H_2} = \frac{1.008 \text{ amu}}{2} = 0.504 \text{ amu} \qquad \mu_{D_2} = \frac{2.014 \text{ amu}}{2} = 1.007 \text{ amu}$$

$$\mu_{DH} = \frac{(2.014 \text{ amu})(1.008 \text{ amu})}{3.022 \text{ amu}} = 0.672 \text{ amu}$$

Now we can use the relationships between μ and Θ_{vib} and μ and Θ_{rot} to find rotational and vibrational temperatures of D_2:

$$\frac{\Theta_{\text{vib},D_2}}{\Theta_{\text{vib},H_2}} = \left(\frac{\mu_{H_2}}{\mu_{D_2}}\right)^{1/2}$$

$$\Theta_{\text{vib},D_2} = \left(\frac{0.504 \text{ amu}}{1.007 \text{ amu}}\right)^{1/2} (6332 \text{ K}) = 4480 \text{ K}$$

$$\frac{\Theta_{\text{rot},D_2}}{\Theta_{\text{rot},H_2}} = \frac{\mu_{H_2}}{\mu_{D_2}}$$

$$\Theta_{\text{rot},D_2} = \frac{0.504 \text{ amu}}{1.007 \text{ amu}}(85.3 \text{ K}) = 42.7 \text{ K}$$

Similarly, for DH

$$\frac{\Theta_{\text{vib},DH}}{\Theta_{\text{vib},H_2}} = \left(\frac{\mu_{H_2}}{\mu_{DH}}\right)^{1/2}$$

$$\Theta_{\text{vib},DH} = \left(\frac{0.504 \text{ amu}}{0.672 \text{ amu}}\right)^{1/2} (6332 \text{ K}) = 5484 \text{ K}$$

$$\frac{\Theta_{\text{rot},DH}}{\Theta_{\text{rot},H_2}} = \frac{\mu_{H_2}}{\mu_{DH}}$$

$$\Theta_{\text{rot},DH} = \frac{0.504 \text{ amu}}{0.672 \text{ amu}}(85.3 \text{ K}) = 64.0 \text{ K}$$

4–35. Using the result for $q_{\text{rot}}(T)$ obtained in Problem 4–14, derive corrections to the expressions $\langle E_{\text{rot}} \rangle = RT$ and $C_{V,\text{rot}} = R$ given in Section 4–5. Express your result in terms of powers of Θ_{rot}/T.

From Problem 4–14, we write q_{rot} as

$$q_{\text{rot}}(T) = \frac{T}{\Theta_{\text{rot}}}\left\{1 + \frac{1}{3}\left(\frac{\Theta_{\text{rot}}}{T}\right) + \frac{1}{15}\left(\frac{\Theta_{\text{rot}}}{T}\right)^2 + O\left[\left(\frac{\Theta_{\text{rot}}}{T}\right)^3\right]\right\}$$

$$\ln q_{\text{rot}} = \ln\frac{T}{\Theta_{\text{rot}}} + \ln\left\{1 + \frac{1}{3}\frac{\Theta_{\text{rot}}}{T} + \frac{1}{15}\left(\frac{\Theta_{\text{rot}}}{T}\right)^2 + O\left[\left(\frac{\Theta_{\text{rot}}}{T}\right)^3\right]\right\}$$

We use the expansion

$$\ln(1+x) = x - \frac{x^2}{2} + \frac{x^3}{3} - O(x^4) \tag{C.12}$$

to write $\ln q_{\text{rot}}$ as

$$\ln q_{\text{rot}} = \ln\frac{T}{\Theta_{\text{rot}}} + \left[\frac{1}{3}\frac{\Theta_{\text{rot}}}{T} + \frac{1}{15}\left(\frac{\Theta_{\text{rot}}}{T}\right)^2\right] - \frac{1}{2}\left(\frac{1}{3}\frac{\Theta_{\text{rot}}}{T}\right)^2 + O\left[\left(\frac{\Theta_{\text{rot}}}{T}\right)^3\right]$$

$$= \ln\frac{T}{\Theta_{\text{rot}}} + \frac{1}{3}\frac{\Theta_{\text{rot}}}{T} + \frac{1}{90}\left(\frac{\Theta_{\text{rot}}}{T}\right)^2 + O\left[\left(\frac{\Theta_{\text{rot}}}{T}\right)^3\right]$$

Now we use Equation 3.21 to write $\langle E\rangle$ as

$$\langle E\rangle = k_{\text{B}}T^2\left(\frac{\partial \ln Q}{\partial T}\right)_{N,V} = Nk_{\text{B}}T^2\left(\frac{\partial \ln q}{\partial T}\right)_{N,V}$$

$$= RT^2\left[\frac{1}{T} - \frac{1}{3}\frac{\Theta_{\text{rot}}}{T^2} - \frac{1}{45}\frac{\Theta_{\text{rot}}^2}{T^3} + O(T^{-4})\right]$$

$$= RT - \frac{R\Theta_{\text{rot}}}{3} - \frac{R\Theta_{\text{rot}}^2}{45T} + O\left[\left(\frac{\Theta_{\text{rot}}}{T}\right)^2\right]$$

Finally, use the definition of constant-volume heat capacity (Equation 3.25) to write

$$C_V = \left(\frac{\partial\langle E\rangle}{\partial T}\right)_{N,V}$$

$$= R + \frac{R}{45}\left(\frac{\Theta_{\text{rot}}}{T}\right)^2 + O\left[\left(\frac{\Theta_{\text{rot}}}{T}\right)^3\right]$$

4–36. Show that the thermodynamic quantities P and C_V are independent of the choice of a zero of energy.

Begin with Equation 3.14,

$$Q = \sum_j e^{-\beta E_j}$$

Now choose E_0 to be the zero of energy and define

$$Q_0 = \sum_j e^{-\beta(E_j - E_0)}$$

such that

$$Q = e^{-\beta E_0}Q_0$$

Now use Equation 3.32 to write

$$P = k_B T \left(\frac{\partial \ln Q}{\partial V}\right)_{N,\beta} = k_B T \left[\frac{\partial}{\partial V}(-\beta E_0 + \ln Q_0)\right]_{N,\beta} = k_B T \left(\frac{\partial \ln Q_0}{\partial V}\right)_{N,\beta}$$

and use Equation 3.21 to write

$$\langle E\rangle = k_B T^2 \left(\frac{\partial \ln Q}{\partial T}\right)_{N,V} = k_B T^2 \left[\frac{\partial}{\partial T}(-\beta E_0 + \ln Q_0)\right]_{N,V} = E_0 + k_B T^2 \left(\frac{\partial \ln Q_0}{\partial T}\right)_{N,V}$$

Because $\overline{C}_V = (\partial\langle E\rangle/\partial T)_{N,V}$ (Equation 3.25), we can write

$$\overline{C}_V = \left\{\frac{\partial}{\partial T}\left[E_0 + k_B T^2 \left(\frac{\partial \ln Q_0}{\partial T}\right)_{N,V}\right]\right\}_{N,V} = \left\{\frac{\partial}{\partial T}\left[k_B T^2 \left(\frac{\partial \ln Q_0}{\partial T}\right)_{N,V}\right]\right\}_{N,V}$$

Therefore, the values of P and $\overline{C}_V$ are independent of the choice of a zero of energy, as they must be.

4–37. Molecular nitrogen is heated in an electric arc. The spectroscopically determined relative populations of excited vibrational levels are listed below.

v	0	1	2	3	4	...
$\frac{f_v}{f_0}$	1.000	0.200	0.040	0.008	0.002	...

Is the nitrogen in thermodynamic equilibrium with respect to vibrational energy? What is the vibrational temperature of the gas? Is this value necessarily the same as the translational temperature? Why or why not?

At thermal equilibrium,

$$\frac{f_v}{f_0} = \frac{e^{-\beta h\nu(v+1/2)}}{q_{\text{vib}}}\left(\frac{e^{-\beta h\nu/2}}{q_{\text{vib}}}\right)^{-1} = e^{\beta h\nu v}$$

Thus, if nitrogen is in thermodynamic equilibrium with respect to vibrational energy, the graph of $\ln(f_v/f_0)$ vs. v will be a straight line with slope $-\beta h\nu$. The following figure shows the plot of $\ln(f_v/f_0)$ versus v.

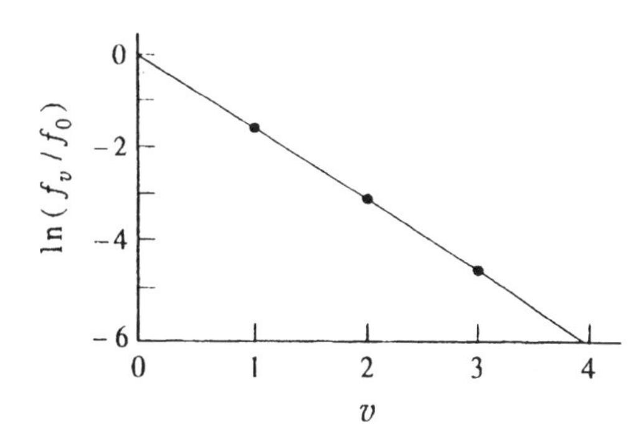

The slope of the line of best fit is -1.5648, which, using $\tilde{\nu}_{N_2} = 2330\ \text{cm}^{-1}$ (Table 1.4), corresponds to a vibrational temperature of 2140 K. The vibrational and translational temperatures need not be equal, because the time scale of the energy transfer between vibrational states and between translational states can be quite different (the energy transfer between vibrational states is usually orders of magnitude slower than that between translational states).

4–38. Consider a system of independent diatomic molecules constrained to move in a plane, that is, a two-dimensional ideal diatomic gas. How many degrees of freedom does a two-dimensional diatomic molecule have? Given that the energy eigenvalues of a two-dimensional rigid rotator are

$$\varepsilon_J = \frac{\hbar^2 J^2}{2I} \qquad J = 0,\ 1,\ 2,\ \ldots$$

(where I is the moment of inertia of the molecule) with a degeneracy $g_J = 2$ for all J except $J = 0$, derive an expression for the rotational partition function. The vibrational partition function is the same as for a three-dimensional diatomic gas. Write out

$$q(T) = q_{\text{trans}}(T)q_{\text{rot}}(T)q_{\text{vib}}(T)$$

and derive an expression for the average energy of this two-dimensional ideal diatomic gas.

A two-dimensional diatomic molecule has two translational degrees of freedom, one vibrational degree of freedom, and one rotational degree of freedom. We know that

$$\epsilon_J = \frac{\hbar^2 J^2}{2I} = \Theta_{\text{rot}} k_{\text{B}} J^2 \qquad J = 0, 1, 2, \cdots$$

So

$$q_{\text{rot}} = \frac{1}{\sigma}\int_0^\infty dJ g_J e^{-\epsilon_J/k_{\text{B}}T}$$

$$q_{\text{rot}} = 1 + 2\int_0^\infty dJ e^{-J^2\Theta_{\text{rot}}/T} \approx \left(\frac{\pi T}{\Theta_{\text{rot}}}\right)^{1/2}$$

We are told that q_{vib} is the same for a two-dimensional gas as it is for a three-dimensional gas, and we know q_{trans} for an ideal two-dimensional gas from Problem 4.3. We can now obtain an expression for the average energy of this gas:

$$\begin{aligned} q(T) &= q_{\text{trans}}(T)q_{\text{rot}}(T)q_{\text{vib}}(T) \\ &= \left(\frac{2a^2\pi m k_{\text{B}} T}{h^2}\right)\left(\frac{\pi T}{\Theta_{\text{rot}}}\right)^{1/2}\left(\frac{e^{-\Theta_{\text{vib}}/2T}}{1 - e^{-\Theta_{\text{vib}}/T}}\right) \end{aligned}$$

We now wish to find the temperature-dependent terms of $\ln q$:

$$\ln q = \ln T + \frac{1}{2}\ln T - \frac{\Theta_{\text{vib}}}{2T} - \ln(1 - e^{-\Theta_{\text{vib}}/2T}) + \text{terms not containing } T$$

Now, as in Example 4.5, we can take

$$\begin{aligned} \langle E\rangle &= Nk_{\text{B}}T^2\left(\frac{\partial \ln q}{\partial T}\right)_V \\ &= RT^2\left(\frac{1}{T} + \frac{1}{2T} + \frac{\Theta_{\text{vib}}}{2T^2} + \frac{\Theta_{\text{vib}}}{T^2}\frac{e^{-\Theta_{\text{vib}}/T}}{1 - e^{-\Theta_{\text{vib}}/T}}\right) \\ &= \frac{3RT}{2} + \frac{R\Theta_{\text{vib}}}{2} + R\Theta_{\text{vib}}\frac{e^{-\Theta_{\text{vib}}/T}}{1 - e^{-\Theta_{\text{vib}}/T}} \end{aligned}$$

4–39. What molar constant-volume heat capacities would you expect under classical conditions for the following gases: (a) Ne, (b) O_2, (c) H_2O, (d) CO_2, and (e) $CHCl_3$?

Each of the gases has a contribution of $3R/2$ to its heat capacity from the translational partition function. In addition, there is a contribution of $R/2$ for each rotational degree of freedom and R for each vibrational degree of freedom. Therefore, the molar heat capacities are

a. $\frac{3}{2}R + 0R + 0R = \frac{3}{2}R$

b. $\frac{3}{2}R + \frac{2}{2}R + R = \frac{7}{2}R$

c. $\frac{3}{2}R + \frac{3}{2}R + 3R = 6R$

d. $\frac{3}{2}R + \frac{2}{2}R + 4R = \frac{13}{2}R$

e. $\frac{3}{2}R + \frac{3}{2}R + 9R = 12R$

4–40. In this problem, we will derive an expression for the number of translational energy states with (translational) energy between ε and $\varepsilon + d\varepsilon$. This expression is essentially the degeneracy of the state whose energy is

$$\varepsilon_{n_x n_y n_z} = \frac{h^2}{8ma^2}(n_x^2 + n_y^2 + n_z^2) \qquad n_x,\ n_y,\ n_z = 1,\ 2,\ 3,\ \ldots \tag{1}$$

The degeneracy is given by the number of ways the integer $M = 8ma^2\varepsilon/h^2$ can be written as the sum of the squares of three positive integers. In general, this is an erratic and discontinuous function of M (the number of ways will be zero for many values of M), but it becomes smooth for large M, and we can derive a simple expression for it. Consider a three-dimensional space spanned by n_x, n_y, and n_z. There is a one-to-one correspondence between energy states given by Equation 1 and the points in this n_x, n_y, n_z space with coordinates given by positive integers. Figure 4.8 shows a two-dimensional version of this space. Equation 1 is an equation for a sphere of radius $R = (8ma^2\varepsilon/h^2)^{1/2}$ in this space

$$n_x^2 + n_y^2 + n_z^2 = \frac{8ma^2\varepsilon}{h^2} = R^2$$

We want to calculate the number of lattice points that lie at some fixed distance from the origin in this space. In general, this is very difficult, but for large R we can proceed as follows. We treat R, or ε, as a continuous variable and ask for the number of lattice points between ε and $\varepsilon + \Delta\varepsilon$. To calculate this quantity, it is convenient to first calculate the number of lattice points consistent with an energy $\leq \varepsilon$. For large ε, an excellent approximation can be made by equating the number of

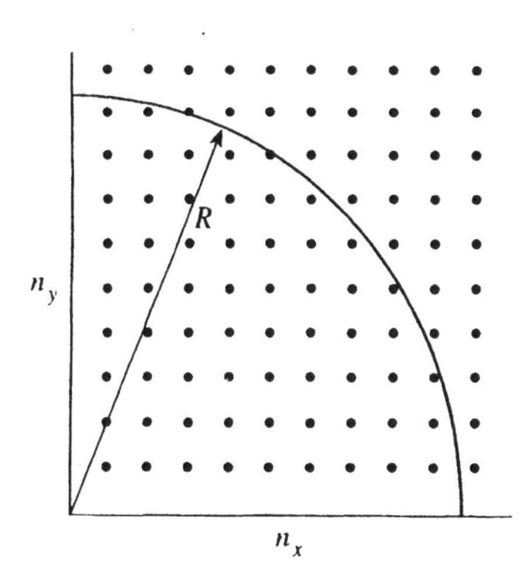

FIGURE 4.8
A two-dimensional version of the (n_x, n_y, n_z) space, the space with the quantum numbers n_x, n_y, and n_z as axes. Each point corresponds to an energy of a particle in a (two-dimensional) box.

lattice points consistent with an energy $\leq \varepsilon$ with the volume of one octant of a sphere of radius R. We take only one octant because n_x, n_y, and n_z are restricted to be positive integers. If we denote the number of such states by $\Phi(\varepsilon)$, we can write

$$\Phi(\varepsilon) = \frac{1}{8}\left(\frac{4\pi R^3}{3}\right) = \frac{\pi}{6}\left(\frac{8ma^2\varepsilon}{h^2}\right)^{3/2}$$

The number of states with energy between ε and $\varepsilon + \Delta\varepsilon$ ($\Delta\varepsilon/\varepsilon \ll 1$) is

$$\omega(\varepsilon, \Delta\varepsilon) = \Phi(\varepsilon + \Delta\varepsilon) - \Phi(\varepsilon)$$

Show that

$$\omega(\varepsilon, \Delta\varepsilon) = \frac{\pi}{4}\left(\frac{8ma^2}{h^2}\right)^{3/2} \varepsilon^{1/2}\Delta\varepsilon + O[(\Delta\varepsilon)^2]$$

Show that if we take $\varepsilon = 3k_BT/2$, $T = 300$ K, $m = 10^{-25}$ kg, $a = 1$ dm, and $\Delta\varepsilon$ to be 0.010ε (in other words 1% of ε), then $\omega(\varepsilon, \Delta\varepsilon)$ is $O(10^{28})$. So, even for a system as simple as a single particle in a box, the degeneracy can be very large at room temperature.

We can express $\Phi(\varepsilon)$ and $\Phi(\varepsilon + \Delta\varepsilon)$, by definition, as

$$\Phi(\varepsilon) = \frac{\pi}{6}\left(\frac{8ma^2\varepsilon}{h^2}\right)^{3/2} \qquad \Phi(\varepsilon + \Delta\varepsilon) = \frac{\pi}{6}\left[\frac{8ma^2(\varepsilon + \Delta\varepsilon)}{h^2}\right]^{3/2}$$

Now we substitute these values into the expression given in the text for $\omega(\varepsilon, \Delta\varepsilon)$ and expand in $\Delta\varepsilon$:

$$\begin{aligned}
\omega(\varepsilon, \Delta\varepsilon) &= \Phi(\varepsilon + \Delta\varepsilon) - \Phi(\varepsilon) \\
&= \frac{\pi}{6}\left(\frac{8ma^2}{h^2}\right)^{3/2}\left[(\varepsilon + \Delta\varepsilon)^{3/2} - \varepsilon^{3/2}\right] \\
&= \frac{\pi}{6}\left(\frac{8ma^2}{h^2}\right)^{3/2}\left\{\varepsilon^{3/2}\left[\left(1 + \frac{\Delta\varepsilon}{\varepsilon}\right)^{3/2} - 1\right]\right\} \\
&= \frac{\pi}{6}\left(\frac{8ma^2}{h^2}\right)^{3/2}\varepsilon^{3/2}\left\{1 + \frac{3}{2}\frac{\Delta\varepsilon}{\varepsilon} + O\left[\left(\frac{\Delta\varepsilon}{\varepsilon}\right)^2\right] - 1\right\} \\
&= \frac{\pi}{4}\left(\frac{8ma^2}{h^2}\right)^{3/2}\varepsilon^{3/2}\frac{\Delta\varepsilon}{\varepsilon} + O\left[(\Delta\varepsilon)^2\right] \\
&= \frac{\pi}{4}\left(\frac{8ma^2}{h^2}\right)^{3/2}\varepsilon^{1/2}\Delta\varepsilon + O[(\Delta\varepsilon)^2]
\end{aligned}$$

Substituting the desired values in the above equations gives $\omega(\varepsilon, 0.01\varepsilon) = 9.5 \times 10^{27} = O(10^{28})$.

4–41. The translational partition function can be written as a single integral over the energy ε if we include the degeneracy

$$q_{\text{trans}}(V, T) = \int_0^\infty \omega(\varepsilon)e^{-\varepsilon/k_BT}d\varepsilon$$

where $\omega(\varepsilon)d\varepsilon$ is the number of states with energy between ε and $\varepsilon + d\varepsilon$. Using the result from the previous problem, show that $q_{\text{trans}}(V, T)$ is the same as that given by Equation 4.6.

As $\Delta\varepsilon \to 0$, $\omega(\varepsilon, \Delta\varepsilon) \to \omega(\varepsilon)d\varepsilon$, so we can use (from the above problem)

$$\omega(\varepsilon)d\varepsilon = \frac{\pi}{4}\left(\frac{8ma^2}{h^2}\right)^{3/2} \varepsilon^{1/2}d\varepsilon$$

Substituting this into the expression given for q_{trans} gives

$$\begin{aligned}
q_{\text{trans}} &= \int_0^\infty \frac{\pi}{4}\left(\frac{8ma^2}{h^2}\right)^{3/2} \varepsilon^{1/2}e^{-\varepsilon/k_BT}d\varepsilon \\
&= \frac{\pi}{4}\left(\frac{8ma^2}{h^2}\right)^{3/2} \int_0^\infty \varepsilon^{1/2}e^{-\varepsilon/k_BT}d\varepsilon \\
&= \frac{\pi}{4}\left(\frac{8ma^2}{h^2}\right)^{3/2} \int_0^\infty xe^{-x^2/k_BT}2xdx \\
&= \frac{\pi}{4}\left(\frac{8ma^2}{h^2}\right)^{3/2} \frac{k_BT}{2}(\pi k_BT)^{1/2} \\
&= \left(\frac{2\pi mk_BT}{h^2}\right)^{3/2} a^3
\end{aligned}$$

which is the same as Equation 4.6.

MATHCHAPTER D

Partial Derivatives

PROBLEMS AND SOLUTIONS

D–1. The isothermal compressibility, κ_T, of a substance is defined as

$$\kappa_T = -\frac{1}{V}\left(\frac{\partial V}{\partial P}\right)_T$$

Obtain an expression for the isothermal compressibility of an ideal gas.

For an ideal gas, $PV = nRT$. Taking the partial derivative of both sides of this equation with respect to P gives

$$V\left(\frac{\partial P}{\partial P}\right)_T + P\left(\frac{\partial V}{\partial P}\right)_T = nR\left(\frac{\partial T}{\partial P}\right)_T$$

$$V + P\left(\frac{\partial V}{\partial P}\right)_T = 0$$

Then

$$\left(\frac{\partial V}{\partial P}\right)_T = -\frac{V}{P}$$

$$-\frac{1}{V}\left(\frac{\partial V}{\partial P}\right)_T = \frac{1}{P}$$

$$-\frac{1}{V}\left(\frac{\partial V}{\partial P}\right)_T = \kappa_T = \frac{1}{P}$$

D–2. The coefficient of thermal expansion, α, of a substance is defined as

$$\alpha = \frac{1}{V}\left(\frac{\partial V}{\partial T}\right)_P$$

Obtain an expression for the coefficient of thermal expansion of an ideal gas.

For an ideal gas, $PV = nRT$. Taking the partial derivative of both sides of this equation with respect to T gives

$$P\left(\frac{\partial V}{\partial T}\right)_P = nR$$
$$\left(\frac{\partial V}{\partial T}\right)_P = \frac{nR}{P}$$
$$\frac{1}{V}\left(\frac{\partial V}{\partial T}\right)_P = \frac{nR}{PV}$$

or

$$\alpha = \frac{1}{T}$$

D–3. Prove that

$$\left(\frac{\partial P}{\partial V}\right)_{n,T} = \frac{1}{\left(\frac{\partial V}{\partial P}\right)_{n,T}}$$

for an ideal gas and for a gas whose equation of state is $P = nRT/(V - nb)$, where b is a constant. This relation is generally true and is called the reciprocal identity. Notice that the same variables must be held fixed on both sides of the identity.

For an ideal gas, $PV = nRT$. The partial of P with respect to V is

$$\left(\frac{\partial P}{\partial V}\right)_{n,T} = -\frac{nRT}{V^2} = -\frac{P}{V}$$

and the partial of V with respect to P is

$$\left(\frac{\partial V}{\partial P}\right)_{n,T} = -\frac{nRT}{P^2} = -\frac{V}{P}$$

Because

$$-\frac{P}{V} = \left(-\frac{V}{P}\right)^{-1}$$

the reciprocal identity given in the text of the problem holds for an ideal gas. Likewise, for a gas with the equation of state $P = nRT/(V - nb)$, the partial of P with respect to V is

$$\left(\frac{\partial P}{\partial V}\right)_{n,T} = -\frac{nRT}{(V - nb)^2} = -\frac{P}{V - nb}$$

and the partial of V with respect to P is

$$\left(\frac{\partial V}{\partial P}\right)_{n,T} = -\frac{nRT}{P^2} = -\frac{V - nb}{P}$$

Because

$$-\frac{P}{(V - nb)} = \left[-\frac{(V - nb)}{P}\right]^{-1}$$

the reciprocal identity also holds for a gas with equation of state $P = nRT/(V - nb)$.

D–4. Given that

$$U = kT^2\left(\frac{\partial \ln Q}{\partial T}\right)_{N,V}$$

where

$$Q(N, V, T) = \frac{1}{N!}\left(\frac{2\pi mk_{\mathrm{B}}T}{h^2}\right)^{3N/2} V^N$$

and k_{B}, m, and h are constants, determine U as a function of T.

We are given

$$U = kT^2\left(\frac{\partial \ln Q}{\partial T}\right)_{N,V}$$

and

$$Q(N, V, T) = \frac{1}{N!}\left(\frac{2\pi mk_{\mathrm{B}}T}{h^2}\right)^{3N/2} V^N$$

Let $K = \frac{V^N}{N!}\left(\frac{2\pi mk_{\mathrm{B}}}{h^2}\right)^{3N/2}$. Then $Q = KT^{3N/2}$, and

$$\ln Q = 3N/2 \ln T + \ln K$$

Taking the partial derivative with respect to T gives

$$\left(\frac{\partial \ln Q}{\partial T}\right)_{N,V} = \frac{3N}{2T}$$

and so

$$U = kT^2\left(\frac{\partial \ln Q}{\partial T}\right)_{N,V} = kT^2\frac{3N}{2T} = \frac{3NkT}{2} = \frac{3nRT}{2}$$

D–5. Show that the total derivative of P for the Redlich-Kwong equation,

$$P = \frac{RT}{\overline{V} - B} - \frac{A}{T^{1/2}\overline{V}(\overline{V} + B)}$$

is given by Equation D.14.

We can write (as in Equation D.11)

$$dP = \left(\frac{\partial P}{\partial T}\right)_{\overline{V}} dT + \left(\frac{\partial P}{\partial \overline{V}}\right)_T d\overline{V}$$

For a Redlich-Kwong gas,

$$\left(\frac{\partial P}{\partial T}\right)_{\overline{V}} = \frac{R}{\overline{V} - B} + \frac{A}{2T^{3/2}\overline{V}\left(\overline{V} + B\right)}$$

$$\left(\frac{\partial P}{\partial \overline{V}}\right)_T = \frac{-RT}{\left(\overline{V} - B\right)^2} + \frac{A}{T^{1/2}\overline{V}^2\left(\overline{V} + B\right)} + \frac{A}{T^{1/2}\overline{V}\left(\overline{V} + B\right)^2}$$

Then

$$dP = \left[\frac{R}{\overline{V} - B} + \frac{A}{2T^{3/2}\overline{V}(\overline{V} + B)}\right] dT + \left[-\frac{RT}{(\overline{V} - B)^2} + \frac{A(2\overline{V} + B)}{T^{1/2}\overline{V}^2(\overline{V} + B)^2}\right] d\overline{V}$$

D–6. Show explicitly that

$$\left(\frac{\partial^2 P}{\partial \overline{V} \partial T}\right) = \left(\frac{\partial^2 P}{\partial T \partial \overline{V}}\right)$$

for the Redlich-Kwong equation (Problem D–5).

We found expressions for $(\partial P/\partial T)_{\overline{V}}$ and $(\partial P/\partial \overline{V})_T$ in Problem D–5. Differentiating $(\partial P/\partial T)_{\overline{V}}$ with respect to $\overline{V}$ and $(\partial P/\partial \overline{V})_T$ with respect to T gives

$$\frac{\partial}{\partial \overline{V}}\left(\frac{\partial P}{\partial T}\right) = \frac{-R}{(\overline{V} - B)^2} - \frac{A}{2T^{3/2}\overline{V}^2(\overline{V} + B)} - \frac{A}{2\overline{V}T^{3/2}(\overline{V} + B)^2}$$

$$\frac{\partial}{\partial T}\left(\frac{\partial P}{\partial \overline{V}}\right) = \frac{-R}{(\overline{V} - B)^2} - \frac{A}{2T^{3/2}\overline{V}^2(\overline{V} + B)} - \frac{A}{2\overline{V}T^{3/2}(\overline{V} + B)^2}$$

So

$$\frac{\partial^2 P}{\partial \overline{V} \partial T} = \frac{\partial^2 P}{\partial T \partial \overline{V}}$$

D–7. We will derive the following equation in Chapter 5:

$$\left(\frac{\partial U}{\partial V}\right)_T = T\left(\frac{\partial P}{\partial T}\right)_V - P$$

Evaluate $(\partial U/\partial V)_T$ for an ideal gas, for a van der Waals gas (Equation D.4), and for a Redlich-Kwong gas (Problem D–5).

For an ideal gas $(\partial P/\partial T)_V = R/V$, so (using the equation given in the problem)

$$\left(\frac{\partial U}{\partial V}\right)_T = T\left(\frac{R}{V}\right) - P = P - P = 0$$

For a van der Waals gas $(\partial P/\partial T)_V = R/(\overline{V} - b)$, so

$$\left(\frac{\partial U}{\partial V}\right)_T = T\frac{R}{\overline{V} - b} - P = \frac{RT}{\overline{V} - b} - \frac{RT}{\overline{V} - b} + \frac{a}{\overline{V}^2} = \frac{a}{\overline{V}^2}$$

For a Redlich-Kwong gas,

$$\left(\frac{\partial P}{\partial T}\right)_{\overline{V}} = \frac{R}{\overline{V} - B} + \frac{A}{2\overline{V}(\overline{V} + B)T^{3/2}}$$

$$\begin{aligned}\left(\frac{\partial U}{\partial V}\right)_T &= T\left[\frac{R}{\overline{V} - B} + \frac{A}{2\overline{V}(\overline{V} + B)T^{3/2}}\right] - P \\ &= \frac{RT}{\overline{V} - B} + \frac{A}{2\overline{V}(\overline{V} + B)T^{1/2}} - \frac{RT}{\overline{V} - B} + \frac{A}{\overline{V}(\overline{V} + B)T^{1/2}} \\ &= \frac{3A}{2\overline{V}(\overline{V} + B)T^{1/2}}\end{aligned}$$

D–8. Given that the heat capacity at constant volume is defined by

$$C_V = \left(\frac{\partial U}{\partial T}\right)_V$$

and given the expression in Problem D–7, derive the equation

$$\left(\frac{\partial C_V}{\partial V}\right)_T = T\left(\frac{\partial^2 P}{\partial T^2}\right)_V$$

We know that

$$C_V = \left(\frac{\partial U}{\partial T}\right)_V \tag{1}$$

and

$$\left(\frac{\partial U}{\partial V}\right)_T = T\left(\frac{\partial P}{\partial T}\right)_V - P \tag{2}$$

Substituting Equation 2 into Equation 1 gives

$$\begin{aligned}\left(\frac{\partial C_V}{\partial V}\right)_T &= \frac{\partial}{\partial V}\left(\frac{\partial U}{\partial T}\right)_V = \frac{\partial^2 U}{\partial V \partial T} = \frac{\partial^2 U}{\partial T \partial V} = \frac{\partial}{\partial T}\left(\frac{\partial U}{\partial V}\right)_T \\ &= \frac{\partial}{\partial T}\left[T\left(\frac{\partial P}{\partial T}\right)_V - P\right] \\ &= \left(\frac{\partial P}{\partial T}\right)_V + T\left(\frac{\partial^2 P}{\partial T^2}\right)_V - \left(\frac{\partial P}{\partial T}\right)_V \\ &= T\left(\frac{\partial^2 P}{\partial T^2}\right)_V\end{aligned}$$

as stated in the text of the problem.

D–9. Use the expression in Problem D–8 to determine $(\partial C_V/\partial V)_T$ for an ideal gas, a van der Waals gas (Equation D.4), and a RedlicD-Kwong gas (see Problem D–5).

The expression in Problem D–8 is

$$\left(\frac{\partial C_V}{\partial V}\right)_T = T\left(\frac{\partial^2 P}{\partial T^2}\right)_V$$

For an ideal gas, $(\partial P/\partial T)_V = R/\overline{V}$, and $(\partial P/\partial T)_V = R/(\overline{V} - b)$ for a van der Waals gas. Since neither of these expressions varies when temperature is held constant, for both ideal and van der Waals gases

$$\left(\frac{\partial C_V}{\partial V}\right)_T = 0$$

For a Redlich-Kwong gas, we know that (Problem D–7)

$$\left(\frac{\partial P}{\partial T}\right)_{\overline{V}} = \frac{R}{\overline{V} - B} + \frac{A}{2\overline{V}(\overline{V} + B)T^{3/2}}$$

Then

$$\left(\frac{\partial^2 P}{\partial T^2}\right)_V = 0 - \frac{3}{4}\left[\frac{A}{\overline{V}(\overline{V}+B)T^{5/2}}\right]$$

Substituting into the expression from Problem D–8 gives

$$\left(\frac{\partial C_V}{\partial V}\right)_T = T\left(\frac{\partial^2 P}{\partial T^2}\right)_V = -\frac{3}{4}\left[\frac{A}{\overline{V}(\overline{V}+B)T^{3/2}}\right]$$

D–10. Is

$$dV = \pi r^2 dh + 2\pi rh dr$$

an exact or inexact differential?

For an exact differential, the cross derivatives are equal. We evaluate the two derivatives

$$\left[\frac{\partial}{\partial r}\pi r^2\right]_h = 2\pi r$$

and

$$\left[\frac{\partial}{\partial h}2\pi rh\right]_r = 2\pi r$$

Because these two derivatives are equal, dV is an exact differential.

D–11. Is

$$dx = C_V(T)dT + \frac{nRT}{V}dV$$

an exact or inexact differential? The quantity $C_V(T)$ is simply an arbitrary function of T. What about dx/T?

We evaluate the two derivatives

$$\left[\frac{\partial}{\partial V}C_V(T)\right]_T = 0$$

and

$$\left[\frac{\partial}{\partial T}\frac{nRT}{V}\right]_V = \frac{nR}{V}$$

Because these derivatives are unequal, dx is an inexact differential. We can express dx/T as

$$\frac{dx}{T} = T^{-1}C_V(T)dT + \frac{nR}{V}dV$$

and evaluate the two derivatives

$$\left[\frac{\partial}{\partial V}T^{-1}C_V(T)\right]_T = 0$$

$$\left[\frac{\partial}{\partial T}\frac{nR}{V}\right]_V = 0$$

Because these derivatives are equal, dx/T is an exact differential.

D–12. Prove that

$$\frac{1}{Y}\left(\frac{\partial Y}{\partial P}\right)_{T,n} = \frac{1}{\overline{Y}}\left(\frac{\partial \overline{Y}}{\partial P}\right)_{T}$$

and that

$$\left(\frac{\partial P}{\partial \overline{Y}}\right)_{T} = n\left(\frac{\partial P}{\partial Y}\right)_{T,n}$$

where $Y = Y(P, T, n)$ is an extensive variable.

We know that $\overline{Y}n = Y$. Then

$$\frac{1}{Y}\left(\frac{\partial Y}{\partial P}\right)_{T,n} = \frac{1}{Y}\left[\frac{\partial(\overline{Y}n)}{\partial P}\right]_{T,n}$$

$$= \frac{n}{Y}\left(\frac{\partial \overline{Y}}{\partial P}\right)_{T,n}$$

Since $\overline{Y} = \overline{Y}(P, T)$, this becomes

$$\frac{1}{Y}\left(\frac{\partial Y}{\partial P}\right)_{T,n} = \frac{1}{\overline{Y}}\left(\frac{\partial \overline{Y}}{\partial P}\right)_{T}$$

We can use the reciprocal identity from Problem D–3 to write this as

$$\frac{1}{Y}\left(\frac{\partial P}{\partial \overline{Y}}\right)_{T} = \frac{1}{\overline{Y}}\left(\frac{\partial P}{\partial Y}\right)_{T,n}$$

$$\left(\frac{\partial P}{\partial \overline{Y}}\right)_{T} = n\left(\frac{\partial P}{\partial Y}\right)_{T,n}$$

D–13. Equation 2.5 gives P for the van der Waals equation as a function of $\overline{V}$ and T. Show that P expressed as a function of V, T, and n is

$$P = \frac{nRT}{V - nb} - \frac{n^2 a}{V^2} \qquad (1)$$

Now evaluate $(\partial P/\partial \overline{V})_T$ from Equation 2.5 and $(\partial P/\partial V)_{T,n}$ from Equation 1 above and show that (see Problem D–12)

$$\left(\frac{\partial P}{\partial \overline{V}}\right)_{T} = n\left(\frac{\partial P}{\partial V}\right)_{T,n}$$

We begin with Equation 2.5:

$$\left(P + \frac{a}{\overline{V}^2}\right)(\overline{V} - b) = RT$$

$$\left(P + \frac{an^2}{V^2}\right) = \frac{nRT}{V - nb}$$

$$P = \frac{nRT}{V - nb} - \frac{n^2 a}{V^2} \qquad (1)$$

Evaluating $(\partial P/\partial \overline{V})_T$ from Equation 2.5 gives

$$P = \frac{RT}{\overline{V} - b} - \frac{a}{\overline{V}^2}$$

$$\left(\frac{\partial P}{\partial \overline{V}}\right)_T = -\frac{RT}{(\overline{V} - b)^2} + \frac{2a}{\overline{V}^3}$$

and evaluating $(\partial P/\partial V)_T$ from Equation 1 gives

$$P = \frac{nRT}{V - nb} - \frac{n^2 a}{V^2}$$

$$\begin{aligned}\left(\frac{\partial P}{\partial V}\right)_T &= -\frac{nRT}{(V - nb)^2} + \frac{2n^2 a}{V^3}\\ &= -\frac{RT}{n(\overline{V} - b)^2} + \frac{2a}{n\overline{V}^3}\\ &= \frac{1}{n}\left(\frac{\partial P}{\partial \overline{V}}\right)_T\end{aligned}$$

D–14. Referring to Problem D–13, show that

$$\left(\frac{\partial P}{\partial T}\right)_{\overline{V}} = \left(\frac{\partial P}{\partial T}\right)_{V,n}$$

and generally that

$$\left[\frac{\partial y(x, \overline{V})}{\partial x}\right]_{\overline{V}} = \left[\frac{\partial y(x, n, V)}{\partial x}\right]_{V,n}$$

where y and x are intensive variables and $y(x, n, V)$ can be written as $y(x, V/n)$.

Evaluating $(\partial P/\partial T)_{\overline{V}}$ from Equation 2.5 gives

$$P = \frac{RT}{\overline{V} - b} - \frac{a}{\overline{V}^2}$$

$$\left(\frac{\partial P}{\partial T}\right)_{\overline{V}} = \frac{R}{\overline{V} - b}$$

and evaluating $(\partial P/\partial T)_V$ from Equation 1 of Problem D–13 gives

$$P = \frac{nRT}{V - nb} - \frac{n^2 a}{V^2}$$

$$\begin{aligned}\left(\frac{\partial P}{\partial T}\right)_V &= \frac{nR}{V - nb} = \frac{R}{\overline{V} - b}\\ &= \left(\frac{\partial P}{\partial T}\right)_{\overline{V}}\end{aligned}$$

Generally, for any $y(x, n, V)$ which can be written as $y(x, \overline{V})$,

$$dy = \left[\frac{\partial y(x, \overline{V})}{\partial x}\right]_{\overline{V}} dx + \left[\frac{\partial y(x, \overline{V})}{\partial \overline{V}}\right]_x d\overline{V}$$

and

$$dy = \left[\frac{\partial y(x, n, V)}{\partial x}\right]_{V,n} dx + \left[\frac{\partial y(x, n, V)}{\partial n}\right]_{x,V} dn + \left[\frac{\partial y(x, n, V)}{\partial V}\right]_{x,n} dV$$

Equating the coefficients of dx in these two expressions gives

$$\left[\frac{\partial y(x, \overline{V})}{\partial x}\right]_{\overline{V}} = \left[\frac{\partial y(x, n, V)}{\partial x}\right]_{V,n}$$

CHAPTER 5

The First Law of Thermodynamics

PROBLEMS AND SOLUTIONS

5–1. Suppose that a 10-kg mass of iron at 20°C is dropped from a height of 100 meters. What is the kinetic energy of the mass just before it hits the ground? What is its speed? What would be the final temperature of the mass if all its kinetic energy at impact is transformed into internal energy? Take the molar heat capacity of iron to be $\overline{C}_P = 25.1\ \mathrm{J\cdot mol^{-1}\cdot K^{-1}}$ and the gravitational acceleration constant to be $9.80\ \mathrm{m\cdot s^{-2}}$.

Just before the mass hits the ground, all of the potential energy that the mass originally had will be converted into kinetic energy. So

$$\mathrm{PE} = mgh = (10\ \mathrm{kg})(9.80\ \mathrm{m\cdot s^{-2}})(100\ \mathrm{m}) = 9.8\ \mathrm{kJ} = \mathrm{KE}$$

Since kinetic energy can be expressed as $mv^2/2$, the speed of the mass just before hitting the ground is

$$v_f = \left(\frac{2\mathrm{KE}}{m}\right)^{1/2} = \left[\frac{2(9.8\ \mathrm{kJ})}{10\ \mathrm{kg}}\right]^{1/2} = 44\ \mathrm{m\cdot s^{-1}}$$

For a solid, the difference between $\overline{C}_V$ and $\overline{C}_P$ is small, so we can write $\Delta U = n\overline{C}_P\Delta T$ (Equation 5.39). Then

$$\Delta T = \frac{9.8\ \mathrm{kJ}}{\left(\dfrac{1\times 10^4\ \mathrm{g}}{55.85\ \mathrm{g\cdot mol^{-1}}}\right)(25.1\ \mathrm{J\cdot mol^{-1}\cdot K^{-1}})} = 2.2\ \mathrm{K}$$

The final temperature of the iron mass is then 22.2°C.

5–2. Consider an ideal gas that occupies $2.50\ \mathrm{dm^3}$ at a pressure of 3.00 bar. If the gas is compressed isothermally at a constant external pressure, P_{ext}, so that the final volume is $0.500\ \mathrm{dm^3}$, calculate the smallest value P_{ext} can have. Calculate the work involved using this value of P_{ext}.

Since the gas is ideal, we can write

$$P_2 = \frac{P_1V_1}{V_2} = \frac{(3.00\ \mathrm{bar})(2.50\ \mathrm{dm^3})}{0.500\ \mathrm{dm^3}} = 15.0\ \mathrm{bar}$$

The smallest possible value of P_{ext} is P_2. The work done in this case is (Equation 5.1)

$$w = -P_{\mathrm{ext}}\Delta V = (-15.0\ \mathrm{bar})(-2.0\ \mathrm{dm^3})\left(\frac{8.3145\ \mathrm{J\cdot mol^{-1}\cdot K^{-1}}}{0.083145\ \mathrm{bar\cdot dm^3\cdot mol^{-1}\cdot K^{-1}}}\right) = 3000\ \mathrm{J}$$

5–3. A one-mole sample of $CO_2(g)$ occupies 2.00 dm^3 at a temperature of 300 K. If the gas is compressed isothermally at a constant external pressure, P_{ext}, so that the final volume is 0.750 dm^3, calculate the smallest value P_{ext} can have, assuming that $CO_2(g)$ satisfies the van der Waals equation of state under these conditions. Calculate the work involved using this value of P_{ext}.

The smallest value P_{ext} can have is P_2, where P_2 is the final pressure of the gas. We can use the van der Waals equation (Equation 2.5) and the constants given in Table 2.3 to find P_2:

$$\begin{aligned} P_2 &= \frac{RT_2}{\overline{V}_2 - b} - \frac{a}{\overline{V}_2^2} \\ &= \frac{(0.083145\ \mathrm{dm^3\cdot bar\cdot mol^{-1}\cdot K^{-1}})(300\ \mathrm{K})}{0.750\ \mathrm{dm^3\cdot mol^{-1}} - 0.042816\ \mathrm{dm^3\cdot mol^{-1}}} - \frac{3.6551\ \mathrm{dm^6\cdot bar\cdot mol^{-2}}}{(0.750\ \mathrm{dm^3\cdot mol^{-1}})^2} \\ &= 28.8\ \mathrm{bar} \end{aligned}$$

The work involved is (Equation 5.1)

$$w = -P\Delta V = -(28.8 \times 10^5\ \mathrm{Pa})(-1.25 \times 10^{-3}\ \mathrm{m^3}) = 3.60\ \mathrm{kJ}$$

5–4. Calculate the work involved when one mole of an ideal gas is compressed reversibly from 1.00 bar to 5.00 bar at a constant temperature of 300 K.

Using the ideal gas equation, we find that

$$V_1 = \frac{nRT}{P_1} \quad \text{and} \quad V_2 = \frac{nRT}{P_2}$$

We can therefore write $V_2/V_1 = P_1/P_2$. Now we substitute into Equation 5.2 to find

$$\begin{aligned} w &= -\int P_{ext} dV = -\int \frac{nRT}{V} dV \\ &= -nRT \ln\left(\frac{V_2}{V_1}\right) = -nRT \ln\left(\frac{P_1}{P_2}\right) \\ &= (-1\ \mathrm{mol})(8.315\ \mathrm{J\cdot mol^{-1}\cdot K^{-1}})(300\ \mathrm{K}) \ln 0.2 = 4.01\ \mathrm{kJ} \end{aligned}$$

5–5. Calculate the work involved when one mole of an ideal gas is expanded reversibly from 20.0 dm^3 to 40.0 dm^3 at a constant temperature of 300 K.

We can integrate Equation 5.2 to find the work involved:

$$\begin{aligned} w &= -nRT \ln\left(\frac{V_2}{V_1}\right) \\ &= (-1\ \mathrm{mol})(8.315\ \mathrm{J\cdot mol^{-1}\cdot K^{-1}})(300\ \mathrm{K}) \ln 2 = -1.73\ \mathrm{kJ} \end{aligned}$$

5–6. Calculate the minimum amount of work required to compress 5.00 moles of an ideal gas isothermally at 300 K from a volume of 100 dm^3 to 40.0 dm^3.

We note that the minimum amount of work required is the amount of work needed to reversibly compress the gas, so we can write Equation 5.2 as

$$w_{\text{min}} = w_{\text{rev}} = -nRT \ln\left(\frac{V_2}{V_1}\right)$$
$$= (-5.00\ \text{mol})(8.315\ \text{J}\cdot\text{mol}^{-1}\cdot\text{K}^{-1})(300\ \text{K}) \ln 0.400 = 11.4\ \text{kJ}$$

5–7. Consider an ideal gas that occupies 2.25 L at 1.33 bar. Calculate the work required to compress the gas isothermally to a volume of 1.50 L at a constant pressure of 2.00 bar followed by another isothermal compression to 0.800 L at a constant pressure of 2.50 bar (Figure 5.4). Compare the result with the work of compressing the gas isothermally and reversibly from 2.25 L to 0.800 L.

We can use Equation 5.2 to describe the work involved with the compressions under different circumstances.

a. Two-step process, each step at constant external pressure

i. From (2.25 L, 1.33 bar) to (1.50 L, 2.00 bar),

$$w = -\int P_{\text{ext}} dV = (-2.00\ \text{bar})(1.50\ \text{L} - 2.25\ \text{L})(100\ \text{J}\cdot\text{bar}^{-1}\cdot\text{dm}^{-3}) = 150\ \text{J}$$

ii. From (1.50 L, 2.00 bar) to (0.800 L, 2.50 bar),

$$w = -\int P_{\text{ext}} dV = (-2.50\ \text{bar})(0.800\ \text{L} - 1.50\ \text{L})(100\ \text{J}\cdot\text{bar}^{-1}\cdot\text{dm}^{-3}) = 175\ \text{J}$$

The total work involved in the two-step process is +325 J.

b. Reversible process

Because the gas is ideal, $PV = nRT$. We can then write

$$PV = (2.25\ \text{L})(1.33\ \text{bar})(100\ \text{J}\cdot\text{bar}^{-1}\cdot\text{dm}^{-3}) = 299.25\ \text{J} = nRT$$

and use Equation 5.2 to find w:

$$w = -\int P dV = -nRT \ln\left(\frac{V_2}{V_1}\right) = -(299.25\ \text{J}) \ln\left(\frac{0.800}{2.25}\right) = 309\ \text{J}$$

The total work involved in the reversible process is +309 J. Note that the work involved in the reversible process is less than the work involved at constant external pressure, as is expected.

5–8. Show that for an isothermal reversible expansion from a molar volume $\overline{V}_1$ to a final molar volume $\overline{V}_2$, the work is given by

$$w = -RT \ln\left(\frac{\overline{V}_2 - B}{\overline{V}_1 - B}\right) - \frac{A}{BT^{1/2}} \ln\left[\frac{(\overline{V}_2 + B)\overline{V}_1}{(\overline{V}_1 + B)\overline{V}_2}\right]$$

for the Redlich-Kwong equation.

For the Redlich-Kwong equation,

$$P = \frac{RT}{\overline{V} - B} - \frac{A}{T^{1/2}\overline{V}(\overline{V} + B)} \tag{2.7}$$

We can then use Equation 5.2 to find w.

$$w = \int_{\overline{V}_1}^{\overline{V}_2} \left[-\frac{RT}{\overline{V} - B} + \frac{A}{T^{1/2}\overline{V}(\overline{V} + B)} \right] d\overline{V}$$

$$= -RT \ln\left(\frac{\overline{V}_2 - B}{\overline{V}_1 - B}\right) - \frac{A}{T^{1/2}B}\left[\ln\left(\frac{\overline{V}_1}{\overline{V}_1 + B}\right) - \ln\left(\frac{\overline{V}_2}{\overline{V}_2 + B}\right)\right]$$

$$= -RT \ln\left(\frac{\overline{V}_2 - B}{\overline{V}_1 - B}\right) - \frac{A}{BT^{1/2}} \ln\left[\frac{(\overline{V}_2 + B)\overline{V}_1}{(\overline{V}_1 + B)\overline{V}_2}\right]$$

5–9. Use the result of Problem 5–8 to calculate the work involved in the isothermal reversible expansion of one mole of $CH_4(g)$ from a volume of 1.00 $dm^3 \cdot mol^{-1}$ to 5.00 $dm^3 \cdot mol^{-1}$ at 300 K. (See Table 2.4 for the values of A and B.)

From Table 2.4, $A = 32.205\ dm^6 \cdot bar \cdot mol^{-2} \cdot K^{1/2}$ and $B = 0.029850\ dm^3 \cdot mol^{-1}$. Then, using the equation for w from the previous problem,

$$w = -RT \ln\left(\frac{\overline{V}_2 - B}{\overline{V}_1 - B}\right) - \frac{A}{BT^{1/2}} \ln\left[\frac{(\overline{V}_2 + B)\overline{V}_1}{(\overline{V}_1 + B)\overline{V}_2}\right]$$

$$= -RT \ln 5.1231 - \frac{A}{BT^{1/2}} \ln 0.97681$$

$$= -(39.3\ dm^3 \cdot bar \cdot mol^{-1})(100\ J \cdot bar^{-1} \cdot dm^{-3}) = -3.93\ kJ \cdot mol^{-1}$$

5–10. Repeat the calculation in Problem 5–9 for a van der Waals gas.

From Equation 2.5,

$$P = \frac{RT}{\overline{V} - b} - \frac{a}{\overline{V}^2}$$

Then (Equation 5.2)

$$w = \int -P d\overline{V} = \int_{\overline{V}_1}^{\overline{V}_2} \left(-\frac{RT}{\overline{V} - b} + \frac{a}{\overline{V}^2} \right) d\overline{V}$$

$$= -RT \ln \frac{\overline{V}_2 - b}{\overline{V}_1 - b} + \frac{a(\overline{V}_2 - \overline{V}_1)}{\overline{V}_2 \overline{V}_1}$$

From Table 2.3, for methane $a = 2.3026\ dm^6 \cdot bar \cdot mol^{-2}$ and $b = 0.043067\ dm^3 \cdot mol^{-1}$. Then

$$w = -39.18\ dm^3 \cdot bar \cdot mol^{-1} = -3.92\ kJ \cdot mol^{-1}$$

5–11. Derive an expression for the reversible isothermal work of an expansion of a gas that obeys the Peng-Robinson equation of state.

The Peng-Robinson equation of state is (Equation 2.8)

$$P = \frac{RT}{\overline{V} - \beta} - \frac{\alpha}{\overline{V}(\overline{V} + \beta) + \beta(\overline{V} - \beta)} \tag{2.7}$$

Substituting into Equation 5.2 gives

$$\begin{aligned}
w_{\text{rev}} &= -\int_1^2 d\overline{V}\left[\frac{RT}{\overline{V} - \beta} - \frac{\alpha}{\overline{V}(\overline{V} + \beta) + \beta(\overline{V} - \beta)}\right] \\
&= -RT \ln\left(\frac{\overline{V}_2 - \beta}{\overline{V}_1 - \beta}\right) - \alpha\int_1^2 d\overline{V}\frac{1}{\overline{V}^2 + 2\overline{V}\beta - \beta^2} \\
&= -RT \ln\left(\frac{\overline{V}_2 - \beta}{\overline{V}_1 - \beta}\right) - \frac{\alpha}{(8\beta^2)^{1/2}} \ln\frac{2\overline{V} + 2\beta - (8\beta^2)^{1/2}}{2\overline{V} + 2\beta + (8\beta^2)^{1/2}}\Bigg|_{\overline{V}_1}^{\overline{V}_2} \\
&= -RT \ln\left(\frac{\overline{V}_2 - \beta}{\overline{V}_1 - \beta}\right) - \frac{\alpha}{(8\beta^2)^{1/2}} \ln\frac{(\overline{V}_2 - 0.4142\beta)(\overline{V}_1 + 2.414\beta)}{(\overline{V}_2 + 2.414\beta)(\overline{V}_1 - 0.4142\beta)}
\end{aligned}$$

5–12. One mole of a monatomic ideal gas initially at a pressure of 2.00 bar and a temperature of 273 K is taken to a final pressure of 4.00 bar by the reversible path defined by P/V = constant. Calculate the values of ΔU, ΔH, q, and w for this process. Take $\overline{C}_V$ to be equal to 12.5 $\text{J}\cdot\text{mol}^{-1}\cdot\text{K}^{-1}$.

Let $P/\overline{V} = C$. Then, since the gas is ideal, we can write

$$T_1 = \frac{P_1V_1}{R} = \frac{P_1^2}{CR}$$

$$C = \frac{4.00 \text{ bar}^2}{(273 \text{ K})R}$$

Since $P/\overline{V}$ is constant throughout the process, we can also write

$$T_2 = \frac{P_2V_2}{R} = \frac{P_2^2}{CR} = \frac{16 \text{ bar}^2}{CR} = \frac{(16.0 \text{ bar}^2)(273 \text{ K})R}{(4.00 \text{ bar}^2)R} = 1092 \text{ K}$$

Because the $\overline{C}_V$ we are given is temperature-independent, we can write (by the definition of molar heat capacity)

$$\begin{aligned}
\Delta U &= n\int_{T_1}^{T_2}\overline{C}_V dT \\
&= (1 \text{ mol})(12.5 \text{ J}\cdot\text{mol}^{-1}\cdot\text{K}^{-1})(1092 \text{ K} - 273 \text{ K}) = 10.2 \text{ kJ}
\end{aligned}$$

Now we can use Equation 5.2 to calculate w, using the equality $P/\overline{V} = C$. Note that $\overline{V} = V$, since we are taking one mole of the gas.

$$\begin{aligned}
w &= -\int_{V_1}^{V_2} P dV = -\int_{V_1}^{V_2} CV dV = -\frac{C}{2}(V_2^2 - V_1^2) \\
&= -\frac{C}{2}\left(\frac{P_2^2}{C^2} - \frac{P_1^2}{C^2}\right) = -\frac{16.0 \text{ bar}^2 - 4.00 \text{ bar}^2}{2C} \\
&= -\frac{(12.0 \text{ bar}^2)(1 \text{ mol})(273 \text{ K})(0.083145 \text{ bar}\cdot\text{dm}^3\cdot\text{mol}^{-1}\cdot\text{K}^{-1})}{2(4.00 \text{ bar}^2)} \\
&= -3.40 \text{ kJ}
\end{aligned}$$

Finally, we can find q from Equation 5.10 and ΔH from Equation 5.35, letting $PV = nRT$.

$$q = \Delta U - w = 13.6 \text{ kJ}$$

$$\Delta H = \Delta U + nR\Delta T = 10.2 \text{ kJ} + (1 \text{ mol})(8.3145 \text{ J}\cdot\text{mol}^{-1}\text{K}^{-1})(819 \text{ K})$$
$$= 17.0 \text{ kJ}$$

5–13. The isothermal compressibility of a substance is given by

$$\beta = -\frac{1}{V}\left(\frac{\partial V}{\partial P}\right)_T \qquad (1)$$

For an ideal gas, $\beta = 1/P$, but for a liquid, β is fairly constant over a moderate pressure range. If β is constant, show that

$$\frac{V}{V_0} = e^{-\beta(P-P_0)} \qquad (2)$$

where V_0 is the volume at a pressure P_0. Use this result to show that the reversible isothermal work of compressing a liquid from a volume V_0 (at a pressure P_0) to a volume V (at a pressure P) is given by

$$\begin{aligned} w &= -P_0(V - V_0) + \beta^{-1}V_0\left(\frac{V}{V_0}\ln\frac{V}{V_0} - \frac{V}{V_0} + 1\right) \\ &= -P_0V_0[e^{-\beta(P-P_0)} - 1] + \beta^{-1}V_0\{1 - [1 + \beta(P - P_0)]e^{-\beta(P-P_0)}\} \end{aligned} \qquad (3)$$

(You need to use the fact that $\int \ln x dx = x\ln x - x$.) The fact that liquids are incompressible is reflected by β being small, so that $\beta(P - P_0) \ll 1$ for moderate pressures. Show that

$$\begin{aligned} w &= \beta P_0V_0(P - P_0) + \frac{\beta V_0(P - P_0)^2}{2} + O(\beta^2) \\ &= \frac{\beta V_0}{2}(P^2 - P_0^2) + O(\beta^2) \end{aligned} \qquad (4)$$

Calculate the work required to compress one mole of toluene reversibly and isothermally from 10 bar to 100 bar at 20°C. Take the value of β to be 8.95×10^{-5} bar^{-1} and the molar volume to be $0.106 \text{ mol}\cdot\text{L}^{-1}$ at 20°C.

We begin with Equation 1 and integrate both sides, letting β be constant with respect to pressure.

$$\int -\beta dP = \int V^{-1}dV$$

$$-\beta(P - P_0) = \ln\left(\frac{V}{V_0}\right)$$

$$\frac{V}{V_0} = e^{-\beta(P-P_0)}$$

$$P = -\beta^{-1}\ln\left(\frac{V}{V_0}\right) + P_0$$

Now we wish to find the reversible isothermal work of compressing a liquid from (P_0, V_0) to (P, V). We know that $\delta w = -PdV$ (Equation 5.2), so we use the expression we found above for P to write

$$\begin{aligned}\delta w &= -\left[-\beta^{-1}\ln\left(\frac{V}{V_0}\right) + P_0\right]dV \\ &= -P_0 dV + \beta^{-1}\ln\left(\frac{V}{V_0}\right)dV\end{aligned}$$

Integrating both sides of this equation gives

$$\begin{aligned}w &= -P_0(V - V_0) + \beta^{-1}\int \ln V dV - \beta^{-1}\int \ln V_0 dV \\ &= -P_0(V - V_0) + \beta^{-1}\left[V\ln V - V - (V_0\ln V_0 - V_0)\right] - \beta^{-1}(V - V_0)\ln V_0 \\ &= -P_0(V - V_0) + \beta^{-1}\left(V\ln V - V - V_0\ln V_0 + V_0 - V\ln V_0 + V_0\ln V_0\right) \\ &= -P_0(V - V_0) + \beta^{-1}V_0\left(\frac{V}{V_0}\ln\frac{V}{V_0} - \frac{V}{V_0} + 1\right)\end{aligned}$$

Substitution for V then yields the result

$$w = -P_0V_0\left[e^{-\beta(P-P_0)} - 1\right] + \beta^{-1}V_0\left\{1 - \left[1 + \beta(P - P_0)\right]e^{-\beta(P-P_0)}\right\}$$

which is Equation 3 in the text of the problem. Now let $x = -\beta(P - P_0)$. Because $-\beta(P - P_0) \ll 1$, $x \ll 1$. We can now write Equation 3 as

$$\begin{aligned}w &= -P_0V_0(e^x - 1) + \beta^{-1}V_0\left\{1 - \left[1 + \beta(P - P_0)\right]e^x\right\} \\ &= -P_0V_0(e^x - 1) + \beta^{-1}V_0 - \beta^{-1}V_0e^x - V_0(P - P_0)e^x\end{aligned}$$

Now, recall from MathChapter C that if x is small, we can write e^x as $1 + x + x^2/2 + O(x^3)$ (Equation C.2). Notice that to find w to $O(\beta^2)$, we must expand e^x to $O(x^3)$, since one of the above terms multiplies e^x by β^{-1}. Expanding the above equation gives

$$\begin{aligned}w &= -P_0V_0\left[x + O(x^2)\right] + \beta^{-1}V_0 - \beta^{-1}V_0\left[1 + x + \frac{x^2}{2} + O(x^3)\right] - V_0(P - P_0)\left[1 + x + O(x^2)\right] \\ &= \beta P_0V_0(P_0 - P_0) + V_0(P - P_0) - \frac{\beta V_0(P - P_0)^2}{2} - V_0(P - P_0) + \beta V_0(P - P_0)^2 + O(\beta^2) \\ &= \beta P_0V_0(P - P_0) + \frac{\beta V_0(P - P_0)^2}{2} + O(\beta^2) \\ &= \frac{\beta V_0}{2}(P^2 - P_0^2) + O(\beta^2)\end{aligned}$$

Now, for one mole of toluene [to $O(\beta^2)$], we use the parameters given in the problem to find

$$w = \frac{(8.95\times10^{-5}\ \text{bar}^{-1})(0.106\ \text{mol}\cdot\text{L}^{-1})^{-1}}{2}\left[(100\ \text{bar})^2 - (10\ \text{bar})^2\right] = 418\ \text{J}$$

5–14. In the previous problem, you derived an expression for the reversible, isothermal work done when a liquid is compressed. Given that β is typically $O(10^{-4})$ bar^{-1}, show that $V/V_0 \approx 1$ for pressures up to about 100 bar. This result, of course, reflects the fact that liquids are not very compressible. We can exploit this result by substituting $dV = -\beta V dP$ from the defining equation of β into $w = -\int PdV$ and then treating V as a constant. Show that this approximation gives Equation 4 of Problem 5–13.

We are given that $\beta \sim O(10^{-4}\ \text{bar}^{-1})$, and the largest pressure differential that can occur under the given conditions is on the order of $O(10^2\ \text{bar})$. Then, using Equation 2 of Problem 5.13, we find

$$\frac{V}{V_0} = e^{-\beta(P-P_0)} = e^{-O(10^{-2})} \approx 0.990$$

Therefore, $V/V_0 \approx 1$ for pressures ranging from 1 to 100 bar. Now $dV = -\beta V dP$, so

$$w = -\int P dV = \beta V_0 \int P dP = \frac{\beta V_0}{2}(P^2 - P_0^2)$$

as in Equation 4 of Problem 5.13.

5–15. Show that

$$\frac{T_2}{T_1} = \left(\frac{V_1}{V_2}\right)^{R/\overline{C}_V}$$

for a reversible adiabatic expansion of an ideal gas.

For an adiabatic expansion $\delta q = 0$, so $dU = \delta w$. By definition, $dU = n\overline{C}_V dT$, and for an ideal gas (Equation 5.2)

$$\delta w = -P dV = -nRTV^{-1}dV$$

We can then write

$$n\overline{C}_V dT = -nRTV^{-1}dV$$

$$\int \frac{\overline{C}_V}{T} dT = \int -\frac{R}{V} dV$$

$$\overline{C}_V \ln\left(\frac{T_2}{T_1}\right) = -R \ln\left(\frac{V_2}{V_1}\right)$$

$$\ln\left(\frac{T_2}{T_1}\right)^{\overline{C}_V} = \ln\left(\frac{V_2}{V_1}\right)^{-R}$$

Finally, exponentiating both sides gives

$$\frac{T_2}{T_1} = \left(\frac{V_2}{V_1}\right)^{-R/\overline{C}_V} = \left(\frac{V_1}{V_2}\right)^{R/\overline{C}_V}$$

5–16. Show that

$$\left(\frac{T_2}{T_1}\right)^{3/2} = \frac{\overline{V}_1 - b}{\overline{V}_2 - b}$$

for a reversible, adiabatic expansion of a monatomic gas that obeys the equation of state $P(\overline{V} - b) = RT$. Extend this result to the case of a diatomic gas.

For an adiabatic expansion $\delta q = 0$, so $dU = \delta w$. By definition, $dU = n\overline{C}_V dT$, and for this gas Equation 5.2 becomes

$$\delta w = -P dV = -n\frac{RT}{\overline{V} - b} d\overline{V}$$

Setting dU and δw equal to one another gives

$$n\overline{C}_V dT = -\frac{nRT}{\overline{V}-b}d\overline{V}$$

$$\int \frac{\overline{C}_V}{T}dT = \int -\frac{R}{\overline{V}-b}d\overline{V}$$

$$\overline{C}_V \ln\left(\frac{T_2}{T_1}\right) = -R\ln\left(\frac{\overline{V}_2-b}{\overline{V}_1-b}\right)$$

$$\left(\frac{T_2}{T_1}\right)^{\overline{C}_V/R} = \frac{\overline{V}_1-b}{\overline{V}_2-b}$$

For a monatomic gas, $\overline{C}_V = 3R/2$, and for a diatomic gas, $\overline{C}_V = 5R/2$. Thus

$$\left(\frac{T_2}{T_1}\right)^{3/2} = \frac{\overline{V}_1-b}{\overline{V}_2-b}$$

for a monatomic gas, and

$$\left(\frac{T_2}{T_1}\right)^{5/2} = \frac{\overline{V}_1-b}{\overline{V}_2-b}$$

for a diatomic gas.

5–17. Show that

$$\frac{T_2}{T_1} = \left(\frac{P_2}{P_1}\right)^{R/\overline{C}_P}$$

for a reversible adiabatic expansion of an ideal gas.

For an ideal gas, $\overline{C}_V + R = \overline{C}_P$ and

$$\frac{P_1V_1}{P_2V_2} = \frac{T_1}{T_2}$$

From Problem 5–15, we can write

$$\left(\frac{T_2}{T_1}\right)^{\overline{C}_V/R} = \left(\frac{V_1}{V_2}\right)$$

$$\frac{P_1}{P_2}\left(\frac{T_2}{T_1}\right)^{\overline{C}_V/R} = \frac{P_1V_1}{P_2V_2} = \frac{T_1}{T_2}$$

Then

$$\frac{P_1}{P_2} = \left(\frac{T_1}{T_2}\right)^{(\overline{C}_V+R)/R} = \left(\frac{T_1}{T_2}\right)^{\overline{C}_P/R}$$

$$\left(\frac{P_1}{P_2}\right)^{R/\overline{C}_P} = \frac{T_1}{T_2}$$

and, finally,

$$\frac{T_2}{T_1} = \left(\frac{P_2}{P_1}\right)^{R/\overline{C}_P}$$

5–18. Show that

$$P_1 V_1^{(\overline{C}_V + R)/\overline{C}_V} = P_2 V_2^{(\overline{C}_V + R)/\overline{C}_V}$$

for an adiabatic expansion of an ideal gas. Show that this formula reduces to Equation 5.23 for a monatomic gas.

For an ideal gas,

$$\frac{P_1 V_1}{P_2 V_2} = \frac{T_1}{T_2}$$

We can substitute this expression into the equation from Problem 5–15 to write

$$\frac{P_1 V_1}{P_2 V_2} = \left(\frac{V_1}{V_2}\right)^{R/\overline{C}_V}$$

Taking the reciprocal gives

$$\frac{P_1 V_1}{P_2 V_2} = \left(\frac{V_2}{V_1}\right)^{R/\overline{C}_V}$$

and rearranging yields

$$P_1 V_1^{\left(1 + R/\overline{C}_V\right)} = P_2 V_2^{\left(1 + R\overline{C}_V\right)}$$

For a monatomic ideal gas, $\overline{C}_V = \frac{3}{2} R$, so

$$P_1 V_1^{5/3} = P_2 V_2^{5/3} \qquad (5.23)$$

5–19. Calculate the work involved when one mole of a monatomic ideal gas at 298 K expands reversibly and adiabatically from a pressure of 10.00 bar to a pressure of 5.00 bar.

Because this process is adiabatic, $\delta q = 0$. This means that

$$\delta w = dU = n\overline{C}_V dT$$

where $\overline{C}_V$ is temperature-independent (since the gas is ideal). We can use the equation from Problem 5–17 to write

$$T_2 = T_1 \left(\frac{P_2}{P_1}\right)^{R/\overline{C}_P}$$

For an ideal gas, $\overline{C}_P = 5R/2$, so

$$T_2 = (298 \text{ K}) \left(\frac{5.00 \text{ bar}}{10.00 \text{ bar}}\right)^{2/5} = 226 \text{ K}$$

Substituting into the expression for δw ($\overline{C}_V = 3R/2$) gives

$$w = n\overline{C}_V \int_{T_1}^{T_2} dT = \frac{3}{2}(8.314\ \text{J}\cdot\text{K}^{-1})(226\ \text{K} - 298\ \text{K}) = -898\ \text{J}$$

5–20. A quantity of $N_2(g)$ at 298 K is compressed reversibly and adiabatically from a volume of 20.0 dm^3 to 5.00 dm^3. Assuming ideal behavior, calculate the final temperature of the $N_2(g)$. Take $\overline{C}_V = 5R/2$.

Using $\overline{C}_V$ given in the problem, we find that (by definition)

$$dU = n\overline{C}_V dT = \frac{5}{2}nRdT$$

and Equation 5.2 gives δw as

$$\delta w = -PdV = -\frac{nRT}{V}dV$$

For a reversible adiabatic compression, $\delta q = 0$, and so $dU = dw$. Then

$$\frac{5}{2}nRdT = -\frac{nRT}{V}dV$$

$$\frac{5}{2}\frac{dT}{T} = -\frac{dV}{V}$$

$$\frac{5}{2}\ln\frac{T_2}{T_1} = -\ln\frac{V_2}{V_1}$$

$$\ln\frac{T_2}{T_1} = -\frac{2}{5}\ln\frac{5.00\ \text{dm}^3}{20.0\ \text{dm}^3}$$

$$T_2 = 519\ \text{K}$$

5–21. A quantity of $CH_4(g)$ at 298 K is compressed reversibly and adiabatically from 50.0 bar to 200 bar. Assuming ideal behavior, calculate the final temperature of the $CH_4(g)$. Take $\overline{C}_V = 3R$.

From Problem 5–17, we have

$$\frac{T_2}{T_1} = \left(\frac{P_2}{P_1}\right)^{R/\overline{C}_P}$$

Assuming ideal behavior, $\overline{C}_P = R + \overline{C}_V = 4R$. Then

$$T_2 = \left(\frac{200\ \text{bar}}{50.0\ \text{bar}}\right)^{1/4}(298\ \text{K}) = 421\ \text{K}$$

5–22. One mole of ethane at 25°C and one atm is heated to 1200°C at constant pressure. Assuming ideal behavior, calculate the values of w, q, ΔU, and ΔH given that the molar heat capacity of ethane is given by

$$\overline{C}_P/R = 0.06436 + (2.137 \times 10^{-2}\ \text{K}^{-1})T$$
$$- (8.263 \times 10^{-6}\ \text{K}^{-2})T^2 + (1.024 \times 10^{-9}\ \text{K}^{-3})T^3$$

over the above temperature range. Repeat the calculation for a constant-volume process.

---

a. For a constant-pressure process, $q_P = \Delta H$ and $d\overline{H} = \overline{C}_P dT$. Then

$$\Delta\overline{H} = \int \overline{C}_P dT$$

$$= R\left[0.06436T + \frac{1}{2}(2.137\times10^{-2})T^2 - \frac{1}{3}(8.263\times10^{-6})T^3 + \frac{1}{4}(1.024\times10^{-9})T^4\right]\Bigg|_{298\text{ K}}^{1473\text{ K}}$$

$$\Delta\overline{H} = 122.9\text{ kJ}\cdot\text{mol}^{-1}$$

We now use Equation 5.36, remembering that the gas is behaving ideally:

$$\Delta\overline{H} = \Delta\overline{U} + P\Delta\overline{V} = \Delta\overline{U} + R\Delta T$$

$$\Delta\overline{U} = 122.9\text{ kJ}\cdot\text{mol}^{-1} + (8.3145\times10^{-3}\text{ kJ}\cdot\text{mol}^{-1}\cdot\text{K}^{-1})(1473\text{ K} - 298\text{ K})$$

$$= 132.7\text{ kJ}\cdot\text{mol}^{-1}$$

Finally, we use the expression $\Delta U = q + w$ to write

$$w = \Delta\overline{U} - q = 132.7\text{ kJ}\cdot\text{mol}^{-1} - 122.9\text{ kJ}\cdot\text{mol}^{-1} = 9.77\text{ kJ}\cdot\text{mol}^{-1}$$

b. For a constant-volume process, $w = 0$, and so $\Delta\overline{U} = q$. $\Delta\overline{H}$ is the same as in the previous situation, so $\Delta\overline{H} = 122.9\text{ kJ}\cdot\text{mol}^{-1}$. We can use Equation 5.36, remembering that the gas behaves ideally, to write

$$\Delta\overline{H} = \Delta\overline{U} + \overline{V}\Delta P = \Delta\overline{U} + R\Delta T$$

$$\Delta U = 122.9\text{ kJ}\cdot\text{mol}^{-1} + (8.3145\times10^{-3}\text{ kJ}\cdot\text{mol}^{-1}\cdot\text{K}^{-1})(1473\text{ K} - 298\text{ K})$$

$$= 132.7\text{ kJ}\cdot\text{mol}^{-1}$$

Note that the value of $\Delta\overline{U}$ is the same as in part a, because U depends only on temperature for an ideal gas.

5–23. The value of $\Delta_r H^\circ$ at 25°C and one bar is +290.8 kJ for the reaction

$$2\,\text{ZnO(s)} + 2\,\text{S(s)} \longrightarrow 2\,\text{ZnS(s)} + \text{O}_2\text{(g)}$$

Assuming ideal behavior, calculate the value of $\Delta_r U^\circ$ for this reaction.

Because both reactants are solid, $V_1 \approx 0$. The final volume will depend only on the amount of oxygen present; assuming it behaves ideally, we write

$$V_2 \approx \frac{nRT}{P} = \frac{(1\text{ mol})(0.08314\text{ dm}^3\cdot\text{bar}\cdot\text{mol}^{-1}\cdot\text{K}^{-1})(298\text{ K})}{1\text{ bar}} = 24.78\text{ dm}^3$$

Then $\Delta V \approx 24.78\text{ dm}^3$ for the reaction. Using Equation 5.36, we write

$$\Delta_r U^\circ = \Delta_r H^\circ - P\Delta V$$

$$= 290.8\text{ kJ} - (1\text{ bar})(24.776\text{ dm}^3)\left(\frac{1\text{ kJ}}{10\text{ dm}^3\cdot\text{bar}}\right)$$

$$= 288.3\text{ kJ}$$

5–24. Liquid sodium is being considered as an engine coolant. How many grams of sodium are needed to absorb 1.0 MJ of heat if the temperature of the sodium is not to increase by more than 10°C. Take $\overline{C}_P = 30.8\ \mathrm{J\cdot K^{-1}\cdot mol^{-1}}$ for Na(l) and $75.2\ \mathrm{J\cdot K^{-1}\cdot mol^{-1}}$ for $H_2O(l)$.

We must have a coolant which can absorb 1.0×10^6 J without changing its temperature by more than 10 K. The smallest amount of sodium required will allow the temperature to change by exactly 10 K. We can consider this a constant-pressure process, because liquids are relatively incompressible. Then, substituting $\Delta T = 10$ K into Equation 5.40, we find

$$\Delta\overline{H} = \overline{C}_P\Delta T = 308\ \mathrm{J\cdot mol^{-1}}$$

We require one mole of sodium to absorb 308 J of heat. Therefore, to absorb 1.0 MJ of heat, we require

$$(1.0 \times 10^6\ \mathrm{J})\left(\frac{1\ \mathrm{mol}}{308\ \mathrm{J}}\right)\left(\frac{22.99\ \mathrm{g}}{1\ \mathrm{mol}}\right) = 74.6\ \mathrm{kg}$$

74.6 kg of liquid sodium is needed.

5–25. A 25.0-g sample of copper at 363 K is placed in 100.0 g of water at 293 K. The copper and water quickly come to the same temperature by the process of heat transfer from copper to water. Calculate the final temperature of the water. The molar heat capacity of copper is $24.5\ \mathrm{J\cdot K^{-1}\cdot mol^{-1}}$.

The heat lost by the copper is gained by the water. Since $\Delta H = n\overline{C}_P\Delta T$ (Equation 5.40), we can let x be the final temperature of the system and write the heat lost by the copper as

$$\left(\frac{25.0\ \mathrm{g}}{63.546\ \mathrm{g\cdot mol^{-1}}}\right)(24.5\ \mathrm{J\cdot mol^{-1}\cdot K^{-1}})(363\ \mathrm{K} - x)$$

and the heat gained by the water as

$$\left(\frac{100.0\ \mathrm{g}}{18.0152\ \mathrm{g\cdot mol^{-1}}}\right)(75.3\ \mathrm{J\cdot mol^{-1}\cdot K^{-1}})(x - 293\ \mathrm{K})$$

Equating these two expressions gives

$$3495\ \mathrm{J} - (9.628\ \mathrm{J\cdot K^{-1}})x = (418.0\ \mathrm{J\cdot K^{-1}})x - 1.224 \times 10^5\ \mathrm{J}$$
$$1.259 \times 10^5\ \mathrm{K} = 427.6x$$
$$295\ \mathrm{K} = x$$

The final temperature of the water is 295 K.

5–26. A 10.0-kg sample of liquid water is used to cool an engine. Calculate the heat removed (in joules) from the engine when the temperature of the water is raised from 293 K to 373 K. Take $\overline{C}_P = 75.2\ \mathrm{J\cdot K^{-1}\cdot mol^{-1}}$ for $H_2O(l)$.

We can use Equation 5.40, where $\Delta T = 373\ \mathrm{K} - 293\ \mathrm{K} = 80$ K. This gives

$$\Delta H = n\overline{C}_P\Delta T = \left(\frac{10.0 \times 10^3\ \mathrm{g}}{18.0152\ \mathrm{g\cdot mol^{-1}}}\right)(75.2\ \mathrm{J\cdot mol^{-1}\cdot K^{-1}})(80\ \mathrm{K}) = 3340\ \mathrm{kJ}$$

3340 kJ of heat is removed by the water.

5–27. In this problem, we will derive a general relation between C_P and C_V. Start with $U = U(P, T)$ and write

$$dU = \left(\frac{\partial U}{\partial P}\right)_T dP + \left(\frac{\partial U}{\partial T}\right)_P dT \tag{1}$$

We could also consider V and T to be the independent variables of U and write

$$dU = \left(\frac{\partial U}{\partial V}\right)_T dV + \left(\frac{\partial U}{\partial T}\right)_V dT \tag{2}$$

Now take $V = V(P, T)$ and substitute its expression for dV into Equation 2 to obtain

$$dU = \left(\frac{\partial U}{\partial V}\right)_T \left(\frac{\partial V}{\partial P}\right)_T dP + \left[\left(\frac{\partial U}{\partial V}\right)_T \left(\frac{\partial V}{\partial T}\right)_P + \left(\frac{\partial U}{\partial T}\right)_V\right] dT$$

Compare this result with Equation 1 to obtain

$$\left(\frac{\partial U}{\partial P}\right)_T = \left(\frac{\partial U}{\partial V}\right)_T \left(\frac{\partial V}{\partial P}\right)_T \tag{3}$$

and

$$\left(\frac{\partial U}{\partial T}\right)_P = \left(\frac{\partial U}{\partial V}\right)_T \left(\frac{\partial V}{\partial T}\right)_P + \left(\frac{\partial U}{\partial T}\right)_V \tag{4}$$

Last, substitute $U = H - PV$ into the left side of Equation (4) and use the definitions of C_P and C_V to obtain

$$C_P - C_V = \left[P + \left(\frac{\partial U}{\partial V}\right)_T\right] \left(\frac{\partial V}{\partial T}\right)_P$$

Show that $C_P - C_V = nR$ if $(\partial U/\partial V)_T = 0$, as it is for an ideal gas.

We can write the total derivatives of $V(P, T)$ and $U(V, T)$ as (MathChapter D)

$$dV = \left(\frac{\partial V}{\partial P}\right)_T dP + \left(\frac{\partial V}{\partial T}\right)_P dT \tag{a}$$

$$dU = \left(\frac{\partial U}{\partial V}\right)_T dV + \left(\frac{\partial U}{\partial T}\right)_V dT \tag{b}$$

Substituting dV from Equation a into Equation b gives

$$\begin{aligned} dU &= \left(\frac{\partial U}{\partial V}\right)_T \left[\left(\frac{\partial V}{\partial P}\right)_T dP + \left(\frac{\partial V}{\partial T}\right)_P dT\right] + \left(\frac{\partial U}{\partial T}\right)_V dT \\ &= \left(\frac{\partial U}{\partial V}\right)_T \left(\frac{\partial V}{\partial P}\right)_T dP + \left[\left(\frac{\partial U}{\partial V}\right)_T \left(\frac{\partial V}{\partial T}\right)_P + \left(\frac{\partial U}{\partial T}\right)_V\right] dT \end{aligned}$$

We can also express U as a function of P and T, in which case the total derivative dU is

$$dU = \left(\frac{\partial U}{\partial P}\right)_T dP + \left(\frac{\partial U}{\partial T}\right)_P dT$$

Because the coefficients of dP and dT in both expressions for dU are equal, we can write

$$\left(\frac{\partial U}{\partial P}\right)_T = \left(\frac{\partial U}{\partial V}\right)_T \left(\frac{\partial V}{\partial P}\right)_T$$

and

$$\left(\frac{\partial U}{\partial T}\right)_P = \left(\frac{\partial U}{\partial V}\right)_T \left(\frac{\partial V}{\partial T}\right)_P + \left(\frac{\partial U}{\partial T}\right)_V \qquad \text{(c)}$$

Substituting $H - PV$ for U into the left side of Equation c gives

$$\left(\frac{\partial [H - PV]}{\partial T}\right)_P = \left(\frac{\partial U}{\partial V}\right)_T \left(\frac{\partial V}{\partial T}\right)_P + \left(\frac{\partial U}{\partial T}\right)_V$$

$$\left(\frac{\partial H}{\partial T}\right)_P - P\left(\frac{\partial V}{\partial T}\right)_P - V\left(\frac{\partial P}{\partial T}\right)_P = \left(\frac{\partial U}{\partial V}\right)_T \left(\frac{\partial V}{\partial T}\right)_P + \left(\frac{\partial U}{\partial T}\right)_V$$

Using the definitions of C_P and C_V (Equations 5.39 and 5.40), this expression becomes

$$C_P - P\left(\frac{\partial V}{\partial T}\right)_P = \left(\frac{\partial U}{\partial V}\right)_T \left(\frac{\partial V}{\partial T}\right)_P + C_V$$

$$C_P - C_V = \left[P + \left(\frac{\partial U}{\partial V}\right)_T\right]\left(\frac{\partial V}{\partial T}\right)_P \qquad \text{(d)}$$

If $(\partial U/\partial V) = 0$, then Equation d becomes

$$C_P - C_V = P\left(\frac{\partial V}{\partial T}\right)_P$$

Using the ideal gas equation to find $P\,(\partial V/\partial T)_P$, we find that

$$PV = nRT$$

$$P\left(\frac{\partial V}{\partial T}\right)_P + V\left(\frac{\partial P}{\partial T}\right)_P = nR\left(\frac{\partial T}{\partial T}\right)_P$$

$$P\left(\frac{\partial V}{\partial T}\right)_P = nR$$

$$C_P - C_V = nR$$

5–28. Following Problem 5–27, show that

$$C_P - C_V = \left[V - \left(\frac{\partial H}{\partial P}\right)_T\right]\left(\frac{\partial P}{\partial T}\right)_V$$

We can write the total derivatives of $P(V, T)$ and $H(P, T)$ as (MathChapter D)

$$dP = \left(\frac{\partial P}{\partial V}\right)_T dV + \left(\frac{\partial P}{\partial T}\right)_V dT \qquad \text{(a)}$$

$$dH = \left(\frac{\partial H}{\partial P}\right)_T dP + \left(\frac{\partial H}{\partial T}\right)_P dT \qquad \text{(b)}$$

Substituting dP from Equation a into Equation b gives

$$\begin{aligned} dH &= \left(\frac{\partial H}{\partial P}\right)_T \left[\left(\frac{\partial P}{\partial V}\right)_T dV + \left(\frac{\partial P}{\partial T}\right)_V dT\right] + \left(\frac{\partial H}{\partial T}\right)_P dT \\ &= \left(\frac{\partial H}{\partial P}\right)_T \left(\frac{\partial P}{\partial V}\right)_T dV + \left[\left(\frac{\partial H}{\partial P}\right)_T \left(\frac{\partial P}{\partial T}\right)_V + \left(\frac{\partial H}{\partial T}\right)_P\right] dT \end{aligned}$$

We can also express H as a function of V and T, in which case the total derivative dH is

$$dH = \left(\frac{\partial H}{\partial V}\right)_T dV + \left(\frac{\partial H}{\partial T}\right)_V dT$$

Because the coefficients of dV and dT in both expressions for dH are equal, we can write

$$\left(\frac{\partial H}{\partial V}\right)_T = \left(\frac{\partial H}{\partial P}\right)_T \left(\frac{\partial P}{\partial V}\right)_T$$

and

$$\left(\frac{\partial H}{\partial T}\right)_V = \left(\frac{\partial H}{\partial P}\right)_T \left(\frac{\partial P}{\partial T}\right)_V + \left(\frac{\partial H}{\partial T}\right)_P \tag{c}$$

Substituting $U + PV$ for H into the left side of Equation c gives

$$\left(\frac{\partial[U + PV]}{\partial T}\right)_V = \left(\frac{\partial H}{\partial P}\right)_T \left(\frac{\partial P}{\partial T}\right)_V + \left(\frac{\partial H}{\partial T}\right)_P$$

$$\left(\frac{\partial U}{\partial T}\right)_V + P\left(\frac{\partial V}{\partial T}\right)_V + V\left(\frac{\partial P}{\partial T}\right)_V = \left(\frac{\partial H}{\partial P}\right)_T \left(\frac{\partial P}{\partial T}\right)_V + \left(\frac{\partial H}{\partial T}\right)_P$$

Using the definitions of C_P and C_V (Equations 5.39 and 5.40), this expression becomes

$$C_V + V\left(\frac{\partial P}{\partial T}\right)_V = \left(\frac{\partial H}{\partial P}\right)_T \left(\frac{\partial P}{\partial T}\right)_V + C_P$$

$$C_P - C_V = \left[V - \left(\frac{\partial H}{\partial P}\right)_T\right]\left(\frac{\partial P}{\partial T}\right)_V$$

which is the desired result.

5–29. Starting with $H = U + PV$, show that

$$\left(\frac{\partial U}{\partial T}\right)_P = C_P - P\left(\frac{\partial V}{\partial T}\right)_P$$

Interpret this result physically.

Take the partial derivative of both sides of this equation with respect to T, holding P constant, and substitute C_P for $(\partial H/\partial T)_P$.

$$H = U + PV$$

$$\left(\frac{\partial H}{\partial T}\right)_P = \left(\frac{\partial U}{\partial T}\right)_P + P\left(\frac{\partial V}{\partial T}\right)_P$$

$$C_P - P\left(\frac{\partial V}{\partial T}\right)_P = \left(\frac{\partial U}{\partial T}\right)_P$$

This expression tells us how the total energy of a constant-pressure system changes with respect to temperature. Recall that for a constant pressure process, $dH = \delta q$. Then $C_P = (\partial q/\partial T)$. Because $dU = \delta q + \delta w$, the work involved in the process must be $-P(\partial V/\partial T)_P$. The equation above is equivalent to the statement

$$\left(\frac{\partial U}{\partial T}\right)_P = \left(\frac{\partial[q + w]}{\partial T}\right)_P$$

5–30. Given that $(\partial U/\partial V)_T = 0$ for an ideal gas, prove that $(\partial H/\partial V)_T = 0$ for an ideal gas.

Begin with Equation 5.35 and use the ideal gas law to write

$$H = U + PV = U + nRT$$

Now take the partial derivative of both sides with respect to volume (note that for an ideal gas, U is dependent only upon temperature) to find

$$\left(\frac{\partial H}{\partial V}\right)_T = \left(\frac{\partial U}{\partial V}\right)_T + nR\left(\frac{\partial T}{\partial V}\right)_T = 0$$

5–31. Given that $(\partial U/\partial V)_T = 0$ for an ideal gas, prove that $(\partial C_V/\partial V)_T = 0$ for an ideal gas.

We define C_V as $(\partial U/\partial V)_T$ (Equation 5.39). Therefore,

$$\left(\frac{\partial C_V}{\partial V}\right)_T = \frac{\partial^2 U}{\partial V \partial T} = \frac{\partial}{\partial T}\left(\frac{\partial U}{\partial V}\right)_T$$

Since $(\partial U/\partial V)_T = 0$ for an ideal gas, $(\partial C_V/\partial V)_T = 0$.

5–32. Show that $C_P - C_V = nR$ if $(\partial H/\partial P)_T = 0$, as is true for an ideal gas.

From Problem 5.28,

$$C_P - C_V = \left[V - \left(\frac{\partial H}{\partial P}\right)_T\right]\left(\frac{\partial P}{\partial T}\right)_V = V\left(\frac{\partial P}{\partial T}\right)_V$$

where, as stated in the problem, $(\partial H/\partial P)_T = 0$. Substituting $P = nRTV^{-1}$ into the above expression gives

$$C_P - C_V = V\left(\frac{\partial[nRTV^{-1}]}{\partial T}\right)_V = nR$$

5–33. Differentiate $H = U + PV$ with respect to V at constant temperature to show that $(\partial H/\partial V)_T = 0$ for an ideal gas.

(Notice that this problem has you prove the same thing as Problem 5.30 without assuming that $(\partial U/\partial V)_T = 0$ for an ideal gas.) We can use the ideal gas equation to write Equation 5.35 as

$$H = U + PV = U + nRT$$

For an ideal gas, U is dependent only on temperature, and the product nRT is also dependent only on temperature. Therefore, H is a function only of temperature, and differentiating H at constant temperature will yield the result

$$\left(\frac{\partial H}{\partial V}\right)_T = \left(\frac{\partial U}{\partial V}\right)_T + nR\left(\frac{\partial T}{\partial V}\right)_T = 0$$

5–34. Given the following data for sodium, plot $\overline{H}(T) - \overline{H}(0)$ against T for sodium: melting point, 361 K; boiling point, 1156 K; $\Delta_{\text{fus}}H^\circ = 2.60\ \text{kJ}\cdot\text{mol}^{-1}$; $\Delta_{\text{vap}}H^\circ = 97.4\ \text{kJ}\cdot\text{mol}^{-1}$; $\overline{C}_P(\text{s}) = 28.2\ \text{J}\cdot\text{mol}^{-1}\cdot\text{K}^{-1}$; $\overline{C}_P(\text{l}) = 32.7\ \text{J}\cdot\text{mol}^{-1}\cdot\text{K}^{-1}$; $\overline{C}_P(\text{g}) = 20.8\ \text{J}\cdot\text{mol}^{-1}\cdot\text{K}^{-1}$.

We can use an extended form of Equation 5.46:

$$\overline{H}(T) - \overline{H}(0) = \int_0^{T_{\text{fus}}} \overline{C}_P(\text{s})dT + \Delta_{\text{fus}}\overline{H} + \int_{T_{\text{fus}}}^{T_{\text{vap}}} \overline{C}_P(\text{l})dT + \Delta_{\text{vap}}\overline{H} + \int_{T_{\text{vap}}}^{T} \overline{C}_P(\text{g})dT$$

Notice the very large jump between the liquid and gaseous phases.

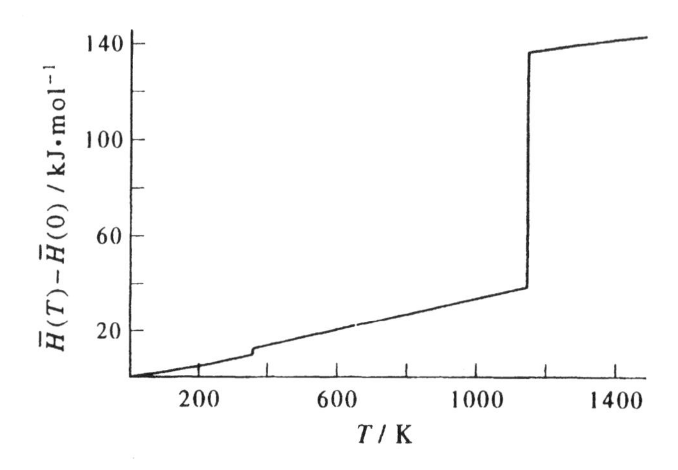

5–35. The $\Delta_r H^\circ$ values for the following equations are

$$2\,\text{Fe(s)} + \tfrac{3}{2}\text{O}_2\text{(g)} \rightarrow \text{Fe}_2\text{O}_3\text{(s)} \quad \Delta_r H^\circ = -206\ \text{kJ}\cdot\text{mol}^{-1}$$

$$3\,\text{Fe(s)} + 2\,\text{O}_2\text{(g)} \rightarrow \text{Fe}_3\text{O}_4\text{(s)} \quad \Delta_r H^\circ = -136\ \text{kJ}\cdot\text{mol}^{-1}$$

Use these data to calculate the value of $\Delta_r H$ for the reaction described by

$$4\,\text{Fe}_2\text{O}_3\text{(s)} + \text{Fe(s)} \longrightarrow 3\,\text{Fe}_3\text{O}_4\text{(s)}$$

Set up the problem so that the summation of two reactions will give the desired reaction:

$$\begin{array}{llr} & 4[\text{Fe}_2\text{O}_3\text{(s)} \rightarrow 2\,\text{Fe(s)} + \tfrac{3}{2}\text{O}_2\text{(g)}] & \Delta_r H = 4(206)\ \text{kJ} \\ + & 3[3\,\text{Fe(s)} + 2\,\text{O}_2\text{(g)} \rightarrow \text{Fe}_3\text{O}_4\text{(s)}] & \Delta_r H = 3(-136)\ \text{kJ} \\ \hline & 4\,\text{Fe}_2\text{O}_3\text{(s)} + \text{Fe(s)} \longrightarrow 3\,\text{Fe}_3\text{O}_4\text{(s)} & \Delta_r H = 416\ \text{kJ} \end{array}$$

5–36. Given the following data,

$$\tfrac{1}{2}\text{H}_2\text{(g)} + \tfrac{1}{2}\text{F}_2\text{(g)} \rightarrow \text{HF(g)} \quad \Delta_r H^\circ = -273.3\ \text{kJ}\cdot\text{mol}^{-1}$$

$$\text{H}_2\text{(g)} + \tfrac{1}{2}\text{O}_2\text{(g)} \rightarrow \text{H}_2\text{O(l)} \quad \Delta_r H^\circ = -285.8\ \text{kJ}\cdot\text{mol}^{-1}$$

calculate the value of $\Delta_r H$ for the reaction described by

$$2\,\text{F}_2\text{(g)} + 2\,\text{H}_2\text{O(l)} \longrightarrow 4\,\text{HF(g)} + \text{O}_2\text{(g)}$$

Set up the problem so that the summation of two reactions will give the desired reaction:

$$\begin{array}{lll} & 4[\tfrac{1}{2}H_2(g) + \tfrac{1}{2}F_2(g) \rightarrow HF(g)] & \Delta_r H = 4(-273.3)\text{ kJ} \\ + & 2[H_2O(l) \rightarrow H_2(g) + \tfrac{1}{2}O_2(g)] & \Delta_r H = 2(285.8)\text{ kJ} \\ \hline & 2\,F_2(g) + 2\,H_2O(l) \longrightarrow 4\,HF(g) + O_2(g) & \Delta_r H = -521.6\text{ kJ} \end{array}$$

5–37. The standard molar heats of combustion of the isomers *m*-xylene and *p*-xylene are $-4553.9\text{ kJ}\cdot\text{mol}^{-1}$ and $-4556.8\text{ kJ}\cdot\text{mol}^{-1}$, respectively. Use these data, together with Hess's Law, to calculate the value of $\Delta_r H°$ for the reaction described by

$$m\text{-xylene} \longrightarrow p\text{-xylene}$$

Because *m*-xylene and *p*-xylene are isomers, their combustion equations are stoichiometrically equivalent. We can therefore write

$$\begin{array}{ll} m\text{-xylene} \rightarrow \text{combustion products} & \Delta_r H = -4553.9\text{ kJ} \\ +\text{combustion products} \rightarrow p\text{-xylene} & \Delta_r H = +4556.8\text{ kJ} \\ \hline m\text{-xylene} \longrightarrow p\text{-xylene} & \Delta_r H = +2.9\text{ kJ} \end{array}$$

5–38. Given that $\Delta_r H° = -2826.7$ kJ for the combustion of 1.00 mol of fructose at 298.15 K,

$$C_6H_{12}O_6(s) + 6\,O_2(g) \longrightarrow 6\,CO_2(g) + 6\,H_2O(l)$$

and the $\Delta_f H°$ data in Table 5.2, calculate the value of $\Delta_f H°$ for fructose at 298.15 K.

We are given $\Delta_r H°$ for the combustion of fructose in the statement of the problem. We use the values given in Table 5.2 for $CO_2(g)$, $H_2O(l)$, and $O_2(g)$:

$$\Delta_f H°[CO_2(g)] = -393.509\text{ kJ}\cdot\text{mol}^{-1} \qquad \Delta_f H°[H_2O(l)] = -285.83\text{ kJ}\cdot\text{mol}^{-1}$$

$$\Delta_f H°[O_2(g)] = 0$$

Now, by Hess's law, we write

$$\Delta_r H° = \sum \Delta_f H°[\text{products}] - \sum \Delta_f H°[\text{reactants}]$$

$$-2826.7\text{ kJ}\cdot\text{mol}^{-1} = 6(-393.509\text{ kJ}\cdot\text{mol}^{-1}) + 6(-285.83\text{ kJ}\cdot\text{mol}^{-1}) - \Delta_r H°[\text{fructose}]$$

$$\Delta_r H°[\text{fructose}] = 1249.3\text{ kJ}\cdot\text{mol}^{-1}$$

5–39. Use the $\Delta_f H°$ data in Table 5.2 to calculate the value of $\Delta_c H°$ for the combustion reactions described by the equations:

a. $CH_3OH(l) + \tfrac{3}{2}\,O_2(g) \longrightarrow CO_2(g) + 2\,H_2O(l)$

b. $N_2H_4(l) + O_2(g) \longrightarrow N_2(g) + 2\,H_2O(l)$

Compare the heat of combustion per gram of the fuels $CH_3OH(l)$ and $N_2H_4(l)$.

We will need the following values from Table 5.2:

$$\Delta_f H°[CO_2(g)] = -393.509 \text{ kJ·mol}^{-1} \qquad \Delta_f H°[H_2O(l)] = -285.83 \text{ kJ·mol}^{-1}$$
$$\Delta_f H°[N_2H_4(l)] = +50.6 \text{ kJ·mol}^{-1} \qquad \Delta_f H°[CH_3OH(l)] = -239.1 \text{ kJ·mol}^{-1}$$
$$\Delta_f H°[N_2(g)] = 0$$

a. Using Hess's law,

$$\begin{aligned}\Delta_r H° &= \sum \Delta_f H°[\text{products}] - \sum \Delta_f H°[\text{reactants}] \\ &= 2(-285.83 \text{ kJ}) + (-393.5 \text{ kJ}) - (-239.1 \text{ kJ}) \\ &= \left(\frac{-726.1 \text{ kJ}}{\text{mol methanol}}\right)\left(\frac{1 \text{ mol}}{32.042 \text{ g}}\right) = -22.7 \text{ kJ·g}^{-1}\end{aligned}$$

b. Again, by Hess's law,

$$\begin{aligned}\Delta_r H° &= \sum \Delta_f H°[\text{products}] - \sum \Delta_f H°[\text{reactants}] \\ &= 2(-285.83 \text{ kJ}) - (+50.6 \text{ kJ}) \\ &= \left(\frac{-622.3 \text{ kJ}}{\text{mol } N_2H_4}\right)\left(\frac{1 \text{ mol}}{32.046 \text{ g}}\right) = -19.4 \text{ kJ·g}^{-1}\end{aligned}$$

More energy per gram is produced by combusting methanol.

5–40. Using Table 5.2, calculate the heat required to vaporize 1.00 mol of $CCl_4(l)$ at 298 K.

$$CCl_4(l) \longrightarrow CCl_4(g)$$

We can subtract $\Delta_f H°[CCl_4(l)]$ from $\Delta_f H°[CCl_4(g)]$ to find the heat required to vaporize CCl_4:

$$\Delta_{vap} H° = -102.9 \text{ kJ} + 135.44 \text{ kJ} = 32.5 \text{ kJ}$$

5–41. Using the $\Delta_f H°$ data in Table 5.2, calculate the values of $\Delta_r H°$ for the following:

a. $C_2H_4(g) + H_2O(l) \longrightarrow C_2H_5OH(l)$

b. $CH_4(g) + 4\,Cl_2(g) \longrightarrow CCl_4(l) + 4\,HCl(g)$

In each case, state whether the reaction is endothermic or exothermic.

---

a. Using Hess's law,

$$\Delta_r H° = -277.69 \text{ kJ} - (-285.83 \text{ kJ} + 52.28 \text{ kJ}) = -44.14 \text{ kJ}$$

This reaction is exothermic.

b. Again, by Hess's law,

$$\Delta_r H° = 4(-92.31 \text{ kJ}) - 135.44 \text{ kJ} - (-74.81 \text{ kJ}) = -429.87 \text{ kJ}$$

This reaction is also exothermic.

5–42. Use the following data to calculate the value of $\Delta_{vap} H°$ of water at 298 K and compare your answer to the one you obtain from Table 5.2: $\Delta_{vap} H°$ at 373 K = 40.7 kJ·mol^{-1}; $\overline{C}_P(l) = 75.2$ J·mol^{-1}·K^{-1}; $\overline{C}_P(g) = 33.6$ J·mol^{-1}·K^{-1}.

We can create a figure similar to Figure 5.10 to illustrate this reaction:

$$\begin{array}{ccc} H_2O(l) & \xrightarrow{\Delta_{vap}H^\circ,373\text{ K}} & H_2O(g) \\ \uparrow \Delta H_2 & & \downarrow \Delta H_3 \\ H_2O(l) & \xrightarrow{\Delta_{vap}H^\circ,298\text{ K}} & H_2O(g) \end{array}$$

Now we use Hess's Law to determine the enthalpy of vaporization.

$$\begin{aligned} \Delta_{vap}H^\circ_{298\text{ K}} &= \Delta H_2 + \Delta H_3 + \Delta_{vap}H^\circ_{373\text{ K}} \\ &= (75\text{ K})(75.2\text{ J}\cdot\text{mol}^{-1}\cdot\text{K}^{-1}) + (-75\text{ K})(33.6\text{ J}\cdot\text{mol}^{-1}\cdot\text{K}^{-1}) + 40.7\text{ kJ}\cdot\text{mol}^{-1} \\ &= 43.8\text{ kJ}\cdot\text{mol}^{-1} \end{aligned}$$

Using Table 5.2, we find

$$\begin{aligned} \Delta_{vap}H^\circ &= \Delta_f H^\circ[H_2O(g)] - \Delta_f H^\circ[H_2O(l)] \\ &= -241.8\text{ kJ}\cdot\text{mol}^{-1} + 285.83\text{ kJ}\cdot\text{mol}^{-1} = 44.0\text{ kJ}\cdot\text{mol}^{-1} \end{aligned}$$

These values are fairly close. (Using values of $\overline{C}_P$ which include temperature-dependent terms may further improve the agreement.)

5–43. Use the following data and the data in Table 5.2 to calculate the standard reaction enthalpy of the water-gas reaction at 1273 K. Assume that the gases behave ideally under these conditions.

$$C(s) + H_2O(g) \longrightarrow CO(g) + H_2(g)$$

$$C^\circ_P[CO(g)]/R = 3.231 + (8.379 \times 10^{-4}\text{ K}^{-1})T - (9.86 \times 10^{-8}\text{ K}^{-2})T^2$$

$$C^\circ_P[H_2(g)]/R = 3.496 + (1.006 \times 10^{-4}\text{ K}^{-1})T + (2.42 \times 10^{-7}\text{ K}^{-2})T^2$$

$$C^\circ_P[H_2O(g)]/R = 3.652 + (1.156 \times 10^{-3}\text{ K}^{-1})T + (1.42 \times 10^{-7}\text{ K}^{-2})T^2$$

$$C^\circ_P[C(s)]/R = -0.6366 + (7.049 \times 10^{-3}\text{ K}^{-1})T - (5.20 \times 10^{-6}\text{ K}^{-2})T^2 + (1.38 \times 10^{-9}\text{ K}^{-3})T^3$$

We can create a figure similar to Figure 5.10 to illustrate this reaction.

$$\begin{array}{ccc} C(s) + H_2O(g) & \xrightarrow{\Delta_r H^\circ,1273\text{ K}} & CO(g) + H_2(g) \\ \downarrow \Delta H_1 & & \uparrow \Delta H_2 \\ C(s) + H_2O(g) & \xrightarrow{\Delta_r H^\circ,298\text{ K}} & CO(g) + H_2(g) \end{array}$$

Now use Hess's Law to calculate the standard reaction enthalpy at 1273 K. To do the integrals, it is helpful to use a program like *Excel* or *Mathematica* (we used *Mathematica*), so that the tedium of adding and multiplying can be avoided.

$$\begin{aligned} \Delta_r H^\circ_{1273} &= \Delta_r H^\circ_{298} + \Delta H_1 + \Delta H_2 \\ &= (-110.5\text{ kJ}\cdot\text{mol}^{-1} + 241.8\text{ kJ}\cdot\text{mol}^{-1}) \\ &\quad + R\int_{298}^{1273} \left\{\overline{C}_P[CO(g)] + \overline{C}_P[H_2(g)] - \overline{C}_P[H_2O(g)] - \overline{C}_P[C(s)]\right\} dT \\ &= 131.3\text{ kJ}\cdot\text{mol}^{-1} + R\,[3725.01\text{ K} + 3649.92\text{ K} - 4542.43\text{ K} - 2151.29\text{ K}] \\ &= 131.3\text{ kJ}\cdot\text{mol}^{-1} + 5.664\text{ kJ}\cdot\text{mol}^{-1} = 136.964\text{ kJ}\cdot\text{mol}^{-1} \end{aligned}$$

5–44. The standard molar enthalpy of formation of $CO_2(g)$ at 298 K is $-393.509\ \text{kJ}\cdot\text{mol}^{-1}$. Use the following data to calculate the value of $\Delta_f H^\circ$ at 1000 K. Assume the gases behave ideally under these conditions.

$$C_P^\circ[CO_2(g)]/R = 2.593 + (7.661\times10^{-3}\ \text{K}^{-1})T - (4.78\times10^{-6}\ \text{K}^{-2})T^2 + (1.16\times10^{-9}\ \text{K}^{-3})T^3$$

$$C_P^\circ[O_2(g)]/R = 3.094 + (1.561\times10^{-3}\ \text{K}^{-1})T - (4.65\times10^{-7}\ \text{K}^{-2})T^2$$

$$C_P^\circ[C(s)]/R = -0.6366 + (7.049\times10^{-3}\ \text{K}^{-1})T - (5.20\times10^{-6}\ \text{K}^{-2})T^2 + (1.38\times10^{-9}\ \text{K}^{-3})T^3$$

We can create a figure similar to Figure 5.10 to illustrate this reaction.

$$\begin{array}{ccc} C(s) + O_2(g) & \xrightarrow{\Delta_r H^\circ,1000\ \text{K}} & CO_2(g) \\ \downarrow \Delta H_1 & & \uparrow \Delta H_2 \\ C(s) + O_2(g) & \xrightarrow{\Delta_r H^\circ,298\ \text{K}} & CO_2(g) \end{array}$$

Now use Hess's Law to calculate the standard reaction enthalpy at 1000 K:

$$\begin{aligned} \Delta_r H^\circ_{1000} &= \Delta_r H^\circ_{298} + \Delta H_1 + \Delta H_2 \\ &= -393.509\ \text{kJ}\cdot\text{mol}^{-1} + R\int_{298}^{1000} \left\{\overline{C}_P[CO_2(g)] - \overline{C}_P[O_2(g)] - \overline{C}_P[C(s)]\right\} dT \\ &= -393.509\ \text{kJ}\cdot\text{mol}^{-1} + R\,[4047.167\ \text{K} - 2732.278\ \text{K} - 1419.433\ \text{K}] \\ &= -393.509\ \text{kJ}\cdot\text{mol}^{-1} - 0.869\ \text{kJ}\cdot\text{mol}^{-1} = -394.378\ \text{kJ}\cdot\text{mol}^{-1} \end{aligned}$$

5–45. The value of the standard molar reaction enthalpy for

$$CH_4(g) + 2\,O_2(g) \longrightarrow CO_2(g) + 2\,H_2O(g)$$

is $-802.2\ \text{kJ}\cdot\text{mol}^{-1}$ at 298 K. Using the heat-capacity data in Problems 5–43 and 5–44 in addition to

$$C_P^\circ[CH_4(g)]/R = 2.099 + (7.272\times10^{-3}\ \text{K}^{-1})T + (1.34\times10^{-7}\ \text{K}^{-2})T^2 - (8.66\times10^{-10}\ \text{K}^{-3})T^3$$

to derive a general equation for the value of $\Delta_r H^\circ$ at any temperature between 300 K and 1500 K. Plot $\Delta_r H^\circ$ versus T. Assume that the gases behave ideally under these conditions.

We can create a figure similar to Figure 5.10 to illustrate this reaction.

$$\begin{array}{ccc} CH_4(s) + 2O_2(g) & \xrightarrow{\Delta_r H^\circ} & CO_2(g) + 2H_2O(g) \\ \downarrow \Delta H_1 & & \uparrow \Delta H_2 \\ CH_4(s) + 2O_2(g) & \xrightarrow{\Delta_r H^\circ,298\ \text{K}} & CO_2(g) + 2H_2O(g) \end{array}$$

Now use Hess's Law:

$$\begin{aligned}
\Delta_r H^\circ &= \Delta_r H^\circ_{298} + \Delta H_1 + \Delta H_2 \\
&= -802.2\ \text{kJ}\cdot\text{mol}^{-1} \\
&\quad + R\int_{298}^{T} \{\overline{C}_P[\text{CO}_2(\text{g})] + 2\overline{C}_P[\text{H}_2\text{O}(\text{g})] - \overline{C}_P[\text{CH}_4(\text{s})] - 2\overline{C}_P[\text{O}_2(\text{g})]\}\, dT \\
&= -802.2\ \text{kJ}\cdot\text{mol}^{-1} + R\int_{298}^{T} \left[1.610 - (4.21\times 10^{-4}\ \text{K}^{-1})T\right] dT \\
&\quad + R\int_{298}^{T} \left[-(3.70\times 10^{-6}\ \text{K}^{-2})T^2 + (2.03\times 10^{-9}\ \text{K}^{-3})T^3\right] dT \\
&= -805.8\ \text{kJ}\cdot\text{mol}^{-1} + (1.339\times 10^{-2}\ \text{kJ}\cdot\text{mol}^{-1}\cdot\text{K}^{-1})T \\
&\quad -(1.750\times 10^{-6}\ \text{kJ}\cdot\text{mol}^{-1}\cdot\text{K}^{-2})T^2 - (1.025\times 10^{-8}\ \text{kJ}\cdot\text{mol}^{-1}\cdot\text{K}^{-3})T^3 \\
&\quad +(4.211\times 10^{-12}\ \text{kJ}\cdot\text{mol}^{-1}\cdot\text{K}^{-4})T^4
\end{aligned}$$

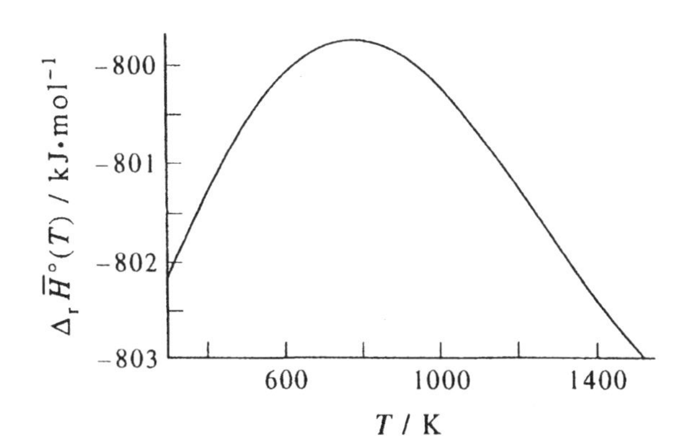

5–46. In all the calculations thus far, we have assumed the reaction takes place at constant temperature, so that any energy evolved as heat is absorbed by the surroundings. Suppose, however, that the reaction takes place under adiabatic conditions, so that all the energy released as heat stays within the system. In this case, the temperature of the system will increase, and the final temperature is called the ***adiabatic flame temperature***. One relatively simple way to estimate this temperature is to suppose the reaction occurs at the initial temperature of the reactants and then determine to what temperature the products can be raised by the quantity $\Delta_r H^\circ$. Calculate the adiabatic flame temperature if one mole of $CH_4(g)$ is burned in two moles of $O_2(g)$ at an initial temperature of 298 K. Use the results of the previous problem.

We know from Problem 5–45 that $802.2\ \text{kJ}\cdot\text{mol}^{-1}$ of energy is produced when one mole of methane is burned in two moles of oxygen at 298 K. Now we determine how much the temperature of the products, one mole of CO_2 and two moles of H_2O, can be raised by this energy:

$$\Delta H(\text{products}) = R\int_{298}^{T} \{\overline{C}_P[\text{CO}_2(\text{g})] + 2\overline{C}_P[\text{H}_2\text{O}(\text{g})]\}\, dT$$

$$\begin{aligned}
802.2\ \text{kJ}\cdot\text{mol}^{-1} &= R\int_{298}^{T} \{\overline{C}_P[\text{CO}_2(\text{g})] + 2\overline{C}_P[\text{H}_2\text{O}(\text{g})]\}\, dT \\
&= R\int_{298}^{T} \left[9.897 + (9.97\times 10^{-3}\ \text{K}^{-1})T - (4.496\times 10^{-6}\ \text{K}^{-2})T^2 \right.\\
&\qquad\qquad \left. +(1.160\times 10^{-9}\ \text{K}^{-3})T^3\right] dT \\
&= -27.89\ \text{kJ}\cdot\text{mol}^{-1} + (8.23\times 10^{-2}\ \text{kJ}\cdot\text{mol}^{-1}\cdot\text{K}^{-1})T + (4.15\times 10^{-5}\ \text{kJ}\cdot\text{mol}^{-1}\cdot\text{K}^{-2})T^2 \\
&\quad -(1.25\times 10^{-8}\ \text{kJ}\cdot\text{mol}^{-1}\cdot\text{K}^{-3})T^3 + (2.41\times 10^{-12}\ \text{kJ}\cdot\text{mol}^{-1}\cdot\text{K}^{-4})T^4
\end{aligned}$$

We can solve this polynomial using Simpson's rule or a numerical software package. Working in *Mathematica*, we find that the final temperature will be 4040 K.

5–47. Explain why the adiabatic flame temperature defined in the previous problem is also called the maximum flame temperature.

The adiabatic flame temperature is the temperature of the system if all the energy released as heat stays within the system. Since we are considering an isolated system, the adiabatic flame temperature is also the maximum temperature which the system can achieve.

5–48. How much energy as heat is required to raise the temperature of 2.00 moles of $O_2(g)$ from 298 K to 1273 K at 1.00 bar? Take

$$\overline{C}_P[O_2(g)]/R = 3.094 + (1.561 \times 10^{-3}\ \text{K}^{-1})T - (4.65 \times 10^{-7}\ \text{K}^{-2})T^2$$

We can use Equation 5.44:

$$\begin{aligned}\Delta H &= \int_{T_1}^{T_2} n\overline{C}_p dT \\ &= (2.00\ \text{mol})R\int_{298}^{1273}\left[3.094 + (1.561 \times 10^{-3}\ \text{K}^{-1})T - (4.65 \times 10^{-7}\ \text{K}^{-2})T^2\right]dT \\ &= 64.795\ \text{kJ}\cdot\text{mol}^{-1}\end{aligned}$$

5–49. When one mole of an ideal gas is compressed adiabatically to one-half of its original volume, the temperature of the gas increases from 273 K to 433 K. Assuming that $\overline{C}_V$ is independent of temperature, calculate the value of $\overline{C}_V$ for this gas.

Equation 5.20 gives an expression for the reversible adiabatic expansion of an ideal gas:

$$\overline{C}_V dT = -\frac{RT}{V}dV$$

Integrating both sides and substituting the temperatures given, we find that

$$\begin{aligned}\int \frac{\overline{C}_V}{T}dT &= -\int \frac{R}{V}dV \\ \overline{C}_V \ln\frac{T_2}{T_1} &= -R\ln\frac{V_2}{V_1} \\ \overline{C}_V \ln\frac{433}{273} &= -R\ln 2 \\ \frac{\overline{C}_V}{R} &= 1.50\end{aligned}$$

5–50. Use the van der Waals equation to calculate the minimum work required to expand one mole of $CO_2(g)$ isothermally from a volume of 0.100 dm^3 to a volume of 100 dm^3 at 273 K. Compare your result with that which you calculate assuming ideal behavior.

In Problem 5–10, we found that the work done by a van der Waals gas was

$$w = -RT \ln \frac{\overline{V}_2 - b}{\overline{V}_1 - b} + \frac{a(\overline{V}_2 - \overline{V}_1)}{\overline{V}_2\overline{V}_1}$$

Substituting $a = 3.6551 \text{ dm}^6\cdot\text{bar}\cdot\text{mol}^{-2}$ and $b = 0.042816 \text{ dm}^3\cdot\text{mol}^{-1}$ from Table 2.3 and using the parameters in the statement of the problem gives

$$\begin{aligned}
w &= -(0.083145 \text{ dm}^3\cdot\text{bar}\cdot\text{mol}^{-1}\cdot\text{K}^{-1})(273 \text{ K}) \ln \frac{100 \text{ dm}^3\cdot\text{mol}^{-1} - 0.042816 \text{ dm}^3\cdot\text{mol}^{-1}}{0.100 \text{ dm}^3\cdot\text{mol}^{-1} - 0.042816 \text{ dm}^3\cdot\text{mol}^{-1}} \\
&\quad +3.6551 \text{ dm}^6\cdot\text{bar}\cdot\text{mol}^{-2}\left[\frac{100 \text{ dm}^3\cdot\text{mol}^{-1} - 0.100 \text{ dm}^3\cdot\text{mol}^{-1}}{(100 \text{ dm}^3\cdot\text{mol}^{-1})(0.100 \text{ dm}^3\cdot\text{mol}^{-1})}\right] \\
&= (-169.5 \text{ dm}^3\cdot\text{bar}\cdot\text{mol}^{-1} + 36.5 \text{ dm}^3\cdot\text{bar}\cdot\text{mol}^{-1})(0.1 \text{ kJ}\cdot\text{dm}^{-3}\cdot\text{bar}^{-1})(1 \text{ mol}) \\
&= -13.3 \text{ kJ}
\end{aligned}$$

For an ideal gas,

$$\begin{aligned}
w &= -\int PdV = -nRT \ln\left(\frac{V_2}{V_1}\right) \\
&= (-156.80 \text{ dm}^3\cdot\text{bar})(0.1 \text{ kJ}\cdot\text{bar}^{-1}) = -15.7 \text{ kJ}
\end{aligned}$$

The work needed to expand the van der Waals gas is greater than that needed for the ideal gas.

5–51. Show that the work involved in a reversible, adiabatic pressure change of one mole of an ideal gas is given by

$$w = \overline{C}_V T_1\left[\left(\frac{P_2}{P_1}\right)^{R/\overline{C}_P} - 1\right]$$

where T_1 is the initial temperature and P_1 and P_2 are the initial and final pressures, respectively.

For a reversible, adiabatic pressure change of an ideal gas, $\delta q = 0$, so $dU = dw$. Since $d\overline{U}$ is defined as $\overline{C}_V dT$,

$$dw = \overline{C}_V dT$$

for one mole of an ideal gas. Integrating, we find

$$w = \overline{C}_V T_2 - \overline{C}_V T_1 = \overline{C}_V(T_2 - T_1) = \overline{C}_V T_1\left(\frac{T_2}{T_1} - 1\right)$$

From Problem 5.17, we know that

$$\frac{T_2}{T_1} = \left(\frac{P_2}{P_1}\right)^{R/\overline{C}_P}$$

and so substituting gives

$$w = \overline{C}_V T_1\left[\left(\frac{P_2}{P_1}\right)^{R/\overline{C}_P} - 1\right]$$

5–52. In this problem, we will discuss a famous experiment called the *Joule-Thomson experiment*. In the first half of the 19th century, Joule tried to measure the temperature change when a gas is expanded into a vacuum. The experimental setup was not sensitive enough, however, and he found that there was no temperature change, within the limits of his error. Soon afterward, Joule and Thomson devised a much more sensitive method for measuring the temperature change upon expansion. In their experiments (see Figure 5.11), a constant applied pressure P_1 causes a quantity of gas to flow slowly from one chamber to another through a porous plug of silk or cotton. If a volume, V_1, of gas is pushed through the porous plug, the work done on the gas is P_1V_1. The pressure on the other side of the plug is maintained at P_2, so if a volume V_2 enters the right-side chamber, then the net work is given by

$$w = P_1V_1 - P_2V_2$$

The apparatus is constructed so that the entire process is adiabatic, so $q = 0$. Use the First Law of Thermodynamics to show that

$$U_2 + P_2V_2 = U_1 + P_1V_1$$

or that $\Delta H = 0$ for a Joule-Thomson expansion. Starting with

$$dH = \left(\frac{\partial H}{\partial P}\right)_T dP + \left(\frac{\partial H}{\partial T}\right)_P dT$$

show that

$$\left(\frac{\partial T}{\partial P}\right)_H = -\frac{1}{C_P}\left(\frac{\partial H}{\partial P}\right)_T$$

Interpret physically the derivative on the left side of this equation. This quantity is called the *Joule-Thomson coefficient* and is denoted by μ_{JT}. In Problem 5–54 you will show that it equals zero for an ideal gas. Nonzero values of $(\partial T/\partial P)_H$ directly reflect intermolecular interactions. Most gases cool upon expansion [a positive value of $(\partial T/\partial P)_H$] and a Joule-Thomson expansion is used to liquefy gases.

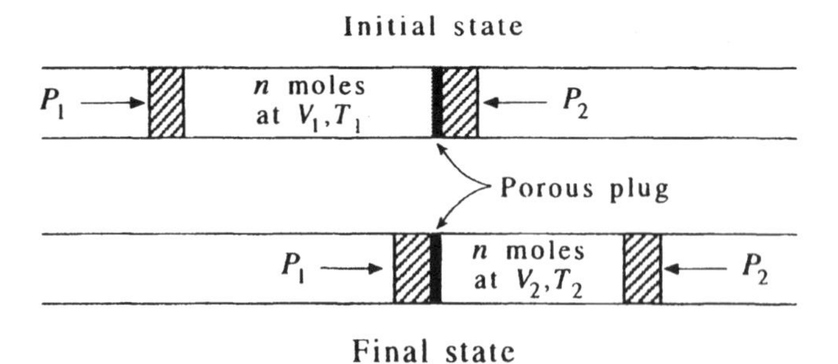

FIGURE 5.11
A schematic description of the Joule-Thomson experiment.

The net work is $w = P_1V_1 - P_2V_2$. Since $q = 0$, $U = w$, so

$$U_2 - U_1 = P_1V_1 - P_2V_2$$
$$U_2 + P_2V_2 = U_1 + P_1V_1$$

Since $\Delta H = U_1 + P_1V_1 - (U_2 + P_2V_2)$, $\Delta H = 0$. Now we write the total derivative of H as a function of P and T:

$$dH = \left(\frac{\partial H}{\partial P}\right)_T dP + \left(\frac{\partial H}{\partial T}\right)_P dT$$

Using the definition of C_P, we write this as

$$dH = \left(\frac{\partial H}{\partial P}\right)_T dP + C_P dT$$

$$-C_P dT = \left(\frac{\partial H}{\partial P}\right)_T dP - dH$$

$$dT = -\frac{1}{C_P}\left(\frac{\partial H}{\partial P}\right)_T dP + \frac{1}{C_P} dH$$

Keep H constant and divide through by dP to obtain

$$\left(\frac{\partial T}{\partial P}\right)_H = -\frac{1}{C_P}\left(\frac{\partial H}{\partial P}\right)_T + 0$$

The Joule-Thomson coefficient is a measure of the change of temperature of a gas with respect to the change in pressure in a Joule-Thomson expansion (or compression).

5–53. The Joule-Thomson coefficient (Problem 5–52) depends upon the temperature and pressure, but assuming an average constant value of 0.15 K·bar^{-1} for $N_2(g)$, calculate the drop in temperature if $N_2(g)$ undergoes a drop in pressure of 200 bar.

$$(0.15\ \text{K}\cdot\text{bar}^{-1})(-200\ \text{bar}) = -30\ \text{K}$$

5–54. Show that the Joule-Thomson coefficient (Problem 5–52) can be written as

$$\mu_{\text{JT}} = \left(\frac{\partial T}{\partial P}\right)_H = -\frac{1}{C_P}\left[\left(\frac{\partial U}{\partial V}\right)_T\left(\frac{\partial V}{\partial P}\right)_T + \left(\frac{\partial (PV)}{\partial P}\right)_T\right]$$

Show that $(\partial T/\partial P)_H = 0$ for an ideal gas.

From Problem 5–52,

$$\mu_{\text{JT}} = -\frac{1}{C_P}\left(\frac{\partial H}{\partial P}\right)_T$$

Since $H = U + PV$,

$$\mu_{\text{JT}} = -\frac{1}{C_P}\left[\left(\frac{\partial U}{\partial P}\right)_T + \left(\frac{\partial (PV)}{\partial P}\right)_T\right]$$

$$= -\frac{1}{C_P}\left[\left(\frac{\partial U}{\partial V}\right)_T\left(\frac{\partial V}{\partial P}\right)_T + \left(\frac{\partial (PV)}{\partial P}\right)_T\right]$$

For an ideal gas, U and PV depend only on temperature, so $\mu_{\text{JT}} = 0$.

5–55. In this problem, we will investigate the pressure dependence of the heat capacity, C_P. Because $C_P = (\partial H/\partial T)_P$, we see that

$$\left(\frac{\partial C_P}{\partial P}\right)_T = \frac{\partial^2 H}{\partial P \partial T}$$

Now show that

$$\left(\frac{\partial C_P}{\partial P}\right)_T = -\mu_{JT}\left(\frac{\partial C_P}{\partial T}\right)_P - C_P\left(\frac{\partial \mu_{JT}}{\partial T}\right)_P$$

where μ_{JT} is the Joule-Thomson coefficient. Show that C_P is independent of pressure for an ideal gas.

From Problem 5–52, we see that

$$\mu_{JT} = -\frac{1}{C_P}\left(\frac{\partial H}{\partial P}\right)_T$$

and so

$$\left(\frac{\partial C_P}{\partial P}\right)_T = \frac{\partial}{\partial T}\left(\frac{\partial H}{\partial P}\right)_T = \frac{\partial}{\partial T}(-\mu_{JT}C_P)$$

$$= -\mu_{JT}\left(\frac{\partial C_P}{\partial T}\right)_P - C_P\left(\frac{\partial \mu_{JT}}{\partial T}\right)_P$$

5–56. Given that $C_P(T)$ for $N_2(g)$ can be represented by

$$C_P(T)/\text{J}\cdot\text{mol}^{-1}\cdot\text{K}^{-1} = 27.296 + (5.230 \times 10^{-3}\ \text{K}^{-1})T - (4.0 \times 10^{-9}\ \text{K}^{-2})T^2$$

for $298\ \text{K} < T < 1500\ \text{K}$ and that μ_{JT} can be represented by

$$\mu_{JT}/\text{K}\cdot\text{atm}^{-1} = 3.722 - (0.02935\ \text{K}^{-1})T + (8.106 \times 10^{-5}\ \text{K}^{-2})T^2 - (7.496 \times 10^{-8}\ \text{K}^{-3})T^3$$

for $150\ \text{K} < T < 500\ \text{K}$ around one atmosphere, estimate the pressure dependence of $C_P(T)$ for $N_2(g)$ at 300 K and one atmosphere pressure.

From Problem 5–55, we have

$$\left(\frac{\partial C_P}{\partial P}\right)_T = -\mu_{JT}\left(\frac{\partial C_P}{\partial T}\right)_P - C_P\left(\frac{\partial \mu_{JT}}{\partial T}\right)_P$$

Using the expressions for C_P and μ_{JT} given in the problem,

$$C_P = 28.86\ \text{J}\cdot\text{mol}^{-1}\cdot\text{K}^{-1}$$

$$\left(\frac{\partial C_P}{\partial T}\right)_P = 5.228 \times 10^{-3}\ \text{J}\cdot\text{mol}^{-1}\cdot\text{K}^{-2}$$

$$\mu_{JT} = 0.188\ \text{K}\cdot\text{atm}^{-1}$$

and

$$\left(\frac{\partial \mu_{JT}}{\partial T}\right)_P = -9.532 \times 10^{-4}\ \text{atm}^{-1}$$

at 300 K. Therefore,

$$\left(\frac{\partial C_P}{\partial P}\right)_T = -(0.188\ \text{K}\cdot\text{atm}^{-1})(5.228\times 10^{-3}\ \text{J}\cdot\text{mol}^{-1}\cdot\text{K}^{-2}) + (28.86\ \text{J}\cdot\text{mol}^{-1}\cdot\text{K}^{-1})(9.532\times 10^{-4}\ \text{atm}^{-1}) = 0.0265\ \text{J}\cdot\text{mol}^{-1}\cdot\text{K}^{-1}\cdot\text{atm}^{-1}$$

5–57. In Chapter 8, we will derive the formula

$$\left(\frac{\partial U}{\partial V}\right)_T = T\left(\frac{\partial P}{\partial T}\right)_V - P \tag{8.22}$$

Show that U is independent of volume for an ideal gas. Show that

$$\Delta U = a\left(\frac{1}{V_1} - \frac{1}{V_2}\right)$$

for an isothermal expansion of a van der Waals gas and that

$$\Delta U = \frac{A}{2T^{1/2}B}\ln\left[\frac{(V_2+B)\,V_1}{(V_1+B)\,V_2}\right]$$

for a Redlich-Kwong gas.

Using the equation $PV = nRT$, we have

$$\left(\frac{\partial U}{\partial V}\right)_T = T\left(\frac{nR}{V}\right) - P = 0$$

for an ideal gas. For the van der Waals equation,

$$\left(\frac{\partial U}{\partial V}\right)_T = T\left(\frac{nR}{V-nb}\right) - \frac{nRT}{V-nb} + \frac{an^2}{V^2} = \frac{an^2}{V^2}$$

and

$$\Delta U = \int_{V_1}^{V_2}\frac{an^2}{V^2}dV = an^2\left(\frac{1}{V_1} - \frac{1}{V_2}\right)$$

For a Redlich-Kwong gas,

$$\left(\frac{\partial U}{\partial V}\right)_T = T\left[\frac{nR}{V-nB} - \frac{An^2}{2T^{3/2}V(V+nB)}\right] - \frac{nRT}{V-nB} + \frac{An^2}{T^{1/2}V(V+nB)} = \frac{An^2}{2T^{1/2}V(V+nB)}$$

and

$$\Delta U = \frac{An^2}{2T^{1/2}B}\ln\left[\frac{(V_2+nB)V_1}{(V_1+nB)V_2}\right]$$

5–58. Use the rigid rotator-harmonic oscillator model and the data in Table 4.2 to plot $\overline{C}_P(T)$ for CO(g) from 300 K to 1000 K. Compare your result with the expression given in Problem 5–43.

From Example 5–8, we know that for an ideal gas

$$\overline{C}_P = \overline{C}_V + R \tag{5.43}$$

And from Chapter 4, we know that for a linear polyatomic ideal gas

$$\frac{\overline{C}_V}{R} = \frac{5}{2} + \left(\frac{\Theta_{\text{vib}}}{T}\right)^2 \frac{e^{-\Theta_{\text{vib}}/T}}{(1 - e^{\Theta_{\text{vib}}/T})^2} \tag{4.41}$$

Therefore, since $\Theta_{\text{vib}}(\text{CO}) = 3103$ K, we wish to graph

$$\frac{\overline{C}_P}{R} = \frac{7}{2} + \left(\frac{3103\text{ K}}{T}\right)^2 \frac{e^{-3103\text{ K}/T}}{(1 - e^{-3103\text{ K}/T})^2}$$

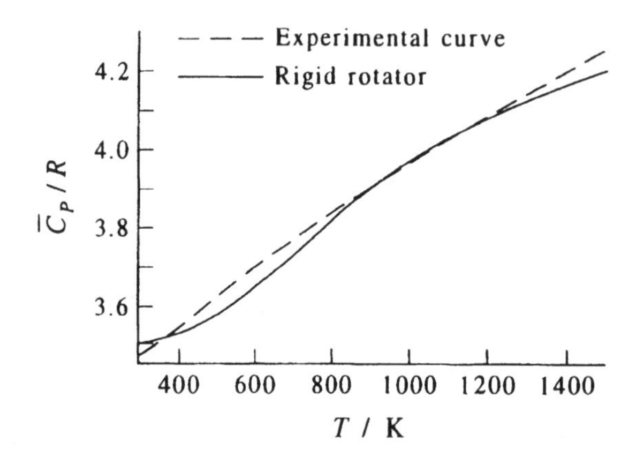

5–59. Use the rigid rotator-harmonic oscillator model and the data in Table 4.4 to plot $\overline{C}_P(T)$ for CH_4(g) from 300 K to 1000 K. Compare your result with the expression given in Problem 5–45.

Again, for an ideal gas

$$\overline{C}_P = \overline{C}_V + R \tag{5.43}$$

And from Chapter 4, we know that for a nonlinear polyatomic ideal gas

$$\frac{\overline{C}_V}{R} = \frac{3}{2} + \frac{3}{2} + \sum_{j=1}^{3n-6} \left(\frac{\Theta_{\text{vib},j}}{T}\right)^2 \frac{e^{-\Theta_{\text{vib},j}/T}}{(1 - e^{-\Theta_{\text{vib},j}/T})^2} \tag{4.62}$$

Using the values given in the problem, we wish to graph

$$\frac{\overline{C}_P}{R} = 4 + \left(\frac{4170\text{ K}}{T}\right)^2 \frac{e^{-4170\text{ K}/T}}{(1 - e^{-4170\text{ K}/T})^2} + 2\left(\frac{2180\text{ K}}{T}\right)^2 \frac{e^{-2180\text{ K}/T}}{(1 - e^{-2180\text{ K}/T})^2}$$
$$+3\left(\frac{4320\text{ K}}{T}\right)^2 \frac{e^{-4320\text{ K}/T}}{(1 - e^{-4320\text{ K}/T})^2} + 3\left(\frac{1870\text{ K}}{T}\right)^2 \frac{e^{-1870\text{ K}/T}}{(1 - e^{-1870\text{ K}/T})^2}$$

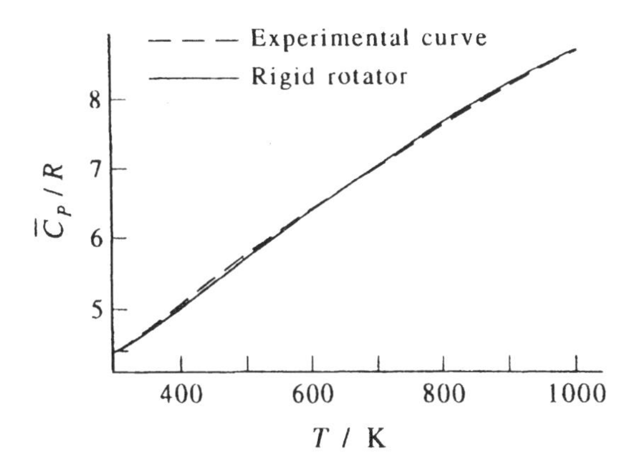

5–60. Why do you think the equations for the dependence of temperature on volume for a reversible adiabatic process (see Equation 5.22 and Example 5.6) depend upon whether the gas is a monatomic gas or a polyatomic gas?

For an adiabatic process, no energy is transferred as heat, so the change in internal energy is equal to the change in work. The internal energy of a monatomic gas (in the electronic ground state) is entirely in the translational degrees of freedom, which is directly related to the temperature of the gas. If the volume of the system increases, work is done by the system, and so the internal energy of the system must decrease. The only way for it to do so is by decreasing the amount of energy in the translational degrees of freedom, which decreases the observed temperature of the gas. The internal energy of a polyatomic gas (in the electronic ground state) is in the rotational, vibrational, and translational degrees of freedom. These vibrational, rotational, and translational energy levels are not necessarily in equilibrium (see Problem 4–37). If the volume of a polyatomic gas is increased, work is done by the system, as before, and the internal energy of the gas must again decrease. However, the gas can lose energy in the rotational or vibrational levels, which will not decrease the observed temperature of the gas. It can also lose energy in the translational levels, which will decrease the observed temperature of the gas, but the other available methods of decreasing the energy of the system will change the dependence of temperature on volume for a polyatomic gas from that observed for a monatomic gas.

MATHCHAPTER E

The Multinominal Distribution and Stirling's Approximation

PROBLEMS AND SOLUTIONS

E–1. Use Equation E.3 to write the expansion of $(1+x)^5$. Use Equation E.4 to do the same thing.

Using Equation E.3, we can write

$$\begin{aligned}(1+x)^5 &= \sum_{N_1=0}^{5} \frac{5!}{N_1!(5-N_1)!} x^{5-N_1} \\ &= 1 + 5x + 10x^2 + 10x^3 + 5x^4 + x^5\end{aligned}$$

Equation E.4 gives the same result.

E–2. Use Equation E.6 to write out the expression for $(x+y+z)^2$. Compare your result to the one that you obtain by multiplying $(x+y+z)$ by $(x+y+z)$.

Using Equation E.6, we can write

$$\begin{aligned}(x+y+z)^2 &= \sum_{N_1=0}^{2}\sum_{N_2=0}^{2}\sum_{N_3=0}^{2}{}^{*} \frac{N!}{N_1!N_2!N_3!} x^{N_1} y^{N_2} z^{N_3} \\ &= \frac{2!}{2!0!0!}x^2 + \frac{2!}{1!0!1!}xz + \frac{2!}{1!1!0!}xy + \frac{2!}{0!1!1!}yz + \frac{2!}{0!2!0!}y^2 + \frac{2!}{0!0!2!}z^2 \\ &= 2yz + 2xz + 2xy + x^2 + y^2 + z^2\end{aligned}$$

We obtain the same result when we multiply $(x+y+z)$ by $(x+y+z)$.

E–3. Use Equation E.6 to write out the expression for $(x+y+z)^4$. Compare your result to the one that you obtain by multiplying $(x+y+z)^2$ from Problem E–2 by itself.

Using Equation E.6, we can write

$$\begin{aligned}(x+y+z)^4 &= \sum_{N_1=0}^{4}\sum_{N_2=0}^{4}\sum_{N_3=0}^{4}{}^{*} \frac{N!}{N_1!N_2!N_3!} x^{N_1} y^{N_2} z^{N_3} \\ &= \frac{4!}{4!0!0!}x^4 + \frac{4!}{3!1!0!}x^3y + \frac{4!}{3!0!1!}x^3z + \frac{4!}{2!2!0!}x^2y^2 + \frac{4!}{2!0!2!}x^2z^2 \\ &\quad + \frac{4!}{2!1!1!}x^2yz + \frac{4!}{1!3!0!}xy^3 + \frac{4!}{1!0!3!}xz^3 + \frac{4!}{1!2!1!}xy^2z + \frac{4!}{1!1!2!}xyz^2\end{aligned}$$

$$+\frac{4!}{0!4!0!}y^4+\frac{4!}{0!0!4!}z^4+\frac{4!}{0!1!3!}yz^3+\frac{4!}{0!3!1!}y^3z+\frac{4!}{0!2!2!}y^2z^2$$
$$=x^4+4x^3y+4x^3z+6x^2y^2+6x^2z^2+12x^2yz+4xy^3+4xz^3$$
$$+12xy^2z+12xyz^2+y^4+z^4+4yz^3+4y^3z+6y^2z^2$$

We obtain the same result when we multiply $(x+y+z)^2$ by $(x+y+z)^2$.

E–4. How many permutations of the letters a, b, c are there?

$3! = 6$

E–5. The coefficients of the expansion of $(1+x)^n$ can be arranged in the following form:

n									
0					1				
1				1		1			
2			1		2		1		
3		1		3		3		1	
4	1		4		6		4		1

Do you see a pattern in going from one row to the next? The triangular arrangement here is called Pascal's triangle.

Each number in a row is the sum of the two numbers above it.

E–6. In how many ways can a committee of three be chosen from nine people?

$$\frac{9!}{3!6!}=3\times 7\times 4=84$$

E–7. Calculate the relative error for $N = 50$ using the formula for Stirling's approximation given in Example E–3, and compare your result with that given in Table E.1 using Equation E.7. Take $\ln N!$ to be 148.47776 (*CRC Handbook of Chemistry and Physics*).

$$\ln N! = N\ln N - N + \ln(2\pi N)^{1/2}$$
$$= 195.6012 - 50 + 2.87495 = 148.4761$$

Compared to $\ln N!$, there is a 1.12×10^{-5} relative error. This can be compared with the 0.0194 relative error obtained when using Equation E.7 for $\ln N!$.

E–8. Prove that $x\ln x \to 0$ as $x \to 0$.

$$\lim_{x\to 0}(x\ln x)=\lim_{x\to 0}\left(\frac{\ln x}{x^{-1}}\right)$$

Since both numerator and denominator are undefined as $x \to 0$, we can apply l'Hôpital's rule:

$$\lim_{x\to 0}(x \ln x) = \lim_{x\to 0}\left(\frac{x^{-1}}{-x^{-2}}\right) = \lim_{x\to 0}\left(\frac{-x^2}{x}\right) = \lim_{x\to 0}(-x) = 0$$

E–9. Prove that the maximum value of $W(N, N_1) = N!/(N - N_1)!N_1!$ is given by $N_1 = N/2$. (*Hint*: Treat N_1 as a continuous variable.)

Let $N_1 = x$, with x being a continuous variable. Then

$$W = \frac{N!}{(N-x)!x!}$$

$$\ln W = \ln\left[\frac{N!}{(N-x)!x!}\right] = \ln N! - \ln(N-x)! - \ln x!$$

We are looking for the maximum value of W, so, at this value, $(dW/dx) = 0$. Also, we know that

$$\frac{1}{W}\frac{dW}{dx} = \frac{d(\ln W)}{dx}$$

At the maximum value of W, then, $d(\ln W)/dx = 0$. We can express $\ln W$ using Stirling's approximation (Equation E.7):

$$\begin{aligned}\ln W &= \ln N! - \ln(N-x)! - \ln x!\\ &= N\ln N - N - [(N-x)\ln(N-x) - (N-x)] - (x\ln x - x)\end{aligned}$$

The derivative of $\ln W$ is then

$$\begin{aligned}\frac{d(\ln W)}{dx} &= 0 + \frac{N-x}{N-x} + \ln(N-x) - 1 - \ln x - \frac{x}{x} + 1\\ &= \ln(N-x) - \ln x\end{aligned}$$

Setting $[d(\ln W)/dx] = 0$ gives

$$\begin{aligned}0 &= \ln(N-x) - \ln x\\ \ln x &= \ln(N-x)\\ 2x &= N\\ x &= \frac{N}{2}\end{aligned}$$

Recall that $N_1 = x$, and that $N_2 = N - N_1$. The maximum value of W then occurs when $N_1 = N/2$.

E–10. Prove that the maximum value of $W(N_1, N_2, \ldots, N_r)$ in Equation E.5 is given by $N_1 = N_2 = \cdots = N_r = N/r$.

Given:

$$W = \frac{N!}{N_1!N_2!\cdots N_r!}$$

Take $N_r = N - N_1 - N_2 - \cdots - N_{r-1}$, because $\sum N_j = N$. Then we can use Stirling's approximation (Equation E.7) to write $\ln W$ as

$$\ln W = N \ln N - N - \sum_{j=1}^{r} N_j \ln N_j - N_j$$

$$= N \ln N - N - \sum_{j=1}^{r-1} N_j \ln N_j - N_j$$

$$-(N - N_1 - N_2 - \cdots - N_{r-1}) \ln(N - N_1 - N_2 - \cdots - N_{r-1}) + N - N_1 - N_2 - \cdots - N_{r-1}$$

Now take the partial derivative of $\ln W$ with respect to N_1. We find that

$$\frac{\partial \ln W}{\partial N_1} = -\ln N_1 + 1 - 1 + \ln(N - N_1 - N_2 - \cdots - N_{r-1}) + 1 - 1$$

As in Problem E–9, we set $d(\ln W)/dN_1 = 0$ to find

$$0 = -\ln N_1 - \ln(N - N_1 - N_2 - \cdots - N_{r-1})$$

$$\ln N_1 = \ln N_r$$

$$N_1 = N_r$$

We get a similar result for N_2 through N_{r-1}, so $N_1 = N_2 = \cdots = N_r$ and all are equal to N/r.

E–11. Prove that

$$\sum_{k=0}^{N} \frac{N!}{k!(N-k)!} = 2^N$$

We know that (Equation E.3)

$$(x + y)^N = \sum_{N_1=0}^{N} \frac{N!}{N_1!(N - N_1)!} x^{N_1} y^{N-N_1}$$

If $x = y = 1$, we can write this as

$$(1 + 1)^N = \sum_{N_1=0}^{N} \frac{N!}{N_1!(N - N_1)!} = \sum_{k=0}^{N} \frac{N!}{k!(N-k)!}$$

Thus

$$\sum_{k=0}^{N} \frac{N!}{k!(N-k)!} = (1 + 1)^N = 2^N$$

E–12. The quantity $n!$ as we have defined it is defined only for positive integer values of n. Consider now the function of x *defined* by

$$\Gamma(x) = \int_0^\infty t^{x-1} e^{-t} dt \tag{1}$$

Integrate by parts (letting $u = t^{x-1}$ and $dv = e^{-t}dt$) to get

$$\Gamma(x) = (x - 1) \int_0^\infty t^{x-2} e^{-t} dt = (x - 1)\Gamma(x - 1) \tag{2}$$

Now use Equation 2 to show that $\Gamma(x) = (x-1)!$ if x is a positive integer. Although Equation 2 provides us with a general function that is equal to $(n-1)!$ when x takes on integer values, it is defined just as well for non-integer values. For example, show that $\Gamma(3/2)$, which in a sense is $(\frac{1}{2})!$, is equal to $\pi^{1/2}/2$. Equation 1 can also be used to explain why $0! = 1$. Let $x = 1$ in Equation 1 to show that $\Gamma(1)$, which we can write as $0!$, is equal to 1. The function $\Gamma(x)$ defined by Equation 1 is called the *gamma function* and was introduced by Euler to generalize the idea of a factorial to general values of n. The gamma function arises in many problems in chemistry and physics.

We can integrate Γ by parts, letting $u = t^{x-1}$ and $v = -e^{-t}$, to find

$$\begin{aligned}
\Gamma(x) &= \int_0^\infty t^{x-1}e^{-t}dt \\
&= -t^{x-1}e^{-t}\Big|_{t=0}^{t=\infty} - \int_0^\infty -e^{-t}(x-1)t^{x-2}dt \\
&= (x-1)\int_0^\infty e^{-t}t^{x-2}dt \\
&= (x-1)\,\Gamma(x-1)
\end{aligned}$$

Now if x is a positive integer, we can write

$$\begin{aligned}
\Gamma(x) &= (x-1)\,\Gamma(x-1) = (x-1)(x-2)\,\Gamma(x-2) \\
&= \prod_{n=2}^{x-1} n\,\Gamma(1) \\
&= \prod_{n=2}^{x-1} n \int_0^\infty e^{-t}dt = \prod_{n=2}^{x-1} n = (x-1)!
\end{aligned}$$

We can let $t = x^2$ and write $\Gamma(3/2)$ as

$$\begin{aligned}
\Gamma(3/2) &= \int_0^\infty t^{1/2}e^{-t}dt = \int_0^\infty 2x^2e^{-x^2}dx \\
&= 2\frac{1}{4}\sqrt{\pi} = \frac{\pi^{1/2}}{2}
\end{aligned}$$

Also, we can express $\Gamma(0)$ as

$$\begin{aligned}
\Gamma(1) &= \int_0^\infty t^0e^{-t}dt = \int_0^\infty e^{-t}dt \\
&= 1
\end{aligned}$$

CHAPTER 6

Entropy and the Second Law of Thermodynamics

PROBLEMS AND SOLUTIONS

6–1. Show that

$$\oint dY = 0$$

if Y is a state function.

If Y is a state function, dY must be an exact differential. This means that $\int_1^2 dY = Y_2 - Y_1$ and $\int_2^1 dY = Y_1 - Y_2$. Then

$$\oint dY = \int_1^2 dY + \int_2^1 dY = Y_2 - Y_1 + (Y_1 - Y_2) = 0$$

6–2. Let $z = z(x, y)$ and $dz = xydx + y^2dy$. Although dz is not an exact differential (why not?), what combination of dz and x and/or y is an exact differential?

The quantity dz is not an exact differential because the coefficient of the dx term is not independent of y. An exact differential would be dz/y, because the coefficient of dx is independent of y and the coefficient of dy is independent of x in

$$\frac{dz}{y} = xdx + ydy$$

6–3. Use the criterion developed in MathChapter D to prove that δq_{rev} in Equation 6.1 is not an exact differential (see also Problem D–11).

We can write δq_{rev} as

$$\delta q_{\text{rev}} = C_V(T)dT + \frac{nRT}{V}dV \tag{6.1}$$

The cross-derivatives of an exact differential are equal, so we will find the cross derivatives of δq_{rev} to determine its nature. These are the coefficient of dT differentiated with respect to V and the coefficient of dV differentiated with respect to T, or

$$\frac{\partial C_V(T)}{\partial V} = 0 \qquad \text{and} \qquad \frac{\partial}{\partial T}\left(\frac{nRT}{V}\right) = \frac{nR}{V}$$

Because these two quantities are not equal, δq_{rev} is an inexact differential.

6–4. Use the criterion developed in MathChapter D to prove that $\delta q_{rev}/T$ in Equation 6.1 is an exact differential.

We use Equation 6.2 to express $\delta q_{rev}/T$ as

$$\frac{\delta q_{rev}}{T} = \frac{C_V(T)}{T}dT + \frac{nR}{V}dV$$

The cross-derivatives of an exact differential are equal, so we will find the cross derivatives of $\delta q_{rev}/T$ to determine its nature. These are the coefficient of dT differentiated with respect to V and the coefficient of dV differentiated with respect to T, or

$$\frac{\partial}{\partial V}\left(\frac{C_V(T)}{T}\right) = 0 \qquad \text{and} \qquad \frac{\partial}{\partial T}\left(\frac{nR}{V}\right) = 0$$

Because these two quantities are equal, $\delta q_{rev}/T$ is an exact differential.

6–5. In this problem, we will prove that Equation 6.5 is valid for an arbitrary system. To do this, consider an isolated system made up of two equilibrium subsystems, A and B, which are in thermal contact with each other; in other words, they can exchange energy as heat between themselves. Let subsystem A be an ideal gas and let subsystem B be arbitrary. Suppose now that an infinitesimal reversible process occurs in A accompanied by an exchange of energy as heat $\delta q_{rev}(\text{ideal})$. Simultaneously, another infinitesimal reversible process takes place in B accompanied by an exchange of energy as heat $\delta q_{rev}(\text{arbitrary})$. Because the composite system is isolated, the First Law requires that

$$\delta q_{rev}(\text{ideal}) = -\delta q_{rev}(\text{arbitrary})$$

Now use Equation 6.4 to prove that

$$\oint \frac{\delta q_{rev}(\text{arbitrary})}{T} = 0$$

Therefore, we can say that the definition given by Equation 6.4 holds for any system.

We use the First Law as suggested in the problem and substitute Equation 6.1 for $\delta q_{rev}(\text{ideal})$ to write

$$\begin{aligned}\delta q_{rev}(\text{arbitrary}) &= -\delta q_{rev}(\text{ideal})\\ &= -C_V(T)dT - \frac{nRT}{V}dV\\ \frac{\delta q_{rev}(\text{arbitrary})}{T} &= \frac{-C_V}{T}dT - \frac{nR}{V}dV\\ &= d\left[\int \frac{-C_V(T)}{T}dT - nR\int \frac{dV}{V} + \text{constant}\right]\end{aligned}$$

Then $\delta q_{rev}(\text{arbitrary})/T$ is the derivative of a state function. We know that the cyclic integral of a state function is equal to 0 (Problem 6–1). Therefore, we can write (as we did in Section 6–2 for ideal gases)

$$\oint \frac{\delta q_{rev}(\text{arbitrary})}{T} = 0$$

and Equation 6.4 holds for any system.

6–6. Calculate q_{rev} and ΔS for a reversible cooling of one mole of an ideal gas at a constant volume V_1 from P_1, V_1, T_1 to P_2, V_1, T_4 followed by a reversible expansion at constant pressure P_2 from P_2, V_1, T_4 to P_2, V_2, T_1 (the final state for all the processes shown in Figure 6.3). Compare your result for ΔS with those for paths A, B + C, and D + E in Figure 6.3.

Step 1. $P_1, V_1, T_1 \rightarrow P_2, V_1, T_4$
Because there is no change in the volume of the ideal gas, $\delta w = 0$, and we can write

$$dq_{rev,1} = dU = C_V(T)dT$$

$$q_{rev,1} = \int_{T_1}^{T_4} C_V(T)dT$$

$$\Delta S_1 = \int_{T_1}^{T_4} \frac{C_V(T)}{T}dT$$

Step 2. $P_2, V_1, T_4 \rightarrow P_2, V_2, T_1$
In this case, we write (by the First Law)

$$\delta q_{rev,2} = dU - \delta w = C_V(T)dT - PdV$$

$$q_{rev,2} = \int_{T_4}^{T_1} C_V(T)dT - \int_{V_1}^{V_2} P_2 dV$$

$$\Delta S_2 = \int_{T_4}^{T_1} \frac{C_V(T)}{T}dT - \int_{V_1}^{V_2} \frac{P_2}{T}dV$$

$$= \int_{T_4}^{T_1} \frac{C_V(T)}{T}dT - \int_{V_1}^{V_2} \frac{R}{V}dV$$

For the entire reaction, $P_1, V_1, T_1 \rightarrow P_2, V_2, T_1$, we have

$$q_{rev} = \int_{T_1}^{T_4} C_V(T)dT + \int_{T_4}^{T_1} C_V(T)dT - \int_{V_1}^{V_2} P_2 dV = -P_2(V_2 - V_1)$$

$$\Delta S = \int_{T_1}^{T_4} \frac{C_V(T)}{T}dT + \int_{T_4}^{T_1} \frac{C_V(T)}{T}dT - \int_{V_1}^{V_2} \frac{R}{V}dV = R \ln \frac{V_2}{V_1}$$

The value of q_{rev} differs from those found for paths A, B + C, and D + E (Section 6–3), but the value of ΔS is the same (because entropy is a path-independent function).

6–7. Derive Equation 6.8 without referring to Chapter 5.

The temperature T_2 is the final temperature resulting from the reversible adiabatic expansion of one mole of an ideal gas. For a reversible expansion, $dw = -PdV = -nRTdV/V$, and for an adiabatic expansion $dU = dw$. Then, because $dU = C_V dT$,

$$\frac{C_V dT}{T} = \frac{-nRdV}{V}$$

$$\int_{T_1}^{T_2} \frac{C_V}{T}dT = -nR \int_{V_1}^{V_2} \frac{dV}{V}$$

$$\int_{T_1}^{T_2} \frac{C_V}{T}dT = -nR \ln \frac{V_2}{V_1}$$

which is Equation 6.8.

6–8. Calculate the value of ΔS if one mole of an ideal gas is expanded reversibly and isothermally from 10.0 dm^3 to 20.0 dm^3. Explain the sign of ΔS.

For an isothermal reaction of an ideal gas, $\delta w = -\delta q$, so $\delta q = PdV$. Then

$$\Delta S = \int \frac{\delta q_{rev}}{T} = \int \frac{P}{T} dV$$

Using T from the ideal gas equation gives

$$\Delta S = \int \frac{nR}{V} dV = nR \ln 2.00 = 5.76 \text{ J}\cdot\text{K}^{-1}$$

The value of ΔS is positive because the gas is expanding.

6–9. Calculate the value of ΔS if one mole of an ideal gas is expanded reversibly and isothermally from 1.00 bar to 0.100 bar. Explain the sign of ΔS.

As in the previous problem, because the reaction is isothermal, $\delta q = PdV$. For an ideal gas,

$$dV = -\frac{nRT}{P^2} dP = -\frac{V}{P} dP$$

so we write ΔS as

$$\Delta S = \int \frac{P}{T} dV = \int -\frac{V}{T} dP = \int -\frac{nR}{P} dP = -nR \ln 0.1 = 19.1 \text{ J}\cdot\text{K}^{-1}$$

The value of ΔS is positive because the gas expands.

6–10. Calculate the values of q_{rev} and ΔS along the path D + E in Figure 6.3 for one mole of a gas whose equation of state is given in Example 6–2. Compare your result with that obtained in Example 6–2.

Path D + E is the path described by $(P_1, V_1, T_1) \rightarrow (P_1, V_2, T_3) \rightarrow (P_2, V_2, T_1)$. For the first step,

$$\delta q_{rev} = dU - \delta w = C_V(T)dT - PdV$$

and for the second step (because the volume remains constant)

$$\delta q_{rev} = dU = C_V(T)dT$$

Then

$$\int \delta q_{rev,D+E} = \int_{T_1}^{T_3} C_V(T)dT - \int_{V_1}^{V_2} P_1 dV + \int_{T_3}^{T_1} C_V(T)dT$$
$$= -\int_{V_1}^{V_2} P_1 dV = -P_1(V_2 - V_1)$$

and

$$\Delta S_{rev,D+E} = -\int_{V_1}^{V_2} \frac{P_1}{T} dV$$

Substituting for T from the equation of state gives

$$\Delta S_{\text{rev,D+E}} = \int_{\overline{V}_1}^{\overline{V}_2} \frac{nP_1 R d\overline{V}}{P_1(\overline{V}-b)} = nR \ln\left(\frac{\overline{V}_2 - b}{\overline{V}_1 - b}\right)$$

The quantity q_{rev} differs from those for the two paths in Example 6–2, but ΔS for all three paths is the same.

6–11. Show that $\Delta S_{\text{D+E}}$ is equal to ΔS_{A} and $\Delta S_{\text{B+C}}$ for the equation of state given in Example 6–2.

From Example 6–3,

$$\Delta S_{\text{A}} = \Delta S_{\text{B+C}} = nR \ln \frac{\overline{V}_2 - b}{\overline{V}_1 - b}$$

and the equation of state used is

$$P = \frac{RT}{\overline{V} - b}$$

In Example 6–1, we calculated $\Delta S_{\text{D+E}}$ for an ideal gas. Without using the ideal gas equation of state, however, we found in Example 6–1 that

$$\Delta S_{\text{D}} = \int_{T_1}^{T_3} \frac{C_V(T)}{T} dT + P_1 \int_{V_1}^{V_2} \frac{dV}{T}$$

$$\Delta S_{\text{E}} = \int_{T_3}^{T_1} \frac{C_V(T)}{T} dT$$

and

$$\Delta S_{\text{D+E}} = P_1 \int_{V_1}^{V_2} \frac{dV}{T}$$

We can substitute T from the equation of state to write

$$\Delta S_{\text{D+E}} = P_1 \int_{\overline{V}_1}^{\overline{V}_2} \frac{nRd\overline{V}}{P_1(\overline{V}-b)} = nR \int_{\overline{V}_1}^{\overline{V}_2} \frac{d\overline{V}}{\overline{V}-b} = nR \ln \frac{\overline{V}_2 - b}{\overline{V}_1 - b}$$

Therefore $\Delta S_{\text{D+E}}$ is equal to ΔS_{A} and $\Delta S_{\text{B+C}}$.

6–12. Calculate the values of q_{rev} and ΔS along the path described in Problem 6–6 for one mole of a gas whose equation of state is given in Example 6–2. Compare your result with that obtained in Example 6–2.

For both steps, because the ideal gas equation was not used in calculating q_{rev}, q_{rev} is the same as it was in Problem 6–6:

$$q_{\text{rev}} = -P_2(V_2 - V_1)$$

Substituting the equation of state from Example 6–2 into the expression for ΔS from Problem 6–6 gives

$$\Delta S = \int_{\overline{V}_1}^{\overline{V}_2} \frac{nP_2}{T} d\overline{V} = \int_{\overline{V}_1}^{\overline{V}_2} \frac{nR}{\overline{V}-b} d\overline{V} = nR \ln\left(\frac{\overline{V}_2 - b}{\overline{V}_1 - b}\right)$$

This is (by no coincidence) the same value as that found for ΔS_A, ΔS_{B+C}, and ΔS_{D+E}.

6–13. Show that

$$\Delta S = C_P \ln \frac{T_2}{T_1}$$

for a constant-pressure process if C_P is independent of temperature. Calculate the change in entropy of 2.00 moles of $H_2O(l)$ ($\overline{C}_P = 75.2\ J\cdot K^{-1}\cdot mol^{-1}$) if it is heated from 10°C to 90°C.

Because ΔS is a state function, we can calculate it using a reversible process. For a constant-pressure reversible process (Equation 5.37), $\delta q_{rev} = dH = C_P dT$, and so

$$\Delta S = \frac{q_{rev}}{T} = \int_{T_1}^{T_2} \frac{C_P}{T} dT = n\overline{C}_P \ln\left(\frac{T_2}{T_1}\right)$$

For 2.00 mol of H_2O,

$$\Delta S = (2.00\ \text{mol})(75.2\ J\cdot K^{-1}\cdot mol^{-1}) \ln \frac{363}{283} = 37.4\ J\cdot K^{-1}$$

6–14. Show that

$$\Delta \overline{S} = \overline{C}_V \ln \frac{T_2}{T_1} + R \ln \frac{V_2}{V_1}$$

if one mole of an ideal gas is taken from T_1, V_1 to T_2, V_2, assuming that $\overline{C}_V$ is independent of temperature. Calculate the value of $\Delta\overline{S}$ if one mole of $N_2(g)$ is expanded from 20.0 dm^3 at 273 K to 300 dm^3 at 400 K. Take $\overline{C}_P = 29.4\ J\cdot K^{-1}\cdot mol^{-1}$.

For the path $(T_1, V_1) \rightarrow (T_2, V_2)$, $\delta w = -PdV$ and $\delta q = dU - \delta w = C_V dT + PdV$. We can then write $\Delta\overline{S}$ as

$$\Delta\overline{S} = \frac{1}{n}\left(\int \frac{C_V}{T} dT + \int \frac{P}{T} dV\right)$$
$$= \int \frac{\overline{C}_V}{T} dT + \int \frac{R}{V} dV$$
$$= \overline{C}_V \ln \frac{T_2}{T_1} + R \ln \frac{V_2}{V_1}$$

Because $\overline{C}_P - \overline{C}_V = R$ for an ideal gas, we can write this as

$$\Delta\overline{S} = (\overline{C}_P - R) \ln \frac{T_2}{T_1} + R \ln \frac{V_2}{V_1}$$

For N_2,

$$\Delta\overline{S} = (29.4\ J\cdot mol^{-1}\cdot K^{-1} - 8.314\ J\cdot K^{-1}\cdot mol^{-1}) \ln \frac{400}{273} + 8.314\ J\cdot K^{-1}\cdot mol^{-1} \ln \frac{300}{20.0}$$
$$= 30.6\ J\cdot K^{-1}\cdot mol^{-1}$$

6–15. In this problem, we will consider a two-compartment system like that in Figure 6.4, except that the two subsystems have the same temperature but different pressures and the wall that separates them is flexible rather than rigid. Show that in this case,

$$dS = \frac{dV_B}{T}(P_B - P_A)$$

Interpret this result with regard to the sign of dV_B when $P_B > P_A$ and when $P_B < P_A$.

We can use the First Law to write δq for each compartment as

$$\delta q_A = dU_A - \delta w_A = dU_A + P_A dV_A$$
$$\delta q_B = dU_B - \delta w_B = dU_B + P_B dV_B$$

We can write the total entropy of the system as

$$dS = dS_A + dS_B = \frac{\delta q_A}{T_A} + \frac{\delta q_B}{T_B}$$
$$= \frac{dU_A}{T_A} + \frac{P_A}{T_A}dV_A + \frac{dU_B}{T_B} + \frac{P_B}{T_B}dV_B$$

Because the two-compartment system is isolated, $dV_A = -dV_B$ and $dU_A = -dU_B$. Also, $T_A = T_B$. The quantity dS above then becomes

$$dS = (P_B - P_A)\frac{dV_B}{T}$$

When $P_B > P_A$, compartment B will expand, so, because dS must be greater than zero, dV_B is positive. Likewise, when $P_B < P_A$, compartment A will expand, so (again, because dS must be greater than zero) dV_B is negative.

6–16. In this problem, we will illustrate the condition $dS_{\text{prod}} \geq 0$ with a concrete example. Consider the two-component system shown in Figure 6.8. Each compartment is in equilibrium with a heat reservoir at different temperatures T_1 and T_2, and the two compartments are separated by a rigid heat-conducting wall. The total change of energy as heat of compartment 1 is

$$dq_1 = d_e q_1 + d_i q_1$$

where $d_e q_1$ is the energy as heat exchanged with the reservoir and $d_i q_1$ is the energy as heat exchanged with compartment 2. Similarly,

$$dq_2 = d_e q_2 + d_i q_2$$

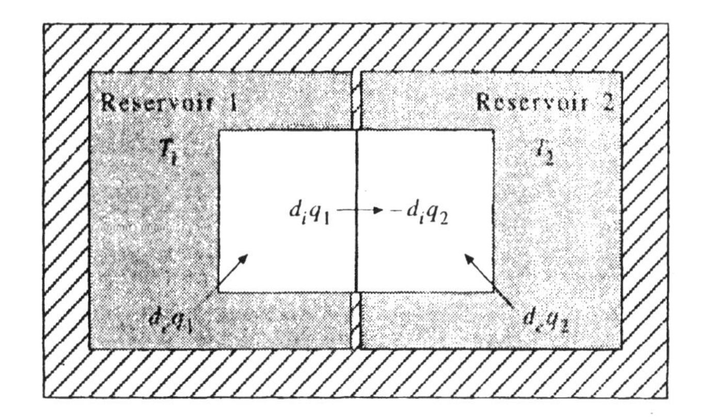

FIGURE 6.8
A two-compartment system with each compartment in contact with an (essentially infinite) heat reservoir, one at temperature T_1 and the other at temperature T_2. The two compartments are separated by a rigid heat-conducting wall.

Clearly,

$$d_i q_1 = -d_i q_2$$

Show that the entropy change for the two-compartment system is given by

$$dS = \frac{d_e q_1}{T_1} + \frac{d_e q_2}{T_2} + d_i q_1 \left(\frac{1}{T_1} - \frac{1}{T_2}\right)$$
$$= dS_{\text{exchange}} + dS_{\text{prod}}$$

where

$$dS_{\text{exchange}} = \frac{d_e q_1}{T_1} + \frac{d_e q_2}{T_2}$$

is the entropy *exchanged* with the reservoirs (surroundings) and

$$dS_{\text{prod}} = d_i q_1 \left(\frac{1}{T_1} - \frac{1}{T_2}\right)$$

is the entropy *produced* within the two-compartment system. Now show that the condition $dS_{\text{prod}} \geq 0$ implies that energy as heat flows spontaneously from a higher temperature to a lower temperature. The value of dS_{exchange}, however, has no restriction and can be positive, negative, or zero.

As stated in the text of the problem, we can write

$$d_i q_1 = -d_i q_2$$

The energy as heat exchanged between compartments 1 and 2 is involved in the entropy transferred between the two compartments. We can therefore express dS_{exchange} as

$$dS_{\text{exchange}} = \frac{d_e q_1}{T_1} + \frac{d_e q_2}{T_2}$$

Similarly, the energy as heat exchanged between the compartments and the reservoirs is involved in the entropy produced within the two-compartment system, so we can write dS_{prod} as

$$dS_{\text{prod}} = \frac{d_i q_1}{T_1} + \frac{d_i q_2}{T_2} = d_i q_1 \left(\frac{1}{T_1} - \frac{1}{T_2}\right)$$

These are the only two means of changing the entropy of the system, so we can find dS_{tot} to be

$$dS_{\text{tot}} = dS_{\text{exchange}} + dS_{\text{prod}} = \frac{d_e q_1}{T_1} + \frac{d_e q_2}{T_2} + d_i q_1 \left(\frac{1}{T_1} - \frac{1}{T_2}\right)$$

Now take the condition $dS_{\text{prod}} \geq 0$. This is the same as saying that

$$d_i q_1 \left(\frac{1}{T_1} - \frac{1}{T_2}\right) \geq 0$$

Arbitrarily, let $T_1 > T_2$. Then $1/T_1 - 1/T_2 < 0$, so $d_i q_1 < 0$ and heat is flowing from compartment 1 to compartment 2. If $T_2 > T_1$, by the same reasoning, $d_i q_1 > 0$ and heat is flowing from compartment 2 to compartment 1.

6–17. Show that

$$\Delta S \geq \frac{q}{T}$$

for an isothermal process. What does this equation say about the sign of ΔS? Can ΔS decrease in a reversible isothermal process? Calculate the entropy change when one mole of an ideal gas is compressed reversibly and isothermally from a volume of 100 dm^3 to 50.0 dm^3 at 300 K.

We defined ΔS as (Equation 6.22)

$$\Delta S \geq \int \frac{\delta q}{T}$$

For an isothermal process T is constant, so we can write this expression as

$$\Delta S \geq \frac{1}{T} \int \delta q$$

and integrate over δq to write

$$\Delta S \geq \frac{q}{T}$$

We know that q can be positive or negative, while T is always positive; therefore, ΔS can be positive or negative for an isothermal process. For one mole of an ideal gas compressed reversibly and isothermally, $dU = 0$, so $\delta q_{\text{rev}} = -\delta w = PdV$. Then

$$\delta q_{\text{rev}} = PdV = \frac{nRT}{V} dV$$

$$q_{\text{rev}} = nRT \ln \frac{V_2}{V_1}$$

and the change in entropy is given by

$$\Delta S = nR \ln \frac{V_2}{V_1} = (1 \text{ mol})(8.3145 \text{ J}\cdot\text{K}^{-1}\cdot\text{mol}^{-1}) \ln 0.5 = -5.76 \text{ J}\cdot\text{K}^{-1}$$

The quantity ΔS is equal to q/T, rather than greater than it, because this is a reversible process.

6–18. Vaporization at the normal boiling point (T_{vap}) of a substance (the boiling point at one atm) can be regarded as a reversible process because if the temperature is decreased infinitesimally below T_{vap}, all the vapor will condense to liquid, whereas if it is increased infinitesimally above T_{vap}, all the liquid will vaporize. Calculate the entropy change when two moles of water vaporize at 100.0°C. The value of $\Delta_{\text{vap}}\overline{H}$ is 40.65 kJ·mol^{-1}. Comment on the sign of $\Delta_{\text{vap}}S$.

At constant pressure and temperature, $q_{\text{rev}} = n\Delta\overline{H}_{\text{rev}}$ (Equation 5.37). We know from the previous problem that, for a reversible isothermal process,

$$\Delta S = \frac{q}{T} = \frac{(2 \text{ mol})(40.65 \text{ kJ}\cdot\text{mol}^{-1})}{373.15 \text{ K}} = 217.9 \text{ J}\cdot\text{K}^{-1}$$

As the water becomes more disordered, changing from liquid to gas, the entropy increases.

6–19. Melting at the normal melting point (T_{fus}) of a substance (the melting point at one atm) can be regarded as a reversible process because if the temperature is changed infinitesimally from exactly T_{fus}, then the substance will either melt or freeze. Calculate the change in entropy when two moles of water melt at 0°C. The value of $\Delta_{\text{fus}}\overline{H}$ is 6.01 kJ·mol^{-1}. Compare your answer with the one you obtained in Problem 6–18. Why is $\Delta_{\text{vap}}S$ much larger than $\Delta_{\text{fus}}S$?

At constant pressure, $q = n\Delta\overline{H}$ (Equation 5.37). For a reversible isothermal process, we can express ΔS as (Problem 6.17)

$$\Delta S = \frac{q}{T} = \frac{(2\text{ mol})(6.01\text{ kJ}\cdot\text{mol}^{-1})}{273.15\text{ K}} = 44.0\text{ J}\cdot\text{K}^{-1}$$

The quantity $\Delta_{\text{fus}}S$ is much less than $\Delta_{\text{vap}}S$ because the difference in disorder between a solid and a liquid is much less than that between a liquid and a gas.

6–20. Consider a simple example of Equation 6.23 in which there are only two states, 1 and 2. Show that $W(a_1, a_2)$ is a maximum when $a_1 = a_2$. *Hint*: Consider ln W, use Stirling's approximation, and treat a_1 and a_2 as continuous variables.

A simplified version of Equation 6.23, for two states only, is

$$W = \frac{(a_1 + a_2)!}{a_1!a_2!}$$

Let $a_1 = x$, with x being a continuous variable, and let $a_1 + a_2 = N$. Then we can express W in terms of N and x:

$$W = \frac{N!}{(N-x)!x!}$$
$$\ln W = \ln N! - \ln(N-x)! - \ln x!$$

Using Stirling's approximation for ln $N!$ (MathChapter E), we can find the first derivative of ln W with respect to x:

$$\begin{aligned}\ln W &= \ln N! - \ln(N-x)! - \ln x! \\ &= N\ln N - N - [(N-x)\ln(N-x) - (N-x)] - (x\ln x - x) \\ \frac{d(\ln W)}{dx} &= 0 + \frac{N-x}{N-x} + \ln(N-x) - 1 - \ln x - \frac{x}{x} + 1 \\ &= \ln(N-x) - \ln x\end{aligned}$$

We are looking for the value of x that produces the maximum value of W, which is where $dW/dx = 0$. Because

$$\frac{1}{W}\frac{dW}{dx} = \frac{d(\ln W)}{dx}$$

the desired value of x will also give $[d(\ln W)/dx] = 0$. Setting this derivative equal to zero, we find

$$\begin{aligned}0 &= \ln(N - x_{\text{max}}) - \ln x_{\text{max}} \\ \ln x_{\text{max}} &= \ln(N - x_{\text{max}}) \\ 2x_{\text{max}} &= N\end{aligned}$$

And clearly $x_{\text{max}} = N/2$. Thus, the maximum value of W occurs when $a_1 = a_2 = N/2$.

6–21. Extend Problem 6–20 to the case of three states. Do you see how to generalize it to any number of states?

For the case of three states, Equation 6.23 becomes

$$W = \frac{(a_1 + a_2 + a_3)!}{a_1! a_2! a_3!}$$

Let $a_1 = x$ and $a_2 = y$, where x and y are continuous variables, and let $a_1 + a_2 + a_3 = A$. Then

$$W = \frac{A!}{(A - x - y)! x! y!}$$
$$\ln W = \ln A! - \ln(A - x - y)! - \ln x! - \ln y!$$

We can use Stirling's approximation for $\ln A!$ (MathChapter E) to write $\ln W$ and differentiate to find an expression for $d(\ln W)/dx$:

$$\ln W = A \ln A - A - [(A - x - y)\ln(A - x - y) - (A - x - y)] - (x \ln x - x) - (y \ln y - y)$$
$$\frac{d(\ln W)}{dx} = \ln(A - x - y) + 1 - 1 - \ln x - 1 + 1 = \ln(A - x - y) - \ln x$$

Similarly,

$$\frac{d(\ln W)}{dy} = \ln(A - x - y) - \ln y$$

As in the previous problem, the values of x and y which give the largest values of W occur where $d(\ln W)/dx = 0$ and $d(\ln W)/dy = 0$. Therefore,

$$0 = \ln(A - x_{max} - y) - \ln x_{max}$$
$$\ln x_{max} = \ln(A - x_{max} - y)$$
$$2x_{max} = A - y$$

Because we want the point at which both x and y are at their maxima, we substitute this value into the expression for $d(\ln W)/dy$ to find

$$0 = \ln(A - x_{max} - y_{max}) - \ln y_{max}$$
$$2y_{max} = \frac{A}{2} + \frac{y_{max}}{2}$$
$$y_{max} = \frac{A}{3}$$

Then (substituting back into the first equality) $x_{max} = A/3$. Thus, the maximum value of W occurs when $a_1 = a_2 = a_3 = A/3$. Problem E.10 generalizes this to any number of states.

6–22. Show that the system partition function can be written as a summation over levels by writing

$$Q(N, V, T) = \sum_E \Omega(N, V, E) e^{-E/k_B T}$$

Now consider the case of an isolated system, for which there is only one term in $Q(N, V, T)$. Now substitute this special case for Q into Equation 6.43 to derive the equation $S = k_B \ln \Omega$.

In Chapter 3, we defined the partition function $Q(N, V, T)$ as (Equation 3.14)

$$Q(N, V, T) = \sum_j e^{-E_j/k_B T}$$

Here a term representing an energy level of degeneracy Ω is written Ω times. We can write this, alternatively, as a sum over energy levels, where a term representing an energy level is written once and multiplied by its degeneracy Ω:

$$Q(N, V, T) = \sum_E \Omega(N, V, E)e^{-E/k_BT}$$

These two expressions are equivalent. For an isolated system, there will be only one term in $Q(N, V, T)$, so

$$Q = \Omega(N, V, E)e^{-E/k_BT}$$
$$\ln Q = \ln \Omega - \left(\frac{E}{k_BT}\right)$$

Applying Equation 6.43 allows us to write

$$\begin{aligned} S &= k_BT\left(\frac{\partial \ln Q}{\partial T}\right)_{N,V} + k_B \ln Q \\ &= k_BT\left(\frac{E}{k_BT^2}\right) + k_B \ln \Omega - \frac{E}{T} \\ &= k_B \ln \Omega \end{aligned}$$

which is Boltzmann's equation.

6–23. In this problem, we will show that $\Omega = c(N)f(E)V^N$ for an ideal gas (Example 6–3). In Problem 4–40 we showed that the number of translational energy states between ε and $\varepsilon + \Delta\varepsilon$ for a particle in a box can be calculated by considering a sphere in n_x, n_y, n_z space,

$$n_x^2 + n_y^2 + n_z^2 = \frac{8ma^2\varepsilon}{h^2} = R^2$$

Show that for an N-particle system, the analogous expression is

$$\sum_{j=1}^{N}(n_{xj}^2 + n_{yj}^2 + n_{zj}^2) = \frac{8ma^2E}{h^2} = R^2$$

or, in more convenient notation

$$\sum_{j=1}^{3N} n_j^2 = \frac{8ma^2E}{h^2} = R^2$$

Thus, instead of dealing with a three-dimensional sphere as we did in Problem 4–40, here we must deal with a $3N$-dimensional sphere. Whatever the formula for the volume of a $3N$-dimensional sphere is (it is known), we can at least say that it is proportional to R^{3N}. Show that this proportionality leads to the following expression for $\Phi(E)$, the number of states with energy $\leq E$,

$$\Phi(E) \propto \left(\frac{8ma^2E}{h^2}\right)^{3N/2} = c(N)E^{3N/2}V^N$$

where $c(N)$ is a constant whose value depends upon N and $V = a^3$. Now, following the argument developed in Problem 4–40, show that the number of states between E and $E + \Delta E$ (which is essentially Ω) is given by

$$\Omega = c(N)f(E)V^N\Delta E$$

where $f(E) = E^{\frac{3N}{2}-1}$.

For an N-particle system, we wish to consider all the N particles in one $3N$-dimensional space, instead of the N particles in N individual three-dimensional spaces. The equation from Problem 4–40 then becomes

$$\sum_{j=1}^{3N} n_j^2 = \frac{8ma^2E}{h^2} = R^2$$

As in Problem 4–40, $\Phi(E) \propto$ the volume of the sphere, so

$$\Phi(E) \propto R^{3N} = \left(\frac{8ma^2E}{h^2}\right)^{3N/2}$$

Letting $c(N)$ be a proportionality constant allows us to write

$$\Phi(E) = c(N)E^{3N/2}V^N$$

Now, as in Problem 4–40, $\Omega = \Phi(E + \Delta E) - \Phi(E)$, so

$$\begin{aligned}
\Omega &= c(N)(E + \Delta E)^{3N/2}V^N - c(N)E^{3N/2}V^N \\
&= c(N)V^N\left[(E + \Delta E)^{3N/2} - E^{3N/2}\right] \\
&= c(N)V^N E^{3N/2}\left[\left(1 + \frac{\Delta E}{E}\right)^{3N/2} - 1\right] \\
&= c(N)V^N E^{3N/2}\left\{1 + \frac{3N}{2}\frac{\Delta E}{E} + O\left[\left(\frac{\Delta E}{E}\right)^2\right] - 1\right\} \\
&= \frac{3N}{2}c(N)V^N E^{3N/2}\frac{\Delta E}{E} + O\left[\left(\frac{\Delta E}{E}\right)^2\right] \\
&= c(N)V^N E^{\frac{3N}{2}-1}\Delta E = c(N)f(E)V^N\Delta E
\end{aligned}$$

where we have incorporated the factor of $3N/2$ into $c(N)$ and defined $f(E) = E^{\frac{3N}{2}-1}$.

6–24. Show that if a process involves only an isothermal transfer of energy as heat (*pure heat transfer*), then

$$dS_{\text{sys}} = \frac{dq}{T} \qquad \text{(pure heat transfer)}$$

The process involves only an isothermal transfer of energy as heat, so $\delta w = 0$ and $dU = \delta q$. Therefore,

$$dS_{\text{sys}} = \frac{\delta q}{T} = \frac{dU}{T} = \frac{dq}{T}$$

where we can write dq instead of δq because $\delta q = dU$, and U is a state function.

6–25. Calculate the change in entropy of the system and of the surroundings and the total change in entropy if one mole of an ideal gas is expanded isothermally and reversibly from a pressure of 10.0 bar to 2.00 bar at 300 K.

Because this is an isothermal reversible expansion, $\delta q = -\delta w = PdV$. We then use the ideal gas equation to write

$$\Delta S_{\text{sys}} = \int \frac{\delta q_{\text{rev}}}{T} = \int \frac{P}{T} dV = \int \frac{nR}{V} dV = nR \ln \frac{V_2}{V_1}$$

For an isothermal expansion of an ideal gas, $P_1 V_1 = P_2 V_2$. We can then write the change of entropy of the gas as

$$\Delta S_{\text{sys}} = (1 \text{ mol})(8.3145 \text{ J}\cdot\text{mol}^{-1}\cdot\text{K}^{-1}) \ln 5.00 = +13.4 \text{ J}\cdot\text{K}^{-1}$$

For a reversible expansion, $\Delta S_{\text{tot}} = 0$, so $\Delta S_{\text{surr}} = -\Delta S_{\text{sys}} = -13.4 \text{ J}\cdot\text{K}^{-1}$.

6–26. Redo Problem 6–25 for an expansion into a vacuum, with an initial pressure of 10.0 bar and a final pressure of 2.00 bar.

As in Problem 6–25, $\Delta S_{\text{sys}} = 13.4 \text{ J}\cdot\text{K}^{-1}$. However, because this is an irreversible expansion into a vacuum, $\Delta S_{\text{surr}} = 0$, so $\Delta S_{\text{tot}} = 13.4 \text{ J}\cdot\text{K}^{-1}$.

6–27. The molar heat capacity of 1-butene can be expressed as

$$\overline{C}_P(T)/R = 0.05641 + (0.04635 \text{ K}^{-1})T - (2.392 \times 10^{-5} \text{ K}^{-2})T^2 + (4.80 \times 10^{-9} \text{ K}^{-3})T^3$$

over the temperature range $300 \text{ K} < T < 1500 \text{ K}$. Calculate the change in entropy when one mole of 1-butene is heated from 300 K to 1000 K at constant pressure.

At constant pressure, $\delta q = dH = n\overline{C}_P dT$. Then Equation 6.22 becomes (assuming a reversible process)

$$\begin{aligned}
\Delta S &= n \int_{300}^{1000} \frac{\overline{C}_P}{T} dT \\
&= (1 \text{ mol})R \int_{300}^{1000} \left[0.05641T^{-1} + (0.04635 \text{ K}^{-1}) - (2.392 \times 10^{-5} \text{ K}^{-2})T \right. \\
&\qquad \left. + (4.80 \times 10^{-9} \text{ K}^{-3})T^2\right] dT \\
&= 192.78 \text{ J}\cdot\text{K}^{-1}
\end{aligned}$$

6–28. Plot $\Delta_{\text{mix}}\overline{S}$ against y_1 for the mixing of two ideal gases. At what value of y_1 is $\Delta_{\text{mix}}\overline{S}$ a maximum? Can you give a physical interpretation of this result?

We can use Equation 6.30 for two gases:

$$\Delta_{\text{mix}}\overline{S} = -R\left(y_1 \ln y_1 - y_2 \ln y_2\right)$$

Because $y_1 + y_2 = 1$, we can write this as

$$\Delta_{\text{mix}}\overline{S}/R = -y_1 \ln y_1 - (1 - y_1) \ln(1 - y_1)$$

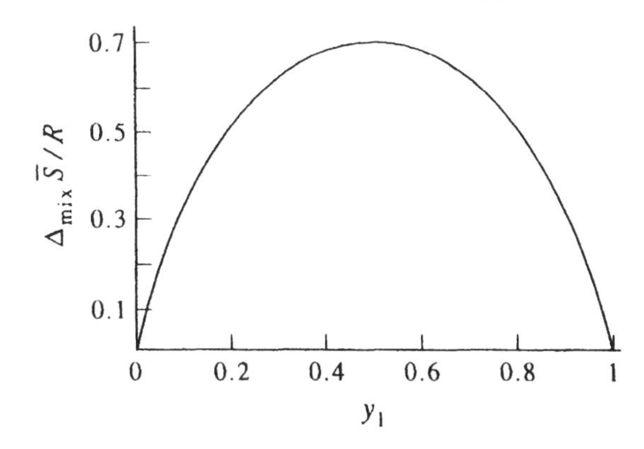

The quantity $\Delta_{mix}\overline{S}$ is a maximum when $y_1 = 0.5$. This means that the gases are most disordered when there are equal amounts of both present in a container.

6–29. Calculate the entropy of mixing if two moles of $N_2(g)$ are mixed with one mole $O_2(g)$ at the same temperature and pressure. Assume ideal behavior.

The mole fractions y are 2/3 for $N_2(g)$ and 1/3 for $O_2(g)$. Therefore,

$$\begin{aligned}\Delta_{mix}\overline{S} &= -Ry_1 \ln y_1 - Ry_2 \ln y_2 \\ &= -\frac{2R}{3}\ln\frac{2}{3} - \frac{R}{3}\ln\frac{1}{3} = 5.29\ \text{J}\cdot\text{K}^{-1}\end{aligned}$$

6–30. Show that $\Delta_{mix}\overline{S} = R\ln 2$ if equal volumes of any two ideal gases under the same conditions are mixed.

Because there are equal volumes of ideal gases under the same conditions, $y_1 = y_2 = 0.5$ (Problem 6–28). Now Equation 6.30 gives

$$\begin{aligned}\Delta_{mix}\overline{S} &= -Ry_1 \ln y_1 - Ry_2 \ln y_2 \\ &= -\frac{R}{2}\ln\frac{1}{2} - \frac{R}{2}\ln\frac{1}{2} \\ &= -R\ln\frac{1}{2} = R\ln 2\end{aligned}$$

6–31. Derive the equation $dU = TdS - PdV$. Show that

$$d\overline{S} = \overline{C}_V\frac{dT}{T} + R\frac{d\overline{V}}{\overline{V}}$$

for one mole of an ideal gas. Assuming that $\overline{C}_V$ is independent of temperature, show that

$$\Delta\overline{S} = \overline{C}_V\ln\frac{T_2}{T_1} + R\ln\frac{\overline{V}_2}{\overline{V}_1}$$

for the change from $T_1, \overline{V}_1$ to $T_2, \overline{V}_2$. Note that this equation is a combination of Equations 6.28 and 6.31.

From the definition of entropy, $\delta q_{\text{rev}} = TdS$, and by definition $\delta w_{\text{rev}} = -PdV$. The first law then gives $dU = \delta q_{\text{rev}} + \delta w_{\text{rev}} = TdS - PdV$. We divide through by T to obtain

$$dS = \frac{dU}{T} - \frac{P}{T}dV$$

Using the relation $dU = n\overline{C}_V dT$ and the ideal gas law, we find

$$dS = n\overline{C}_V \frac{dT}{T} - nR\frac{dV}{V}$$

or

$$d\overline{S} = \overline{C}_V \frac{dT}{T} - R\frac{d\overline{V}}{\overline{V}}$$

If $\overline{C}_V$ is temperature-independent, integrating gives

$$\int d\overline{S} = \overline{C}_V \int \frac{dT}{T} + R\int \frac{d\overline{V}}{\overline{V}}$$

$$\Delta\overline{S} = \overline{C}_V \ln\frac{T_2}{T_1} + R\ln\frac{\overline{V}_2}{\overline{V}_1}$$

6–32. Derive the equation $dH = TdS + VdP$. Show that

$$\Delta\overline{S} = \overline{C}_P \ln\frac{T_2}{T_1} - R\ln\frac{P_2}{P_1}$$

for the change of one mole of an ideal gas from T_1, P_1 to T_2, P_2, assuming that $\overline{C}_P$ is independent of temperature.

We derived the equation $dU = TdS - PdV$ in Problem 6–31. Now add $d(PV)$ to both sides of this equation to obtain

$$dU + d(PV) = dH = TdS + VdP$$

Now divide both sides by T to write

$$dS = \frac{dH}{T} - \frac{V}{T}dP$$

and use the relation $dH = n\overline{C}_P dT$ and the ideal gas law to obtain

$$dS = \frac{n\overline{C}_P}{T}dT - \frac{nR}{P}dP$$

or

$$d\overline{S} = \frac{\overline{C}_P}{T}dT - \frac{R}{P}dP$$

Assuming that $\overline{C}_P$ is temperature-independent, integrating gives

$$\Delta\overline{S} = \overline{C}_P \ln\frac{T_2}{T_1} - R\ln\frac{P_2}{P_1}$$

6–33. Calculate the change in entropy if one mole of $SO_2(g)$ at 300 K and 1.00 bar is heated to 1000 K and its pressure is decreased to 0.010 bar. Take the molar heat capacity of $SO_2(g)$ to be

$$\overline{C}_P(T)/R = 7.871 - \frac{1454.6\text{ K}}{T} + \frac{160\,351\text{ K}^2}{T^2}$$

We can use the result of the previous problem,

$$dS = \frac{n\overline{C}_P dT}{T} - \frac{nR\,dP}{P}$$

Then

$$\int dS = nR\int \frac{\overline{C}_P}{T}dT - nR\int \frac{dP}{P}$$

$$= (1.00\text{ mol})(8.314\text{ J}\cdot\text{mol}^{-1}\cdot\text{K}^{-1})\left[\int_{300\text{ K}}^{1000\text{ K}}\left(\frac{7.871}{T} - \frac{1454.6\text{ K}}{T^2} + \frac{160\,351\text{ K}^2}{T^3}\right)dT - \ln\frac{0.010}{1.00}\right]$$

$$\Delta S = 95.6\text{ J}\cdot\text{K}^{-1}$$

6–34. In the derivation of Equation 6.32, argue that $\Delta S_c > 0$ and $\Delta S_h < 0$. Now show that

$$\Delta S = \Delta S_c + \Delta S_h > 0$$

by showing that

$$\Delta S_c - |\Delta S_h| > 0$$

The two quantities $\Delta S_c > 0$ and $\Delta S_h < 0$ because the colder piece will become hotter and the hotter piece will become colder. Using the expressions for ΔS_c and ΔS_h in Section 6–6,

$$\Delta S_c = C_V \ln\frac{T_h + T_c}{2T_c} > 0 \quad \text{and} \quad \Delta S_h = C_V \ln\frac{T_h + T_c}{2T_h} < 0$$

Now, because $\Delta S_h < 0$, $|\Delta S_h| = -\Delta S_h$, and

$$|\Delta S_h| = C_V \ln\frac{2T_h}{T_h + T_c} > 0$$

The total change in entropy is given by

$$\Delta S = \Delta S_c + \Delta S_h = \Delta S_c - |\Delta S_h|$$

$$= C_V \ln\frac{(T_h + T_c)^2}{4T_hT_c} > 0$$

where we proved that $(T_h + T_c)^2 > 4T_hT_c$ in Section 6–6.

6–35. We can use the equation $S = k\ln W$ to derive Equation 6.28. First, argue that the probability that an ideal-gas molecule is found in a subvolume V_s of some larger volume V is V_s/V. Because the molecules of an ideal gas are independent, the probability that N ideal-gas molecules are found in V_s is $(V_s/V)^N$. Now show that the change in entropy when the volume of one mole of an ideal gas changes isothermally from V_1 to V_2 is

$$\Delta S = R \ln \frac{V_2}{V_1}$$

The probability that the molecule is in subvolume V_s is V_s/V, because the numerator represents the situations where the molecule is in V_s and the denominator represents all positions available to the molecule. Now we can write (using Boltzmann's equation)

$$S = k_B \ln W = k_B \ln \left(\frac{V_s}{V}\right)^N = R \ln \frac{V_s}{V}$$

Now take $V_s = V_1$ and an arbitrary V_2 and V. The change in entropy when an ideal gas goes from V_1 to V_2 isothermally is then

$$\Delta S = R\left(\ln \frac{V_2}{V} - \ln \frac{V_1}{V}\right) = R \ln \frac{V_2}{V_1}$$

6–36. The relation $n_j \propto e^{-\varepsilon_j/k_BT}$ can be derived by starting with $S = k \ln W$. Consider a gas with n_0 molecules in the ground state and n_j in the jth state. Now add an energy $\varepsilon_j - \varepsilon_0$ to this system so that a molecule is promoted from the ground state to the jth state. If the volume of the gas is kept constant, then no work is done, so $dU = dq$,

$$dS = \frac{dq}{T} = \frac{dU}{T} = \frac{\varepsilon_j - \varepsilon_0}{T}$$

Now, assuming that n_0 and n_j are large, show that

$$dS = k_B \ln \left\{\frac{N!}{(n_0 - 1)!n_1!\cdots(n_j + 1)!\cdots}\right\} - k_B \ln \left\{\frac{N!}{n_0!n_1!\cdots n_j!\cdots}\right\}$$

$$= k_B \ln \left\{\frac{n_j!}{(n_j + 1)!}\frac{n_0!}{(n_0 - 1)!}\right\} = k \ln \frac{n_0}{n_j}$$

Equating the two expressions for dS, show that

$$\frac{n_j}{n_0} = e^{-(\varepsilon_j - \varepsilon_0)/k_BT}$$

From the problem,

$$dS = \frac{\varepsilon_j - \varepsilon_0}{T}$$

Recall that (Equation 6.24) $S = k_B \ln W$. For the initial state,

$$S_{\text{initial}} = k_B \ln W = k_B \ln \left[\frac{N!}{n_0!n_1!\cdots n_j!\cdots}\right]$$

and for the final state

$$S_{\text{final}} = k_B \ln W = k_B \ln \left[\frac{N!}{(n_0 - 1)!n_1!\cdots(n_j + 1)!\cdots}\right]$$

Then

$$\begin{aligned} dS &= S_{\text{final}} - S_{\text{initial}} \\ &= k_B \ln\left[\frac{N!}{(n_0-1)!n_1!\cdots(n_j+1)!\cdots}\right] - k_B \ln\left[\frac{N!}{n_0!n_1!\cdots n_j!\cdots}\right] \\ &= k_B \ln\left[\frac{n_j!}{(n_j+1)!}\frac{n_0!}{(n_0-1)!}\right] = k_B \ln\frac{n_0}{n_j+1} \\ &= k_B \ln\frac{n_0}{n_j} \end{aligned}$$

where the last equality holds because $n_j \gg 1$. Equating the two expressions for dS, we find that

$$\begin{aligned} k_B \ln\frac{n_0}{n_j} &= \frac{\varepsilon_j - \varepsilon_0}{T} \\ \ln\frac{n_j}{n_0} &= -\frac{\varepsilon_j - \varepsilon_0}{k_B T} \\ \frac{n_j}{n_0} &= e^{-(\varepsilon_j-\varepsilon_0)/k_B T} \end{aligned}$$

6–37. We can use Equation 6.24 to calculate the probability of observing fluctuations from the equilibrium state. Show that

$$\frac{W}{W_{\text{eq}}} = e^{-\Delta S/k_B}$$

where W represents the nonequilibrium state and ΔS is the entropy difference between the two states. We can interpret the ratio W/W_{eq} as the probability of observing the nonequilibrium state. Given that the entropy of one mole of oxygen is $205.0\ \text{J}\cdot\text{K}^{-1}\cdot\text{mol}^{-1}$ at 25°C and one bar, calculate the probability of observing a decrease in entropy that is one millionth of a percent of this amount.

We can use Equation 6.24 to write

$$S_{\text{eq}} = k_B \ln W_{\text{eq}} \qquad \text{and} \qquad S = k_B \ln W$$

Then ΔS is

$$\begin{aligned} \Delta S = S - S_{\text{eq}} &= k_B(\ln W - \ln W_{\text{eq}}) \\ &= k_B \ln\frac{W}{W_{\text{eq}}} \\ -\frac{\Delta S}{k_B} &= \ln\frac{W}{W_{\text{eq}}} \\ e^{-\Delta S/k_B} &= \frac{W}{W_{\text{eq}}} \end{aligned}$$

The probability of observing a ΔS which is one millionth of one percent of $205\ \text{J}\cdot\text{K}^{-1}\cdot\text{mol}^{-1}$ is

$$\exp\left[-\frac{(1\ \text{mol})(1.00\times10^{-8})(205.0\ \text{J}\cdot\text{K}^{-1}\cdot\text{mol}^{-1})}{1.381\times10^{-23}\ \text{J}\cdot\text{K}^{-1}}\right] = \exp[-1.485\times10^{17}] \approx 0$$

6–38. Consider one mole of an ideal gas confined to a volume V. Calculate the probability that all the N_A molecules of this ideal gas will be found to occupy one half of this volume, leaving the other half empty.

From Problem 6–35, we can write the probability as

$$\left(\frac{\frac{1}{2}V}{V}\right)^{N_A} = \left(\frac{1}{2}\right)^{N_A} \approx 0$$

6–39. Show that S_{system} given by Equation 6.40 is a maximum when all the p_j are equal. Remember that $\sum p_j = 1$, so that

$$\sum_j p_j \ln p_j = p_1 \ln p_1 + p_2 \ln p_2 + \cdots + p_{n-1} \ln p_{n-1} + (1 - p_1 - p_2 - \cdots - p_{n-1}) \ln(1 - p_1 - p_2 - \cdots - p_{n-1})$$

See also Problem E–10.

Begin with Equation 6.40,

$$S_{system} = -k_B \sum_j p_j \ln p_j$$

Substituting the expression given for $\sum_j p_j \ln p_j$, we find

$$S_{system} = -k_B \left[p_1 \ln p_1 + p_2 \ln p_2 + \cdots + p_{n-1} \ln p_{n-1} + (1 - p_1 - p_2 - \cdots - p_{n-1}) \ln(1 - p_1 - p_2 - \cdots - p_{n-1})\right]$$

$$\frac{\partial S_{system}}{\partial p_j} = \ln p_j + 1 - \ln(1 - p_1 - \cdots - p_{j-1} - p_{j+1} - \cdots - p_{n-1}) - 1$$

$$0 = \ln p_j - \ln(1 - p_1 - \cdots - p_{j-1} - p_{j+1} - \cdots - p_{n-1})$$

$$p_j = 1 - p_1 - \cdots - p_{j-1} - p_{j+1} - \cdots - p_{n-1}$$

Because p_j can be any of p_1 to p_{n-1}, and the above equality holds for all p_j, all the p_j must be equal.

6–40. Use Equation 6.45 to calculate the molar entropy of krypton at 298.2 K and one bar, and compare your result with the experimental value of $164.1\ \text{J}\cdot\text{K}^{-1}\cdot\text{mol}^{-1}$.

This problem is like Example 6–6. We use Equation 6.45,

$$\overline{S} = \frac{5}{2}R + R \ln\left[\left(\frac{2\pi m k_B T}{h^2}\right)^{3/2} \frac{\overline{V}}{N_A}\right]$$

Assuming ideal behavior, at 298.2 K and one bar

$$\frac{N_A}{\overline{V}} = \frac{N_A P}{RT}$$

$$= \frac{(6.022 \times 10^{23}\ \text{mol}^{-1})(1\ \text{bar})}{(0.08314\ \text{dm}^3\cdot\text{bar}\cdot\text{mol}^{-1}\cdot\text{K}^{-1})(298.2\ \text{K})}$$

$$= 2.429 \times 10^{22}\ \text{dm}^{-3} = 2.429 \times 10^{25}\ \text{m}^{-3}$$

and

$$\left(\frac{2\pi mk_BT}{h^2}\right)^{3/2} = \left[\frac{2\pi(0.08380\text{ kg}\cdot\text{mol}^{-1})(1.3806\times10^{-23}\text{ J}\cdot\text{K}^{-1})(298.2\text{ K})}{(6.022\times10^{23}\text{ mol}^{-1})(6.626\times10^{-34}\text{ J}\cdot\text{s})^2}\right]^{3/2}$$
$$= (8.199\times10^{21}\text{ m}^{-2})^{3/2} = 7.424\times10^{32}\text{ m}^{-3}$$

Then

$$\overline{S} = \frac{5}{2}R + 17.235R$$
$$= 164.1\text{ J}\cdot\text{K}^{-1}\cdot\text{mol}^{-1}$$

This value is the same as the experimental value.

6–41. Use Equation 4.39 and the data in Table 4.2 to calculate the entropy of nitrogen at 298.2 K and one bar. Compare your result with the experimental value of 191.6 $\text{J}\cdot\text{K}^{-1}\cdot\text{mol}^{-1}$.

Recall from Chapter 4 that

$$Q = \frac{q^N}{N!}$$

Substituting into Equation 6.43 gives

$$S = Nk_B\ln q - k_B\ln N! + Nk_BT\left(\frac{\partial \ln q}{\partial T}\right)_V$$
$$= Nk_B\ln q - Nk_B\ln N + Nk_B + Nk_BT\left(\frac{\partial \ln q}{\partial T}\right)_V$$
$$= Nk_B + Nk_B\ln\frac{q}{N} + Nk_BT\left(\frac{\partial \ln q}{\partial T}\right)_V \qquad (1)$$

For a diatomic ideal gas,

$$q = \left(\frac{2\pi Mk_BT}{h^2}\right)^{3/2} V\frac{T}{\sigma\Theta_{\text{rot}}}\frac{e^{-\Theta_{\text{vib}}/2T}}{1-e^{-\Theta_{\text{vib}}/T}}g_{e1}e^{D_e/k_BT} \qquad (4.39)$$

Then

$$\ln\frac{q}{N} = \ln\left[\left(\frac{2\pi Mk_BT}{h^2}\right)^{3/2}\frac{\overline{V}}{N_A}\right] + \ln\frac{T}{\sigma\Theta_{\text{rot}}} - \ln(1-e^{-\Theta_{\text{vib}}/T})$$
$$-\frac{\Theta_{\text{vib}}}{2T} + \ln g_{e1} + \frac{D_e}{k_BT}$$

and

$$\left(\frac{\partial \ln q}{\partial T}\right) = \frac{3}{2T} + \frac{1}{T} + \frac{\Theta_{\text{vib}}}{2T^2} + \frac{(\Theta_{\text{vib}}/T^2)e^{-\Theta_{\text{vib}}/T}}{1-e^{-\Theta_{\text{vib}}/T}} - \frac{D_e}{k_BT^2}$$

Substituting into Equation 1 above, we find that

$$\frac{\overline{S}}{R} = \frac{7}{2} + \ln\left[\left(\frac{2\pi Mk_BT}{h^2}\right)^{3/2}\frac{\overline{V}}{N_A}\right] + \ln\frac{T}{\sigma\Theta_{\text{rot}}} - \ln(1-e^{-\Theta_{\text{vib}}/T})$$
$$+\frac{\Theta_{\text{vib}}/T}{e^{\Theta_{\text{vib}}/T}-1} + \ln g_{e1}$$

For N_2, $\Theta_{vib} = 3374$ K, $\Theta_{rot} = 2.88$ K, $\sigma = 2$, and $g_{e1} = 1$. The various factors are as follows:

$$\left(\frac{2\pi M k_B T}{h^2}\right)^{3/2} = \left[\frac{2\pi(4.653\times10^{-26}\text{ kg})(1.381\times10^{-23}\text{ J}\cdot\text{K}^{-1})(298.2\text{ K})}{(6.626\times10^{34}\text{ J}\cdot\text{s})^2}\right]^{3/2}$$
$$= 1.435\times10^{32}\text{ m}^{-3}$$

$$\frac{\overline{V}}{N_A} = \frac{RT}{N_A P} = \frac{(0.08314\text{ dm}^3\cdot\text{bar}\cdot\text{mol}^{-1}\cdot\text{K}^{-1})(298.2\text{ K})}{(6.022\times10^{23}\text{ mol}^{-1})(1\text{ bar})}$$
$$= 4.117\times10^{-23}\text{ dm}^3 = 4.117\times10^{-26}\text{ m}^3$$

$$\ln\frac{T}{\sigma\Theta_{rot}} = \ln\frac{298.2\text{ K}}{2(2.88\text{ K})} = 3.947$$

$$\ln(1-e^{-\Theta_{vib}/T}) = -1.22\times10^{-5}$$

$$\frac{\Theta_{vib}/T}{e^{\Theta_{vib}/T}-1} = \frac{11.31}{e^{11.31}-1} = 1.380\times10^{-4}$$

The standard molar entropy is then

$$\frac{\overline{S}}{R} = 3.5 + 15.59 + 3.947 + 1.22\times10^{-5} + 1.380\times10^{-4}$$
$$= 23.04$$

This is 191.6 $\text{J}\cdot\text{K}^{-1}\cdot\text{mol}^{-1}$, which is also the experimental value.

6–42. Use Equation 4.57 and the data in Table 4.4 to calculate the entropy of CO_2(g) at 298.2 K and one bar. Compare your result with the experimental value of 213.8 $\text{J}\cdot\text{K}^{-1}\cdot\text{mol}^{-1}$.

For a linear polyatomic ideal gas having three atoms,

$$q = \left(\frac{2\pi M k_B T}{h^2}\right)^{3/2} V \frac{T}{\sigma\Theta_{rot}}\left(\prod_{j=1}^{4}\frac{e^{-\Theta_{vib,j}/2T}}{1-e^{-\Theta_{vib,j}/T}}\right) g_{e1} e^{D_e/k_BT} \qquad (4.57)$$

Then

$$\ln\frac{q}{N} = \ln\left[\left(\frac{2\pi M k_B T}{h^2}\right)^{3/2}\frac{\overline{V}}{N_A}\right] + \ln\frac{T}{\sigma\Theta_{rot}} - \sum_{j=1}^{4}\ln(1-e^{-\Theta_{vib,j}/T})$$
$$-\sum_{j=1}^{4}\frac{\Theta_{vib,j}}{2T} + \ln g_{e1} + \frac{D_e}{k_BT}$$

and

$$\left(\frac{\partial \ln q}{\partial T}\right) = \frac{3}{2T} + \frac{1}{T} + \sum_{j=1}^{4}\frac{\Theta_{vib,j}}{2T^2} + \sum_{j=1}^{4}\frac{(\Theta_{vib,j}/T^2)e^{-\Theta_{vib,j}/T}}{1-e^{-\Theta_{vib,j}/T}} - \frac{D_e}{k_BT^2}$$

Substituting into Equation 1 from Problem 6–41, we find

$$\frac{\overline{S}}{R} = \frac{7}{2} + \ln\left[\left(\frac{2\pi M k_B T}{h^2}\right)^{3/2}\frac{\overline{V}}{N_A}\right] + \ln\frac{T}{\sigma\Theta_{rot}} - \sum_{j=1}^{4}\ln(1-e^{-\Theta_{vib,j}/T})$$
$$+\sum_{j=1}^{4}\left[\frac{(\Theta_{vib,j}/T)e^{-\Theta_{vib,j}/T}}{1-e^{-\Theta_{vib,j}/T}}\right] + \ln g_{e1}$$

For CO_2, $\Theta_{vib,1} = 3360$ K, $\Theta_{vib,2} = \Theta_{vib,3} = 954$ K, $\Theta_{vib,4} = 1890$ K, $\Theta_{rot} = 0.561$ K, $\sigma = 2$, and $g_{e1} = 1$. The various factors are as follows:

$$\left(\frac{2\pi Mk_BT}{h^2}\right)^{3/2} = \left[\frac{2\pi(7.308\times10^{-26}\text{ kg})(1.381\times10^{-23}\text{ J}\cdot\text{K}^{-1})(298.2\text{ K})}{(6.626\times10^{34}\text{ J}\cdot\text{s})^2}\right]^{3/2}$$
$$= 2.825\times10^{32}\text{ m}^{-3}$$

$$\frac{\overline{V}}{N_A} = \frac{RT}{N_AP} = \frac{(0.08314\text{ dm}^3\cdot\text{bar}\cdot\text{mol}^{-1}\cdot\text{K}^{-1})(298.2\text{ K})}{(6.022\times10^{23}\text{ mol}^{-1})(1\text{ bar})}$$
$$= 4.117\times10^{-23}\text{ dm}^3 = 4.117\times10^{-26}\text{ m}^3$$

$$\ln\frac{T}{\sigma\Theta_{rot}} = \ln\left[\frac{298.2\text{ K}}{2(0.561\text{ K})}\right] = 5.583$$

$$\sum_{j=1}^{4}\ln(1-e^{-\Theta_{vib,j}/T}) = -0.08508$$

$$\sum_{j=1}^{4}\frac{\Theta_{vib,j}}{T}\left(\frac{e^{-\Theta_{vib,j}/T}}{1-e^{-\Theta_{vib,j}/T}}\right) = 0.2835$$

The standard molar entropy is then

$$\frac{\overline{S}}{R} = 3.5 + 16.269 + 5.583 + 0.08508 + 0.2835$$
$$= 25.72$$

This is 213.8 $\text{J}\cdot\text{K}^{-1}\cdot\text{mol}^{-1}$, which is also the experimental value.

6–43. Use Equation 4.60 and the data in Table 4.4 to calculate the entropy of $NH_3(g)$ at 298.2 K and one bar. Compare your result with the experimental value of 192.8 $\text{J}\cdot\text{K}^{-1}\cdot\text{mol}^{-1}$.

For a nonlinear polyatomic ideal gas having four atoms,

$$q = \left(\frac{2\pi Mk_BT}{h^2}\right)^{3/2} V\frac{\pi^{1/2}}{\sigma}\left(\frac{T}{\Theta_{rot,A}\Theta_{rot,B}\Theta_{rot,C}}\right)^{1/2}\left(\prod_{j=1}^{6}\frac{e^{-\Theta_{vib,j}/2T}}{1-e^{-\Theta_{vib,j}/T}}\right)g_{e1}e^{D_e/k_BT} \qquad (4.60)$$

Then

$$\ln\frac{q}{N} = \ln\left[\left(\frac{2\pi Mk_BT}{h^2}\right)^{3/2}\frac{\overline{V}}{N_A}\right] + \ln\frac{\pi^{1/2}}{\sigma} + \frac{1}{2}\ln\left(\frac{T^3}{\Theta_{rot,A}\Theta_{rot,B}\Theta_{rot,C}}\right)$$
$$-\sum_{j=1}^{6}\ln(1-e^{-\Theta_{vib,j}/T}) - \sum_{j=1}^{6}\frac{\Theta_{vib,j}}{2T} + \ln g_{e1} + \frac{D_e}{k_BT}$$

and

$$\left(\frac{\partial\ln q}{\partial T}\right) = \frac{3}{2T} + \frac{3}{2T} + \sum_{j=1}^{6}\frac{\Theta_{vib,j}}{2T^2} + \sum_{j=1}^{6}\frac{(\Theta_{vib,j}/T^2)e^{-\Theta_{vib,j}/T}}{1-e^{-\Theta_{vib,j}/T}} - \frac{D_e}{k_BT^2}$$

Substituting into Equation 1 from Problem 6–41, we find

$$\frac{\overline{S}}{R} = 4 + \ln\left[\left(\frac{2\pi Mk_BT}{h^2}\right)^{3/2}\frac{\overline{V}}{N_A}\right] + \ln\frac{\pi^{1/2}}{\sigma} + \frac{1}{2}\ln\frac{T^3}{\Theta_{rot,A}\Theta_{rot,B}\Theta_{rot,C}}$$

$$-\sum_{j=1}^{6}\ln(1-e^{-\Theta_{\text{vib},j}/T})+\sum_{j=1}^{6}\left[\frac{(\Theta_{\text{vib},j}/T)e^{-\Theta_{\text{vib},j}/T}}{1-e^{-\Theta_{\text{vib},j}/T}}\right]+\ln g_{el}$$

For NH_3, $\Theta_{\text{vib},1} = 4800$ K, $\Theta_{\text{vib},2} = 1360$ K, $\Theta_{\text{vib},3} = \Theta_{\text{vib},4} = 4880$ K, $\Theta_{\text{vib},5} = \Theta_{\text{vib},6} = 2330$ K, $\Theta_{\text{rot,A}} = \Theta_{\text{rot,B}} = 13.6$ K, $\Theta_{\text{rot,C}} = 8.92$ K, $\sigma = 3$, and $g_{el} = 1$. The various factors are as follows:

$$\left(\frac{2\pi Mk_BT}{h^2}\right)^{3/2} = \left[\frac{2\pi(2.828\times10^{-26}\text{ kg})(1.381\times10^{-23}\text{ J}\cdot\text{K}^{-1})(298.2\text{ K})}{(6.626\times10^{34}\text{ J}\cdot\text{s})^2}\right]^{3/2}$$
$$= 6.801\times10^{31}\text{ m}^{-3}$$

$$\frac{\overline{V}}{N_A} = \frac{RT}{N_AP} = \frac{(0.08314\text{ dm}^3\cdot\text{bar}\cdot\text{mol}^{-1}\cdot\text{K}^{-1})(298.2\text{ K})}{(6.022\times10^{23}\text{ mol}^{-1})(1\text{ bar})}$$
$$= 4.117\times10^{-23}\text{ dm}^3 = 4.117\times10^{-26}\text{ m}^3$$

$$\frac{1}{2}\ln\frac{T^3}{\Theta_{\text{rot,A}}\Theta_{\text{rot,B}}\Theta_{\text{rot,C}}} = 4.842$$

$$\sum_{j=1}^{6}\ln(1-e^{-\Theta_{\text{vib},j}/T}) = -0.01132$$

$$\sum_{j=1}^{6}\frac{\Theta_{\text{vib},j}}{T}\left(\frac{e^{-\Theta_{\text{vib},j}/T}}{1-e^{-\Theta_{\text{vib},j}/T}}\right) = 0.05451$$

The standard molar entropy is then

$$\frac{\overline{S}}{R} = 4 + 14.845 - 0.5262 + 4.842 + 0.01132 + 0.05451$$
$$= 23.23$$

This is 193.1 $\text{J}\cdot\text{K}^{-1}\cdot\text{mol}^{-1}$. The slight disagreement with the experimental value is due to our use of the rigid rotator-harmonic oscillator model.

6–44. Derive Equation 6.35.

The maximum efficiency is defined as

$$\text{maxmimum efficiency} = \frac{-w}{q_{\text{rev,h}}} = \frac{q_{\text{rev,h}}+q_{\text{rev,c}}}{q_{\text{rev,h}}} = 1 + \frac{q_{\text{rev,c}}}{q_{\text{rev,h}}}$$

Because the process is cyclic and reversible, $\Delta S_{\text{engine}} = 0$, and so (as in Equation 6.34)

$$q_{\text{rev,c}} = -q_{\text{rev,h}}\frac{T_c}{T_h}$$

The efficiency becomes

$$\text{maximum efficiency} = 1 - \frac{T_c}{T_h} = \frac{T_h - T_c}{T_h} \tag{6.35}$$

6–45. The boiling point of water at a pressure of 25 atm is 223°C. Compare the theoretical efficiencies of a steam engine operating between 20°C and the boiling point of water at 1 atm and at 25 atm.

At one atmosphere, using Equation 6.35 gives an efficiency of

$$\text{efficiency} = 1 - \frac{293}{373} = 21\%$$

At 25 atm, the same engine will give an efficiency of

$$\text{efficiency} = 1 - \frac{293}{496} = 41\%$$

CHAPTER 7

Entropy and the Third Law of Thermodynamics

PROBLEMS AND SOLUTIONS

7–1. Form the total derivative of H as a function of T and P and equate the result to dH in Equation 7.6 to derive Equations 7.7 and 7.8.

The total derivative of $S(T, P)$ is

$$dS = \left(\frac{\partial S}{\partial T}\right)_P dT + \left(\frac{\partial S}{\partial P}\right)_T dP$$

We can substitute this in Equation 7.6 to obtain

$$dH = TdS + VdP$$

$$dH = T\left(\frac{\partial S}{\partial T}\right)_P dT + T\left(\frac{\partial S}{\partial P}\right)_T dP + VdP$$

$$= T\left(\frac{\partial S}{\partial T}\right)_P dT + \left[V + T\left(\frac{\partial S}{\partial P}\right)_T\right] dP \tag{1}$$

We now write the total derivative of $H(T, P)$ as

$$dH = \left(\frac{\partial H}{\partial T}\right)_P dT + \left(\frac{\partial H}{\partial P}\right)_T dP$$

$$dH = C_P dT + \left(\frac{\partial H}{\partial P}\right)_T dP \tag{2}$$

Set the coefficients of dT in Equations 1 and 2 equal to each other to find Equation 7.7

$$\left(\frac{\partial S}{\partial T}\right)_P = \frac{C_P}{T}$$

and set the coefficients of dP equal to each other to obtain Equation 7.8:

$$\left[V + T\left(\frac{\partial S}{\partial P}\right)_T\right] = \left(\frac{\partial H}{\partial P}\right)_T$$

$$\left(\frac{\partial S}{\partial P}\right)_T = \frac{1}{T}\left[\left(\frac{\partial H}{\partial P}\right)_T - V\right] \tag{7.8}$$

7–2. The molar heat capacity of $H_2O(l)$ has an approximately constant value of $\overline{C}_P = 75.4\ \mathrm{J \cdot K^{-1} \cdot mol^{-1}}$ from 0°C to 100°C. Calculate ΔS if two moles of $H_2O(l)$ are heated from 10°C to 90°C at constant pressure.

We use Equation 7.9 to write

$$\Delta S = \int_{T_1}^{T_2} \frac{n\overline{C}_P}{T} dT = \int_{283\text{ K}}^{363\text{ K}} \frac{(2\text{ mol})(75.4\text{ J}\cdot\text{K}^{-1}\cdot\text{mol}^{-1})}{T} dT$$

$$= (150.8\text{ J}\cdot\text{K}^{-1}) \ln \frac{363}{283} = 37.5\text{ J}\cdot\text{K}^{-1}$$

7–3. The molar heat capacity of butane can be expressed by

$$\overline{C}_P/R = 0.05641 + (0.04631\text{ K}^{-1})T - (2.392 \times 10^{-5}\text{ K}^{-2})T^2 + (4.807 \times 10^{-9}\text{ K}^{-3})T^3$$

over the temperature range $300\text{ K} \le T \le 1500\text{ K}$. Calculate ΔS if one mole of butane is heated from 300 K to 1000 K at constant pressure.

We can use Equation 7.9 to write

$$\Delta S = \int_{T_1}^{T_2} \frac{n\overline{C}_P}{T} dT$$

$$= nR \int_{300\text{ K}}^{1000\text{ K}} \left[\frac{0.05641}{T} + 0.04631\text{ K}^{-1} - (2.392 \times 10^{-5}\text{ K}^{-2})T + (4.807 \times 10^{-9}\text{ K}^{-3})T^2\right] dT$$

$$= (23.16R)(1\text{ mol}) = 192.6\text{ J}\cdot\text{K}^{-1}$$

7–4 . The molar heat capacity of $C_2H_4(g)$ can be expressed by

$$\overline{C}_V(T)/R = 16.4105 - \frac{6085.929\text{ K}}{T} + \frac{822\,826\text{ K}^2}{T^2}$$

over the temperature range $300\text{ K} < T < 1000\text{ K}$. Calculate ΔS if one mole of ethene is heated from 300 K to 600 K at constant volume.

We can use Equation 7.5 to write

$$\Delta S = \int_{T_1}^{T_2} \frac{n\overline{C}_V}{T} dT$$

$$= nR \int_{300\text{ K}}^{600\text{ K}} \left[\frac{16.4105}{T} - \frac{6085.929\text{ K}}{T^2} + \frac{822\,826\text{ K}^2}{T^3}\right] dT$$

$$= (4.660R)(1\text{ mol}) = 38.75\text{ J}\cdot\text{K}^{-1}$$

7–5. Use the data in Problem 7–4 to calculate ΔS if one mole of ethene is heated from 300 K to 600 K at constant pressure. Assume ethene behaves ideally.

For an ideal gas, $\overline{C}_P - \overline{C}_V = R$, so we can express $\overline{C}_P$ as

$$\overline{C}_P/R = 1 + 16.4105 - \frac{6085.929\text{ K}}{T} + \frac{822\,826\text{ K}^2}{T^2}$$

Then we can use Equation 7.9 to calculate ΔS:

$$\begin{aligned}\Delta S &= \int_{T_1}^{T_2} \frac{n\overline{C}_P}{T} dT \\ &= nR \int_{300\text{ K}}^{600\text{ K}} \left[\frac{17.4105}{T} - \frac{6085.929\text{ K}}{T^2} + \frac{822\,826\text{ K}^2}{T^3} \right] dT \\ &= (5.353R)(1\text{ mol}) = 44.51\text{ J}\cdot\text{K}^{-1}\end{aligned}$$

7–6. We can calculate the difference in the results of Problems 7–4 and 7–5 in the following way. First, show that because $\overline{C}_P - \overline{C}_V = R$ for an ideal gas,

$$\Delta\overline{S}_P = \Delta\overline{S}_V + R \ln \frac{T_2}{T_1}$$

Check to see numerically that your answers to Problems 7–4 and 7–5 differ by $R \ln 2 = 0.693R = 5.76\text{ J}\cdot\text{K}^{-1}\cdot\text{mol}^{-1}$.

For an ideal gas, $\overline{C}_P - \overline{C}_V = R$. Equations 7.6 and 7.9 state that

$$\Delta\overline{S}_V = \int_{T_1}^{T_2} \frac{\overline{C}_V}{T} dT \qquad \text{and} \qquad \Delta\overline{S}_P = \int_{T_1}^{T_2} \frac{\overline{C}_P}{T} dT$$

Subtracting $\Delta\overline{S}_P$ from $\Delta\overline{S}_V$ gives

$$\Delta\overline{S}_P - \Delta\overline{S}_V = \int_{T_1}^{T_2} \frac{\overline{C}_P - \overline{C}_V}{T} dT = \int_{T_1}^{T_2} \frac{R}{T} dT = R \ln \frac{T_2}{T_1}$$

and so $\Delta\overline{S}_P$ can be written as

$$\Delta\overline{S}_P = \Delta\overline{S}_V + R \ln \frac{T_2}{T_1}$$

The answers to Problems 24–4 and 24–5 differ by $R \ln 2$, as required.

7–7. The results of Problems 7–4 and 7–5 must be connected in the following way. Show that the two processes can be represented by the diagram

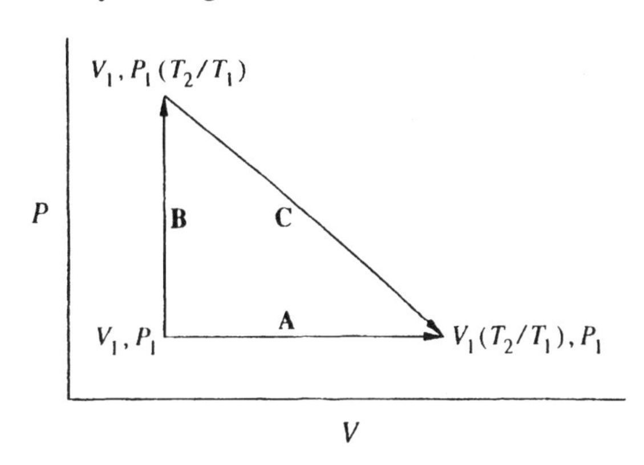

where paths A and B represent the processes in Problems 7–5 and 7–4, respectively.

Now, path A is equivalent to the sum of paths B and C. Show that ΔS_C is given by

$$\Delta S_C = nR \ln \frac{V_1\left(\frac{T_2}{T_1}\right)}{V_1} = nR \ln \frac{P_1\left(\frac{T_2}{T_1}\right)}{P_1} = nR \ln \frac{T_2}{T_1}$$

and that the result given in Problem 7–6 follows.

In Problem 7–4, ethane is heated at constant volume, so (assuming ideal behavior)

$$P_2 = P_1\left(\frac{T_2}{T_1}\right)$$

Likewise, in Problem 7–5, the system is kept at constant pressure and so (assuming ideal behavior)

$$V_2 = V_1\left(\frac{T_2}{T_1}\right)$$

These values correspond to those shown in the diagram. Now, path C represents an isothermal process. Because we are assuming ideal behavior, $dU = 0$, which means that $\delta q_{rev} = -\delta w = nRT/V dV$. Then we can write ΔS_C as (Equation 6.22)

$$\Delta S_C = \int \frac{\delta q_{rev}}{T} = nR \ln \frac{V_2}{V_1} = R \ln \frac{V_1(T_2/T_1)}{V_1} = R \ln \frac{T_2}{T_1}$$

Note that path A is equivalent to the sums of paths B and C, so $\Delta S_A = \Delta S_B + \Delta S_C$. Because $\Delta S_A = \Delta S_P$ and $\Delta S_B = \Delta S_V$, we can write

$$\Delta S_C = \Delta S_P - \Delta S_V$$

and the result given in Problem 7–6 follows.

7–8. Use Equations 6.23 and 6.24 to show that $S = 0$ at 0 K, where every system will be in its ground state.

We begin with Equations 6.23 and 6.24,

$$W = \frac{\mathcal{A}}{a_1!a_2!\ldots} \quad \text{and} \quad S = k_B \ln W$$

Let a_1 represent the ground state, so all other $a_j = 0$ when the system is in the ground state. Then $\mathcal{A} = \sum_j a_j = a_1 + 0 = a_1$, and Equation 6.23 becomes

$$W = \frac{a_1!}{a_1!} = 1$$

Substitute this into Equation 6.24 for S to find

$$S = k_B \ln 1 = 0$$

7–9. Prove that $S = -k \sum p_j \ln p_j = 0$ when $p_1 = 1$ and all the other $p_j = 0$. In other words, prove that $x \ln x \to 0$ as $x \to 0$.

Let $p_1 = 1$ and all other $p_j = 0$. Then Equation 6.40 becomes

$$S = -k_B \sum p_j \ln p_j = 0 - k_B \sum x \ln x$$

where $x \to 0$. In Problem E–8, we proved that $x \ln x \to 0$ as $x \to 0$, so $S = 0 - 0 = 0$ under the conditions given.

7–10. It has been found experimentally that $\Delta_{vap}\overline{S} \approx 88\ \mathrm{J\cdot K^{-1}\cdot mol^{-1}}$ for many nonassociated liquids. This rough rule of thumb is called *Trouton's rule*. Use the following data to test the validity of Trouton's rule.

Substance	t_{fus}/°C	t_{vap}/°C	$\Delta_{fus}\overline{H}/\mathrm{kJ\cdot mol^{-1}}$	$\Delta_{vap}\overline{H}/\mathrm{kJ\cdot mol^{-1}}$
Pentane	−129.7	36.06	8.42	25.79
Hexane	−95.3	68.73	13.08	28.85
Heptane	−90.6	98.5	14.16	31.77
Ethylene oxide	−111.7	10.6	5.17	25.52
Benzene	5.53	80.09	9.95	30.72
Diethyl ether	−116.3	34.5	7.27	26.52
Tetrachloromethane	−23	76.8	3.28	29.82
Mercury	−38.83	356.7	2.29	59.11
Bromine	−7.2	58.8	10.57	29.96

Use Equation 7.16,

$$\Delta_{vap}\overline{S} = \frac{\Delta_{vap}\overline{H}}{T_{vap}}$$

to construct a table of values of $\Delta_{vap}\overline{S}$.

Substance	$\Delta_{vap}\overline{S}/\mathrm{J\cdot mol^{-1}\cdot K^{-1}}$
Pentane	83.41
Hexane	84.39
Heptane	85.5
Ethylene oxide	89.9
Benzene	86.97
Diethyl ether	86.2
Tetrachloromethane	85.2
Mercury	93.85
Bromine	90.3

7–11. Use the data in Problem 7–10 to calculate the value of $\Delta_{fus}\overline{S}$ for each substance.

Use Equation 7.16,

$$\Delta_{\text{fus}}\overline{S} = \frac{\Delta_{\text{fus}}\overline{H}}{T_{\text{fus}}}$$

to construct a table of values of $\Delta_{\text{fus}}\overline{S}$.

Substance	$\Delta_{\text{fus}}\overline{S}/\text{J}\cdot\text{mol}^{-1}\cdot\text{K}^{-1}$
Pentane	58.7
Hexane	73.5
Heptane	77.6
Ethylene oxide	32.0
Benzene	35.7
Diethyl ether	46.3
Tetrachloromethane	13
Mercury	9.77
Bromine	40

7–12. Why is $\Delta_{\text{vap}}\overline{S} > \Delta_{\text{fus}}\overline{S}$?

$\Delta_{\text{vap}}\overline{S} \gg \Delta_{\text{fus}}\overline{S}$ because gases are essentially completely unordered; the molecules of a gas travel more or less randomly within the gas's container. Liquids, however, are much more cohesive and structured, and solids are very structured. The difference between the entropy of a liquid and that of a solid is less than the difference between the entropy of a liquid and that of a gas.

7–13. Show that if $C_P^s(T) \to T^\alpha$ as $T \to 0$, where α is a positive constant, then $S(T) \to 0$ as $T \to 0$.

We assume in the statement of the problem that

$$\lim_{T\to 0} C_P^s(T) = T^\alpha$$

where α is a positive constant. Then express S using Equation 7.10 and take the limit of S as $T \to 0$:

$$\lim_{T\to 0} S(T) = S(0\,\text{K}) + \lim_{T\to 0}\int_0^T \frac{C_P(T)}{T}dT = \lim_{T\to 0}\int_0^T \frac{T^\alpha}{T}dT = \lim_{T\to 0}\frac{T^\alpha}{T} = 0$$

as long as $S(0\,\text{K}) = 0$ and $\alpha > 0$ (as stipulated in the statement of the problem).

7–14. Use the following data to calculate the standard molar entropy of $N_2(g)$ at 298.15 K.

$$C_P^\circ[N_2(s_1)]/R = -0.03165 + (0.05460\ K^{-1})T + (3.520 \times 10^{-3}\ K^{-2})T^2 - (2.064 \times 10^{-5}\ K^{-3})T^3$$
$$10\ K \leq T \leq 35.61\ K$$

$$C_P^\circ[N_2(s_2)]/R = -0.1696 + (0.2379\ K^{-1})T - (4.214 \times 10^{-3}\ K^{-2})T^2 + (3.036 \times 10^{-5}\ K^{-3})T^3$$
$$35.61\ K \leq T \leq 63.15\ K$$

$$C_P^\circ[N_2(l)]/R = -18.44 + (1.053\ K^{-1})T - (0.0148\ K^{-2})T^2 + (7.064 \times 10^{-5}\ K^{-3})T^3$$
$$63.15\ K \leq T \leq 77.36\ K$$

$C_P^\circ[N_2(g)]/R = 3.500$ from $77.36\ K \leq T \leq 1000\ K$, $C_P(T = 10.0\ K) = 6.15\ J\cdot K^{-1}\cdot mol^{-1}$, $T_{trs} = 35.61\ K$, $\Delta_{trs}\overline{H} = 0.2289\ kJ\cdot mol^{-1}$, $T_{fus} = 63.15\ K$, $\Delta_{fus}\overline{H} = 0.71\ kJ\cdot mol^{-1}$, $T_{vap} = 77.36\ K$, and $\Delta_{vap}\overline{H} = 5.57\ kJ\cdot mol^{-1}$. The correction for nonideality (Problem 8–20) $= 0.02\ J\cdot K^{-1}\cdot mol^{-1}$.

The easiest way to do this series of problems is to input the given data into a program like *Mathematica* and use Equation 7.17. For temperatures below the minimum value for which the formulae provided are valid, we can use the expression from Example 7–3, $\overline{S}(T) = \overline{C}_P(T)/3$. Here, we solve the formula

$$\begin{aligned}\overline{S}(T) &= \frac{\overline{C}_P(10\ K)}{3} + \int_{10}^{35.61} \frac{\overline{C}_P[N_2(s_1)]}{T}dT + \frac{\Delta_{trs}\overline{H}}{35.61\ K} + \int_{35.61}^{63.15} \frac{\overline{C}_P[N_2(s_2)]}{T}dT \\ &+ \frac{\Delta_{fus}\overline{H}}{63.15\ K} + \int_{63.15}^{77.36} \frac{\overline{C}_P[N_2(l)]}{T}dT + \frac{\Delta_{vap}\overline{H}}{77.36\ K} \\ &+ \int_{77.36}^{298.15} \frac{\overline{C}_P[N_2(g)]}{T}dT + \text{correction} \\ &= 2.05\ J\cdot mol^{-1}\cdot K^{-1} + 25.86\ J\cdot mol^{-1}\cdot K^{-1} + 6.428\ J\cdot mol^{-1}\cdot K^{-1} \\ &\quad +23.41\ J\cdot mol^{-1}\cdot K^{-1} + 11.24\ J\cdot mol^{-1}\cdot K^{-1} + 11.78\ J\cdot mol^{-1}\cdot K^{-1} \\ &\quad +72.00\ J\cdot mol^{-1}\cdot K^{-1} + 39.26\ J\cdot mol^{-1}\cdot K^{-1} + 0.02\ J\cdot mol^{-1}\cdot K^{-1} \\ &= 192.05\ J\cdot K^{-1}\cdot mol^{-1}\end{aligned}$$

The literature value of the standard molar entropy is $191.6\ J\cdot K^{-1}\cdot mol^{-1}$. The slight discrepancy between these two values reflects the use of the ideal expression for $\overline{C}_P[N_2(g)]$. (Using the $\overline{C}_P[N_2(g)]$ that is given in the next problem, which is linear in T, gives a standard molar entropy of $191.04\ J\cdot mol^{-1}\cdot K^{-1}$.)

7–15. Use the data in Problem 7–14 and $\overline{C}_P[N_2(g)]/R = 3.307 + (6.29 \times 10^{-4}\ K^{-1})T$ for $T \geq 77.36$ K to plot the standard molar entropy of nitrogen as a function of temperature from 0 K to 1000 K.

The function which describes the standard molar entropy of nitrogen must be defined differently for each phase and phase transition. Notice that the correction factor must be added to all functions to correct for nonideality.

From 0 K to 10 K,

$$\overline{S}(T) = \frac{\overline{C}_P(10\ K)}{3} + \text{corr}$$

From 10 K to 35.61 K,

$$\overline{S}(T) = \frac{\overline{C}_P(10\text{ K})}{3} + \int_{10}^{T} \frac{\overline{C}_P[\text{N}_2(\text{s}_1)]}{T} dT + \text{corr}$$

At 35.61 K,

$$\overline{S}(T) = \frac{\overline{C}_P(10\text{ K})}{3} + \int_{10}^{35.61} \frac{\overline{C}_P[\text{N}_2(\text{s}_1)]}{T} dT + \frac{\Delta_{\text{trs}}\overline{H}}{35.61\text{ K}} + \text{corr}$$

From 35.61 K to 63.15 K,

$$\overline{S}(T) = \frac{\overline{C}_P(10\text{ K})}{3} + \int_{10}^{35.61} \frac{\overline{C}_P[\text{N}_2(\text{s}_1)]}{T} dT + \frac{\Delta_{\text{trs}}\overline{H}}{35.61\text{ K}} + \int_{35.61}^{T} \frac{\overline{C}_P[\text{N}_2(\text{s}_2)]}{T} dT + \text{corr}$$

At 63.15 K,

$$\overline{S}(T) = \frac{\overline{C}_P(10\text{ K})}{3} + \int_{10}^{35.61} \frac{\overline{C}_P[\text{N}_2(\text{s}_1)]}{T} dT + \frac{\Delta_{\text{trs}}\overline{H}}{35.61\text{ K}} + \int_{35.61}^{63.15} \frac{\overline{C}_P[\text{N}_2(\text{s}_2)]}{T} dT + \frac{\Delta_{\text{fus}}\overline{H}}{63.15\text{ K}} + \text{corr}$$

From 63.15 K to 77.36 K,

$$\overline{S}(T) = \frac{\overline{C}_P(10\text{ K})}{3} + \int_{10}^{35.61} \frac{\overline{C}_P[\text{N}_2(\text{s}_1)]}{T} dT + \frac{\Delta_{\text{trs}}\overline{H}}{35.61\text{ K}} + \int_{35.61}^{63.15} \frac{\overline{C}_P[\text{N}_2(\text{s}_2)]}{T} dT + \frac{\Delta_{\text{fus}}\overline{H}}{63.15\text{ K}} + \int_{63.15}^{T} \frac{\overline{C}_P[\text{N}_2(\text{l})]}{T} dT + \text{corr}$$

At 77.36 K,

$$\overline{S}(T) = \frac{\overline{C}_P(10\text{ K})}{3} + \int_{10}^{35.61} \frac{\overline{C}_P[\text{N}_2(\text{s}_1)]}{T} dT + \frac{\Delta_{\text{trs}}\overline{H}}{35.61\text{ K}} + \int_{35.61}^{63.15} \frac{\overline{C}_P[\text{N}_2(\text{s}_2)]}{T} dT + \frac{\Delta_{\text{fus}}\overline{H}}{63.15\text{ K}} + \int_{63.15}^{77.36} \frac{\overline{C}_P[\text{N}_2(\text{l})]}{T} dT + \frac{\Delta_{\text{vap}}\overline{H}}{77.36\text{ K}} + \text{corr}$$

From 77.36 K to 1000 K,

$$\overline{S}(T) = \frac{\overline{C}_P(10\text{ K})}{3} + \int_{10}^{35.61} \frac{\overline{C}_P[\text{N}_2(\text{s}_1)]}{T} dT + \frac{\Delta_{\text{trs}}\overline{H}}{35.61\text{ K}} + \int_{35.61}^{63.15} \frac{\overline{C}_P[\text{N}_2(\text{s}_2)]}{T} dT + \frac{\Delta_{\text{fus}}\overline{H}}{63.15\text{ K}} + \int_{63.15}^{77.36} \frac{\overline{C}_P[\text{N}_2(\text{l})]}{T} dT + \frac{\Delta_{\text{vap}}\overline{H}}{77.36\text{ K}} + \int_{77.36}^{T} \frac{\overline{C}_P[\text{N}_2(\text{g})]}{T} dT + \text{corr}$$

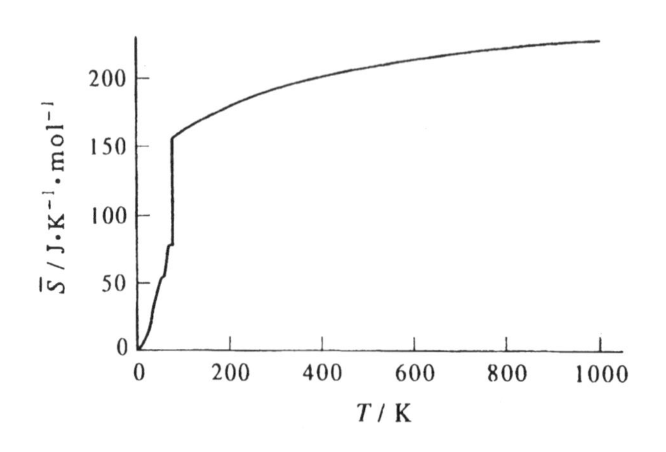

7–16. The molar heat capacities of solid, liquid, and gaseous chlorine can be expressed as

$$C_P^\circ[\mathrm{Cl_2(s)}]/R = -1.545 + (0.1502\ \mathrm{K^{-1}})T - (1.179\times10^{-3}\ \mathrm{K^{-2}})T^2 + (3.441\times10^{-6}\ \mathrm{K^{-3}})T^3$$
$$15\ \mathrm{K} \le T \le 172.12\ \mathrm{K}$$

$$C_P^\circ[\mathrm{Cl_2(l)}]/R = 7.689 + (5.582\times10^{-3}\ \mathrm{K^{-1}})T - (1.954\times10^{-5}\ \mathrm{K^{-2}})T^2$$
$$172.12\ \mathrm{K} \le T \le 239.0\ \mathrm{K}$$

$$C_P^\circ[\mathrm{Cl_2(g)}]/R = 3.812 + (1.220\times10^{-3}\ \mathrm{K^{-1}})T - (4.856\times10^{-7}\ \mathrm{K^{-2}})T^2$$
$$239.0\ \mathrm{K} \le T \le 1000\ \mathrm{K}$$

Use the above molar heat capacities and $T_{\mathrm{fus}} = 172.12$ K, $\Delta_{\mathrm{fus}}\overline{H} = 6.406\ \mathrm{kJ\cdot mol^{-1}}$, $T_{\mathrm{vap}} = 239.0$ K, $\Delta_{\mathrm{vap}}\overline{H} = 20.40\ \mathrm{kJ\cdot mol^{-1}}$, and $\Theta_{\mathrm{D}} = 116$ K. The correction for nonideality $= 0.502\ \mathrm{J\cdot K^{-1}\cdot mol^{-1}}$ to calculate the standard molar entropy of chlorine at 298.15 K. Compare your result with the value given in Table 7.2.

$$\overline{S}(T) = \int_0^{15} \frac{12\pi^4}{5T} R\left(\frac{T}{\Theta_{\mathrm{D}}}\right)^3 dT + \int_{15}^{172.12} \frac{\overline{C}_P[\mathrm{Cl_2(s)}]}{T} dT + \frac{\Delta_{\mathrm{fus}}\overline{H}}{172.12\ \mathrm{K}}$$
$$+ \int_{172.12}^{239.0} \frac{\overline{C}_P[\mathrm{Cl_2(l)}]}{T} dT + \frac{\Delta_{\mathrm{vap}}\overline{H}}{239.0\ \mathrm{K}} + \int_{239.0}^{298} \frac{\overline{C}_P[\mathrm{Cl_2(g)}]}{T} dT + \text{correction}$$
$$= 1.401\ \mathrm{J\cdot K^{-1}\cdot mol^{-1}} + 69.37\ \mathrm{J\cdot K^{-1}\cdot mol^{-1}} + 37.22\ \mathrm{J\cdot K^{-1}\cdot mol^{-1}}$$
$$+21.86\ \mathrm{J\cdot K^{-1}\cdot mol^{-1}} + 85.36\ \mathrm{J\cdot K^{-1}\cdot mol^{-1}} + 7.54\ \mathrm{J\cdot K^{-1}\cdot mol^{-1}}$$
$$+0.502\ \mathrm{J\cdot K^{-1}\cdot mol^{-1}}$$
$$= 223.2\ \mathrm{J\cdot K^{-1}\cdot mol^{-1}}$$

The result is extremely close to the value of 223.1 $\mathrm{J\cdot K^{-1}\cdot mol^{-1}}$ found in Table 7.2.

7–17. Use the data in Problem 7–16 to plot the standard molar entropy of chlorine as a function of temperature from 0 K to 1000 K.

Do this in the same manner as Problem 7–15, using the appropriate values from Problem 7–16 and changing the limits of integration as required.

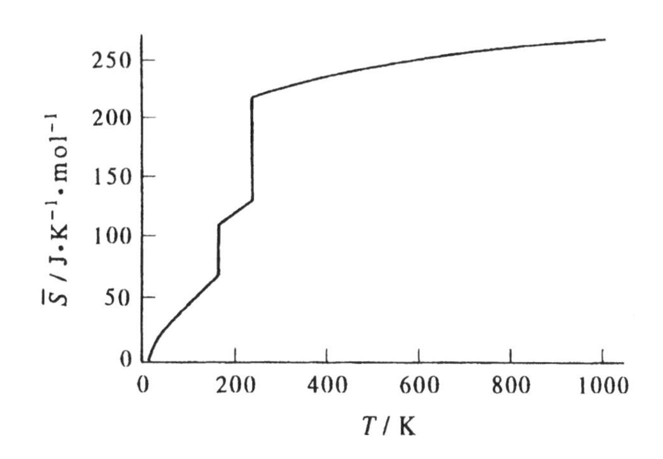

7–18. Use the following data to calculate the standard molar entropy of cyclopropane at 298.1 K.

$$C_P^\circ[\text{C}_3\text{H}_6(\text{s})]/R = -1.921 + (0.1508\ \text{K}^{-1})T - (9.670 \times 10^{-4}\ \text{K}^{-2})T^2 + (2.694 \times 10^{-6}\ \text{K}^{-3})T^3$$
$$15\ \text{K} \le T \le 145.5\ \text{K}$$

$$C_P^\circ[\text{C}_3\text{H}_6(\text{l})]/R = 5.624 + (4.493 \times 10^{-2}\ \text{K}^{-1})T - (1.340 \times 10^{-4}\ \text{K}^{-2})T^2$$
$$145.5\ \text{K} \le T \le 240.3\ \text{K}$$

$$C_P^\circ[\text{C}_3\text{H}_6(\text{g})]/R = -1.793 + (3.277 \times 10^{-2}\ \text{K}^{-1})T - (1.326 \times 10^{-5}\ \text{K}^{-2})T^2$$
$$240.3\ \text{K} \le T \le 1000\ \text{K}$$

$T_{\text{fus}} = 145.5$ K, $T_{\text{vap}} = 240.3$ K, $\Delta_{\text{fus}}\overline{H} = 5.44\ \text{kJ}\cdot\text{mol}^{-1}$, $\Delta_{\text{vap}}\overline{H} = 20.05\ \text{kJ}\cdot\text{mol}^{-1}$, and $\Theta_D = 130$ K. The correction for nonideality $= 0.54\ \text{J}\cdot\text{K}^{-1}\cdot\text{mol}^{-1}$.

$$\overline{S}(T) = \int_0^{15} \frac{12\pi^4}{5T} R\left(\frac{T}{\Theta_D}\right)^3 dT + \int_{15}^{145.5} \frac{\overline{C}_P[\text{C}_3\text{H}_6(\text{s})]}{T} dT + \frac{\Delta_{\text{fus}}\overline{H}}{145.5\ \text{K}}$$
$$+ \int_{145.5}^{240.3} \frac{\overline{C}_P[\text{C}_3\text{H}_6(\text{l})]}{T} dT + \frac{\Delta_{\text{vap}}\overline{H}}{240.3\ \text{K}} + \int_{240.3}^{298.1} \frac{\overline{C}_P[\text{C}_3\text{H}_6(\text{g})]}{T} dT$$
$$+ \text{correction}$$
$$= 0.995\ \text{J}\cdot\text{K}^{-1}\cdot\text{mol}^{-1} + 66.1\ \text{J}\cdot\text{K}^{-1}\cdot\text{mol}^{-1} + 37.4\ \text{J}\cdot\text{K}^{-1}\cdot\text{mol}^{-1}$$
$$+38.5\ \text{J}\cdot\text{K}^{-1}\cdot\text{mol}^{-1} + 83.4\ \text{J}\cdot\text{K}^{-1}\cdot\text{mol}^{-1} + 10.8\ \text{J}\cdot\text{K}^{-1}\cdot\text{mol}^{-1}$$
$$+0.54\ \text{J}\cdot\text{K}^{-1}\cdot\text{mol}^{-1}$$
$$= 237.8\ \text{J}\cdot\text{K}^{-1}\cdot\text{mol}^{-1}$$

This compares very well with the literature value of 237.5 $\text{J}\cdot\text{K}^{-1}\cdot\text{mol}^{-1}$.

7–19. Use the data in Problem 7–18 to plot the standard molar entropy of cyclopropane from 0 K to 1000 K.

Do this in the same manner as Problem 7–15, using the appropriate values from Problem 7–18 and changing the limits of integration as required.

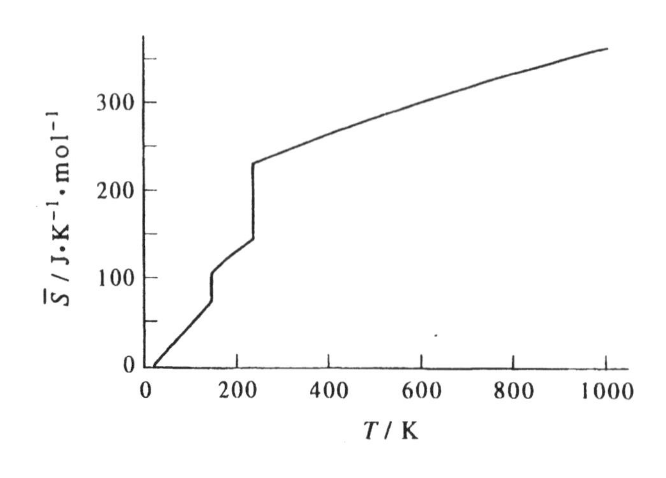

7–20. The constant-pressure molar heat capacity of N_2O as a function of temperature is tabulated below. Dinitrogen oxide melts at 182.26 K with $\Delta_{fus}\overline{H} = 6.54\ \text{kJ}\cdot\text{mol}^{-1}$, and boils at 184.67 K with $\Delta_{vap}\overline{H} = 16.53\ \text{kJ}\cdot\text{mol}^{-1}$ at one bar. Assuming the heat capacity of solid dinitrogen oxide can be described by the Debye theory up to 15 K, calculate the molar entropy of $N_2O(g)$ at its boiling point.

T/K	$\overline{C}_P$/J·K^{-1}·mol^{-1}	T/K	$\overline{C}_P$/J·K^{-1}·mol^{-1}
15.17	2.90	120.29	45.10
19.95	6.19	130.44	47.32
25.81	10.89	141.07	48.91
33.38	16.98	154.71	52.17
42.61	23.13	164.82	54.02
52.02	28.56	174.90	56.99
57.35	30.75	180.75	58.83
68.05	34.18	182.26	Melting point
76.67	36.57	183.55	77.70
87.06	38.87	183.71	77.45
98.34	41.13	184.67	Boiling point
109.12	42.84		

We can do this problem in the same way we did Problems 7–14, 7–16, and 7–18. Because we are not given equations for the molar heat capacity, we can graph the heat capacity of the solid and liquid dinitrogen oxide, find a best-fit line, and use this to calculate the molar entropy of N_2O at the boiling point.

For solid dinitrogen oxide, a best-fit line gives the equation

$$\overline{C}_P[\text{N}_2\text{O(s)}]/\text{J}\cdot\text{K}^{-1}\cdot\text{mol}^{-1} = -13.153 + (1.1556\ \text{K}^{-1})T - (8.3372 \times 10^{-3}\ \text{K}^{-2})T^2 + (2.3026 \times 10^{-5}\ \text{K}^{-3})T^3$$

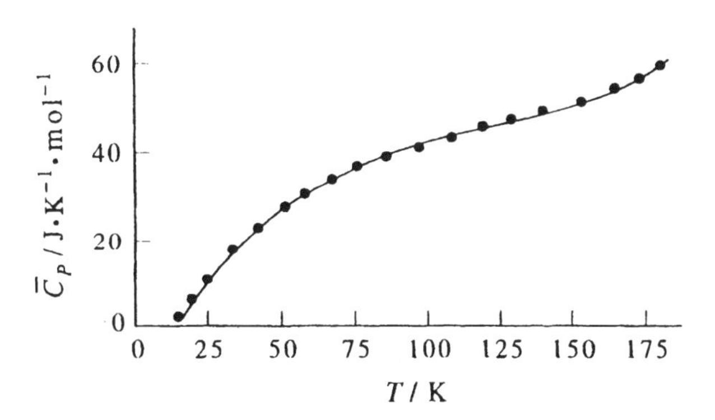

And for liquid dinitrogen oxide (with only two points), a line drawn between those two points has the equation

$$\overline{C}_P[\text{N}_2\text{O(l)}]/\text{J}\cdot\text{K}^{-1}\cdot\text{mol}^{-1} = 364.49 - (1.5625\ \text{K}^{-1})T$$

Note that, although we are given only two data points, the temperature varies by only 2° for dinitrogen oxide. From the Debye theory (Example 7–3) we can write the low temperature entropy as

$$\overline{S}(15\ \text{K}) = \frac{\overline{C}_P(15\ \text{K})}{3} \qquad 0 < T \le 15\ \text{K}$$

Now we can substitute into Equation 7.17, as before.

$$\overline{S}(T) = \frac{\overline{C}_P(15\text{ K})}{3} + \int_{15}^{182.26} \frac{\overline{C}_P[\text{N}_2\text{O(s)}]}{T} dT + \frac{\Delta_{\text{fus}}\overline{H}}{182.26\text{ K}}$$

$$+ \int_{182.26}^{184.67} \frac{\overline{C}_P[\text{N}_2\text{O(l)}]}{T} dT + \frac{\Delta_{\text{vap}}\overline{H}}{184.67\text{ K}}$$

$$= 0.967\text{ J}\cdot\text{K}^{-1}\cdot\text{mol}^{-1} + 69.34\text{ J}\cdot\text{K}^{-1}\cdot\text{mol}^{-1} + 35.9\text{ J}\cdot\text{K}^{-1}\cdot\text{mol}^{-1}$$

$$+1.02\text{ J}\cdot\text{K}^{-1}\cdot\text{mol}^{-1} + 89.5\text{ J}\cdot\text{K}^{-1}\cdot\text{mol}^{-1}$$

$$= 196.7\text{ J}\cdot\text{K}^{-1}\cdot\text{mol}^{-1}$$

7–21. Methylammonium chloride occurs as three crystalline forms, called β, γ, and α, between 0 K and 298.15 K. The constant-pressure molar heat capacity of methylammonium chloride as a function of temperature is tabulated below. The $\beta \rightarrow \gamma$ transition occurs at 220.4 K with $\Delta_{\text{trs}}\overline{H} = 1.779\text{ kJ}\cdot\text{mol}^{-1}$ and the $\gamma \rightarrow \alpha$ transition occurs at 264.5 K with $\Delta_{\text{trs}}\overline{H} = 2.818\text{ kJ}\cdot\text{mol}^{-1}$. Assuming the heat capacity of solid methylammonium chloride can be described by the Debye theory up to 12 K, calculate the molar entropy of methylammonium chloride at 298.15 K.

T/K	$\overline{C}_P$/J·K^{-1}·mol^{-1}	T/K	$\overline{C}_P$/J·K^{-1}·mol^{-1}
12	0.837	180	73.72
15	1.59	200	77.95
20	3.92	210	79.71
30	10.53	220.4	$\beta \rightarrow \gamma$ transition
40	18.28	222	82.01
50	25.92	230	82.84
60	32.76	240	84.27
70	38.95	260	87.03
80	44.35	264.5	$\gamma \rightarrow \alpha$ transition
90	49.08	270	88.16
100	53.18	280	89.20
120	59.50	290	90.16
140	64.81	295	90.63
160	69.45		

We can this problem in the same way as Problem 7–20, graphing the molar heat capacities of the β, α, and γ crystalline forms versus temperature:

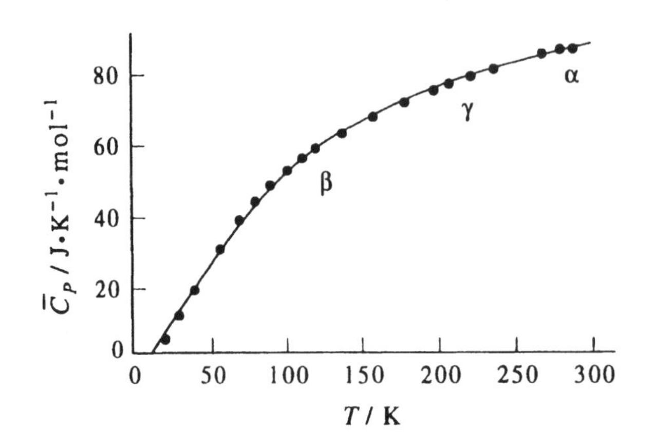

Fitting the curves to a polynomial, we find the following expressions for $\overline{C}_P$:

$$\overline{C}_P[\beta]/\text{J}\cdot\text{K}^{-1}\cdot\text{mol}^{-1} = -12.432 + (0.93892\ \text{K}^{-1})T - (3.4126\times10^{-3}\ \text{K}^{-2})T^2 + (4.8562\times10^{-6}\ \text{K}^{-3})T^3$$

$$12\ \text{K} \le T \le 220.4\ \text{K}$$

$$\overline{C}_P[\gamma]/\text{J}\cdot\text{K}^{-1}\cdot\text{mol}^{-1} = 78.265 - (8.2955\times10^{-2}\ \text{K}^{-1})T + (4.4885\times10^{-4}\ \text{K}^{-2})T^2$$

$$220.4\ \text{K} \le T \le 264.5\ \text{K}$$

$$\overline{C}_P[\alpha]/\text{J}\cdot\text{K}^{-1}\cdot\text{mol}^{-1} = 35.757 + (0.28147\ \text{K}^{-1})T - (3.2362\times10^{-4}\ \text{K}^{-2})T^2$$

$$264.5\ \text{K} \le T$$

From the Debye theory (Example 7–3),

$$\overline{S}(12\ \text{K}) = \frac{\overline{C}_P(12\ \text{K})}{3} \qquad 0 < T \le 12\ \text{K}$$

Now we can write the molar entropy of methylammonium chloride as

$$\begin{aligned}\overline{S}(298.15\ \text{K}) &= \frac{\overline{C}_P(12\ \text{K})}{3} + \int_{12}^{220.4}\frac{\overline{C}_P[\beta]}{T}dT + \frac{\Delta_{\beta\to\gamma}\overline{H}}{220.4\ \text{K}} + \int_{220.4}^{264.5}\frac{\overline{C}_P[\gamma]}{T}dT \\ &\quad + \frac{\Delta_{\gamma\to\alpha}\overline{H}}{264.5\ \text{K}} + \int_{264.5}^{298.15}\frac{\overline{C}_P[\alpha]}{T}dT \\ &= 0.279\ \text{J}\cdot\text{K}^{-1}\cdot\text{mol}^{-1} + 94.17\ \text{J}\cdot\text{K}^{-1}\cdot\text{mol}^{-1} + 8.07\ \text{J}\cdot\text{K}^{-1}\cdot\text{mol}^{-1} \\ &\quad + 15.42\ \text{J}\cdot\text{K}^{-1}\cdot\text{mol}^{-1} + 10.65\ \text{J}\cdot\text{K}^{-1}\cdot\text{mol}^{-1} + 10.69\ \text{J}\cdot\text{K}^{-1}\cdot\text{mol}^{-1} \\ &= 139.3\ \text{J}\cdot\text{K}^{-1}\cdot\text{mol}^{-1}\end{aligned}$$

7–22. The constant-pressure molar heat capacity of chloroethane as a function of temperature is tabulated below. Chloroethane melts at 134.4 K with $\Delta_{\text{fus}}\overline{H} = 4.45\ \text{kJ}\cdot\text{mol}^{-1}$, and boils at 286.2 K with $\Delta_{\text{vap}}\overline{H} = 24.65\ \text{kJ}\cdot\text{mol}^{-1}$ at one bar. Furthermore, the heat capacity of solid chloroethane can be described by the Debye theory up to 15 K. Use these data to calculate the molar entropy of chloroethane at its boiling point.

T/K	$\overline{C}_P/\text{J}\cdot\text{K}^{-1}\cdot\text{mol}^{-1}$	T/K	$\overline{C}_P/\text{J}\cdot\text{K}^{-1}\cdot\text{mol}^{-1}$
15	5.65	130	84.60
20	11.42	134.4	90.83 (solid)
25	16.53		97.19 (liquid)
30	21.21	140	96.86
35	25.52	150	96.40
40	29.62	160	96.02
50	36.53	180	95.65
60	42.47	200	95.77
70	47.53	220	96.04
80	52.63	240	97.78
90	55.23	260	99.79
100	59.66	280	102.09
110	65.48	286.2	102.13
120	73.55		

Do this problem in the same way as Problem 7–20, graphing the molar heat capacities of solid and liquid chloroethane versus temperature:

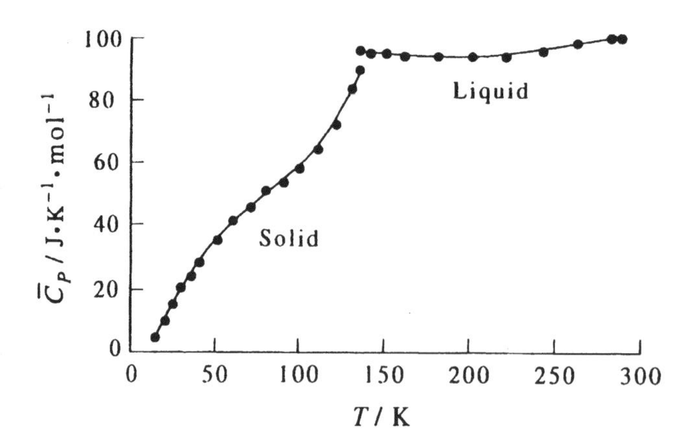

Fitting the curves to a polynomial, we find the following expressions for $\overline{C}_P$:

$$\overline{C}_P[\text{solid}]/\text{ J}\cdot\text{K}^{-1}\cdot\text{mol}^{-1} = -19.195 + (1.863\text{ K}^{-1})T - (1.8997\times10^{-2}\text{ K}^{-2})T^2 + (8.3132\times10^{-5}\text{ K}^{-3})T^3$$

$$15\text{ K} \leq T \leq 134.4\text{ K}$$

$$\overline{C}_P[\text{liquid}]/\text{ J}\cdot\text{K}^{-1}\cdot\text{mol}^{-1} = 118.15 - (0.24544\text{ K}^{-1})T + (6.675\times10^{-4}\text{ K}^{-2})T^2$$

$$134.4\text{ K} \leq T \leq 298.15\text{ K}$$

From the Debye theory (Example 7–3),

$$\overline{S}(15\text{ K}) = \frac{\overline{C}_P(15\text{ K})}{3} \qquad 0 < T \leq 15\text{ K}$$

Now

$$\begin{aligned}\overline{S}(T) &= \frac{\overline{C}_P(15\text{ K})}{3} + \int_{15}^{134.4}\frac{\overline{C}_P[\text{solid}]}{T}dT + \frac{\Delta_{\text{fus}}\overline{H}}{134.4\text{ K}} + \int_{134.4}^{286.2}\frac{\overline{C}_P[\text{liquid}]}{T}dT + \frac{\Delta_{\text{vap}}\overline{H}}{286.2\text{ K}}\\ &= 1.88\text{ J}\cdot\text{K}^{-1}\cdot\text{mol}^{-1} + 78.1\text{ J}\cdot\text{K}^{-1}\cdot\text{mol}^{-1} + 33.1\text{ J}\cdot\text{K}^{-1}\cdot\text{mol}^{-1}\\ &\quad +73.4\text{ J}\cdot\text{K}^{-1}\cdot\text{mol}^{-1} + 86.1\text{ J}\cdot\text{K}^{-1}\cdot\text{mol}^{-1}\\ &= 272.6\text{ J}\cdot\text{K}^{-1}\cdot\text{mol}^{-1}\end{aligned}$$

7–23. The constant-pressure molar heat capacity of nitromethane as a function of temperature is tabulated below. Nitromethane melts at 244.60 K with $\Delta_{\text{fus}}\overline{H} = 9.70\text{ kJ}\cdot\text{mol}^{-1}$, and boils at 374.34 K at one bar with $\Delta_{\text{vap}}\overline{H} = 38.27\text{ kJ}\cdot\text{mol}^{-1}$ at 298.15 K. Furthermore, the heat capacity of solid nitromethane can be described by the Debye theory up to 15 K. Use these data to calculate the molar entropy of nitromethane at 298.15 K and one bar. The vapor pressure of nitromethane is 36.66 torr at 298.15 K. (Be sure to take into account ΔS for the isothermal compression of nitromethane from its vapor pressure to one bar at 298.15 K).

T/K	$\overline{C}_P$/J·K^{-1}·mol^{-1}	T/K	$\overline{C}_P$/J·K^{-1}·mol^{-1}
15	3.72	200	71.46
20	8.66	220	75.23
30	19.20	240	78.99
40	28.87	244.60	melting point
60	40.84	250	104.43
80	47.99	260	104.64
100	52.80	270	104.93
120	56.74	280	105.31
140	60.46	290	105.69
160	64.06	300	106.06
180	67.74		

Do this problem in the same way as Problems 7–20, but include ΔS for the isothermal compression of nitromethane. Graph the molar heat capacities of solid and liquid chloroethane versus temperature:

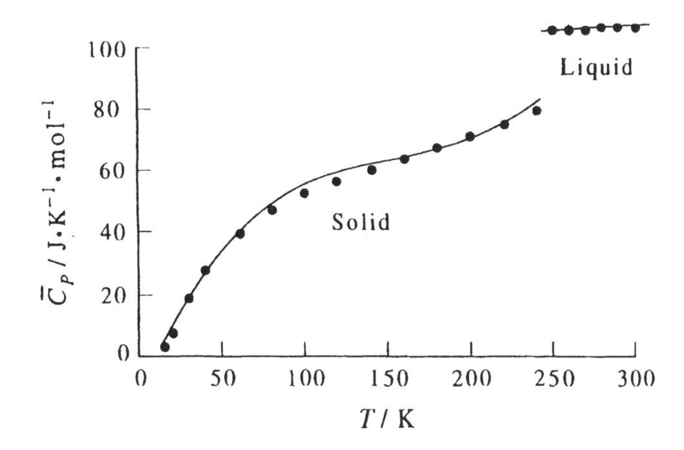

Fitting the curves to a polynomial, we find the following expressions for $\overline{C}_P$:

$$\overline{C}_P[\text{solid}]/\text{J·K}^{-1}\text{·mol}^{-1} = -11.177 + (1.1831\ \text{K}^{-1})T - (6.6826 \times 10^{-3}\ \text{K}^{-2})T^2 + (1.3948 \times 10^{-5}\ \text{K}^{-3})T^3$$

$$15\ \text{K} \leq T \leq 244.60\ \text{K}$$

$$\overline{C}_P[\text{liquid}]/\text{J·K}^{-1}\text{·mol}^{-1} = 111.6 - (8.0557 \times 10^{-2}\ \text{K}^{-1})T + (2.0714 \times 10^{-4}\ \text{K}^{-2})T^2$$

$$244.60\ \text{K} \leq T \leq 300\ \text{K}$$

From the Debye theory (Example 7–3),

$$\overline{S}(T) = \frac{\overline{C}_P(15\ \text{K})}{3} \qquad 0 < T \leq 15\ \text{K}$$

Assuming that nitromethane behaves ideally, $dU = 0$ for the isothermal compression and so $\delta q = PdV$. Then we can express the change in entropy for the isothermal compression as

$$\Delta\overline{S} = \int Pd\overline{V} = \int_{P_1}^{P_2} \frac{R}{P} dP$$

We can now write the molar entropy of nitromethane at the given conditions as

$$\overline{S}(T) = \frac{\overline{C}_P(15\text{ K})}{3} + \int_{15}^{244.60} \frac{\overline{C}_P[\text{solid}]}{T} dT + \frac{\Delta_{\text{fus}}\overline{H}}{244.60\text{ K}} + \int_{244.60}^{374.34} \frac{\overline{C}_P[\text{liquid}]}{T} dT$$

$$+ \frac{\Delta_{\text{vap}}\overline{H}}{374.34\text{ K}} - \int_{0.0489\text{ bar}}^{1\text{ bar}} \frac{R}{P} dP$$

$$= 1.24\text{ J}\cdot\text{K}^{-1}\cdot\text{mol}^{-1} + 109.3\text{ J}\cdot\text{K}^{-1}\cdot\text{mol}^{-1} + 39.66\text{ J}\cdot\text{K}^{-1}\cdot\text{mol}^{-1}$$

$$+20.79\text{ J}\cdot\text{K}^{-1}\cdot\text{mol}^{-1} + 128.4\text{ J}\cdot\text{K}^{-1}\cdot\text{mol}^{-1} - 25.1\text{ J}\cdot\text{K}^{-1}\cdot\text{mol}^{-1}$$

$$= 274.3\text{ J}\cdot\text{K}^{-1}\cdot\text{mol}^{-1}$$

7–24. Use the following data to calculate the standard molar entropy of CO(g) at its normal boiling point. Carbon monoxide undergoes a solid-solid phase transition at 61.6 K. Compare your result with the calculated value of 160.3 $\text{J}\cdot\text{K}^{-1}\cdot\text{mol}^{-1}$. Why is there a discrepancy between the calculated value and the experimental value?

$$\overline{C}_P[\text{CO(s}_1)]/R = -2.820 + (0.3317\text{ K}^{-1})T - (6.408\times10^{-3}\text{ K}^{-2})T^2 + (6.002\times10^{-5}\text{ K}^{-3})T^3$$
$$10\text{ K} \le T \le 61.6\text{ K}$$

$$\overline{C}_P[\text{CO(s}_2)]/R = 2.436 + (0.05694\text{ K}^{-1})T$$
$$61.6\text{ K} \le T \le 68.1\text{ K}$$

$$\overline{C}_P[\text{CO(l)}]/R = 5.967 + (0.0330\text{ K}^{-1})T - (2.088\times10^{-4}\text{ K}^{-2})T^2$$
$$68.1\text{ K} \le T \le 81.6\text{ K}$$

and $T_{\text{trs}}(\text{s}_1 \rightarrow \text{s}_2) = 61.6$ K, $T_{\text{fus}} = 68.1$ K, $T_{\text{vap}} = 81.6$ K, $\Delta_{\text{fus}}\overline{H} = 0.836\text{ kJ}\cdot\text{mol}^{-1}$, $\Delta_{\text{trs}}\overline{H} = 0.633\text{ kJ}\cdot\text{mol}^{-1}$, $\Delta_{\text{vap}}\overline{H} = 6.04\text{ kJ}\cdot\text{mol}^{-1}$, $\Theta_{\text{D}} = 79.5$ K, and the correction for nonideality $= 0.879\text{ J}\cdot\text{K}^{-1}\cdot\text{mol}^{-1}$.

$$\overline{S}(T) = \int_0^{10} \frac{12\pi^4}{5T} R\left(\frac{T}{\Theta_{\text{D}}}\right)^3 dT + \int_{10}^{61.6} \frac{\overline{C}_P[\text{CO(s}_1)]}{T} dT + \frac{\Delta_{\text{trs}}\overline{H}}{61.6\text{ K}}$$

$$+ \int_{61.6}^{68.1} \frac{\overline{C}_P[\text{CO(s}_2)]}{T} dT + \frac{\Delta_{\text{fus}}\overline{H}}{68.1\text{ K}} + \int_{68.1}^{81.6} \frac{\overline{C}_P[\text{CO(l)}]}{T} dT + \frac{\Delta_{\text{vap}}\overline{H}}{81.6\text{ K}} + \text{correction}$$

$$= 1.29\text{ J}\cdot\text{K}^{-1}\cdot\text{mol}^{-1} + 40.0\text{ J}\cdot\text{K}^{-1}\cdot\text{mol}^{-1} + 10.3\text{ J}\cdot\text{K}^{-1}\cdot\text{mol}^{-1} + 5.11\text{ J}\cdot\text{K}^{-1}\cdot\text{mol}^{-1}$$

$$+12.3\text{ J}\cdot\text{K}^{-1}\cdot\text{mol}^{-1} + 10.9\text{ J}\cdot\text{K}^{-1}\cdot\text{mol}^{-1} + 74.0\text{ J}\cdot\text{K}^{-1}\cdot\text{mol}^{-1} + 0.879\text{ J}\cdot\text{K}^{-1}\cdot\text{mol}^{-1}$$

$$= 154.7\text{ J}\cdot\text{K}^{-1}\cdot\text{mol}^{-1}$$

We have found an experimental value for $\overline{S}$ of 154.7 $\text{J}\cdot\text{K}^{-1}\cdot\text{mol}^{-1}$. The difference between this and the calculated value is the residual entropy of the crystal, which is approximately $R \ln 2$, or 5.8 $\text{J}\cdot\text{K}^{-1}\cdot\text{mol}^{-1}$ (in agreement with the difference calculated here of 5.6 $\text{J}\cdot\text{K}^{-1}\cdot\text{mol}^{-1}$).

7–25. The molar heat capacities of solid and liquid water can be expressed by

$$\overline{C}_P[H_2O(s)]/R = -0.2985 + (2.896 \times 10^{-2}\ K^{-1})T - (8.6714 \times 10^{-5}\ K^{-2})T^2 + (1.703 \times 10^{-7}\ K^{-3})T^3$$

$$10\ K \leq T \leq 273.15\ K$$

$$\overline{C}_P[H_2O(l)]/R = 22.447 - (0.11639\ K^{-1})T + (3.3312 \times 10^{-4}\ K^{-2})T^2 - (3.1314 \times 10^{-7}\ K^{-3})T^3$$

$$273.15\ K \leq T \leq 298.15\ K$$

and $T_{fus} = 273.15$ K, $\Delta_{fus}\overline{H} = 6.007\ kJ\cdot mol^{-1}$, $\Delta_{vap}\overline{H}(T = 298.15\ K) = 43.93\ kJ\cdot mol^{-1}$, $\Theta_D = 192$ K, the correction for nonideality $= 0.32\ J\cdot K^{-1}\cdot mol^{-1}$, and the vapor pressure of H_2O at 298.15 K = 23.8 torr. Use these data to calculate the standard molar entropy of $H_2O(g)$ at 298.15 K. You need the vapor pressure of water at 298.15 K because that is the equilibrium pressure of $H_2O(g)$ when it is vaporized at 298.15 K. You must include the value of ΔS that results when you compress the $H_2O(g)$ from 23.8 torr to its standard value of one bar. Your answer should come out to be $185.6\ J\cdot K^{-1}\cdot mol^{-1}$, which does not agree exactly with the value in Table 7.2. There is a residual entropy associated with ice, which a detailed analysis of the structure of ice gives as $\Delta S_{residual} = R\ln(3/2) = 3.4\ J\cdot K^{-1}\cdot mol^{-1}$, which is in good agreement with $\overline{S}_{calc} - \overline{S}_{exp}$.

We can do this problem in the same way as Problems 7–14, 7–16, and 7–18, taking into account ΔS for the isothermal compression of water. For an isothermal reaction, $\delta w = -\delta q$, so $\delta q = PdV$ and we assume that the gas is ideal. Then

$$\begin{aligned}\overline{S}(T) = &\int_0^{10} \frac{12\pi^4}{5T}R\left(\frac{T}{\Theta_D}\right)^3 dT + \int_{10}^{273.15} \frac{\overline{C}_P[H_2O(s)]}{T}dT + \frac{\Delta_{fus}\overline{H}}{273.15\ K} \\ &+ \int_{273.15}^{298.15} \frac{\overline{C}_P[H_2O(l)]}{T}dT + \frac{\Delta_{vap}\overline{H}}{298.15\ K} \\ &- \int_{0.0317}^{1} \frac{R}{P}dP + \text{correction} \\ = &\ 0.0915\ J\cdot K^{-1}\cdot mol^{-1} + 37.9\ J\cdot K^{-1}\cdot mol^{-1} + 22.0\ J\cdot K^{-1}\cdot mol^{-1} \\ &+ 6.62\ J\cdot K^{-1}\cdot mol^{-1} + 147.3\ J\cdot K^{-1}\cdot mol^{-1} - 28.69\ J\cdot K^{-1}\cdot mol^{-1} \\ &+ 0.32\ J\cdot K^{-1}\cdot mol^{-1} \\ = &\ 185.6\ J\cdot K^{-1}\cdot mol^{-1}\end{aligned}$$

Adding in the residual entropy gives a molar entropy of $189\ J\cdot K^{-1}\cdot mol^{-1}$.

7–26. Use the data in Problem 7–25 and the empirical expression

$$\overline{C}_P[H_2O(g)]/R = 3.652 + (1.156 \times 10^{-3}\ K^{-1})T - (1.424 \times 10^{-7}\ K^{-2})T^2$$

$$300\ K \leq T \leq 1000\ K$$

to plot the standard molar entropy of water from 0 K to 500 K.

Do this in the same manner as Problem 7–15, using the appropriate values from Problem 7–25 and changing the limits of integration as required.

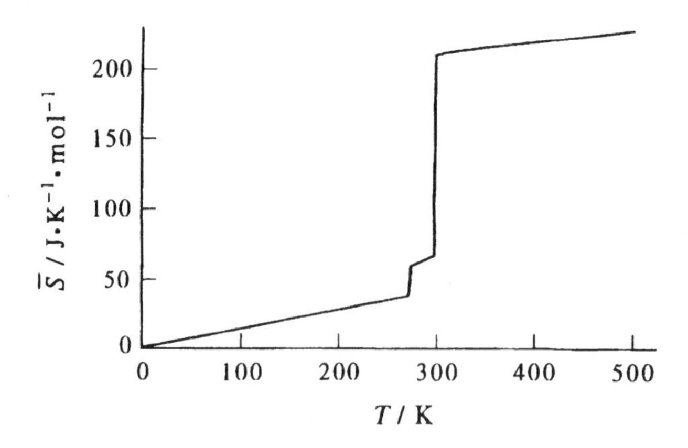

7–27. Show that

$$\overline{S} = R \ln \frac{qe}{N} + RT \left(\frac{\partial \ln q}{\partial T} \right)_V$$

We express the partition function of an ideal gas as (Equation 3.38)

$$Q(N, V, T) = \frac{q(V, T)^N}{N!}$$

and so

$$\ln Q(N, V, T) = N \ln q(V, T) - \ln N!$$

We substitute into Equation 7.19 to write

$$S = Nk_B \ln q - k_B \ln N + Nk_B T \left(\frac{\partial \ln q}{\partial T} \right)_V$$

We use Stirling's approximation and divide both sides of the equation by n to find

$$\overline{S} = N_A k_B \ln q - N_A k_B \ln N + k_B N_A + N_A k_B T \left(\frac{\partial \ln q}{\partial T} \right)_V$$

$$= R \ln \frac{q}{N} + R + RT \left(\frac{\partial \ln q}{\partial T} \right)_V$$

$$= R \ln \frac{qe}{N} + RT \left(\frac{\partial \ln q}{\partial T} \right)_V$$

7–28. Show that Equations 3.21 and 7.19 are consistent with Equations 7.2 and 7.3.

We begin with Equations 7.19 and 3.21,

$$S = k_B \ln Q + k_B T \left(\frac{\partial \ln Q}{\partial T} \right)_{N,V} \tag{7.19}$$

$$\langle E \rangle = k_B T^2 \left(\frac{\partial \ln Q}{\partial T} \right)_{N,V} \tag{3.21}$$

We can substitute $k_B T(\partial Q/\partial T)$ from Equation 3.21 into Equation 7.19 to find

$$S = k_B \ln Q + \frac{\langle E \rangle}{T} \tag{1}$$

$$\left(\frac{\partial S}{\partial T}\right)_V = k_B \left(\frac{\partial \ln Q}{\partial T}\right)_{N,V} + \frac{1}{T}\left(\frac{\partial \langle E \rangle}{\partial T}\right)_V - \frac{\langle E \rangle}{T^2}$$

Because $(\partial \langle E \rangle/\partial T)_V = C_V$, this becomes

$$\left(\frac{\partial S}{\partial T}\right)_V = k_B \left(\frac{\partial \ln Q}{\partial T}\right) + \frac{C_V}{T} - \frac{1}{T^2} k_B T^2 \left(\frac{\partial \ln Q}{\partial T}\right) = \frac{C_V}{T}$$

which is Equation 7.2. Now differentiate Equation 1 with respect to V:

$$S = k_B \ln Q + \frac{\langle E \rangle}{T}$$

$$\left(\frac{\partial S}{\partial V}\right)_T = k_B \left(\frac{\partial \ln Q}{\partial V}\right)_T + \frac{1}{T}\left(\frac{\partial \langle E \rangle}{\partial V}\right)_V$$

$$= \frac{1}{T}\left[k_B T \left(\frac{\partial \ln Q}{\partial V}\right)_T + \left(\frac{\partial U}{\partial V}\right)_T\right]$$

For an ideal gas, $k_B T = PV/N$, and $(\partial \ln Q/\partial V)_T = N/V$. Thus we have

$$\left(\frac{\partial S}{\partial V}\right)_T = \frac{1}{T}\left[P + \left(\frac{\partial U}{\partial V}\right)_T\right]$$

which is Equation 7.3.

7–29. Substitute Equation 7.23 into Equation 7.19 and derive the equation (Problem 6–31)

$$\Delta \overline{S} = \overline{C}_V \ln \frac{T_2}{T_1} + R \ln \frac{V_2}{V_1}$$

for one mole of a monatomic ideal gas.

We know that $Q = q^N/N!$ for an ideal gas (Equation 7.22), so substituting Equation 7.23 into this equation gives

$$Q = \frac{1}{N!}\left[\left(\frac{2\pi m k_B T}{h^2}\right)^{3/2} V \cdot g_{e1}\right]^N$$

Then

$$\ln Q = N \ln \left[\left(\frac{2\pi m k_B T}{h^2}\right)^{3/2} V \cdot g_{e1}\right] - N!$$

so, if only temperature and volume vary,

$$\ln Q_2 - \ln Q_1 = \frac{3}{2} N \ln \frac{T_2}{T_1} + \ln \frac{V_2}{V_1}$$

Also, we can find $(\partial \ln Q/\partial T)_V$:

$$\left(\frac{\partial \ln Q}{\partial T}\right)_{N,V} = \frac{3N}{2T}$$

Equation 7.19 states that

$$S = k_B \ln Q + k_B T \left(\frac{\partial \ln Q}{\partial T}\right)_{N,V}$$

Let $\Delta\overline{S} = \overline{S}_2 - \overline{S}_1$. Then

$$\begin{aligned}
\Delta\overline{S} &= \frac{1}{n}\left[k_B \ln Q_2 + k_B T_2 \left(\frac{\partial \ln Q_2}{\partial T}\right)_{N,V} - k_B \ln Q_1 - k_B T_1 \left(\frac{\partial \ln Q_1}{\partial T}\right)_{N,V}\right] \\
&= \frac{3}{2} R \ln \frac{T_2}{T_1} + R \ln \frac{V_2}{V_1} + \frac{3R}{2T_2} T_2 - \frac{3R}{2T_1} T_1 \\
&= \frac{3}{2} R \ln \frac{T_2}{T_1} + R \ln \frac{V_2}{V_1}
\end{aligned}$$

where $3R/2$ is equal to $\overline{C}_V$ for a monatomic ideal gas (Equation 3.26).

7–30. Use Equation 7.24 and the data in Chapter 4 to calculate the standard molar entropy of $Cl_2(g)$ at 298.15 K. Compare your answer with the experimental value of 223.1 $J \cdot K^{-1} \cdot mol^{-1}$.

$$q = \left(\frac{2\pi M k_B T}{h^2}\right) V \frac{T}{\sigma \Theta_{rot}} \frac{e^{-\Theta_{vib}/2T}}{1 - e^{-\Theta_{vib}/T}} g_{el} e^{D_e/k_B T} \tag{7.24}$$

Equation 7.28 (which is also for a diatomic ideal gas) can be written as

$$\frac{\overline{S}}{R} = \frac{7}{2} + \ln\left[\left(\frac{2\pi M k_B T}{h^2}\right)^{3/2} \frac{\overline{V}}{N_A}\right] + \ln \frac{T}{\sigma \Theta_{rot}} - \ln(1 - e^{-\Theta_{vib}/T}) + \frac{\Theta_{vib}/T}{e^{\Theta_{vib}/T} - 1} + \ln g_{el}$$

(see also Problem 20–41).

For chlorine, $\Theta_{vib} = 805$ K, $\Theta_{rot} = 0.351$ K, $\sigma = 2$, and $g_{el} = 1$. Then

$$\left(\frac{2\pi M k_B T}{h^2}\right)^{3/2} = \left[\frac{2\pi(1.1774 \times 10^{-25}\text{ kg})(1.381 \times 10^{-23}\text{ J}\cdot\text{K}^{-1})(298.15\text{ K})}{(6.626 \times 10^{34}\text{ J}\cdot\text{s})^2}\right]^{3/2} = 5.777 \times 10^{32}\text{ m}^{-3}$$

$$\frac{\overline{V}}{N_A} = \frac{RT}{N_A P} = \frac{(0.08314\text{ dm}^3\cdot\text{bar}\cdot\text{mol}^{-1}\cdot\text{K}^{-1})(298.15\text{ K})}{(6.022 \times 10^{23}\text{ mol}^{-1})(1\text{ bar})} = 4.116 \times 10^{-26}\text{ m}^3$$

$$\ln\left(\frac{T}{\sigma \Theta_{rot}}\right) = \ln\left[\frac{298.15\text{ K}}{2(0.351\text{ K})}\right] = 6.051$$

$$\ln(1 - e^{-\Theta_{vib}/T}) = -0.06957$$

$$\frac{\Theta_{vib}/T}{e^{\Theta_{vib}/T} - 1} = 0.1945$$

The standard molar entropy is then

$$\frac{\overline{S}}{R} = 3.5 + 16.98 + 6.051 + 0.06954 + 0.1945 = 26.80$$

This is 222.8 $J \cdot K^{-1} \cdot mol^{-1}$, which is very close to the experimental value.

7–31. Use Equation 7.24 and the data in Chapter 4 to calculate the standard molar entropy of CO(g) at its standard boiling point, 81.6 K. Compare your answer with the experimental value of 155.6 $J \cdot K^{-1} \cdot mol^{-1}$. Why is there a discrepancy of about 5 $J \cdot K^{-1} \cdot mol^{-1}$?

As in Problem 7–30,

$$\frac{\overline{S}}{R} = \frac{7}{2} + \ln\left[\left(\frac{2\pi M k_{\mathrm{B}} T}{h^2}\right)^{3/2} \frac{\overline{V}}{N_{\mathrm{A}}}\right] + \ln\frac{T}{\sigma\Theta_{\mathrm{rot}}} - \ln(1 - e^{-\Theta_{\mathrm{vib}}/T}) + \frac{\Theta_{\mathrm{vib}}/T}{e^{\Theta_{\mathrm{vib}}/T} - 1} + \ln g_{e1}$$

For carbon monoxide, $\Theta_{\mathrm{vib}} = 3103$ K, $\Theta_{\mathrm{rot}} = 2.77$ K, $\sigma = 1$, and $g_{e1} = 1$. Then

$$\left(\frac{2\pi M k_{\mathrm{B}} T}{h^2}\right)^{3/2} = \left[\frac{2\pi(4.651\times10^{-26}\ \mathrm{kg})(1.381\times10^{-23}\ \mathrm{J\cdot K^{-1}})(81.6\ \mathrm{K})}{(6.626\times10^{34}\ \mathrm{J\cdot s})^2}\right]^{3/2} = 2.054\times10^{31}\ \mathrm{m^{-3}}$$

$$\frac{\overline{V}}{N_{\mathrm{A}}} = \frac{RT}{N_{\mathrm{A}}P} = \frac{(0.08314\ \mathrm{dm^3\cdot bar\cdot mol^{-1}\cdot K^{-1}})(81.6\ \mathrm{K})}{(6.022\times10^{23}\ \mathrm{mol^{-1}})(1\ \mathrm{bar})} = 1.127\times10^{-26}\ \mathrm{m^3}$$

$$\ln\left(\frac{T}{\sigma\Theta_{\mathrm{rot}}}\right) = \ln\left[\frac{81.6\ \mathrm{K}}{(2.77\ \mathrm{K})}\right] = 3.383$$

$$\ln(1 - e^{-\Theta_{\mathrm{vib}}/T}) = -3.057\times10^{-17}$$

$$\frac{\Theta_{\mathrm{vib}}/T}{e^{\Theta_{\mathrm{vib}}/T} - 1} = 1.162\times10^{-15}$$

The standard molar entropy is then

$$\frac{\overline{S}}{R} = 3.5 + 12.352 + 3.383 + 3.057\times10^{-17} + 1.162\times10^{-15} = 19.23$$

This is 159.9 $\mathrm{J\cdot K^{-1}\cdot mol^{-1}}$, which is about 4 $\mathrm{J\cdot K^{-1}\cdot mol^{-1}}$ larger than the experimental value. The discrepancy is due to residual entropy.

7–32. Use Equation 7.26 and the data in Chapter 4 to calculate the standard molar entropy of $NH_3(g)$ at 298.15 K. Compare your answer with the experimental value of 192.8 $\mathrm{J\cdot K^{-1}\cdot mol^{-1}}$.

See Problem 6–43. The value calculated is 193.1 $\mathrm{J\cdot K^{-1}\cdot mol^{-1}}$.

7–33. Use Equation 7.24 and the data in Chapter 4 to calculate the standard molar entropy of $Br_2(g)$ at 298.15 K. Compare your answer with the experimental value of 245.5 $\mathrm{J\cdot K^{-1}\cdot mol^{-1}}$.

As in Problem 7–30,

$$\frac{\overline{S}}{R} = \frac{7}{2} + \ln\left[\left(\frac{2\pi M k_{\mathrm{B}} T}{h^2}\right)^{3/2} \frac{\overline{V}}{N_{\mathrm{A}}}\right] + \ln\frac{T}{\sigma\Theta_{\mathrm{rot}}} - \ln(1 - e^{-\Theta_{\mathrm{vib}}/T}) + \frac{\Theta_{\mathrm{vib}}/T}{e^{\Theta_{\mathrm{vib}}/T} - 1} + \ln g_{e1}$$

For bromine, $\Theta_{\mathrm{vib}} = 463$ K, $\Theta_{\mathrm{rot}} = 0.116$ K, $\sigma = 2$, and $g_{e1} = 1$. Then

$$\left(\frac{2\pi M k_{\mathrm{B}} T}{h^2}\right)^{3/2} = \left[\frac{2\pi(2.654\times10^{-25}\ \mathrm{kg})(1.381\times10^{-23}\ \mathrm{J\cdot K^{-1}})(298.15\ \mathrm{K})}{(6.626\times10^{34}\ \mathrm{J\cdot s})^2}\right]^{3/2} = 1.955\times10^{33}\ \mathrm{m^{-3}}$$

$$\frac{\overline{V}}{N_{\mathrm{A}}} = \frac{RT}{N_{\mathrm{A}}P} = \frac{(0.08314\ \mathrm{dm^3\cdot bar\cdot mol^{-1}\cdot K^{-1}})(298.15\ \mathrm{K})}{(6.022\times10^{23}\ \mathrm{mol^{-1}})(1\ \mathrm{bar})} = 4.116\times10^{-26}\ \mathrm{m^3}$$

$$\ln\left(\frac{T}{\sigma\Theta_{\mathrm{rot}}}\right) = \ln\left[\frac{298.15\ \mathrm{K}}{2(0.116\ \mathrm{K})}\right] = 7.159$$

$$\ln(1 - e^{-\Theta_{\mathrm{vib}}/T}) = -0.2378$$

$$\frac{\Theta_{\mathrm{vib}}/T}{e^{\Theta_{\mathrm{vib}}/T} - 1} = 0.417$$

The standard molar entropy is then

$$\frac{\overline{S}}{R} = 3.5 + 18.203 + 7.158 + 0.238 + 0.417 = 29.52$$

This is $245.4\ \mathrm{J{\cdot}K^{-1}{\cdot}mol^{-1}}$, almost identical to the experimental value.

7–34. The vibrational and rotational constants for HF(g) within the harmonic oscillator-rigid rotator model are $\tilde{\nu}_0 = 3959\ \mathrm{cm^{-1}}$ and $\tilde{B}_0 = 20.56\ \mathrm{cm^{-1}}$. Calculate the standard molar entropy of HF(g) at 298.15 K. How does this value compare with that in Table 7.3?

We can use the equalities (given in Chapter 4)

$$\Theta_{\text{vib}} = \frac{hc\tilde{\nu}_0}{k_B} \qquad \Theta_{\text{rot}} = \frac{hc\tilde{B}_0}{k_B}$$

to find $\Theta_{\text{vib}} = 5696$ K and $\Theta_{\text{rot}} = 29.58$ K, and then solve for entropy as we did in the previous problems.

As in Problem 7–30,

$$\frac{\overline{S}}{R} = \frac{7}{2} + \ln\left[\left(\frac{2\pi M k_B T}{h^2}\right)^{3/2}\frac{\overline{V}}{N_A}\right] + \ln\frac{T}{\sigma\Theta_{\text{rot}}} - \ln(1 - e^{-\Theta_{\text{vib}}/T}) + \frac{\Theta_{\text{vib}}/T}{e^{\Theta_{\text{vib}}/T} - 1} + \ln g_{e1}$$

Substitute in to find the values of the various components of entropy. Note that $\sigma = 1$ and $g_{e1} = 1$.

$$\left(\frac{2\pi M k_B T}{h^2}\right)^{3/2} = \left[\frac{2\pi(3.322\times10^{-26}\ \text{kg})(1.381\times10^{-23}\ \mathrm{J{\cdot}K^{-1}})(298.15\ \text{K})}{(6.626\times10^{34}\ \mathrm{J{\cdot}s})^2}\right]^{3/2} = 8.658\times10^{31}\ \mathrm{m^{-3}}$$

$$\frac{\overline{V}}{N_A} = \frac{RT}{N_A P} = \frac{(0.08314\ \mathrm{dm^3{\cdot}bar{\cdot}mol^{-1}{\cdot}K^{-1}})(298.15\ \text{K})}{(6.022\times10^{23}\ \mathrm{mol^{-1}})(1\ \text{bar})} = 4.116\times10^{-26}\ \mathrm{m^3}$$

$$\ln\left(\frac{T}{\sigma\Theta_{\text{rot}}}\right) = 2.310$$

$$\ln(1 - e^{-\Theta_{\text{vib}}/T}) = -5.046\times10^{-9}$$

$$\frac{\Theta_{\text{vib}}/T}{e^{\Theta_{\text{vib}}/T} - 1} = 9.639\times10^{-8}$$

$$\frac{\overline{S}}{R} = 3.5 + 15.09 + 2.310 + 5.046\times10^{-9} + 9.639\times10^{-8} = 20.90$$

The standard molar entropy is $173.7\ \mathrm{J{\cdot}K^{-1}{\cdot}mol^{-1}}$, which is very close to the value in Table 7.3.

7–35. Calculate the standard molar entropy of H_2(g) and D_2(g) at 298.15 K given that the bond length of both diatomic molecules is 74.16 pm and the vibrational temperatures of H_2(g) and D_2(g) are 6332 K and 4480 K, respectively. Calculate the standard molar entropy of HD(g) at 298.15 K ($R_e = 74.13$ pm and $\Theta_{\text{vib}} = 5496$ K).

We can use the relation $\Theta_{\text{rot}} = \hbar^2/2Ik_B$ (Equation 4.32) to find Θ_{rot} for HD, H_2, and D_2. Then we can solve for molar entropy as in Problem 7–30. For both H_2 and D_2, $\sigma = 2$ and $g_{e1} = 1$; for HD, $\sigma = 1$ and $g_{e1} = 1$. For H_2,

$$\Theta_{\text{rot}} = \frac{\hbar^2}{2\mu R_e^2 k_B} = \frac{\hbar^2}{2(8.368 \times 10^{-28}\text{ kg})(74.16 \times 10^{-12}\text{ m})^2 k_B} = 87.51\text{ K}$$

and

$$\frac{\overline{S}}{R} = \frac{7}{2} + \ln\left[\left(\frac{2\pi M k_B T}{h^2}\right)^{3/2} \frac{\overline{V}}{N_A}\right] + \ln\frac{T}{\sigma\Theta_{\text{rot}}} - \ln(1 - e^{-\Theta_{\text{vib}}/T}) + \frac{\Theta_{\text{vib}}/T}{e^{\Theta_{\text{vib}}/T} - 1} + \ln g_{e1}$$

Substitute in to find the values of the various components of entropy.

$$\left(\frac{2\pi M k_B T}{h^2}\right)^{3/2} = \left[\frac{2\pi(3.348 \times 10^{-27}\text{ kg})(1.381 \times 10^{-23}\text{ J}\cdot\text{K}^{-1})(298.15\text{ K})}{(6.626 \times 10^{34}\text{ J}\cdot\text{s})^2}\right]^{3/2} = 2.769 \times 10^{30}\text{ m}^{-3}$$

$$\frac{\overline{V}}{N_A} = \frac{RT}{N_A P} = \frac{(0.08314\text{ dm}^3\cdot\text{bar}\cdot\text{mol}^{-1}\cdot\text{K}^{-1})(298.15\text{ K})}{(6.022 \times 10^{23}\text{ mol}^{-1})(1\text{ bar})} = 4.116 \times 10^{-26}\text{ m}^3$$

$$\ln\left(\frac{T}{\sigma\Theta_{\text{rot}}}\right) = 0.533$$

$$\ln(1 - e^{-\Theta_{\text{vib}}/T}) = -8.852 \times 10^{-10}$$

$$\frac{\Theta_{\text{vib}}/T}{e^{\Theta_{\text{vib}}/T} - 1} = 1.845 \times 10^{-8}$$

$$\frac{\overline{S}}{R} = 3.55 + 11.64 + 0.533 + 8.852 \times 10^{-10} + 1.845 \times 10^{-8} = 15.72$$

The standard molar entropy of H_2(g) at 298.15 K is 130.7 $\text{J}\cdot\text{K}^{-1}\cdot\text{mol}^{-1}$.

For D_2,

$$\Theta_{\text{rot}} = \frac{\hbar^2}{2\mu R_e^2 k_B} = \frac{\hbar^2}{2(1.672 \times 10^{-27}\text{ kg})(74.16 \times 10^{-12}\text{ m})^2 k_B} = 43.79\text{ K}$$

and

$$\frac{\overline{S}}{R} = \frac{7}{2} + \ln\left[\left(\frac{2\pi M k_B T}{h^2}\right)^{3/2} \frac{\overline{V}}{N_A}\right] + \ln\frac{T}{\sigma\Theta_{\text{rot}}} - \ln(1 - e^{-\Theta_{\text{vib}}/T}) + \frac{\Theta_{\text{vib}}/T}{e^{\Theta_{\text{vib}}/T} - 1} + \ln g_{e1}$$

Substitute in to find the values of the various components of entropy.

$$\left(\frac{2\pi M k_B T}{h^2}\right)^{3/2} = \left[\frac{2\pi(6.689 \times 10^{-27}\text{ kg})(1.381 \times 10^{-23}\text{ J}\cdot\text{K}^{-1})(298.15\text{ K})}{(6.626 \times 10^{34}\text{ J}\cdot\text{s})^2}\right]^{3/2} = 7.822 \times 10^{30}\text{ m}^{-3}$$

$$\frac{\overline{V}}{N_A} = \frac{RT}{N_A P} = \frac{(0.08314\text{ dm}^3\cdot\text{bar}\cdot\text{mol}^{-1}\cdot\text{K}^{-1})(298.15\text{ K})}{(6.022 \times 10^{23}\text{ mol}^{-1})(1\text{ bar})} = 4.116 \times 10^{-26}\text{ m}^3$$

$$\ln\left(\frac{T}{\sigma\Theta_{\text{rot}}}\right) = 1.225$$

$$\ln(1 - e^{-\Theta_{\text{vib}}/T}) = -3.977 \times 10^{-7}$$

$$\frac{\Theta_{\text{vib}}/T}{e^{\Theta_{\text{vib}}/T} - 1} = 5.861 \times 10^{-6}$$

$$\frac{\overline{S}}{R} = 3.52 + 12.682 + 1.226 + 3.977 \times 10^{-7} + 5.861 \times 10^{-6} = 17.43$$

The standard molar entropy of D_2(g) at 298.15 K is 144.9 $\text{J}\cdot\text{K}^{-1}\cdot\text{mol}^{-1}$.

For HD,

$$\Theta_{\text{rot}} = \frac{\hbar^2}{2\mu R_e^2 k_B} = \frac{\hbar^2}{2(1.115 \times 10^{-27}\ \text{kg})(74.13 \times 10^{-12}\ \text{m})^2 k_B} = 65.71\ \text{K}$$

and

$$\frac{\overline{S}}{R} = \frac{7}{2} + \ln\left[\left(\frac{2\pi M k_B T}{h^2}\right)^{3/2} \frac{\overline{V}}{N_A}\right] + \ln\frac{T}{\sigma\Theta_{\text{rot}}} - \ln(1 - e^{-\Theta_{\text{vib}}/T}) + \frac{\Theta_{\text{vib}}/T}{e^{\Theta_{\text{vib}}/T} - 1} + \ln g_{e1}$$

Substitute in to find the values of the various components of entropy.

$$\left(\frac{2\pi M k_B T}{h^2}\right)^{3/2} = \left[\frac{2\pi(5.018 \times 10^{-27}\ \text{kg})(1.381 \times 10^{-23}\ \text{J}\cdot\text{K}^{-1})(298.15\ \text{K})}{(6.626 \times 10^{34}\ \text{J}\cdot\text{s})^2}\right]^{3/2} = 5.082 \times 10^{30}\ \text{m}^{-3}$$

$$\frac{\overline{V}}{N_A} = \frac{RT}{N_A P} = \frac{(0.08314\ \text{dm}^3\cdot\text{bar}\cdot\text{mol}^{-1}\cdot\text{K}^{-1})(298.15\ \text{K})}{(6.022 \times 10^{23}\ \text{mol}^{-1})(1\ \text{bar})} = 4.116 \times 10^{-26}\ \text{m}^3$$

$$\ln\left(\frac{T}{\sigma\Theta_{\text{rot}}}\right) = 1.512$$

$$\ln(1 - e^{-\Theta_{\text{vib}}/T}) = -9.871 \times 10^{-9}$$

$$\frac{\Theta_{\text{vib}}/T}{e^{\Theta_{\text{vib}}/T} - 1} = 1.820 \times 10^{-7}$$

$$\frac{\overline{S}}{R} = 3.53 + 12.251 + 1.512 + 9.871 \times 10^{-9} + 1.820 \times 10^{-7} = 17.30$$

The standard molar entropy of HD(g) at 298.15 K is 143.8 $\text{J}\cdot\text{K}^{-1}\cdot\text{mol}^{-1}$.

7–36. Calculate the standard molar entropy of HCN(g) at 1000 K given that $I = 1.8816 \times 10^{-46}\ \text{kg}\cdot\text{m}^2$, $\tilde{\nu}_1 = 2096.70\ \text{cm}^{-1}$, $\tilde{\nu}_2 = 713.46\ \text{cm}^{-1}$, and $\tilde{\nu}_3 = 3311.47\ \text{cm}^{-1}$. Recall that HCN(g) is a linear triatomic molecule and therefore the bending mode, ν_2, is doubly degenerate.

In Problem 4–24, we found $\Theta_{\text{vib},j}$ and Θ_{rot} of HCN to be

$$\Theta_{\text{vib},1} = 3016\ \text{K} \qquad \Theta_{\text{vib},4} = 4764\ \text{K}$$
$$\Theta_{\text{vib},2,3} = 1026\ \text{K} \qquad \Theta_{\text{rot}} = 2.135\ \text{K}$$

For a linear polyatomic ideal gas having three atoms,

$$q = \left(\frac{2\pi M k_B T}{h^2}\right)^{3/2} V \frac{T}{\sigma\Theta_{\text{rot}}} \left(\prod_{j=1}^{4} \frac{e^{-\Theta_{\text{vib},j}/2T}}{1 - e^{-\Theta_{\text{vib},j}/T}}\right) g_{e1} e^{D_e/k_B T} \tag{7.25}$$

Substituting into Equation 7.27, we find

$$\frac{\overline{S}}{R} = \frac{7}{2} + \ln\left[\left(\frac{2\pi M k_B T}{h^2}\right)^{3/2} \frac{\overline{V}}{N_A}\right] + \ln\frac{T}{\sigma\Theta_{\text{rot}}} - \sum_{j=1}^{4} \ln(1 - e^{-\Theta_{\text{vib},j}/T}) + \sum_{j=1}^{4} \left[\frac{(\Theta_{\text{vib},j}/T)e^{-\Theta_{\text{vib},j}/T}}{1 - e^{-\Theta_{\text{vib},j}/T}}\right] + \ln g_{e1}$$

Because HCN is asymmetrical, its symmetry number is unity. Then

$$\left(\frac{2\pi M k_B T}{h^2}\right)^{3/2} = \left[\frac{2\pi(4.488 \times 10^{-26}\ \text{kg})(1.381 \times 10^{-23}\ \text{J}\cdot\text{K}^{-1})(1000\ \text{K})}{(6.626 \times 10^{34}\ \text{J}\cdot\text{s})^2}\right]^{3/2} = 8.350 \times 10^{32}\ \text{m}^{-3}$$

$$\frac{\overline{V}}{N_A} = \frac{RT}{N_A P} = \frac{(0.08314\ \text{dm}^3\cdot\text{bar}\cdot\text{mol}^{-1}\cdot\text{K}^{-1})(1000\ \text{K})}{(6.022 \times 10^{23}\ \text{mol}^{-1})(1\ \text{bar})} = 1.381 \times 10^{-25}\ \text{m}^3$$

$$\ln\left(\frac{T}{\sigma\Theta_{\text{rot}}}\right) = \ln\left[\frac{1000\text{ K}}{(2.135\text{ K})}\right] = 6.149$$

$$\sum_{j=1}^{4} \ln(1 - e^{-\Theta_{\text{vib},j}/T}) = -0.9465$$

$$\sum_{j=1}^{4} \frac{\Theta_{\text{vib},j}}{T}\left(\frac{e^{-\Theta_{\text{vib},j}/T}}{1 - e^{-\Theta_{\text{vib},j}/T}}\right) = 1.343$$

The standard molar entropy is then

$$\frac{\overline{S}}{R} = 3.5 + 18.563 + 6.149 + 0.9465 + 1.343 = 30.5$$

The standard molar entropy of HCN(g) at 1000 K is 253.6 $\text{J}\cdot\text{K}^{-1}\cdot\text{mol}^{-1}$. The experimentally observed value is 253.7 $\text{J}\cdot\text{K}^{-1}\cdot\text{mol}^{-1}$.

7–37. Given that $\tilde{\nu}_1 = 1321.3\text{ cm}^{-1}$, $\tilde{\nu}_2 = 750.8\text{ cm}^{-1}$, $\tilde{\nu}_3 = 1620.3\text{ cm}^{-1}$, $\tilde{A}_0 = 7.9971\text{ cm}^{-1}$, $\tilde{B}_0 = 0.4339\text{ cm}^{-1}$, and $\tilde{C}_0 = 0.4103\text{ cm}^{-1}$, calculate the standard molar entropy of NO_2(g) at 298.15 K. (Note that NO_2(g) is a bent triatomic molecule.) How does your value compare with that in Table 7.2?

In Problem 4–29, we found that

$$\Theta_{\text{vib},1} = 1898.7\text{ K} \qquad \Theta_{\text{rot},A} = 11.512\text{ K}$$
$$\Theta_{\text{vib},2} = 1078.8\text{ K} \qquad \Theta_{\text{rot},B} = 0.62304\text{ K}$$
$$\Theta_{\text{vib},3} = 2327.6\text{ K} \qquad \Theta_{\text{rot},C} = 0.59047\text{ K}$$

For a nonlinear polyatomic ideal gas having three atoms,

$$q = \left(\frac{2\pi M k_B T}{h^2}\right)^{3/2} V \frac{\pi^{1/2}}{\sigma}\left(\frac{T^3}{\Theta_{\text{rot,A}}\Theta_{\text{rot,B}}\Theta_{\text{rot,C}}}\right)^{1/2}\left(\prod_{j=1}^{3}\frac{e^{-\Theta_{\text{vib},j}/2T}}{1 - e^{-\Theta_{\text{vib},j}/T}}\right) g_{e1} e^{D_e/k_B T} \tag{7.26}$$

Substituting into Equation 7.27, we find

$$\frac{\overline{S}}{R} = 4 + \ln\left[\left(\frac{2\pi M k_B T}{h^2}\right)^{3/2}\frac{\overline{V}}{N_A}\right] + \ln\left(\frac{\pi^{1/2}}{\sigma}\right) + \frac{1}{2}\ln\frac{T^3}{\Theta_{\text{rot,A}}\Theta_{\text{rot,B}}\Theta_{\text{rot,C}}}$$
$$- \sum_{j=1}^{3}\ln(1 - e^{-\Theta_{\text{vib},j}/T}) + \sum_{j=1}^{3}\left[\frac{(\Theta_{\text{vib},j}/T)e^{-\Theta_{\text{vib},j}/T}}{1 - e^{-\Theta_{\text{vib},j}/T}}\right] + \ln g_{e1}$$

From Table 4.4, $\sigma = 2$, and $g_{e1} = 1$. Then

$$\left(\frac{2\pi M k_B T}{h^2}\right)^{3/2} = \left[\frac{2\pi(7.639\times10^{-26}\text{ kg})(1.381\times10^{-23}\text{ J}\cdot\text{K}^{-1})(298.15\text{ K})}{(6.626\times10^{34}\text{ J}\cdot\text{s})^2}\right]^{3/2} = 3.019\times10^{32}\text{ m}^{-3}$$

$$\frac{\overline{V}}{N_A} = \frac{RT}{N_A P} = \frac{(0.08314\text{ dm}^3\cdot\text{bar}\cdot\text{mol}^{-1}\cdot\text{K}^{-1})(298.15\text{ K})}{(6.022\times10^{23}\text{ mol}^{-1})(1\text{ bar})} = 4.116\times10^{-26}\text{ m}^3$$

$$\frac{1}{2}\ln\left(\frac{T^3}{\Theta_{\text{rot,A}}\Theta_{\text{rot,B}}\Theta_{\text{rot,C}}}\right) = 7.825$$

$$\sum_{j=1}^{3}\ln(1-e^{-\Theta_{\text{vib},j}/T}) = -0.0293$$

$$\sum_{j=1}^{3}\frac{\Theta_{\text{vib},j}}{T}\left(\frac{e^{-\Theta_{\text{vib},j}/T}}{1-e^{-\Theta_{\text{vib},j}/T}}\right) = 0.114$$

The standard molar entropy is then

$$\frac{\overline{S}}{R} = 4 + 16.335 - 0.121 + 7.825 + 0.0293 + 0.114 = 28.18$$

This is 234.3 $\text{J}\cdot\text{K}^{-1}\cdot\text{mol}^{-1}$. The experimental value is 240.1 $\text{J}\cdot\text{K}^{-1}\cdot\text{mol}^{-1}$. The difference is due to residual entropy.

7–38. In Problem 7–48, you are asked to calculate the value of $\Delta_r S^\circ$ at 298.15 K using the data in Table 7.2 for the reaction described by

$$2\,\text{CO(g)} + \text{O}_2\text{(g)} \longrightarrow 2\,\text{CO}_2\text{(g)}$$

Use the data in Table 4.2 to calculate the standard molar entropy of each of the reagents in this reaction [see Example 7–5 for the calculation of the standard molar entropy of CO_2(g)]. Then use these results to calculate the standard entropy change for the above reaction. How does your answer compare with what you obtained in Problem 7–48?

From Example 7–5, $S^\circ[\text{CO}_2(g)] = 213.8\ \text{J}\cdot\text{K}^{-1}\cdot\text{mol}^{-1}$. Because both CO and O_2 are diatomic molecules, we can write (as in Problem 7–30)

$$\frac{\overline{S}}{R} = \frac{7}{2} + \ln\left[\left(\frac{2\pi M k_B T}{h^2}\right)^{3/2}\frac{\overline{V}}{N_A}\right] + \ln\frac{T}{\sigma\Theta_{\text{rot}}} - \ln(1-e^{-\Theta_{\text{vib}}/T}) + \frac{\Theta_{\text{vib}}/T}{e^{\Theta_{\text{vib}}/T}-1} + \ln g_{e1}$$

Because CO is a heteronuclear diatomic molecule, $\sigma = 1$; because O_2 is homonuclear, $\sigma = 2$. For CO $\Theta_{\text{vib}} = 3103$ K and $\Theta_{\text{rot}} = 2.77$ K. Then

$$\left(\frac{2\pi M k_B T}{h^2}\right)^{3/2} = \left[\frac{2\pi(4.651\times10^{-26}\ \text{kg})(1.381\times10^{-23}\ \text{J}\cdot\text{K}^{-1})(298.15\ \text{K})}{(6.626\times10^{34}\ \text{J}\cdot\text{s})^2}\right]^{3/2} = 1.434\times10^{32}\ \text{m}^{-3}$$

$$\frac{\overline{V}}{N_A} = \frac{RT}{N_A P} = \frac{(0.08314\ \text{dm}^3\cdot\text{bar}\cdot\text{mol}^{-1}\cdot\text{K}^{-1})(298.15\ \text{K})}{(6.022\times10^{23}\ \text{mol}^{-1})(1\ \text{bar})} = 4.116\times10^{-26}\ \text{m}^3$$

$$\ln\left(\frac{T}{\sigma\Theta_{\text{rot}}}\right) = 4.679$$

$$\ln(1-e^{-\Theta_{\text{vib}}/T}) = -3.02\times10^{-5}$$

$$\frac{\Theta_{\text{vib}}/T}{e^{\Theta_{\text{vib}}/T}-1} = 3.14\times10^{-4}$$

$$\frac{\overline{S}}{R} = 3.5 + 15.591 + 4.679 + 3.02\times10^{-5} + 3.14\times10^{-4} = 23.77$$

The standard molar entropy of CO(g) at 298.15 K is 197.6 $\text{J}\cdot\text{K}^{-1}\cdot\text{mol}^{-1}$.

We follow the same procedure for O_2, with $\Theta_{vib} = 2256$ K and $\Theta_{rot} = 2.07$ K. Note that $g_{el} = 3$ for O_2, so we cannot neglect the $\ln g_{el}$ term!

$$\left(\frac{2\pi M k_B T}{h^2}\right)^{3/2} = \left[\frac{2\pi(5.313\times10^{-26}\text{ kg})(1.381\times10^{-23}\text{ J}\cdot\text{K}^{-1})(298.15\text{ K})}{(6.626\times10^{34}\text{ J}\cdot\text{s})^2}\right]^{3/2} = 1.751\times10^{32}\text{ m}^{-3}$$

$$\frac{\overline{V}}{N_A} = \frac{RT}{N_A P} = \frac{(0.08314\text{ dm}^3\cdot\text{bar}\cdot\text{mol}^{-1}\cdot\text{K}^{-1})(298.15\text{ K})}{(6.022\times10^{23}\text{ mol}^{-1})(1\text{ bar})} = 4.116\times10^{-26}\text{ m}^3$$

$$\ln\left(\frac{T}{\sigma\Theta_{rot}}\right) = 4.277$$

$$\ln(1-e^{-\Theta_{vib}/T}) = -5.18\times10^{-4}$$

$$\frac{\Theta_{vib}/T}{e^{\Theta_{vib}/T}-1} = 3.92\times10^{-3}$$

$$\ln g_{el} = \ln 3 = 1.099$$

$$\frac{\overline{S}}{R} = 3.5 + 15.79 + 4.277 + 5.18\times10^{-4} + 3.92\times10^{-3} + 1.099 = 24.67$$

The standard molar entropy of $O_2(g)$ at 298.15 K is 205.1 $\text{J}\cdot\text{K}^{-1}\cdot\text{mol}^{-1}$.

We can calculate the entropy change for the above reaction easily using the method described in Section 7–9:

$$\begin{aligned}\Delta_r S^\circ &= 2S^\circ[CO_2] - S^\circ[O_2] - 2S^\circ[CO]\\ &= 2(213.8\text{ J}\cdot\text{K}^{-1}\cdot\text{mol}^{-1}) - (205.1\text{ J}\cdot\text{K}^{-1}\cdot\text{mol}^{-1}) - 2(197.6\text{ J}\cdot\text{K}^{-1}\cdot\text{mol}^{-1})\\ &= -172.7\text{ J}\cdot\text{K}^{-1}\cdot\text{mol}^{-1}\end{aligned}$$

This value is very close to that found in Problem 7–48.

7–39. Calculate the value of $\Delta_r S^\circ$ for the reaction described by

$$H_2(g) + \tfrac{1}{2}O_2(g) \longrightarrow H_2O(g)$$

at 500 K using the data in Tables 4.2 and 4.4.

Because both H_2 and O_2 are diatomic molecules, we can write (as in the previous problem)

$$\frac{\overline{S}}{R} = \frac{7}{2} + \ln\left[\left(\frac{2\pi M k_B T}{h^2}\right)^{3/2}\frac{\overline{V}}{N_A}\right] + \ln\frac{T}{\sigma\Theta_{rot}} - \ln(1-e^{-\Theta_{vib}/T}) + \frac{\Theta_{vib}/T}{e^{\Theta_{vib}/T}-1} + \ln g_{el}$$

Because both are homonuclear, $\sigma = 2$ for both H_2 and O_2. For H_2 $\Theta_{vib} = 6332$ K and $\Theta_{rot} = 85.3$ K. Then

$$\left(\frac{2\pi M k_B T}{h^2}\right)^{3/2} = \left[\frac{2\pi(3.347\times10^{-27}\text{ kg})(1.381\times10^{-23}\text{ J}\cdot\text{K}^{-1})(500\text{ K})}{(6.626\times10^{34}\text{ J}\cdot\text{s})^2}\right]^{3/2} = 6.014\times10^{30}\text{ m}^{-3}$$

$$\frac{\overline{V}}{N_A} = \frac{RT}{N_A P} = \frac{(0.08314\text{ dm}^3\cdot\text{bar}\cdot\text{mol}^{-1}\cdot\text{K}^{-1})(500\text{ K})}{(6.022\times10^{23}\text{ mol}^{-1})(1\text{ bar})} = 6.903\times10^{-26}\text{ m}^3$$

$$\ln\left(\frac{T}{\sigma\Theta_{rot}}\right) = 1.08$$

$$\ln(1-e^{-\Theta_{vib}/T}) = -4.00\times10^{-6}$$

$$\frac{\Theta_{vib}/T}{e^{\Theta_{vib}/T}-1} = 4.97\times10^{-5}$$

$$\frac{\overline{S}}{R} = 3.5 + 12.94 + 1.08 + 4.00\times10^{-6} + 4.97\times10^{-5} = 17.51$$

The standard molar entropy of H_2 at 500 K is 145.6 $J\cdot K^{-1}\cdot mol^{-1}$.

We can do the same for O_2 (with $\Theta_{vib} = 2256$ K and $\Theta_{rot} = 2.07$ K), keeping in mind that $g_{el} = 3$:

$$\left(\frac{2\pi M k_B T}{h^2}\right)^{3/2} = \left[\frac{2\pi(5.313\times10^{-26}\text{ kg})(1.381\times10^{-23}\text{ J}\cdot\text{K}^{-1})(500\text{ K})}{(6.626\times10^{34}\text{ J}\cdot\text{s})^2}\right]^{3/2} = 3.803\times10^{32}\text{ m}^{-3}$$

$$\frac{\overline{V}}{N_A} = \frac{RT}{N_A P} = \frac{(0.08314\text{ dm}^3\cdot\text{bar}\cdot\text{mol}^{-1}\cdot\text{K}^{-1})(500\text{ K})}{(6.022\times10^{23}\text{ mol}^{-1})(1\text{ bar})} = 6.903\times10^{-26}\text{ m}^3$$

$$\ln\left(\frac{T}{\sigma\Theta_{rot}}\right) = 4.79$$

$$\ln(1-e^{-\Theta_{vib}/T}) = -0.0110$$

$$\frac{\Theta_{vib}/T}{e^{\Theta_{vib}/T}-1} = 0.0501$$

$$\ln g_{el} = \ln 3 = 1.099$$

$$\frac{\overline{S}}{R} = 3.5 + 17.08 + 4.79 + 0.0110 + 0.0501 + 1.099 = 26.54$$

The standard molar entropy of O_2 at 500 K is 220.6 $J\cdot K^{-1}\cdot mol^{-1}$.

Because H_2O is a bent polyatomic molecule, we treat it as we did NO_2 in Problem 7–37. From Table 4.4, $\sigma = 2$, $\Theta_{rot,A} = 40.1$ K, $\Theta_{rot,B} = 20.9$ K, $\Theta_{rot,C} = 13.4$ K, $\Theta_{vib,1} = 5360$ K, $\Theta_{vib,2} = 5160$ K, and $\Theta_{vib,3} = 2290$ K. Then

$$\left(\frac{2\pi M k_B T}{h^2}\right)^{3/2} = \left[\frac{2\pi(2.991\times10^{-26}\text{ kg})(1.381\times10^{-23}\text{ J}\cdot\text{K}^{-1})(500\text{ K})}{(6.626\times10^{34}\text{ J}\cdot\text{s})^2}\right]^{3/2} = 1.607\times10^{32}\text{ m}^{-3}$$

$$\frac{\overline{V}}{N_A} = \frac{RT}{N_A P} = \frac{(0.08314\text{ dm}^3\cdot\text{bar}\cdot\text{mol}^{-1}\cdot\text{K}^{-1})(500\text{ K})}{(6.022\times10^{23}\text{ mol}^{-1})(1\text{ bar})} = 6.903\times10^{-26}\text{ m}^3$$

$$\frac{1}{2}\ln\left(\frac{T^3}{\Theta_{rot,A}\Theta_{rot,B}\Theta_{rot,C}}\right) = 4.66$$

$$\sum_{j=1}^{6}\ln(1-e^{-\Theta_{vib,j}/T}) = -0.0104$$

$$\sum_{j=1}^{6}\frac{\Theta_{vib,j}}{T}\left(\frac{e^{-\Theta_{vib,j}/T}}{1-e^{-\Theta_{vib,j}/T}}\right) = 0.048$$

The standard molar entropy is then

$$\frac{\overline{S}}{R} = 4 + \ln\left[\left(\frac{2\pi M k_B T}{h^2}\right)^{3/2}\frac{\overline{V}}{N_A}\right] + \ln\left(\frac{\pi^{1/2}}{\sigma}\right) + \frac{1}{2}\ln\frac{T^3}{\Theta_{rot,A}\Theta_{rot,B}\Theta_{rot,C}}$$

$$-\sum_{j=1}^{3}\ln(1-e^{-\Theta_{vib,j}/T}) + \sum_{j=1}^{3}\left[\frac{(\Theta_{vib,j}/T)e^{-\Theta_{vib,j}/T}}{1-e^{-\Theta_{vib,j}/T}}\right] + \ln g_{el}$$

$$= 4 + 16.22 - 0.121 + 4.66 + 0.0104 + 0.048 = 24.82$$

which gives a value of $\overline{S} = 206.3$ $J\cdot K^{-1}\cdot mol^{-1}$.

Finally, we can calculate the value of $\Delta_r S^\circ$ for the reaction above, as we did in the previous problem.

$$\begin{aligned}\Delta_r S^\circ &= S^\circ[H_2O] - \tfrac{1}{2}S^\circ[O_2] - S^\circ[H_2] \\ &= (206.3\ J\cdot K^{-1}\cdot mol^{-1}) - \tfrac{1}{2}(220.6\ J\cdot K^{-1}\cdot mol^{-1}) - (145.6\ J\cdot K^{-1}\cdot mol^{-1}) \\ &= -49.6\ J\cdot K^{-1}\cdot mol^{-1}\end{aligned}$$

7–40. In each case below, predict which molecule of the pair has the greater molar entropy under the same conditions (assume gaseous species).

a. CO $\quad CO_2$

b. $CH_3CH_2CH_3$ $\quad$ cyclopropane (ring of H_2C—CH_2—CH_2)

c. $CH_3CH_2CH_2CH_2CH_3$ $\quad$ $C(CH_3)_4$ (H_3C—C(CH_3)(CH_3)—CH_3)

---

a. CO_2 (more atoms)

b. $CH_3CH_2CH_3$ (more flexibility)

c. $CH_3CH_2CH_2CH_2CH_3$ (more flexibility)

7–41. In each case below, predict which molecule of the pair has the greater molar entropy under the same conditions (assume gaseous species).

a. H_2O $\quad D_2O$

b. CH_3CH_2OH $\quad$ ethylene oxide (ring of H_2C—CH_2—O)

c. $CH_3CH_2CH_2CH_2NH_2$ $\quad$ pyrrolidine (ring of N(H), H_2C, CH_2, H_2C, CH_2)

---

a. D_2O (larger mass)

b. CH_3CH_2OH (more flexibility)

c. $CH_3CH_2CH_2CH_2NH_2$ (more flexibility)

7–42. Arrange the following reactions according to increasing values of $\Delta_r S^\circ$ (do not consult any references).

a. $S(s) + O_2(g) \longrightarrow SO_2(g)$

b. $H_2(g) + O_2(g) \longrightarrow H_2O_2(l)$

c. $CO(g) + 3\,H_2(g) \longrightarrow CH_4(g) + H_2O(l)$

d. $C(s) + H_2O(g) \longrightarrow CO(g) + H_2(g)$

Recall that molar entropies of solids and liquids are much smaller than those of gases, so we can ignore the contribution of the solids and liquids to $\Delta_r S^\circ$ when we order these reactions. Considering only the gaseous products and reactants, we can find Δn for each reaction to be

a. $\Delta n = 0$ **b.** $\Delta n = -2$ **c.** $\Delta n = -3$ **d.** $\Delta n = +1$

The correct ordering of the reactions is therefore **d** > **a** > **b** > **c**.

7–43. Arrange the following reactions according to increasing values of $\Delta_r S^\circ$ (do not consult any references).

a. $2\,H_2(g) + O_2(g) \longrightarrow 2\,H_2O(l)$

b. $NH_3(g) + HCl(g) \longrightarrow NH_4Cl(s)$

c. $K(s) + O_2(g) \longrightarrow KO_2(s)$

d. $N_2(g) + 3\,H_2(g) \longrightarrow 2\,NH_3(g)$

Again, calculating Δn for each reaction for the gaseous products and reactants gives

a. $\Delta n = -3$ **b.** $\Delta n = -2$ **c.** $\Delta n = -1$ **d.** $\Delta n = -2$

The correct ordering of the reactions is therefore **c** > **b** $\approx$ **d** > **a**.

7–44. In Problem 7–40, you are asked to predict which molecule, CO(g) or $CO_2(g)$, has the greater molar entropy. Use the data in Tables 4.2 and 4.4 to calculate the standard molar entropy of CO(g) and $CO_2(g)$ at 298.15 K. Does this calculation confirm your intuition? Which degree of freedom makes the dominant contribution to the molar entropy of CO? Of CO_2?

In Problem 7–38 and Example 7–5, we used the data in Tables 4.2 and 4.4 to find that the standard molar entropy of CO(g) is 197.6 $J\cdot K^{-1}\cdot mol^{-1}$ and that of $CO_2(g)$ is 213.8 $J\cdot K^{-1}\cdot mol^{-1}$. In both cases, the translational degrees of freedom make the dominant contribution to the molar entropy.

7–45. Table 7.2 gives $S^\circ[CH_3OH(l)] = 126.8\ J\cdot K^{-1}\cdot mol^{-1}$ at 298.15 K. Given that $T_{vap} = 337.7$ K, $\Delta_{vap}\overline{H}(T_b) = 36.5\ kJ\cdot mol^{-1}$, $\overline{C}_P[CH_3OH(l)] = 81.12\ J\cdot K^{-1}\cdot mol^{-1}$, and $\overline{C}_P[CH_3OH(g)] = 43.8\ J\cdot K^{-1}\cdot mol^{-1}$, calculate the value of $S^\circ[CH_3OH(g)]$ at 298.15 K and compare your answer with the experimental value of 239.8 $J\cdot K^{-1}\cdot mol^{-1}$.

This is done in the same way as $S^\circ[Br_2(g)]$ was found in Section 7–7. First, we heat the methanol to its boiling point:

$$\Delta\overline{S}_1 = \overline{S}^l(337.7\text{ K}) - \overline{S}^l(298.15\text{ K}) = \overline{C}^l_P \ln\frac{T_2}{T_1}$$

$$= (81.12\ J\cdot K^{-1}\cdot mol^{-1})\ln\frac{337.7}{298.15} = 10.10\ J\cdot K^{-1}\cdot mol^{-1}$$

Then vaporize the methanol at its normal boiling point:

$$\Delta\overline{S}_2 = \overline{S}^g(337.7\text{ K}) - \overline{S}^l(337.7\text{ K}) = \frac{\Delta_{vap}\overline{H}}{T_{vap}}$$

$$= \frac{36\,500\ J\cdot mol^{-1}}{337.7\text{ K}} = 108.1\ J\cdot K^{-1}\cdot mol^{-1}$$

Finally, cool the gas back down to 298.15 K:

$$\Delta\overline{S}_3 = \overline{S}^g(298.15\text{ K}) - \overline{S}^g(337.7\text{ K}) = \overline{C}^g_P \ln\frac{T_2}{T_1}$$

$$= (43.8\ J\cdot K^{-1}\cdot mol^{-1})\ln\frac{298.15}{337.7} = -5.456\ J\cdot K^{-1}\cdot mol^{-1}$$

The sum of these three steps plus $S^\circ_{298}[\text{CH}_3\text{OH(l)}] = 126.8\ \text{J}\cdot\text{K}^{-1}\cdot\text{mol}^{-1}$ will be the desired value:

$$\begin{aligned} S^\circ_{298}[\text{CH}_3\text{OH(g)}] &= S^\circ_{298}[\text{CH}_3\text{OH(l)}] + \Delta\overline{S}_1 + \Delta\overline{S}_2 + \Delta\overline{S}_3 \\ &= 239.5\ \text{J}\cdot\text{K}^{-1}\cdot\text{mol}^{-1} \end{aligned}$$

which is within 0.1% of the experimental value.

7–46. Given the following data, $T_{\text{vap}} = 373.15$ K, $\Delta\overline{H}_{\text{vap}}(T_{\text{vap}}) = 40.65\ \text{kJ}\cdot\text{mol}^{-1}$, $\overline{C}_P[\text{H}_2\text{O(l)}] = 75.3\ \text{J}\cdot\text{K}^{-1}\cdot\text{mol}^{-1}$, and $\overline{C}_P[\text{H}_2\text{O(g)}] = 33.8\ \text{J}\cdot\text{K}^{-1}\cdot\text{mol}^{-1}$, show that the values of $S^\circ[\text{H}_2\text{O(l)}]$ and $S^\circ[\text{H}_2\text{O(g)}]$ in Table 7.2 are consistent.

This is done in the same way as Problem 7–45. First, we heat the water to its boiling point:

$$\begin{aligned} \Delta\overline{S}_1 &= \overline{S}^{\text{l}}(373.15\ \text{K}) - \overline{S}^{\text{l}}(298.15\ \text{K}) = \overline{C}^{\text{l}}_P \ln\frac{T_2}{T_1} \\ &= (75.3\ \text{J}\cdot\text{K}^{-1}\cdot\text{mol}^{-1}) \ln\frac{373.15}{298.15} = 16.90\ \text{J}\cdot\text{K}^{-1}\cdot\text{mol}^{-1} \end{aligned}$$

Then vaporize the water at its normal boiling point:

$$\begin{aligned} \Delta\overline{S}_2 &= \overline{S}^{\text{g}}(373.15\ \text{K}) - \overline{S}^{\text{l}}(373.15\ \text{K}) = \frac{\Delta_{\text{vap}}\overline{H}}{T_{\text{vap}}} \\ &= \frac{40\,650\ \text{J}\cdot\text{mol}^{-1}}{373.15\ \text{K}} = 108.9\ \text{J}\cdot\text{K}^{-1}\cdot\text{mol}^{-1} \end{aligned}$$

Finally, cool the gas back down to 298.15 K:

$$\begin{aligned} \Delta\overline{S}_3 &= \overline{S}^{\text{g}}(298.15\ \text{K}) - \overline{S}^{\text{g}}(373.15\ \text{K}) = \overline{C}^{\text{g}}_P \ln\frac{T_2}{T_1} \\ &= (33.8\ \text{J}\cdot\text{K}^{-1}\cdot\text{mol}^{-1}) \ln\frac{298.15}{373.15} = -7.584\ \text{J}\cdot\text{K}^{-1}\cdot\text{mol}^{-1} \end{aligned}$$

The sum of these three steps plus $S^\circ_{298}[\text{H}_2\text{O(l)}] = 70.0\ \text{J}\cdot\text{K}^{-1}\cdot\text{mol}^{-1}$ will be the desired value:

$$\begin{aligned} S^\circ_{298}[\text{H}_2\text{O(g)}] &= S^\circ_{298}[\text{H}_2\text{O(l)}] + \Delta\overline{S}_1 + \Delta\overline{S}_2 + \Delta\overline{S}_3 \\ &= 188.2\ \text{J}\cdot\text{K}^{-1}\cdot\text{mol}^{-1} \end{aligned}$$

which is within 0.4% of the value in Table 7.2.

7–47. Use the data in Table 7.2 to calculate the value of $\Delta_r S^\circ$ for the following reactions at 25°C and one bar.

a. $\text{C(s, graphite)} + \text{O}_2\text{(g)} \longrightarrow \text{CO}_2\text{(g)}$

b. $\text{CH}_4\text{(g)} + 2\,\text{O}_2\text{(g)} \longrightarrow \text{CO}_2\text{(g)} + 2\,\text{H}_2\text{O(l)}$

c. $\text{C}_2\text{H}_2\text{(g)} + \text{H}_2\text{(g)} \longrightarrow \text{C}_2\text{H}_4\text{(g)}$

---

a.
$$\begin{aligned} \Delta_r S^\circ &= S^\circ[\text{products}] - S^\circ[\text{reactants}] \\ &= 213.8\ \text{J}\cdot\text{K}^{-1}\cdot\text{mol}^{-1} - 205.2\ \text{J}\cdot\text{K}^{-1}\cdot\text{mol}^{-1} - 5.74\ \text{J}\cdot\text{K}^{-1}\cdot\text{mol}^{-1} \\ &= 2.86\ \text{J}\cdot\text{K}^{-1}\cdot\text{mol}^{-1} \end{aligned}$$

b. $\Delta_r S^\circ = S^\circ[\text{products}] - S^\circ[\text{reactants}]$

$$= 2(70.0\ \text{J·K}^{-1}\text{·mol}^{-1}) + 213.8\ \text{J·K}^{-1}\text{·mol}^{-1} - 2(205.2\ \text{J·K}^{-1}\text{·mol}^{-1}) - 186.3\ \text{J·K}^{-1}\text{·mol}^{-1}$$

$$= -242.9\ \text{J·K}^{-1}\text{·mol}^{-1}$$

c. $\Delta_r S^\circ = S^\circ[\text{products}] - S^\circ[\text{reactants}]$

$$= 219.6\ \text{J·K}^{-1}\text{·mol}^{-1} - 130.7\ \text{J·K}^{-1}\text{·mol}^{-1} - 200.9\ \text{J·K}^{-1}\text{·mol}^{-1}$$

$$= -112.0\ \text{J·K}^{-1}\text{·mol}^{-1}$$

7–48. Use the data in Table 7.2 to calculate the value of $\Delta_r S^\circ$ for the following reactions at 25°C and one bar.

a. $CO(g) + 2\,H_2(g) \longrightarrow CH_3OH(l)$

b. $C(s, \text{graphite}) + H_2O(l) \longrightarrow CO(g) + H_2(g)$

c. $2\,CO(g) + O_2(g) \longrightarrow 2\,CO_2(g)$

---

a. $\Delta_r S^\circ = S^\circ[\text{products}] - S^\circ[\text{reactants}]$

$$= 126.8\ \text{J·K}^{-1}\text{·mol}^{-1} - 197.7\ \text{J·K}^{-1}\text{·mol}^{-1} - 2(130.7\ \text{J·K}^{-1}\text{·mol}^{-1})$$

$$= -332.3\ \text{J·K}^{-1}\text{·mol}^{-1}$$

b. $\Delta_r S^\circ = S^\circ[\text{products}] - S^\circ[\text{reactants}]$

$$= 130.7\ \text{J·K}^{-1}\text{·mol}^{-1} + 197.7\ \text{J·K}^{-1}\text{·mol}^{-1} - 70.0\ \text{J·K}^{-1}\text{·mol}^{-1} - 5.74\ \text{J·K}^{-1}\text{·mol}^{-1}$$

$$= 252.66\ \text{J·K}^{-1}\text{·mol}^{-1}$$

c. $\Delta_r S^\circ = S^\circ[\text{products}] - S^\circ[\text{reactants}]$

$$= 2(213.8\ \text{J·K}^{-1}\text{·mol}^{-1}) - 205.2\ \text{J·K}^{-1}\text{·mol}^{-1} - 2(197.7\ \text{J·K}^{-1}\text{·mol}^{-1})$$

$$= -173.0\ \text{J·K}^{-1}\text{·mol}^{-1}$$

CHAPTER 8

Helmholtz and Gibbs Energies

PROBLEMS AND SOLUTIONS

8–1. The molar enthalpy of vaporization of benzene at its normal boiling point (80.09°C) is $30.72\ \text{kJ}\cdot\text{mol}^{-1}$. Assuming that $\Delta_{\text{vap}}\overline{H}$ and $\Delta_{\text{vap}}\overline{S}$ stay constant at their values at 80.09°C, calculate the value of $\Delta_{\text{vap}}\overline{G}$ at 75.0°C, 80.09°C, and 85.0°C. Interpret these results physically.

We can write (as in Section 8–2)

$$\Delta_{\text{vap}}\overline{G} = \Delta_{\text{vap}}\overline{H} - T\Delta_{\text{vap}}\overline{S}$$

At the boiling point of benzene, the liquid and vapor phases are in equilibrium, so $\Delta_{\text{vap}}\overline{G} = 0$. Thus, at 80.09°C,

$$0 = 30.72\ \text{kJ}\cdot\text{mol}^{-1} - (353.24\ \text{K})\Delta_{\text{vap}}\overline{S}$$

$$\Delta_{\text{vap}}\overline{S} = 86.97\ \text{J}\cdot\text{K}^{-1}\cdot\text{mol}^{-1}$$

Since $\Delta_{\text{vap}}\overline{H}$ and $\Delta_{\text{vap}}\overline{S}$ are assumed to stay constant at their boiling-point values, we know their numerical values and can substitute into our first equation:

$$\Delta_{\text{vap}}\overline{G}(75.0^\circ\text{C}) = 30.72\ \text{kJ}\cdot\text{mol}^{-1} - (348.15\ \text{K})(86.97\ \text{J}\cdot\text{K}^{-1}\cdot\text{mol}^{-1}) = 441.4\ \text{J}\cdot\text{mol}^{-1}$$

$$\Delta_{\text{vap}}\overline{G}(85.0^\circ\text{C}) = 30.72\ \text{kJ}\cdot\text{mol}^{-1} - (358.15\ \text{K})(86.97\ \text{J}\cdot\text{K}^{-1}\cdot\text{mol}^{-1}) = -428.3\ \text{J}\cdot\text{mol}^{-1}$$

From these values, we can see that at 75.0°C benzene will spontaneously condense, whereas at 85.0°C it will spontaneously evaporate (just as we would expect).

8–2. Redo Problem 8–1 without assuming that $\Delta_{\text{vap}}\overline{H}$ and $\Delta_{\text{vap}}\overline{S}$ do not vary with temperature. Take the molar heat capacities of liquid and gaseous benzene to be $136.3\ \text{J}\cdot\text{K}^{-1}\cdot\text{mol}^{-1}$ and $82.4\ \text{J}\cdot\text{K}^{-1}\cdot\text{mol}^{-1}$, respectively. Compare your results with those you obtained in Problem 8–1. Are any of your physical interpretations different?

We wish to consider the temperature variation of $\Delta_{\text{vap}}\overline{G}$, so we must use Equation 8.31*a*,

$$\left(\frac{\partial \Delta_{\text{vap}}\overline{G}}{\partial T}\right)_P = -\Delta_{\text{vap}}\overline{S}(T)$$

where (as in Example 6–5)

$$\Delta_{\text{vap}}\overline{S}(T) = \Delta_{\text{vap}}\overline{S}(80.09^\circ\text{ C}) + \int_{353.24\text{ K}}^{T} \frac{\Delta\overline{C}_P}{T} dT$$

$$= 86.97\text{ J}\cdot\text{K}^{-1}\cdot\text{mol}^{-1} - (53.9\text{ J}\cdot\text{K}^{-1}\cdot\text{mol}^{-1}) \ln \frac{T}{353.24\text{ K}}$$

$$= 403.2\text{ J}\cdot\text{K}^{-1}\cdot\text{mol}^{-1} - (53.9\text{ J}\cdot\text{K}^{-1}\cdot\text{mol}^{-1}) \ln(T/\text{K})$$

Substituting into Equation 8.31a, we can write

$$\Delta_{\text{vap}}\overline{G}(T) - \Delta_{\text{vap}}\overline{G}(353.24\text{ K}) = -\int_{353.24\text{ K}}^{T} \big[403.2\text{ J}\cdot\text{K}^{-1}\cdot\text{mol}^{-1} - (53.9\text{ J}\cdot\text{K}^{-1}\cdot\text{mol}^{-1}) \ln(T/\text{K})\big] dT$$

$$= -(403.2\text{ J}\cdot\text{K}^{-1}\cdot\text{mol}^{-1})(T - 353.24\text{ K}) + (53.9\text{ J}\cdot\text{K}^{-1}\cdot\text{mol}^{-1})\,[T\ln(T/\text{K}) - T - 1719.3\text{ K}]$$

Letting $T = 348.15$ K gives

$$\Delta_{\text{vap}}\overline{G}(T) = 2052\text{ J}\cdot\text{mol}^{-1} - 1608\text{ J}\cdot\text{mol}^{-1} = +\,444\text{ J}\cdot\text{mol}^{-1}$$

and letting $T = 358.15$ K gives

$$\Delta_{\text{vap}}\overline{G}(T) = -1980\text{ J}\cdot\text{mol}^{-1} + 1555\text{ J}\cdot\text{mol}^{-1} = -425\text{ J}\cdot\text{mol}^{-1}$$

Notice that taking the temperature variation of $\Delta_{\text{vap}}\overline{H}$ and $\Delta_{\text{vap}}\overline{S}$ into account made little difference over such a small temperature range.

8–3. Substitute $(\partial P/\partial T)_{\overline{V}}$ from the van der Waals equation into Equation 8.19 and integrate from $\overline{V}^{\text{id}}$ to $\overline{V}$ to obtain

$$\overline{S}(T, \overline{V}) - \overline{S}^{\text{id}}(T) = R \ln \frac{\overline{V} - b}{\overline{V}^{\text{id}} - b}$$

Now let $\overline{V}^{\text{id}} = RT/P^{\text{id}}$, $P^{\text{id}} = P^\circ = 1$ bar, and $\overline{V}^{\text{id}} \gg b$ to obtain

$$\overline{S}(T, \overline{V}) - \overline{S}^{\text{id}}(T) = -R \ln \frac{RT/P^\circ}{\overline{V} - b}$$

Given that $\overline{S}^{\text{id}} = 246.35\text{ J}\cdot\text{mol}^{-1}\cdot\text{K}^{-1}$ for ethane at 400 K, show that

$$\overline{S}(\overline{V})/\text{J}\cdot\text{mol}^{-1}\cdot\text{K}^{-1} = 246.35 - 8.3145 \ln \frac{33.258\text{ L}\cdot\text{mol}^{-1}}{\overline{V} - 0.065144\text{ L}\cdot\text{mol}^{-1}}$$

Calculate $\overline{S}$ as a function of $\rho = 1/\overline{V}$ for ethane at 400 K and compare your results with the experimental results shown in Figure 8.2.

Show that

$$\overline{S}(\overline{V})/\text{J}\cdot\text{mol}^{-1}\cdot\text{K}^{-1} = 246.35 - 8.3145 \ln \frac{33.258\text{ L}\cdot\text{mol}^{-1}}{\overline{V} - 0.045153\text{ L}\cdot\text{mol}^{-1}} + 13.68 \ln \frac{\overline{V} + 0.045153\text{ L}\cdot\text{mol}^{-1}}{\overline{V}}$$

for the Redlich-Kwong equation for ethane at 400 K. Calculate $\overline{S}$ as a function of $\rho = 1/\overline{V}$ and compare your results with the experimental results shown in Figure 8.2.

From the van der Waals equation,

$$\left(\frac{\partial P}{\partial T}\right)_V = \frac{R}{\overline{V} - b}$$

We can substitute into Equation 8.19 to find an expression for $d\overline{S}$:

$$\left(\frac{\partial \overline{S}}{\partial \overline{V}}\right)_T = \left(\frac{\partial P}{\partial T}\right)_V$$

$$d\overline{S} = \frac{R}{\overline{V} - b} d\overline{V}$$

Integrating both sides of this equation gives

$$\overline{S}(T, \overline{V}) - \overline{S}^{\text{id}}(T) = R \ln \frac{\overline{V} - b}{\overline{V}^{\text{id}} - b}$$

Since $\overline{V}^{\text{id}}$ is quite large compared to b, we can neglect b in $\overline{V}^{\text{id}} - b$. Letting $\overline{V}^{\text{id}} = RT/P^{\text{id}}$ and $P^{\text{id}} = P^\circ = 1$ bar, we find

$$\begin{aligned}\overline{S}(T, \overline{V}) - \overline{S}^{\text{id}}(T) &= -R \ln \frac{\overline{V}^{\text{id}}}{\overline{V} - b} \\ &= -R \ln \frac{RT/P^\circ}{\overline{V} - b} \\ \overline{S}(T, \overline{V}) &= \overline{S}^{\text{id}}(T) - R \ln \frac{RT}{\overline{V} - b}\end{aligned}$$

where, in the last equality, the pressure units of R are in bars. For ethane, $a = 5.5818\ \text{dm}^6\cdot\text{bar}\cdot\text{mol}^{-2}$ and $b = 0.065144\ \text{dm}^3\cdot\text{mol}^{-1}$ (Table 2.3), so

$$\overline{S}(\overline{V}) = 246.35\ \text{J}\cdot\text{K}^{-1}\cdot\text{mol}^{-1} - (8.3145\ \text{J}\cdot\text{K}^{-1}\cdot\text{mol}^{-1}) \ln \frac{33.258\ \text{L}\cdot\text{mol}^{-1}}{\overline{V} - 0.065144\ \text{L}\cdot\text{mol}^{-1}}$$

To make graphing entropy vs. density easier, we can break up the logarithmic term and then graph, as shown:

$$\begin{aligned}\overline{S}/\text{J}\cdot\text{K}^{-1}\cdot\text{mol}^{-1} &= 246.35 - (8.3145) \ln 33.258 \\ &\quad + (8.3145) \ln \left(\frac{\text{mol}\cdot\text{dm}^{-3}}{\rho} - 0.065144\right)\end{aligned}$$

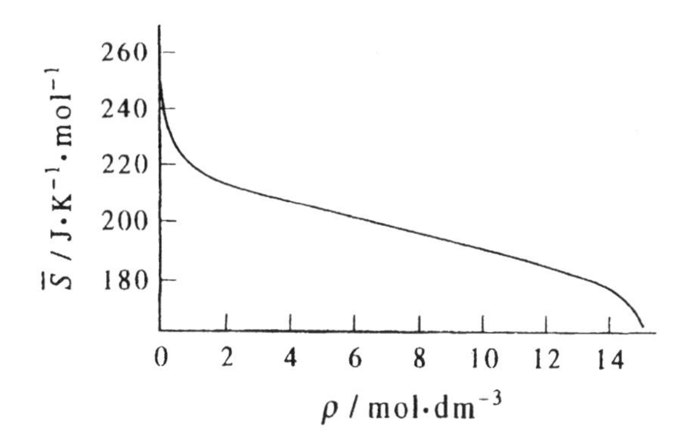

For a Redlich-Kwong gas,

$$P = \frac{RT}{\overline{V} - B} - \frac{A}{T^{1/2}\overline{V}(\overline{V} + B)}$$

and

$$\left(\frac{\partial P}{\partial T}\right)_V = \frac{R}{\overline{V} - B} + \frac{A}{2\overline{V}(\overline{V} + B)T^{3/2}}$$

Using Equation 8.19 and integrating (keeping temperature constant) gives

$$\left(\frac{\partial \overline{S}}{\partial \overline{V}}\right)_T = \frac{R}{\overline{V} - B} + \frac{A}{2\overline{V}(\overline{V} + B)T^{3/2}}$$

$$\int d\overline{S} = \int \left[\frac{R}{\overline{V} - B} + \frac{A}{2\overline{V}(\overline{V} + B)T^{3/2}}\right] d\overline{V}$$

$$\overline{S} - \overline{S}^{\text{id}} = R \ln \frac{\overline{V} - B}{\overline{V}^{\text{id}} - B} - \frac{A}{2T^{3/2}}\left[-\frac{1}{B}\ln\left(\frac{B + \overline{V}}{\overline{V}}\right) + \frac{1}{B}\ln\left(\frac{B + \overline{V}^{\text{id}}}{\overline{V}^{\text{id}}}\right)\right]$$

Since for an ideal gas B is negligible compared to $\overline{V}$,

$$\overline{S} - \overline{S}^{\text{id}} = R \ln \frac{\overline{V} - B}{\overline{V}^{\text{id}}} + \frac{A}{2BT^{3/2}} \ln \frac{\overline{V}}{\overline{V} + B}$$

Then, for ethane, since $A = 98.831 \text{ dm}^6\cdot\text{bar}\cdot\text{mol}^{-2}\cdot\text{K}^{1/2}$ and $B = 0.045153 \text{ dm}^3\cdot\text{mol}^{-1}$ (Table 2.4),

$$\overline{S} = \overline{S}^{\text{id}} - R \ln \frac{\overline{V}^{\text{id}}}{\overline{V} - B} + \frac{A}{2BT^{3/2}} \ln \frac{\overline{V} + B}{\overline{V}}$$

$$\overline{S}/\text{J}\cdot\text{K}^{-1}\cdot\text{mol}^{-1} = 246.35 - 8.3145 \ln \frac{33.258 \text{ dm}^3\cdot\text{mol}^{-1}}{\overline{V} - 0.045153 \text{ dm}^3\cdot\text{mol}^{-1}} + 13.68 \ln \frac{\overline{V} + 0.045153 \text{ dm}^3\cdot\text{mol}^{-1}}{\overline{V}}$$

To make graphing entropy vs. density easier, we can break up the logarithmic term and then graph, as shown. We have divided both numerator and denominator of the logarithmic terms in the previous expression by $\text{dm}^3\cdot\text{mol}^{-1}$.

$$\overline{S}/\text{J}\cdot\text{K}^{-1}\cdot\text{mol}^{-1} = 246.35 - 8.3145\left[\ln 33.258 - \ln\left(\frac{1}{\rho} - 0.045153\right)\right] + 13.68\left[\ln\left(\frac{1}{\rho} + 0.045153\right) - \ln\frac{1}{\rho}\right]$$

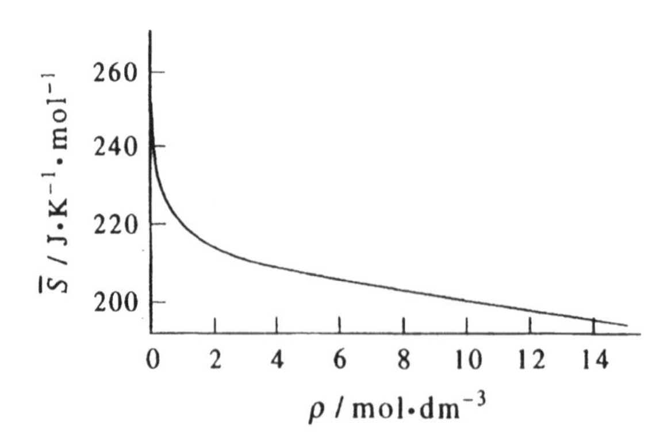

8–4. Use the van der Waals equation to derive

$$\overline{U}(T, \overline{V}) - \overline{U}^{\text{id}}(T) = -\frac{a}{\overline{V}}$$

Use this result along with the van der Waals equation to calculate the value of $\overline{U}$ as a function of $\overline{V}$ for ethane at 400 K, given that $\overline{U}^{\text{id}} = 14.55\ \text{kJ}\cdot\text{mol}^{-1}$. To do this, specify $\overline{V}$ (from 0.0700 L·mol^{-1} to 7.00 L·mol^{-1}, see Figure 8.2), calculate both $\overline{U}(\overline{V})$ and $P(\overline{V})$, and plot $\overline{U}(\overline{V})$ versus $P(\overline{V})$. Compare your result with the experimental data in Figure 8.3. Use the Redlich-Kwong equation to derive

$$\overline{U}(T,\overline{V}) - \overline{U}^{\text{id}}(T) = -\frac{3A}{2BT^{1/2}}\ln\frac{\overline{V}+B}{\overline{V}}$$

Repeat the above calculation for ethane at 400 K.

Begin with Equation 8.22,

$$\left(\frac{\partial U}{\partial V}\right)_T = -P + T\left(\frac{\partial P}{\partial T}\right)_V$$

In Problem 8–3, we found $(\partial P/\partial T)_V$ for the van der Waals and Redlich-Kwong equations. For the van der Waals equation, Equation 8.22 becomes

$$\left(\frac{\partial U}{\partial V}\right)_T = -\frac{RT}{\overline{V}-b} + \frac{a}{\overline{V}^2} + \frac{RT}{\overline{V}-b} = \frac{a}{\overline{V}^2}$$

$$\int_{\overline{U}^{\text{id}}}^{\overline{U}} dU = \int_{\overline{V}^{\text{id}}}^{\overline{V}} \frac{a}{\overline{V}^2} d\overline{V}$$

$$\overline{U} - \overline{U}^{\text{id}} = -\frac{a}{\overline{V}}$$

The van der Waals constants for ethane are listed in the previous problem. Substituting, we find that

$$\overline{U} = 14.55\ \text{kJ}\cdot\text{mol}^{-1} - \frac{0.55818\ \text{kJ}\cdot\text{dm}^3\cdot\text{mol}^{-2}}{\overline{V}}$$

$$P = \frac{(0.083145\ \text{dm}^3\cdot\text{bar}\cdot\text{mol}^{-1}\cdot\text{K}^{-1})(400\ \text{K})}{\overline{V} - 0.065144\ \text{dm}^3\cdot\text{mol}^{-1}} - \frac{5.5818\ \text{dm}^6\cdot\text{bar}\cdot\text{mol}^{-1}}{\overline{V}^2}$$

We can use a parametric plot to plot $\overline{U}(\overline{V})$ vs. $P(\overline{V})$:

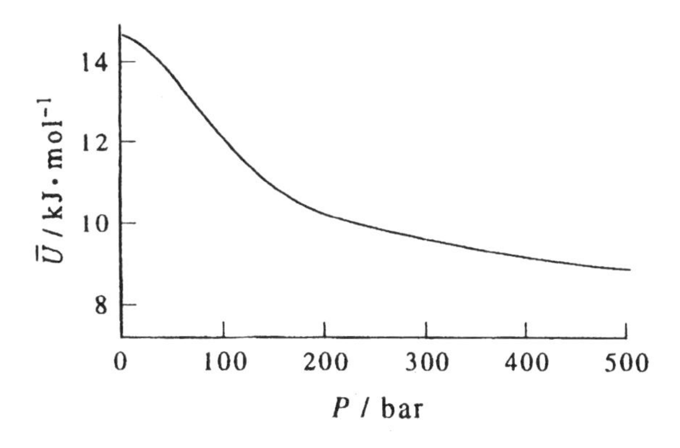

For the Redlich-Kwong equation, Equation 8.22 becomes

$$\left(\frac{\partial U}{\partial V}\right)_T = -\frac{RT}{\overline{V}-B} + \frac{A}{T^{1/2}\overline{V}(\overline{V}+B)} + \frac{RT}{\overline{V}-B} + \frac{A}{2T^{1/2}\overline{V}(\overline{V}+B)}$$

$$\int_{\overline{U}^{\text{id}}}^{\overline{U}} dU = \int_{\overline{V}^{\text{id}}}^{\overline{V}} \frac{3A}{2T^{1/2}\overline{V}(\overline{V}+B)} d\overline{V}$$

$$\overline{U} - \overline{U}^{\text{id}} = \frac{3A}{2T^{1/2}}\left(-\frac{1}{B}\ln\frac{\overline{V}+B}{\overline{V}} + \frac{1}{B}\ln\frac{\overline{V}^{\text{id}}+B}{\overline{V}^{\text{id}}}\right)$$

$$\overline{U} = \overline{U}^{\text{id}} - \frac{3A}{2BT^{1/2}}\ln\frac{\overline{V}+B}{\overline{V}}$$

The Redlich-Kwong constants for ethane are listed in the previous problem. Substituting, we find that

$$\overline{U} = 14.55\ \text{kJ}\cdot\text{mol}^{-1} - \frac{3(9.8831\ \text{kJ}\cdot\text{dm}^3\cdot\text{mol}^{-2}\cdot\text{K}^{1/2})}{2(0.045153\ \text{dm}^3\cdot\text{mol}^{-1})(400\ \text{K})^{1/2}}\ln\frac{\overline{V}+0.045153\ \text{dm}^3\cdot\text{mol}^{-1}}{\overline{V}}$$

$$P = \frac{(0.083145\ \text{dm}^3\cdot\text{bar}\cdot\text{mol}^{-1}\cdot\text{K}^{-1})(400\ \text{K})}{\overline{V} - 0.045153\ \text{dm}^3\cdot\text{mol}^{-1}} - \frac{98.831\ \text{dm}^6\cdot\text{bar}\cdot\text{mol}^{-2}\cdot\text{K}^{1/2}}{(400\ \text{K})^{1/2}\overline{V}(\overline{V}+0.045153\ \text{dm}^3\cdot\text{mol}^{-1})}$$

Again, use a parametric plot to plot $\overline{U}(\overline{V})$ vs. $P(\overline{V})$:

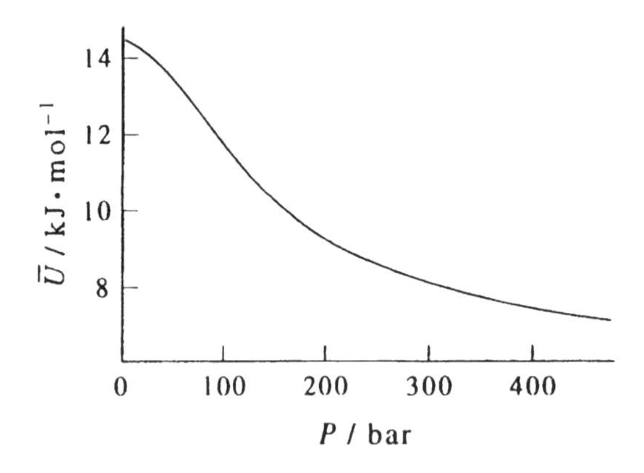

8–5. Show that $(\partial U/\partial V)_T = 0$ for a gas that obeys an equation of state of the form $Pf(V) = RT$. Give two examples of such equations of state that appear in the text.

We first take the partial derivative of P with respect to T, keeping V constant:

$$Pf(V) = RT$$

$$\left(\frac{\partial P}{\partial T}\right)_V = \frac{R}{f(V)}$$

From Equation 8.22, we can write

$$\left(\frac{\partial U}{\partial V}\right)_T = -P + T\left(\frac{\partial P}{\partial T}\right)_V = -P + \frac{RT}{f(V)} = -P + P = 0$$

Two such equations of state are the ideal gas equation and $P(\overline{V} - b) = RT$.

8–6. Show that

$$\left(\frac{\partial \overline{U}}{\partial \overline{V}}\right)_T = \frac{RT^2}{\overline{V}^2}\frac{dB_{2V}}{dT} + \frac{RT^2}{\overline{V}^3}\frac{dB_{3V}}{dT} + \cdots$$

Begin with Equation 8.22,

$$\left(\frac{\partial \overline{U}}{\partial \overline{V}}\right)_T = -P + T\left(\frac{\partial P}{\partial T}\right)_V$$

We can use the virial expansion in the volume to express Z as (Equation 2.22)

$$Z = 1 + B_{2V}\overline{V}^{-1} + B_{3V}\overline{V}^{-2} + O(\overline{V}^{-3})$$
$$P = RT\overline{V}^{-1} + B_{2V}RT\overline{V}^{-2} + B_{3V}RT\overline{V}^{-3} + O(\overline{V}^{-4})$$
$$\left(\frac{\partial P}{\partial T}\right)_V = \frac{R}{\overline{V}} + B_{2V}\frac{R}{\overline{V}^2} + \frac{dB_{2V}}{dT}\frac{RT}{\overline{V}^2} + B_{3V}\frac{R}{\overline{V}^3} + \frac{dB_{3V}}{dT}\frac{RT}{\overline{V}^3} + O(\overline{V}^{-4})$$

Substituting into Equation 8.22 gives

$$\left(\frac{\partial \overline{U}}{\partial \overline{V}}\right)_T = -\left(\frac{RT}{\overline{V}} + B_{2V}\frac{RT}{\overline{V}^2} + B_{3V}\frac{RT}{\overline{V}^3}\right) + \frac{RT}{\overline{V}} + O(\overline{V}^{-4})$$
$$+B_{2V}\frac{RT}{\overline{V}^2} + \frac{dB_{2V}}{dT}\frac{RT^2}{\overline{V}^2} + B_{3V}\frac{RT}{\overline{V}^3} + \frac{dB_{3V}}{dT}\frac{RT^2}{\overline{V}^3} + O(\overline{V}^{-4})$$
$$= \frac{RT^2}{\overline{V}^2}\frac{dB_{2V}}{dT} + \frac{RT^2}{\overline{V}^3}\frac{dB_{3V}}{dT} + O(\overline{V}^{-4})$$

8–7. Use the result of the previous problem to show that

$$\Delta\overline{U} = -T\frac{dB_{2V}}{dT}(P_2 - P_1) + \cdots$$

Use Equation 2.41 for the square-well potential to show that

$$\Delta\overline{U} = -\frac{2\pi\sigma^3 N_A}{3}(\lambda^3 - 1)\frac{\varepsilon}{k_B T}e^{\varepsilon/k_B T}(P_2 - P_1) + \cdots$$

Given that $\sigma = 327.7$ pm, $\varepsilon/k_B = 95.2$ K, and $\lambda = 1.58$ for $N_2(g)$, calculate the value of $\Delta\overline{U}$ for a pressure increase from 1.00 bar to 10.0 bar at 300 K.

We integrate the equation we found in the previous problem (keeping T constant):

$$\left(\frac{\partial \overline{U}}{\partial \overline{V}}\right)_T = \frac{RT^2}{\overline{V}^2}\frac{dB_{2V}}{dT} + \frac{RT^2}{\overline{V}^3}\frac{dB_{3V}}{dT} + O(\overline{V}^{-4})$$
$$d\overline{U} = \frac{RT^2}{\overline{V}^2}\frac{dB_{2V}}{dT}d\overline{V} + O(\overline{V}^{-3})$$
$$\Delta\overline{U} = -\frac{RT^2}{\overline{V}}\frac{dB_{2V}}{dT}\Bigg|_{\overline{V}_1}^{\overline{V}_2}$$

Substitute in for $\overline{V} = RT/P$ to get

$$\Delta\overline{U} = -T(P_2 - P_1)\frac{dB_{2V}}{dT} + \cdots$$

Using Equation 2.41,

$$B_{2V}(T) = \frac{2\pi\sigma^3 N_A}{3}\left[1 - (\lambda^3 - 1)\left(e^{\varepsilon/k_BT} - 1\right)\right]$$

$$\frac{dB_{2V}}{dT} = \frac{2\pi\sigma^3 N_A}{3}(\lambda^3 - 1)e^{\varepsilon/k_BT}\left(\frac{\varepsilon}{k_BT^2}\right)$$

Substitute into the equation for $\Delta\overline{U}$ to find

$$\Delta\overline{U} = -\frac{2\pi\sigma^3 N_A}{3}(\lambda^3 - 1)\frac{\varepsilon}{k_BT}e^{\varepsilon/k_BT}(P_2 - P_1) + \cdots$$

For N_2(g) under the conditions specified,

$$\begin{aligned}\Delta\overline{U} &= -\frac{2\pi\sigma^3 N_A}{3}(\lambda^3 - 1)\frac{\varepsilon}{k_BT}e^{\varepsilon/k_BT}(P_2 - P_1) + \cdots \\ &= -\frac{2\pi(327.7\times10^{-12}\text{ m})^3(6.022\times10^{23}\text{ mol}^{-1})}{3}(1.58^3 - 1)\left(\frac{95.2}{300}\right)e^{95.2/300}(10.0 - 1.00)\text{ bar} \\ &= -5.13\times10^{-4}\text{ bar}\cdot\text{m}^3\cdot\text{mol}^{-1} \\ &= -51.3\text{ J}\cdot\text{mol}^{-1}\end{aligned}$$

8–8. Determine $\overline{C}_P - \overline{C}_V$ for a gas that obeys the equation of state $P(\overline{V} - b) = RT$.

We can write, from the equation of state,

$$P(\overline{V} - b) = RT$$

$$\left(\frac{\partial P}{\partial T}\right)_V = \frac{R}{\overline{V} - b}$$

$$\left(\frac{\partial \overline{V}}{\partial T}\right)_P = \frac{R}{P}$$

Now we substitute into Equation 8.23:

$$\overline{C}_P - \overline{C}_V = T\left(\frac{\partial P}{\partial T}\right)_V\left(\frac{\partial \overline{V}}{\partial T}\right)_P = \frac{RT}{\overline{V} - b}\left(\frac{R}{P}\right) = R$$

8–9. The coefficient of thermal expansion of water at 25°C is 2.572×10^{-4} K^{-1}, and its thermal compressibility is 4.525×10^{-5} bar^{-1}. Calculate the value of $C_P - C_V$ for one mole of water at 25°C. The density of water at 25°C is 0.99705 g·mL^{-1}.

The molar volume of water is

$$\overline{V} = \left(\frac{1}{0.99705\text{ g}\cdot\text{mL}^{-1}}\right)\left(\frac{18.015\text{ g}}{1\text{ mol}}\right)\left(\frac{1\text{ dm}^3}{1000\text{ mL}}\right) = 0.018068\text{ dm}^3\cdot\text{mol}^{-1}$$

We can now substitute into Equation 8.27 to find $C_P - C_V$. For one mole,

$$\begin{aligned}C_P - C_V &= \frac{\alpha^2 TV}{\kappa} \\ &= \frac{(2.572\times10^{-4}\text{ K}^{-1})^2(298.15\text{ K})(0.018068\text{ dm}^3)}{4.525\times10^{-5}\text{ bar}^{-1}} \\ &= 7.875\times10^{-3}\text{ dm}^3\cdot\text{bar}^{-1}\cdot\text{K}^{-1}\end{aligned}$$

8–10. Use Equation 8.22 to show that

$$\left(\frac{\partial C_V}{\partial V}\right)_T = T\left(\frac{\partial^2 P}{\partial T^2}\right)_V$$

Show that $(\partial C_V/\partial V)_T = 0$ for an ideal gas and a van der Waals gas, and that

$$\left(\frac{\partial C_V}{\partial V}\right)_T = -\frac{3A}{4T^{3/2}\overline{V}(\overline{V}+B)}$$

for a Redlich-Kwong gas.

Recall that, by definition, $C_V = (\partial U/\partial T)_V$, so

$$\left(\frac{\partial C_V}{\partial V}\right)_T = \frac{\partial^2 U}{\partial V \partial T} = \frac{\partial^2 U}{\partial T \partial V} = \frac{\partial}{\partial T}\left(\frac{\partial U}{\partial V}\right)_T$$

Express $(\partial U/\partial V)$ using Equation 8.22 and write

$$\left(\frac{\partial U}{\partial V}\right)_T = -P + T\left(\frac{\partial P}{\partial T}\right)_V$$

$$\frac{\partial}{\partial T}\left(\frac{\partial U}{\partial V}\right)_T = -\left(\frac{\partial P}{\partial T}\right)_V + \left(\frac{\partial P}{\partial T}\right)_V + T\left(\frac{\partial^2 P}{\partial T^2}\right)_V$$

$$\left(\frac{\partial C_V}{\partial V}\right)_T = T\left(\frac{\partial^2 P}{\partial T^2}\right)_V$$

For an ideal gas and for a van der Waals gas,

$$P\overline{V} = RT \qquad\qquad P = \frac{RT}{\overline{V}-b} + \frac{a}{\overline{V}^2}$$

$$\left(\frac{\partial^2 P}{\partial T^2}\right)_V = 0 \qquad\qquad \left(\frac{\partial^2 P}{\partial T^2}\right)_V = 0$$

$$T\left(\frac{\partial^2 P}{\partial T^2}\right)_V = 0 = \left(\frac{\partial C_V}{\partial V}\right)_T \qquad\qquad T\left(\frac{\partial^2 P}{\partial T^2}\right)_V = 0 = \left(\frac{\partial C_V}{\partial V}\right)_T$$

For a Redlich-Kwong gas,

$$P = \frac{RT}{\overline{V}-B} - \frac{A}{T^{1/2}\overline{V}(\overline{V}+B)}$$

$$\left(\frac{\partial P}{\partial T}\right)_V = \frac{R}{\overline{V}-B} + \frac{A}{2\overline{V}(\overline{V}+B)T^{3/2}}$$

$$\left(\frac{\partial^2 P}{\partial T^2}\right)_V = -\frac{3}{4}\frac{A}{T^{5/2}\overline{V}(\overline{V}+B)}$$

$$T\left(\frac{\partial^2 P}{\partial T^2}\right)_V = -\frac{3A}{4T^{3/2}\overline{V}(\overline{V}+B)} = \left(\frac{\partial C_V}{\partial V}\right)_T$$

8–11. In this problem you will derive the equation (Equation 8.24)

$$C_P - C_V = -T\left(\frac{\partial V}{\partial T}\right)_P^2\left(\frac{\partial P}{\partial V}\right)_T$$

To start, consider V to be a function of T and P and write out dV. Now divide through by dT at constant volume ($dV = 0$) and then substitute the expression for $(\partial P/\partial T)_V$ that you obtain into Equation 8.23 to get the above expression.

The total derivative of $V(T, P)$ is

$$dV(T, P) = \left(\frac{\partial V}{\partial T}\right)_P dT + \left(\frac{\partial V}{\partial P}\right)_T dP$$

Dividing through by dT at constant volume gives

$$0 = \left(\frac{\partial V}{\partial T}\right)_P + \left(\frac{\partial V}{\partial P}\right)_T \left(\frac{\partial P}{\partial T}\right)_V$$

$$\left(\frac{\partial P}{\partial T}\right)_V = -\left(\frac{\partial V}{\partial T}\right)_P \left(\frac{\partial P}{\partial V}\right)_T$$

Now substitute for $(\partial P/\partial T)_V$ into Equation 8.23:

$$C_P - C_V = T\left(\frac{\partial P}{\partial T}\right)_V \left(\frac{\partial V}{\partial T}\right)_P = -T\left(\frac{\partial V}{\partial T}\right)_P^2 \left(\frac{\partial P}{\partial V}\right)_T$$

which is Equation 8.24.

8–12. The quantity $(\partial U/\partial V)_T$ has units of pressure and is called the *internal pressure*, which is a measure of the intermolecular forces within the body of a substance. It is equal to zero for an ideal gas, is nonzero but relatively small for dense gases, and is relatively large for liquids, particularly those whose molecular interactions are strong. Use the following data to calculate the internal pressure of ethane as a function of pressure at 280 K. Compare your values with the values you obtain from the van der Waals equation and the Redlich-Kwong equation.

P/bar	(dP/dT)/bar·K^{-1}	$\overline{V}$/dm^3·mol^{-1}	P/bar	(dP/dT)/bar·K^{-1}	$\overline{V}$/dm^3·mol^{-1}
4.458	0.01740	5.000	307.14	6.9933	0.06410
47.343	4.1673	0.07526	437.40	7.9029	0.06173
98.790	4.9840	0.07143	545.33	8.5653	0.06024
157.45	5.6736	0.06849	672.92	9.2770	0.05882

Use Equation 8.22 to write

$$\left(\frac{\partial U}{\partial V}\right)_T = -P + T\left(\frac{\partial P}{\partial T}\right)_V$$

To find the experimental values of internal pressure, we can substitute the data given into the above equation. We expressed $(\partial U/\partial V)_T$ for the van der Waals equation in Problem 8–4 as

$$\left(\frac{\partial U}{\partial V}\right)_T = \frac{a}{\overline{V}^2}$$

and $(\partial U/\partial V)_T$ for the Redlich-Kwong equation as

$$\left(\frac{\partial U}{\partial V}\right)_T = \frac{3A}{2T^{1/2}\overline{V}(\overline{V}+B)}$$

We can use the molar volumes given in the statement of the problem and the constants from Tables 2.3 and 2.4 ($a = 5.5818$ dm^6·bar·mol^{-2}, $b = 0.065144$ dm^3·mol^{-1}, $A = 98.831$ dm^6·bar·mol^{-2}·K$^{1/2}$,

$B = 0.045153\ \text{dm}^3\cdot\text{mol}^{-1}$) to create a table of values of internal pressures for each experimental pressure.

	$(\partial U/\partial V)_T$/bar		
P/bar	Experimental	van der Waals	Redlich-Kwong
4.458	0.4140	0.2233	0.3512
47.343	1119.5	972.5	967.1
98.790	1296.7	1094	1064
157.45	1431.2	1190	1138
307.14	1651.0	1359	1265
437.40	1775.4	1465	1343
545.33	1853.0	1538	1395
672.92	1924.6	1613	1449

8–13. Show that

$$\left(\frac{\partial \overline{H}}{\partial P}\right)_T = -RT^2\left(\frac{dB_{2P}}{dT} + \frac{dB_{3P}}{dT}P + \cdots\right)$$

$$= B_{2V}(T) - T\frac{dB_{2V}}{dT} + O(P)$$

Use Equation 2.41 for the square-well potential to obtain

$$\left(\frac{\partial \overline{H}}{\partial P}\right)_T = \frac{2\pi\sigma^3 N_A}{3}\left[\lambda^3 - (\lambda^3 - 1)\left(1 + \frac{\varepsilon}{k_B T}\right)e^{\varepsilon/k_B T}\right]$$

Given that $\sigma = 327.7$ pm, $\varepsilon/k_B = 95.2$ K, and $\lambda = 1.58$ for $N_2(g)$, calculate the value of $(\partial\overline{H}/\partial P)_T$ at 300 K. Evaluate $\Delta\overline{H} = \overline{H}(P{=}10.0\text{ bar}) - \overline{H}(P{=}1.0\text{ bar})$. Compare your result with $8.724\ \text{J}\cdot\text{mol}^{-1}$, the value of $\overline{H}(T) - \overline{H}(0)$ for nitrogen at 300 K.

Use the virial expansion in the pressure (Equation 2.23):

$$Z = 1 + B_{2P}P + B_{3P}P^2 + O(P^3)$$

$$\overline{V} = \frac{RT}{P} + RTB_{2P} + PRTB_{3P} + O(P^2)$$

$$\left(\frac{\partial \overline{V}}{\partial T}\right)_P = \frac{R}{P} + RB_{2P} + RT\frac{dB_{2P}}{dT} + PRB_{3P} + PRT\frac{dB_{3P}}{dT} + O(P^2)$$

Substitute into Equation 8.34:

$$\left(\frac{\partial \overline{H}}{\partial P}\right)_T = \overline{V} - T\left(\frac{\partial \overline{V}}{\partial T}\right)_P$$

$$\left(\frac{\partial \overline{H}}{\partial P}\right)_T = \left[\frac{RT}{P} + RTB_{2P} + PRTB_{3P}\right] - \left[\frac{RT}{P} + RTB_{2P} + RT^2\frac{dB_{2P}}{dT} + PRTB_{3P} + PRT^2\frac{dB_{3P}}{dT}\right] + O(P^2)$$

$$= -RT^2\left[\frac{dB_{2P}}{dT} + \frac{dB_{3P}}{dT}P + O(P^2)\right] \qquad (1)$$

Since $B_{2V} = RT B_{2P}$ (Equation 2.24),

$$\frac{dB_{2V}}{dT} = RB_{2P} + RT\frac{dB_{2P}}{dT}$$

$$\frac{dB_{2V}}{dT}\frac{1}{RT} = \frac{B_{2V}}{RT^2} + \frac{dB_{2P}}{dT}$$

$$\frac{dB_{2P}}{dT} = \frac{1}{RT}\frac{dB_{2V}}{dT} - \frac{B_{2V}}{RT^2}$$

Then Equation 1 becomes

$$\left(\frac{\partial \overline{H}}{\partial P}\right)_T = -RT^2\left[\frac{dB_{2P}}{dT} + O(P)\right]$$

$$= B_{2V} - T\frac{dB_{2V}}{dT} + O(P) \qquad (2)$$

Start with Equation 2.41 and find $-TdB_{2V}/dT$:

$$B_{2V}(T) = \frac{2\pi\sigma^3 N_A}{3}\left[1 - (\lambda^3 - 1)\left(e^{\varepsilon/k_BT} - 1\right)\right]$$

$$\frac{dB_{2V}}{dT} = \frac{2\pi\sigma^3 N_A}{3}(\lambda^3 - 1)e^{\varepsilon/k_BT}\left(\frac{\varepsilon}{k_BT^2}\right)$$

$$-T\frac{dB_{2V}}{dT} = -\frac{2\pi\sigma^3 N_A}{3}(\lambda^3 - 1)e^{\varepsilon/k_BT}\left(\frac{\varepsilon}{k_BT}\right)$$

Substituting this value into Equation 2 and ignoring terms of P or higher, we find

$$\left(\frac{\partial \overline{H}}{\partial P}\right)_T = \frac{2\pi\sigma^3 N_A}{3}\left[1 - (\lambda^3 - 1)\left(e^{\varepsilon/k_BT} - 1\right) - e^{\varepsilon/k_BT}\left(\frac{\varepsilon}{k_BT}\right)(\lambda^3 - 1)\right]$$

$$= \frac{2\pi\sigma^3 N_A}{3}\left[1 - \left(\lambda^3 e^{\varepsilon/k_BT} + 1 - \lambda^3 - e^{\varepsilon/k_BT}\right) - \lambda^3\frac{\varepsilon}{k_BT}e^{\varepsilon/k_BT} + \frac{\varepsilon}{k_BT}e^{\varepsilon/k_BT}\right]$$

$$= \frac{2\pi\sigma^3 N_A}{3}\left[\lambda^3 - e^{\varepsilon/k_BT}\left(\lambda^3 - 1 - \frac{\varepsilon}{k_BT} + \lambda^3\frac{\varepsilon}{k_BT}\right)\right]$$

$$= \frac{2\pi\sigma^3 N_A}{3}\left[\lambda^3 - (\lambda^3 - 1)\left(1 + \frac{\varepsilon}{k_BT}\right)e^{\varepsilon/k_BT}\right]$$

Using the parameters provided for nitrogen, this expression becomes

$$\left(\frac{\partial \overline{H}}{\partial P}\right)_T = \frac{2\pi(327.7\times10^{-12}\ \text{m})^3(6.022\times10^{23}\ \text{mol}^{-1})}{3}\left[1.58^3 - (1.58^3 - 1)\left(1 + \frac{95.2}{300}\right)e^{95.2/300}\right]$$

$$= -6.138\times10^{-5}\ \text{m}^3\cdot\text{mol}^{-1}$$

Then

$$\Delta\overline{H} = (-6.138\times10^{-5}\ \text{m}^3\cdot\text{mol}^{-1})\Delta P = -5.52\times10^{-4}\ \text{m}^3\cdot\text{bar}\cdot\text{mol}^{-1} = -55.2\ \text{J}\cdot\text{mol}^{-1}$$

8–14. Show that the enthalpy is a function of only the temperature for a gas that obeys the equation of state $P(\overline{V} - bT) = RT$.

$$\left(\frac{\partial \overline{H}}{\partial P}\right)_T = \overline{V} - T\left(\frac{\partial \overline{V}}{\partial T}\right)_P \tag{8.34}$$

For a gas obeying the equation of state $P(\overline{V} - bT) = RT$,

$$\overline{V} = \frac{RT}{P} + bT$$

$$\left(\frac{\partial \overline{V}}{\partial T}\right)_P = \frac{R}{P} + b$$

Substituting into Equation 8.34, we find

$$\left(\frac{\partial \overline{H}}{\partial P}\right)_T = \frac{RT}{P} + bT - T\left(\frac{R}{P} + b\right) = 0$$

and so enthalpy does not depend on pressure for a gas with this equation of state.

8–15. Use your results for the van der Waals equation and the Redlich-Kwong equation in Problem 8–4 to calculate $\overline{H}(T, \overline{V})$ as a function of volume for ethane at 400 K. In each case, use the equation $\overline{H} = \overline{U} + P\overline{V}$. Compare your results with the experimental data shown in Figure 8.5.

$$\overline{H} = \overline{U}(T, \overline{V}) + P\overline{V}$$

From Problem 8–4, for a van der Waals gas

$$\overline{U} = 14.55\ \text{kJ}\cdot\text{mol}^{-1} - \frac{0.55818\ \text{kJ}\cdot\text{dm}^3\cdot\text{mol}^{-2}}{\overline{V}}$$

$$P = \frac{(0.083145\ \text{dm}^3\cdot\text{bar}\cdot\text{mol}^{-1}\cdot\text{K}^{-1})(400\ \text{K})}{\overline{V} - 0.065144\ \text{dm}^3\cdot\text{mol}^{-1}} - \frac{5.5818\ \text{dm}^6\cdot\text{bar}\cdot\text{mol}^{-1}}{\overline{V}^2}$$

and for a Redlich-Kwong gas

$$\overline{U} = 14.55\ \text{kJ}\cdot\text{mol}^{-1} - \frac{3(9.8831\ \text{kJ}\cdot\text{dm}^3\cdot\text{mol}^{-2}\cdot\text{K}^{1/2})}{2(0.045153\ \text{dm}^3\cdot\text{mol}^{-1})(400\ \text{K})^{1/2}} \ln\frac{\overline{V} + 0.045153\ \text{dm}^3\cdot\text{mol}^{-1}}{\overline{V}}$$

$$P = \frac{(0.083145\ \text{dm}^3\cdot\text{bar}\cdot\text{mol}^{-1}\cdot\text{K}^{-1})(400\ \text{K})}{\overline{V} - 0.045153\ \text{dm}^3\cdot\text{mol}^{-1}} - \frac{98.831\ \text{dm}^6\cdot\text{bar}\cdot\text{mol}^{-2}\cdot\text{K}^{1/2}}{(400\ \text{K})^{1/2}\overline{V}(\overline{V} + 0.045153\ \text{dm}^3\cdot\text{mol}^{-1})}$$

Note that in using these values, we find $\overline{U}$ in terms of $\text{kJ}\cdot\text{mol}^{-1}$ and $P\overline{V}$ in terms of $\text{dm}^3\cdot\text{bar}\cdot\text{mol}^{-1}$. Dividing $P\overline{V}$ by 10 will result in values of enthalpy given in $\text{kJ}\cdot\text{mol}^{-1}$. Using these values, we can produce plots of $\overline{H}$ vs. P.

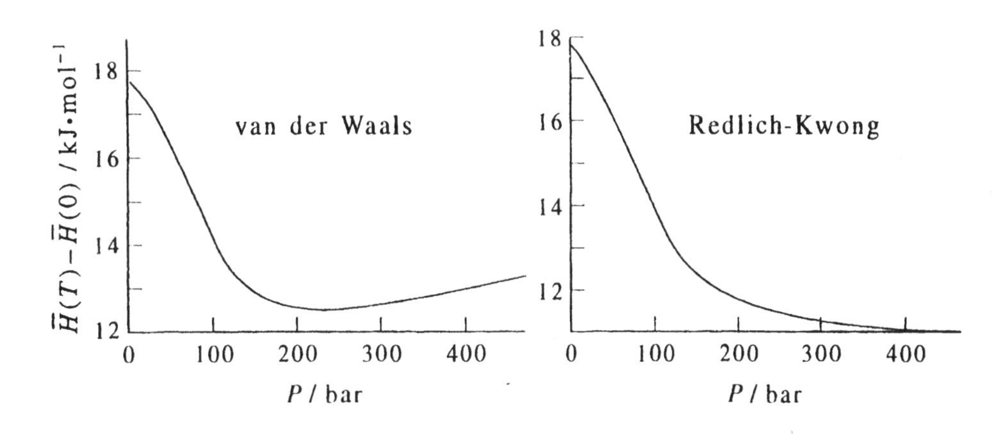

8–16. Use Equation 8.34 to show that

$$\left(\frac{\partial C_P}{\partial P}\right)_T = -T\left(\frac{\partial^2 V}{\partial T^2}\right)_P$$

Use a virial expansion in P to show that

$$\left(\frac{\partial \overline{C}_P}{\partial P}\right)_T = -T\frac{d^2 B_{2V}}{dT^2} + O(P)$$

Use the square-well second virial coefficient (Equation 2.41) and the parameters given in Problem 8–13 to calculate the value of $(\partial \overline{C}_P/\partial P)_T$ for $N_2(g)$ at 0°C. Now calculate $\overline{C}_P$ at 100 atm and 0°C, using $\overline{C}_P^{\text{id}} = 5R/2$.

We define C_P as $(\partial H/\partial T)_P$ (Equation 5.40). Starting with Equation 8.34,

$$\left(\frac{\partial H}{\partial P}\right)_T = V - T\left(\frac{\partial V}{\partial T}\right)_P$$

$$\frac{\partial}{\partial T}\left(\frac{\partial H}{\partial P}\right)_T = \left(\frac{\partial V}{\partial T}\right)_P - \left(\frac{\partial V}{\partial T}\right)_P - T\frac{\partial}{\partial T}\left(\frac{\partial V}{\partial T^2}\right)_P$$

$$\frac{\partial}{\partial P}\left(\frac{\partial H}{\partial T}\right)_P = -T\left(\frac{\partial^2 V}{\partial T^2}\right)_P$$

$$\left(\frac{\partial C_P}{\partial P}\right)_T = -T\left(\frac{\partial^2 V}{\partial T^2}\right)_P \tag{1}$$

Using a virial expansion in P, we find

$$\overline{V} = \frac{RT}{P} + RTB_{2P} + PRTB_{3P} + O(P^2)$$

$$\left(\frac{\partial \overline{V}}{\partial T}\right)_P = \frac{R}{P} + RB_{2P} + RT\frac{dB_{2P}}{dT} + O(P)$$

$$\left(\frac{\partial^2 \overline{V}}{\partial T^2}\right)_P = R\frac{dB_{2P}}{dT} + R\frac{dB_{2P}}{dT} + RT\frac{d^2 B_{2P}}{dT^2} + O(P)$$

$$= 2R\frac{dB_{2P}}{dT} + RT\frac{d^2 B_{2P}}{dT^2} + O(P)$$

Now, since $B_{2V} = RTB_{2P}$ (Equation 2.24),

$$B_{2P} = \frac{B_{2V}}{RT}$$

$$\frac{dB_{2P}}{dT} = \frac{1}{RT}\frac{dB_{2V}}{dT} - \frac{B_{2V}}{RT^2}$$

$$\frac{d^2 B_{2P}}{dT^2} = \frac{1}{RT}\frac{d^2 B_{2V}}{dT^2} - \frac{1}{RT^2}\frac{dB_{2V}}{dT} - \frac{1}{RT^2}\frac{dB_{2V}}{dT} + \frac{2B_{2V}}{RT^3}$$

$$= \frac{1}{RT}\frac{d^2 B_{2V}}{dT^2} - \frac{2}{RT^2}\frac{dB_{2V}}{dT} + \frac{2B_{2V}}{RT^3}$$

Now we solve Equation 1 for $(\partial \overline{C}_P/\partial P)_T$ in terms of B_{2V}:

$$\begin{aligned}
\left(\frac{\partial \overline{C}_P}{\partial P}\right)_T &= -T\left(\frac{\partial^2 \overline{V}}{\partial T^2}\right)_P \\
&= -T\left[2R\frac{dB_{2P}}{dT} + RT\frac{d^2B_{2P}}{dT^2} + O(P)\right] \\
&= -T\left[2R\left(\frac{1}{RT}\frac{dB_{2V}}{dT} - \frac{B_{2V}}{RT^2}\right) + RT\left(\frac{1}{RT}\frac{d^2B_{2V}}{dT^2} - \frac{2}{RT^2}\frac{dB_{2V}}{dT} + \frac{2B_{2V}}{RT^3}\right)\right] + O(P) \\
&= -\frac{2dB_{2V}}{dT} + \frac{2B_{2V}}{T} - T\frac{d^2B_{2V}}{dT^2} + \frac{2dB_{2V}}{dT} - \frac{2B_{2V}}{T} + O(P) \\
&= -T\frac{d^2B_{2V}}{dT^2} + O(P)
\end{aligned}$$

Using Equation 2.41,

$$\begin{aligned}
B_{2V}(T) &= \frac{2\pi\sigma^3 N_A}{3}\left[1 - \left(\lambda^3 - 1\right)\left(e^{\varepsilon/k_BT} - 1\right)\right] \\
\frac{dB_{2V}}{dT} &= \frac{2\pi\sigma^3 N_A}{3}\left(\lambda^3 - 1\right)e^{\varepsilon/k_BT}\left(\frac{\varepsilon}{k_BT^2}\right) \\
\frac{d^2B_{2V}}{dT^2} &= -\frac{2\pi\sigma^3 N_A}{3}\left(\lambda^3 - 1\right)e^{\varepsilon/k_BT}\frac{\varepsilon}{k_BT^3}\left(\frac{\varepsilon}{k_BT} + 2\right) \\
\left(\frac{\partial \overline{C}_P}{\partial P}\right)_T &= -T\frac{d^2B_{2V}}{dT} = \frac{2\pi\sigma^3 N_A}{3}\left(\lambda^3 - 1\right)e^{\varepsilon/k_BT}\frac{\varepsilon}{k_BT^2}\left(\frac{\varepsilon}{k_BT} + 2\right)
\end{aligned}$$

For nitrogen at 298.15 K,

$$\begin{aligned}
\left(\frac{\partial \overline{C}_P}{\partial P}\right)_T &= \frac{2\pi(327.7 \times 10^{-12}\text{ m})^3(6.022 \times 10^{23}\text{ mol}^{-1})}{3}\left(1.58^3 - 1\right)e^{95.2/298.15}\frac{95.2}{(298.15)^2\text{ K}}\left(\frac{95.2}{298.15} + 2\right) \\
&= 4.467 \times 10^{-7}\text{ m}^3\cdot\text{mol}^{-1}\cdot\text{K}^{-1} = 4.47 \times 10^{-4}\text{ dm}^3\cdot\text{mol}^{-1}\cdot\text{K}^{-1}
\end{aligned}$$

Finally,

$$\begin{aligned}
\overline{C}_P - \overline{C}_P^{\text{id}} &= (4.467 \times 10^{-4}\text{ dm}^3\cdot\text{mol}^{-1}\cdot\text{K}^{-1})(P - P^{\text{id}}) \\
\overline{C}_P &= \frac{5R}{2} + (4.467 \times 10^{-4}\text{ dm}^3\cdot\text{mol}^{-1}\cdot\text{K}^{-1})(99\text{ atm}) \\
&= (2.5)(8.3145\text{ J}\cdot\text{mol}^{-1}\cdot\text{K}^{-1}) + 4.42\text{ J}\cdot\text{mol}^{-1}\cdot\text{K}^{-1} \\
&= 25.21\text{ J}\cdot\text{mol}^{-1}\cdot\text{K}^{-1}
\end{aligned}$$

8–17. Show that the molar enthalpy of a substance at pressure P relative to its value at one bar is given by

$$\overline{H}(T, P) = \overline{H}(T, P=1\text{ bar}) + \int_1^P \left[\overline{V} - T\left(\frac{\partial \overline{V}}{\partial T}\right)_P\right] dP'$$

Calculate the value of $\overline{H}(T, P) - \overline{H}(T, P=1\text{ bar})$ at 0°C and 100 bar for mercury given that the molar volume of mercury varies with temperature according to

$$\overline{V}(t) = (14.75\text{ mL}\cdot\text{mol}^{-1})(1 + 0.182 \times 10^{-3}t + 2.95 \times 10^{-9}t^2 + 1.15 \times 10^{-10}t^3)$$

where t is the Celsius temperature. Assume that $\overline{V}(0)$ does not vary with pressure over this range and express your answer in units of $\text{kJ}\cdot\text{mol}^{-1}$.

Begin with Equation 8.34:

$$\left(\frac{\partial \overline{H}}{\partial P}\right)_T = \overline{V} - T\left(\frac{\partial \overline{V}}{\partial T}\right)_P$$

$$d\overline{H} = \left[\overline{V} - T\left(\frac{\partial \overline{V}}{\partial T}\right)_P\right] dP$$

$$\overline{H}(T, P) - \overline{H}(T, 1\text{ bar}) = \int_{1\text{ bar}}^{P} \left[\overline{V} - T\left(\frac{\partial \overline{V}}{\partial T}\right)_P\right] dP'$$

where we have begun using P' as the quantity integrated over in order to distinguish it from P, the final pressure of the substance.
Using the values given for mercury,

$$\left(\frac{\partial \overline{V}}{\partial T}\right)_P = (14.75\text{ mL}\cdot\text{mol}^{-1})(0.182 \times 10^{-3} + 5.90 \times 10^{-9}t + 3.45 \times 10^{-10}t^2)$$

Then at 0° C and 100 bar,

$$\begin{aligned}\overline{H}(T, 100\text{ bar}) - \overline{H}(T, 1\text{ bar}) &= (14.75\text{ mL}\cdot\text{mol}^{-1}) \int_{1\text{ bar}}^{100\text{ bar}} \left[1 - T(0.182 \times 10^{-3})\right] dP' \\ &= (14.75\text{ mL}\cdot\text{mol}^{-1})(99\text{ bar})[1 - (298.15)(0.182 \times 10^{-3})] \\ &= 1381\text{ mL}\cdot\text{bar}\cdot\text{mol}^{-1} = 138.1\text{ J}\cdot\text{mol}^{-1}\end{aligned}$$

8–18. Show that

$$dH = \left[V - T\left(\frac{\partial V}{\partial T}\right)_P\right] dP + C_p dT$$

What does this equation tell you about the natural variables of H?

Write the total derivative of $H(P, T)$:

$$dH = \left(\frac{\partial H}{\partial P}\right)_T dP + \left(\frac{\partial H}{\partial T}\right)_P dT$$

We can now use Equation 8–34 and the definition of C_p to write this as

$$dH = \left[V - T\left(\frac{\partial V}{\partial T}\right)_P\right] dP + C_p dT$$

Since the coefficients of dP and dT are not simple, this tells us that the natural variables of H are not P and T.

8–19. What are the natural variables of the entropy?

$$dS = PdV + \frac{1}{T}dU \tag{8.39}$$

Because the coefficients of dV and dU are simple thermodynamic quantities, we say that V and U are the natural variables of entropy.

8–20. Experimentally determined entropies are commonly adjusted for nonideality by using an equation of state called the (modified) Berthelot equation:

$$\frac{P\overline{V}}{RT} = 1 + \frac{9}{128}\frac{PT_c}{P_cT}\left(1 - 6\frac{T_c^2}{T^2}\right)$$

Show that this equation leads to the correction

$$S^\circ(\text{at one bar}) = \overline{S}(\text{at one bar}) + \frac{27}{32}\frac{RT_c^3}{P_cT^3}(1\text{ bar})$$

This result needs only the critical data for the substance. Use this equation along with the critical data in Table 2.5 to calculate the nonideality correction for $N_2(g)$ at 298.15 K. Compare your result with the value used in Table 7.1.

$$S^\circ(1\text{ bar}) - \overline{S}(1\text{ bar}) = \int_{P^{id}}^{1\text{ bar}}\left[\left(\frac{\partial\overline{V}}{\partial T}\right)_P - \frac{R}{P}\right]dP \tag{8.54}$$

We find $(\partial\overline{V}/\partial T)$ from the modified Bethelot equation:

$$\frac{P\overline{V}}{RT} = 1 + \frac{9}{128}\frac{PT_c}{P_cT}\left(1 - 6\frac{T_c^2}{T^2}\right)$$

$$\overline{V} = \frac{RT}{P} + \frac{9R}{128}\frac{T_c}{P_c} - \frac{9\cdot 6}{128}\frac{RT_c^3}{P_cT^2}$$

$$\left(\frac{\partial\overline{V}}{\partial T}\right)_P = \frac{R}{P} + \frac{9\cdot 6\cdot 2}{128}\frac{RT_c^3}{P_cT^3}$$

Now substitute into Equation 8.54 to find $S^\circ(1\text{ bar}) - \overline{S}(1\text{ bar})$, neglecting P^{id} with respect to 1 bar:

$$S^\circ(1\text{ bar}) - \overline{S}(1\text{ bar}) = \int_{P^{id}}^{1\text{ bar}}\left[\frac{R}{P} + \frac{27}{32}\frac{RT_c^3}{P_cT^3} - \frac{R}{P}\right]dP$$

$$= \frac{27}{32}\frac{RT_c^3}{P_cT^3}(1\text{ bar})$$

For N_2 at 298.15 K, $T_c = 126.2$ K and $P_c = 34.00$ bar. Then the nonideality correction (the difference between the two values of S) is

$$S^\circ(1\text{ bar}) - \overline{S}(1\text{ bar}) = \frac{27}{32}\frac{(8.3145\text{ J}\cdot\text{mol}^{-1}\cdot\text{K}^{-1})(126.2\text{ K})^3}{(34.00\text{ bar})(298.15\text{ K})^3}(1\text{ bar})$$

$$= 0.0156\text{ J}\cdot\text{mol}^{-1}\cdot\text{K}^{-1}$$

This is essentially the value used in Table 7.1 ($0.02\text{ J}\cdot\text{K}^{-1}\cdot\text{mol}^{-1}$).

8–21. Use the result of Problem 8–20 along with the critical data in Table 2.5 to determine the nonideality correction for CO(g) at its normal boiling point, 81.6 K. Compare your result with the value used in Problem 7–24.

$$S^\circ(1\text{ bar}) - \overline{S}(1\text{ bar}) = \frac{27}{32}\frac{RT_c^3}{P_cT^3}(1\text{ bar})$$
$$= \frac{27}{32}\frac{(8.3145\text{ J}\cdot\text{mol}^{-1}\cdot\text{K}^{-1})(132.85\text{ K})^3}{(34.935\text{ bar})(81.6\text{ K})^3}(1\text{ bar})$$
$$= 0.867\text{ J}\cdot\text{mol}^{-1}\cdot\text{K}^{-1}$$

This is comparable to the value used in Problem 7–24 ($0.879\text{ J}\cdot\text{K}^{-1}\cdot\text{mol}^{-1}$).

8–22. Use the result of Problem 8–20 along with the critical data in Table 2.5 to determine the nonideality correction for $Cl_2(g)$ at its normal boiling point, 239 K. Compare your result with the value used in Problem 7–16.

$$S^\circ(1\text{ bar}) - \overline{S}(1\text{ bar}) = \frac{27}{32}\frac{RT_c^3}{P_cT^3}(1\text{ bar})$$
$$= \frac{27}{32}\frac{(8.3145\text{ J}\cdot\text{mol}^{-1}\cdot\text{K}^{-1})(416.9\text{ K})^3}{(79.91\text{ bar})(239\text{ K})^3}(1\text{ bar})$$
$$= 0.466\text{ J}\cdot\text{mol}^{-1}\cdot\text{K}^{-1}$$

This is comparable to the value of $0.502\text{ J}\cdot\text{K}^{-1}\cdot\text{mol}^{-1}$ used in Problem 21–16.

8–23. Derive the equation

$$\left(\frac{\partial(A/T)}{\partial T}\right)_V = -\frac{U}{T^2}$$

which is a Gibbs-Helmholtz equation for A.

Begin with the definition of A (Equation 8.4):

$$A = U - TS$$
$$\frac{\partial}{\partial T}\left(\frac{A}{T}\right) = \frac{\partial}{\partial T}\left[\frac{U}{T} - S\right]$$
$$\left[\frac{\partial(A/T)}{\partial T}\right]_V = -\frac{U}{T^2} + \frac{1}{T}\left(\frac{\partial U}{\partial T}\right)_V - \left(\frac{\partial S}{\partial T}\right)_V$$

Now, by the definition of C_V,

$$\frac{C_V}{T} = \frac{1}{T}\left(\frac{\partial U}{\partial T}\right)_V$$

From Equation 7.2, we also know that

$$\frac{C_V}{T} = \left(\frac{\partial S}{\partial T}\right)_V$$

Therefore,

$$\left(\frac{\partial(A/T)}{\partial T}\right)_V = -\frac{U}{T^2} + \frac{C_V}{T} - \frac{C_V}{T} = -\frac{U}{T^2}$$

8–24. We can derive the Gibbs-Helmholtz equation directly from Equation 8.31a in the following way. Start with $(\partial G/\partial T)_P = -S$ and substitute for S from $G = H - TS$ to obtain

$$\frac{1}{T}\left(\frac{\partial G}{\partial T}\right)_P - \frac{G}{T^2} = -\frac{H}{T^2}$$

Now show that the left side is equal to $(\partial[G/T]/\partial T)_P$ to get the Gibbs-Helmholtz equation.

Begin with the equality in the statement of the problem and substitute for S as suggested (from Equation 8.13):

$$\left(\frac{\partial G}{\partial T}\right)_P = -S = \frac{G}{T} - \frac{H}{T}$$

$$\frac{1}{T}\left(\frac{\partial G}{\partial T}\right)_P - \frac{G}{T^2} = -\frac{H}{T^2} \qquad (1)$$

Taking the partial derivative of G/T with respect to T gives

$$\frac{\partial}{\partial T}\left(\frac{G}{T}\right) = \frac{1}{T}\left(\frac{\partial G}{\partial T}\right)_P - \frac{G}{T^2}$$

so we can write Equation 1 as

$$\frac{\partial}{\partial T}\left(\frac{G}{T}\right) = -\frac{H}{T^2}$$

which is the Gibbs-Helmholtz equation.

8–25. Use the following data for benzene to plot $\overline{G}(T) - \overline{H}(0)$ versus T. [In this case we will ignore the (usually small) corrections due to nonideality of the gas phase.]

$$\overline{C}_P^{\rm s}(T)/R = \frac{12\pi^4}{5}\left(\frac{T}{\Theta_{\rm D}}\right)^3 \qquad \Theta_{\rm D} = 130.5\text{ K} \qquad 0\text{ K} < T < 13\text{ K}$$

$$\overline{C}_P^{\rm s}(T)/R = -0.6077 + (0.1088\text{ K}^{-1})T - (5.345\times10^{-4}\text{ K}^{-2})T^2 + (1.275\times10^{-6}\text{ K}^{-3})T^3$$
$$13\text{ K} < T < 278.6\text{ K}$$

$$\overline{C}_P^{\rm l}(T)/R = 12.713 + (1.974\times10^{-3}\text{ K}^{-1})T - (4.766\times10^{-5}\text{ K}^{-2})T^2$$
$$278.6\text{ K} < T < 353.2\text{ K}$$

$$\overline{C}_P^{\rm g}(T)/R = -4.077 + (0.05676\text{ K}^{-1})T - (3.588\times10^{-5}\text{ K}^{-2})T^2 + (8.520\times10^{-9}\text{ K}^{-3})T^3$$
$$353.2\text{ K} < T < 1000\text{ K}$$

$$T_{\rm fus} = 278.68\text{ K} \qquad \Delta_{\rm fus}\overline{H} = 9.95\text{ kJ}\cdot\text{mol}^{-1}$$

$$T_{\rm vap} = 353.24\text{ K} \qquad \Delta_{\rm vap}\overline{H} = 30.72\text{ kJ}\cdot\text{mol}^{-1}$$

Use the Equation $\overline{G}(T) - \overline{H}(0) = \overline{H}(T) - \overline{H}(0) - T\overline{S}(T)$ (as in Section 8–7) and Equations 8.62 and 8.63 for $\overline{H}(T) - \overline{H}(0)$ and $\overline{S}(T)$ to plot. (See Problem 7.15 for an explanation of how to assign the values of entropy and enthalpy as functions of temperature.)

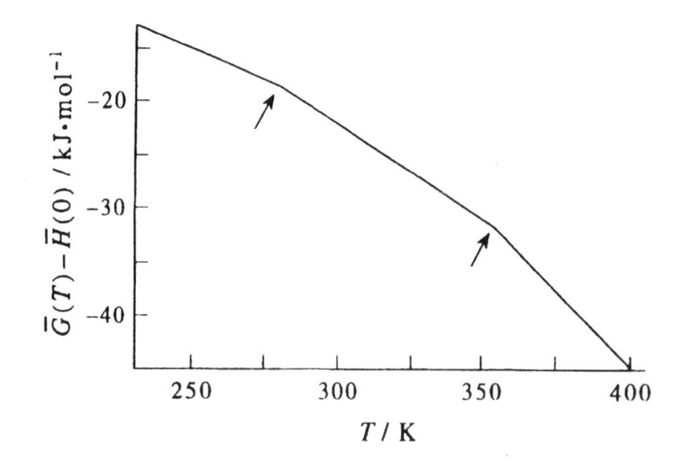

The discontinuities in the slope at the melting and boiling points are difficult to see and so are highlighted with arrows.

8–26. Use the following data for propene to plot $\overline{G}(T) - \overline{H}(0)$ versus T. [In this case we will ignore the (usually small) corrections due to nonideality of the gas phase.]

$$\overline{C}_P^{\rm s}(T)/R = \frac{12\pi^4}{5}\left(\frac{T}{\Theta_{\rm D}}\right)^3 \qquad \Theta_{\rm D} = 100\ \text{K} \qquad 0\ \text{K} < T < 15\ \text{K}$$

$$\overline{C}_P^{\rm s}(T)/R = -1.616 + (0.08677\ \text{K}^{-1})T - (9.791 \times 10^{-4}\ \text{K}^{-2})T^2 + (2.611 \times 10^{-6}\ \text{K}^{-3})T^3$$

$$15\ \text{K} < T < 87.90\ \text{K}$$

$$\overline{C}_P^{\rm l}(T)/R = 15.935 - (0.08677\ \text{K}^{-1})T + (4.294 \times 10^{-4}\ \text{K}^{-2})T^2 - (6.276 \times 10^{-7}\ \text{K}^{-3})T^3$$

$$87.90\ \text{K} < T < 225.46\ \text{K}$$

$$\overline{C}_P^{\rm g}(T)/R = 1.4970 + (2.266 \times 10^{-2}\ \text{K}^{-1})T - (5.725 \times 10^{-6}\ \text{K}^{-2})T^2$$

$$225.46\ \text{K} < T < 1000\ \text{K}$$

$$T_{\rm fus} = 87.90\ \text{K} \qquad \Delta_{\rm fus}\overline{H} = 3.00\ \text{kJ}\cdot\text{mol}^{-1}$$

$$T_{\rm vap} = 225.46\ \text{K} \qquad \Delta_{\rm vap}\overline{H} = 18.42\ \text{kJ}\cdot\text{mol}^{-1}$$

This is done in the same way as Problem 8–25.

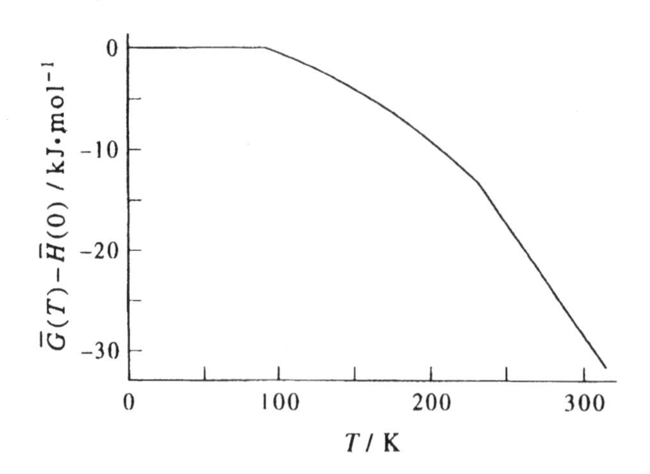

8–27. Use a virial expansion for Z to prove (a) that the integrand in Equation 8.74 is finite as $P \to 0$, and (b) that $(Z-1)/P = 0$ for an ideal gas.

---

a. Use the virial expansion $Z = 1 + B_{2P}P + B_{3P}P^2 + O(P^3)$. Then

$$\frac{Z-1}{P} = \frac{1 + B_{2P}P + B_{3P}P^2 + O(P^3) - 1}{P}$$
$$= B_{2P} + B_{3P}P + O(P^3)$$

As $P \to 0$, then, the integrand approaches B_{2P}, wich is finite.

b. For an ideal gas, $Z = P\overline{V}/RT$ and $P\overline{V} = RT$, so

$$\frac{Z-1}{P} = \frac{1-1}{P} = 0$$

8–28. Derive a virial expansion in the pressure for $\ln \gamma$.

Begin with Equation 8.74 and expand Z as in Problem 8-27(a):

$$\ln \gamma = \int_0^P \frac{Z-1}{P'} dP'$$
$$= \int_0^P (B_{2P} + B_{3P}P' + O(P'^2)\, dP'$$
$$= B_{2P}P + \frac{B_{3P}P^2}{2} + O(P^3)$$

8–29. The compressibility factor for ethane at 600 K can be fit to the expression

$$Z = 1.0000 - 0.000612(P/\text{bar}) + 2.661 \times 10^{-6}(P/\text{bar})^2$$
$$- 1.390 \times 10^{-9}(P/\text{bar})^3 - 1.077 \times 10^{-13}(P/\text{bar})^4$$

for $0 \leq P/\text{bar} \leq 600$. Use this expression to determine the fugacity coefficient of ethane as a function of pressure at 600 K.

Substitute into Equation 8.74:

$$\ln \gamma = \int_0^P \frac{Z-1}{P'} dP'$$
$$= \int_0^P \left[-6.12 \times 10^{-4}\ \text{bar}^{-1} + (2.661 \times 10^{-6}\ \text{bar}^{-2})P'\right.$$
$$\left. -(1.390 \times 10^{-9}\ \text{bar}^{-3})P'^2 - (1.077 \times 10^{-13}\ \text{bar}^{-4})P'^3\right] dP'$$
$$= -(6.12 \times 10^{-4}\ \text{bar}^{-1})P + \tfrac{1}{2}(2.661 \times 10^{-6}\ \text{bar}^{-2})P^2$$
$$-\tfrac{1}{3}(1.390 \times 10^{-9}\ \text{bar}^{-3})P^3 - \tfrac{1}{4}(1.077 \times 10^{-13}\ \text{bar}^{-4})P^4$$
$$\gamma = \exp\left[-(6.12 \times 10^{-4}\ \text{bar}^{-1})P + (1.3305 \times 10^{-6}\ \text{bar}^{-2})P^2\right.$$
$$\left. -(4.633 \times 10^{-10}\ \text{bar}^{-3})P^3 - (2.693 \times 10^{-14}\ \text{bar}^{-4})P^4\right]$$

8–30. Use Figure 8.11 and the data in Table 2.5 to estimate the fugacity of ethane at 360 K and 1000 atm.

From Table 2.5, $T_c = 305.34$ K and $P_c = 48.077$ atm. Then at 360 K and 1000 atm, $T/T_c = 1.18$ and $P/P_c = 20.8$. Using Figure 8.11, it appears that $\gamma \approx 0.63$.

8–31. Use the following data for ethane at 360 K to plot the fugacity coefficient against pressure.

ρ/mol·dm^{-3}	P/bar	ρ/mol·dm^{-3}	P/bar	ρ/mol·dm^{-3}	P/bar
1.20	31.031	6.00	97.767	10.80	197.643
2.40	53.940	7.20	112.115	12.00	266.858
3.60	71.099	8.40	130.149	13.00	381.344
4.80	84.892	9.60	156.078	14.40	566.335

Compare your result with the result you obtained in Problem 8–30.

By definition,

$$\frac{(Z-1)}{P} = \frac{1}{\rho RT} - \frac{1}{P} = y$$

Now we can plot $(Z-1)/P$ vs. P:

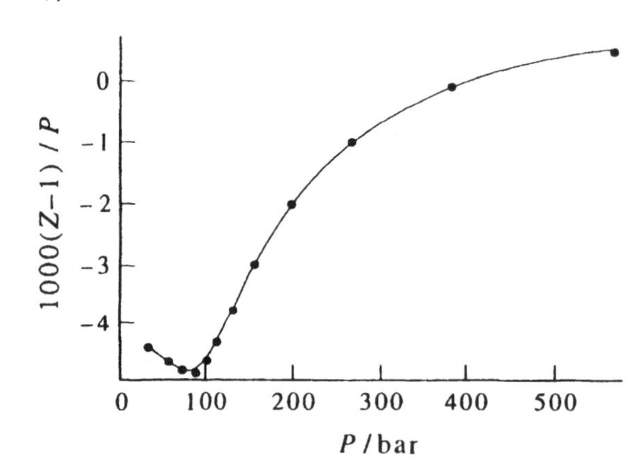

Numerical integration using the trapezoidal approximation allows us to graph f/P vs. P, in the same way that Figure 8.10 was created from Figure 8.9.

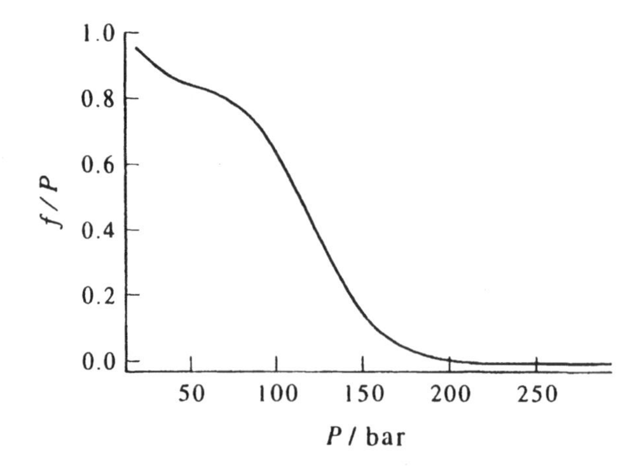

8–32. Use the following data for $N_2(g)$ at 0°C to plot the fugacity coefficient as a function of pressure.

P/atm	$Z = P\overline{V}/RT$	P/atm	$Z = P\overline{V}/RT$	P/atm	$Z = P\overline{V}/RT$
200	1.0390	1000	2.0700	1800	3.0861
400	1.2570	1200	2.3352	2000	3.3270
600	1.5260	1400	2.5942	2200	3.5640
800	1.8016	1600	2.8456	2400	3.8004

Again, plot $(Z-1)/P$ vs. P:

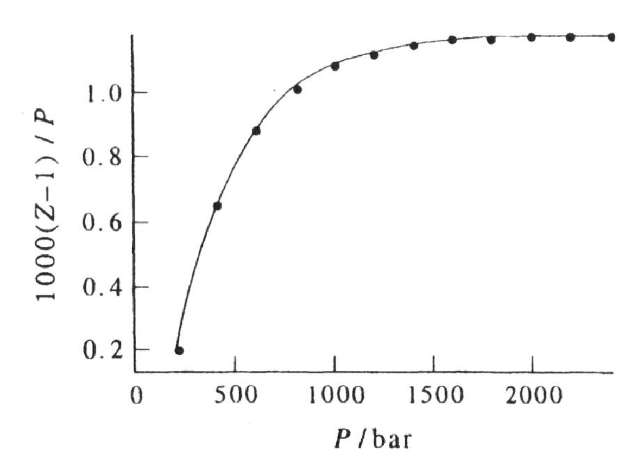

Then do numerical integration with the trapezoidal approximation to graph f/P vs. P:

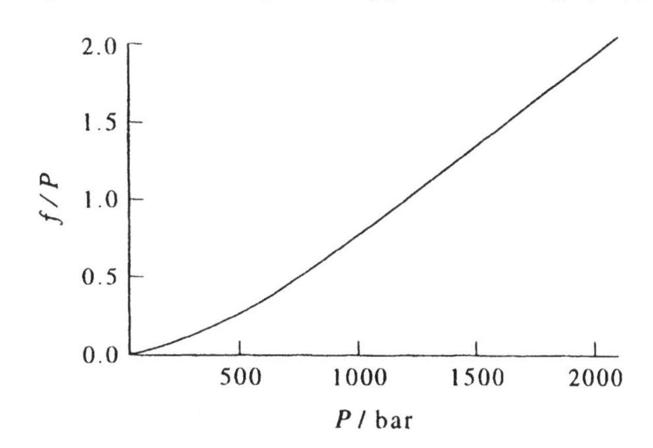

8–33. It might appear that we can't use Equation 8.72 to determine the fugacity of a van der Waals gas because the van der Waals equation is a cubic equation in $\overline{V}$, so we can't solve it analytically for $\overline{V}$ to carry out the integration in Equation 8.72. We can get around this problem, however, by integrating Equation 8.72 by parts. First show that

$$RT\ln\gamma = P\overline{V} - RT - \int_{\overline{V}^{\text{id}}}^{\overline{V}} Pd\overline{V}' - RT\ln\frac{P}{P^{\text{id}}}$$

where $P^{\text{id}} \to 0$, $\overline{V}^{\text{id}} \to \infty$, and $P^{\text{id}}\overline{V}^{\text{id}} \to RT$. Substitute P from the van der Waals equation into the first term and the integral on the right side of the above equation and integrate to obtain

$$RT\ln\gamma = \frac{RT\overline{V}}{\overline{V}-b} - \frac{a}{\overline{V}} - RT - RT\ln\frac{\overline{V}-b}{\overline{V}^{\text{id}}-b} - \frac{a}{\overline{V}} - RT\ln\frac{P}{P^{\text{id}}}$$

Now use the fact that $\overline{V}^{\text{id}} \to \infty$ and that $P^{\text{id}}\overline{V}^{\text{id}} = RT$ to show that

$$\ln \gamma = -\ln\left[1 - \frac{a(\overline{V}-b)}{RT\overline{V}^2}\right] + \frac{b}{\overline{V}-b} - \frac{2a}{RT\overline{V}}$$

This equation gives the fugacity of a van der Waals gas as a function of $\overline{V}$. You can use the van der Waals equation itself to calculate P from $\overline{V}$, so the above equation, in conjunction with the van der Waals equation, gives $\ln \gamma$ as a function of pressure.

First we integrate Equation 8.72 by parts:

$$\ln \gamma = \int_{P^{\text{id}}}^{P}\left(\frac{\overline{V}}{RT} - \frac{1}{P'}\right)dP'$$

$$RT\ln\gamma = \int_{P^{\text{id}}}^{P}\overline{V}dP' - \int_{P^{\text{id}}}^{P}\frac{RT}{P'}dP'$$

$$= P'\overline{V}\Big|_{P^{\text{id}}}^{P} - \int_{\overline{V}^{\text{id}}}^{\overline{V}}P'd\overline{V} - RT\ln\frac{P}{P^{\text{id}}}$$

$$= P\overline{V} - (P\overline{V})^{\text{id}} - \int_{\overline{V}^{\text{id}}}^{\overline{V}}P'd\overline{V}' - RT\ln\frac{P}{P^{\text{id}}}$$

$$= P\overline{V} - RT - \int_{\overline{V}^{\text{id}}}^{\overline{V}}P'd\overline{V}' - RT\ln\frac{P}{P^{\text{id}}}$$

Substituting P from the van der Waals equation, we find that this equation becomes

$$RT\ln\gamma = \overline{V}\left[\frac{RT}{\overline{V}-b} - \frac{a}{\overline{V}^2}\right] - RT - \int_{\overline{V}^{\text{id}}}^{\overline{V}}\left[\frac{RT}{\overline{V}'-b} - \frac{a}{\overline{V}'^2}\right]d\overline{V}' - RT\ln\frac{P}{P^{\text{id}}}$$

$$= \frac{RT\overline{V}}{\overline{V}-b} - \frac{a}{\overline{V}} - RT - RT\ln\frac{\overline{V}-b}{\overline{V}^{\text{id}}-b} - \frac{a}{\overline{V}} + \frac{a}{\overline{V}^{\text{id}}} - RT\ln\frac{P}{P^{\text{id}}}$$

$$= \frac{RT\overline{V}}{\overline{V}-b} - \frac{a}{\overline{V}} - RT - RT\ln\frac{\overline{V}-b}{\overline{V}^{\text{id}}} - \frac{a}{\overline{V}} + \frac{a}{\overline{V}^{\text{id}}} - RT\ln\frac{P}{P^{\text{id}}}$$

Now, since $\overline{V}^{\text{id}} \longrightarrow \infty$, we can neglect b in the denominator of the logarithmic term and consider the $a/\overline{V}^{\text{id}}$ term negligible. Also, since $P^{\text{id}}\overline{V}^{\text{id}} = RT$, we write the above expression as

$$RT\ln\gamma = \frac{RT\overline{V}}{\overline{V}-b} - \frac{2a}{\overline{V}} - RT - RT\left(\ln\frac{\overline{V}-b}{\overline{V}^{\text{id}}} + \ln\frac{P}{P^{\text{id}}}\right)$$

$$\ln\gamma = \frac{\overline{V}}{\overline{V}-b} - \frac{2a}{\overline{V}RT} - 1 + \ln\overline{V}^{\text{id}}P^{\text{id}} - \ln P(\overline{V}-b)$$

$$= \frac{\overline{V}-(\overline{V}-b)}{\overline{V}-b} - \frac{2a}{RT\overline{V}} - \ln\frac{P(\overline{V}-b)}{RT}$$

$$= \frac{b}{\overline{V}-b} - \frac{2a}{RT\overline{V}} - \ln\left[\frac{(\overline{V}-b)}{RT}\left(\frac{RT}{\overline{V}-b} - \frac{a}{\overline{V}^2}\right)\right]$$

$$= \frac{b}{\overline{V}-b} - \frac{2a}{RT\overline{V}} - \ln\left[1 - \frac{a(\overline{V}-b)}{RT\overline{V}^2}\right]$$

8–34. Use the final equation in Problem 8–33 along with the van der Waals equation to plot $\ln\gamma$ against pressure for CO(g) at 200 K. Compare your result with Figure 8.10.

From Table 2.3, $a = 1.4734$ dm^6·bar·mol^{-1}·K^{-1} and $b = 0.039523$ dm^3·mol^{-1} for CO. We are given $\ln\gamma$ as a function of $\overline{V}$ in Problem 8–33, and the van der Waals equation gives P as a function of $\overline{V}$. Therefore, we can choose values of $\overline{V}$ and calculate the corresponding values of $\ln\gamma$ and P. Then we can plot $\ln\gamma$ against P. The result is

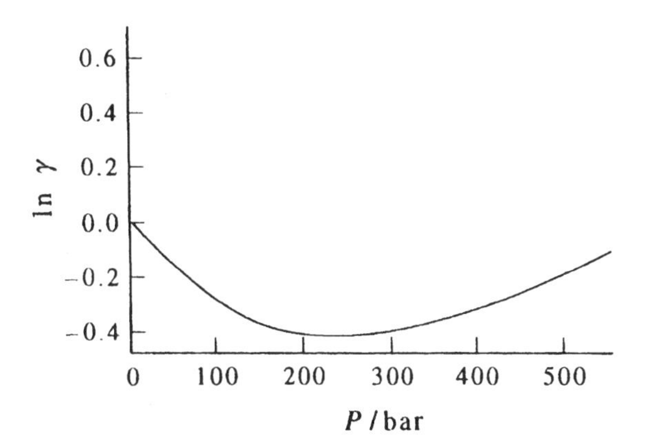

8–35. Show that the expression for $\ln\gamma$ for the van der Waals equation (Problem 8–33) can be written in the reduced form

$$\ln\gamma = \frac{1}{3V_R - 1} - \frac{9}{4V_R T_R} - \ln\left[1 - \frac{3(3V_R - 1)}{8T_R V_R^2}\right]$$

Use this equation along with the van der Waals equation in reduced form (Equation 2.19) to plot γ against P_R for $T_R = 1.00$ and 2.00 and compare your results with Figure 8.11.

We can use Equations 2.12 to express a and b in terms of T_c and $\overline{V}_c$:

$$3\overline{V}_c = b + \frac{RT_c}{P_c} \qquad 3\overline{V}_c^2 = \frac{a}{P_c} \qquad \overline{V}_c^3 = \frac{ab}{P_c}$$

Combining the first and the second, and the second and the third, equations gives

$$3\overline{V}_c = b + \frac{\overline{V}_c^3 RT_c}{ab} \qquad 3\overline{V}_c^2 = \frac{a\overline{V}_c^3}{ab}$$

$$3\overline{V}_c = \frac{\overline{V}_c}{3} + \frac{3\overline{V}_c^2 RT_c}{a} \qquad b = \frac{\overline{V}_c}{3}$$

$$a = \frac{9}{8}\overline{V}_c RT_c$$

Now we can substitute into the expression for $\ln\gamma$ we found in the previous problem:

$$\begin{aligned}\ln\gamma &= \frac{b}{\overline{V} - b} - \frac{2a}{RT\overline{V}} - \ln\left[1 - \frac{a(\overline{V} - b)}{RT\overline{V}^2}\right] \\ &= \frac{\overline{V}_c}{3\overline{V} - \overline{V}_c} - \frac{2}{RT\overline{V}}\left(\frac{9}{8}\overline{V}_c RT_c\right) - \ln\left[1 - \frac{9}{8}\overline{V}_c RT_c \frac{(3\overline{V} - \overline{V}_c)}{3RT\overline{V}^2}\right] \\ &= \frac{1}{3\overline{V}_R - 1} - \frac{9}{4T_R\overline{V}_R} - \ln\left[1 - \frac{3(3\overline{V}_R - 1)}{8T_R\overline{V}_R^2}\right]\end{aligned}$$

Equation 2.19 gives

$$P_R = \frac{8}{3}T_R\left(\overline{V}_R - \frac{1}{3}\right)^{-1} - \frac{3}{\overline{V}_R^2}$$

We have now found expressions for both $\ln\gamma$ and P as functions of $\overline{V}_R$ and T_R. We can therefore choose values of $\overline{V}_R$ for a specific value of T_R and calculate the corresponding values of $\ln\gamma$ and P, and plot γ against P. The result is

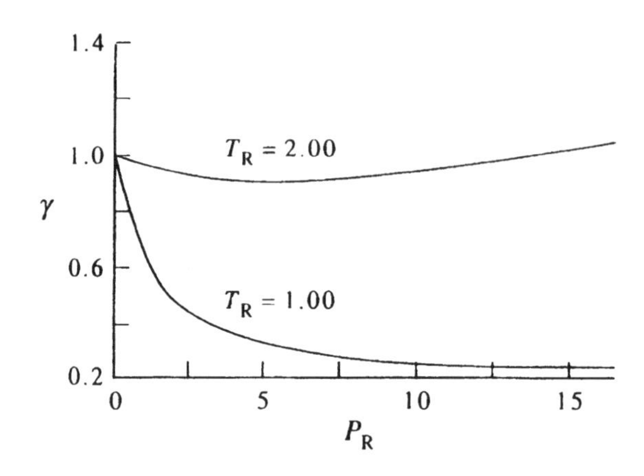

This looks very much like the experimental curves plotted in Figure 8.11.

8–36. Use the method outlined in Problem 8–33 to show that

$$\ln\gamma = \frac{B}{\overline{V}-B} - \frac{A}{RT^{3/2}(\overline{V}+B)} - \frac{A}{BRT^{3/2}}\ln\frac{\overline{V}+B}{\overline{V}} - \ln\left[1 - \frac{A(\overline{V}-B)}{RT^{3/2}\overline{V}(\overline{V}+B)}\right]$$

for the Redlich-Kwong equation. You need to use the standard integral

$$\int\frac{dx}{x(a+bx)} = -\frac{1}{a}\ln\frac{a+bx}{x}$$

For a Redlich-Kwong gas,

$$P = \frac{RT}{\overline{V}-B} - \frac{A}{T^{1/2}\overline{V}(\overline{V}+B)}$$

We can still use the first equation in Problem 8–43, since it was independent of the equation of state chosen. Also, realize that as $\overline{V}^{\text{id}} \to \infty$, B becomes negligible with respect to $\overline{V}^{\text{id}}$.

$$RT\ln\gamma = P\overline{V} - RT - \int_{\overline{V}^{\text{id}}}^{\overline{V}} P'd\overline{V} - RT\ln\frac{P}{P^{\text{id}}}$$

$$= \frac{RT\overline{V}}{\overline{V}-B} - \frac{A}{T^{1/2}(\overline{V}+B)} - RT - \int_{\overline{V}^{\text{id}}}^{\overline{V}}\left[\frac{RT}{\overline{V}-B} - \frac{A}{T^{1/2}\overline{V}(\overline{V}+B)}\right]d\overline{V} - RT\ln\frac{P}{P^{\text{id}}}$$

$$= RT\left(\frac{\overline{V}}{\overline{V}-B} - 1\right) - \frac{A}{T^{1/2}(\overline{V}+B)} - RT\ln\frac{\overline{V}-B}{\overline{V}^{\text{id}}-B}$$

$$- \frac{A}{T^{1/2}B}\left[\ln\frac{\overline{V}+B}{\overline{V}} - \ln\frac{\overline{V}^{\text{id}}+B}{\overline{V}^{\text{id}}}\right] - RT\ln\frac{P}{P^{\text{id}}}$$

$$\ln\gamma = \frac{B}{\overline{V}-B} - \frac{A}{RT^{3/2}(\overline{V}+B)} - \frac{A}{BRT^{3/2}}\ln\frac{\overline{V}+B}{\overline{V}} - \ln\frac{P(\overline{V}-B)}{P^{\text{id}}\overline{V}^{\text{id}}}$$

$$= \frac{B}{\overline{V}-B} - \frac{A}{RT^{3/2}(\overline{V}+B)} - \frac{A}{BRT^{3/2}}\ln\frac{\overline{V}+B}{\overline{V}} - \ln\left\{\frac{(\overline{V}-B)}{RT}\left[\frac{RT}{\overline{V}-B} - \frac{A}{T^{1/2}\overline{V}(\overline{V}+B)}\right]\right\}$$

$$= \frac{B}{\overline{V}-B} - \frac{A}{RT^{3/2}(\overline{V}+B)} - \frac{A}{BRT^{3/2}}\ln\frac{\overline{V}+B}{\overline{V}} - \ln\left[1 - \frac{A(\overline{V}-B)}{RT^{3/2}\overline{V}(\overline{V}+B)}\right]$$

8–37. Show that $\ln\gamma$ for the Redlich-Kwong equation (see Problem 8–36) can be written in the reduced form

$$\ln\gamma = \frac{0.25992}{\overline{V}_R - 0.25992} - \frac{1.2824}{T_R^{3/2}(\overline{V}_R + 0.25992)} - \frac{4.9340}{T_R^{3/2}}\ln\frac{\overline{V}_R + 0.25992}{\overline{V}_R} - \ln\left[1 - \frac{1.2824(\overline{V}_R - 0.25992)}{T_R^{3/2}\overline{V}_R(\overline{V}_R + 0.25992)}\right]$$

From Problem 2–26, we can express A and B in terms of T_c and $\overline{V}_c$:

$$3\overline{V}_c = \frac{RT_c}{P_c} \qquad B = 0.25992\overline{V}_c \qquad A = 0.42748\frac{R^2T_c^{5/2}}{P_c}$$

Then

$$A = 0.42748\frac{R^2T_c^{5/2}3\overline{V}_c}{RT_c} = 1.2824R\overline{V}_cT_c^{3/2}$$

Now we can substitute into our expression for $\ln\gamma$ in the previous problem:

$$\ln\gamma = \frac{B}{\overline{V}-B} - \frac{A}{RT^{3/2}(\overline{V}+B)} - \frac{A}{BRT^{3/2}}\ln\frac{\overline{V}+B}{\overline{V}} - \ln\left[1 - \frac{A(\overline{V}-B)}{RT^{3/2}\overline{V}(\overline{V}+B)}\right]$$

$$= \frac{0.25992\overline{V}_c}{\overline{V}-0.25992\overline{V}_c} - \frac{1.2824R\overline{V}_cT_c^{3/2}}{RT^{3/2}(\overline{V}+0.25992\overline{V}_c)} - \frac{1.2824R\overline{V}_cT_c^{3/2}}{RT^{3/2}(0.25992\overline{V}_c)}\ln\frac{\overline{V}+0.25992\overline{V}_c}{\overline{V}} - \ln\left[1 - \frac{1.2824R\overline{V}_cT_c^{3/2}(\overline{V}-0.25992\overline{V}_c)}{RT^{3/2}(\overline{V}+0.25992\overline{V}_c)\overline{V}}\right]$$

$$= \frac{0.25992}{\overline{V}_R - 0.25992} - \frac{1.2824}{T_R^{3/2}(\overline{V}_R + 0.25992)} - \frac{4.9340}{T_R^{3/2}}\ln\frac{\overline{V}_R + 0.25992}{\overline{V}_R} - \ln\left[1 - \frac{1.2824(\overline{V}_R - 0.25992)}{T_R^{3/2}\overline{V}_R(\overline{V}_R + 0.25992)}\right]$$

8–38. Use the expression for $\ln\gamma$ in reduced form given in Problem 8–37 along with the Redlich-Kwong equation in reduced form (Example 2–7) to plot $\ln\gamma$ versus P_R for $T_R = 1.00$ and 2.00 and compare your results with those you obtained in Problem 8–35 for the van der Waals equation.

From Example 2–7, we have an expression for P_R as a function of T_R and $\overline{V}_R$:

$$P_R = \frac{3T_R}{\overline{V}_R - 0.25992} - \frac{3.8473}{T_R^{1/2}\overline{V}_R(\overline{V}_R + 0.25992)}$$

and from the previous problem we have an expression for $\ln \gamma$ as a function of T_R and $\overline{V}_R$. We can therefore choose values of $\overline{V}_R$ for a specific value of T_R and calculate the corresponding values of $\ln \gamma$ and P, and then plot γ against P. The result is

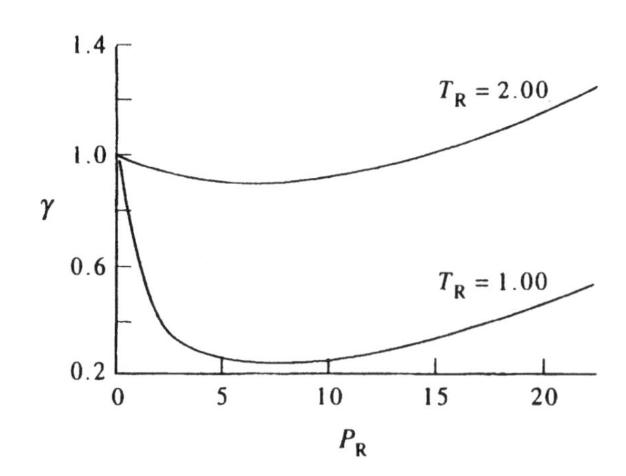

The upward curvature observed here is more marked than that in the plot we obtained from the van der Waals equation.

8–39. Compare $\ln \gamma$ for the van der Waals equation (Problem 8–33) with the values of $\ln \gamma$ for ethane at 600 K (Problem 8–29).

We can graph both the experimental and van der Waals $\ln \gamma$ vs. P, using $a = 5.5818\ \text{dm}^3\cdot\text{bar}\cdot\text{mol}^{-2}$ and $b = 0.065144\ \text{dm}^3\cdot\text{mol}^{-1}$. This is only a good fit at extremely low pressures.

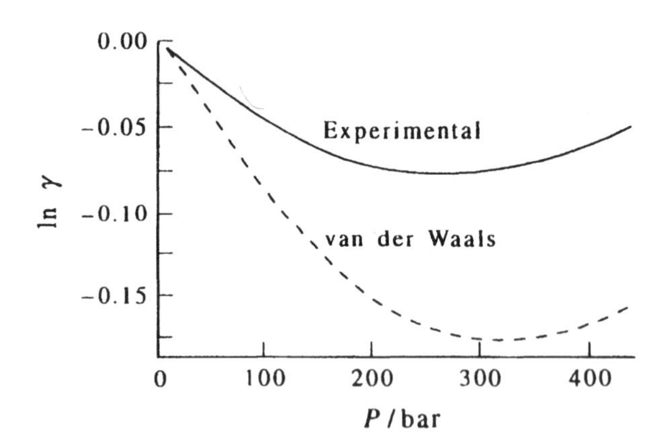

8–40. Compare $\ln \gamma$ for the Redlich-Kwong equation (Problem 8–36) with the values of $\ln \gamma$ for ethane at 600 K (Problem 8–29).

We can graph both the experimental and Redlich-Kwong $\ln \gamma$ vs. P, using $A = 98.831\ \text{dm}^3\cdot\text{bar}\cdot\text{K}^{-1/2}\cdot\text{mol}^{-2}$ and $B = 0.045153\ \text{dm}^3\cdot\text{mol}^{-1}$. The Redlich-Kwong equation provides a markedly better description of the behavior of $\ln \gamma$ than the van der Waals equation does.

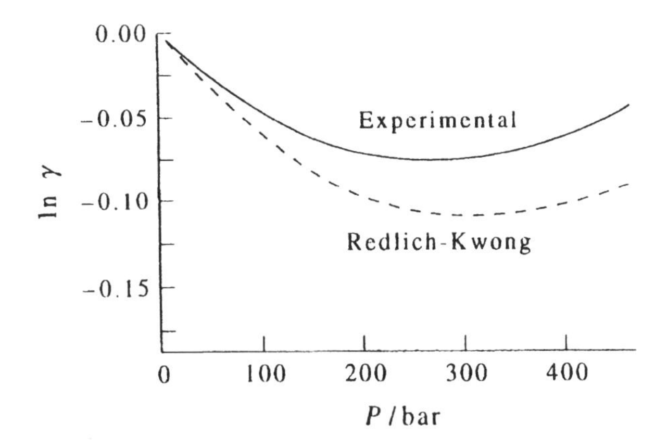

8–41. We can use the equation $(\partial S/\partial U)_V = 1/T$ to illustrate the consequence of the fact that entropy always increases during an irreversible adiabatic process. Consider a two-compartment system enclosed by rigid adiabatic walls, and let the two compartments be separated by a rigid heat-conducting wall. We assume that each compartment is at equilibrium but that they are not in equilibrium with each other. Because no work can be done by this two-compartment system (rigid walls) and no energy as heat can be exchanged with the surroundings (adiabatic walls),

$$U = U_1 + U_2 = \text{constant}$$

Show that

$$dS = \left(\frac{\partial S_1}{\partial U_1}\right) dU_1 + \left(\frac{\partial S_2}{\partial U_2}\right) dU_2$$

because the entropy of each compartment can change only as a result of a change in energy. Now show that

$$dS = dU_1 \left(\frac{1}{T_1} - \frac{1}{T_2}\right) \geq 0$$

Use this result to discuss the direction of the flow of energy as heat from one temperature to another.

We know that $U_1 + U_2$ is a constant, so $dU_1 = -dU_2$. Since the change of entropy of each compartment is dependent only on the energy change of each compartment (using the fact that we can express dS in terms of dU and dV, its natural variables), we can write

$$dS(U_1, U_2) = \left(\frac{\partial S_1}{\partial U_1}\right) dU_1 + \left(\frac{\partial S_2}{\partial U_2}\right) dU_2$$

Notice that this is a constant-volume process, so we use the expression $(\partial S/\partial U)_V = 1/T$ to write

$$\left(\frac{\partial S_1}{\partial U_1}\right)_V = \frac{1}{T_1} \quad \text{and} \quad \left(\frac{\partial S_2}{\partial U_2}\right)_V = \frac{1}{T_2}$$

Substituting into the expression for dS gives

$$dS = \frac{dU_1}{T_1} + \frac{dU_2}{T_2} = dU_1\left(\frac{1}{T_1} - \frac{1}{T_2}\right) \geq 0$$

where the inequality holds because entropy always increases in an irreversible adiabatic process. If the energy flows from compartment 2 into compartment 1, dU_1 must be positive, and so $T_1 < T_2$ in order for the inequality to hold. Likewise, if the energy flows from compartment 2 into compartment 1, dU_1 is negative and so $T_1 > T_2$. The energy always flows from the higher temperature to the lower temperature.

8–42. Modify the argument in Problem 8–41 to the case in which the two compartments are separated by a nonrigid, insulating wall. Derive the result

$$dS = \left(\frac{P_1}{T_1} - \frac{P_2}{T_2}\right) dV_1$$

Use this result to discuss the direction of a volume change under an isothermal pressure difference.

Since the entire system is isolated, we know

$$U_1 + U_2 = \text{constant} \qquad \text{and} \qquad V_1 + V_2 = \text{constant}$$

This means that $dU_1 = -dU_2$ and $dV_1 = -dV_2$. Now entropy depends on the energy and the volume, so

$$dS = \left(\frac{\partial S_1}{\partial U_1}\right) dU_1 + \left(\frac{\partial S_1}{\partial V_1}\right) dV_1 + \left(\frac{\partial S_2}{\partial U_2}\right) dU_2 + \left(\frac{\partial S_2}{\partial V_2}\right) dV_1$$

From Equation 8.40, $(\partial S/\partial U)_V = 1/T$ and $(\partial S/\partial V)_U = P/T$, so

$$dS = \frac{dU_1}{T_1} - \frac{dU_1}{T_2} + \frac{P_1}{T_1} dV_1 - \frac{P_2}{T_2} dV_1$$

$$= dU\left(\frac{1}{T_1} - \frac{1}{T_2}\right) + dV_1\left(\frac{P_1}{T_1} - \frac{P_2}{T_2}\right)$$

For an isothermal process, this expression becomes

$$dS = \frac{dV_1}{T}(P_1 - P_2)$$

If the volume of compartment 1 increases, dV_1 is positive and so $P_1 > P_2$ in order for dS to be positive. If the volume of compartment 1 decreases, dV_1 is negative and $P_2 > P_1$. The higher pressure compartment will expand under an isothermal pressure difference.

8–43. In this problem, we will derive virial expansions for $\overline{U}$, $\overline{H}$, $\overline{S}$, $\overline{A}$, and $\overline{G}$. Substitute

$$Z = 1 + B_{2P}P + B_{3P}P^2 + \cdots$$

into Equation 8.65 and integrate from a small pressure, P^{id}, to P to obtain

$$\overline{G}(T, P) - \overline{G}(T, P^{\text{id}}) = RT \ln \frac{P}{P^{\text{id}}} + RTB_{2P}P + \frac{RTB_{3P}}{2}P^2 + \cdots$$

Now use Equation 8.64 (realize that $P = P^{\text{id}}$ in Equation 8.64) to get

$$\overline{G}(T, P) - G^\circ(T) = RT \ln P + RTB_{2P}P + \frac{RTB_{3P}}{2}P^2 + \cdots \tag{1}$$

at $P^\circ = 1$ bar. Now use Equation 8.31a to get

$$\overline{S}(T, P) - S^\circ(T) = -R \ln P - \frac{d(RTB_{2P})}{dT}P - \frac{1}{2}\frac{d(RTB_{3P})}{dT}P^2 + \cdots \tag{2}$$

at $P^\circ = 1$ bar. Now use $\overline{G} = \overline{H} - T\overline{S}$ to get

$$\overline{H}(T, P) - H^\circ(T) = -RT^2 \frac{dB_{2P}}{dT} P - \frac{RT^2}{2} \frac{dB_{3P}}{dT} P^2 + \cdots \tag{3}$$

Now use the fact that $\overline{C}_P = (\partial \overline{H}/\partial T)_P$ to get

$$\overline{C}_P(T, P) - C_P^\circ(T) = -RT\left[2\frac{dB_{2P}}{dT} + T\frac{d^2B_{2P}}{dT^2}\right] P - \frac{RT}{2}\left[2\frac{dB_{3P}}{dT} + T\frac{d^2B_{3P}}{dT^2}\right] P^2 + \cdots \tag{4}$$

We can obtain expansions for $\overline{U}$ and $\overline{A}$ by using the equation $\overline{H} = \overline{U} + P\overline{V} = \overline{U} + RTZ$ and $\overline{G} = \overline{A} + P\overline{V} = \overline{A} + RTZ$. Show that

$$\overline{U} - U^\circ = -RT\left(B_{2P} + T\frac{dB_{2P}}{dT}\right) P - RT\left(B_{3P} + \frac{T}{2}\frac{dB_{3P}}{dT}\right) P^2 + \cdots \tag{5}$$

and

$$\overline{A} - A^\circ = RT \ln P - \frac{RTB_{3P}}{2} P^2 + \cdots \tag{6}$$

at $P^\circ = 1$ bar.

We can use the virial expansion in pressure to write

$$Z = 1 + B_{2P}P + B_{3P}P^2 + \cdots$$

$$\overline{V} = \frac{RT}{P} + B_{2P}RT + B_{3P}RTP + \cdots$$

Now substitute into Equation 8.65:

$$\left(\frac{\partial \overline{G}}{\partial P}\right)_T = \overline{V}$$

$$\int_{P^{\text{id}}}^{P} d\overline{G} = \int_{P^{\text{id}}}^{P} \left[\frac{RT}{P} + B_{2P}RT + B_{3P}RTP + O(P^2)\right] dP$$

$$\overline{G}(T, P) - \overline{G}(T, P^{\text{id}}) = RT \ln \frac{P}{P^{\text{id}}} + RTB_{2P}(P - P^{\text{id}}) + \frac{RTB_{3P}}{2}(P^2 - P^{\text{id}\,2}) + O(P^3)$$

Since P^{id} is very small, we can neglect it with respect to P in the last two terms and find

$$\overline{G}(T, P) - \overline{G}(T, P^{\text{id}}) = RT \ln \frac{P}{P^{\text{id}}} + RTB_{2P}P + \frac{RTB_{3P}}{2} P^2 + O(P^3)$$

Equation 8.64 states that

$$\overline{G}(T, P^{\text{id}}) = G^\circ(T) + RT \ln \frac{P^{\text{id}}}{P^\circ}$$

Substituting,

$$\overline{G}(T, P) - G^\circ(T) = RT \ln \frac{P}{P^\circ} + RTB_{2P}P + \frac{RTB_{3P}}{2} P^2 + O(P^3)$$

which is Equation 1 when $P^\circ = 1$ bar. Now

$$\overline{S} = -\left(\frac{\partial \overline{G}}{\partial T}\right)_P$$

$$\overline{S}(T,P) - S^\circ(T) = -\left(\frac{\partial[\overline{G}(T,P) - G^\circ(T)]}{\partial T}\right)_P$$

$$= -R\ln P - \frac{d(RTB_{2P})}{dT}P - \frac{1}{2}\frac{d(TRB_{3P})}{dT}P^2 + O(P^3)$$

which is Equation 2. Since $\overline{G} = \overline{H} - T\overline{S}$, we can now write

$$\overline{G}(T,P) - G^\circ(T) = \overline{H}(T,P) - H^\circ(T) - T\left[\overline{S}(T,P) - S^\circ(T)\right]$$

$$\overline{H}(T,P) - H^\circ(T) = RT\ln P + RTB_{2P}P + \frac{RTB_{3P}}{2}P^2 + T\left[-R\ln P - \frac{d(RTB_{2P})}{dT}P - \frac{1}{2}\frac{d(TRB_{3P})}{dT}P^2 + O(P^3)\right]$$

$$= -RT^2\frac{dB_{2P}}{dT}P - \frac{RT^2}{2}\frac{dB_{3P}}{dT}P^2 + O(P^3)$$

This is Equation 3. Now we can take the partial derivative of enthalpy with respect to temperature to find $\overline{C}_P$:

$$\overline{C}_P = \left(\frac{\partial \overline{H}}{\partial T}\right)_P$$

$$\overline{C}_P(T,P) - C_P^\circ(T) = \left(\frac{\partial[\overline{H}(T,P) - H^\circ(T)]}{\partial T}\right)_P$$

$$= \left(-2RT\frac{dB_{2P}}{dT} - RT^2\frac{d^2B_{2P}}{dT^2}\right)P - \left(RT\frac{dB_{3P}}{dT} + \frac{RT^2}{2}\frac{d^2B_{3P}}{dT^2}\right)P^2 + O(P^3)$$

$$= -RT\left(2\frac{dB_{2P}}{dT} + T\frac{d^2B_{2P}}{dT^2}\right)P - \frac{RT}{2}\left(2\frac{dB_{3P}}{dT} + T\frac{d^2B_{3P}}{dT^2}\right)P^2 + O(P^3)$$

This is Equation 4. Now use the fact that $\overline{U} = \overline{H} - P\overline{V}$:

$$\overline{U}(T,P) - U^\circ(T) = \overline{H}(T,P) - H^\circ(T) - P\left[\overline{V}(T,P) - V^\circ(T)\right]$$

$$= \left[-RT^2\frac{dB_{2P}}{dT}P - \frac{RT^2}{2}\frac{dB_{3P}}{dT}P^2\right] - P\left[\frac{ZRT}{P} - \frac{RT}{P}\right] + O(P^3)$$

$$= -RT^2\frac{dB_{2P}}{dT}P - \frac{RT^2}{2}\frac{dB_{3P}}{dT}P^2 - ZRT - RT + O(P^3)$$

$$= -RT^2\frac{dB_{2P}}{dT}P - \frac{RT^2}{2}\frac{dB_{3P}}{dT}P^2 - \left[1 + B_{2P}P + B_{3P}P^2\right]RT - RT + O(P^3)$$

$$= -RT\left(B_{2P} + T\frac{dB_{2P}}{dT}\right)P - RT\left(B_{3P} + \frac{T}{2}\frac{dB_{3P}}{dT}\right)P^2 + O(P^3)$$

In the above derivation, we realized that $PV^\circ = RT$, or $Z = 1$ for these conditions. We can similarly use the fact that $\overline{G} = \overline{A} + RTZ$, and write

$$\overline{A} - A^\circ = \overline{G} - G^\circ - RT(Z-1)$$

$$= \left[RT\ln P + RTB_{2P}P + \frac{RTB_{3P}}{2}P^2\right] - RT\left(1 + B_{2P}P + B_{3P}P^2 - 1\right) + O(P^3)$$

$$= RT \ln P + PRTB_{2P} + \frac{RTB_{3P}}{2}P^2 - PRTB_{2P} - RTB_{3P}P^2 + O(P^3)$$

$$= RT \ln P - \frac{RTB_{3P}}{2}P^2 + O(P^3)$$

8–44. In this problem, we will derive the equation

$$\overline{H}(T, P) - H^\circ(T) = RT(Z - 1) + \int_{\overline{V}^{\text{id}}}^{\overline{V}} \left[T\left(\frac{\partial P}{\partial T}\right)_V - P\right] d\overline{V}'$$

where $\overline{V}^{\text{id}}$ is a very large (molar) volume, where the gas is sure to behave ideally. Start with $dH = TdS + VdP$ to derive

$$\left(\frac{\partial H}{\partial V}\right)_T = T\left(\frac{\partial S}{\partial V}\right)_T + V\left(\frac{\partial P}{\partial V}\right)_T$$

and use one of the Maxwell relations for $(\partial S/\partial V)_T$ to obtain

$$\left(\frac{\partial H}{\partial V}\right)_T = T\left(\frac{\partial P}{\partial T}\right)_V + V\left(\frac{\partial P}{\partial V}\right)_T$$

Now integrate by parts from an ideal-gas limit to an arbitrary limit to obtain the desired equation.

Start with Equation 8.49 and take the partial derivative of both sides with respect to V:

$$dH = TdS + VdP$$

$$\left(\frac{\partial H}{\partial V}\right)_T = T\left(\frac{\partial S}{\partial V}\right)_T + V\left(\frac{\partial P}{\partial V}\right)_T$$

Now use Equation 8.19 to write this as

$$\left(\frac{\partial H}{\partial V}\right)_T = T\left(\frac{\partial P}{\partial T}\right)_V + V\left(\frac{\partial P}{\partial V}\right)_T$$

We now integrate the above equation. Recall that $PV = nZRT$, and for an ideal gas $Z = 1$.

$$\int_{H^{\text{id}}}^{H} dH = \int_{V^{\text{id}}}^{V} \left[T\left(\frac{\partial P}{\partial T}\right)_V + V\left(\frac{\partial P}{\partial V}\right)_T\right] dV'$$

$$H - H^\circ = \int_{V^{\text{id}}}^{V} \left[T\left(\frac{\partial P}{\partial T}\right)_V - V\left(\frac{nZRT}{V^2}\right) + nRT\left(\frac{\partial Z}{\partial V}\right)_T\right] dV'$$

$$H - H^\circ = \int_{V^{\text{id}}}^{V} \left[T\left(\frac{\partial P}{\partial T}\right)_V - \frac{nZRT}{V}\right] dV' + \int_1^Z nRT\,dZ'$$

$$= nRT\int_1^Z dZ' + \int_{V^{\text{id}}}^{V} \left[T\left(\frac{\partial P}{\partial T}\right)_V - P\right] dV'$$

$$= nRT(Z - 1) + \int_{V^{\text{id}}}^{V} \left[T\left(\frac{\partial P}{\partial T}\right)_V - P\right] dV'$$

Dividing both sides of the above equation by n gives the desired equation.

8–45. Using the result of Problem 8–44, show that H is independent of volume for an ideal gas. What about a gas whose equation of state is $P(\overline{V} - b) = RT$? Does U depend upon volume for this equation of state? Account for any difference.

For an ideal gas,

$$\left(\frac{\partial P}{\partial T}\right)_V = \frac{nR}{V} \quad \text{and} \quad \left(\frac{\partial P}{\partial V}\right)_T = -\frac{nRT}{V^2}$$

Substituting into the equation from the previous problem,

$$\left(\frac{\partial H}{\partial V}\right)_T = T\left(\frac{\partial P}{\partial T}\right)_V + V\left(\frac{\partial P}{\partial V}\right)_T = \frac{nRT}{V} - \frac{nRTV}{V^2} = 0$$

For the second equation of state given in the problem,

$$\left(\frac{\partial P}{\partial T}\right)_{\overline{V}} = \frac{R}{\overline{V} - b} \quad \text{and} \quad \left(\frac{\partial P}{\partial \overline{V}}\right)_T = -\frac{RT}{(\overline{V} - b)^2}$$

$$\begin{aligned}\left(\frac{\partial \overline{H}}{\partial \overline{V}}\right)_T &= T\left(\frac{\partial P}{\partial T}\right)_{\overline{V}} + \overline{V}\left(\frac{\partial P}{\partial \overline{V}}\right)_T \\ &= \frac{RT}{\overline{V} - b} - \frac{\overline{V}RT}{(\overline{V} - b)^2} = \frac{RT}{\overline{V} - b}\left[1 - \frac{\overline{V}}{\overline{V} - b}\right]\end{aligned}$$

Remember that $\overline{U} = \overline{H} - P\overline{V}$, so

$$\begin{aligned}\left(\frac{\partial \overline{U}}{\partial \overline{V}}\right)_T &= \left(\frac{\partial \overline{H}}{\partial \overline{V}}\right)_T - P - \overline{V}\left(\frac{\partial P}{\partial \overline{V}}\right)_T \\ &= \frac{RT}{\overline{V} - b}\left[1 - \frac{\overline{V}}{\overline{V} - b}\right] - \frac{RT}{\overline{V} - b} + \overline{V}\frac{RT}{(\overline{V} - b)^2} = 0\end{aligned}$$

Therefore $\overline{U}$ does not depend on volume for a gas that obeys the equation of state $P(\overline{V} - b) = RT$.

8–46. Using the result of Problem 8–44, show that

$$\overline{H} - H^{\circ} = \frac{RTb}{\overline{V} - b} - \frac{2a}{\overline{V}}$$

for the van der Waals equation.

For the van der Waals equation of state,

$$P = \frac{RT}{\overline{V} - b} - \frac{a}{\overline{V}^2} \quad \text{and} \quad \left(\frac{\partial P}{\partial T}\right)_{\overline{V}} = \frac{R}{\overline{V} - b}$$

Also,

$$Z = \frac{P\overline{V}}{RT} = \left[\frac{RT}{\overline{V} - b} - \frac{a}{\overline{V}^2}\right]\frac{\overline{V}}{RT}$$

Now we substitute these values into the equation from Problem 8–44:

$$\begin{aligned}\overline{H} - \overline{H}^{\circ} &= ZRT - RT + \int_{\overline{V}^{\text{id}}}^{\overline{V}}\left[T\left(\frac{\partial P}{\partial T}\right)_{\overline{V}} - P\right]d\overline{V}' \\ &= \overline{V}\left[\frac{RT}{\overline{V} - b} - \frac{a}{\overline{V}^2}\right] - RT + \int_{\overline{V}^{\text{id}}}^{\overline{V}}\left[\frac{RT}{\overline{V} - b} - \frac{RT}{\overline{V} - b} + \frac{a}{\overline{V}^2}\right]d\overline{V}'\end{aligned}$$

$$= RT\frac{\overline{V} - (\overline{V} - b)}{\overline{V} - b} - \frac{a}{\overline{V}} - \frac{a}{\overline{V}}\bigg|_{\overline{V}^{\text{id}}}^{\overline{V}}$$

$$= \frac{RTb}{\overline{V} - b} - \frac{2a}{\overline{V}} + \frac{a}{\overline{V}^{\text{id}}}$$

$$= \frac{RTb}{\overline{V} - b} - \frac{2a}{\overline{V}}$$

because $\overline{V}^{\text{id}}$ is very large compared to $\overline{V}$.

8–47. Using the result of Problem 8–44, show that

$$\overline{H} - H^\circ = \frac{RTB}{\overline{V} - B} - \frac{A}{T^{1/2}(\overline{V} + B)} - \frac{3A}{2BT^{1/2}}\ln\frac{\overline{V} + B}{\overline{V}}$$

for the Redlich-Kwong equation.

For the Redlich-Kwong equation of state,

$$P = \frac{RT}{\overline{V} - B} - \frac{A}{T^{1/2}\overline{V}(\overline{V} + B)} \qquad \left(\frac{\partial P}{\partial T}\right)_{\overline{V}} = \frac{R}{\overline{V} - B} + \frac{A}{2T^{3/2}\overline{V}(\overline{V} + B)}$$

Also, we write Z as

$$Z = \left[\frac{RT}{\overline{V} - B} - \frac{A}{T^{1/2}\overline{V}(\overline{V} + B)}\right]\frac{\overline{V}}{RT}$$

Now we substitute these values into the equation from Problem 8–44:

$$\overline{H} - \overline{H}^\circ = ZRT - RT + \int_{\overline{V}^{\text{id}}}^{\overline{V}}\left[T\left(\frac{\partial P}{\partial T}\right)_{\overline{V}} - P\right]d\overline{V}'$$

$$= \overline{V}\left[\frac{RT}{\overline{V} - B} - \frac{A}{T^{1/2}\overline{V}(\overline{V} + B)}\right] - RT + \int_{\overline{V}^{\text{id}}}^{\overline{V}}\left[\frac{RT}{\overline{V}' - B} + \frac{A}{2T^{1/2}\overline{V}'(\overline{V}' + B)} - \frac{RT}{\overline{V}' - B} + \frac{A}{T^{1/2}\overline{V}'(\overline{V}' + B)}\right]d\overline{V}'$$

$$= \frac{\overline{V}RT - RT(\overline{V} - B)}{\overline{V} - B} - \frac{A}{T^{1/2}(\overline{V} + B)} + \int_{\overline{V}^{\text{id}}}^{\overline{V}}\frac{3A}{2T^{1/2}\overline{V}'(\overline{V}' + B)}d\overline{V}'$$

$$= \frac{BRT}{\overline{V} - B} - \frac{A}{T^{1/2}(\overline{V} + B)} - \frac{3A}{2BT^{1/2}}\left(\ln\frac{\overline{V} + B}{\overline{V}} - \ln\frac{\overline{V}^{\text{id}} + B^{\text{id}}}{\overline{V}^{\text{id}}}\right)$$

$$= \frac{BRT}{\overline{V} - B} - \frac{A}{T^{1/2}(\overline{V} + B)} - \frac{3A}{2BT^{1/2}}\ln\frac{\overline{V} + B}{\overline{V}}$$

because $\overline{V}^{\text{id}}$ is very large compared to $\overline{V}$.

The following six problems involve the Joule-Thomson coefficient.

8–48. We introduced the Joule-Thomson effect and the Joule-Thomson coefficient in Problems 5–52 through 5–54. The Joule-Thomson coefficient is defined by

$$\mu_{JT} = \left(\frac{\partial T}{\partial P}\right)_H = -\frac{1}{C_P}\left(\frac{\partial H}{\partial P}\right)_T$$

and is a direct measure of the expected temperature change when a gas is expanded through a throttle. We can use one of the equations derived in this chapter to obtain a convenient working equation for μ_{JT}. Show that

$$\mu_{JT} = \frac{1}{C_P}\left[T\left(\frac{\partial V}{\partial T}\right)_P - V\right]$$

Use this result to show that $\mu_{JT} = 0$ for an ideal gas.

Start with

$$\left(\frac{\partial H}{\partial P}\right)_T = V - T\left(\frac{\partial V}{\partial T}\right)_P \tag{8.34}$$

Substitute this into the expression for μ_{JT} to obtain

$$\mu_{JT} = \frac{1}{C_P}\left[T\left(\frac{\partial V}{\partial T}\right)_P - V\right]$$

For an ideal gas, $(\partial V/\partial T)_P = nR/P$, so

$$\mu_{JT} = \frac{1}{C_P}\left[\frac{nRT}{P} - V\right] = 0$$

since $PV = nRT$.

8–49. Use the virial equation of state of the form

$$\frac{P\overline{V}}{RT} = 1 + \frac{B_{2V}(T)}{RT}P + \cdots$$

to show that

$$\mu_{JT} = \frac{1}{C_P^{\text{id}}}\left[T\frac{dB_{2V}}{dT} - B_{2V}\right] + O(P)$$

It so happens that B_{2V} is negative and dB_{2V}/dT is positive for $T^* < 3.5$ (see Figure 2.15) so that μ_{JT} is positive for low temperatures. Therefore, the gas will cool upon expansion under these conditions. (See Problem 8–48.)

Use the virial equation of state to express $\overline{V}$:

$$\frac{P\overline{V}}{RT} = 1 + \frac{B_{2V}(T)}{RT}P + O(P^2)$$

$$\overline{V} = \frac{RT}{P} + B_{2V}(T) + O(P)$$

$$\left(\frac{\partial \overline{V}}{\partial T}\right)_P = \frac{R}{P} + \frac{dB_{2V}}{dT} + O(P)$$

Substituting into the equation for μ_{JT} from Problem 8-48,

$$\begin{aligned}\mu_{JT} &= \frac{1}{C_P}\left[T\left(\frac{\partial V}{\partial T}\right)_P - V\right] \\ &= \frac{1}{C_P}\left[\frac{RT}{P} + T\frac{dB_{2V}}{dT} - \frac{RT}{P} - B_{2V} + O(P)\right] \\ &= \frac{1}{C_P}\left[T\frac{dB_{2V}}{dT} - B_{2V}\right] + O(P)\end{aligned}$$

8–50. Show that

$$\mu_{JT} = -\frac{b}{C_P}$$

for a gas that obeys the equation of state $P(\overline{V} - b) = RT$. (See Problem 8–48.)

For such a gas,

$$\left(\frac{\partial \overline{V}}{\partial T}\right)_P = \frac{R}{P} \qquad \text{and} \qquad \overline{V} = \frac{RT}{P} + b$$

Substituting into the equation for μ_{JT} from Problem 8-48,

$$\mu_{JT} = \frac{1}{C_P}\left[T\left(\frac{\partial V}{\partial T}\right)_P - V\right] = \frac{1}{C_P}\left[\frac{RT}{P} - \frac{RT}{P} - b\right] = -\frac{b}{C_P}$$

8–51. The second virial coefficient for a square-well potential is (Equation 2.41)

$$B_{2V}(T) = b_0[1 - (\lambda^3 - 1)(e^{\varepsilon/k_BT} - 1)]$$

Show that

$$\mu_{JT} = \frac{b_0}{C_P}\left[(\lambda^3 - 1)\left(1 + \frac{\varepsilon}{k_BT}\right)e^{\varepsilon/k_BT} - \lambda^3\right]$$

where $b_0 = 2\pi\sigma^3 N_A/3$. Given the following square-well parameters, calculate μ_{JT} at 0°C and compare your values with the given experimental values. Take $C_P = 5R/2$ for Ar and $7R/2$ for N_2 and CO_2.

Gas	$b_0/\text{mL}\cdot\text{mol}^{-1}$	λ	ε/k_B	$\mu_{JT}(\text{exptl})/\text{K}\cdot\text{atm}^{-1}$
Ar	39.87	1.85	69.4	0.43
N_2	45.29	1.87	53.7	0.26
CO_2	75.79	1.83	119	1.3

From Problem 8–49, we have

$$\mu_{JT} = \frac{1}{C_P}\left[T\frac{dB_{2V}}{dT} - B_{2V}\right] + O(P)$$

Now we find dB_{2V}/dT from Equation 2.41:

$$B_{2V} = b_0\left[1 - (\lambda^3 - 1)(e^{\varepsilon/k_BT} - 1)\right]$$

$$\frac{dB_{2V}}{dT} = \frac{\varepsilon b_0}{k_BT^2}(\lambda^3 - 1)e^{\varepsilon/k_BT}$$

Substituting into the expression for μ_{JT}, we find

$$\begin{aligned}
\mu_{JT} &= \frac{1}{C_P}\left[T\frac{dB_{2V}}{dT} - B_{2V}\right] + O(P) \\
&= \frac{1}{C_P}\left[\frac{b_0\varepsilon}{k_BT}(\lambda^3 - 1)e^{\varepsilon/k_BT} - b_0 + b_0(\lambda^3 - 1)(e^{\varepsilon/k_BT} - 1)\right] \\
&= \frac{b_0}{C_P}\left\{e^{\varepsilon/k_BT}\left[\frac{\varepsilon}{k_BT}(\lambda^3 - 1)\right] - 1 + \lambda^3 e^{\varepsilon/k_BT} - e^{\varepsilon/k_BT} - \lambda^3 + 1\right\} \\
&= \frac{b_0}{C_P}\left\{e^{\varepsilon/k_BT}\left[\frac{\varepsilon}{k_BT}(\lambda^3 - 1) + (\lambda^3 - 1)\right] - \lambda^3\right\} \\
&= \frac{b_0}{C_P}\left[(\lambda^3 - 1)\left(1 + \frac{\varepsilon}{k_BT}\right)e^{\varepsilon/k_BT} - \lambda^3\right]
\end{aligned}$$

We can now use the given values of λ, b_0, and ε/k_B to calculate μ_{JT}(theoretical) for Ar, N_2, and CO_2. We use $\overline{C}_P = 5R/2$ for argon and $\overline{C}_P = 7R/2$ for N_2 and CO_2.

Gas	Ar	N_2	CO_2
μ_{JT}(theor.)/K·atm^{-1}	0.44	0.24	1.39
μ_{JT}(exp.)/K·atm^{-1}	0.43	0.26	1.3
Percent Difference	3.4	7.3	6.6

8–52. The temperature at which the Joule-Thomson coefficient changes sign is called the *Joule-Thomson inversion temperature*, T_i. The low-pressure Joule-Thomson inversion temperature for the square-well potential is obtained by setting $\mu_{JT} = 0$ in Problem 8–51. This procedure leads to an equation for k_BT/ε in terms of λ^3 that cannot be solved analytically. Solve the equation numerically to calculate T_i for the three gases given in the previous problem. The experimental values are 794 K, 621 K, and 1500 K for Ar, N_2, and CO_2, respectively.

$$0 = \frac{b_0}{C_P}\left[(\lambda^3 - 1)\left(1 + \frac{\varepsilon}{k_BT}\right)e^{\varepsilon/k_BT} - \lambda^3\right]$$

$$\lambda^3 = (\lambda^3 - 1)\left(1 + \frac{\varepsilon}{k_BT}\right)e^{\varepsilon/k_BT}$$

We can use the experimental values as initial values and then use the Newton-Raphson method to find T_i. The inversion temperatures found are tabulated below.

Gas	Ar	N_2	CO_2
T_i(theor.)/K	791	634	1310
T_i(exp.)/K	794	621	1500
Percent Difference	0.378	2.09	12.7

8–53. Use the data in Problem 8–51 to estimate the temperature drop when each of the gases undergoes an expansion for 100 atm to one atm.

By definition, $\mu_{JT} = (\partial T/\partial P)_H$, so

$$\frac{1}{\mu_{JT}} = \left(\frac{\partial P}{\partial T}\right)_H$$

$$\int \mu_{JT} dP = \int dT$$

Let us assume that μ_{JT} does not change significantly over the pressure range. Then

$$\Delta T = \mu_{JT} \Delta P$$

Using the experimental values of μ_{JT}, we see that Ar(g) experiences a temperature drop of 42.6 K, N_2(g) has a temperature drop of 25.7 K, and CO_2(g) drops in temperature by 129 K.

8–54. When a rubber band is stretched, it exerts a restoring force, f, which is a function of its length L and its temperature T. The work involved is given by

$$w = \int f(L, T) dL \tag{1}$$

Why is there no negative sign in front of the integral, as there is in Equation 5.2 for P-V work? Given that the volume change upon stretching a rubber band is negligible, show that

$$dU = TdS + fdL \tag{2}$$

and that

$$\left(\frac{\partial U}{\partial L}\right)_T = T\left(\frac{\partial S}{\partial L}\right)_T + f \tag{3}$$

Using the definition $A = U - TS$, show that Equation 2 becomes

$$dA = -SdT + fdL \tag{4}$$

and derive the Maxwell relation

$$\left(\frac{\partial f}{\partial T}\right)_L = -\left(\frac{\partial S}{\partial L}\right)_T \tag{5}$$

Substitute Equation 5 into Equation 3 to obtain the analog of Equation 8.22

$$\left(\frac{\partial U}{\partial L}\right)_T = f - T\left(\frac{\partial f}{\partial T}\right)_L$$

For many elastic systems, the observed temperature-dependence of the force is linear. We define an *ideal rubber band* by

$$f = T\phi(L) \qquad \text{(ideal rubber band)} \tag{6}$$

Show that $(\partial U/\partial L)_T = 0$ for an ideal rubber band. Compare this result with $(\partial U/\partial V)_T = 0$ for an ideal gas.

Now let's consider what happens when we stretch a rubber band quickly (and, hence, adiabatically). In this case, $dU = dw = fdL$. Use the fact that U depends upon only the temperature for an ideal rubber band to show that

$$dU = \left(\frac{\partial U}{\partial T}\right)_L dT = fdL \tag{7}$$

The quantity $(\partial U/\partial T)_L$ is a heat capacity, so Equation 7 becomes

$$C_L dT = f dL \tag{8}$$

Argue now that if a rubber band is suddenly stretched, then its temperature will rise. Verify this result by holding a rubber band against your upper lip and stretching it quickly.

There is no negative sign in front of the integral because the force the rubber band exerts is a restoring force, which means that it is acting to contract the rubber band.

Since $dU = \delta q + \delta w$ and $\delta q = TdS$,

$$dU = TdS + f dL \tag{2}$$

Taking the partial derivative of both sides with respect to L at constant T gives

$$\left(\frac{\partial U}{\partial L}\right)_T = T\left(\frac{\partial S}{\partial L}\right)_T + f \tag{3}$$

Then, since $A = U - TS$, $U = A + TS$ and

$$\begin{aligned} dU &= TdS + f dL \\ dA + TdS + SdT &= TdS + f dL \\ dA &= -SdT + f dL \end{aligned} \tag{4}$$

We can also write dA as the total derivative of $A(T, L)$:

$$dA = \left(\frac{\partial A}{\partial T}\right)_L dT + \left(\frac{\partial A}{\partial L}\right)_T dL$$

Comparing the above equation and Equation 4, we see that

$$\left(\frac{\partial A}{\partial T}\right)_L = -S \qquad \text{and} \qquad \left(\frac{\partial A}{\partial L}\right)_T = f$$

and equating the second cross partial derivatives gives

$$-\left(\frac{\partial S}{\partial L}\right)_T = \left(\frac{\partial f}{\partial T}\right)_L \tag{5}$$

Substituting into Equation 3 gives

$$\left(\frac{\partial U}{\partial L}\right)_T = -T\left(\frac{\partial f}{\partial T}\right)_L + f$$

For an ideal rubber band,

$$f = T\phi(L) \qquad \text{and so} \qquad \left(\frac{\partial f}{\partial T}\right)_L = \phi_L$$

Then

$$\left(\frac{\partial U}{\partial L}\right)_T = -T\left(\frac{\partial f}{\partial T}\right)_L + f = -T\phi + T\phi = 0$$

Both this result and the result $(\partial U/\partial V)_T = 0$ essentially state that the energy of the system is independent of the length of the rubber band or the volume of the gas at constant temperature.

Now define $C_L = (\partial U/\partial T)_L$. For the ideal rubber band, U depends only on temperature, so we can write

$$U = \int \left(\frac{\partial U}{\partial T}\right)_L dT$$

$$dU = \left(\frac{\partial U}{\partial L}\right)_T dT = C_L dT$$

We know that $dU = f dL$ from the problem text, so we can now write

$$C_L dT = f dL$$

If we suddenly stretch a rubber band, we are applying force f over the distance we stretch the rubber band. Then

$$\int f dL = C_L \int dT$$

which is approximately

$$f \Delta L = C_L \Delta T$$

If ΔL is positive (we are stretching the rubber band), then ΔT must also be positive, and the rubber band heats up when we stretch it.

8–55. Derive an expression for ΔS for the reversible, isothermal expansion of one mole of a gas that obeys van der Waals equation. Use your result to calculate ΔS for the isothermal compression of ethane from 10.0 $dm^3 \cdot mol^{-1}$ to 1.00 $dm^3 \cdot mol^{-1}$ at 400 K. Compare your result to what you would get using the ideal-gas equation.

We can use the Maxwell relation (Equation 8.19)

$$\left(\frac{\partial \overline{S}}{\partial \overline{V}}\right)_T = \left(\frac{\partial P}{\partial T}\right)_{\overline{V}}$$

For the van der Waals equation,

$$P = \frac{RT}{\overline{V} - b} - \frac{a}{\overline{V}^2}$$

$$\left(\frac{\partial P}{\partial T}\right)_{\overline{V}} = \frac{R}{\overline{V} - b}$$

Substituting into the Maxwell equation above, we find that

$$\left(\frac{\partial \overline{S}}{\partial \overline{V}}\right)_T = \frac{R}{\overline{V} - b}$$

or

$$\Delta \overline{S} = R \ln \frac{\overline{V}_2 - b}{\overline{V}_1 - b}$$

For ethane, $b = 0.065144$ $dm^3 \cdot mol^{-1}$, so

$$\Delta \overline{S} = (8.3145\ \text{J} \cdot \text{K}^{-1} \cdot \text{mol}^{-1}) \ln \frac{1.00\ \text{dm}^3 \cdot \text{mol}^{-1} - 0.065144\ \text{dm}^3 \cdot \text{mol}^{-1}}{10.0\ \text{dm}^3 \cdot \text{mol}^{-1} - 0.065144\ \text{dm}^3 \cdot \text{mol}^{-1}}$$

$$= -19.7\ \text{J} \cdot \text{K}^{-1} \cdot \text{mol}^{-1}$$

Using the ideal gas equation, we find

$$\left(\frac{\partial \overline{S}}{\partial \overline{V}}\right)_T = \left(\frac{\partial P}{\partial T}\right)_{\overline{V}} = \frac{R}{\overline{V}}$$

$$\Delta S = R \ln \frac{V_2}{V_1} = (8.3145 \text{ J}\cdot\text{K}^{-1}\cdot\text{mol}^{-1}) \ln \frac{1.00}{10.0} = -19.1 \text{ J}\cdot\text{K}^{-1}\cdot\text{mol}^{-1}$$

The van der Waals result is smaller than the value obtained with the ideal gas equation.

8–56. Derive an expression for ΔS for the reversible, isothermal expansion of one mole of a gas that obeys the Redlich-Kwong equation (Equation 2.7). Use your result to calculate ΔS for the isothermal compression of ethane from $10.0 \text{ dm}^3\cdot\text{mol}^{-1}$ to $1.00 \text{ dm}^3\cdot\text{mol}^{-1}$ at 400 K. Compare your result with the result you would get using the ideal-gas equation.

Because these are the same parameters as those used in the previous problem, the ideal gas equation of state gives a value of $-19.1 \text{ J}\cdot\text{K}^{-1}\cdot\text{mol}^{-1}$ for $\Delta\overline{S}$.

We can use the Maxwell relation (Equation 8.19)

$$\left(\frac{\partial \overline{S}}{\partial \overline{V}}\right)_T = \left(\frac{\partial P}{\partial T}\right)_{\overline{V}}$$

For the Redlich-Kwong equation,

$$P = \frac{RT}{\overline{V} - B} - \frac{A}{T^{1/2}\overline{V}(\overline{V} + B)}$$

$$\left(\frac{\partial P}{\partial T}\right)_{\overline{V}} = \frac{R}{\overline{V} - B} + \frac{A}{2T^{3/2}\overline{V}(\overline{V} + B)}$$

Substituting into the Maxwell equation (Equation 8.19), we find that

$$\left(\frac{\partial \overline{S}}{\partial \overline{V}}\right)_T = \left(\frac{\partial P}{\partial T}\right)_{\overline{V}} = \frac{R}{\overline{V} - B} + \frac{A}{2T^{3/2}\overline{V}(\overline{V} + B)}$$

or

$$\Delta\overline{S} = R \ln \frac{\overline{V}_2 - B}{\overline{V}_1 - B} - \frac{A}{2BT^{3/2}} \ln \frac{\overline{V}_1(\overline{V}_2 + B)}{\overline{V}_2(\overline{V}_1 + B)}$$

For ethane, $A = 98.831 \text{ dm}^6\cdot\text{bar}\cdot\text{mol}^{-2}\cdot\text{K}^{1/2}$ and $B = 0.045153 \text{ dm}^3\cdot\text{mol}^{-1}$, so

$$\Delta\overline{S} = (0.083145 \text{ dm}^3\cdot\text{bar}\cdot\text{mol}^{-1}\cdot\text{K}^{-1}) \ln \frac{1.00 \text{ dm}^3\cdot\text{mol}^{-1} - 0.045153 \text{ dm}^3\cdot\text{mol}^{-1}}{10.0 \text{ dm}^3\cdot\text{mol}^{-1} - 0.045153 \text{ dm}^3\cdot\text{mol}^{-1}}$$

$$- \frac{98.831 \text{ dm}^6\cdot\text{bar}\cdot\text{mol}^{-2}\cdot\text{K}^{1/2}}{2(0.045153 \text{ dm}^3\cdot\text{mol}^{-1})(400 \text{ K})^{3/2}} \ln \frac{(10.0 \text{ dm}^3\cdot\text{mol}^{-1})(1.045153 \text{ dm}^3\cdot\text{mol}^{-1})}{(1.00 \text{ dm}^3\cdot\text{mol}^{-1})(10.045153 \text{ dm}^3\cdot\text{mol}^{-1})}$$

$$= -0.200 \text{ dm}^3\cdot\text{bar} = -20.0 \text{ J}\cdot\text{K}^{-1}\cdot\text{mol}^{-1}$$

This is smaller than the value obtained with the ideal gas equation.

8–57. Equation 8.2 says that $dU \leq 0$ for any spontaneous process that occurs at constant S and V. Thus, the energy will always decrease as a result of a spontaneous process and will attain a minimum value at equilibrium. Consider now a two-compartment system like that shown in the figure below.

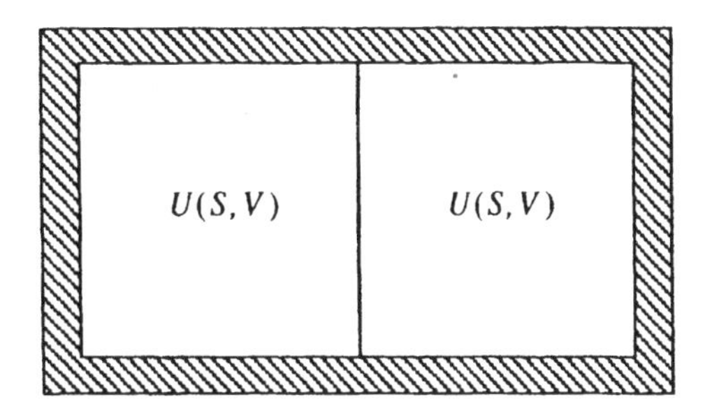

The entire system is enclosed by a rigid, adiabatic wall (so that it is isolated). The wall that separates the system into two compartments is totally restrictive (rigid, nonheat-conducting, and nonpermeable). The two identical compartments are at equilibrium with $U = U(S, V)$ at constant N.

Now let's partition the volume unequally between the two compartments (by moving the wall slightly), so that one compartment has a volume $V + \Delta V$ and the other has a volume $V - \Delta V$. Argue that

$$U(S, V + \Delta V) + U(S, V - \Delta V) > 2U(S, V) \tag{1}$$

Expand both $U(S, V + \Delta V)$ and $U(S, V - \Delta V)$ in a Taylor series (MathChapter C) in ΔV to obtain

$$\left(\frac{\partial^2 U}{\partial V^2}\right)_S > 0 \tag{2}$$

Equation 2 is one of several conditions for an equilibrium state to be stable with respect to small changes in its extensive variables.

Starting with $dU = TdS - PdV$, show that Equation 2 implies that the so-called adiabatic compressibility, κ_S, must obey the inequality

$$\kappa_S = -\frac{1}{V}\left(\frac{\partial V}{\partial P}\right)_S > 0$$

or that κ_S must be a positive quantity.

The energy at equilibrium is a minimum, so the energy when the volume is repartitioned into $V + \Delta V$ and $V - \Delta V$ must be greater than when the two volumes are both equal to V. Therefore,

$$U(S, V + \Delta V) + U(S, V - \Delta V) > 2U(S, V) \tag{1}$$

The Taylor expansions of $U(S, V + \Delta V)$ and $U(S, V - \Delta V)$ are

$$U(S, V + \Delta V) = U(S, V) + \left(\frac{\partial U}{\partial V}\right)_S \Delta V + \frac{1}{2}\left(\frac{\partial^2 U}{\partial V^2}\right)_S (\Delta V)^2 + \cdots$$

and

$$U(S, V - \Delta V) = U(S, V) - \left(\frac{\partial U}{\partial V}\right)_S \Delta V + \frac{1}{2}\left(\frac{\partial^2 U}{\partial V^2}\right)_S (\Delta V)^2 + \cdots$$

Substitute these two expansions into Equation 1 to obtain

$$\left(\frac{\partial^2 U}{\partial V^2}\right)_S (\Delta V)^2 > 0$$

But $(\Delta V)^2 > 0$, so

$$\left(\frac{\partial^2 U}{\partial V^2}\right)_S > 0$$

The equation $dU = TdS - PdV$ (Equation 8.35) gives $(\partial U/\partial V)_S = -P$, so

$$\left(\frac{\partial^2 U}{\partial V^2}\right)_S = -\left(\frac{\partial P}{\partial V}\right)_S > 0$$

or

$$\left(\frac{\partial P}{\partial V}\right)_S < 0 \qquad \text{and} \qquad \left(\frac{\partial V}{\partial P}\right)_S < 0$$

and so

$$\kappa_S = -\frac{1}{V}\left(\frac{\partial V}{\partial P}\right)_S > 0$$

8–58. In the previous problem, we showed that one of the stability conditions of an isolated system is that $(\partial^2 U/\partial V^2)_S > 0$. We did this by repartitioning the volume between the two compartments. Now consider the effect of repartitioning the entropy between the two compartments keeping their volumes equal. Show that

$$\left(\frac{\partial^2 U}{\partial S^2}\right)_V > 0$$

Now use the equation $dU = TdS - PdV$ to show that this inequality implies that

$$\left(\frac{\partial T}{\partial S}\right)_V = \frac{T}{C_V} > 0$$

or that $C_V > 0$.

Repartitioning the entropy between the two compartments gives

$$U(S + \Delta S, V) + U(S - \Delta S, V) > 2U(S, V) \tag{1}$$

The Taylor expansions of $U(S + \Delta S, V)$ and $U(S - \Delta S, V)$ are

$$U(S + \Delta S, V) = U(S, V) + \left(\frac{\partial U}{\partial S}\right)_V \Delta S + \frac{1}{2}\left(\frac{\partial^2 U}{\partial S^2}\right)_V (\Delta S)^2 + \cdots$$

and

$$U(S - \Delta S, V) = U(S, V) - \left(\frac{\partial U}{\partial S}\right)_V \Delta S + \frac{1}{2}\left(\frac{\partial^2 U}{\partial S^2}\right)_V (\Delta S)^2 + \cdots$$

Substitute these two expansions into Equation 1 to obtain

$$\left(\frac{\partial^2 U}{\partial S^2}\right)_V (\Delta S)^2 > 0$$

But $(\Delta S)^2 > 0$, so

$$\left(\frac{\partial^2 U}{\partial S^2}\right)_V > 0$$

The equation $dU = TdS - PdV$ (Equation 8.35) gives $(\partial U/\partial S)_V = T$, so

$$\left(\frac{\partial^2 U}{\partial S^2}\right)_V = \left(\frac{\partial T}{\partial S}\right)_V > 0$$

But (Equation 6.31)

$$\left(\frac{\partial S}{\partial T}\right)_V = \frac{C_V}{T}$$

so

$$\frac{T}{C_V} > 0 \qquad \text{and} \qquad \frac{C_V}{T} > 0$$

and $C_V > 0$.

8–59. The enthalpy, H, (at constant N) has natural variables S and P, one an extensive variable (S) and the other an intensive variable (P). To determine the sign of $(\partial^2 H/\partial S^2)_P$, we use the repartitioning approach that we described in Problem 8–57. Show that

$$\left(\frac{\partial^2 H}{\partial S^2}\right)_P > 0$$

Now show that this inequality implies that

$$\left(\frac{\partial T}{\partial S}\right)_P = \frac{T}{C_P} > 0$$

or that $C_P > 0$.

The pressure is an intensive variable, so we are unable to repartition it. To determine the sign of $(\partial^2 H/\partial P^2)_S$, we first determine $(\partial H/\partial P)_S$. Show that

$$\left(\frac{\partial H}{\partial P}\right)_S = V$$

Now differentiate once again with respect to P keeping S constant.

$$\left(\frac{\partial V}{\partial P}\right)_S = \left(\frac{\partial^2 H}{\partial P^2}\right)_S$$

Now show that the reciprocal of $(\partial V/\partial P)_S$ is given by

$$\left(\frac{\partial P}{\partial V}\right)_S = -\left(\frac{\partial^2 U}{\partial V^2}\right)_S$$

Put this all together to show that

$$\left(\frac{\partial V}{\partial P}\right)_S = \left(\frac{\partial^2 H}{\partial P^2}\right)_S = -\frac{1}{\left(\frac{\partial^2 U}{\partial V^2}\right)_S} < 0$$

Thus, we see once again (see Problem 8–57) that

$$\kappa_S = -\frac{1}{V}\left(\frac{\partial V}{\partial P}\right)_S > 0$$

Notice that we determine the sign of the second derivative of H with respect to the extensive variable, S, by using the repartitioning that we used in Problem 8–57 and the sign of the second derivative with respect to the intensive variable, P, by first determining $(\partial H/\partial P)_S$, differentiating it, and then relating its reciprocal to $(\partial^2 U/\partial V^2)_S$, whose sign we already knew.

Repartition the entropy between the two compartments to obtain

$$H(S+\Delta S, P) + H(S-\Delta S, P) > 2H(S, P)$$

Using a Taylor expansion as we do in Problems 8–57 and 8–58 gives

$$\left(\frac{\partial^2 H}{\partial S^2}\right)_P (\Delta S)^2 > 0$$

or simply

$$\left(\frac{\partial^2 H}{\partial S^2}\right)_P > 0$$

The equation $dH = TdS + VdP$ (Equation 8.41) gives $(\partial H/\partial S)_P = T$, so

$$\left(\frac{\partial^2 H}{\partial S^2}\right)_P = \left(\frac{\partial T}{\partial S}\right)_P > 0$$

But (Problem 6–13)

$$\left(\frac{\partial S}{\partial T}\right)_P = \frac{C_P}{T}$$

so

$$\frac{T}{C_P} > 0 \qquad \text{and} \qquad \frac{C_P}{T} > 0$$

and $C_P > 0$.

Using the equation $dH = TdS + VdP$ (Equation 8.41), we see that $(\partial H/\partial P)_S = V$, and so

$$\left(\frac{\partial^2 H}{\partial P^2}\right)_S = \left(\frac{\partial V}{\partial P}\right)_S$$

Now use $dU = TdS - PdV$ to see that $(\partial U/\partial V)_S = -P$ and that

$$\left(\frac{\partial^2 U}{\partial V^2}\right)_S = -\left(\frac{\partial P}{\partial V}\right)_S = -\frac{1}{\left(\frac{\partial V}{\partial P}\right)_S}$$

Therefore,

$$\left(\frac{\partial^2 H}{\partial P^2}\right)_S = \left(\frac{\partial V}{\partial P}\right)_S = \frac{1}{\left(\frac{\partial P}{\partial V}\right)_S} = -\frac{1}{\left(\frac{\partial^2 U}{\partial V^2}\right)_S}$$

But Problem 8–57 shows that $(\partial^2 U/\partial V^2)_S > 0$, so

$$\left(\frac{\partial^2 H}{\partial P^2}\right)_S = \left(\frac{\partial V}{\partial P}\right)_S < 0$$

and so

$$\kappa_S = -\frac{1}{V}\left(\frac{\partial V}{\partial P}\right)_S > 0$$

8–60. Of the two natural variables of the Helmholtz energy (at constant N), one is an extensive variable, V, and the other is an intensive variable, T. Using the repartitioning approach shown in Problem 8–57, show that

$$\left(\frac{\partial^2 A}{\partial V^2}\right)_T > 0$$

and that

$$\left(\frac{\partial P}{\partial V}\right)_T < 0$$

Now show that the isothermal compressibility, κ_T, is always positive, or that

$$\kappa_T = -\frac{1}{V}\left(\frac{\partial P}{\partial V}\right)_T > 0$$

The Helmholtz energy is also a function of one intensive natural variable, T. Show that (see problem 8–59)

$$\left(\frac{\partial S}{\partial T}\right)_V = -\left(\frac{\partial^2 A}{\partial T^2}\right)_V = \frac{1}{\left(\frac{\partial T}{\partial S}\right)_V} = \frac{1}{\left(\frac{\partial^2 U}{\partial S^2}\right)_V} > 0$$

Thus, we see that the entropy increases monotonically with temperature at constant volume.

The proof that $(\partial^2 A/\partial V^2)_T > 0$ is completely analogous to the proof that $(\partial^2 U/\partial V^2)_S > 0$ in Problem 8–57. Using the equation $dA = -SdT - PdV$, we see that $(\partial A/\partial V)_T = -P$ and that

$$\left(\frac{\partial^2 A}{\partial V^2}\right)_T = -\left(\frac{\partial P}{\partial V}\right)_T > 0$$

Therefore,

$$\left(\frac{\partial P}{\partial V}\right)_T < 0 \quad \text{and} \quad \left(\frac{\partial V}{\partial P}\right)_T < 0$$

and

$$\kappa_T = -\frac{1}{V}\left(\frac{\partial V}{\partial P}\right)_T > 0$$

Using the equation $dA = -SdT - PdV$ (Equation 8.42), we see that $(\partial A/\partial T)_V = -S$ and that

$$\left(\frac{\partial^2 A}{\partial T^2}\right)_V = -\left(\frac{\partial S}{\partial T}\right)_V$$

But the equation $dU = TdS - PdV$ gives $(\partial U/\partial S)_V = T$, and so we see that

$$\left(\frac{\partial T}{\partial S}\right)_V = \left(\frac{\partial^2 U}{\partial S^2}\right)_V$$

and that

$$\left(\frac{\partial^2 A}{\partial T^2}\right)_V = -\left(\frac{\partial S}{\partial T}\right)_V = -\frac{1}{\left(\frac{\partial T}{\partial S}\right)_V} = -\frac{1}{\left(\frac{\partial^2 U}{\partial S^2}\right)_V} < 0$$

where we have used the fact that $(\partial^2 U/\partial S^2)_V > 0$ (Problem 8–57). Thus we see that $(\partial S/\partial T)_V > 0$.

8–61. The two natural variables of the Gibbs energy (at constant $\tilde{N}$) are both intensive variables, T and P. First show that

$$\left(\frac{\partial S}{\partial T}\right)_P = -\left(\frac{\partial^2 G}{\partial T^2}\right)_P = \frac{1}{\left(\frac{\partial T}{\partial S}\right)_P} = \frac{1}{\left(\frac{\partial^2 H}{\partial S^2}\right)_P} > 0$$

Thus, we see that the entropy increases monotonically with temperature at constant pressure.

Now show that

$$\left(\frac{\partial V}{\partial P}\right)_T = \left(\frac{\partial^2 G}{\partial P^2}\right)_T = \frac{1}{\left(\frac{\partial P}{\partial V}\right)_T} = \frac{1}{\left(\frac{\partial^2 A}{\partial V^2}\right)_T} < 0$$

or that

$$\kappa_T = -\frac{1}{V}\left(\frac{\partial V}{\partial P}\right)_T > 0$$

Using the equation $dG = -SdT + VdP$ (Equation 8.45), we see that $(\partial G/\partial T)_P = -S$ and that

$$\left(\frac{\partial^2 G}{\partial T^2}\right)_P = -\left(\frac{\partial S}{\partial T}\right)_P$$

Furthermore, the equation $dH = TdS + VdP$ (Equation 8.41) shows that $(\partial H/\partial S)_P = T$ and that

$$\left(\frac{\partial T}{\partial S}\right)_P = \left(\frac{\partial^2 H}{\partial S^2}\right)_P$$

Therefore, using the fact that $(\partial^2 H/\partial S^2)_P > 0$ (Problem 8–59), we see that

$$\left(\frac{\partial^2 G}{\partial T^2}\right)_P = -\left(\frac{\partial S}{\partial T}\right)_P = -\frac{1}{\left(\frac{\partial T}{\partial S}\right)_P} = -\frac{1}{\left(\frac{\partial^2 H}{\partial S^2}\right)_P} < 0$$

and so $(\partial S/\partial T)_P > 0$.

The equation $dG = -SdT + VdP$ gives $(\partial G/\partial P)_T = V$ and that

$$\left(\frac{\partial^2 G}{\partial P^2}\right)_T = \left(\frac{\partial V}{\partial P}\right)_T$$

Furthermore, the equation $dA = -SdT - PdV$ gives $(\partial A/\partial V)_T = -P$ and that

$$\left(\frac{\partial P}{\partial V}\right)_T = -\left(\frac{\partial^2 A}{\partial V^2}\right)_T$$

Therefore, using the fact that $(\partial^2 A/\partial V^2)_T > 0$ (Problem 8–60), we see that

$$\left(\frac{\partial^2 G}{\partial P^2}\right)_T = \left(\frac{\partial V}{\partial P}\right)_T = \frac{1}{\left(\frac{\partial P}{\partial V}\right)_T} = -\frac{1}{\left(\frac{\partial^2 A}{\partial V^2}\right)_T} < 0$$

or that

$$\kappa_T = -\frac{1}{V}\left(\frac{\partial V}{\partial P}\right)_T > 0$$

8–62. Use the results of the previous problem and Equation 8.27 to prove that $C_P > C_V$.

Equation 8.27 is

$$C_P - C_V = \frac{\alpha^2 T V}{\kappa_T}$$

But all the quantities on the right side are positive, so $C_P > C_V$.

8–63. In Problems 8–57 and 8–58, we repartitioned the volume of each compartment without changing the entropy of each compartment, or we repartitioned the entropy without changing the volume. Let's now look at the case in which both V and S change. Show that in this case,

$$U(S + \Delta S, V + \Delta V) + U(S - \Delta S, V - \Delta V) > 2U(S, V) \tag{1}$$

The Taylor expansion of a function of two variables is

$$\begin{aligned} f(x + \Delta x, y + \Delta y) = f(x, y) &+ \left(\frac{\partial f}{\partial x}\right)_y \Delta x + \left(\frac{\partial f}{\partial y}\right)_x \Delta x \\ &+ \frac{1}{2}\left(\frac{\partial^2 f}{\partial x^2}\right)_y (\Delta x)^2 + \left(\frac{\partial^2 f}{\partial x \partial y}\right)(\Delta x)(\Delta y) \\ &+ \frac{1}{2}\left(\frac{\partial^2 f}{\partial y^2}\right)_x (\Delta y)^2 \\ &+ \text{cubic terms in } \Delta x, \Delta y, \text{and their products} \end{aligned}$$

Using this result, show that Equation 1 becomes

$$\left(\frac{\partial^2 U}{\partial S^2}\right)_V (\Delta S)^2 + 2\left(\frac{\partial^2 U}{\partial S \partial V}\right)(\Delta S)(\Delta V) + \left(\frac{\partial^2 U}{\partial V^2}\right)_S (\Delta V)^2 > 0 \tag{2}$$

Equation 2 is valid for any values of ΔS and ΔV. Let $\Delta V = 0$ and $\Delta S = 0$ in turn to show that

$$\left(\frac{\partial^2 U}{\partial S^2}\right)_V > 0 \qquad \left(\frac{\partial^2 U}{\partial V^2}\right)_S > 0 \tag{3}$$

Now let

$$\Delta S = \frac{(\partial^2 U/\partial S \partial V)}{\partial^2 U/\partial S^2)_V} \qquad \text{and} \qquad \Delta V = 1$$

to obtain

$$\left(\frac{\partial^2 U}{\partial S^2}\right)_V \left(\frac{\partial^2 U}{\partial V^2}\right)_S - \left(\frac{\partial^2 U}{\partial S \partial V}\right) > 0 \tag{4}$$

It turns out that Equations 3 and 4 are the mathematical conditions that the extremum given by $dU = 0$ is actually a minimum rather than a maximum or a saddle point.

Equation 1 simply says that the energy is a minimum at equilibrium. The Taylor expansion of $U(S + \Delta S, V + \Delta V)$ and $U(S - \Delta S, V - \Delta V)$ are

$$\begin{aligned} U(S + \Delta S, V + \Delta V) = U(S, V) &+ \left(\frac{\partial U}{\partial S}\right)_V \Delta S + \left(\frac{\partial U}{\partial V}\right)_S \Delta V + \frac{1}{2}\left(\frac{\partial^2 U}{\partial S^2}\right)_V (\Delta S)^2 \\ &+ \left(\frac{\partial^2 U}{\partial S \partial V}\right)(\Delta S)(\Delta V) + \frac{1}{2}\left(\frac{\partial^2 U}{\partial V^2}\right)_S (\Delta V)^2 + \cdots \end{aligned}$$

and

$$U(S-\Delta S, V-\Delta V) = U(S,V) - \left(\frac{\partial U}{\partial S}\right)_V \Delta S - \left(\frac{\partial U}{\partial V}\right)_S \Delta V + \frac{1}{2}\left(\frac{\partial^2 U}{\partial S^2}\right)_V (\Delta S)^2$$
$$+ \left(\frac{\partial^2 U}{\partial S \partial V}\right)(\Delta S)(\Delta V) + \frac{1}{2}\left(\frac{\partial^2 U}{\partial V^2}\right)_S (\Delta V)^2 + \cdots$$

Substitute these two expansions into Equation 1 and get

$$\left(\frac{\partial^2 U}{\partial S^2}\right)_V (\Delta S)^2 + 2\left(\frac{\partial^2 U}{\partial S \partial V}\right)(\Delta S)(\Delta V) + \left(\frac{\partial^2 U}{\partial V^2}\right)_S (\Delta V)^2 > 0 \tag{2}$$

Letting $\Delta V = 0$ gives $(\partial^2 U/\partial S^2)_V > 0$ and $\Delta S = 0$ gives $(\partial^2 U/\partial V^2)_S > 0$. Now let

$$\Delta S = -\frac{(\partial^2 U/\partial S \partial V)}{(\partial^2 U/\partial S^2)_V} \qquad \text{and} \qquad \Delta V = 1$$

in Equation 2 to get

$$\frac{(\partial^2 U/\partial S \partial V)^2}{(\partial^2 U/\partial S^2)_V} - 2\frac{(\partial^2 U/\partial S \partial V)^2}{(\partial^2 U/\partial S^2)_V} + \left(\frac{\partial^2 U}{\partial V^2}\right)_S > 0$$

or

$$\left(\frac{\partial^2 U}{\partial S^2}\right)_V \left(\frac{\partial^2 U}{\partial V^2}\right)_S - \left(\frac{\partial^2 U}{\partial S \partial V}\right)^2 > 0 \tag{3}$$

8–64. Equation 8.45 gives $\overline{S}$ for a monatomic ideal gas in terms of T and V. Use the fact that $\overline{U} = 3RT/2$ to show that Equation 8.45 can be written as

$$\overline{U} = \overline{U}(\overline{S}, \overline{V}) = \frac{\alpha e^{2\overline{S}/3R}}{\overline{V}^{2/3}} \tag{5}$$

where α is a constant independent of $\overline{S}$ and $\overline{V}$. Show that $\overline{U}$ given by Equation 5 satisfies Equations 3 and 4 of Problem 8–63. Show also that

$$\left(\frac{\partial \overline{U}}{\partial \overline{S}}\right)_{\overline{V}} = T \qquad \left(\frac{\partial \overline{U}}{\partial \overline{V}}\right)_{\overline{S}} = -P$$

Equation 8.45 is

$$\overline{S} = \frac{5}{2}R + R\ln\left[\left(\frac{2\pi m k_B T}{h^2}\right)^{3/2} \frac{g_{e1}\overline{V}}{N_A}\right]$$

The molar energy of a monatomic ideal gas is $\overline{U} = 3N_A k_B T/2$, so $\overline{S}$ can be written as

$$\overline{S} = R\ln\left[\left(\frac{4\pi m}{3h^2}\right)^{3/2} \frac{g_{e1}e^{5/2}}{N_A^{5/2}} \overline{U}^{3/2}\overline{V}\right]$$
$$= R\ln\left[c\overline{U}^{3/2}\overline{V}\right]$$

where c is a constant. Solving for $\overline{U}$ gives

$$\overline{U} = \frac{\alpha e^{2\overline{S}/3R}}{\overline{V}^{2/3}}$$

where $\alpha = c^{-2/3}$ is a constant. Therefore,

$$\left(\frac{\partial \overline{U}}{\partial \overline{V}}\right)_{\overline{S}} = -\frac{2\alpha}{3}\frac{e^{2\overline{S}/3R}}{\overline{V}^{5/3}} = -\frac{2}{3}\frac{\overline{U}}{\overline{V}} = -\frac{2}{3\overline{V}}\left(\frac{3}{2}RT\right) = -\frac{RT}{\overline{V}} = -P$$

$$\left(\frac{\partial \overline{U}}{\partial \overline{S}}\right)_{\overline{V}} = \frac{2\alpha}{3R}\frac{e^{2\overline{S}/3R}}{\overline{V}^{2/3}} = \frac{2}{3R}\overline{U} = \frac{2}{3R}\left(\frac{3}{2}RT\right) = T$$

$$\left(\frac{\partial^2 \overline{U}}{\partial \overline{V}^2}\right)_{\overline{S}} = \frac{10\alpha}{9}\frac{e^{2\overline{S}/3R}}{\overline{V}^{8/3}}$$

$$\left(\frac{\partial^2 \overline{U}}{\partial \overline{S}^2}\right)_{\overline{V}} = \frac{4\alpha}{9R^2}\frac{e^{2\overline{S}/3R}}{\overline{V}^{2/3}}$$

$$\left(\frac{\partial^2 \overline{U}}{\partial \overline{S}\partial \overline{V}}\right) = -\frac{4\alpha}{9R}\frac{e^{2\overline{S}/3R}}{\overline{V}^{5/3}}$$

Substitute these results into Equation 4 to get

$$\frac{40\alpha^2}{9^2R^2}\frac{e^{4\overline{S}/3R}}{\overline{V}^{10/3}} > \frac{16\alpha^2}{9^2R^2}\frac{e^{4\overline{S}/3R}}{\overline{V}^{10/3}}$$

or $40 > 16$.

CHAPTER 9

Phase Equilibria

PROBLEMS AND SOLUTIONS

9–1. Sketch the phase diagram for oxygen using the following data: triple point, 54.3 K and 1.14 torr; critical point, 154.6 K and 37 828 torr; normal melting point, −218.4°C; and normal boiling point, −182.9°C. Does oxygen melt under an applied pressure as water does?

We can use the triple point and the normal melting point to construct the liquid-solid line and the triple point, normal boiling point, and critical point to construct the liquid-gas line. The liquid-gas line stops at the critical point. We produce the diagram

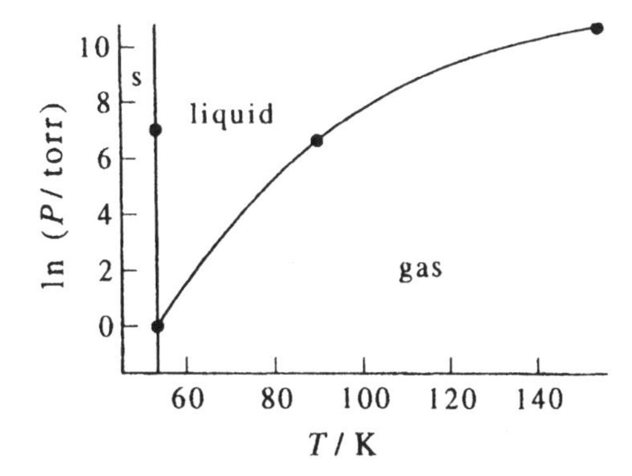

We can see that oxygen does not melt under an applied pressure, because its normal melting point temperature is higher than the triple point temperature.

9–2. Sketch the phase diagram for I_2 given the following data: triple point, 113°C and 0.12 atm; critical point, 512°C and 116 atm; normal melting point, 114°C; normal boiling point, 184°C; and density of liquid > density of solid.

We use the triple point and normal melting point to construct the liquid-solid line and the triple point, normal boiling point, and critical point to construct the liquid-gas line. Because the density of the liquid is greater than the density of the solid, the solid-liquid line has a positive slope.

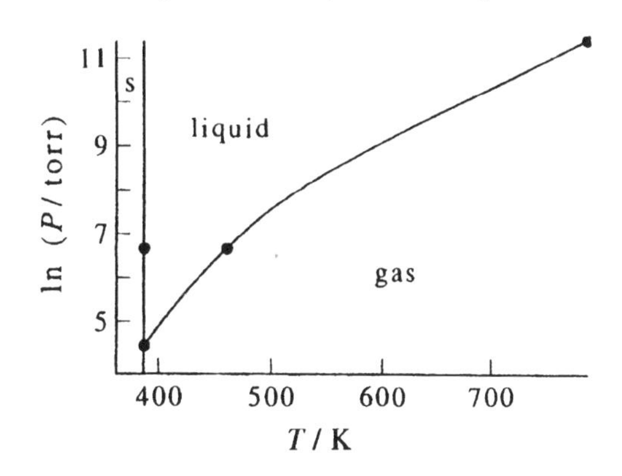

9–3. Figure 9.14 shows a density-temperature phase diagram for benzene. Using the following data for the triple point and the critical point, interpret this phase diagram. Why is the triple point indicated by a line in this type of phase diagram?

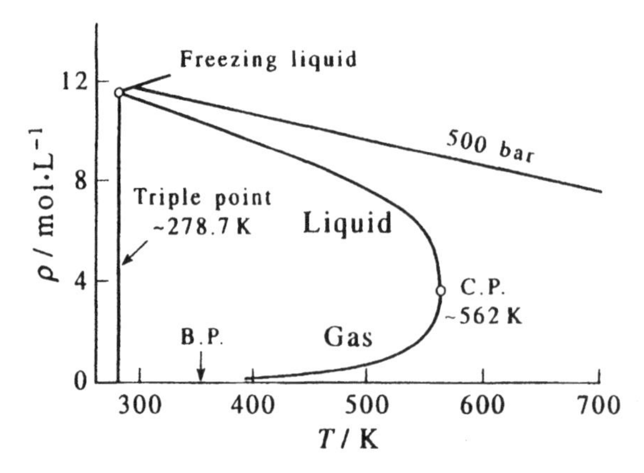

FIGURE 9.14
A density-temperature phase diagram of benzene.

	T/K	P/bar	ρ/mol·L^{-1} Vapor	Liquid
Triple point	278.680	0.04785	0.002074	11.4766
Critical point	561.75	48.7575	3.90	3.90
Normal freezing point	278.68	1.01325		
Normal boiling point	353.240	1.01325	0.035687	10.4075

The triple point is indicated by a line because it represents a temperature at which the solid, liquid, and gas phases all coexist at equilibrium. The line labelled triple point connects the densities of the liquid and vapor in equilibrium with each other. Notice that the liquid and gaseous densities become equal at the critical point. The line labelled 500 bar represents the density of benzene at 500 bar as a function of temperature. Below the information conveyed by the density-temperature phase diagram is represented in a pressure-temperature phase diagram.

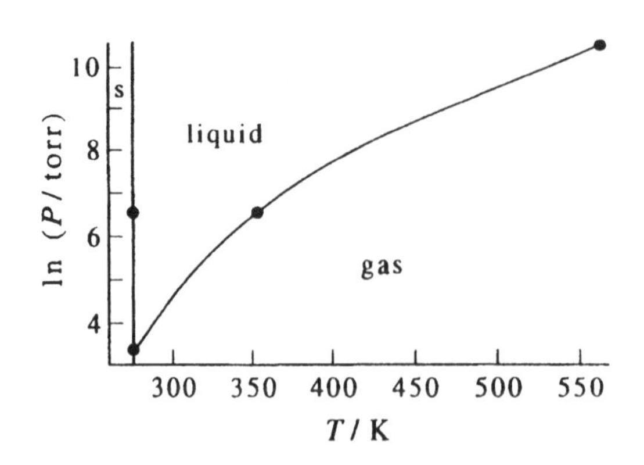

9–4. The vapor pressures of solid and liquid chlorine are given by

$$\ln(P^{s}/\text{torr}) = 24.320 - \frac{3777\ \text{K}}{T}$$
$$\ln(P^{l}/\text{torr}) = 17.892 - \frac{2669\ \text{K}}{T}$$

where T is the absolute temperature. Calculate the temperature and pressure at the triple point of chlorine.

This problem is done in the same way as Example 9–1. At the triple point, the two equations for the vapor pressure must be equivalent, since the solid and liquid coexist. Then

$$24.320 - \frac{3777\ \text{K}}{T_{\text{tp}}} = 17.892 - \frac{2669\ \text{K}}{T_{\text{tp}}}$$
$$(24.320 - 17.892)T_{\text{tp}} = -2669\ \text{K} + 3777\ \text{K}$$
$$T_{\text{tp}} = 172.4\ \text{K}$$

We can check this by substituting back into both expressions, and we find $\ln(P^{s}) = \ln(P^{l}) = 2.41$ torr and so $P_{\text{tp}} = 11.1$ torr.

9–5. The pressure along the melting curve from the triple-point temperature to an arbitrary temperature can be fit empirically by the Simon equation, which is

$$(P - P_{\text{tp}})/\text{bar} = a\left[\left(\frac{T}{T_{\text{tp}}}\right)^{\alpha} - 1\right]$$

where a and α are constants whose values depend upon the substance. Given that $P_{\text{tp}} = 0.04785$ bar, $T_{\text{tp}} = 278.68$ K, $a = 4237$, and $\alpha = 2.3$ for benzene, plot P against T and compare your result with that given in Figure 9.2.

Substituting into this expression, we find that we must plot

$$P/\text{bar} = 0.04785 + 4237\left[\left(\frac{T/\text{K}}{278.68\ \text{K}}\right)^{2.3} - 1\right]$$
$$P/\text{bar} = 0.04785 - 4237 + \frac{4237}{(278.68)^{2.3}}T^{2.3}$$

where T is on the y-axis and P is on the x-axis. The result looks very much like Figure 9.2.

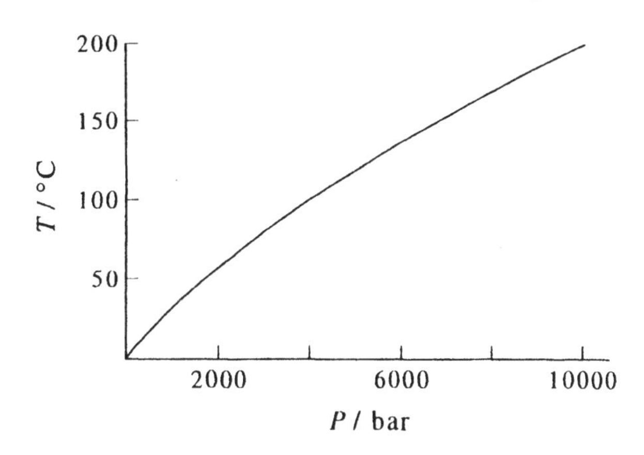

9–6. The slope of the melting curve of methane is given by

$$\frac{dP}{dT} = (0.08446\ \text{bar}\cdot\text{K}^{-1.85})T^{0.85}$$

from the triple point to arbitrary temperatures. Using the fact that the temperature and pressure at the triple point are 90.68 K and 0.1174 bar, calculate the melting pressure of methane at 300 K.

Integrating from the triple point to 300 K gives

$$\int_{0.1174\ \text{bar}}^{P_2} dP = \int_{90.68\ \text{K}}^{300\ \text{K}} 0.08446\ \text{bar}\cdot\text{K}^{-1.85}T^{0.85}dT$$

$$P_2 - 0.1174\ \text{bar} = \frac{0.08446\ \text{bar}\cdot\text{K}^{-1.85}}{1.85}\left[(300\ \text{K})^{1.85} - (90.68\ \text{K})^{1.85}\right]$$

$$P_2 = 1556\ \text{bar}$$

This is the melting pressure of methane at 300 K.

9–7. The vapor pressure of methanol along the entire liquid-vapor coexistence curve can be expressed very accurately by the empirical equation

$$\ln(P/\text{bar}) = -\frac{10.752849}{x} + 16.758207 - 3.603425x + 4.373232x^2 - 2.381377x^3 + 4.572199(1-x)^{1.70}$$

where $x = T/T_c$, and $T_c = 512.60$ K. Use this formula to show that the normal boiling point of methanol is 337.67 K.

At the normal boiling point, $P = 1$ atm $= 1.01325$ bar. If the normal boiling point of methanol is 337.67 K, then the equality below should hold when $x = 337.67/512.60$:

$$\ln(1.01325) \stackrel{?}{=} -\frac{10.752849}{x} + 16.758207 - 3.603425x + 4.373232x^2 - 2.381377x^3 + 4.572199(1-x)^{1.70}$$

$$0.013163 \stackrel{?}{=} -16.323364 + 16.758207 - 2.373719 + 1.897712 - 0.6807220 + 0.735141$$

$$0.013163 \approx 0.0132546$$

9–8. The standard boiling point of a liquid is the temperature at which the vapor pressure is exactly one bar. Use the empirical formula given in the previous problem to show that the standard boiling point of methanol is 337.33 K.

We do this in the same way as the previous problem, but substitute $x = 337.33/512.60$ into

$$\ln(1) \stackrel{?}{=} -\frac{10.752849}{x} + 16.758207 - 3.603425x + 4.373232x^2 - 2.381377x^3 + 4.572199(1-x)^{1.70}$$

$$0 \stackrel{?}{=} -16.339820 + 16.758207 - 2.371329 + 1.893892 - 0.678668 + 0.737572$$

$$0 \approx -0.000143$$

9–9. The vapor pressure of benzene along the liquid-vapor coexistence curve can be accurately expressed by the empirical expression

$$\ln(P/\text{bar}) = -\frac{10.655375}{x} + 23.941912 - 22.388714x + 20.2085593x^2 - 7.219556x^3 + 4.84728(1-x)^{1.70}$$

where $x = T/T_c$, and $T_c = 561.75$ K. Use this formula to show that the normal boiling point of benzene is 353.24 K. Use the above expression to calculate the standard boiling point of benzene.

This problem is essentially the same as Problem 9–7. We must substitute $x = 353.24/561.75$ into the equation

$$\ln(1.01325) \stackrel{?}{=} -\frac{10.655375}{x} + 23.941912 - 22.388714x + 20.2085593x^2 - 7.219556x^3 + 4.84728(1-x)^{1.70}$$

$$0.013163 \stackrel{?}{=} -16.945014 + 23.941912 - 14.078486 + 7.990776 - 1.795109 + 0.899064$$

$$0.013163 \approx 0.0131423$$

To calculate the standard boiling point of benzene, we must solve the polynomial equation for x when $P = 1$ bar, or when $\ln P/\text{bar} = 0$:

$$0 = -\frac{10.655375}{x} + 23.941912 - 22.388714x + 20.2085593x^2 - 7.219556x^3 + 4.84728(1-x)^{1.70}$$

Inputting this formula into a computational mathematics program such as *Mathematica* (or using the Newton-Raphson method) gives $x = 0.62806$, so the standard boiling point is $T = (561.75\text{ K})x = 352.8$ K.

9–10. Plot the following data for the densities of liquid and gaseous ethane in equilibrium with each other as a function of temperature, and determine the critical temperature of ethane.

T/K	$\rho^l/\text{mol}\cdot\text{dm}^{-3}$	$\rho^g/\text{mol}\cdot\text{dm}^{-3}$	T/K	$\rho^l/\text{mol}\cdot\text{dm}^{-3}$	$\rho^g/\text{mol}\cdot\text{dm}^{-3}$
100.00	21.341	1.336×10^{-3}	283.15	12.458	2.067
140.00	19.857	0.03303	293.15	11.297	2.880
180.00	18.279	0.05413	298.15	10.499	3.502
220.00	16.499	0.2999	302.15	9.544	4.307
240.00	15.464	0.5799	304.15	8.737	5.030
260.00	14.261	1.051	304.65	8.387	5.328
270.00	13.549	1.401	305.15	7.830	5.866

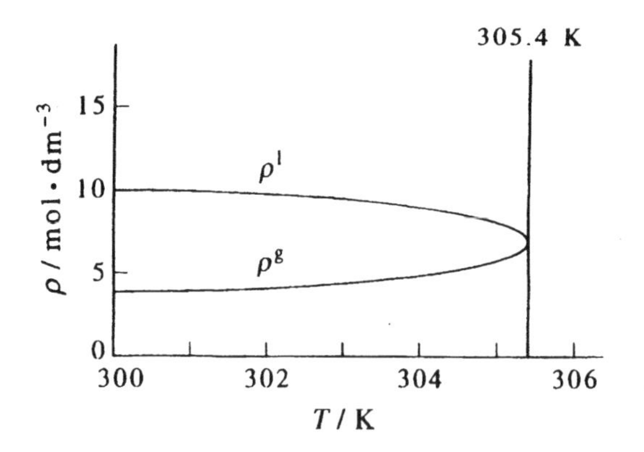

The critical temperature of ethane is about 305.4 K.

9–11. Use the data in the preceding problem to plot $(\rho^l + \rho^g)/2$ against $T_c - T$, with $T_c = 305.4$ K. The resulting straight line is an empirical law called the *law of rectilinear diameters*. If this curve is plotted on the same figure as in the preceding problem, the intersection of the two curves gives the critical density, ρ_c.

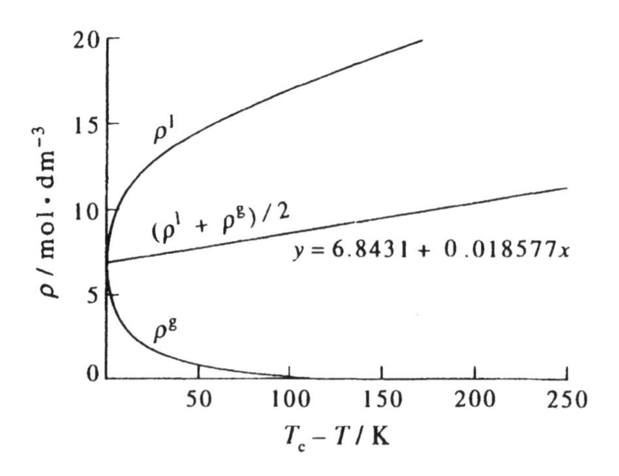

The critical density ρ_c is about 6.84 mol·dm^{-3}.

9–12. Use the data in Problem 9–10 to plot $(\rho^l - \rho^g)$ against $(T_c - T)^{1/3}$ with $T_c = 305.4$ K. What does this plot tell you?

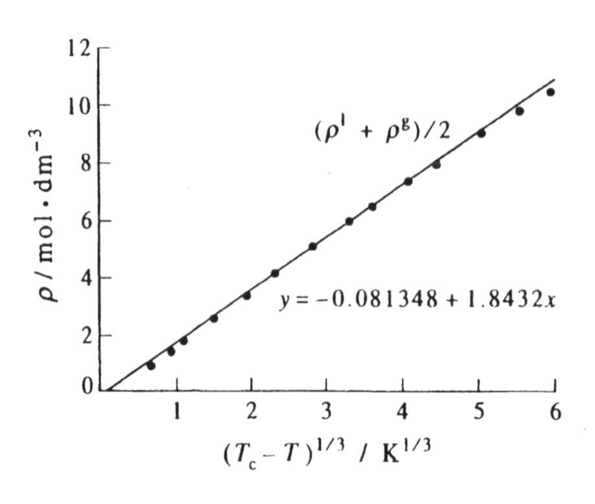

The linear nature of this plot tells us that $(\rho^{\text{l}} - \rho^{\text{g}})$ varies as $(T - T_{\text{c}})^{1/3}$ near the critical point.

9–13. The densities of the coexisting liquid and vapor phases of methanol from the triple point to the critical point are accurately given by the empirical expressions

$$\frac{\rho^{\text{l}}}{\rho_{\text{c}}} - 1 = 2.51709(1-x)^{0.350} + 2.466694(1-x) - 3.066818(1-x^2) + 1.325077(1-x^3)$$

and

$$\ln\frac{\rho^{\text{g}}}{\rho_{\text{c}}} = -10.619689\frac{1-x}{x} - 2.556682(1-x)^{0.350} + 3.881454(1-x) + 4.795568(1-x)^2$$

where $\rho_{\text{c}} = 8.40\ \text{mol}\cdot\text{L}^{-1}$ and $x = T/T_{\text{c}}$, where $T_{\text{c}} = 512.60$ K. Use these expressions to plot ρ^{l} and ρ^{g} against temperature, as in Figure 9.7. Now plot $(\rho^{\text{l}} + \rho^{\text{g}})/2$ against T. Show that this line intersects the ρ^{l} and ρ^{g} curves at $T = T_{\text{c}}$.

In this graph, the highest line represents ρ^{l}, the lowest line represents ρ^{g}, and the dashed line which comes between the two represents $(\rho^{\text{l}} + \rho^{\text{g}})/2$. At $T = T_{\text{c}}$, $\rho^{\text{l}} = \rho^{\text{g}}$, and so $(\rho^{\text{l}} + \rho^{\text{g}})/2 = \rho^{\text{l}} = \rho^{\text{g}}$. Therefore, the lines all meet at $T = T_{\text{c}}$.

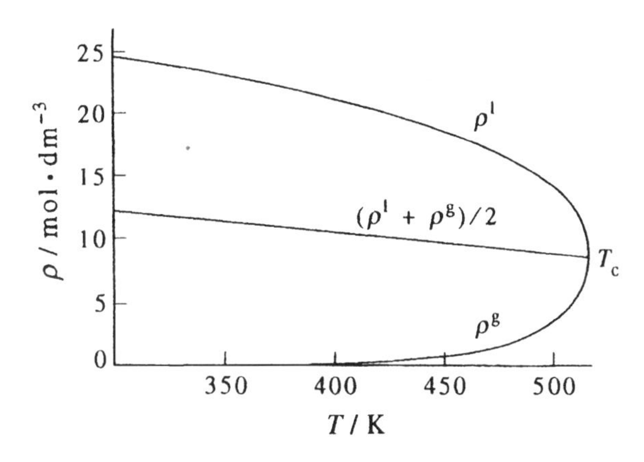

9–14. Use the expressions given in the previous problem to plot $(\rho^{\text{l}} - \rho^{\text{g}})/2$ against $(T_{\text{c}} - T)^{1/3}$. Do you get a reasonably straight line? If not, determine the value of the exponent of $(T_{\text{c}} - T)$ that gives the best straight line.

We find a line which is reasonably straight (although the curvature shown here is marked, note the scale of the y-axis).

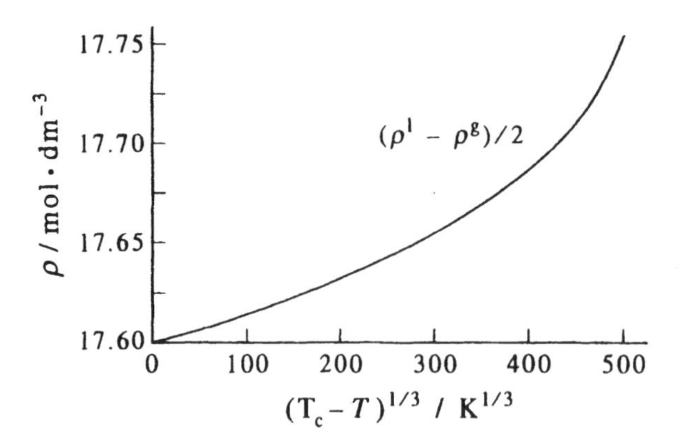

9–15. The molar enthalpy of vaporization of ethane can be expressed as

$$\Delta_{\text{vap}}\overline{H}(T)/\text{kJ}\cdot\text{mol}^{-1} = \sum_{j=1}^{6} A_j x^j$$

where $A_1 = 12.857$, $A_2 = 5.409$, $A_3 = 33.835$, $A_4 = -97.520$, $A_5 = 100.849$, $A_6 = -37.933$, and $x = (T_c - T)^{1/3}/(T_c - T_{\text{tp}})^{1/3}$ where the critical temperature $T_c = 305.4$ K and the triple point temperature $T_{\text{tp}} = 90.35$ K. Plot $\Delta_{\text{vap}}\overline{H}(T)$ versus T and show that the curve is similar to that of Figure 9.8.

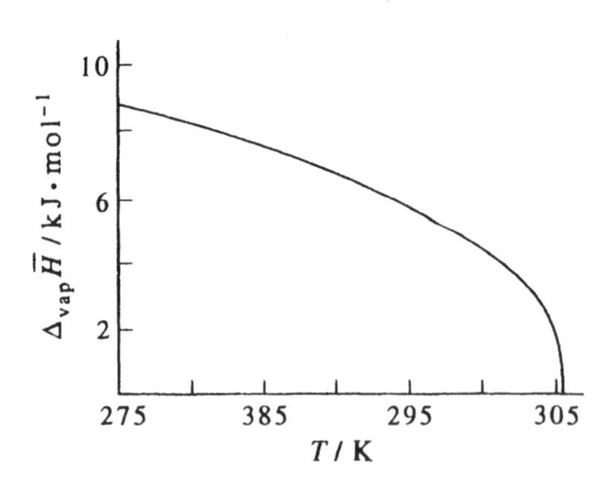

9–16. Fit the following data for argon to a cubic polynomial in T. Use your result to determine the critical temperature.

T/K	$\Delta_{\text{vap}}\overline{H}/\text{J}\cdot\text{mol}^{-1}$	T/K	$\Delta_{\text{vap}}\overline{H}/\text{J}\cdot\text{mol}^{-1}$
83.80	6573.8	122.0	4928.7
86.0	6508.4	126.0	4665.0
90.0	6381.8	130.0	4367.7
94.0	6245.2	134.0	4024.7
98.0	6097.7	138.0	3618.8
102.0	5938.8	142.0	3118.2
106.0	5767.6	146.0	2436.3
110.0	5583.0	148.0	1944.5
114.0	5383.5	149.0	1610.2
118.0	5166.5	150.0	1131.5

Fitting the data to a cubic polynomial in T gives the expression

$$\Delta_{\text{vap}}\overline{H}/\text{J}\cdot\text{mol}^{-1} = 39458.8 - (912.758\ \text{K}^{-1})T + (8.53681\ \text{K}^{-2})T^2 - (0.0276089\ \text{K}^{-3})T^3$$

Solving for T when $\Delta_{\text{vap}}\overline{H} = 0$ (at the critical temperature) gives a critical temperature of $T_c = 156.0\ \text{K}$. A better fit is to a fifth-order polynomial in T, which gives the expression

$$\Delta_{\text{vap}}\overline{H}/\text{J}\cdot\text{mol}^{-1} = 474232 - (21594.6\ \text{K}^{-1})T + (396.54\ \text{K}^{-2})T^2 - (3.61587\ \text{K}^{-3})T^3$$
$$+(1.63603\times 10^{-2}\ \text{K}^{-4})T^4 - (2.94294\times 10^{-5}\ \text{K}^{-5})T^5$$

We can solve this fifth-order equation for T when $\Delta_{\text{vap}}\overline{H} = 0$ by using a computational mathematics program or the Newton-Raphson method, which both give a critical temperature of $T_c = 152\ \text{K}$.

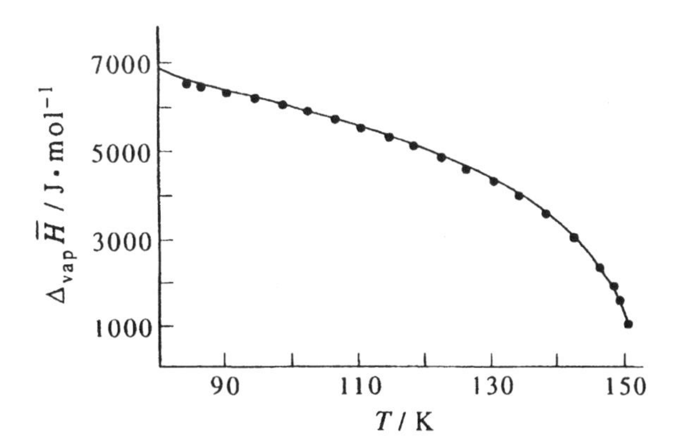

9–17. Use the following data for methanol at one atm to plot $\overline{G}$ versus T around the normal boiling point (337.668 K). What is the value of $\Delta_{\text{vap}}\overline{H}$?

T/K	$\overline{H}$/kJ·mol^{-1}	$\overline{S}$/J·mol^{-1}·K^{-1}
240	4.7183	112.259
280	7.7071	123.870
300	9.3082	129.375
320	10.9933	134.756
330	11.8671	137.412
337.668	12.5509	139.437
337.668	47.8100	243.856
350	48.5113	245.937
360	49.0631	247.492
380	50.1458	250.419
400	51.2257	253.189

We can use the formula $\overline{G} = \overline{H} - T\overline{S}$ to find $\overline{G}$ from this data and then plot $\overline{G}$ vs. T:

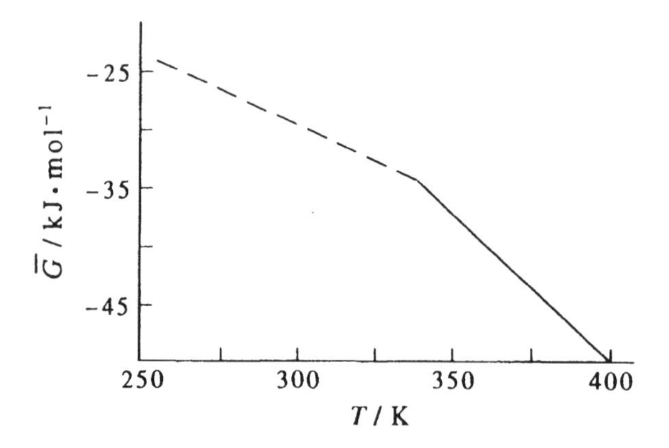

The line of best fit for the gaseous phase is $\overline{G}^{\text{g}}$/kJ·mol^{-1} = 49.57 − 0249T and the line of best fit for the liquid phase is $\overline{G}^{\text{l}}$/kJ·mol^{-1} = 8.1913 − 0.126T. $\Delta_{\text{vap}}\overline{H}$ will simply be the change in enthalpy when going from a liquid to a gas:

$$\Delta_{\text{vap}}\overline{H}/\text{kJ}\cdot\text{mol}^{-1} = 47.8100 - 12.5509 = 35.2591$$

9–18. In this problem, we will sketch $\overline{G}$ versus P for the solid, liquid, and gaseous phases for a generic ideal substance as in Figure 9.11. Let $\overline{V}^{\text{s}} = 0.600$, $\overline{V}^{\text{l}} = 0.850$, and $RT = 2.5$, in arbitrary units. Now show that

$$\overline{G}^{\text{s}} = 0.600(P - P_0) + \overline{G}_0^{\text{s}}$$
$$\overline{G}^{\text{l}} = 0.850(P - P_0) + \overline{G}_0^{\text{l}}$$

and

$$\overline{G}^{\text{g}} = 2.5\ln(P - P_0) + \overline{G}_0^{\text{g}}$$

where $P_0 = 1$ and $\overline{G}_0^{\text{s}}$, $\overline{G}_0^{\text{l}}$, and $\overline{G}_0^{\text{g}}$ are the respective zeros of energy. Show that if we (arbitrarily) choose the solid and liquid phases to be in equilibrium at $P = 2.00$ and the liquid and gaseous phases to be in equilibrium at $P = 1.00$, then we obtain

$$\overline{G}_0^{\text{s}} - \overline{G}_0^{\text{l}} = 0.250$$

and

$$\overline{G}_0^{\mathrm{l}} = \overline{G}_0^{\mathrm{g}}$$

from which we obtain

$$\overline{G}_0^{\mathrm{s}} - \overline{G}_0^{\mathrm{g}} = 0.250$$

Now we can express $\overline{G}^{\mathrm{s}}$, $\overline{G}^{\mathrm{l}}$, and $\overline{G}^{\mathrm{g}}$ in terms of a common zero of energy, $\overline{G}_0^{\mathrm{g}}$, which we must do to compare them with each other and to plot them on the same graph. Show that

$$\overline{G}^{\mathrm{s}} - \overline{G}_0^{\mathrm{g}} = 0.600(P-1) + 0.250$$
$$\overline{G}^{\mathrm{l}} - \overline{G}_0^{\mathrm{g}} = 0.850(P-1)$$
$$\overline{G}^{\mathrm{g}} - \overline{G}_0^{\mathrm{g}} = 2.5 \ln P$$

Plot these on the same graph from $P = 0.100$ to 3.00 and compare your result with Figure 9.11.

We know from Chapter 8 that $(\partial \overline{G}/\partial P)_T = \overline{V}$. This means that (for an ideal gas)

$$\overline{G} - \overline{G}_0 = \int_{P_0}^{P} \overline{V}\, dP = RT \ln \frac{P}{P_0}$$

For the solid and liquid phases, $\overline{V}$ is essentially constant with respect to pressure, and so $\overline{G} - \overline{G}_0 = \overline{V}(P - P_0)$. Therefore, we have

$$\overline{G}^{\mathrm{s}} = 0.600(P-1) + \overline{G}_0^{\mathrm{s}}$$
$$\overline{G}^{\mathrm{l}} = 0.850(P-1) + \overline{G}_0^{\mathrm{l}}$$
$$\overline{G}^{\mathrm{g}} = RT \ln \frac{P}{P_0} + \overline{G}_0^{\mathrm{g}} = 2.5 \ln P + \overline{G}_0^{\mathrm{g}}$$

where the units are arbitrary. Now, at equilibrium, $\overline{G}^{1} = \overline{G}^{2}$. Since the solid and liquid are in equilibrium at $P = 2.00$ and the liquid and gas are in equilibrium at $P = 1.00$,

$$\overline{G}^{\mathrm{s}}(P = 2.00) = \overline{G}^{\mathrm{l}}(P = 2.00)$$
$$0.600 + \overline{G}_0^{\mathrm{s}} = 0.850 + \overline{G}_0^{\mathrm{l}}$$
$$\overline{G}_0^{\mathrm{s}} - \overline{G}_0^{\mathrm{l}} = 0.250$$
$$\overline{G}^{\mathrm{l}}(P = 1.00) = \overline{G}^{\mathrm{g}}(P = 1.00)$$
$$\overline{G}_0^{\mathrm{l}} = \overline{G}_0^{\mathrm{g}}$$

and so $\overline{G}_0^{\mathrm{s}} - \overline{G}_0^{\mathrm{g}} = 0.250$. Now substitute into the first equations we found:

$$\overline{G}^{\mathrm{s}} = 0.600(P-1) + \overline{G}_0^{\mathrm{s}} = 0.600(P-1) + 0.250 + \overline{G}_0^{\mathrm{g}}$$
$$\overline{G}^{\mathrm{s}} - \overline{G}_0^{\mathrm{g}} = 0.600(P-1) + 0.250$$

Also

$$\overline{G}^{\mathrm{l}} - \overline{G}_0^{\mathrm{g}} = 0.0850(P-1)$$

and

$$\overline{G}^{\mathrm{g}} - \overline{G}_0^{\mathrm{g}} = 2.5 \ln P$$

A plot of the Gibbs energies of the gas, liquid, and solid using $\overline{G}_0^{\mathrm{g}}$ as the zero of energy is shown:

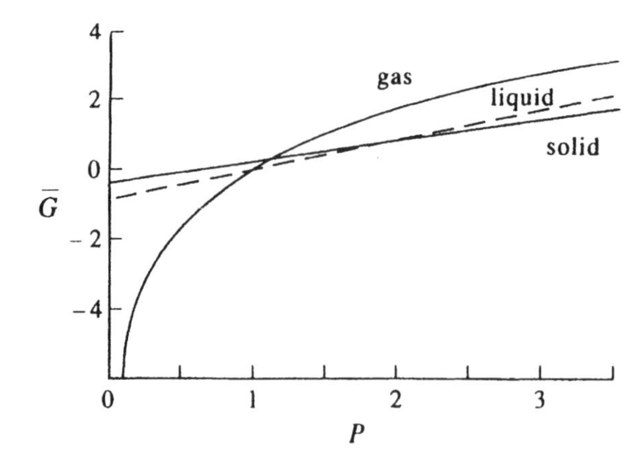

Since we see a gas-liquid-solid progression, we are looking at a temperature less than the triple point temperature (as explained in the caption of Figure 9.11).

9–19. In this problem, we will demonstrate that entropy always increases when there is a material flow from a region of higher concentration to one of lower concentration. (Compare with Problems 8–41 and 8–42.) Consider a two-compartment system enclosed by rigid, impermeable, adiabatic walls, and let the two compartments be separated by a rigid, insulating, but permeable wall. We assume that the two compartments are in equilibrium but that they are not in equilibrium with each other. Show that

$$U_1 = \text{constant}, \quad U_2 = \text{constant}, \quad V_1 = \text{constant}, \quad V_2 = \text{constant},$$

and

$$n_1 + n_2 = \text{constant}$$

for this system. Now show that

$$dS = \frac{dU}{T} + \frac{P}{T}dV - \frac{\mu}{T}dn$$

in general, and that

$$dS = \left(\frac{\partial S_1}{\partial n_1}\right) dn_1 + \left(\frac{\partial S_2}{\partial n_2}\right) dn_2$$
$$= dn_1 \left(\frac{\mu_2}{T} - \frac{\mu_1}{T}\right) \geq 0$$

for this system. Use this result to discuss the direction of a (isothermal) material flow under a chemical potential difference.

The volume of each compartment cannot change, since the walls of the compartments are rigid. Thus V_1 = constant and V_2 = constant. Since the walls are adiabatic, $\delta q = 0$ for the gases in each compartment. For both components of the system $\delta w = 0$, since there is no change in volume, so $dU = 0$ for both compartments. Thus U_1 = constant and U_2 = constant. Finally, since the entire system is surrounded by impermeable walls, the total number of moles of gas in the system must remain constant, so $n_1 + n_2$ = constant.

We have defined $\mu = (\partial G/\partial n)_{P,T}$ (Equation 9.3), so $\mu dn = dG$. Now recall (Equation 8.13) that

$$G = U - TS + PV$$
$$dG = dU - TdS + PdV$$
$$\frac{\mu dn}{T} = \frac{dU}{T} - dS + \frac{P}{T}dV$$
$$dS = \frac{dU}{T} + \frac{P}{T}dV - \frac{\mu}{T}dn$$

For this system, since $dU = dV = 0$,

$$dS_1 = -\frac{\mu_1}{T}dn_1 \qquad dS_2 = -\frac{\mu_2}{T}dn_2$$

Then

$$\begin{aligned} dS_{\text{system}} &= dS_1 + dS_2 \\ &= \left(\frac{\partial S_1}{\partial n_1}\right) dn_1 + \left(\frac{\partial S_2}{\partial n_2}\right) dn_2 \\ &= -\frac{\mu_1}{T}dn_1 - \frac{\mu_2}{T}(-dn_1) \\ &= dn_1\left(\frac{\mu_2}{T} - \frac{\mu_1}{T}\right) \end{aligned}$$

If molecules are flowing into compartment 1, then dn_1 is positive and $\mu_2 > \mu_1$ (since transfer occurs from the system with higher chemical potential to the system with lower chemical potential). Then both terms in the expression above are positive and $dS_{\text{system}} > 0$. If molecules are flowing into compartment 2, then dn_1 is negative and $\mu_2 < \mu_1$, making both terms negative and $dS_{\text{system}} > 0$. If dn_1 is 0 (no transfer occurs), then the two compartments are in equilibrium with respect to material flow, and $dS_{\text{system}} = 0$.

9–20. Determine the value of dT/dP for water at its normal boiling point of 373.15 K given that the molar enthalpy of vaporization is 40.65 $kJ \cdot mol^{-1}$, and the densities of the liquid and vapor are 0.9584 $g \cdot L^{-1}$ and 0.6010 $g \cdot mL^{-1}$, respectively. Estimate the boiling point of water at 2 atm.

First find $\overline{V}^g - \overline{V}^l$:

$$\begin{aligned} \overline{V}^g - \overline{V}^l &= \left(\frac{1}{0.6010\ g \cdot dm^{-3}} - \frac{1}{958.4\ g \cdot dm^{-3}}\right)(18.015\ g \cdot mol^{-1}) \\ &= 29.96\ dm^3 \cdot mol^{-1} \end{aligned}$$

Now use Equation 9.10 to write

$$\begin{aligned} \frac{dT}{dP} &= \frac{T\Delta_{\text{vap}}\overline{V}}{\Delta_{\text{vap}}\overline{H}} \\ &= \frac{(373.15\ K)(\overline{V}^g - \overline{V}^l)}{40\,650\ J \cdot mol^{-1}} \\ &= \left[\frac{(373.15\ K)(29.96\ dm^3 \cdot mol^{-1})}{40\,650\ J \cdot mol^{-1}}\right]\left(\frac{8.314\ J}{0.08206\ dm^3 \cdot atm}\right) \\ &= 27.9\ K \cdot atm^{-1} \end{aligned}$$

To estimate the boiling point of water at 2 atm, we can find the change in temperature which accompanies a change in pressure of 1 atm (since we know the boiling point of water at 1 atm). That is $\Delta T = (27.9\ \text{K}\cdot\text{atm}^{-1})(1\ \text{atm}) = 27.9\ \text{K}$. Therefore, the boiling point of water at 2 atm is about 127.9°C.

9–21. The orthobaric densities of liquid and gaseous ethyl acetate are $0.826\ \text{g}\cdot\text{mL}^{-1}$ and $0.00319\ \text{g}\cdot\text{mL}^{-1}$, respectively, at its normal boiling point (77.11°C). The rate of change of vapor pressure with temperature is $23.0\ \text{torr}\cdot\text{K}^{-1}$ at the normal boiling point. Estimate the molar enthalpy of vaporization of ethyl acetate at its normal boiling point.

First find $\overline{V}^{\text{g}} - \overline{V}^{\text{l}}$:

$$\overline{V}^{\text{g}} - \overline{V}^{\text{l}} = \left(\frac{1}{0.00319\ \text{g}\cdot\text{mL}^{-1}} - \frac{1}{0.826\ \text{g}\cdot\text{mL}^{-1}}\right)(88.102\ \text{g}\cdot\text{mol}^{-1})$$

$$\Delta_{\text{vap}}\overline{V} = 27\,510\ \text{cm}^3\cdot\text{mol}^{-1} = 275.10\ \text{dm}^3\cdot\text{mol}^{-1}$$

Now use Equation 9.10 to write

$$\begin{aligned}\Delta_{\text{vap}}\overline{H} &= T\Delta_{\text{vap}}\overline{V}\left(\frac{dP}{dT}\right)\\ &= (350.26\ \text{K})(27.51\ \text{dm}^3\cdot\text{mol}^{-1})(23.0\ \text{torr}\cdot\text{K}^{-1})\left(\frac{1\ \text{atm}}{760\ \text{torr}}\right)\left(\frac{8.314\ \text{J}}{0.08206\ \text{L}\cdot\text{atm}}\right)\\ &= 29.5\ \text{kJ}\cdot\text{mol}^{-1}\end{aligned}$$

9–22. The vapor pressure of mercury from 400°C to 1300°C can be expressed by

$$\ln(P/\text{torr}) = -\frac{7060.7\ \text{K}}{T} + 17.85$$

The density of the vapor at its normal boiling point is $3.82\ \text{g}\cdot\text{L}^{-1}$ and that of the liquid is $12.7\ \text{g}\cdot\text{mL}^{-1}$. Estimate the molar enthalpy of vaporization of mercury at its normal boiling point.

If we express P using the above equation, we find that

$$\frac{dP}{dT} = P\left(\frac{7060.7\ \text{K}}{T^2}\right)$$

At the boiling point and one atmosphere of pressure,

$$\frac{dP}{dT} = (760\ \text{torr})\left[\frac{7060.7\ \text{K}}{(629.88\ \text{K})^2}\right] = 13.52\ \text{torr}\cdot\text{K}^{-1}$$

We find $\Delta_{\text{vap}}\overline{V}$ by subtracting $\overline{V}^{\text{l}}$ from $\overline{V}^{\text{g}}$:

$$\overline{V}^{\text{g}} - \overline{V}^{\text{l}} = \left(\frac{1}{3.82\ \text{g}\cdot\text{dm}^{-3}} - \frac{1}{12700\ \text{g}\cdot\text{dm}^{-3}}\right)(200.59\ \text{g}\cdot\text{mol}^{-1})$$

$$\Delta_{\text{vap}}\overline{V} = 52.49\ \text{dm}^3\cdot\text{mol}^{-1}$$

Now we use Equation 9.10 to estimate $\Delta_{\text{vap}}\overline{H}$.

$$\begin{aligned}\Delta_{\text{vap}}\overline{H} &= T\Delta_{\text{vap}}\overline{V}\frac{dP}{dT}\\ &= (629.88\text{ K})(52.49\text{ dm}^3\cdot\text{mol}^{-1})(13.52\text{ torr}\cdot\text{K}^{-1})\\ &= 447\ 200\text{ dm}^3\cdot\text{torr}\cdot\text{mol}^{-1}\left(\frac{1\text{ atm}}{760\text{ torr}}\right)\left(\frac{8.314\text{ J}}{0.08206\text{ dm}^3\cdot\text{atm}}\right)\\ &= 59.62\text{ kJ}\cdot\text{mol}^{-1}\end{aligned}$$

9–23. The pressures at the solid-liquid coexistence boundary of propane are given by the empirical equation

$$P = -718 + 2.38565T^{1.283}$$

where P is in bars and T is in kelvins. Given that $T_{\text{fus}} = 85.46$ K and $\Delta_{\text{fus}}\overline{H} = 3.53\text{ kJ}\cdot\text{mol}^{-1}$, calculate $\Delta_{\text{fus}}\overline{V}$ at 85.46 K.

At 85.46 K, the empirical equation gives

$$\frac{dP}{dT} = 3.06079(85.46)^{0.283} = 10.778\text{ bar}\cdot\text{K}^{-1}$$

We substitute into Equation 9.10 to find

$$\begin{aligned}\Delta_{\text{fus}}\overline{V} &= \frac{\Delta_{\text{fus}}\overline{H}}{T}\left(\frac{dP}{dT}\right)^{-1}\\ &= \frac{35\ 300\text{ J}\cdot\text{mol}^{-1}}{85.46\text{ K}}(10.778\text{ bar}\cdot\text{K}^{-1})\left(\frac{10\text{ bar}\cdot\text{cm}^3}{1\text{ J}}\right)\\ &= 383\text{ cm}^3\cdot\text{mol}^{-1}\end{aligned}$$

9–24. Use the vapor pressure data given in Problem 9–7 and the density data given in Problem 9–13 to calculate $\Delta_{\text{vap}}\overline{H}$ for methanol from the triple point (175.6 K) to the critical point (512.6 K). Plot your result.

We are given ρ in units of $\text{mol}\cdot\text{dm}^{-3}$ in Problem 9–13 and P in units of bars in Problem 9–7. We want to find $\Delta_{\text{vap}}\overline{H}$ using Equation 9.10:

$$\Delta_{\text{vap}}\overline{H} = T\Delta_{\text{vap}}\overline{V}\frac{dP}{dT}$$

Taking the derivative of the expression for P given in Problem 9–7 gives us

$$\begin{aligned}\frac{dP}{dT}/\text{bar}\cdot\text{K}^{-1} = P\Big[&-0.0070297 - 0.0151634(1 - 0.00195084T)^{0.7}\\ &+\frac{5511.91}{T^2} + 3.32871\times10^{-5}T - 5.30412\times10^{-8}T^2\Big]\end{aligned}$$

Using $1/\rho^{\text{l}}$ for $\overline{V}^{\text{l}}$ and $1/\rho^{\text{g}}$ for $\overline{V}^{\text{g}}$ gives

$$\Delta_{\text{vap}}\overline{V}/\text{dm}^3\cdot\text{mol}^{-1} = \frac{1}{\rho^{\text{g}}} - \frac{1}{\rho^{\text{l}}}$$

or (where $\Delta_{vap}\overline{V}$ is in units of $dm^3\cdot mol^{-1}$

$$\Delta_{vap}\overline{V} = 0.119048\exp\Big[2.55668(1-0.0019508T)^{0.35} - 3.88145(1-0.0019508T)$$
$$-4.79557(1-0.0019508T)^2 + \frac{5443.65(1-0.0019508)}{T}\Big]$$
$$-0.119048\big[1 + 2.5171(1-0.0019508T)^{0.35} + 2.46669(1-0.0019508T)$$
$$-3.06682(1-0.0019508T)^2 + 1.32508(1-0.0019508T)^3\big]^{-1}$$

Substituting these expressions into Equation 9.10 gives $\Delta_{vap}\overline{H}$ in units of $dm^3\cdot bar\cdot mol^{-1}$. To convert this to $kJ\cdot mol^{-1}$ we must divide by 10.

Now graph

$$\Delta_{vap}\overline{H}/kJ\cdot mol^{-1} = \frac{\Delta_{vap}\overline{V}}{10}\frac{dP}{dT}T$$

using the expressions found above for $\Delta_{vap}\overline{V}$ and dP/dT:

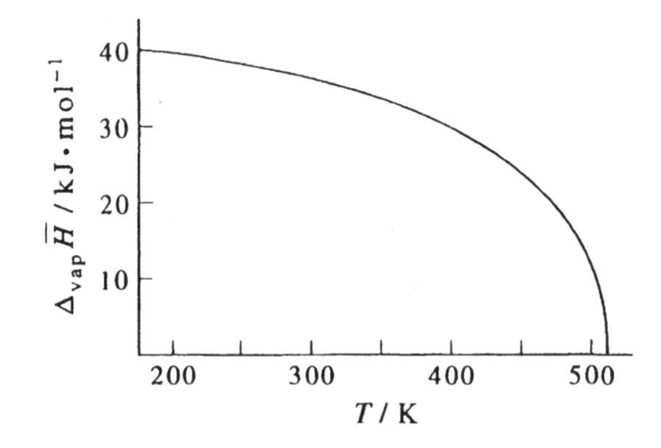

9–25. Use the result of the previous problem to plot $\Delta_{vap}\overline{S}$ of methanol from the triple point to the critical point.

Since at a transition point $\Delta_{trs}\overline{G} = 0$,

$$\frac{\Delta_{vap}\overline{H}}{T} = \Delta_{vap}\overline{S}$$

We can use the expression for $\Delta_{vap}\overline{H}$ given in Problem 9–24 (converting it to $J\cdot mol^{-1}$, since these are the usual units of entropy) to graph $\Delta_{vap}\overline{S}$.

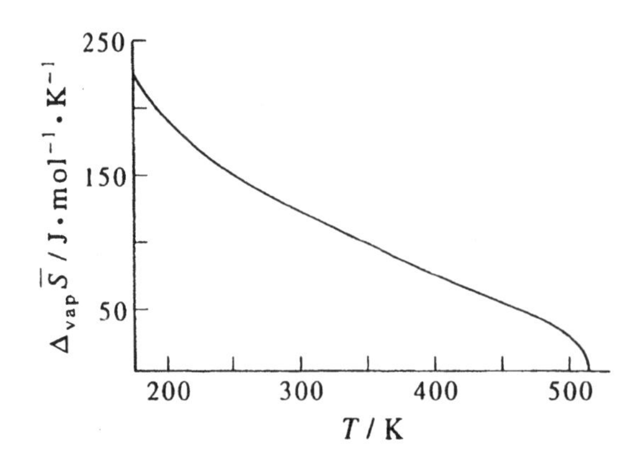

Notice that $\Delta_{\text{vap}}\overline{S} \to 0$ as $T \to T_c$.

9–26. Use the vapor pressure data for methanol given in Problem 9–7 to plot $\ln P$ against $1/T$. Using your calculations from Problem 9–24, over what temperature range do you think the Clausius-Clapeyron equation will be valid?

Use the formula for $\ln P$ given in Problem 9–7 to plot $\ln P$ vs. $1/T$.

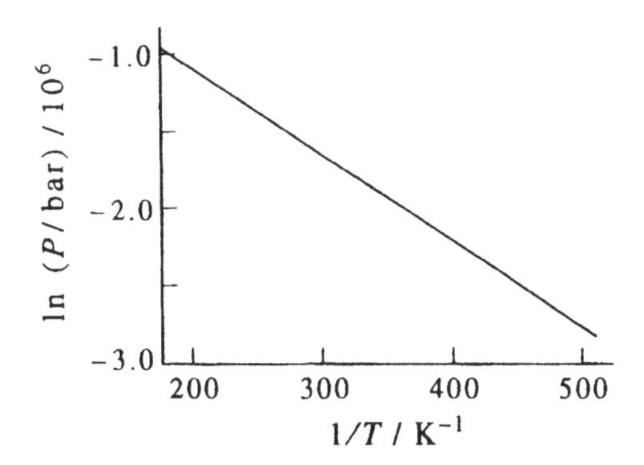

The slope of the line plotted should be constant for the Clausius-Clapeyron equation to be valid. It appears that the Clausius-Clapeyron equation is valid over the range plotted in Problem 9–31.

9–27. The molar enthalpy of vaporization of water is $40.65\ \text{kJ}\cdot\text{mol}^{-1}$ at its normal boiling point. Use the Clausius-Clapeyron equation to calculate the vapor pressure of water at 110°C. The experimental value is 1075 torr.

Assuming $\Delta_{\text{vap}}\overline{H}$ remains constant with respect to temperature over this ten-degree temperature range, we can use Equation 9.13:

$$\ln\frac{P_2}{P_1} = \frac{\Delta_{\text{vap}}\overline{H}}{R}\left(\frac{T_2 - T_1}{T_1T_2}\right)$$

$$\ln\frac{P_2}{1\ \text{atm}} = \frac{40\,650\ \text{J}\cdot\text{mol}^{-1}}{8.3145\ \text{J}\cdot\text{mol}^{-1}\cdot\text{K}^{-1}}\left[\frac{10\ \text{K}}{(373.15\ \text{K})(383.15\ \text{K})}\right]$$

$$\ln(P_2/\ \text{atm}) = 0.342$$

$$P_2 = 1.408\ \text{atm} = 1070\ \text{torr}$$

9–28. The vapor pressure of benzaldehyde is 400 torr at 154°C and its normal boiling point is 179°C. Estimate its molar enthalpy of vaporization. The experimental value is $42.50\ \text{kJ}\cdot\text{mol}^{-1}$.

Again, assuming $\Delta_{\text{vap}}\overline{H}$ does not vary over this temperature range, we can use Equation 9.13.

$$\ln\frac{P_2}{P_1} = \frac{\Delta_{\text{vap}}\overline{H}}{R}\left(\frac{T_2 - T_1}{T_1T_2}\right)$$

$$\Delta_{\text{vap}}\overline{H} = \frac{RT_1T_2}{T_2 - T_1}\ln\frac{P_2}{P_1}$$

$$= \frac{(8.3145\ \text{J}\cdot\text{mol}^{-1}\cdot\text{K}^{-1})(427.15\ \text{K})(452.15\ \text{K})}{25\ \text{K}}\ln\frac{760}{400}$$

$$= 41.2\ \text{kJ}\cdot\text{mol}^{-1}$$

9–29. Use the following data to estimate the normal boiling point and the molar enthalpy of vaporization of lead.

T/K	1500	1600	1700	1800	1900
P/torr	19.72	48.48	107.2	217.7	408.2

Plot $\ln P$ vs. $1/T$:

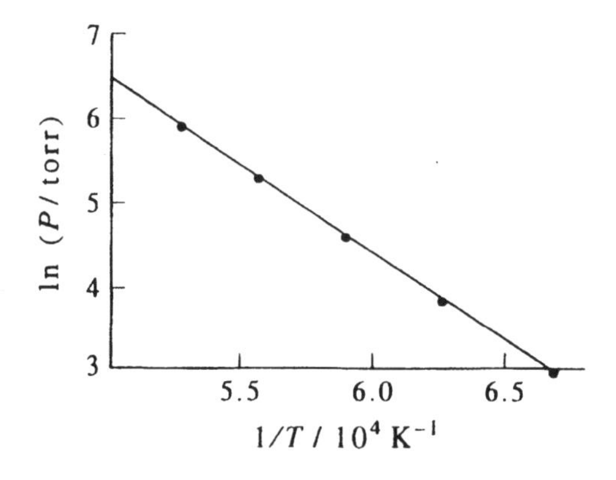

The equation for the line of best fit is $y = 17.3799 - 21597.6x$. At 1 atm (the normal boiling point pressure), $\ln 760 = y$ and so

$$17.3799 = 6.6333 + \frac{21597.6}{T}$$
$$T = 2010 \text{ K}$$

The normal boiling point is about 2010 K. Now recall that the slope of the plot we have created should be $-\Delta_{\text{vap}}\overline{H}/R$ (Equation 9.14). Then

$$\Delta_{\text{vap}}\overline{H} = (8.314 \text{ J}\cdot\text{mol}^{-1}\cdot\text{K}^{-1})(21597.6 \text{ K}) = 179.6 \text{ kJ}\cdot\text{mol}^{-1}$$

9–30. The vapor pressure of solid iodine is given by

$$\ln(P/\text{atm}) = -\frac{8090.0 \text{ K}}{T} - 2.013\ln(T/\text{K}) + 32.908$$

Use this equation to calculate the normal sublimation temperature and the molar enthalpy of sublimation of I_2(s) at 25°C. The experimental value of $\Delta_{\text{sub}}\overline{H}$ is 62.23 kJ·mol^{-1}.

The sublimation temperature is found by setting $P = 1$ atm in the above equation:

$$0 = -\frac{8090.0 \text{ K}}{T_{\text{sub}}} - 2.013\ln(T_{\text{sub}}/\text{K}) + 32.908$$

We can solve this equation for T using the Newton-Raphson method, and we find that $T_{\text{sub}} = 386.8$ K.

We can now use the equation provided and Equation 9.12 to find the molar enthalpy of sublimation:

$$\frac{\Delta_{\text{sub}}\overline{H}}{RT^2} = \frac{d\ln P}{dT} = \frac{8090.0 \text{ K}}{T^2} - \frac{2.013}{T}$$
$$\Delta_{\text{sub}}\overline{H} = R(8090.0 \text{ K} - 2.013T)$$

At 25°C,

$$\Delta_{\text{sub}}\overline{H} = (8.314\ \text{J}\cdot\text{mol}^{-1}\cdot\text{K}^{-1})\ [8090.0\ \text{K} - 2.013(298.15\ \text{K})] = 62.27\ \text{kJ}\cdot\text{mol}^{-1}$$

9–31. Fit the following vapor pressure data of ice to an equation of the form

$$\ln P = -\frac{a}{T} + b\ln T + cT + d$$

where T is temperature in kelvins. Use your result to determine the molar enthalpy of sublimation of ice at 0°C.

t/°C	P/torr	t/°C	P/torr
−10.0	1.950	−4.8	3.065
− 9.6	2.021	−4.4	3.171
− 9.2	2.093	−4.0	3.280
− 8.8	2.168	−3.6	3.393
− 8.4	2.246	−3.2	3.509
− 8.0	2.326	−2.8	3.360
− 7.6	2.408	−2.4	3.753
− 7.2	2.493	−2.0	3.880
− 6.8	2.581	−1.6	4.012
− 6.4	2.672	−1.2	4.147
− 6.0	2.765	−0.8	4.287
− 5.6	2.862	−0.4	4.431
− 5.2	2.962	0.0	4.579

Fitting the data to an equation of this form gives

$$\ln P = -\frac{5924.3}{T} + 2.7221\ln T - 0.007221T + 9.9118$$

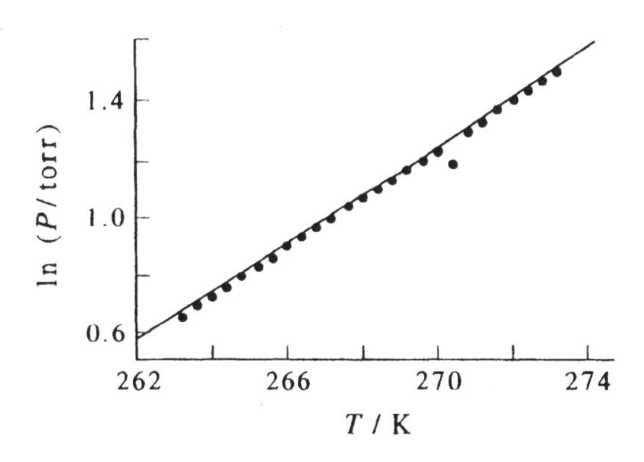

Using this equation and Equation 9.12, we find that

$$\frac{\Delta_{\text{sub}}\overline{H}}{RT^2} = \frac{d\ln P}{dT} = \frac{5924.3\ \text{K}}{T^2} + \frac{2.7221}{T} - 0.007221\ \text{K}^{-1}$$

$$\Delta_{\text{sub}}\overline{H} = R\left[5924.3\ \text{K} + 2.7221T - (0.007221\ \text{K}^{-1})T^2\right]$$

At 0°C,

$$\begin{aligned}\Delta_{\text{sub}}\overline{H} &= (8.314\ \text{J}\cdot\text{mol}^{-1}\cdot\text{K}^{-1})\ [5924.3\ \text{K} + 2.7221(298.15\ \text{K}) \\ &\quad - (0.007221\ \text{K}^{-1})(298.15\ \text{K})^2] \\ &= 50.96\ \text{kJ}\cdot\text{mol}^{-1}\end{aligned}$$

9–32. The following table gives the vapor pressure data for liquid palladium as a function of temperature:

T/K	P/bar
1587	1.002×10^{-9}
1624	2.152×10^{-9}
1841	7.499×10^{-8}

Estimate the molar enthalpy of vaporization of palladium.

Plot ln P vs. $1/T$:

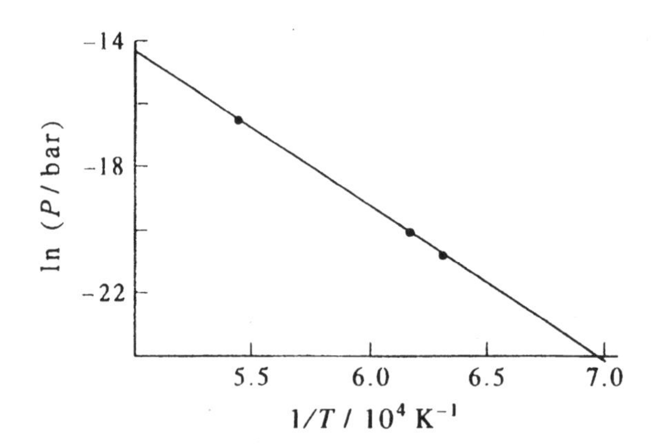

The line of best fit is $y = 10.4359 - 49407x$. Since the slope of this line is equal to $-\Delta_{vap}\overline{H}/R$,

$$\Delta_{vap}\overline{H} = (49407\text{ K})(8.31451\text{ J}\cdot\text{mol}^{-1}\cdot\text{K}^{-1}) = 410.8\text{ kJ}\cdot\text{mol}^{-1}$$

9–33. The sublimation pressure of CO_2 at 138.85 K and 158.75 K is 1.33×10^{-3} bar and 2.66×10^{-2} bar, respectively. Estimate the molar enthalpy of sublimation of CO_2.

Substitute into Equation 9.13:

$$\ln\frac{P_2}{P_1} = \frac{\Delta_{sub}\overline{H}}{R}\left(\frac{T_2 - T_1}{T_2 T_1}\right)$$

$$\Delta_{sub}\overline{H} = (8.3145\text{ J}\cdot\text{mol}^{-1}\cdot\text{K}^{-1})\ln\left(\frac{2.66 \times 10^{-2}}{1.33 \times 10^{-3}}\right)\left[\frac{(138.85\text{ K})(158.75\text{ K})}{19.9\text{ K}}\right]$$

$$= 27.6\text{ kJ}\cdot\text{mol}^{-1}$$

9–34. The vapor pressures of solid and liquid hydrogen iodide can be expressed empirically as

$$\ln(P^s/\text{torr}) = -\frac{2906.2\text{ K}}{T} + 19.020$$

and

$$\ln(P^l/\text{torr}) = -\frac{2595.7\text{ K}}{T} + 17.572$$

Calculate the ratio of the slopes of the solid-gas curve and the liquid-gas curve at the triple point.

We can write the slopes of the solid-gas and liquid-gas curves as

$$\frac{dP^{\mathrm{s}}}{dT} = P^{\mathrm{s}}\left(\frac{2906.2\ \mathrm{K}}{T^2}\right) \qquad \text{and} \qquad \frac{dP^{\mathrm{l}}}{dT} = P^{\mathrm{l}}\left(\frac{2595.7\ \mathrm{K}}{T^2}\right)$$

where pressures are in units of torr. Since $P^{\mathrm{s}} = P^{\mathrm{l}}$ at the triple point, the ratio of the slopes at the triple point is

$$\frac{dP^{\mathrm{s}}/dT}{dP^{\mathrm{l}}/dT} = \frac{2906.2\ \mathrm{K}}{2595.7\ \mathrm{K}} = 1.120$$

9–35. Given that the normal melting point, the critical temperature, and the critical pressure of hydrogen iodide are 222 K, 424 K and 82.0 atm, respectively, use the data in the previous problem to sketch the phase diagram of hydrogen iodide.

The triple point is located where the solid and liquid vapor pressures are the same, so at the triple point

$$19.020 - \frac{2906.2\ \mathrm{K}}{T_{\mathrm{tp}}} = 17.572 - \frac{2595.7\ \mathrm{K}}{T_{\mathrm{tp}}}$$

$$1.448T_{\mathrm{tp}} = 310.5\ \mathrm{K}$$

$$T_{\mathrm{tp}} = 214.43\ \mathrm{K}$$

Substituting to solve for P_{tp}, we find that $P_{\mathrm{tp}} = 236.8$ torr. We also know that the normal melting point of HI is 222 K, so we can produce a phase diagram of hydrogen iodide by plotting the line between the solid and gas, the line between the vapor and gas, the critical point, the triple point, and the normal melting point. Note that the equation for the liquid-gas line is not completely accurate at high temperatures and pressures (it does not intersect the critical point).

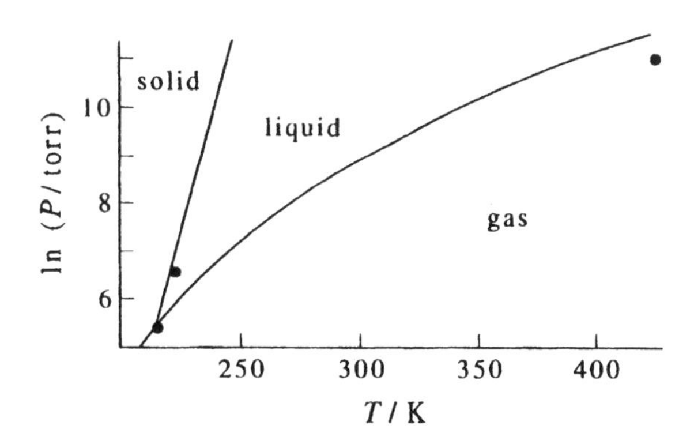

9–36. Consider the phase change

$$\mathrm{C(graphite)} \rightleftharpoons \mathrm{C(diamond)}$$

Given that $\Delta_r G°/\mathrm{J{\cdot}mol^{-1}} = 1895 + 3.363T$, calculate $\Delta_r H°$ and $\Delta_r S°$. Calculate the pressure at which diamond and graphite are in equilibrium with each other at 25°C. Take the density of diamond and graphite to be 3.51 $\mathrm{g{\cdot}cm^{-3}}$ and 2.25 $\mathrm{g{\cdot}cm^{-3}}$, respectively. Assume that both diamond and graphite are incompressible.

To find the standard molar Gibbs entropy, we use the Maxwell relation (Equation 8.46)

$$\Delta_r \overline{S}^\circ = -\left(\frac{\partial \Delta_r \overline{G}^\circ}{\partial T}\right)_P = -3.363\ \text{J}\cdot\text{mol}^{-1}$$

Substituting into $\Delta_r \overline{G}^\circ = \Delta_r \overline{H}^\circ - T\Delta_r \overline{S}^\circ$ (Equation 8.13), we find that $\Delta_r \overline{H}^\circ = 1895\ \text{J}\cdot\text{mol}^{-1}$. Because both graphite and diamond are incompressible, we can write (as in Problem 9–18)

$$\overline{G}_{\text{graph}} = \overline{G}^\circ_{\text{graph}} + \overline{V}_{\text{graph}}(P - P_0)$$

and

$$\overline{G}_{\text{diam}} = \overline{G}^\circ_{\text{diam}} + \overline{V}_{\text{diam}}(P - P_0)$$

Combining these two equations gives

$$\Delta_r \overline{G} = \Delta_r \overline{G}^\circ + (\overline{V}_{\text{diam}} - \overline{V}_{\text{graph}})(P - P_0)$$

When graphite and diamond are in equilibrium, $\Delta_r \overline{G} = 0$. Substituting into the equation given in the problem, we see that at 25°C, $\Delta_r G^\circ = 2898\ \text{J}\cdot\text{mol}^{-1}$. Then

$$\begin{aligned}
0 &= \Delta_r \overline{G}^\circ + \left(\frac{1}{3510\ \text{g}\cdot\text{dm}^{-3}} - \frac{1}{2250\ \text{g}\cdot\text{dm}^{-3}}\right)(12.01\ \text{g}\cdot\text{mol}^{-1})(P - 1\ \text{bar}) \\
&= 2898\ \text{J}\cdot\text{mol}^{-1} - (1.916 \times 10^{-3}\ \text{dm}^3\cdot\text{mol}^{-1})(P - 1\ \text{bar}) \\
P &= \frac{1}{1.916 \times 10^{-3}\ \text{dm}^3\cdot\text{mol}^{-1}}\left(2898\ \text{J}\cdot\text{mol}^{-1}\right)\left(\frac{0.08206\ \text{dm}^3\cdot\text{bar}}{8.3145\ \text{J}}\right) + 1\ \text{bar} \\
&= 15\,000\ \text{bar}
\end{aligned}$$

9–37. Use Equation 9.36 to calculate $\mu^\circ - E_0$ for Kr(g) at 298.15 K. The literature value is $-42.72\ \text{kJ}\cdot\text{mol}^{-1}$.

$$\mu^\circ - E_0 = -RT \ln\left[\left(\frac{q^0}{V}\right)\frac{k_B T}{P^\circ}\right] \tag{9.36}$$

We do this in the same way we found $\mu^\circ - E_0$ for Ar(g) in Section 9–5. First,

$$\begin{aligned}
\frac{q^0(V,T)}{V} &= \left(\frac{2\pi m k_B T}{h^2}\right)^{3/2} \\
&= \left[\frac{(2\pi)(1.391 \times 10^{-25}\ \text{kg}\cdot\text{mol}^{-1})(1.381 \times 10^{-23}\ \text{J}\cdot\text{K}^{-1})(298.15\ \text{K})}{(6.626 \times 10^{-34}\ \text{J}\cdot\text{s})^2}\right]^{3/2} \\
&= 7.422 \times 10^{32}\ \text{m}^{-3} \\
\frac{k_B T}{P^\circ} &= \frac{(1.381 \times 10^{-23}\ \text{J}\cdot\text{K}^{-1})(298.15\ \text{K})}{1.0 \times 10^5\ \text{Pa}} = 4.116 \times 10^{-26}\ \text{m}^{-3}
\end{aligned}$$

Substituting into Equation 9.34 gives

$$\begin{aligned}
\mu^\circ - E_0 &= -RT \ln\left[\left(\frac{q^0}{V}\right)\frac{k_B T}{P^\circ}\right] \\
&= -R(298.15\ \text{K}) \ln[(7.422 \times 10^{32}\ \text{m}^{-3})(4.116 \times 10^{-26}\ \text{m}^{-3})] \\
&= -4.272 \times 10^4\ \text{J}\cdot\text{mol}^{-1} = -42.72\ \text{kJ}\cdot\text{mol}^{-1}
\end{aligned}$$

9–38. Show that Equations 9.30 and 9.32 for $\mu(T, P)$ for a monatomic ideal gas are equivalent to using the relation $\overline{G} = \overline{H} - T\overline{S}$ with $\overline{H} = 5RT/2$ and S given by Equation 6.45.

Recall that μ for a pure substance is $\overline{G}$. Equation 6.45 is

$$\overline{S} = \frac{5}{2}R + R\ln\left[\left(\frac{2\pi mk_BT}{h^2}\right)^{3/2}\frac{\overline{V}}{N_A}\right]$$

Therefore,

$$\begin{aligned}
\overline{G} = \overline{H} - T\overline{S} &= \frac{5RT}{2} - TS \\
&= \frac{5RT}{2} - \frac{5RT}{2} - RT\ln\left[\left(\frac{2\pi mk_BT}{h^2}\right)^{3/2}\frac{k_BT}{P}\right] \\
&= -RT\ln\left[\left(\frac{2\pi mk_BT}{h^2}\right)^{3/2}\frac{k_BT}{P}\right] + RT\ln P \\
&= -RT\ln\left[\left(\frac{q}{V}\right)k_BT\right] + RT\ln P
\end{aligned}$$

This is Equation 9.30. Equation 9.32 appears when we substitute $P^\circ = 1$ bar into this equation.

9–39. Use Equation 9.37 and the molecular parameters in Table 4.2 to calculate $\mu^\circ - E_0$ for $N_2(g)$ at 298.15 K. The literature value is $-48.46\ \text{kJ}\cdot\text{mol}^{-1}$.

$$\begin{aligned}
\frac{q^0}{V} &= \left(\frac{2\pi mk_BT}{h^2}\right)^{3/2}\frac{T}{\sigma\Theta_{\text{rot}}}\frac{1}{1 - e^{-\Theta_{\text{vib}}/T}} \\
&= \left[\frac{2\pi(4.65\times10^{-26}\ \text{kg}\cdot\text{mol}^{-1})k_B(298.15\ \text{K})}{h^2}\right]^{3/2}\frac{298.15\ \text{K}}{2(2.88\ \text{K})}\frac{1}{1 - e^{-3374/298.15}} \\
&= 7.42\times10^{33}\ \text{m}^{-3}
\end{aligned}$$

$$\frac{RT}{N_AP^\circ} = 4.116\times10^{-26}\ \text{m}^3$$

where use the value of RT/N_AP° from Problem 9–36, since the fraction RT/N_AP° is independent of the substance. Now, substituting, we see that

$$\begin{aligned}
\mu^\circ - E_0 &= -RT\ln\left[\left(\frac{q^0}{V}\right)\frac{RT}{N_AP^\circ}\right] \\
&= -R(298.15\ \text{K})\ln[(7.42\times10^{33}\ \text{m}^{-3})(4.116\times10^{-26}\ \text{m}^3)] \\
&= -48.43\ \text{kJ}\cdot\text{mol}^{-1}
\end{aligned}$$

9–40. Use Equation 9.37 and the molecular parameters in Table 4.2 to calculate $\mu^\circ - E_0$ for CO(g) at 298.15 K. The literature value is $-50.26\ \text{kJ}\cdot\text{mol}^{-1}$.

$$\frac{q^0}{V} = \left(\frac{2\pi m k_B T}{h^2}\right)^{3/2} \frac{T}{\sigma \Theta_{rot}} \frac{1}{1 - e^{-\Theta_{vib}/T}}$$

$$= \left[\frac{2\pi(4.65 \times 10^{-26}\ \text{kg}\cdot\text{mol}^{-1}) k_B (298.15\ \text{K})}{h^2}\right]^{3/2} \frac{298.15\ \text{K}}{2(2.77\ \text{K})} \frac{1}{1 - e^{-3103/298.15}}$$

$$= 1.54 \times 10^{34}\ \text{m}^{-3}$$

$$\frac{RT}{N_A P^\circ} = 4.116 \times 10^{-26}\ \text{m}^3$$

Now, substituting, we see that

$$\mu^\circ - E_0 = -RT \ln\left[\left(\frac{q^0}{V}\right) \frac{RT}{N_A P^\circ}\right]$$

$$= -R(298.15\ \text{K}) \ln[(1.54 \times 10^{34}\ \text{m}^{-3})(4.116 \times 10^{-26}\ \text{m}^3)]$$

$$= -50.25\ \text{kJ}\cdot\text{mol}^{-1}$$

9–41. Use Equation 4.60 [without the factor of $\exp(D_e/k_B T)$] and the molecular parameters in Table 4.4 to calculate $\mu^\circ - E_0$ for $CH_4(g)$ at 298.15 K. The literature value is $-45.51\ \text{kJ}\cdot\text{mol}^{-1}$.

$$\frac{q}{V} = \left(\frac{2\pi M k_B T}{h^2}\right)^{3/2} \frac{\pi^{1/2}}{\sigma} \left(\frac{T^3}{\Theta_{rot,A}\Theta_{rot,B}\Theta_{rot,C}}\right)^{1/2} \left[\prod_{j=1}^{9} \frac{e^{-\Theta_{vib,j}/2T}}{(1 - e^{-\Theta_{vib,j}/T})}\right] \tag{4.60}$$

The ground-state energy must be considered for each vibrational state, so, in analogy to the derivation of q^0 in Section 9–5,

$$\frac{q^0}{V} = \left(\frac{2\pi M k_B T}{h^2}\right)^{3/2} \frac{\pi^{1/2}}{\sigma} \left(\frac{T^3}{\Theta_{rot,A}\Theta_{rot,B}\Theta_{rot,C}}\right)^{1/2} \left[\prod_{j=1}^{9} \frac{1}{(1 - e^{-\Theta_{vib,j}/T})}\right]$$

$$= 2.30 \times 10^{33}\ \text{m}^{-3}$$

$$\frac{RT}{N_A P^\circ} = 4.116 \times 10^{-26}\ \text{m}^3$$

$$\mu^\circ - E_0 = -RT \ln\left[\left(\frac{q^0}{V}\right) \frac{RT}{N_A P^\circ}\right]$$

$$= -R(298.15\ \text{K}) \ln[(2.30 \times 10^{33}\ \text{m}^{-3})(4.116 \times 10^{-26}\ \text{m}^3)]$$

$$= -45.53\ \text{kJ}\cdot\text{mol}^{-1}$$

9–42. When we refer to the equilibrium vapor pressure of a liquid, we tacitly assume that some of the liquid has evaporated into a vacuum and that equilibrium is then achieved. Suppose, however, that we are able by some means to exert an additional pressure on the surface of the liquid. One way to do this is to introduce an insoluble, inert gas into the space above the liquid. In this problem, we will investigate how the equilibrium vapor pressure of a liquid depends upon the total pressure exerted on it.

Consider a liquid and a vapor in equilibrium with each other, so that $\mu^l = \mu^g$. Show that

$$\overline{V}^l dP^l = \overline{V}^g dP^g$$

because the two phases are at the same temperature. Assuming that the vapor may be treated as an ideal gas and that $\overline{V}^{\mathrm{l}}$ does not vary appreciably with pressure, show that

$$\ln \frac{P^{\mathrm{g}}(\text{at } P^{\mathrm{l}} = P)}{P^{\mathrm{g}}(\text{at } P^{\mathrm{l}} = 0)} = \frac{\overline{V}^{\mathrm{l}} P^{\mathrm{l}}}{RT}$$

Use this equation to calculate the vapor pressure of water at a total pressure of 10.0 atm at 25°C. Take P^{g} (at $P^{\mathrm{l}} = 0$) $= 0.313$ atm.

We start with the fact that $\mu^{\mathrm{l}} = \mu^{\mathrm{g}}$. Since μ can be written as a function of T and P, and since the temperature does not change, we can write

$$d\mu^{\mathrm{l}} = \left(\frac{\partial \overline{G}}{\partial P}\right) dP^{\mathrm{l}} = \overline{V}^{\mathrm{l}} dP^{\mathrm{l}}$$

Likewise, $d\mu^{\mathrm{g}} = \overline{V}^{\mathrm{g}} dP^{\mathrm{g}}$, and so

$$\overline{V}^{\mathrm{l}} dP^{\mathrm{l}} = \overline{V}^{\mathrm{g}} dP^{\mathrm{g}}$$

follows naturally from the inital assumption. Now we assume that the vapor can be treated as an ideal gas and that $\overline{V}^{\mathrm{l}}$ does not vary with respect to pressure, so

$$\overline{V}^{\mathrm{l}} dP^{\mathrm{l}} = \frac{RT}{P^{\mathrm{g}}} dP^{\mathrm{g}}$$

$$\frac{\overline{V}^{\mathrm{l}}}{RT} dP^{\mathrm{l}} = \frac{dP^{\mathrm{g}}}{P^{\mathrm{g}}}$$

$$\frac{\overline{V}^{\mathrm{l}}}{RT} \int_0^P dP^{\mathrm{l}} = \int_{P^{\mathrm{g}}(P^{\mathrm{l}}=0)}^{P^{\mathrm{g}}(P^{\mathrm{l}}=P)} \frac{dP^{\mathrm{g}}}{P^{\mathrm{g}}}$$

$$\frac{\overline{V}^{\mathrm{l}} P}{RT} = \ln \frac{P^{\mathrm{g}}(P^{\mathrm{l}} = P)}{P^{\mathrm{g}}(P^{\mathrm{l}} = 0)}$$

The specific density of water is 1 g·cm^{-3}, so $\overline{V} = 0.018$ dm^3·mol^{-1}. For water at a total pressure of 10.0 atm at 298.15 K, since the vapor pressure of water expanding into a vacuum is 0.0313 atm at 298.15 K,

$$\ln \frac{P^{\mathrm{g}}(P^{\mathrm{l}} = P)}{0.0313 \text{ atm}} = \frac{\overline{V}^{\mathrm{l}} P^{\mathrm{l}}}{R(298.15 \text{ K})}$$

$$\ln \frac{P^{\mathrm{g}}(P^{\mathrm{l}} = P)}{0.0313 \text{ atm}} = \frac{(0.018 \text{ dm}^3\cdot\text{mol}^{-1})(10.0 \text{ atm})}{(0.082058 \text{ dm}^3\cdot\text{atm}\cdot\text{mol}^{-1}\cdot\text{K}^{-1})(298.15 \text{ K})}$$

$$\frac{P^{\mathrm{g}}}{0.0313 \text{ atm}} = 1.007$$

$$P^{\mathrm{g}} = 0.0315 \text{ atm}$$

The vapor pressure of water at a total pressure 10.0 atm at 298.15 K is 0.0315 atm, or a change of 2×10^{-4} atm.

9–43. Using the fact that the vapor pressure of a liquid does not vary appreciably with the total pressure, show that the final result of the previous problem can be written as

$$\frac{\Delta P^{\mathrm{g}}}{P^{\mathrm{g}}} = \frac{\overline{V}^{\mathrm{l}} P^{\mathrm{l}}}{RT}$$

Hint: Let $P^g(\text{at } P = P^l) = P^g(\text{at } P = 0) + \Delta P$ and use the fact that ΔP is small. Calculate ΔP for water at a total pressure of 10.0 atm at 25°C. Compare your answer with the one you obtained in the previous problem.

$$\ln \frac{P^g(P^l = P)}{P^g(P^l = 0)} = \frac{\overline{V}^l P^l}{RT}$$

$$\ln \frac{P^g(P^l = 0) + \Delta P^g}{P^g(P^l = 0)} = \frac{\overline{V}^l P^l}{RT}$$

Use the relation $\ln(1 + x) \approx x$ when x is small to obtain

$$\ln\left(1 + \frac{\Delta P^g}{P^g}\right) = \frac{\overline{V}^l P^l}{RT}$$

$$\ln \frac{\Delta P^g}{P^g} = \frac{\overline{V}^l P^l}{RT}$$

Again, the vapor pressure of water expanding into a vacuum is 0.0313 atm at 298.15 K and $\overline{V} = 0.018\ \text{dm}^3\cdot\text{mol}^{-1}$, so

$$\frac{\Delta P^g}{P^g} = \frac{\overline{V}^l P^l}{RT}$$

$$\Delta P^g = \frac{(0.0313\ \text{atm})(0.018\ \text{dm}^3\cdot\text{mol}^{-1})(10.0\ \text{atm})}{(0.082058\ \text{dm}^3\cdot\text{atm}\cdot\text{mol}^{-1}\cdot\text{K}^{-1})(298.15\ \text{K})}$$

$$= 2.30 \times 10^{-4}\ \text{atm}$$

This would give a vapor pressure of 0.0315 atm, as in the previous problem.

9–44. In this problem, we will show that the vapor pressure of a droplet is not the same as the vapor pressure of a relatively large body of liquid. Consider a spherical droplet of liquid of radius r in equilibrium with a vapor at a pressure P, and a flat surface of the same liquid in equilibrium with a vapor at a pressure P_0. Show that the change in Gibbs energy for the isothermal transfer of dn moles of the liquid from the flat surface to the droplet is

$$dG = dnRT \ln \frac{P}{P_0}$$

This change in Gibbs energy is due to the change in surface energy of the droplet (the change in surface energy of the large, flat surface is negligible). Show that

$$dnRT \ln \frac{P}{P_0} = \gamma dA$$

where γ is the surface tension of the liquid and dA is the change in the surface area of a droplet. Assuming the droplet is spherical, show that

$$dn = \frac{4\pi r^2 dr}{\overline{V}^l}$$

$$dA = 8\pi r dr$$

and finally that

$$\ln \frac{P}{P_0} = \frac{2\gamma \overline{V}^l}{rRT} \qquad (1)$$

Because the right side is positive, we see that the vapor pressure of a droplet is greater than that of a planar surface. What if $r \to \infty$?

For an isothermal process involving an ideal gas,

$$\Delta G = nRT \ln \frac{P_2}{P_1} \tag{8.58}$$

When a small amount of Gibbs energy goes from the flat surface (of vapor pressure P_0) to the droplet (with vapor pressure P), corresponding to adding dn moles to the droplet from the flat surface, the change in Gibbs energy dG is

$$G(\text{droplet}) - G(\text{surface}) = dnRT \ln \frac{P}{P_1} - dnRT \ln \frac{P_0}{P_1}$$

where P_1 is an arbitrary reference pressure. Therefore, we have

$$dG = dnRT \ln \frac{P}{P_0}$$

This change in surface energy is equal to γdA, so

$$dnRT \ln \frac{P}{P_0} = \gamma dA$$

If a spherical droplet contains n moles, then

$$n\overline{V}^{\mathrm{l}} = \frac{4}{3}\pi r^3$$

$$dn = \frac{4\pi r^2 dr}{\overline{V}^{\mathrm{l}}}$$

$$A = 4\pi r^2$$

$$dA = 8\pi r dr$$

Substituting these expressions back into $dnRT \ln(P/P_0) = \gamma dA$, we find that

$$\frac{4\pi r^2 dr}{\overline{V}^{\mathrm{l}}} RT \ln \frac{P}{P_0} = \gamma 8\pi r dr$$

$$\ln \frac{P}{P_0} = \frac{2\gamma \overline{V}^{\mathrm{l}}}{rRT}$$

If $r \to \infty$, then $\ln(P/P_0) \to 0$: the spherical droplet becomes more and more like the flat surface.

9–45. Use Equation 1 of Problem 9–44 to calculate the vapor pressure at 25°C of droplets of water of radius 1.0×10^{-5} cm. Take the surface tension of water to be $7.20 \times 10^{-4}\ \mathrm{J \cdot m^{-2}}$.

$$\ln \frac{P}{P_0} = \frac{2\gamma \overline{V}^{\mathrm{l}}}{rRT}$$

$$\ln \frac{P}{0.0313\ \mathrm{atm}} = \frac{2(7.20 \times 10^{-4}\ \mathrm{J \cdot m^{-2}})(18.0 \times 10^{-6}\ \mathrm{m^3 \cdot mol^{-1}})}{(1.0 \times 10^{-7}\ \mathrm{m})(8.3145\ \mathrm{J \cdot mol^{-1} \cdot K^{-1}})(298.15\ \mathrm{K})} = 1.046 \times 10^{-4}$$

Solving for P, we find that

$$P = (0.0313\ \mathrm{atm})e^{1.046 \times 10^{-4}} = 0.0313\ \mathrm{atm}$$

The vapor pressure of these droplets of water is not significantly different from the vapor pressure of water in a surface.

9–46. Figure 9.15 shows reduced pressure, P_R, plotted against reduced volume, $\overline{V}_R$, for the van der Waals equation at a reduced temperature, T_R, of 0.85. The so-called van der Waals loop apparent in the figure will occur for any reduced temperature less than unity and is a consequence of the simplified form of the van der Waals equation. It turns out that any analytic equation of state (one that can be written as a Maclaurin expansion in the reduced density, $1/\overline{V}_R$) will give loops for subcritical temperatures ($T_R < 1$). The correct behavior as the pressure is increased is given by the path abdfg in Figure 9.15. The horizontal region bdf, not given by the van der Waals equation, represents the condensation of the gas to a liquid at a fixed pressure. We can draw the horizontal line (called a *tie line*) at the correct position by recognizing that the chemical potentials of the liquid and the vapor must be equal at the points b and f. Using this requirement, Maxwell showed that the horizontal line representing condensation should be drawn such that the areas of the loops above and below the line must be equal. To prove *Maxwell's equal-area construction rule*, integrate $(\partial\mu/\partial P)_T = \overline{V}$ by parts along the path bcdef and use the fact that μ^l (the value of μ at point f) $= \mu^g$ (the value of μ at point b) to obtain

$$\mu^l - \mu^g = P_0(\overline{V}^l - \overline{V}^g) - \int_{\text{bcdef}} P d\overline{V}$$
$$= \int_{\text{bcdef}} (P_0 - P) d\overline{V}$$

where P_0 is the pressure corresponding to the tie line. Interpret this result.

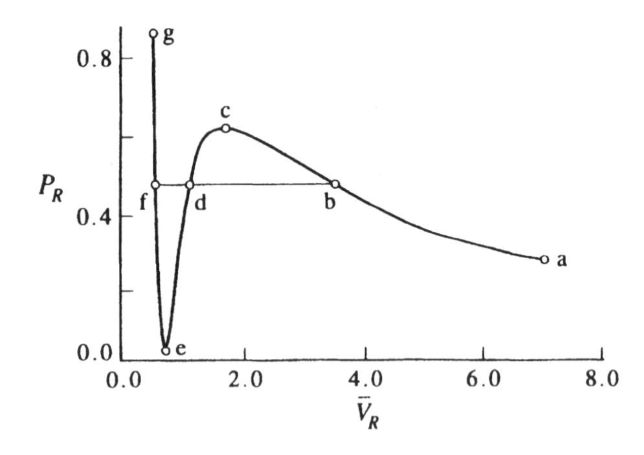

FIGURE 9.15
A plot of reduced pressure, P_R, versus reduced volume, $\overline{V}_R$, for the van der Waals equation at a reduced temperature, T_R, of 0.85.

Start with the equation

$$\int_{\text{bcdef}} d\mu = \int_{\text{bcdef}} \overline{V} dP$$

and integrate by parts to obtain

$$\mu^l - \mu^g = P_0(\overline{V}^l - \overline{V}^g) - \int_{\text{bcdef}} P d\overline{V}$$

Now combine the two terms on the right to get

$$\mu^l - \mu^g = \int_{\text{bcdef}} (P_0 - P) d\overline{V}$$

Now $\mu^{\text{l}} = \mu^{\text{g}}$, so

$$\int_{\text{bcdef}} (P_0 - P)d\overline{V} = 0$$

or the specified area in a plot of P against $\overline{V}$ is equal to zero.

9–47. The isothermal compressibility, κ_T, is defined by

$$\kappa_T = -\frac{1}{V}\left(\frac{\partial V}{\partial P}\right)_T$$

Because $(\partial P/\partial V)_T = 0$ at the critical point, κ_T diverges there. A question that has generated a great deal of experimental and theoretical research is the question of the manner in which κ_T diverges as T approaches T_c. Does it diverge as $\ln(T - T_c)$ or perhaps as $(T - T_c)^{-\gamma}$ where γ is some *critical exponent*? An early theory of the behavior of thermodynamic functions such as κ_T very near the critical point was proposed by van der Waals, who predicted that κ_T diverges as $(T - T_c)^{-1}$. To see how van der Waals arrived at this prediction, we consider the (double) Taylor expansion of the pressure $P(\overline{V}, T)$ about T_c and V_c:

$$P(\overline{V}, T) = P(\overline{V}_c, T_c) + (T - T_c)\left(\frac{\partial P}{\partial T}\right)_c + \frac{1}{2}(T - T_c)^2\left(\frac{\partial^2 P}{\partial T^2}\right)_c$$
$$+ (T - T_c)(\overline{V} - \overline{V}_c)\left(\frac{\partial^2 P}{\partial V \partial T}\right)_c + \frac{1}{6}(\overline{V} - \overline{V}_c)^3\left(\frac{\partial^3 P}{\partial \overline{V}^3}\right)_c + \cdots$$

Why are there no terms in $(\overline{V} - \overline{V}_c)$ or $(\overline{V} - \overline{V}_c)^2$? Write this Taylor series as

$$P = P_c + a(T - T_c) + b(T - T_c)^2 + c(T - T_c)(\overline{V} - \overline{V}_c) + d(\overline{V} - \overline{V}_c)^3 + \cdots$$

Now show that

$$\left(\frac{\partial P}{\partial \overline{V}}\right)_T = c(T - T_c) + 3d(\overline{V} - \overline{V}_c)^2 + \cdots \qquad \begin{pmatrix} T \to T_c \\ \overline{V} \to \overline{V}_c \end{pmatrix}$$

and that

$$\kappa_T = \frac{-1/\overline{V}}{c(T - T_c) + 3d(\overline{V} - \overline{V}_c)^2 + \cdots}$$

Now let $\overline{V} = \overline{V}_c$ to obtain

$$\kappa_T \propto \frac{1}{T - T_c} \qquad T \to (T_c)$$

Accurate experimental measurements of κ_T as $T \to T_c$ suggest that κ_T diverges a little more strongly than $(T - T_c)^{-1}$. In particular, it is found that $\kappa_T \to (T - T_c)^{-\gamma}$ where $\gamma = 1.24$. Thus, the theory of van der Waals, although qualitatively correct, is not quantitatively correct.

The Taylor expansion of $P(\overline{V}, T)$ about T_c and $\overline{V}_c$ is

$$P(\overline{V}, T) = P(\overline{V}_c, T_c) + (T - T_c)\left(\frac{\partial P}{\partial T}\right)_c + (\overline{V} - \overline{V}_c)\left(\frac{\partial P}{\partial \overline{V}}\right)_c$$
$$+\frac{1}{2}(T - T_c)^2\left(\frac{\partial^2 P}{\partial T^2}\right)_c + (T - T_c)(\overline{V} - \overline{V}_c)\left(\frac{\partial^2 P}{\partial \overline{V} \partial T}\right)_c$$
$$+\frac{1}{2}(\overline{V} - \overline{V}_c)^2\left(\frac{\partial^2 P}{\partial \overline{V}^2}\right)_c + \frac{1}{6}(\overline{V} - \overline{V}_c)^3\left(\frac{\partial^3 P}{\partial \overline{V}^3}\right)_c + \cdots$$

However, recall from Section 2–3 that, at the critical point, $(\partial P/\partial \overline{V})_c = (\partial^2 P/\partial \overline{V}^2)_c = 0$. Thus, the Taylor expansion becomes

$$\begin{aligned}P &= P(\overline{V}_c, T_c) + (T - T_c)\left(\frac{\partial P}{\partial T}\right)_c + \frac{1}{2}(T - T_c)^2\left(\frac{\partial^2 P}{\partial T^2}\right)_c \\ &\quad + (T - T_c)(\overline{V} - \overline{V}_c)\left(\frac{\partial^2 P}{\partial \overline{V}\partial T}\right)_c + \frac{1}{6}(\overline{V} - \overline{V}_c)^3\left(\frac{\partial^3 P}{\partial \overline{V}^3}\right)_c + \cdots \\ &= P_c + a(T - T_c) + b(T - T_c)^2 + c(T - T_c)(\overline{V} - \overline{V}_c) \\ &\qquad + d(\overline{V} - \overline{V}_c)^3 + \cdots \\ \left(\frac{\partial P}{\partial \overline{V}}\right)_T &= c(T - T_c) + 3d(\overline{V} - \overline{V}_c)^2 + \cdots\end{aligned}$$

Note that in differentiating, we truncated our expansion and dropped terms of $O[(\overline{V} - \overline{V}_c)^4]$, $O[(T - T_c)^2]$, and third-order terms of $O[(\overline{V} - \overline{V}_c)^x(T - T_c)^y]$. Our partial derivative is thus accurate to $O[(\overline{V} - \overline{V}_c)^4]$ and $O[(T - T_c)^2]$. We can truncate these terms because when $\overline{V} \to \overline{V}_c$ and $T \to T_c$ the higher-order terms become negligible, so we find

$$\begin{aligned}\kappa_T &= -\frac{1}{\overline{V}}\left(\frac{\partial P}{\partial \overline{V}}\right)_T^{-1} \\ &= -\frac{1}{\overline{V}}\left[\frac{1}{c(T - T_c) + 3d(\overline{V} - \overline{V}_c)^2 + \cdots}\right] \qquad T \to T_c, \quad \overline{V} \to \overline{V}_c\end{aligned}$$

Letting $\overline{V} = \overline{V}_c$, we find that

$$\kappa_T = -\frac{1}{\overline{V}_c c(T - T_c)}$$

$$\kappa_T \propto \frac{1}{(T - T_c)} \qquad T \to T_c$$

Again, this expression is only accurate to $O[(T - T_c)^2]$.

9–48. We can use the ideas of the previous problem to predict how the difference in the densities (ρ^l and ρ^g) of the coexisting liquid and vapor states (*orthobaric densities*) behave as $T \to T_c$. Substitute

$$P = P_c + a(T - T_c) + b(T - T_c)^2 + c(T - T_c)(\overline{V} - \overline{V}_c) + d(\overline{V} - \overline{V}_c)^3 + \cdots \tag{1}$$

into the Maxwell equal-area construction (Problem 9–46) to get

$$\begin{aligned}P_0 &= P_c + a(T - T_c) + b(T - T_c)^2 + \frac{c}{2}(T - T_c)(\overline{V}^l + \overline{V}^g - 2\overline{V}_c) \\ &\quad + \frac{d}{4}[(\overline{V}^g - \overline{V}_c)^2 + (\overline{V}^l - \overline{V}_c)^2](\overline{V}^l + \overline{V}^g - 2\overline{V}_c) + \cdots\end{aligned} \tag{2}$$

For $P < P_c$, Equation 1 gives loops and so has three roots, $\overline{V}^l$, $\overline{V}_c$, and $\overline{V}^g$ for $P = P_0$. We can obtain a first approximation to these roots by assuming that $\overline{V}_c \approx \frac{1}{2}(\overline{V}^l + \overline{V}^g)$ in Equation 2 and writing

$$P_0 = P_c + a(T - T_c) + b(T - T_c)^2$$

To this approximation, the three roots to Equation 1 are obtained from

$$d(\overline{V} - \overline{V}_c)^3 + c(T - T_c)(\overline{V} - \overline{V}_c) = 0$$

Show that the three roots are

$$\overline{V}_1 = \overline{V}^{\rm l} = \overline{V}_{\rm c} - \left(\frac{c}{d}\right)^{1/2}(T_{\rm c} - T)^{1/2}$$
$$\overline{V}_2 = \overline{V}_{\rm c}$$
$$\overline{V}_3 = \overline{V}^{\rm g} = \overline{V}_{\rm c} + \left(\frac{c}{d}\right)^{1/2}(T_{\rm c} - T)^{1/2}$$

Now show that

$$\overline{V}^{\rm g} - \overline{V}^{\rm l} = 2\left(\frac{c}{d}\right)^{1/2}(T_{\rm c} - T)^{1/2} \qquad \begin{pmatrix} T < T_{\rm c} \\ T \to T_{\rm c} \end{pmatrix}$$

and that this equation is equivalent to

$$\rho^{\rm l} - \rho^{\rm g} \longrightarrow A(T_{\rm c} - T)^{1/2} \qquad \begin{pmatrix} T < T_{\rm c} \\ T \to T_{\rm c} \end{pmatrix}$$

Thus, the van der Waals theory predicts that the critical exponent in this case is 1/2. It has been shown experimentally that

$$\rho^{\rm l} - \rho^{\rm g} \longrightarrow A(T_{\rm c} - T)^{\beta}$$

where $\beta = 0.324$. Thus, as in the previous problem, although qualitatively correct, the van der Waals theory is not quantitatively correct.

We start with the result of Problem 9–46,

$$\mu^{\rm l} - \mu^{\rm g} = \int_{\rm bcdef} (P_0 - P)d\overline{V}$$

$$0 = \int \left[P_0 - P_{\rm c} - a(T - T_{\rm c}) - b(T - T_{\rm c})^2 - c(T - T_{\rm c})(\overline{V} - \overline{V}_{\rm c}) - d(\overline{V} - \overline{V}_{\rm c})^3 - \cdots\right] d\overline{V}$$

$$= \left[P_0 - P_{\rm c} - a(T - T_{\rm c}) - b(T - T_{\rm c})^2\right](\overline{V}^{\rm l} - \overline{V}^{\rm g}) - \frac{1}{2}c(T - T_{\rm c})\left[(\overline{V}^{\rm l} - \overline{V}_{\rm c})^2 - (\overline{V}^{\rm g} - \overline{V}_{\rm c})^2\right] - \frac{1}{4}d\left[(\overline{V}^{\rm l} - \overline{V}_{\rm c})^4 - (\overline{V}^{\rm g} - \overline{V}_{\rm c})^4\right]$$

$$P_0 = P_{\rm c} + a(T - T_{\rm c}) + b(T - T_{\rm c})^2 + \frac{c}{2}\left[\frac{(\overline{V}^{\rm l})^2 - 2\overline{V}_{\rm c}\overline{V}^{\rm l} + \overline{V}_{\rm c}^2 - (\overline{V}^{\rm g})^2 + 2\overline{V}^{\rm g}\overline{V}_{\rm c} - \overline{V}_{\rm c}^2}{\overline{V}^{\rm l} - \overline{V}^{\rm g}}\right] + \frac{d}{4}\left[\frac{(\overline{V}^{\rm l} - \overline{V}_{\rm c})^2 - (\overline{V}^{\rm g} - \overline{V}_{\rm c})^2}{\overline{V}^{\rm l} - \overline{V}^{\rm g}}\right]\left[(\overline{V}^{\rm l} - \overline{V}_{\rm c})^2 + (\overline{V}^{\rm g} - \overline{V}_{\rm c})^2\right]$$

$$= P_{\rm c} + a(T - T_{\rm c}) + b(T - T_{\rm c})^2 + \frac{c}{2}\left[\frac{(\overline{V}^{\rm l} + \overline{V}^{\rm g})(\overline{V}^{\rm l} - \overline{V}^{\rm g}) - 2\overline{V}_{\rm c}(\overline{V}^{\rm l} - \overline{V}^{\rm g})}{\overline{V}^{\rm l} - \overline{V}^{\rm g}}\right] + \frac{d}{4}\left[\frac{(\overline{V}^{\rm l} + \overline{V}^{\rm g})(\overline{V}^{\rm l} - \overline{V}^{\rm g}) - 2\overline{V}_{\rm c}(\overline{V}^{\rm l} - \overline{V}^{\rm g})}{\overline{V}^{\rm l} - \overline{V}^{\rm g}}\right]\left[(\overline{V}^{\rm l} - \overline{V}_{\rm c})^2 + (\overline{V}^{\rm g} - \overline{V}_{\rm c})^2\right]$$

$$= P_{\rm c} + a(T - T_{\rm c}) + b(T - T_{\rm c})^2 + \frac{c}{2}\left[\overline{V}^{\rm l} + \overline{V}^{\rm g} - 2\overline{V}_{\rm c}\right] + \frac{d}{4}\left[\overline{V}^{\rm l} + \overline{V}^{\rm g} - 2\overline{V}_{\rm c}\right]\left[(\overline{V}^{\rm l} - \overline{V}_{\rm c})^2 + (\overline{V}^{\rm g} - \overline{V}_{\rm c})^2\right]$$

Now assume that $\overline{V}_c \approx \frac{1}{2}(\overline{V}^g + \overline{V}^l)$ to get

$$P_0 = P_c + a(T - T_c) + b(T - T_c)^2$$

To this approximation, the three roots to Equation 1 are given by

$$d(\overline{V} - \overline{V}_c)^3 + c(T - T_c)(V - V_c) = 0$$
$$(\overline{V} - \overline{V}_c)\left[d(\overline{V} - \overline{V}_c)^2 + c(T - T_c)\right] = 0$$

This expression is accurate only to $O(T - T_c)$. We then find the three roots

$$\overline{V} - \overline{V}_c = 0$$
$$\overline{V}_2 = \overline{V}_c$$
$$\overline{V} - \overline{V}_c = \left[-\frac{c(T - T_c)}{d}\right]^{1/2}$$
$$\overline{V}_3 = \overline{V}_c + \left(\frac{c}{d}\right)^{1/2}(T_c - T)^{1/2}$$
$$\overline{V} - \overline{V}_c = -\left[-\frac{c(T - T_c)}{d}\right]^{1/2}$$
$$\overline{V}_1 = \overline{V}_c - \left(\frac{c}{d}\right)^{1/2}(T_c - T)^{1/2}$$

These values are only accurate to $O[(T - T_c)^{1/2}]$. We know that the largest root is the value of $\overline{V}^g$ and that the smallest root is the value of $\overline{V}^l$. Then, to the correct accuracy,

$$\overline{V}^g - \overline{V}^l = \overline{V}_c + \left(\frac{c}{d}\right)^{1/2}(T_c - T)^{1/2} - \left[\overline{V}_c - \left(\frac{c}{d}\right)^{1/2}(T_c - T)^{1/2}\right]$$
$$= 2\left(\frac{c}{d}\right)^{1/2}(T_c - T)^{1/2}$$

where $T \to T_c$, but $T < T_c$ (in order for the quantity $T_c - T$ to be real). Notice that the $(T - T_c)$ term in the product of the molar volumes drops out, since we have been truncating our expressions. The difference in densities is then

$$\rho^l - \rho^g = \frac{1}{\overline{V}^l} - \frac{1}{\overline{V}^g}$$
$$= \frac{\overline{V}^g - \overline{V}^l}{\overline{V}^l\overline{V}^g}$$
$$= 2\left(\frac{c}{d}\right)^{1/2}(T_c - T)^{1/2}\overline{V}_c^{-2}$$
$$= A(T_c - T)^{1/2}$$

9–49. The following data give the temperature, the vapor pressure, and the density of the coexisting vapor phase of butane. Use the van der Waals equation and the Redlich-Kwong equation to calculate the vapor pressure and compare your result with the experimental values given below.

T/K	P/bar	ρ^g/mol·L^{-1}
200	0.0195	0.00117
210	0.0405	0.00233
220	0.0781	0.00430
230	0.1410	0.00746
240	0.2408	0.01225
250	0.3915	0.01924
260	0.6099	0.02905
270	0.9155	0.04239
280	1.330	0.06008

For butane, from Tables 2.3 and 2.4, $a = 13.888$ dm^6·bar·mol^{-2}, $b = 0.11641$ dm^3·mol^{-1}, $A = 290.16$ dm^6·bar·mol^{-2}·K$^{1/2}$, and $B = 0.080683$ dm^3·mol^{-1}. We can substitute the given density and temperature into the van der Waals and Redlich-Kwong equations and thus find the vapor pressure P that is given by each equation.

For the van der Waals approximation,

$$P = \frac{RT}{\overline{V} - b} - \frac{a}{\overline{V}^2} \tag{2.5}$$

and for the Redlich-Kwong approximation

$$P = \frac{RT}{\overline{V} - B} - \frac{A}{T^{1/2}\overline{V}(\overline{V} + B)} \tag{2.8}$$

T/K	P(van der Waals)/bar	P(Redlich-Kwong)/bar
200	0.0194	0.0194
210	0.0406	0.0406
220	0.0784	0.0783
230	0.1420	0.1417
240	0.2427	0.2419
250	0.3957	0.3938
260	0.6184	0.6143
270	0.9314	0.9233
280	1.3584	1.3432

9–50. The following data give the temperature, the vapor pressure, and the density of the coexisting vapor phase of benzene. Use the van der Waals equation and the Redlich-Kwong equation to calculate the vapor pressure and compare your result with the experimental values given below. Use

Equations 2.17 and 2.18 with $T_c = 561.75$ K and $P_c = 48.7575$ bar to calculate the van der Waals parameters and the Redlich-Kwong parameters.

T/K	P/bar	ρ^g/mol·L^{-1}
290.0	0.0860	0.00359
300.0	0.1381	0.00558
310.0	0.2139	0.00839
320.0	0.3205	0.01223
330.0	0.4666	0.01734
340.0	0.6615	0.02399
350.0	0.9161	0.03248

We can do this in the same way as the previous problem after finding a, b, A, and B using Equations 2.17 and 2.18:

$$a = \frac{27(RT_c)^2}{64P_c} = \frac{27[(0.083145\ \text{dm}^3\cdot\text{bar}\cdot\text{mol}^{-1})(561.75)]^2}{64(48.7575\ \text{bar})} = 18.876\ \text{dm}^6\cdot\text{bar}\cdot\text{mol}^{-2}$$

$$b = \frac{RT_c}{8P_c} = \frac{(0.083145\ \text{dm}^3\cdot\text{bar}\cdot\text{mol}^{-1})(561.75)}{8(48.7575\ \text{bar})} = 0.11974\ \text{dm}^3\cdot\text{mol}^{-1}$$

$$A = 0.42748\frac{R^2T_c^{5/2}}{P_c} = 0.42748\frac{(0.083145\ \text{dm}^3\cdot\text{bar}\cdot\text{mol}^{-1}\cdot\text{K}^{-1})^2(561.75\ \text{K})^{5/2}}{48.7575\ \text{bar}}$$

$$= 453.21\ \text{dm}^6\cdot\text{bar}\cdot\text{mol}^{-1}\cdot\text{K}^{1/2}$$

$$B = 0.086640\frac{RT_c}{P_c} = 0.086640\frac{(0.083145\ \text{dm}^3\cdot\text{bar}\cdot\text{mol}^{-1})(561.75)}{48.7575\ \text{bar}} = 0.082996\ \text{dm}^3\cdot\text{mol}^{-1}$$

Now substitute into the appropriate equations to find the vapor pressure P for each temperature and density:

T/K	P(van der Waals)/bar	P(Redlich-Kwong)/bar
290	0.00842	0.00757
300	0.13869	0.13843
310	0.21514	0.21459
320	0.32305	0.32194
330	0.47109	0.46897
340	0.66927	0.66541
350	0.92897	0.92225

9–51. In Problems 8–57 to 8–63, we considered only cases with constant n. Use the approach of Problem 8–57 to repartition n between two initially identical compartments to show that $(\partial^2 A/\partial n^2)_{V,T} > 0$ and that $(\partial \mu/\partial n)_{V,T} > 0$.

The equation corresponding to Equation 1 of Problem 8–57 is

$$A(V, T, n + \Delta n) + A(V, T, n - \Delta n) > 2A(V, T, n) \qquad (1)$$

The Taylor expansions of $A(V, T, n + \Delta n)$ and $A(V, T, n - \Delta n)$ are

$$A(V, T, n + \Delta n) = A(V, T, n) + \left(\frac{\partial A}{\partial n}\right)_{V,T} \Delta n + \frac{1}{2}\left(\frac{\partial^2 A}{\partial n^2}\right)_{V,T} (\Delta n)^2 + \cdots$$

$$A(V, T, n - \Delta n) = A(V, T, n) - \left(\frac{\partial A}{\partial n}\right)_{V,T} \Delta n + \frac{1}{2}\left(\frac{\partial^2 A}{\partial n^2}\right)_{V,T} (\Delta n)^2 + \cdots$$

Substitute these expansions into Equation 1 to obtain

$$\left(\frac{\partial^2 A}{\partial n^2}\right)_{V,T} (\Delta n)^2 > 0$$

or simply

$$\left(\frac{\partial^2 A}{\partial n^2}\right)_{V,T} > 0$$

Using the equation $dA = -SdT - PdV + \mu dn$ (Equation 8.35), we see that $(\partial A/\partial n)_{V,T} = \mu$. Therefore,

$$\left(\frac{\partial^2 A}{\partial n^2}\right)_{V,T} = \left(\frac{\partial \mu}{\partial n}\right)_{V,T} > 0$$

CHAPTER 10

Liquid-Liquid Solutions

PROBLEMS AND SOLUTIONS

10–1. In the text, we went from Equation 10.5 to 10.6 using a physical argument involving varying the size of the system while keeping T and P fixed. We could also have used a mathematical process called *Euler's theorem*. Before we can learn about Euler's theorem, we must first define a *homogeneous function*. A function $f(z_1, z_2, \ldots, z_N)$ is said to be homogeneous if

$$f(\lambda z_1, \lambda z_2, \ldots, \lambda z_N) = \lambda f(z_1, z_2, \ldots, z_N)$$

Argue that extensive thermodynamic quantities are homogeneous functions of their extensive variables.

If we change extensive variables by a factor of λ, then we change an extensive function of these variables by a factor of λ.

10–2. Euler's theorem says that if $f(z_1, z_2, \ldots, z_N)$ is homogeneous, then

$$f(z_1, z_2, \ldots, z_N) = z_1 \frac{\partial f}{\partial \lambda_1} + z_2 \frac{\partial f}{\partial z_2} + \cdots + z_N \frac{\partial f}{\partial \lambda_N}$$

Prove Euler's theorem by differentiating the equation in Problem 10–1 with respect to λ and then setting $\lambda = 1$.

Apply Euler's theorem to $G = G(n_1, n_2, T, P)$ to derive Equation 10.6. (*Hint*: Because T and P are intensive variables, they are simply irrevelant variables in this case.)

Start with

$$\lambda f(z_1, z_2, \ldots, z_N) = f(\lambda z_1, \lambda z_2, \ldots, \lambda z_N)$$

differentiate with respect to λ to obtain

$$\begin{aligned} f(z_1, z_2, \ldots, z_N) &= \frac{\partial f(\lambda z_1, \lambda z_2, \ldots, \lambda z_N)}{\partial \lambda z_1} \frac{\partial \lambda z_1}{\partial \lambda} + \cdots + \frac{\partial f(\lambda z_1, \lambda z_2, \ldots, \lambda z_N)}{\partial \lambda z_N} \frac{\partial \lambda z_N}{\partial \lambda} \\ &= z_1 \frac{\partial f(\lambda z_1, \lambda z_2, \ldots, \lambda z_N)}{\partial \lambda z_1} + \cdots + z_N \frac{\partial f(\lambda z_1, \lambda z_2, \ldots, \lambda z_N)}{\partial \lambda z_N} \end{aligned}$$

Now set $\lambda = 1$

$$f(z_1, z_2, \ldots, z_N) = z_1 \frac{\partial f}{\partial z_1} + \cdots + z_N \frac{\partial f}{\partial z_N}$$

To apply this result to $G = G(n_1, n_2, T, P)$, we let $f = G$, $z_1 = n_1$, and $z_2 = n_2$ to write

$$\begin{aligned} G(n_1, n_2, T, P) &= n_1 \left(\frac{\partial G}{\partial n_1}\right)_{T,P,n_2} + n_2 \left(\frac{\partial G}{\partial n_2}\right)_{T,P,n_1} \\ &= n_1\mu_1 + n_2\mu_2 \end{aligned}$$

10–3. Use Euler's theorem (Problem 10–2) to prove that

$$Y(n_1, n_2, \ldots, T, P) = \sum n_j \overline{Y}_j$$

for any extensive quantity Y.

Simply let $Y = f$ and $n_j = z_j$ in Problem 10–2 to write

$$\begin{aligned} Y(n_1, n_2, \ldots, T, P) &= n_1 \left(\frac{\partial Y}{\partial n_1}\right)_{T,P,n_{k\neq 1}} + n_2 \left(\frac{\partial Y}{\partial n_2}\right)_{T,P,n_{k\neq 2}} + \cdots \\ &= n_1\overline{Y}_1 + n_2\overline{Y}_2 + \cdots \end{aligned}$$

10–4. Apply Euler's theorem to $U = U(S, V, n)$. Do you recognize the resulting equation?

All three variables, S, V and n, are extensive. Using Euler's theorem (Problem 10–2) gives

$$\begin{aligned} U &= S\left(\frac{\partial U}{\partial S}\right)_{V,n} + V\left(\frac{\partial U}{\partial V}\right)_{S,n} + n\left(\frac{\partial U}{\partial n}\right)_{S,V} \\ &= S(T) + V(-P) + n(\mu) \\ &= TS - PV + \mu n \end{aligned}$$

or

$$G = \mu n = U - TS + PV = H - TS = A + PV$$

This is the defining equation for the Gibbs energy.

10–5. Apply Euler's theorem to $A = A(T, V, n)$. Do you recognize the resulting equation?

The extensive variables are V and n. Using Euler's theorem (Problem 10–2) gives

$$\begin{aligned} A &= V\left(\frac{\partial A}{\partial V}\right)_{T,n} + n\left(\frac{\partial A}{\partial n}\right)_{T,V} \\ &= V(-P) + n(\mu) \end{aligned}$$

or

$$G = A + PV$$

This is the defining equation for the Gibbs energy.

10–6. Apply Euler's theorem to $V = V(T, P, n_1, n_2)$ to derive Equation 10.7.

Use the result of Problem 10–3 with $V = Y$.

10–7. The properties of many solutions are given as a function of the mass percent of the components. If we let the mass percent of component-2 be A_2, then derive a relation between A_2 and the mole fractions, x_1 and x_2.

$$A_2 = \frac{m_2}{m_1 + m_2} \times 100 = \frac{M_2 n_2}{M_1 n_1 + M_2 n_2} \times 100 \qquad (1)$$

where the number of moles of component j is $n_j = m_j/M_j$ where M_j is its molar mass. Now divide numerator and denominator of Equation 1 by $n_1 + n_2$ to write

$$A_2 = \frac{M_2 x_2}{M_1 x_1 + M_2 x_2} \times 100$$

10–8. The *CRC Handbook of Chemistry and Physics* gives the densities of many aqueous solutions as a function of the mass percentage of solute. If we denote the density by ρ and the mass percentage of component-2 by A_2, the *Handbook* gives $\rho = \rho(A_2)$ (in $\text{g}\cdot\text{mL}^{-1}$). Show that the quantity $V = (n_1 M_1 + n_2 M_2)/\rho(A_2)$ is the volume of the solution containing n_1 moles of component 1 and n_2 moles of component-2. Now show that

$$\overline{V}_1 = \frac{M_1}{\rho(A_2)}\left[1 + \frac{A_2}{\rho(A_2)}\frac{d\rho(A_2)}{dA_2}\right]$$

and

$$\overline{V}_2 = \frac{M_2}{\rho(A_2)}\left[1 + \frac{(A_2 - 100)}{\rho(A_2)}\frac{d\rho(A_2)}{dA_2}\right]$$

Show that

$$V = n_1 \overline{V}_1 + n_2 \overline{V}_2$$

in agreement with Equation 10.7.

The mass of component j in the solution is $m_j = n_j M_j$, so the total mass is $n_1 M_1 + n_2 M_2$. Therefore, the volume is the mass divided by the density, or $V = (n_1 M_1 + n_2 M_2)/\rho(A_2)$. Now

$$\overline{V}_1 = \left(\frac{\partial V}{\partial n_1}\right)_{n_2} = \frac{M_1}{\rho(A_2)} - \frac{n_1 M_1 + n_2 M_2}{\rho^2(A_2)}\left[\frac{\partial \rho(A_2)}{\partial n_1}\right]_{n_2}$$

But, using Equation 1 of Problem 10–7,

$$\left[\frac{\partial \rho(A_2)}{\partial n_1}\right]_{n_2} = \left[\frac{d\rho(A_2)}{dA_2}\right]\left(\frac{\partial A_2}{\partial n_1}\right)_{n_2} = \left[\frac{d\rho(A_2)}{dA_2}\right]\left[-\frac{M_2 n_2 M_1}{(n_1 M_1 + n_2 M_2)^2} \times 100\right]$$

$$= \left[\frac{d\rho(A_2)}{dA_2}\right]\left(-\frac{A_2 M_1}{n_1 M_1 + n_2 M_2}\right)$$

Substitute this into $\overline{V}_1$ above to get

$$\overline{V}_1 = \frac{M_1}{\rho(A_2)} + \frac{A_2 M_1}{\rho^2(A_2)}\frac{d\rho(A_2)}{dA_2} = \frac{M_1}{\rho(A_2)}\left[1 + \frac{A_2}{\rho(A_2)}\frac{d\rho(A_2)}{dA_2}\right]$$

Similarly

$$\overline{V}_2 = \left(\frac{\partial V}{\partial n_2}\right)_{n_1} = \frac{M_2}{\rho(A_2)} - \frac{n_1 M_1 + n_2 M_2}{\rho^2(A_2)}\left[\frac{\partial \rho(A_2)}{\partial n_2}\right]_{n_1}$$

But

$$\begin{aligned}\left[\frac{\partial \rho(A_2)}{\partial n_2}\right]_{n_1} &= \left[\frac{d\rho(A_2)}{dA_2}\right]\left(\frac{\partial A_2}{\partial n_2}\right)_{n_1} \\ &= \left[\frac{d\rho(A_2)}{dA_2}\right]\left[\frac{100M_2}{n_1M_1 + n_2M_2} - \frac{100n_2M_2^2}{(n_1M_1 + n_2M_2)^2}\right] \\ &= \left[\frac{d\rho(A_2)}{dA_2}\right]\left[\frac{100M_2 n_1 M_1}{(n_1M_1 + n_2M_2)^2}\right]\end{aligned}$$

Substituting this into $\overline{V}_2$ gives

$$\overline{V}_2 = \frac{M_2}{\rho(A_2)}\left[1 - \frac{A_1}{\rho(A_2)}\frac{d\rho(A_2)}{dA_2}\right] = \frac{M_2}{\rho(A_2)}\left[1 + \frac{(A_2 - 100)}{\rho(A_2)}\frac{d\rho(A_2)}{dA_2}\right]$$

because $A_1 + A_2 = 100$. Finally,

$$n_1\overline{V}_1 + n_2\overline{V}_2 = \frac{n_1M_1 + n_2M_2}{\rho(A_2)} + \frac{(n_1M_1A_2 - n_2M_2A_1)}{\rho^2(A_2)}\frac{d\rho(A_2)}{dA_2}$$

But

$$n_1M_1A_2 - n_2M_2A_1 = \frac{n_1M_1M_2n_2 - n_2M_2M_1n_1}{n_1M_1 + n_2M_2} \times 100 = 0$$

so

$$n_1\overline{V}_1 + n_2\overline{V}_2 = \frac{n_1M_1 + n_2M_2}{\rho(A_2)} = V$$

10–9. The density (in $\text{g}\cdot\text{mol}^{-1}$) of a 1-propanol-water solution at 20°C as a function of A_2, the mass percentage of 1-propanol, can be expressed as

$$\rho(A_2) = \sum_{j=0}^{7} \alpha_j A_2^j$$

where

$$\begin{aligned} \alpha_0 &= 0.99823 & \alpha_4 &= 1.5312 \times 10^{-7} \\ \alpha_1 &= -0.0020577 & \alpha_5 &= -2.0365 \times 10^{-9} \\ \alpha_2 &= 1.0021 \times 10^{-4} & \alpha_6 &= 1.3741 \times 10^{-11} \\ \alpha_3 &= -5.9518 \times 10^{-6} & \alpha_7 &= -3.7278 \times 10^{-14} \end{aligned}$$

Use this expression to plot $\overline{V}_{H_2O}$ and $\overline{V}_{1\text{-propanol}}$ versus A_2 and compare your values with those in Figure 10.1.

Substitute $\rho(A_2)$ into $\overline{V}_1 = \overline{V}_{\text{1-propanol}}$ in Problem 10–8 and into $\overline{V}_2 = \overline{V}_{H_2O}$ to obtain

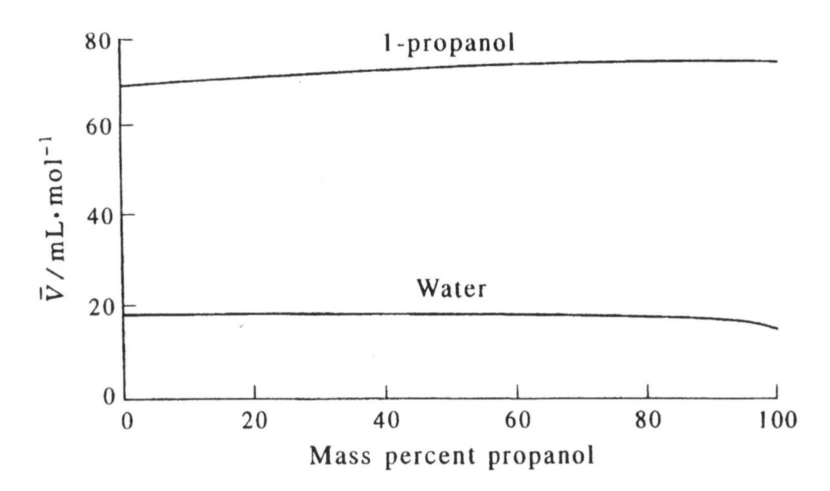

10–10. Given the density of a binary solution as a function of the mole fraction of component 2 $[\rho = \rho(x_2)]$, show that the volume of the solution containing n_1 moles of component 1 and n_2 moles of component 2 is given by $V = (n_1M_1 + n_2M_2)/\rho(x_2)$. Now show that

$$\overline{V}_1 = \frac{M_1}{\rho(x_2)}\left[1 + \left(\frac{x_2(M_2 - M_1) + M_1}{M_1}\right)\frac{x_2}{\rho(x_2)}\frac{d\rho(x_2)}{dx_2}\right]$$

and

$$\overline{V}_2 = \frac{M_2}{\rho(x_2)}\left[1 - \left(\frac{x_2(M_2 - M_1) + M_1}{M_2}\right)\frac{1 - x_2}{\rho(x_2)}\frac{d\rho(x_2)}{dx_2}\right]$$

Show that

$$V = n_1\overline{V}_1 + n_2\overline{V}_2$$

in agreement with Equation 10.7.

The total mass of the solution is $n_1M_1 + n_2M_2$, so its volume (mass/density) is $V = (n_1M_1 + n_2M_2)/\rho(x_2)$.

$$\overline{V}_1 = \left(\frac{\partial V}{\partial n_1}\right)_{n_2} = \frac{M_1}{\rho(x_2)} - \frac{n_1M_1 + n_2M_2}{\rho^2(x_2)}\left[\frac{\partial\rho(x_2)}{\partial n_1}\right]_{n_2}$$

But,

$$\left[\frac{\partial\rho(x_2)}{\partial n_1}\right]_{n_2} = \left[\frac{d\rho(x_2)}{dx_2}\right]\left(\frac{\partial x_2}{\partial n_1}\right)_{n_2} = \left[\frac{d\rho(x_2)}{dx_2}\right]\left[-\frac{n_2}{(n_1 + n_2)^2}\right]$$

$$= -\left[\frac{d\rho(x_2)}{dx_2}\right]\left(\frac{x_2}{n_1 + n_2}\right)$$

Substitute this result into $\overline{V}_1$ to get

$$\overline{V}_1 = \frac{M_1}{\rho(x_2)} + \frac{(n_1M_1 + n_2M_2)x_2}{(n_1+n_2)\rho^2(x_2)}\frac{d\rho(x_2)}{dx_2} = \frac{M_1}{\rho(x_2)} + \frac{(x_1M_1 + x_2M_2)x_2}{\rho^2(x_2)}\frac{d\rho(x_2)}{dx_2}$$

$$= \frac{M_1}{\rho(x_2)}\left[1 + \frac{x_1M_1 + x_2M_2}{M_1}\frac{x_2}{\rho(x_2)}\frac{d\rho(x_2)}{dx_2}\right]$$

$$= \frac{M_1}{\rho(x_2)}\left[1 + \frac{M_1 + x_2(M_2 - M_1)}{M_1}\frac{x_2}{\rho(x_2)}\frac{d\rho(x_2)}{dx_2}\right]$$

Similarly

$$\overline{V}_2 = \left(\frac{\partial V}{\partial n_2}\right)_{n_1} = \frac{M_2}{\rho(x_2)} - \frac{n_1M_1 + n_2M_2}{\rho^2(x_2)}\left[\frac{\partial\rho(x_2)}{\partial n_2}\right]_{n_1}$$

But

$$\left[\frac{\partial\rho(x_2)}{\partial n_2}\right]_{n_1} = \frac{d\rho(x_2)}{dx_2}\left[\frac{\partial x_2}{\partial n_2}\right]_{n_1} = \frac{d\rho(x_2)}{dx_2}\left[\frac{1}{n_1+n_2} - \frac{n_2}{(n_1+n_2)^2}\right]$$

$$= \frac{d\rho(x_2)}{dx_2}\left[\frac{n_1}{(n_1+n_2)^2}\right] = \frac{d\rho(x_2)}{dx_2}\frac{x_1}{n_1+n_2}$$

Substitute this result into $\overline{V}_2$ to get

$$\overline{V}_2 = \frac{M_2}{\rho(A_2)} - \frac{(x_1M_1 + x_2M_2)x_1}{\rho^2(x_2)}\frac{d\rho(x_2)}{dx_2}$$

$$= \frac{M_2}{\rho(A_2)}\left\{1 - \left[\frac{M_1 + x_2(M_2 - M_1)}{M_2}\right]\frac{1-x_2}{\rho(x_2)}\frac{d\rho(x_2)}{dx_2}\right\}$$

Finally,

$$n_1\overline{V}_1 + n_2\overline{V}_2 = \frac{n_1M_1 + n_2M_2}{\rho(x_2)} + \frac{M_1 + x_2(M_2 - M_1)}{\rho^2(A_2)}\frac{d\rho(A_2)}{dA_2}(n_1x_2 - n_2x_1)$$

But

$$n_1x_2 - n_2x_1 = \frac{n_1n_2}{n_1+n_2} - \frac{n_2n_1}{n_1+n_2} = 0$$

so

$$n_1\overline{V}_1 + n_2\overline{V}_2 = \frac{n_1M_1 + n_2M_2}{\rho(x_2)} = V$$

10–11. The density (in $\text{g}\cdot\text{mol}^{-1}$) of a 1-propanol/water solution at 20°C as a function of x_2, the mole fraction of 1-propanol, can be expressed as

$$\rho(x_2) = \sum_{j=0}^{4}\alpha_j x_2^j$$

where

$$\alpha_0 = 0.99823 \qquad \alpha_3 = -0.17163$$

$$\alpha_1 = -0.48503 \qquad \alpha_4 = -0.01387$$

$$\alpha_2 = 0.47518$$

Use this expression to calculate the values of $\overline{V}_{H_2O}$ and $\overline{V}_{1\text{-propanol}}$ as a function of x_2 according to the equation in Problem 10–10.

Substitute $\rho(x_2)$ into $\overline{V}_1 = \overline{V}_{1\text{-propanol}}$ and $\overline{V}_2 = \overline{V}_{H_2O}$ in Problem 10–10 to obtain

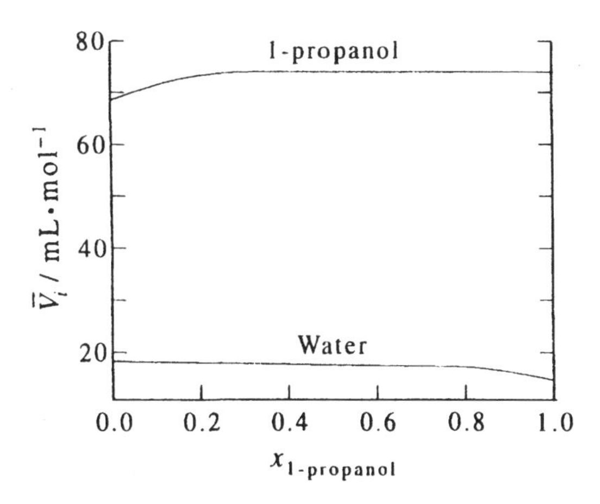

10–12. Use the data in the *CRC Handbook of Chemistry and Physics* to curve fit the density of a water/glycerol solution to a fifth-order polynomial in the mole fraction of glycerol, and then determine the partial molar volumes of water and glycerol as a function of mole fraction. Plot your result.

The curve fit of the density-mole fraction data gives (see the accompanying figure)

$$\rho(x_2) = 0.99849 + 1.1328x_2 - 2.7605x_2^2 + 4.1281x_2^3 - 3.2887x_2^4 + 1.0512x_2^5$$

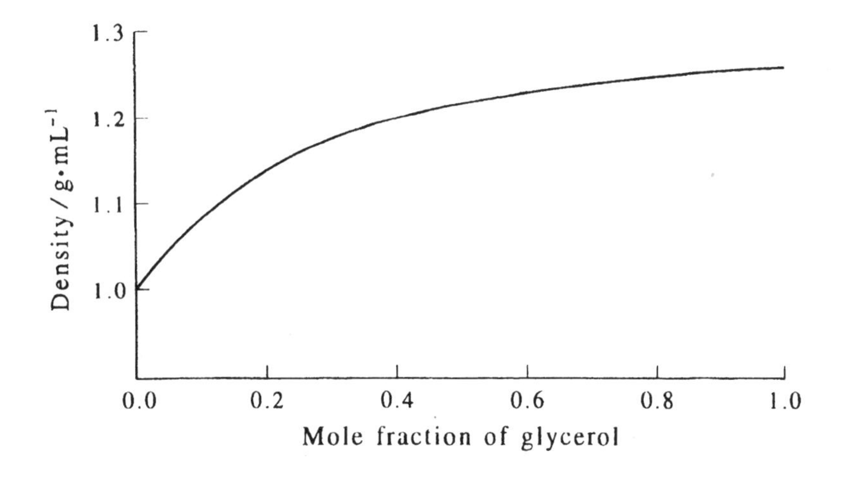

Substitute this result into the equations for $\overline{V}_1$ and $\overline{V}_2$ given in Problem 10–10 with $M_1 = 18.02$ and $M_2 = 92.09$ to get the following result:

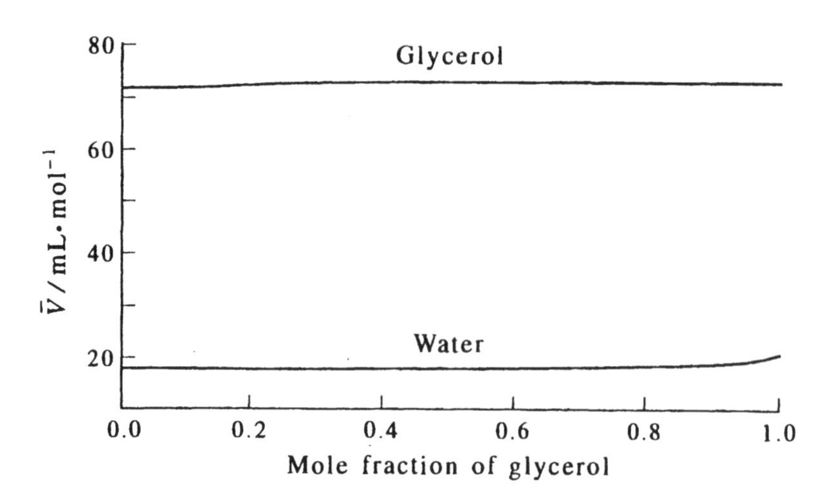

10–13. Just before Example 10–2, we showed that if one component of a binary solution obeys Raoult's law over the entire composition range, the other component does also. Now show that if $\mu_2 = \mu_2' + RT \ln x_2$ for $x_{2,\text{min}} \leq x_2 \leq 1$, then $\mu_1 = \mu_1' + RT \ln x_1$ for $0 < x_1 < 1 - x_{2,\text{min}}$). Notice that for the range over which μ_2 obeys the simple form given, μ_1 obeys a similarly simple form. If we let $x_{2,\text{min}} = 0$, we obtain $\mu_1 = \mu_1^* + RT \ln x_1$ $(0 \leq x_1 \leq 1)$.

Start with the Gibbs-Duhem equation

$$x_1 d\mu_1 + x_2 d\mu_2 = 0$$

Solve for $d\mu_1$

$$d\mu_1 = -\frac{x_2}{x_1} d\mu_1 = -\frac{x_2}{x_1}\frac{RT}{x_1} dx_2 \qquad x_{2,\text{min}} \leq x_2 \leq 1$$

$$= \frac{RT}{x_1} dx_1 \qquad 0 \leq x_1 \leq 1 - x_{2,\text{min}}$$

Integrate to obtain

$$\mu_1 = \mu_1' + RT \ln x_1 \qquad 0 \leq x_1 \leq 1 - x_{2,\text{min}}$$

10–14. Continue the calculations in Example 10–3 to obtain y_2 as a function of x_2 by varying x_2 from 0 to 1. Plot your result.

We use the equation

$$y_2 = \frac{P_2}{P_{\text{total}}} = \frac{x_2 P_2^*}{x_1 P_1^* + x_2 P_2^*} = \frac{x_2(45.2 \text{ torr})}{(1 - x_2)(20.9 \text{ torr}) + x_2(45.2 \text{ torr})}$$

A plot of y_2 against x_2 is

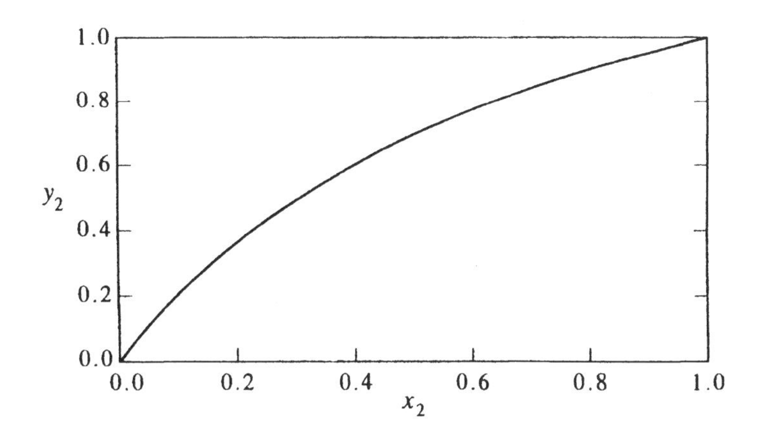

10–15. Use your results from Problem 10–14 to construct the pressure-composition diagram in Figure 10.4.

$$P_{\text{total}} = (1 - x_2)(20.9 \text{ torr}) + x_2(45.2 \text{ torr})$$

Solve the equation given in Problem 10–14 for x_2 in terms of y_2

$$x_2 = \frac{(20.9 \text{ torr})y_2}{45.2 \text{ torr} + (20.9 \text{ torr} - 45.2 \text{ torr})y_2}$$

Let x_2 vary from 0 to 1 in the first equation to calculate P_{total} as a function of x_2. Now let y_2 vary from 0 to 1 to calculate x_2 and then P_{total} to give P_{total} as a function of y_2. A plot of P_{total} against x_2 and y_2 is

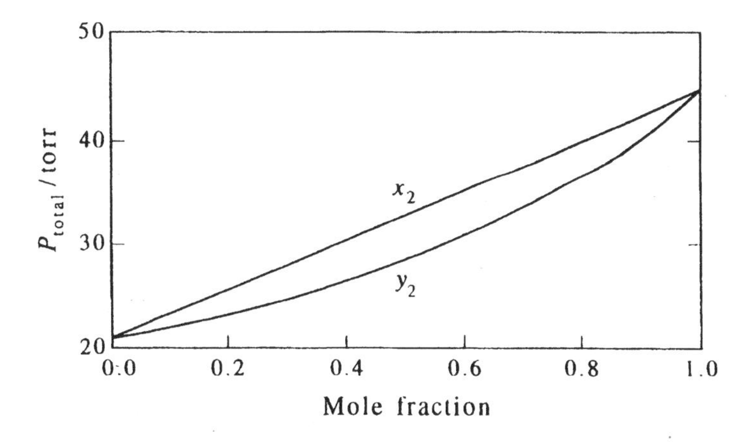

10–16. Calculate the relative amounts of liquid and vapor phases at an overall composition of 0.50 for one of the pair of values, $x_2 = 0.38$ and $y_2 = 0.57$, that you obtained in Problem 10–14.

We use Equation 10.19

$$\frac{n^{\mathrm{l}}}{n^{\mathrm{vap}}} = \frac{y_2 - x_a}{x_a - x_2} = \frac{0.57 - 0.50}{0.50 - 0.38} = 0.58$$

10–17. In this problem, we will derive analytic expressions for the pressure-composition curves in Figure 10.4. The liquid (upper) curve is just

$$P_{\mathrm{total}} = x_1 P_1^* + x_2 P_2^* = (1 - x_2) P_1^* + x_2 P_2^* = P_1^* + x_2 (P_2^* - P_1^*) \qquad (1)$$

which is a straight line, as seen in Figure 10.4. Solve the equation

$$y_2 = \frac{x_2 P_2^*}{P_{\mathrm{total}}} = \frac{x_2 P_2^*}{P_1^* + x_2 (P_2^* - P_1^*)} \qquad (2)$$

for x_2 in terms of y_2 and substitute into Equation (1) to obtain

$$P_{\mathrm{total}} = \frac{P_1^* P_2^*}{P_2^* - y_2 (P_2^* - P_1^*)}$$

Plot this result versus y_2 and show that it gives the vapor (lower) curve in Figure 10.4.

We solve Equation 2 for x_2 to obtain

$$x_2 = \frac{y_2 P_1^*}{P_2^* - y_2 (P_2^* - P_1^*)}$$

Substitute this result into

$$P_{\mathrm{total}} = P_1^* + x_2 (P_2^* - P_1^*)$$

to get

$$P_{\mathrm{total}} = \frac{P_1^* P_2^*}{P_2^* - y_2 (P_2^* - P_1^*)}$$

The plots of P_{total} against x_2 and y_2 for $P_1^* = 20.9$ torr and $P_2^* = 45.2$ torr are

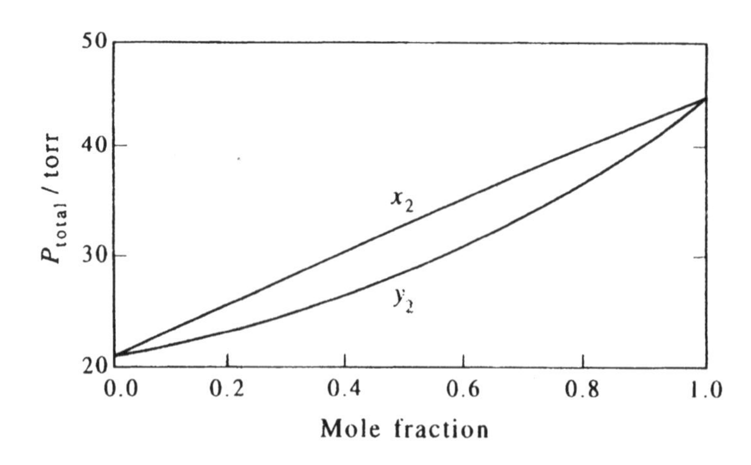

10–18. Prove that $y_2 > x_2$ if $P_2^* > P_1^*$ and that $y_2 < x_2$ if $P_2^* < P_1^*$. Interpret this result physically.

Start with

$$y_2 = \frac{x_2 P_2^*}{P_1^* + x_2(P_2^* - P_1^*)}$$

Divide both sides by x_2 and the numerator and denominator of the right side by P_1^* to obtain

$$\frac{y_2}{x_2} = \frac{P_2^*/P_1^*}{1 + x_2(P_2^*/P_1^* - 1)} = \frac{R}{1 + x_2(R - 1)}$$

where $R = P_2^*/P_1^*$. Now subtract 1 from both sides

$$\frac{y_2}{x_2} - 1 = \frac{R - 1 - x_2(R - 1)}{1 + x_2(R - 1)} = \frac{x_1(R - 1)}{1 + x_2(R - 1)}$$

If $R > 1$ ($P_2^* > P_1^*$), then the right side is always positive because $0 \le x_1 \le 1$ and $0 \le x_2 \le 1$ and so $y_2 > x_2$. If $R < 1$ ($P_2^* < P_1^*$), then the right side is always negative.

This result simply says that the mole fraction of a given component in the vapor phase will be greater than that of the other component if it is more volatile.

10–19. Tetrachloromethane and trichloroethylene form essentially an ideal solution at 40°C at all concentrations. Given that the vapor pressure of tetrachloromethane and trichloroethylene at 40°C are 214 torr and 138 torr, respectively, plot the pressure-composition diagram for this system (see Problem 10–17).

Plot

$$P_{\text{total}} = P_1^* + x_2(P_2^* - P_1^*) = 214 \text{ torr} - x_2(76 \text{ torr})$$

and

$$P_{\text{total}} = \frac{P_1^* P_2^*}{P_2^* - y_2(P_2^* - P_1^*)} = \frac{(214 \text{ torr})(138 \text{ torr})}{138 \text{ torr} + y_2(76 \text{ torr})}$$

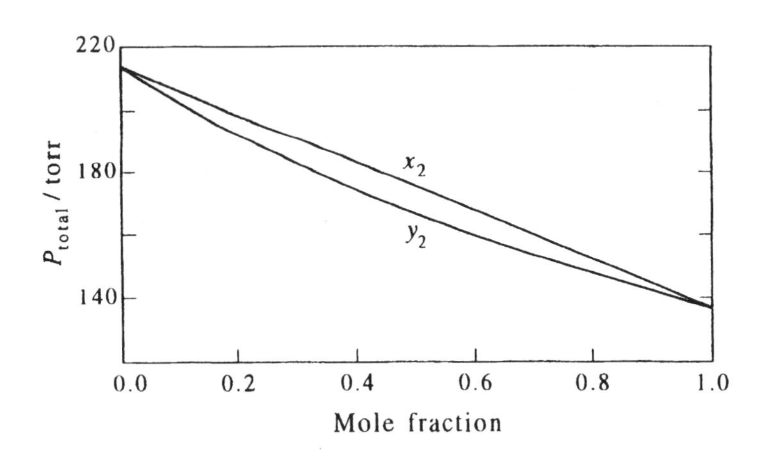

10–20. The vapor pressures of tetrachloromethane (1) and trichloroethylene (2) between 76.8°C and 87.2°C can be expressed empirically by the formulas

$$\ln(P_1^*/\text{torr}) = 15.8401 - \frac{2790.78}{t + 226.4}$$

and

$$\ln(P_2^*/\text{torr}) = 15.0124 - \frac{2345.4}{t + 192.7}$$

where t is the Celsius temperature. Assuming that tetrachloromethane and trichloroethylene form an ideal solution between 76.8°C and 87.2°C at all compositions, calculate the values of x_1 and y_1 at 82.0°C (at an ambient pressure of 760 torr).

Let 1 denote tetrachloromethane and 2 denote trichloroethylene.

$$\ln(P_1^*/\text{torr}) = 15.8401 - \frac{2790.78}{82.0 + 226.4} = 6.7919$$

or $P_1^* = 890$ torr. Similarly, $P_2^* = 648$ torr. Therefore, (see Example 10–5)

$$x_1 = \frac{P_2^* - 760\text{ torr}}{P_2^* - P_1^*} = \frac{648\text{ torr} - 760\text{ torr}}{648\text{ torr} - 890\text{ torr}} = 0.463$$

$$y_1 = \frac{P_1}{760\text{ torr}} = \frac{x_1 P_1^*}{760\text{ torr}} = \frac{(0463)(890\text{ torr})}{760\text{ torr}} = 0.542$$

10–21. Use the data in Problem 10–20 to construct the entire temperature-composition diagram of a tetrachloromethane/trichlororethylene solution.

The vapor pressures of tetrachloromethane (1) and trichloroethylene (2) between 76.8°C and 87.2°C are given by

$$\ln(P_1^*/\text{torr}) = 15.8401 - \frac{2790.84}{t + 226.4}$$

$$\ln(P_2^*/\text{torr}) = 15.0124 - \frac{2345.4}{t + 192.7}$$

where t is the Celsius temperature. The mole fractions of tetrachloromethane (1) in the liquid and vapor phases at temperature t are given by

$$x_1 = \frac{P_2^* - 760\text{ torr}}{P_2^* - P_1^*} \quad \text{and} \quad y_1 = \frac{x_1 P_1^*}{760\text{ torr}}$$

Some data and the plot are given below.

t/°C	P_1^*/torr	P_2^*/torr	x_1	y_1
76.8	761.7	549.8	0.992	0.994
77.6	780.3	564.2	0.906	0.930
78.4	799.3	578.8	0.822	0.864
79.2	818.7	593.7	0.739	0.796
80.0	838.5	608.9	0.658	0.726
80.8	858.6	624.5	0.579	0.654
81.6	879.1	640.3	0.501	0.580
82.4	900.0	656.4	0.425	0.504
83.2	921.3	672.8	0.360	0.435
84.0	942.9	706.6	0.207	0.262
85.6	987.4	724.0	0.137	0.178
86.4	1010	741.7	0.068	0.091
87.2	1033	759.7	0.001	0.001

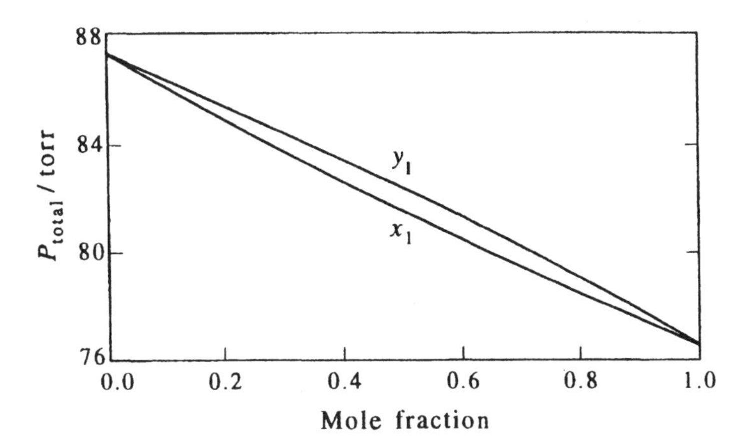

10–22. The vapor pressures of benzene and toluene between 80°C and 110°C as a function of the Kelvin temperature are given by the empirical formulas

$$\ln(P^*_{\text{benz}}/\text{torr}) = -\frac{3856.6\text{ K}}{T} + 17.551$$

and

$$\ln(P^*_{\text{tol}}/\text{torr}) = -\frac{4514.6\text{ K}}{T} + 18.397$$

Assuming that benzene and toluene form an ideal solution, use these formulas to construct a temperature-composition diagram of this system at an ambient pressure of 760 torr.

This problem is very similar to Problems 10–20 and 10–21. Some data and the plot are given below

T/K	t/°C	P_1^*/torr	P_2^*/torr	x_1	y_1
353.0	79.85	754.3	272.5	1.000	1.000
355.0	81.85	802.2	292.9	0.917	0.968
357.0	83.85	852.5	314.5	0.828	0.929
359.0	85.85	905.4	337.5	0.744	0.886
361.0	87.85	960.9	361.8	0.665	0.840
363.0	89.85	1019.2	387.6	0.590	0.791
365.0	91.85	1080.3	415.0	0.519	0.737
367.0	93.85	1144.3	443.9	0.451	0.680
369.0	95.85	1211.4	474.5	0.387	0.618
371.0	97.85	1281.6	506.9	0.327	0.551
373.0	99.85	1355.0	541.1	0.269	0.480
375.0	101.8	1431.9	577.1	0.214	0.403
377.0	103.8	1512.2	615.2	0.161	0.321
379.0	105.8	1596.0	655.3	0.111	0.234
381.0	107.8	1683.6	697.6	0.063	0.140
383.0	109.8	1775.0	742.1	0.017	0.040

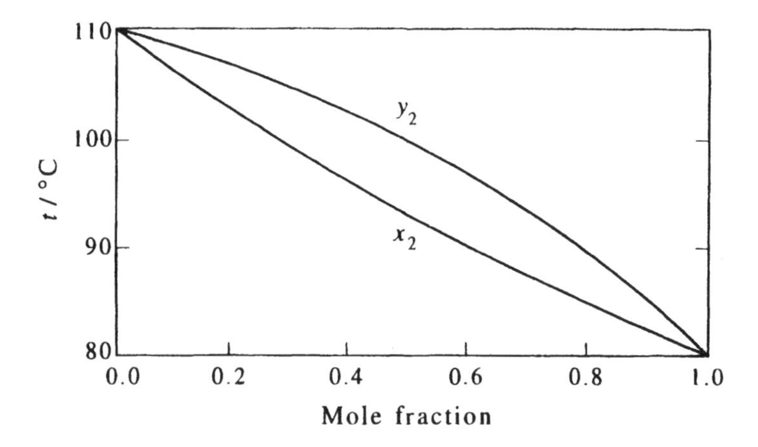

10–23. Construct the temperature-composition diagram for 1-propanol and 2-propanol in Figure 10.5 by varying t from 82.3°C (the boiling point of 2-propanol) to 97.2°C (the boiling point of 1-propanol), calculating (1) P_1^* and P_2^* at each temperature (see Example 10–5), (2) x_1 according to $x_1 = (P_2^* - 760)/(P_2^* - P_1^*)$, and (3) y_1 according to $y_1 = x_1 P_1^*/760$. Now plot t versus x_1 and y_1 on the same graph to obtain the temperature-composition diagram.

This problem is very similar to Problem 10–21. Some data and the plot are

$t/°C$	P_1^*/torr	P_2^*/torr	x_1	y_1
82.3	419.6	760.9	0.003	0.001
84.3	456.0	823.7	0.173	0.104
85.3	475.2	856.7	0.254	0.159
86.3	495.1	890.8	0.331	0.215
87.3	515.6	926.1	0.405	0.274
88.3	536.8	962.4	0.476	0.336
89.3	558.8	1000.0	0.544	0.400
90.3	581.5	1038.7	0.610	0.466
91.3	605.0	1078.7	0.673	0.536
92.3	629.2	1120.0	0.733	0.607
93.3	654.2	1162.5	0.792	0.682
94.3	680.1	1206.4	0.848	0.759
95.3	706.8	1251.7	0.902	0.839
96.3	734.3	1298.3	0.955	0.922
97.2	759.9	1341.5	1.000	1.000

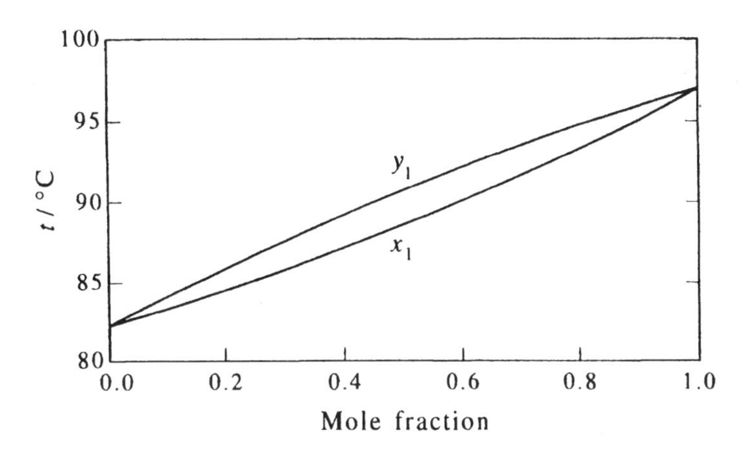

10–24. Prove that $\overline{V}_j = \overline{V}_j^*$ for an ideal solution, where $\overline{V}_j^*$ is the molar volume of pure component j.

The chemical potentials of compounds 1 and 2 of an ideal solution are given by

$$\mu_j = \mu_j^* + RT \ln x_j \qquad j = 1 \text{ and } 2$$

The Gibbs energy is given by

$$G = n_1\mu_1 + n_2\mu_2 = n_1\mu_1^* + n_2\mu_2^* + n_1RT \ln x_1 + n_2RT \ln x_2$$

The volume is given by

$$V = \left(\frac{\partial G}{\partial P}\right)_{T,n_1,n_2} = n_1\left(\frac{\partial \mu_1^*}{\partial P}\right)_T + n_2\left(\frac{\partial \mu_2^*}{\partial P}\right)_T$$
$$= n_1\overline{V}_1^* + n_2\overline{V}_2^*$$

By comparing this result to Equation 10.7, we see that $\overline{V}_j^* = \overline{V}_j$.

10–25. The volume of mixing of miscible liquids is defined as the volume of the solution minus the volume of the individual pure components. Show that

$$\Delta_{\text{mix}}\overline{V} = \sum x_1(\overline{V}_i - \overline{V}_i^*)$$

at constant P and T, where $\overline{V}_i^*$ is the molar volume of pure component i. Show that $\Delta_{\text{mix}}\overline{V} = 0$ for an ideal solution (see Problem 10–24).

Problem 10–24 shows that $\overline{V}_j^* = \overline{V}_j$ for an ideal solution, so $\Delta_{\text{mix}}\overline{V} = 0$.

10–26. Suppose the vapor pressures of the two components of a binary solution are given by

$$P_1 = x_1 P_1^* e^{x_2^2/2}$$

and

$$P_2 = x_2 P_2^* e^{x_1^2/2}$$

Given that $P_1^* = 75.0$ torr and $P_2^* = 160$ torr, calculate the total vapor pressure and the composition of the vapor phase at $x_1 = 0.40$.

$$\begin{aligned}
P_{\text{total}} &= P_1 + P_2 = x_1 P_1^* e^{x_2^2/2} + x_2 P_2^* e^{x_1^2/2} \\
&= (0.40)(75.0\text{ torr})e^{(0.60)^2/2} + (0.60)(160\text{ torr})e^{(0.40)^2/2} \\
&= 35.9\text{ torr} + 104\text{ torr} = 140\text{ torr}
\end{aligned}$$

$$y_1 = \frac{P_1}{P_{\text{total}}} = \frac{35.9\text{ torr}}{140\text{ torr}} = 0.26$$

10–27. Plot y_1 versus x_1 for the system described in the previous problem. Why does the curve lie below the straight line connecting the origin with the point $x_1 = 1$, $y_1 = 1$? Describe a system for which the curve would lie above the diagonal line.

We simply use

$$\begin{aligned}
y_1 = \frac{P_1}{P_{\text{total}}} &= \frac{x_1 P_1^* e^{x_2^2/2}}{x_1 P_1^* e^{x_2^2/2} + x_2 P_2^* e^{x_1^2/2}} \\
&= \frac{x_1(75.0\text{ torr})e^{(1-x_1)^2/2}}{x_1(75.0\text{ torr})e^{(1-x_1)^2/2} + (1-x_1)(160\text{ torr})e^{x_1^2/2}}
\end{aligned}$$

A plot of y_1 against x_1 is the curved line shown below.

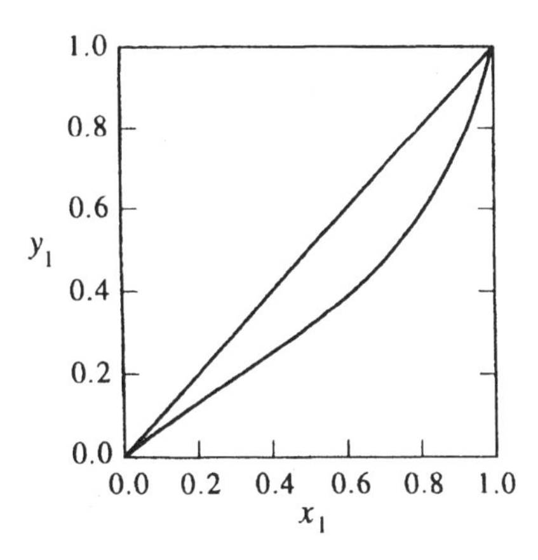

The straight line connecting the origin with the point $x_1 = y_1 = 1$ is given by $y_1 = x_1$. Therefore, component 1 is less volatile than component 2, so the vapor is richer in component 2; $y_1 < x_1$ because $P_1^* < P_2^*$.

10–28. Use the expressions for P_1 and P_2 given in Problem 10–26 to construct a pressure-composition diagram.

Start with

$$P_{\text{total}} = P_1 + P_2 = x_1 P_1^* e^{x_2^2/2} + x_2 P_2^* e^{x_1^2/2}$$
$$= x_1(75.0 \text{ torr})e^{(1-x_1)^2/2} + (1 - x_1)(160 \text{ torr})e^{x_1^2/2}$$

and

$$y_1 = \frac{P_1}{P_{\text{total}}} = \frac{x_1 P_1^* e^{x_2^2/2}}{x_1 P_1^* e^{x_2^2/2} + x_2 P_2^* e^{x_1^2/2}}$$
$$= \frac{x_1(75.0 \text{ torr})e^{(1-x_1)^2/2}}{x_1(75.0 \text{ torr})e^{(1-x_1)^2/2} + (1 - x_1)(160 \text{ torr})e^{x_1^2/2}}$$

Now calculate P_{total} and y_1 as a function of x_1 and then plot P_{total} against x_1 and y_1.

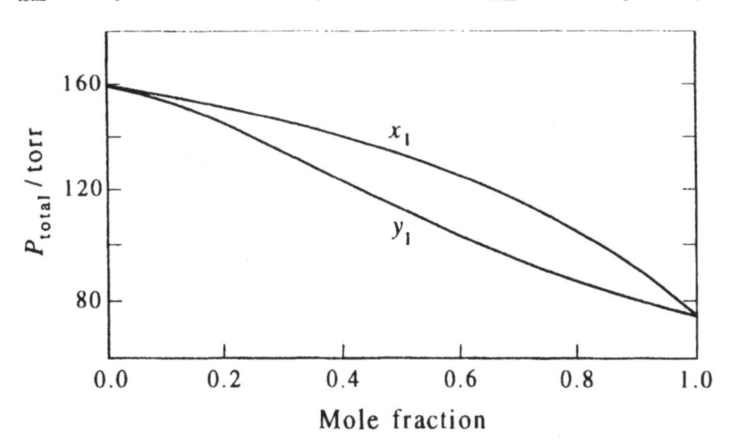

10–29. The vapor pressure (in torr) of the two components in a binary solution are given by

$$P_1 = 120x_1 e^{0.20x_2^2+0.10x_2^3}$$

and

$$P_2 = 140x_2 e^{0.35x_1^2-0.10x_1^3}$$

Determine the values of P_1^*, P_2^*, $k_{H,1}$, and $k_{H,2}$.

Use the fact that $P_j \to x_j P_j^*$ as $x_j \to 1$ to obtain

$$P_1 \longrightarrow 120x_1 \quad \text{as} \quad x_1 \longrightarrow 1 \quad \text{and} \quad P_2 \longrightarrow 140x_2 \quad \text{as} \quad x_2 \longrightarrow 1$$

or $P_1^* = 120$ torr and $P_2^* = 140$ torr. Now use the fact that $P_j \to k_{H,j}x_j$ as $x_j \to 0$ to obtain

$$P_1 \longrightarrow 120x_1 e^{0.30} = 162 \text{ torr} = k_{H,1}$$

and

$$P_2 \longrightarrow 140x_1 e^{0.25} = 180 \text{ torr} = k_{H,2}$$

10–30. Suppose the vapor pressure of the two components of a binary solution are given by

$$P_1 = x_1 P_1^* e^{\alpha x_2^2+\beta x_2^3}$$

and

$$P_2 = x_2 P_2^* e^{(\alpha+3\beta/2)x_1^2-\beta x_1^3}$$

Show that $k_{H,1} = P_1^* e^{\alpha+\beta}$ and $k_{H,2} = P_2^* e^{\alpha+\beta/2}$.

Use the fact that $P_j \to k_{H,j}x_j$ as $x_j \to 0$ to obtain

$$P_1 \longrightarrow x_1 P_1^* e^{\alpha+\beta} \quad \text{and} \quad P_2 \longrightarrow x_2 P_2^* e^{\alpha+\beta/2}$$

or

$$k_{H,1} = P_1^* e^{\alpha+\beta} \quad \text{and} \quad k_{H,2} = P_2^* e^{\alpha+\beta/2}$$

10–31. The empirical expression for the vapor pressure that we used in Examples 10–6 and 10–7, for example,

$$P_1 = x_1 P_1^* e^{\alpha x_2^2+\beta x_2^3+\cdots}$$

is sometimes called the *Margules equation*. Use Equation 10.29 to prove that there can be no linear term in the exponential factor in P_1, for otherwise P_2 will not satisfy Henry's law as $x_2 \to 0$.

Assume that there is a linear term in the exponent of P_1.

$$P_1 = x_1 P_1^* e^{\alpha x_2}$$

Then

$$\frac{\partial \ln P_1}{\partial x_1} = \frac{1}{x_1} - \alpha$$

According to Equation 10.29,

$$\begin{aligned}\frac{\partial \ln P_2}{\partial x_2} &= \frac{x_1}{x_2}\frac{\partial \ln P_1}{\partial x_1}\\ &= \frac{1}{x_2} - \alpha\frac{x_1}{x_2} = \frac{1}{x_2} - \alpha\frac{1-x_2}{x_2}\\ &= \frac{1-\alpha}{x_2} + \alpha\end{aligned}$$

Integration with respect to x_2 gives

$$\ln P_2 = (1-\alpha)\ln x_2 + \alpha x_2 + \ln A$$

where $\ln A$ is an integration constant. Then

$$P_2 = A x_2^{1-\alpha} e^{\alpha x_2}$$

As $x_2 \to 0$, $P_2 \to A x_2^{1-\alpha}$. But according to Henry's law, $P_2 \to k_{H,2}x_2$ as $x_2 \to 0$, so α must equal zero.

10–32. In the text, we showed that the Henry's law behavior of component-2 as $x_2 \to 0$ is a direct consequence of the Raoult's law behavior of component 1 as $x_1 \to 1$. In this problem, we will prove the converse: the Raoult's law behavior of component 1 as $x_1 \to 1$ is a direct consequence of the Henry's law behavior of component-2 as $x_2 \to 0$. Show that the chemical potential of component-2 as $x_2 \to 0$ is

$$\mu_2(T,P) = \mu_2^\circ(T) + RT\ln k_{H,2} + RT\ln x_2 \qquad x_2 \to 0$$

Differentiate μ_2 with respect to x_2 and substitute the result into the Gibbs-Duhem equation to obtain

$$d\mu_1 = RT\frac{dx_1}{x_1} \qquad x_2 \longrightarrow 0$$

Integrate this expression from $x_1 = 1$ to $x_1 \approx 1$ and use the fact that $\mu_1(x_1 = 1) = \mu_1^*$ to obtain

$$\mu_1(T,P) = \mu_1^*(T) + RT\ln x_1 \qquad x_1 \to 1$$

which is the Raoult's law expression for chemical potential. Show that this result follows directly from Equation 10.29.

Start with

$$\mu_2 = \mu_2^* + RT\ln\frac{P_2}{P_2^*} = \mu_2^* - RT\ln P_2^* + RT\ln P_2$$

As $x_2 \to 0$, $P_2 \to k_{H,2}x_2$, so

$$\begin{aligned}\mu_2 &= \mu_2^* - RT\ln P_2^* + RT\ln k_{H,2} + RT\ln x_2 \qquad (x_2 \longrightarrow 0)\\ &= \mu^\circ(T) + RT\ln k_{H,2} + RT\ln x_2 \qquad (x_2 \longrightarrow 0)\end{aligned}$$

Now

$$\frac{d\mu_2}{dx_2} = \frac{RT}{x_2}$$

and so according to the Gibbs-Duhem equation

$$d\mu_1 = -\frac{x_2}{x_1}d\mu_2 = -\frac{x_2}{x_1}\frac{RT}{x_2}dx_2 = RT\frac{dx_1}{x_1} \qquad (x_2 \longrightarrow 0)$$

Now integrate this expression from $x_1 = 1$ to $x_1 \approx 1$ (because the expression is valid only for $x_2 \approx 0$, or $x_1 \approx 1$) to obtain

$$\mu_1 = \mu_1^* + RT \ln x_1 \qquad (x_1 \longrightarrow 1)$$

10–33. In Example 10–7, we saw that if

$$P_1 = x_1 P_1^* e^{\alpha x_2^2 + \beta x_2^3}$$

then

$$P_2 = x_2 P_2^* e^{(\alpha + 3\beta/2)x_1^2 - \beta x_1^3} \qquad (x_2 \longrightarrow 0)$$

Show that this result follows directly from Equation 10.29.

Start with

$$\ln P_1 = \ln x_1 + \ln P_1^* + \alpha(1 - x_1)^2 + \beta(1 - x_1)^3$$

and differentiate with respect to x_1 to obtain

$$\frac{\partial \ln P_1}{\partial x_1} = \frac{1}{x_1} - 2\alpha(1 - x_1) - 3\beta(1 - x_1)^2$$

According to Equation 10.29

$$\frac{\partial \ln P_2}{\partial x_2} = \frac{x_1}{x_2}\frac{\partial \ln P_1}{\partial x_1} = \frac{1}{x_2} - 2\alpha x_1 - 3\beta x_1 x_2$$

$$= \frac{1}{x_2} - 2\alpha + 2\alpha x_2 - 3\beta x_2 + 3\beta x_2^2$$

Now integrate with respect to x_2 to obtain

$$\ln P_2 = \ln x_2 - 2\alpha x_2 + (2\alpha - 3\beta)\frac{x_2^2}{2} + \beta x_2^3 + A$$

where A is an integration constant. Substituting $x_2 = 1 - x_1$ in the last three terms gives

$$\ln P_2 = \ln x_2 - 2\alpha(1 - x_1) + \left(\frac{2\alpha - 3\beta}{2}\right)(1 - x_1)^2 + \beta(1 - x_1)^3 + A$$

$$= \ln x_2 - 2\alpha + 2\alpha x_1 + \alpha - 2\alpha x_1 + \alpha x_1^2 - \frac{3}{2}\beta + 3\beta x_1 - \frac{3}{2}\beta x_1^2$$

$$+ \beta - 3\beta x_1 + 3\beta x_1^2 - \beta x_1^3 + A$$

$$= \ln x_2 + \alpha x_1^2 + \frac{3}{2}\beta x_1^2 - \beta x_1^3 + \left(A - \alpha - \frac{\beta}{2}\right)$$

Rewrite this expression as

$$P_2 = x_2 B e^{\alpha x_1^2 + \frac{3}{2}\beta x_1^2 - \beta x_1^3}$$

where $B = A - \alpha - \beta/2$. Note that $B = P_2^*$ because $P_2 \rightarrow x_2 P_2^*$ as $x_2 \rightarrow 1$.

10–34. Suppose we express the vapor pressures of the components of a binary solution by

$$P_1 = x_1 P_1^* e^{\alpha x_2^2}$$

and

$$P_2 = x_2 P_2^* e^{\beta x_1^2}$$

Use the Gibbs-Duhem equation or Equation 10.29 to prove that α must equal β.

Start with

$$\ln P_1 = \ln x_1 + \ln P_1^* + \alpha x_2^2$$

Differentiate with respect to x_1 to obtain

$$\frac{\partial \ln P_1}{\partial x_1} = \frac{1}{x_1} - 2\alpha(1 - x_1)$$

Use Equation 10.29 to get

$$\frac{\partial \ln P_2}{\partial x_2} = \frac{x_1}{x_2}\frac{\partial \ln P_1}{\partial x_1} = \frac{1}{x_2} - 2\alpha x_1 = \frac{1}{x_2} - 2\alpha + 2\alpha x_2$$

Integrate with respect to x_2 to get

$$\begin{aligned}
\ln P_2 &= \ln x_2 - 2\alpha x_2 + \alpha x_2^2 + A \\
&= \ln x_2 - 2\alpha + 2\alpha x_1 + \alpha - 2\alpha x_1 + \alpha x_1^2 + A \\
&= \ln x_2 + \alpha x_1^2 + (A - \alpha)
\end{aligned}$$

where A is an integration constant. Therefore,

$$P_2 = B x_2 e^{\alpha x_1^2}$$

where $B = A - \alpha$. Clearly $B = P_2^*$ because $P_2 \rightarrow P_2^* x_2$ as $x_2 \rightarrow 1$.

10–35. Use Equation 10.29 to show that if one component of a binary solution obeys Raoult's law for all concentrations, then the other component also obeys Raoult's law for all concentrations.

According to Raoult's law

$$P_1 = x_1 P_1^*$$

Therefore,

$$\frac{\partial \ln P_1}{\partial x_1} = \frac{1}{x_1}$$

and Equation 10.29 gives

$$\frac{\partial \ln P_2}{\partial x_2} = \frac{1}{x_2}$$

which upon integration gives

$$\ln P_2 = \ln x_2 + A$$

$$P_2 = Ax_2 = P_2^* x_2$$

where A is an integration constant.

10–36. Use Equation 10.29 to show that if one component of a binary solution has positive deviations from Raoult's law, then the other component must also.

Equation 10.29 says that

$$x_1 \left(\frac{\partial \ln P_1}{\partial x_1} \right) = x_2 \left(\frac{\partial \ln P_2}{\partial x_2} \right)$$

or

$$\frac{\partial \ln P_1}{\partial \ln x_1} = \frac{\partial \ln P_2}{\partial \ln x_2}$$

For an ideal solution, $P_j^{\text{id}} = x_j P_j^*$ and so

$$\frac{\partial \ln P_1^{\text{id}}}{\partial \ln x_1} = \frac{\partial \ln P_2^{\text{id}}}{\partial \ln x_2} = 1$$

If $P_1 > P_1^{\text{id}} = x_1 P_1^*$ (positive deviation from ideality), then

$$\frac{\partial \ln P_1}{\partial \ln x_1} > 1$$

and so

$$\frac{\partial \ln P_2}{\partial \ln x_2} > 1$$

Conversely, if one component has negative deviations from ideality $(\partial \ln P_1/\partial \ln x_1) < 1$, then the other must also.

10–37. If the vapor pressures of the two components in a binary solution are given by

$$P_1 = x_1 P_1^* e^{ux_2^2/RT} \quad \text{and} \quad P_2 = x_2 P_2^* e^{ux_1^2/RT}$$

show that

$$\Delta_{\text{mix}} \overline{G}/u = \Delta_{\text{mix}} G/(n_1 + n_2)u = \frac{RT}{u}(x_1 \ln x_1 + x_2 \ln x_2) + x_1 x_2$$

$$\Delta_{\text{mix}} \overline{S}/R = \Delta_{\text{mix}} S/(n_1 + n_2)R = -(x_1 \ln x_1 + x_2 \ln x_2)$$

and

$$\Delta_{\text{mix}}\overline{H}/u = \Delta_{\text{mix}}H/(n_1 + n_2)u = x_1x_2$$

A solution that satisfies these equations is called a *regular solution*. A statistical thermodynamic model of binary solutions shows that u is proportional to $2\varepsilon_{12} - \varepsilon_{11} - \varepsilon_{22}$, where ε_{ij} is the interaction energy between molecules of components i and j. Note that $u = 0$ if $\varepsilon_{12} = (\varepsilon_{11} + \varepsilon_{22})/2$, which means that energetically, molecules of components 1 and 2 "like" the opposite molecules as well as their own.

Use the equations

$$G^{\text{sln}} = n_1\mu_1 + n_2\mu_2$$

and

$$\mu_j = \mu_j^* + RT \ln \frac{P_j}{P_j^*}$$

to write

$$G^{\text{sln}} = n_1\mu_1^* + n_2\mu_2^* + n_1RT \ln(x_1e^{ux_2^2/RT}) + n_2RT \ln(x_2e^{ux_1^2/RT})$$

But $n_1\mu_1^* + n_2\mu_2^*$ is the Gibbs energy of the two pure liquid components, so

$$\Delta_{\text{mix}}G = G^{\text{sln}} - n_1\mu_1^* + n_2\mu_2^* = n_1RT \ln x_1 + n_2RT \ln x_2 + u(n_1x_2^2 + n_2x_1^2)$$

Divide by the total number of moles, $n_1 + n_2$, to get

$$\Delta_{\text{mix}}\overline{G} = RT(x_1 \ln x_1 + x_2 \ln x_2) + ux_1x_2(x_2 + x_1)$$

Now divide by u and use the fact that $x_1 + x_2 = 1$ to get

$$\Delta_{\text{mix}}\overline{G}/u = \frac{RT}{u}(x_1 \ln x_1 + x_2 \ln x_2) + x_1x_2$$

Use the equation $\Delta_{\text{mix}}\overline{S} = -(\partial\Delta_{\text{mix}}\overline{G}/\partial T)$ to obtain

$$\Delta_{\text{mix}}\overline{S} = -R(x_1 \ln x_1 + x_2 \ln x_2)$$

Now use $\Delta_{\text{mix}}\overline{G} = \Delta_{\text{mix}}\overline{H} - T\Delta_{\text{mix}}\overline{S}$ to get

$$\Delta_{\text{mix}}\overline{H}/u = x_1x_2$$

10–38. Prove that $\Delta_{\text{mix}}\overline{G}$, $\Delta_{\text{mix}}\overline{S}$, and $\Delta_{\text{mix}}\overline{H}$ in the previous problem are symmetric about the point $x_1 = x_2 = 1/2$.

Each expression is symmetric in x_1 and x_2. Therefore, they must be symmetric about $x_1 = x_2 = 1/2$.

10–39. Plot $P_1/P_1^* = x_1e^{ux_2^2/RT}$ versus x_1 for $RT/u = 0.60, 0.50, 0.45, 0.40$, and 0.35. Note that some of the curves have regions where the slope is negative. The following problem has you show that this behavior occurs when $RT/u < 0.50$. These regions are similar to the loops of the van der Waals equation or the Redlich-Kwong equation when $T < T_c$ (Figure 2.8), and in this case correspond to regions in which the two liquids are not miscible. The critical value $RT/u = 0.50$ corresponds to a solution critical temperature of $0.50u/R$.

See plot

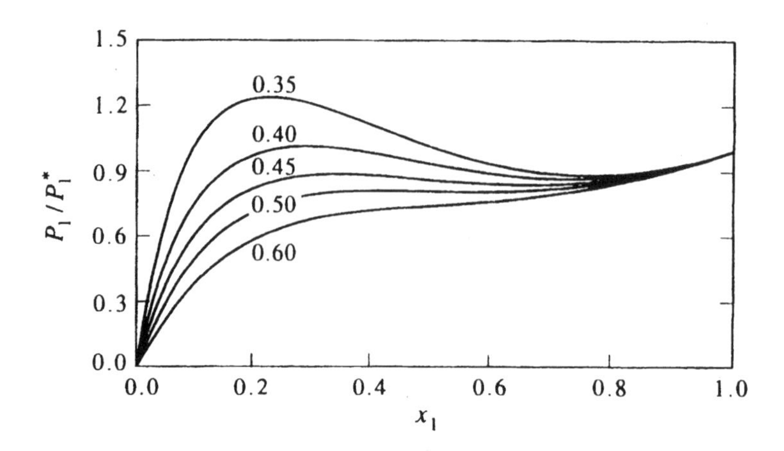

10–40. Differentiate $P_1 = x_1 P_1^* e^{u(1-x_1)^2/RT}$ with respect to x_1 to prove that P_1 has a maximum or a minimum at the points $x_1 = \frac{1}{2} \pm \frac{1}{2}(1 - \frac{2RT}{u})^{1/2}$. Show that $RT/u < 0.50$ for either a maximum or a minimum to occur. Do the positions of these extremes when $RT/u = 0.35$ correspond to the plot you obtained in the previous problem?

$$\frac{dP_1}{dx_1} = 0 = P_1^* e^{u(1-x_1)^2/RT} - \frac{2u}{RT} x_1 (1 - x_1) P_1^* e^{u(1-x_1)^2/RT} = 0$$

Cancelling several factors and rearranging gives

$$x_1^2 - x_1 + \frac{RT}{2u} = 0$$

or

$$x_1 = \frac{1}{2} \pm \frac{1}{2}\left(1 - \frac{2RT}{u}\right)^{1/2}$$

The values of x_1 will not be real unless $2RT/u < 1$, or unless $RT/u < 0.50$. When $RT/u = 0.35$, $x_1 = 0.226$ (a maximum) and 0.774 (a minimum).

10–41. Plot $\Delta_{\text{mix}}\overline{G}/u$ in Problem 10–37 versus x_1 for $RT/u = 0.60$, 0.50, 0.45, 0.40, and 0.35. Note that some of the curves have regions where $\partial^2 \Delta_{\text{mix}}\overline{G}/\partial x_1^2 < 0$. These regions correspond to regions in which the two liquids are not miscible. Show that $RT/u = 0.50$ is a critical value, in the sense that unstable regions occur only when $RT/u < 0.50$. (See the previous problem.)

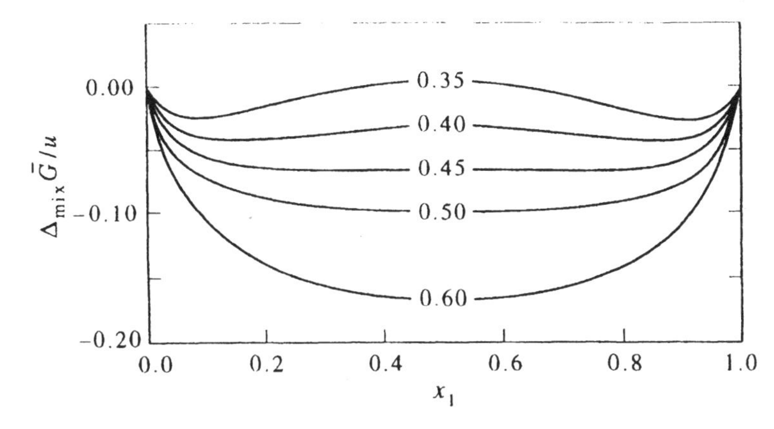

Differentiate $\Delta_{mix}\overline{G}$ with respect to x_1 to obtain

$$\frac{\partial \Delta_{mix}\overline{G}}{\partial x_1} = \frac{RT}{u}[\ln x_1 + 1 - \ln(1 - x_1) - 1] + 1 - 2x_1 = 0$$

or

$$\frac{RT}{u}\ln\frac{x_1}{1 - x_1} = 2x_1 - 1$$

Note that both sides of this equation equal zero when $x_1 = 1/2$. The unstable regions occur when $\partial^2\Delta_{mix}\overline{G}/\partial x_1^2 < 0$:

$$\frac{\partial^2 \Delta_{mix}\overline{G}}{\partial x_1^2} = \frac{RT}{u}\left(\frac{1}{x_1} + \frac{1}{1 - x_1}\right) - 2$$
$$= \frac{RT}{u}\left[\frac{1}{x_1(1 - x_1)}\right] - 2$$

The unstable regions are centered at $x_1 = x_2 = 1/2$, so substituting $x_1 = x_2 = 1/2$ into $\partial^2\Delta_{mix}\overline{G}/\partial x_1^2$ gives the inequality

$$\frac{4RT}{u} - 2 < 0$$

or

$$\frac{RT}{u} < \frac{1}{2}$$

10–42. Plot both $P_1/P_1^* = x_1 e^{ux_2^2/RT}$ and $P_2/P_2^* = x_2 e^{ux_1^2/RT}$ for $RT/u = 0.60, 0.50, 0.45, 0.40$, and 0.35. Prove that the loops occur for values of $RT/u < 0.50$.

See Problem 10–40 for proof that unstable regions occur only for $RT/u < 0.50$.

10–43. Plot both $P_1/P_1^* = x_1 e^{ux_2^2/RT}$ and $P_2/P_2^* = x_2 e^{ux_1^2/RT}$ against x_1 for $RT/u = 0.40$. The loops indicate regions in which the two liquids are not miscible, as explained in Problem 10–39. Draw a horizontal line connecting the left-side and the right-side intersections of the two curves. This line, which connects states in which the vapor pressure (or chemical potential) of each component is the same in the two solutions of different composition, corresponds to one of the horizontal lines in Figure 10.12. Now set $P_1/P_1^* = x_1 e^{ux_2^2/RT}$ equal to $P_2/P_2^* = x_2 e^{ux_1^2/RT}$ and solve for RT/u in terms of x_1. Plot RT/u against x_1 and obtain a coexistence curve like the one in Figure 10.13.

The plot of P_1/P_1^* and P_2/P_2^* against x_1 for $RT/u = 0.40$ is shown below.

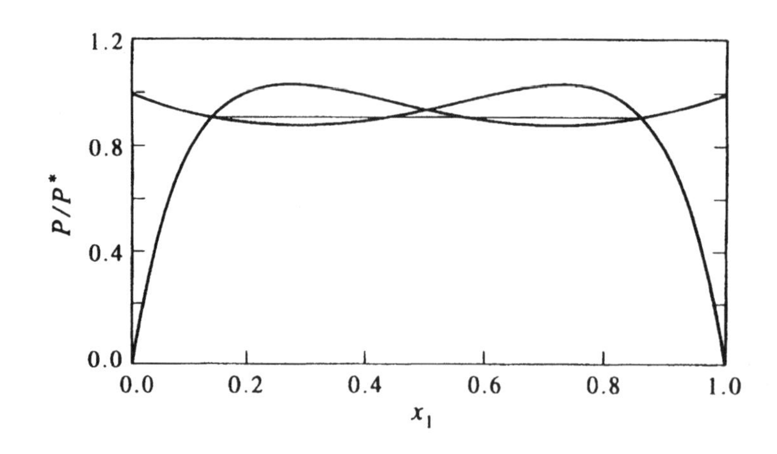

Write

$$x_1 e^{ux_2^2/RT} = x_2 e^{ux_1^2/RT}$$

as

$$e^{u(x_2^2 - x_1^2)/RT} = \frac{x_2}{x_1}$$

and take logarithms to get

$$\frac{RT}{u} = \frac{x_2^2 - x_1^2}{\ln(x_2/x_1)} = \frac{1 - 2x_1}{\ln[(1 - x_1)/x_1]}$$

A plot of RT/u against x_1 follows.

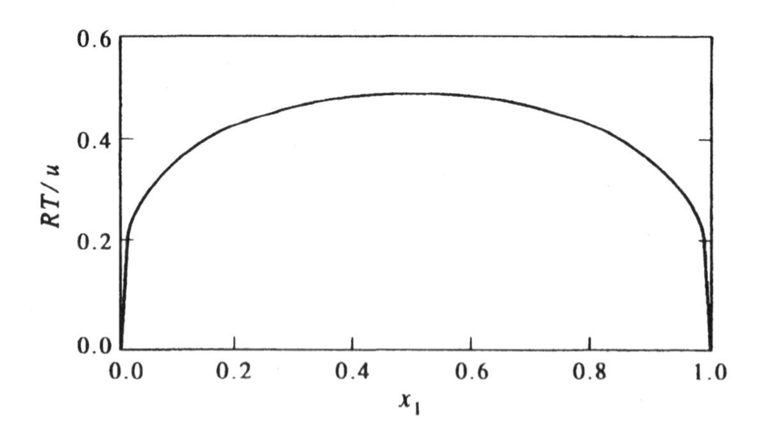

10–44. The molar enthalpies of mixing of solutions of tetrachloromethane (1) and cyclohexane (2) at 25°C are listed below.

x_1	$\Delta_{\text{mix}}\overline{H}/\text{J}\cdot\text{mol}^{-1}$
0.0657	37.8
0.2335	107.9
0.3495	134.9
0.4745	146.7
0.5955	141.6
0.7213	118.6
0.8529	73.6

Plot $\Delta_{\text{mix}}\overline{H}$ against x_1x_2 according to Problem 10–37. Do tetrachloromethane and cyclohexane form a regular solution?

If tetrachloromethane and cyclohexane form a regular solution at 25°C, then a plot of $\Delta_{\text{mix}}\overline{H}/x_2$ against x_1 should be linear. The linearity of the following plot shows that they form a regular solution.

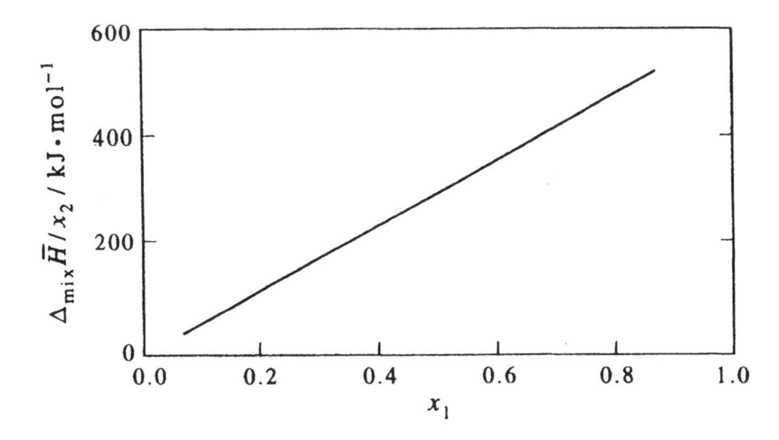

10–45. The molar enthalpies of mixing of solutions of tetrahydrofuran and trichloromethane at 25°C are listed below.

x_{THF}	$\Delta_{\text{mix}}\overline{H}/\text{J}\cdot\text{mol}^{-1}$
0.0568	−0.469
0.1802	−1.374
0.3301	−2.118
0.4508	−2.398
0.5702	−2.383
0.7432	−1.888
0.8231	−1.465
0.9162	−0.802

Do tetrahydrofuran and trichloromethane form a regular solution?

If tetrahydrofuran and trichloromethane form a regular solution at 25°C, then a plot of $\Delta_{\text{mix}}\overline{H}/x_2$ against x_1 should be linear. The nonlinearity of the following plot shows that they do not quite form a regular solution.

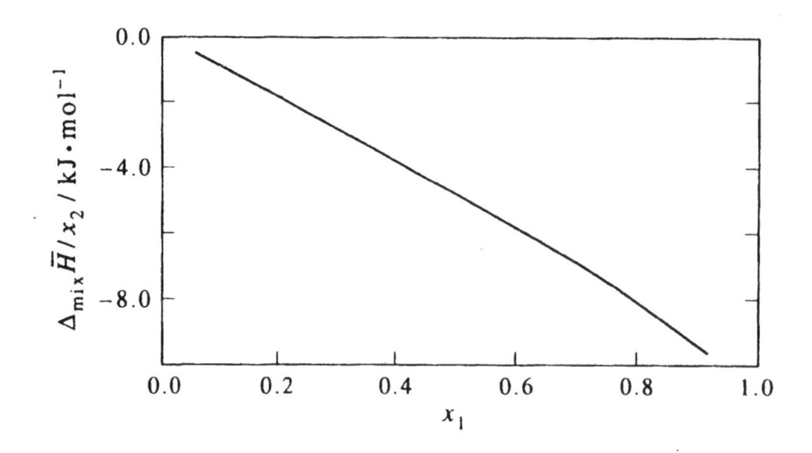

10–46. Derive the equation

$$x_1 d\ln\gamma_1 + x_2 d\ln\gamma_2 = 0$$

by starting with Equation 10.11. Use this equation to obtain the same result as in Example 10–8.

Equation 10.11 is

$$x_1 d\mu_1 + x_2 d\mu_2 = 0$$

Substitute $\mu_j = \mu_j^* + RT\ln\gamma_j x_j$ for μ_1 and μ_2 to obtain

$$x_1 RT\left(\frac{dx_1}{x_1} + d\ln\gamma_1\right) + x_2 RT\left(\frac{dx_2}{x_2} + d\ln\gamma_2\right) = 0$$

or

$$RT(dx_1 + dx_2) + RTx_1 d\ln\gamma_1 + RTx_2 d\ln\gamma_2 = 0$$

But $dx_1 + dx_2 = 0$ because $x_1 + x_2 = 1$, and so we have

$$x_1 d\ln\gamma_1 + x_2 d\ln\gamma_2 = 0$$

According to Example 10–8,

$$\gamma_1 = e^{\alpha x_2^2}$$

Therefore,

$$\begin{aligned} d\ln\gamma_2 &= -\frac{x_1}{x_2}d\ln\gamma_1 = -\frac{x_1}{x_2}(2\alpha x_2 dx_2) \\ &= -2\alpha(1 - x_2)dx_2 \end{aligned}$$

Integration from $x_2 = 1$ (where $\gamma_2 = 1$) to arbitrary x_2 gives

$$\ln \gamma_2 = -2\alpha \int_1^{x_2} (1 - x_2')dx_2'$$

$$= -2\alpha \left(x_2 - 1 - \frac{x_2^2 - 1}{2} \right) = \alpha(1 - 2x_2 + x_2^2)$$

$$= \alpha(1 - x_2)^2 = \alpha x_1^2$$

in agreement with Example 10–8.

10–47. The vapor pressure data for carbon disulfide in Table 10.1 can be curve fit by

$$P_1 = x_1(514.5 \text{ torr})e^{1.4967x_2^2 - 0.68175x_2^3}$$

Using the results of Example 10–7, show that the vapor pressure of dimethoxymethane is given by

$$P_2 = x_2(587.7 \text{ torr})e^{0.4741x_1^2 + 0.68175x_1^3}$$

Now plot P_2 versus x_2 and compare the result with the data in Table 10.1. Plot $\overline{G}^E$ against x_1. Is the plot symmetric about a vertical line at $x_1 = 1/2$? Do carbon disulfide and dimethoxymethane form a regular solution at 35.2°C?

According to Example 10–7, if

$$P_1 = x_2 P_1^* e^{\alpha x_2^2 + \beta x_2^3}$$

then

$$P_2 = x_2 P_2^* e^{(\alpha + 3\beta/2)x_1^2 - \beta x_1^3}$$

Therefore, since $\alpha = 1.4967$ and $\beta = -0.68175$,

$$P_2 = x_2 P_2^* e^{0.4741x_1^2 + 0.68175x_1^3}$$

A comparison of P_2 from Table 10.1 with that calculated from the above equation is shown below. The solid curve is the calculated curve and the dots represent the experimental data. The agreement is very good.

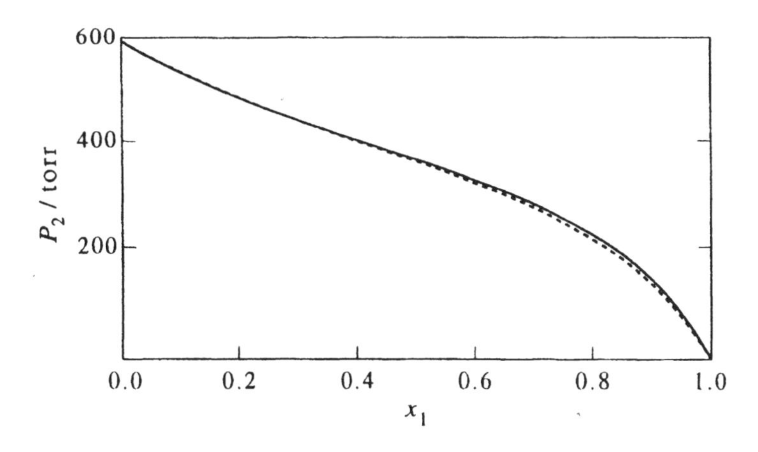

Now use Equation 10.51:

$$\begin{aligned}\overline{G}^{E}/R &= x_1 \ln \gamma_1 + x_2 \ln \gamma_2 \\ &= x_1(1.4967x_2^2 - 0.68175x_2^3 + 0.4741x_1^2 + 0.68175x_1^3) \\ &= 0.8149x_1x_2(1 + 0.4183x_1)\end{aligned}$$

The following plot shows that $\overline{G}^{E}$ is not symmetric about $x_1 = x_2 = 1/2$. Carbon disulfide and dimethoxymethane do not form a regular solution under the given conditions.

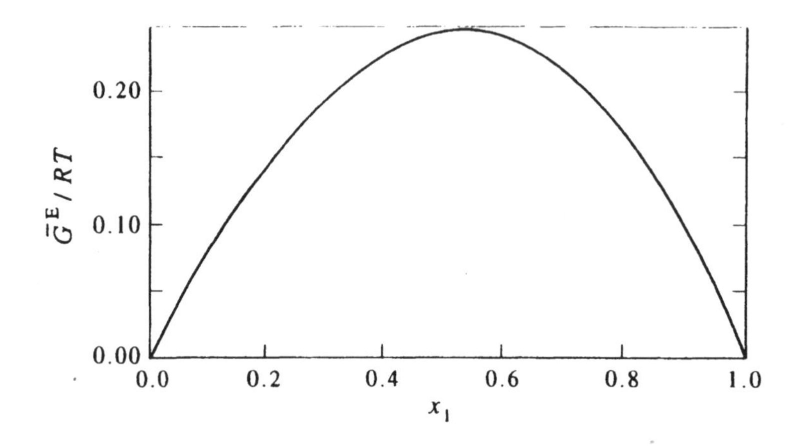

10–48. A mixture of trichloromethane and acetone with $x_{acet} = 0.713$ has a total vapor pressure of 220.5 torr at 28.2°C, and the mole fraction of acetone in the vapor is $y_{acet} = 0.818$. Given that the vapor pressure of pure trichloromethane at 28.2°C is 221.8 torr, calculate the activity and the activity coefficient (based upon a Raoult's law standard state) of trichloromethane in the mixture. Assume the vapor behaves ideally.

We have $x_{acet} = 0.713$, $y_{acet} = 0.818$, and $P_{total} = 220.5$ torr. Therefore,

$$P_{tri} = (1.000 - 0.818)(220.5 \text{ torr}) = 40.13 \text{ torr}$$

and

$$a_{tri}^{(R)} = \frac{P_{tri}}{P_{tri}^*} = \frac{40.13 \text{ torr}}{221.8 \text{ torr}} = 0.181$$

and

$$\gamma_{tri}^{(R)} = \frac{a_{tri}^{(R)}}{x_{tri}} = \frac{0.181}{1.000 - 0.713} = 0.631$$

10–49. Consider a binary solution for which the vapor pressure (in torr) of one of the components (say component 1) is given empirically by

$$P_1 = 78.8x_1e^{0.65x_2^2 + 0.18x_2^3}$$

Calculate the activity and the activity coefficient of component 1 when $x_1 = 0.25$ based on a solvent and a solute standard state.

$$a_1^{(R)} = \frac{P_1}{P_1^*} = x_1 e^{0.65x_2^2+0.18x_2^3}$$

When $x_1 = 0.25$, $a_1^{(R)} = 0.25e^{0.4416} = 0.39$ and $\gamma_1^{(R)} = a_1^{(R)}/0.25 = 1.6$. The activity based upon a Henry's law standard state is given by

$$a_1^{(H)} = \frac{P_1}{k_{H,1}} = \frac{x_1 P_1^* e^{0.65x_2^2+0.18x_2^3}}{P_1^* e^{0.65+0.18}} = \frac{0.39}{2.29} = 0.17$$

and $\gamma_1^{(H)} = 0.17/0.25 = 0.68$

10–50. Some vapor pressure data for ethanol/water solutions at 25°C are listed below.

x_{ethanol}	P_{ethanol}/torr	P_{water}/torr
0.00	0.00	23.78
0.02	4.28	23.31
0.05	9.96	22.67
0.08	14.84	22.07
0.10	17.65	21.70
0.20	27.02	20.25
0.30	31.23	19.34
0.40	33.93	18.50
0.50	36.86	17.29
0.60	40.23	15.53
0.70	43.94	13.16
0.80	48.24	9.89
0.90	53.45	5.38
0.93	55.14	3.83
0.96	56.87	2.23
0.98	58.02	1.13
1.00	59.20	0.00

Plot these data to determine the Henry's law constant for ethanol in water and for water in ethanol at 25°C.

Henry's law constant of component j is given by the limiting slope of the vapor pressure of component j as $x_j \to 0$. The straight lines are shown in the following figure. The slopes of these lines give $k_{H,\text{water}} \approx 20$ torr$/0.35 = 57$ torr and $k_{H,\text{eth}} \approx 25$ torr$/0.10 = 250$ torr

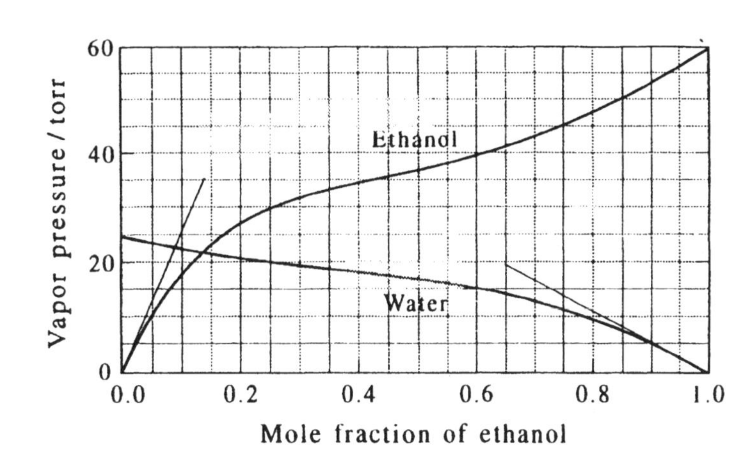

10–51. Using the data in Problem 10–50, plot the activity coefficients (based upon Raoult's law) of both ethanol and water against the mole fraction of ethanol.

The activity coefficients based upon Raoult's law are given by $\gamma_j = P_j/x_j P_j^*$, where all these quantities are given in the problem. The activities are shown in the following figure.

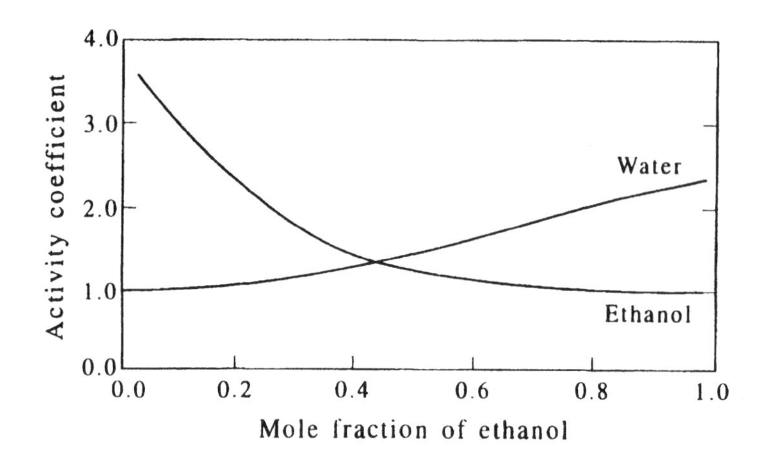

10–52. Using the data in Problem 10–50, plot $\overline{G}^{E}/RT$ against x_{H_2O}. Is a water/ethanol solution at 25°C a regular solution?

According to Equation 10.51,

$$\frac{\overline{G}^{E}}{RT} = x_1 \ln \gamma_1 + x_2 \ln \gamma_2$$

The activity coefficients are calculated in Problem 10–51, and $\overline{G}^{E}/RT$ is plotted against the mole fraction of water below. The plot is not symmetric about $x_1 = x_2 = 1/2$, and so water and ethanol do not form a regular solution under these conditions.

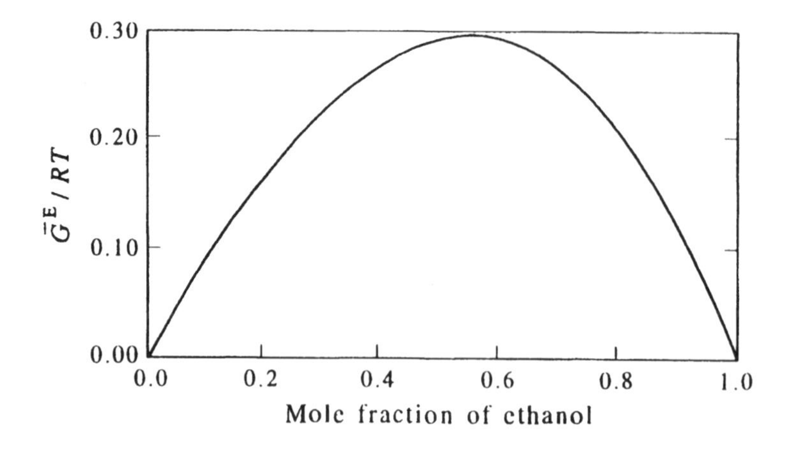

10–53. Some vapor pressure data for a 2-propanol/benzene solution at 25°C are

$x_{\text{2-propanol}}$	$P_{\text{2-propanol}}$/torr	P_{total}/torr
0.000	0.0	94.4
0.059	12.9	104.5
0.146	22.4	109.0
0.362	27.6	108.4
0.521	30.4	105.8
0.700	36.4	99.8
0.836	39.5	84.0
0.924	42.2	66.4
1.000	44.0	44.0

Plot the activities and the activity coefficients of 2-propanol and benzene relative to a Raoult's law standard state versus the mole fraction of 2-propanol.

The activity coefficients based upon Raoult's law are given by $\gamma_j = P_j/x_jP_j^*$, where all these quantities are given above. The activities and activity coefficients are shown in the following figures.

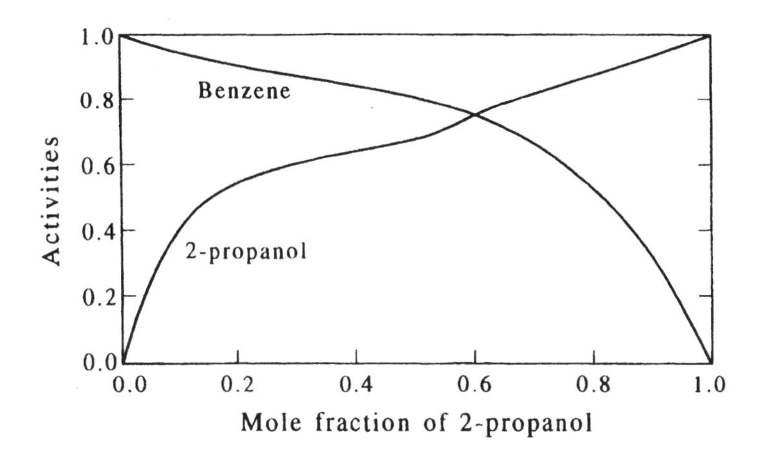

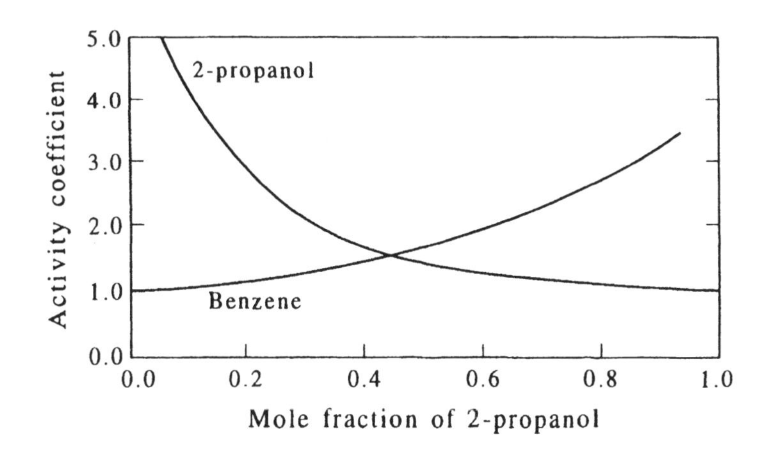

10–54. Using the data in Problem 10–53, plot $\overline{G}^{E}/RT$ versus $x_{\text{2-propanol}}$.

According to Equation 10.51,

$$\frac{\overline{G}^{E}}{RT} = x_1 \ln \gamma_1 + x_2 \ln \gamma_2$$

The activity coefficients are calculated in Problem 10–53, and $\overline{G}^{E}/RT$ is plotted against the mole fraction of 2-propanol below. The plot is not symmetric about $x_1 = x_2 = 1/2$, and so 2-propanol and benzene do not form a regular solution under these conditions.

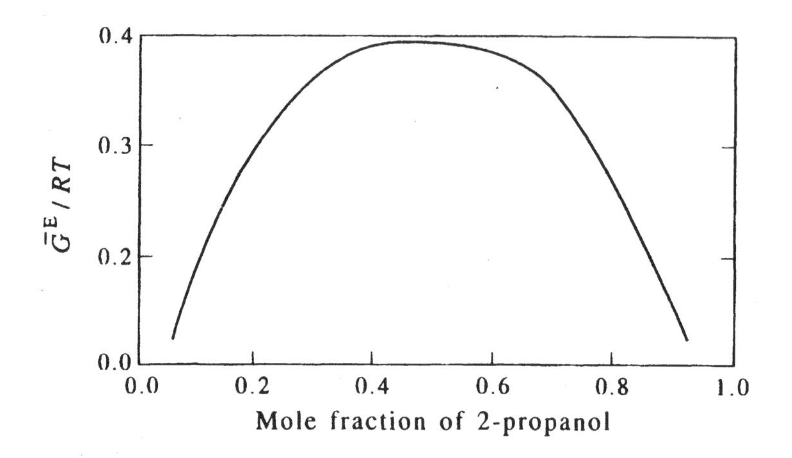

10–55. *Excess thermodynamic quantities* are defined relative to the values the quantities would have if the pure components formed an ideal solution at the same given temperature and pressure. For example, we saw that (Equation 10.51)

$$\frac{\overline{G}^{E}}{RT} = \frac{G^{E}}{(n_1 + n_2)RT} = x_1 \ln \gamma_1 + x_2 \ln \gamma_2$$

Show that

$$\frac{\overline{S}^{E}}{R} = \frac{\Delta S^{E}}{(n_1 + n_2)R} = -(x_1 \ln \gamma_1 + x_2 \ln \gamma_2) - T\left(x_1 \frac{\partial \ln \gamma_1}{\partial T} + x_2 \frac{\partial \ln \gamma_2}{\partial T}\right)$$

Use the relation $S = -(\partial G/\partial T)_{P,n_1,n_2}$ to write

$$\overline{S}^{E} = -\left(\frac{\partial \overline{G}^{E}}{\partial T}\right)_{P,x_1} = -\frac{\partial}{\partial T}[RT(x_1 \ln \gamma_1 + x_2 \ln \gamma_2)]$$

$$= -R(x_1 \ln \gamma_1 + x_2 \ln \gamma_2) - RT\left(x_1 \frac{\partial \ln \gamma_1}{\partial T} + x_2 \frac{\partial \ln \gamma_2}{\partial T}\right)_P$$

10–56. Show that

$$\overline{G}^{\mathrm{E}} = \frac{G^{\mathrm{E}}}{n_1 + n_2} = ux_1x_2$$

$$\overline{S}^{\mathrm{E}} = \frac{S^{\mathrm{E}}}{n_1 + n_2} = 0$$

and

$$\overline{H}^{\mathrm{E}} = \frac{H^{\mathrm{E}}}{n_1 + n_2} = ux_1x_2$$

for a regular solution (see Problem 10–37).

A regular solution is defined in Problem 10–37. For example,

$$\Delta_{\mathrm{mix}}\overline{G} = RT(x_1 \ln x_1 + x_2 \ln x_2) + ux_1x_2$$

But the first terms on the right side of this equation are $\Delta_{\mathrm{mix}}\overline{G}^{\mathrm{id}}$, and so

$$\overline{G}^{\mathrm{E}} = \Delta_{\mathrm{mix}}\overline{G} - \Delta_{\mathrm{mix}}\overline{G}^{\mathrm{id}} = ux_1x_2$$

Similarly,

$$\overline{S}^{\mathrm{E}} = \Delta_{\mathrm{mix}}\overline{S} - \Delta_{\mathrm{mix}}\overline{S}^{\mathrm{id}} = -R(x_1 \ln x_1 + x_2 \ln x_2) + R(x_1 \ln x_1 + x_2 \ln x_2)$$
$$= 0$$

and $\overline{G}^{\mathrm{E}} = \overline{H}^{\mathrm{E}} - T\overline{S}^{\mathrm{E}}$ gives

$$\overline{H}^{\mathrm{E}} = ux_1x_2$$

10–57. Example 10–7 expresses the vapor pressures of the two components of a binary solution as

$$P_1 = x_1 P_1^* e^{\alpha x_2^2 + \beta x_2^3}$$

and

$$P_2 = x_2 P_2^* e^{(\alpha + 3\beta/2)x_1^2 - \beta x_1^3}$$

Show that these expressions are equivalent to

$$\gamma_1 = e^{\alpha x_2^2 + \beta x_2^3} \quad \text{and} \quad \gamma_2 = e^{(\alpha + 3\beta/2)x_1^2 - \beta x_1^3}$$

Using these expressions for the activity coefficients, derive an expression for $\overline{G}^{\mathrm{E}}$ in terms of α and β. Show that your expression reduces to that for $\overline{G}^{\mathrm{E}}$ for a regular solution.

Start with Equation 10.51

$$\overline{G}^{\mathrm{E}}/RT = x_1 \ln \gamma_1 + x_2 \ln \gamma_2$$

and $\gamma_j = P_j/x_jP_j^*$. For P_1 and P_2 given in the problem

$$\gamma_1 = \frac{P_1}{x_1P_1^*} = e^{\alpha x_2^2+\beta x_2^3} \quad \text{and} \quad \gamma_2 = \frac{P_2}{x_1P_2^*} = e^{(\alpha+3\beta/2)x_1^2-\beta x_1^3}$$

and so

$$\begin{aligned}
\overline{G}^{\mathrm{E}}/RT &= x_1(\alpha x_2^2 + \beta x_2^3) + x_2\left[\left(\alpha + \frac{3\beta}{2}\right)x_1^2 - \beta x_1^3\right] \\
&= x_1x_2\left(\alpha x_2 + \beta x_2^2 + \alpha x_1 + \frac{3\beta}{2}x_1 - \beta x_1^2\right) \\
&= x_1x_2\left[\alpha(x_1 + x_2) + \frac{3\beta}{2}x_1 + \beta(x_2^2 - x_1^2)\right] \\
&= x_1x_2\left[\alpha + \frac{3\beta}{2}x_1 + \beta(x_2 - x_1)(x_2 + x_1)\right] \\
&= x_1x_2\left[\alpha + \beta\left(1 - \frac{x_1}{2}\right)\right]
\end{aligned}$$

This expression reduces to that of a regular solution when $\beta = 0$.

10–58. Prove that the maxima or minima of $\Delta_{\text{mix}}\overline{G}$ defined in Problem 10–37 occur at $x_1 = x_2 = 1/2$ for any value of RT/u. Now prove that

$$\frac{\partial^2\Delta_{\text{mix}}\overline{G}}{\partial x_1^2} \begin{cases} > 0 & \text{for } RT/u > 0.50 \\ = 0 & \text{for } RT/u = 0.50 \\ < 0 & \text{for } RT/u < 0.50 \end{cases}$$

at $x_1 = x_2 = 1/2$. Is this result consistent with the graphs you obtained in Problem 10–41?

Start with

$$\frac{\Delta_{\text{mix}}\overline{G}}{u} = \frac{RT}{u}(x_1 \ln x_1 + x_2 \ln x_2) + x_1x_2$$

The maxima or minima are given by

$$\left(\frac{\partial\Delta_{\text{mix}}\overline{G}/u}{\partial x_1}\right) = \frac{RT}{u}[\ln x_1 + 1 - \ln(1 - x_2) - 1] + 1 - 2x_1 = 0$$

or by

$$\frac{RT}{u}\ln\frac{x_1}{1 - x_1} = 1 - 2x_1$$

Note that this equation is satisfied by $x_1 = x_2 = 1/2$ for any value of RT/u.

$$\begin{aligned}
\left(\frac{\partial^2\Delta_{\text{mix}}\overline{G}/u}{\partial x_1^2}\right) &= \frac{RT}{u}\left(\frac{1}{x_1} + \frac{1}{1 - x_1}\right) - 2 \\
&= \frac{RT}{u}\left[\frac{1}{x_1(1 - x_1)}\right] - 2 \\
&= \frac{4RT}{u} - 2
\end{aligned}$$

at $x_1 = x_2 = 1/2$. This expression is greater than zero when $RT/u > 0.50$, less than zero when $RT/u < 0.50$, and equal to zero when $RT/w = 0.50$.

10–59. Use the data in Table 10.1 to plot Figures 10.15 through 10.17 and 10.20.

Use the relations $a_j^{(R)} = P_j/P_j^*$ and $\gamma_j^{(R)} = P_j/x_jP_j^*$. The results of the calculations are given below.

x_1	P_{CS_2}/torr	P_{dimeth}/torr	$a_{CS_2}^{(R)}$	$a_{dimeth}^{(R)}$	$\gamma_{CS_2}^{(R)}$	$\gamma_{dimeth}^{(R)}$	ΔG^E/kJ·mol^{-1}
0.0000	0.0	587.7	0.000	1.000	2.22	1.00	0.000
0.0489	54.5	558.3	0.106	0.950	2.17	1.00	0.037
0.1030	109.3	529.1	0.212	0.900	2.06	1.00	0.078
0.1640	159.5	500.4	0.310	0.851	1.89	1.02	0.120
0.2710	234.8	451.2	0.456	0.768	1.68	1.05	0.179
0.3470	277.6	412.7	0.540	0.706	1.55	1.08	0.204
0.4536	324.8	378.0	0.631	0.643	1.39	1.18	0.239
0.4946	340.2	360.8	0.661	0.614	1.34	1.21	0.242
0.5393	357.2	342.2	0.694	0.582	1.29	1.26	0.244
0.6071	381.9	313.3	0.742	0.533	1.22	1.36	0.242
0.6827	407.0	277.8	0.791	0.473	1.16	1.49	0.227
0.7377	424.3	250.1	0.825	0.426	1.12	1.62	0.209
0.7950	442.6	217.4	0.860	0.370	1.08	1.80	0.184
0.8445	458.1	184.9	0.890	0.315	1.05	2.02	0.154
0.9108	481.8	124.2	0.936	0.211	1.03	2.37	0.102
0.9554	501.0	65.1	0.974	0.111	1.02	2.48	0.059
1.0000	514.5	0.0	1.000	0.000	1.00	2.50	0.000

10–60. Use Equation 10.62 to show that the slopes of all the curves in Figure 10.20 are equal to zero when $x_1 = x_2 = 1/2$.

Differentiate Equation 10.62 (with $x_2 = 1 - x_1$) with respect to x_1

$$\frac{\partial \Delta_{\text{mix}}\overline{G}/u}{\partial x_1} = \frac{RT}{u}[\ln x_1 + 1 - \ln(1 - x_1) - 1] + 1 - 2x_1$$

$$= \frac{RT}{u}\ln\frac{x_1}{1 - x_1} + 1 - 2x_1$$

This quantity is equal to zero when $x_1 = x_2 = 1/2$.

10–61. Derive Equations 10.59 through 10.61.

Start with Equation 10.57

$$\Delta_{\text{mix}}\overline{G} = RT(x_1 \ln x_1 + x_2 \ln x_2) + ux_1x_2$$

Use the relation $\Delta_{\text{mix}}\overline{S} = -(\partial \Delta_{\text{mix}}\overline{G}/\partial T)_P$ to obtain

$$\Delta_{\text{mix}}\overline{S} = -R(x_1 \ln x_1 + x_2 \ln x_2)$$

But this expression for $\Delta_{\text{mix}}\overline{S}$ is also equal to $\Delta_{\text{mix}}\overline{S}^{\text{id}}$, so

$$\overline{S}^{\text{E}} = \Delta_{\text{mix}}\overline{S} - \Delta_{\text{mix}}\overline{S}^{\text{id}} = 0$$

Now use the relation $G = H - TS$ (along with Equation 10.58) and get

$$\Delta_{\text{mix}}\overline{H} = \Delta_{\text{mix}}\overline{G} + T\Delta_{\text{mix}}\overline{S} = ux_1x_2$$

and

$$\overline{H}^{\text{E}} = \overline{G}^{\text{E}} + T\overline{S}^{\text{E}} = ux_1x_2$$

10–62. In this problem, we will prove that $(\partial^2\Delta_{\text{mix}}\overline{G}/\partial x_1^2)$ is greater than zero in a stable region of a binary solution. First choose some point x_1^0 in Figure 10.20 and draw a straight line at x_1^0 tangent to the curve of $\Delta_{\text{mix}}\overline{G}$ against x_1. Now, the system will be stable if $\Delta_{\text{mix}}\overline{G}$ is a minimum. Because $\Delta_{\text{mix}}\overline{G}$ is a minimum, $\Delta_{\text{mix}}\overline{G}$ around the point $x_1 = x_1^0$ will necessarily lie above the tangent line at $x_1 = x_1^0$. Therefore, we can write

$$\Delta_{\text{mix}}\overline{G} \text{ at } x_1 \text{ around } x_1^0 > \text{tangent line at } x_1^0$$

Show that the equation for the tangent line is

$$y = (x_1 - x_1^0)\left(\frac{\partial\Delta_{\text{mix}}\overline{G}}{\partial x_1}\right)_{x_1=x_1^0} + \Delta_{\text{mix}}\overline{G}(x_1^0)$$

Now expand $\Delta_{\text{mix}}\overline{G}$ in a Taylor series (MathChapter C) about the point x_1^0 and show that

$$\left(\frac{\partial^2\Delta_{\text{mix}}\overline{G}}{\partial x_1^2}\right)_{x_1=x_1^0} > 0$$

for the binary solution to be stable.

The slope of the tangent line at $x_1 = x_1^0$ is given by

$$\text{slope} = \left(\frac{\partial\Delta_{\text{mix}}\overline{G}}{\partial x_1}\right)_{x_1=x_1^0}$$

and the following figure shows that its intercept is

$$\text{intercept} = \Delta_{\text{mix}}\overline{G}(x_1^0) - x_1^0\left(\frac{\partial\Delta_{\text{mix}}\overline{G}}{\partial x_1}\right)_{x_1=x_1^0}$$

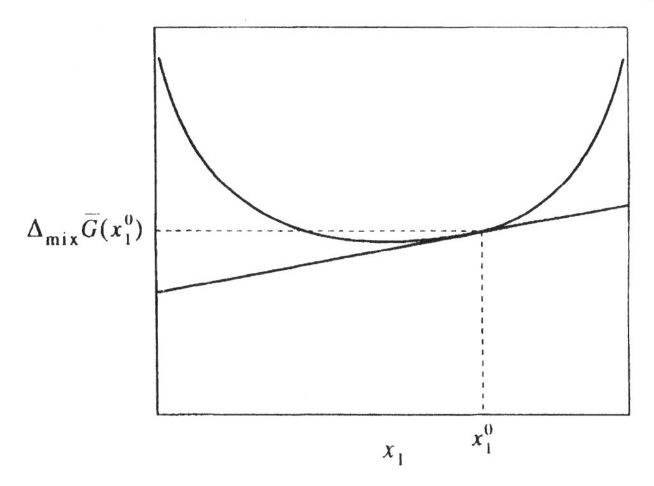

Therefore, the equation of the tangent line is

$$\begin{aligned} y &= (\text{slope})x + \text{intercept} \\ &= x_1 \left(\frac{\partial \Delta_{mix}\overline{G}}{\partial x_1} \right)_{x_1 = x_1^0} + \Delta_{mix}\overline{G}(x_1^0) - x_1^0 \left(\frac{\partial \Delta_{mix}\overline{G}}{\partial x_1} \right)_{x_1 = x_1^0} \\ &= (x_1 - x_1^0) \left(\frac{\partial \Delta_{mix}\overline{G}}{\partial x_1} \right)_{x_1 = x_1^0} + \Delta_{mix}\overline{G}(x_1^0) \end{aligned}$$

The Taylor expansion of $\Delta_{mix}\overline{G}(x_1)$ about x_1^0 is

$$\begin{aligned} \Delta_{mix}\overline{G}(x_1) = \Delta_{mix}\overline{G}(x_1^0) &+ (x_1 - x_1^0) \left(\frac{\partial \Delta_{mix}\overline{G}}{\partial x_1} \right)_{x_1 = x_1^0} \\ &+ \frac{(x_1 - x_1^0)^2}{2} \left(\frac{\partial^2 \Delta_{mix}\overline{G}}{\partial x_1^2} \right)_{x_1 = x_1^0} + \cdots \end{aligned}$$

If we substitute this result into $\Delta_{mix}\overline{G}$ at x_1 around $x_1^0 >$ tangent line at x_1^0, the constant terms and linear terms cancel and we are left with

$$\frac{(x_1 - x_1^0)^2}{2} \left(\frac{\partial^2 \Delta_{mix}\overline{G}}{\partial x_1^2} \right)_{x_1 = x_1^0} > 0$$

or simply

$$\left(\frac{\partial^2 \Delta_{mix}\overline{G}}{\partial x_1^2} \right)_{x_1 = x_1^0} > 0$$

CHAPTER 11

Solutions II
Solid-Liquid Solutions

PROBLEMS AND SOLUTIONS

11–1. The density of a glycerol/water solution that is 40.0% glycerol by mass is $1.101\ \text{g}\cdot\text{mL}^{-1}$ at 20°C. Calculate the molality and the molarity of glycerol in the solution at 20°C. Calculate the molality at 0°C.

The mass of glycerol per millimeter of solution is

$$\text{g glycerol per mL} = (0.400)(1.101\ \text{g}\cdot\text{mL}^{-1}) = 0.4404\ \text{g}\cdot\text{mL}^{-1}$$

The number of moles of glycerol per liter of solution is

$$\text{molarity} = \frac{440.4\ \text{g}\cdot\text{L}^{-1}}{92.093\ \text{g}\cdot\text{mol}^{-1}} = 4.78\ \text{mol}\cdot\text{L}^{-1}$$

The number of grams of water per 0.4404 grams of glycerol is given by

$$1.101\ \text{g} - 0.4404\ \text{g} = 0.6606\ \text{g}\ H_2O$$

or 0.4404 g glycerol per 0.6606 g H_2O, or 0.6666 g glycerol per g H_2O. Therefore,

$$\text{molality} = \frac{666.6\ \text{g}\cdot\text{kg}^{-1}}{92.094\ \text{g}\cdot\text{mol}^{-1}} = 7.24\ \text{mol}\cdot\text{kg}^{-1}$$

11–2. Concentrated sulfuric acid is sold as a solution that is 98.0% sulfuric acid and 2.0% water by mass. Given that the density is $1.84\ \text{g}\cdot\text{mL}^{-1}$, calculate the molarity of concentrated sulfuric acid.

$$\text{g}\ H_2SO_4\ \text{per mL solution} = (0.980)(1.84\ \text{g}\cdot\text{mL}^{-1}) = 1.80\ \text{g}\cdot\text{mL}^{-1})$$

$$\text{molarity} = \frac{1800\ \text{g}\cdot\text{L}^{-1}}{98.08\ \text{g}\cdot\text{mol}^{-1}} = 18.4\ \text{mol}\cdot\text{L}^{-1}$$

11–3. Concentrated phosphoric acid is sold as a solution that is 85% phosphoric acid and 15% water by mass. Given that the molarity is 15 M, calculate the density of concentrated phosphoric acid.

A 15 molar solution implies that there are

$$(15\ \text{mol}\cdot\text{L}^{-1})(97.998\ \text{g}\cdot\text{mol}^{-1}) = 1470\ \text{g of phosphoric acid per liter of solution}$$

Therefore, the density of the solution is

$$\text{density} = \frac{1470\ \text{g}\cdot\text{L}^{-1}}{0.85} = 1700\ \text{g}\cdot\text{L}^{-1} = 1.7\ \text{g}\cdot\text{mL}^{-1}$$

11–4. Calculate the mole fraction of glucose in an aqueous solution that is 0.500 mol·kg^{-1} glucose.

There are 0.500 mol glucose per kg H_2O. so

$$x_2 = \frac{0.500\ \text{mol}}{0.500\ \text{mol} + \dfrac{1000\ \text{g H}_2\text{O}}{18.02\ \text{g}\cdot\text{mol}^{-1}\ \text{H}_2\text{O}}} = 0.00893$$

11–5. Show that the relation between molarity and molality for a solution with a single solute is

$$c = \frac{(1000\ \text{mL}\cdot\text{L}^{-1})\rho m}{1000\ \text{g}\cdot\text{kg}^{-1} + mM_2}$$

where c is the molarity, m is the molality, ρ is the density of the solution in g·mL^{-1}, and M_2 is the molar mass (g·mol^{-1}) of the solute.

Consider a solution of a certain molality, m, containing 1000 g of solvent. The total mass of the solution is $1000\ \text{g}\cdot\text{kg}^{-1} + mM_2$ and its volume (in mL) is $(1000\ \text{g}\cdot\text{kg}^{-1} + mM_2)/\rho$, where ρ is the density of the the solution in g·mL^{-1}. The volume of the solution in liters is $(1000\ \text{g}\cdot\text{kg}^{-1} + mM_2)/\rho(1000\ \text{mL}\cdot\text{L}^{-1})$ liters. There are m moles of solute per $(1000\ \text{g}\cdot\text{kg}^{-1} + mM_2)/\rho(1000\ \text{mL}\cdot\text{L}^{-1})$ liters, so the molarity is

$$c = \frac{(1000\ \text{mL}\cdot\text{L}^{-1})\rho m}{1000\ \text{g}\cdot\text{kg}^{-1} + mM_2}$$

11–6. The *CRC Handbook of Chemistry and Physics* has tables of "concentrative properties of aqueous solutions" for many solutions. Some entries for CsCl(s) are

$A/\%$	$\rho/\text{g}\cdot\text{mL}^{-1}$	$c/\text{mol}\cdot\text{L}^{-1}$
1.00	1.0058	0.060
5.00	1.0374	0.308
10.00	1.0798	0.641
20.00	1.1756	1.396
40.00	1.4226	3.380

where A is the mass percent of the solute, ρ is the density of the solution, and c is the molarity. Using these data, calculate the molality at each concentration.

Using the result of the previous problem,

$$m = \frac{(1000\ \text{g}\cdot\text{kg}^{-1})c}{(1000\ \text{mL}\cdot\text{L}^{-1})\rho - M_2 c}$$

We have then ($M_2 = 168.36\ \text{g}\cdot\text{mol}^{-1}$)

$c/\text{mol}\cdot\text{L}^{-1}$	$m/\text{mol}\cdot\text{kg}^{-1}$
0.060	0.060
0.308	0.313
0.641	0.660
1.396	1.484
3.380	3.960

11–7. Derive a relation between the mass percentage (A) of a solute in a solution and its molality (m). Calculate the molality of an aqueous sucrose solution that is 18% sucrose ($C_{12}H_{22}O_{11}$) by mass.

Mass percentage of solute, A_2, is given by

$$A_2 = \frac{\text{mass}_2}{\text{mass}_1 + \text{mass}_2} \times 100$$

If we take a solution containing 1000 g of solvent, then $\text{mass}_2 = mM_2$ and $\text{mass}_1 = 1000\ \text{g}\cdot\text{kg}^{-1}$, so

$$A_2 = \frac{mM_2}{1000\ \text{g}\cdot\text{kg}^{-1} + mM_2} \times 100$$

Solve for m to get

$$m = \frac{(1000\ \text{g}\cdot\text{kg}^{-1})A_2}{(100 - A_2)M_2}$$

For an aqueous sucrose solution that is 18% sucrose by mass,

$$m = \frac{(1000\ \text{g}\cdot\text{kg}^{-1})(18)}{(100 - 18)(342.3\ \text{g}\cdot\text{mol}^{-1})} = 0.73\ \text{mol}\cdot\text{kg}^{-1}$$

11–8. Derive a relation between the mole fraction of the solvent and the molality of a solution.

Start with

$$x_2 = \frac{n_2}{n_1 + n_2}$$

Now take a solution containing 1000 g of solvent, so that $n_2 = m$ and $n_1 = (1000\ \text{g})/M_1$, where M_1 is the molar mass of the solvent. Therefore,

$$x_2 = \frac{m}{\dfrac{1000\ \text{g}\cdot\text{kg}^{-1}}{M_1} + m} = \frac{mM_1}{1000\ \text{g}\cdot\text{kg}^{-1} + mM_1}$$

and

$$x_1 = 1 - x_2 = \frac{1000\ \text{g}\cdot\text{kg}^{-1}}{1000\ \text{g}\cdot\text{kg}^{-1} + mM_1}$$

11–9. The volume of an aqueous sodium chloride solution at 25°C can be expressed as

$$V/\text{mL} = 1001.70 + (17.298\ \text{kg}\cdot\text{mol}^{-1})m + (0.9777\ \text{kg}^2\cdot\text{mol}^{-2})m^2$$
$$- (0.0569\ \text{kg}^3\cdot\text{mol}^{-3})m^3$$
$$0 \le m \le 6\ \text{mol}\cdot\text{kg}^{-1}$$

where m is the molality. Calculate the molarity of a solution that is 3.00 $\text{mol}\cdot\text{kg}^{-1}$ of sodium chloride.

The volume of the solution at a 3.00 $\text{mol}\cdot\text{kg}^{-1}$ concentration is

$$V/\text{mL} = 1060.86$$

The mass of a 3.00 $\text{mol}\cdot\text{kg}^{-1}$ NaCl(aq) solution that contains 1000 g of solvent is

$$\begin{aligned}\text{mass} &= 1000\ \text{g}\cdot\text{kg}^{-1} + (3.00\ \text{mol}\cdot\text{kg}^{-1})(58.444\ \text{g}\cdot\text{mol}^{-1})\\ &= 1175.33\ \text{g}\end{aligned}$$

The density of the solution is

$$\rho^{\text{sln}} = \frac{1175.33\ \text{g}}{1060.86\ \text{mL}} = 1.108\ \text{g}\cdot\text{mL}^{-1}$$

and so the molarity is (see Problem 11–5)

$$\begin{aligned}c &= \frac{(1000\ \text{mL}\cdot\text{L}^{-1})\rho m}{1000\ \text{g}\cdot\text{kg}^{-1} + mM_2}\\ &= \frac{(1000\ \text{mL}\cdot\text{L}^{-1})(1.108\ \text{g}\cdot\text{mL}^{-1})(3.00\ \text{mol}\cdot\text{kg}^{-1})}{1000\ \text{g}\cdot\text{kg}^{-1} + (3.00\ \text{mol}\cdot\text{kg}^{-1})(58.444\ \text{g}\cdot\text{mol}^{-1})}\\ &= 2.83\ \text{mol}\cdot\text{L}^{-1}\end{aligned}$$

11–10. If x_2^∞, m^∞, and c^∞ are the mole fraction, molality, and molarity, respectively, of a solute at infinite dilution, show that

$$x_2^\infty = \frac{m^\infty M_1}{1000\ \text{g}\cdot\text{kg}^{-1}} = \frac{c^\infty M_1}{(1000\ \text{mL}\cdot\text{L}^{-1})\rho_1}$$

where M_1 is the molar mass ($\text{g}\cdot\text{mol}^{-1}$) and ρ_1 is the density ($\text{g}\cdot\text{mL}^{-1}$) of the solvent. Note that mole fraction, molality, and molarity are all directly proportional to each other at low concentrations.

Start with $x_2 = n_2/(n_1 + n_2)$. At infinite dilution, $n_2 \to 0$, and so

$$x_2 = \frac{n_2}{n_1 + n_2} \longrightarrow \frac{n_2}{n_1} \quad \text{as} \quad n_2 \longrightarrow 0$$

Consider a solution containing 1000 g of solvent. In this case, $n_2 = m$ and $n_1 = (1000\ \text{g}\cdot\text{kg}^{-1})/M_1$, where M_1 is the molar mass of the solvent. Then

$$x_2^\infty = \frac{m^\infty}{\dfrac{1000\ \text{g}\cdot\text{kg}^{-1}}{M_1}} = \frac{m^\infty M_1}{1000\ \text{g}\cdot\text{kg}^{-1}}$$

According to Problem 11–5, $c^\infty \to (\text{mL}\cdot\text{L}^{-1})\rho m^\infty/(\text{g}\cdot\text{kg}^{-1})$, so

$$x_2^\infty = \frac{c^\infty}{\left(\dfrac{1000\ \text{mL}\cdot\text{L}^{-1}}{M_1}\right)\rho} = \frac{c^\infty M_1}{(1000\ \text{mL}\cdot\text{L}^{-1})\rho}$$

11–11. Consider two solutions whose solute activities are a_2' and a_2'', referred to the same standard state. Show that the difference in the chemical potentials of these two solutions is independent of the standard state and depends only upon the ratio a_2'/a_2''. Now choose one of these solutions to be at an arbitrary concentration and the other at a very dilute concentration (essentially infinitely dilute) and argue that

$$\frac{a_2'}{a_2''} = \frac{\gamma_{2x}x_2}{x_2^\infty} = \frac{\gamma_{2m}m}{m^\infty} = \frac{\gamma_{2c}c}{c^\infty}$$

Let

$$\mu_2' = (\mu_2^\circ)' + RT\ln a_2'$$
$$\mu_2'' = (\mu_2^\circ)'' + RT\ln a_2'' = (\mu_2^\circ)' + RT\ln a_2''$$

Therefore,

$$\Delta\mu = \mu_2' - \mu_2'' = RT\ln\frac{a_2'}{a_2''}$$

If the solution denoted by the double prime is very dilute, then $a_2'' = x_2^\infty$, m^∞, or c^∞. Therefore,

$$\frac{a_2'}{a_2''} = \frac{\gamma_{2x}x_2}{x_2^\infty} = \frac{\gamma_{2m}m}{m^\infty} = \frac{\gamma_{2c}c}{c^\infty}$$

11–12. Use Equations 11.4, 11.11, and the results of the previous two problems to show that

$$\gamma_{2x} = \gamma_{2m}\left(1 + \frac{mM_1}{1000\ \text{g}\cdot\text{kg}^{-1}}\right) = \gamma_{2c}\left(\frac{\rho}{\rho_1} + \frac{c[M_1 - M_2]}{\rho_1[1000\ \text{mL}\cdot\text{L}^{-1}]}\right)$$

where ρ is the density of the solution. Thus, we see that the three different activity coefficients are related to one another.

Using the result of the previous problem,

$$\gamma_{2x} = \frac{x_2^\infty}{m^\infty}\frac{m}{x_2}\gamma_{2m}$$

Using the result of Equation 11.4 and Problem 11–10, we have

$$\gamma_{2x} = \gamma_{2m}\left(\frac{M_1}{1000\ \text{g}\cdot\text{kg}^{-1}}\right)\left(\frac{1000\ \text{g}\cdot\text{kg}^{-1}}{M_1} + m\right) = \gamma_{2m}\left(1 + \frac{mM_1}{1000\ \text{g}\cdot\text{kg}^{-1}}\right)$$

Similarly, Problem 11–11 gives us

$$\gamma_{2x} = \gamma_{2c} \frac{x_2^\infty}{c^\infty} \frac{c}{x_2}$$

Using Equation 11.11 and the result of Problem 11–11, we have

$$\begin{aligned}\gamma_{2x} &= \gamma_{2c} \left[\frac{M_1}{(1000\ \mathrm{mL \cdot L^{-1}})\rho_1} \right] \left[\frac{(1000\ \mathrm{mL \cdot L^{-1}})\rho + c(M_1 - M_2)}{M_1} \right] \\ &= \gamma_{2c} \left[\frac{\rho}{\rho_1} + \frac{c(M_1 - M_2)}{(1000\ \mathrm{mL \cdot L^{-1}})\rho_1} \right]\end{aligned}$$

11–13. Use Equations 11.4, 11.11, and the results of Problem 11–12 to derive

$$\gamma_{2m} = \gamma_{2c} \left(\frac{\rho}{\rho_1} - \frac{cM_2}{\rho_1 [1000\ \mathrm{mL \cdot L^{-1}}]} \right)$$

Given that the density of an aqueous citric acid ($M_2 = 192.12\ \mathrm{g \cdot mol^{-1}}$) solution at 20°C is given by

$$\rho / \mathrm{g \cdot mL^{-1}} = 0.99823 + (0.077102\ \mathrm{L \cdot mol^{-1}})c$$

$$0 \le c < 1.772\ \mathrm{mol \cdot L^{-1}}$$

plot γ_{2m}/γ_{2c} versus c. Up to what concentration do γ_{2m} and γ_{2c} differ by 2%?

From Problem 11–11,

$$\gamma_{2m} = \gamma_{2c} \frac{m^\infty}{c^\infty} \frac{c}{m}$$

Using the results from Problems 11–5 and 11–10,

$$\gamma_{2m} = \gamma_{2c} \frac{1}{\rho_1} \left[\frac{(1000\ \mathrm{mL \cdot L^{-1}})\rho - cM_2}{1000\ \mathrm{mL \cdot L^{-1}}} \right] = \gamma_{2c} \left[\frac{\rho}{\rho_1} - \frac{cM_2}{\rho_1 (1000\ \mathrm{mL \cdot L^{-1}})} \right]$$

The ratio γ_{2m}/γ_{2c} is plotted below.

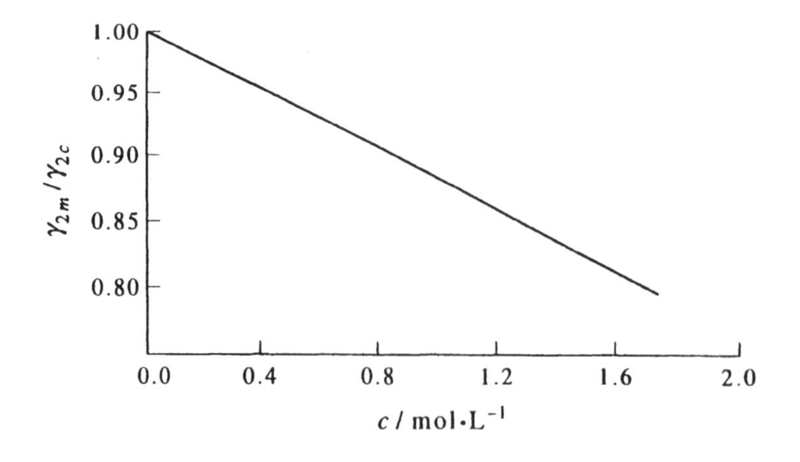

11–14. The *CRC Handbook of Chemistry and Physics* gives a table of mass percent of sucrose in an aqueous solution and its corresponding molarity at 25°C. Use these data to plot molality versus molarity for an aqueous sucrose solution.

Use the relation between mass percentage and molality that is derived in Problem 11–7 to calculate the molality at each mass percentage. Some representative values of A, c, and m and the plot of m against c are given below.

A	$c/\text{mol}\cdot\text{L}^{-1}$	$m/\text{mol}\cdot\text{kg}^{-1}$	A	$c/\text{mol}\cdot\text{L}^{-1}$	$m/\text{mol}\cdot\text{kg}^{-1}$
1.00	0.029	0.030	24.00	0.771	0.923
2.00	0.059	0.060	28.00	0.914	1.136
3.00	0.089	0.090	32.00	1.063	1.375
4.00	0.118	0.122	36.00	1.216	1.643
5.00	0.149	0.154	40.00	1.375	1.948
6.00	0.179	0.186	44.00	1.539	2.295
7.00	0.210	0.220	48.00	1.709	2.697
8.00	0.241	0.254	52.00	1.885	3.165
9.00	0.272	0.289	56.00	2.067	3.718
10.00	0.303	0.325	60.00	2.255	4.382
12.00	0.367	0.398	64.00	2.450	5.194
14.00	0.431	0.476	68.00	2.652	6.208
16.00	0.497	0.556	72.00	2.860	7.512
18.00	0.564	0.641	76.00	3.076	9.251
20.00	0.632	0.730	80.00	3.299	11.686

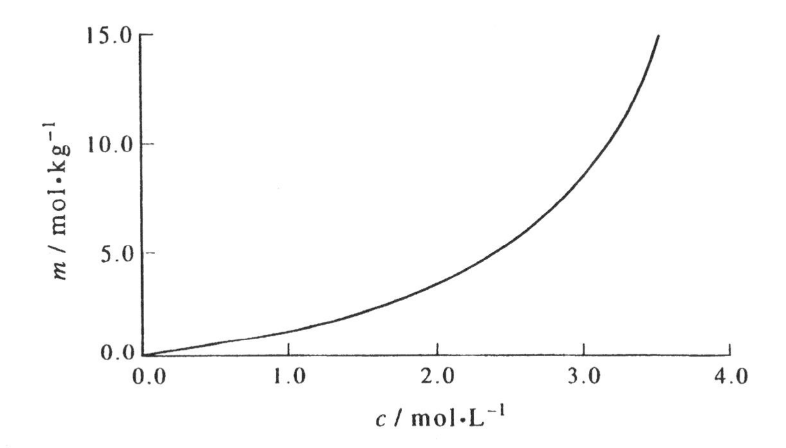

11–15. Using the data in Table 11.2, calculate the activity coefficient of water (on a mole fraction basis) at a sucrose concentration of 3.00 molal.

We use the equation $\gamma_1 = a_1/x_1$. The relation between molality and mole fraction is given by Equation 11.4:

$$x_1 = 1 - x_2 = \frac{1}{1 + \dfrac{mM_1}{1000\ \text{g}\cdot\text{kg}^{-1}}} = \frac{1000\ \text{g}\cdot\text{kg}^{-1}}{1000\ \text{g}\cdot\text{kg}^{-1} + mM_1}$$

At $m = 3.00 \text{ mol}\cdot\text{kg}^{-1}$ (with $M_1 = 18.02 \text{ g}\cdot\text{mol}^{-1}$), we have $x_1 = 0.9487$. Therefore,

$$\gamma_1 = \frac{0.93276}{0.9487} = 0.983$$

11–16. Using the data in Table 11.2, plot the activity coefficient of water (on a mole fraction basis) against the mole fraction of water.

Calculate the mole fraction from the molality according to Problem 11–15, use the relation $\gamma_{1x} = a_1/x_1$, and plot the results to get

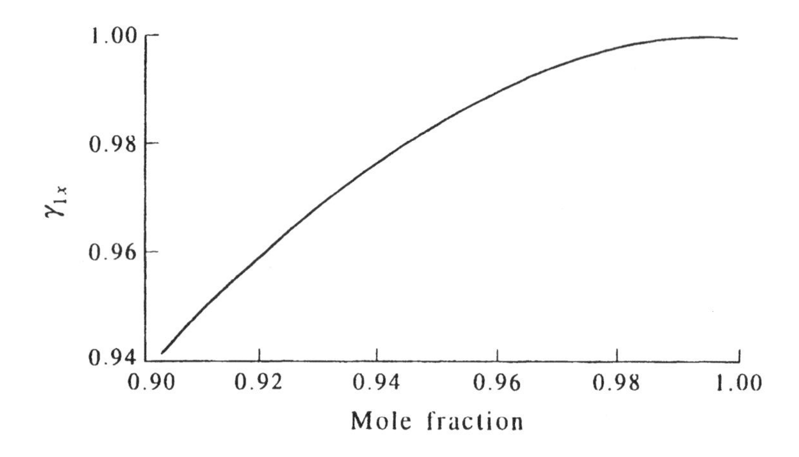

11–17. Using the data in Table 11.2, calculate ϕ at each value of m and reproduce Figure 11.2.

Use Equation 11.13,

$$\ln a_1 = -\frac{m\phi}{55.506 \text{ mol}\cdot\text{kg}^{-1}}$$

11–18. Fit the data for the osmotic coefficient of sucrose in Table 11.2 to a fourth-degree polynomial and calculate γ_{2m} for a 1.00-molal solution. Compare your result with the one obtained in Example 11–5.

Suppressing the units of the coefficients, we get

$$\phi = 1 + 0.075329m + 0.016554m^2 - 0.0039647m^3 + 0.00024694m^4$$

Use Equation 11.15 to write

$$\ln \gamma_{2m} = \phi - 1 + \int_0^{1.00} \frac{\phi - 1}{m} dm = 0.08816 + 0.08235 = 0.1705$$

and so $\gamma_{2m} = 1.186$.

11–19. Using the data for sucrose given in Table 11.2, determine $\ln \gamma_{2m}$ at 3.00 molal by plotting $(\phi - 1)/m$ versus m and determining the area under the curve by numerical integration (Mathchapter A) rather than by curve fitting ϕ first. Compare your result with the value given in Table 11.2.

Using Kaleidagraph, we obtain

$$\int_0^{3.00} \frac{\phi - 1}{m} dm = 0.272$$

From Table 11.2, $\phi - 1 = 0.2879$, and so $\ln \gamma_{2m} = 0.560$, and $\gamma_{2m} = 1.75$.

11–20. Equation 11.18 can be used to determine the activity of the solvent at its freezing point. Assuming that ΔC_P^* is independent of temperature, show that

$$\Delta_{\text{fus}}\overline{H}(T) = \Delta_{\text{fus}}\overline{H}(T_{\text{fus}}^*) + \Delta\overline{C}_P^*(T - T_{\text{fus}}^*)$$

where $\Delta_{\text{fus}}\overline{H}(T_{\text{fus}}^*)$ is the molar enthalpy of fusion at the freezing point of the pure solvent (T_{fus}^*) and $\Delta\overline{C}_P^*$ is the difference in the molar heat capacities of liquid and solid solvent. Using Equation 11.18, show that

$$-\ln a_1 = \frac{\Delta_{\text{fus}}\overline{H}(T_{\text{fus}}^*)}{R(T_{\text{fus}}^*)^2}\theta + \frac{1}{R(T_{\text{fus}}^*)^2}\left(\frac{\Delta_{\text{fus}}\overline{H}(T_{\text{fus}}^*)}{T_{\text{fus}}^*} - \frac{\Delta\overline{C}_P^*}{2}\right)\theta^2 + \cdots$$

where $\theta = T_{\text{fus}}^* - T_{\text{fus}}$.

Use

$$\left(\frac{\partial \Delta_{\text{fus}}\overline{H}}{\partial T}\right)_P = \Delta\overline{C}_P^*$$

to derive

$$\Delta_{\text{fus}}\overline{H}(T) - \Delta_{\text{fus}}\overline{H}(T^*) = \Delta\overline{C}_P^*(T - T^*)$$

Using Equation 11.18

$$\begin{aligned}
\ln a_1 &= \int_{T_{\text{fus}}^*}^{T_{\text{fus}}} \frac{\Delta_{\text{fus}}\overline{H}(T)}{RT^2} dT \\
&= \frac{\Delta_{\text{fus}}\overline{H}(T_{\text{fus}}^*)}{R}\int_{T_{\text{fus}}^*}^{T_{\text{fus}}} \frac{dT}{T^2} + \frac{\Delta\overline{C}_P^*}{R}\int_{T_{\text{fus}}^*}^{T_{\text{fus}}} \frac{dT(T - T_{\text{fus}}^*)}{T^2} \\
&= \frac{\Delta_{\text{fus}}\overline{H}(T_{\text{fus}}^*)}{R}\left(-\frac{1}{T_{\text{fus}}} + \frac{1}{T_{\text{fus}}^*}\right) + \frac{\Delta\overline{C}_P^*}{R}\left(\ln\frac{T_{\text{fus}}}{T_{\text{fus}}^*} + \frac{T_{\text{fus}}^*}{T_{\text{fus}}} - \frac{T_{\text{fus}}^*}{T_{\text{fus}}^*}\right) \\
&= \frac{\Delta_{\text{fus}}\overline{H}(T_{\text{fus}}^*)}{RT_{\text{fus}}^*}\left(\frac{T_{\text{fus}} - T_{\text{fus}}^*}{T_{\text{fus}}}\right) + \frac{\Delta\overline{C}_P^*}{R}\left(\ln\frac{T_{\text{fus}}}{T_{\text{fus}}^*} + \frac{T_{\text{fus}}^* - T_{\text{fus}}}{T_{\text{fus}}}\right)
\end{aligned}$$

Now let $T_{\text{fus}} = T_{\text{fus}}^* - \theta$ and use $1/(1 - x) = 1 + x + x^2 + \cdots$ to get

$$\frac{1}{T_{\text{fus}}} = \frac{1}{T_{\text{fus}}^* - \theta} = \frac{1}{T_{\text{fus}}^*}\left[1 + \frac{\theta}{T_{\text{fus}}^*} + \frac{\theta^2}{(T_{\text{fus}}^*)^2} + \cdots\right]$$

and use $\ln(1-x) = -x - x^2/2 + \cdots$ to get

$$\ln \frac{T_{\text{fus}}}{T^*_{\text{fus}}} = \ln\left(1 - \frac{\theta}{T^*_{\text{fus}}}\right) = -\frac{\theta}{T^*_{\text{fus}}} - \frac{1}{2}\frac{\theta^2}{(T^*_{\text{fus}})^2} + \cdots$$

Finally, then

$$\begin{aligned} -\ln a_1 &= -\frac{\Delta_{\text{fus}}\overline{H}(T^*_{\text{fus}})}{R(T^*_{\text{fus}})^2}\left(-\theta - \frac{\theta^2}{T^*_{\text{fus}}} + \cdots\right) \\ &\quad - \frac{\Delta\overline{C}^*_P}{RT^*_{\text{fus}}}\left[-\theta - \frac{\theta^2}{2T^*_{\text{fus}}} + \cdots + \theta\left(1 + \frac{\theta}{T^*_{\text{fus}}} + \cdots\right)\right] \\ &= \frac{\Delta_{\text{fus}}\overline{H}(T^*_{\text{fus}})}{R(T^*_{\text{fus}})^2}\theta + \frac{1}{R(T^*_{\text{fus}})^2}\left(\frac{\Delta_{\text{fus}}\overline{H}(T^*_{\text{fus}})}{T^*_{\text{fus}}} - \frac{\Delta\overline{C}^*_P}{2}\right)\theta^2 + \cdots \end{aligned}$$

11–21. Take $\Delta_{\text{fus}}\overline{H}(T^*_{\text{fus}}) = 6.01\ \text{kJ}\cdot\text{mol}^{-1}$, $\overline{C}^{\text{l}}_P = 75.2\ \text{J}\cdot\text{K}^{-1}\cdot\text{mol}^{-1}$, and $\overline{C}^{\text{s}}_P = 37.6\ \text{J}\cdot\text{K}^{-1}\cdot\text{mol}^{-1}$ to show that the equation for $-\ln a_1$ in the previous problem becomes

$$-\ln a_1 = (0.00969\ \text{K}^{-1})\theta + (5.2 \times 10^{-6}\ \text{K}^{-2})\theta^2 + \cdots$$

for an aqueous solution. The freezing point depression of a 1.95-molal aqueous sucrose solution is 4.45°C. Calculate the value of a_1 at this concentration. Compare your result with the value in Table 11.2. The value you calculated in this problem is for 0°C, whereas the value in Table 11.2 is for 25°C, but the difference is fairly small because a_1 does not vary greatly with temperature (Problem 11–61).

Using the final equation in Problem 11–20, we have

$$\begin{aligned} -\ln a_1 &= \frac{(6.01\ \text{kJ}\cdot\text{mol}^{-1})\theta}{(8.314\ \text{J}\cdot\text{mol}^{-1}\cdot\text{K}^{-1})(273.2\ \text{K})^2} \\ &\quad + \frac{1}{(8.314\ \text{J}\cdot\text{mol}^{-1}\cdot\text{K}^{-1})(273.2\ \text{K})^2}\left(\frac{6.01\ \text{kJ}\cdot\text{mol}^{-1}}{273.2\ \text{K}} - 18.8\ \text{J}\cdot\text{mol}^{-1}\cdot\text{K}^{-1}\right) \\ &= (0.00969\ \text{K}^{-1})\theta + (5.2 \times 10^{-6}\ \text{K}^{-2})\theta^2 + \cdots \end{aligned}$$

If $\theta = 4.45$ K, then

$$\begin{aligned} \ln a_1 &= -(0.00969\ \text{K}^{-1})(4.45\ \text{K}) - (5.2 \times 10^{-6}\ \text{K}^{-2})(4.45\ \text{K})^2 \\ &= -0.0432 \end{aligned}$$

and so $a_1 = 0.958$.

11–22. The freezing point of a 5.0-molal aqueous glycerol (1,2,3-propanetriol) solution is -10.6°C. Calculate the activity of water at 0°C in this solution. (See Problems 11–20 and 11–21.)

Use the equation derived in Problem 11–21

$$\ln a_1 = -(0.00969\ \text{K}^{-1})\theta - (5.2 \times 10^{-6}\ \text{K}^{-2})\theta^2$$

with $\theta = 10.6$ K to get

$$\ln a_1 = -(0.00969\ \mathrm{K^{-1}})(10.6\ \mathrm{K}) - (5.2\times10^{-6}\ \mathrm{K^{-2}})(10.6\ \mathrm{K})^2 = -0.103$$

and so $a_1 = 0.902$.

11–23. Show that replacing T_{fus} by T^*_{fus} in the denominator of $(T_{\mathrm{fus}} - T^*_{\mathrm{fus}})/T^*_{\mathrm{fus}}T_{\mathrm{fus}}$ (see Equation 11.20) gives $-\theta/(T^*_{\mathrm{fus}})^2 - \theta^2/(T^*_{\mathrm{fus}})^3 + \cdots$ where $\theta = T^*_{\mathrm{fus}} - T_{\mathrm{fus}}$.

$$\frac{T_{\mathrm{fus}} - T^*_{\mathrm{fus}}}{T_{\mathrm{fus}}T^*_{\mathrm{fus}}} = \frac{-\theta}{T^*_{\mathrm{fus}}(T^*_{\mathrm{fus}} - \theta)} = -\frac{\theta}{(T^*_{\mathrm{fus}})^2\left(1 - \dfrac{\theta}{T^*_{\mathrm{fus}}}\right)}$$

Now use the expansion $1/(1-x) = x + x^2 + \cdots$ to write

$$\frac{T_{\mathrm{fus}} - T^*_{\mathrm{fus}}}{T_{\mathrm{fus}}T^*_{\mathrm{fus}}} = -\frac{\theta}{(T^*_{\mathrm{fus}})^2}\left[1 + \frac{\theta}{T^*_{\mathrm{fus}}} + \frac{\theta^2}{(T^*_{\mathrm{fus}})^2} + \cdots\right] = -\frac{\theta}{(T^*_{\mathrm{fus}})^2} - \frac{\theta^2}{(T^*_{\mathrm{fus}})^3} + \cdots$$

11–24. Calculate the value of the freezing point depression constant for nitrobenzene, whose freezing point is 5.7°C and whose molar enthalpy of fusion is 11.59 kJ·mol^{-1}.

Using Equation 11.23, we write

$$K_{\mathrm{f}} = \left(\frac{M_1}{1000\ \mathrm{g\cdot kg^{-1}}}\right)\frac{R(T^*_{\mathrm{fus}})^2}{\Delta_{\mathrm{fus}}\overline{H}} = \left(\frac{123.11\ \mathrm{g\cdot mol^{-1}}}{1000\ \mathrm{g\cdot kg^{-1}}}\right)\left[\frac{(8.314\ \mathrm{J\cdot mol^{-1}\cdot K^{-1}})(278.9\ \mathrm{K})^2}{11.59\times10^3\ \mathrm{J\cdot mol^{-1}}}\right] = 6.87\ \mathrm{K\cdot kg\cdot mol^{-1}}$$

11–25. Use an argument similar to the one we used to derive Equations 11.22 and 11.23 to derive Equations 11.24 and 11.25.

The condition for equilibrium at a temperature T is

$$\mu_1^{\mathrm{g}}(T, P) = \mu_1^{\mathrm{sln}}(T, P) = \mu_1^*(T, P) + RT\ln a_1 = \mu^{\mathrm{l}} + RT\ln a_1$$

Solving for $\ln a_1$ gives

$$\ln a_1 = \frac{\mu_1^{\mathrm{g}} - \mu_1^{l}}{RT}$$

Use the Gibbs-Helmholtz equation (Example 10–1) to get

$$\left(\frac{\partial \ln a_1}{\partial T}\right)_{P,x_1} = \frac{\overline{H}_1^{\mathrm{l}} - \overline{H}_1^{\mathrm{g}}}{RT^2} = -\frac{\Delta_{\mathrm{vap}}\overline{H}}{RT^2}$$

This equation is similar to Equation 11.17 except for the negative sign, which occurs because boiling points of solutions are elevated whereas freezing points are lowered. The rest of the derivation follows Equations 11.18 through 11.23.

11–26. Calculate the boiling point elevation constant for cyclohexane given that $T_{\text{vap}} = 354$ K and that $\Delta_{\text{vap}}\overline{H} = 29.97\ \text{kJ}\cdot\text{mol}^{-1}$.

Using the analog of Equation 11.23, we have

$$\begin{aligned} K_{\text{b}} &= \frac{(84.161\ \text{g}\cdot\text{mol}^{-1})(8.314\ \text{J}\cdot\text{mol}^{-1}\cdot\text{K}^{-1})(354\ \text{K})^2}{(1000\ \text{g}\cdot\text{kg}^{-1})(29.97\times 10^3\ \text{J}\cdot\text{mol}^{-1})} \\ &= 2.93\ \text{K}\cdot\text{kg}\cdot\text{mol}^{-1} \end{aligned}$$

11–27. A solution containing 1.470 g of dichlorobenzene in 50.00 g of benzene boils at 80.60°C at a pressure of 1.00 bar. The boiling point of pure benzene is 80.09°C and the molar enthalpy of vaporization of pure benzene is 32.0 kJ·mol^{-1}. Determine the molecular mass of dichlorobenzene from these data.

The value of $\Delta_{\text{vap}}T$ is

$$\Delta_{\text{vap}}T = 80.60^\circ\text{C} - 80.09^\circ\text{C} = 0.51^\circ\text{C} = 0.51\ \text{K}$$

Using the analog of Equation 11.23, we have

$$\begin{aligned} K_{\text{b}} &= \frac{(78.108\ \text{g}\cdot\text{mol}^{-1})(8.314\ \text{J}\cdot\text{mol}^{-1}\cdot\text{K}^{-1})(353.2\ \text{K})^2}{(1000\ \text{g}\cdot\text{kg}^{-1})(32.0\times 10^3\ \text{J}\cdot\text{mol}^{-1})} \\ &= 2.53\ \text{K}\cdot\text{kg}\cdot\text{mol}^{-1} \end{aligned}$$

The molality is given by

$$m = \frac{\Delta_{\text{vap}}T}{K_{\text{b}}} = \frac{0.51\ \text{K}}{2.53\ \text{K}\cdot\text{kg}^{-1}\cdot\text{mol}^{-1}} = 0.20\ \text{mol}\cdot\text{kg}^{-1}$$

Therefore,

$$\begin{aligned} &1.470\ \text{g}\ C_6H_4Cl_2 \longleftrightarrow 50.0\ \text{g}\ C_6H_6 \\ &29.4\ \text{g}\ C_6H_4Cl_2 \longleftrightarrow 1000\ \text{g}\ C_6H_6 \longleftrightarrow 0.20\ \text{mol} \end{aligned}$$

and so the molecular mass is 147.

11–28. Consider the following phase diagram for a typical pure substance. Label the region corresponding to each phase. Illustrate how this diagram changes for a dilute solution of a nonvolatile solute.

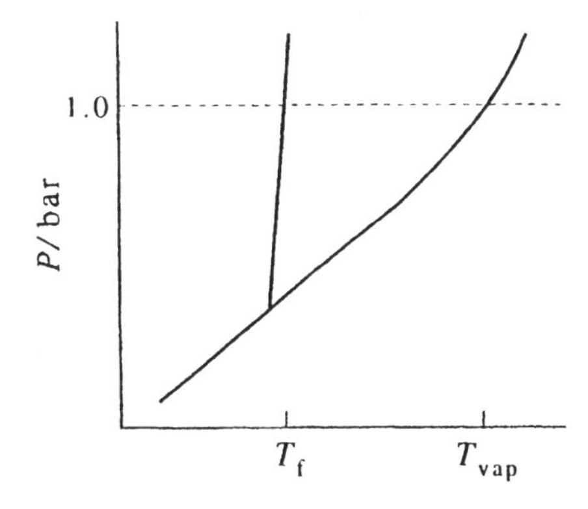

Now demonstrate that the boiling point increases and the freezing point decreases as a result of the dissolution of the solute.

Use the following figure for water, which is self-explanatory

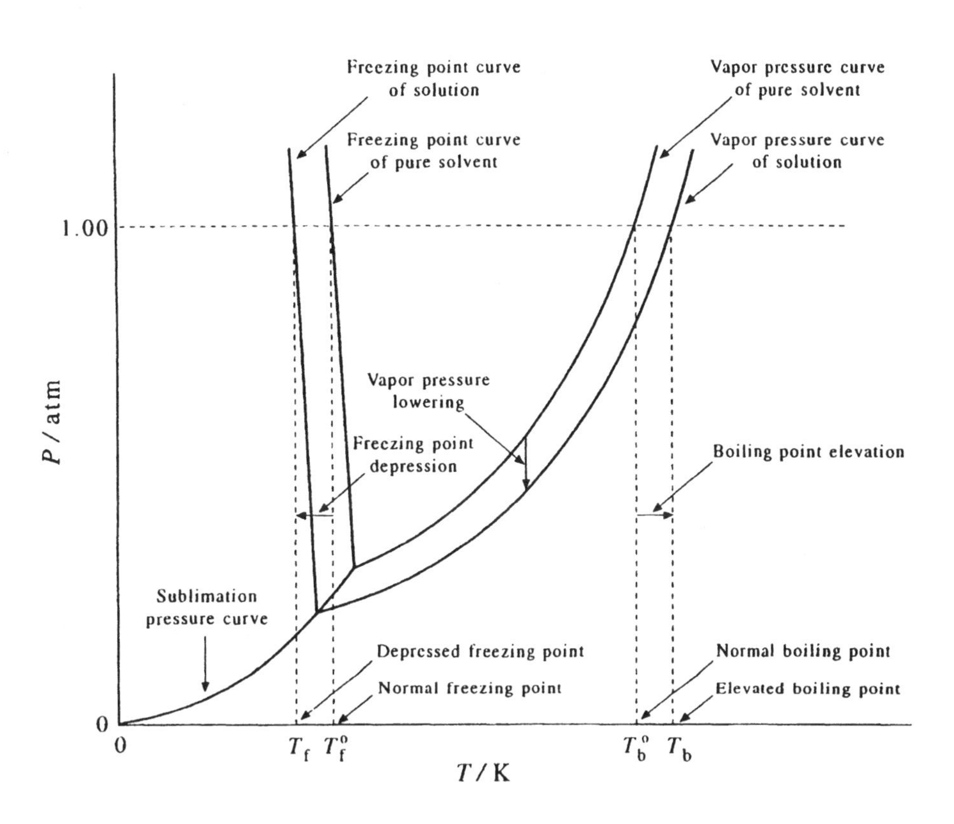

11–29. A solution containing 0.80 grams of a protein in 100 mL of a solution has an osmotic pressure of 2.06 torr at 25.0°C. What is the molecular mass of the protein?

We use Equation 11.31,

$$c = \frac{\Pi}{RT} = \frac{(2.06 \text{ torr})/(760 \text{ torr}\cdot\text{atm}^{-1})}{(0.08206 \text{ L}\cdot\text{atm}\cdot\text{mol}^{-1}\cdot\text{K}^{-1})(298.2 \text{ K})}$$
$$= 1.11 \times 10^{-4} \text{ mol}\cdot\text{L}^{-1} = 1.11 \times 10^{-5} \text{ mol}/100 \text{ mL}$$

Therefore, 1.11×10^{-5} mol corresponds to 0.80 g protein, and so the molecular mass of the protein is 72 000.

11–30. Show that the osmotic pressure of an aqueous solution can be written as

$$\Pi = \frac{RT}{\overline{V}^*}\left(\frac{m}{55.506\ \mathrm{mol\cdot kg^{-1}}}\right)\phi$$

Simply substitute Equation 11.13 into Equation 11.30.

11–31. According to Table 11.2, the activity of the water in a 2.00 molal sucrose solution is 0.95807. What external pressure must be applied to the solution at 25.0°C to make the activity of the water in the solution the same as that in pure water at 25.0°C and 1 atm? Take the density of water to be $0.997\ \mathrm{g\cdot mL^{-1}}$.

Using Equation 11.30, we have

$$\Pi = -\frac{RT\ln a_1}{\overline{V}_1^*} = -\frac{(0.08206\ \mathrm{L\cdot atm\cdot mol^{-1}\cdot K^{-1}})(298.2\ \mathrm{K})(\ln 0.95807)}{0.01807\ \mathrm{L\cdot mol^{-1}}}$$
$$= 58.0\ \mathrm{atm}$$

11–32. Show that $a_2 = a_\pm^2 = m^2\gamma_\pm^2$ for a 2–2 salt such as $CuSO_4$ and that $a_2 = a_\pm^4 = 27m^4\gamma_\pm^4$ for a 1–3 salt such as $LaCl_3$.

Equation 11.40 gives us $a_2 = a_\pm^\nu = m_\pm^\nu\gamma_\pm^\nu$. For a 2–2 salt, such as $MgSO_4$, $\nu_+ = 1$, $\nu_- = 1$, $m_+ = m$, and $m_- = m$, and so $a_2 = a_\pm^2 = m^2\gamma_\pm^2$, or $a_\pm = m\gamma_\pm$.

For a 1–3 salt, such as $LaCl_3$, $\nu_+ = 1$, $\nu_- = 3$, $m_+ = m$, and $m_- = 3m$, and so $a_2 = a_\pm^4 = [m^1(3m)^3]\gamma_\pm^4$, or $a_\pm = 27^{1/4}m\gamma_\pm$.

11–33. Verify the following table:

Type of salt	Example	I_m
1 – 1	KCl	m
1 – 2	$CaCl_2$	$3m$
2 – 1	K_2SO_4	$3m$
2 – 2	$MgSO_4$	$4m$
1 – 3	$LaCl_3$	$6m$
3 – 1	Na_3PO_4	$6m$

Show that the general result for I_m is $|z_+z_-|(\nu_1 + \nu_2)m/2$.

We use Equation 11.52 in terms of molality.

$$I_m = \frac{1}{2}\sum_{j=1}^{s} z_j^2 m_j = \frac{1}{2}(z_+^2 m_+ + z_-^2 m_-) \qquad (1)$$

for a binary salt. Therefore, we have the following table.

type	I_m
1–1	$\frac{1}{2}(1m_+ + 1m_-) = \frac{1}{2}(2m) = m$
1–2	$\frac{1}{2}(4m_+ + 1m_-) = \frac{1}{2}(4m + 2m) = 3m$
2–1	$\frac{1}{2}(m_+ + 4m_-) = \frac{1}{2}(2m + 4m) = 3m$
2–2	$\frac{1}{2}(4m_+ + 4m_-) = \frac{1}{2}(8m) = 4m$
1–3	$\frac{1}{2}(9m_+ + 1m_-) = \frac{1}{2}(9m + 3m) = 6m$
3–1	$\frac{1}{2}(m_+ + 9m_-) = \frac{1}{2}(3m + 9m) = 6m$

To prove the general result, substitute $m_+ = \nu_+ m$ and $m_- = \nu_- m$ into Equation 1 to get

$$I_m = \frac{1}{2}(z_+^2\nu_+ + z_-^2\nu_-)m$$

Now use the electroneutrality condition $z_+\nu_+ = |z_-|\nu_-$ to get

$$\begin{aligned} I_m &= \frac{1}{2}(z_+|z_-|\nu_- + |z_-|^2\nu_-)m \\ &= \frac{z_+|z_-|}{2}\left(\nu_- + \frac{|z_-|\nu_-}{z_+}\right)m = \frac{|z_-z_+|}{2}(\nu_+ + \nu_-)m \end{aligned}$$

11–34. Show that the inclusion of the factor ν in Equation 11.41 allows $\phi \to 1$ as $m \to 0$ for solutions of electrolytes as well as nonelectrolytes. *Hint*: Realize that x_2 involves the total number of moles of solute particles (see Equation 11.44).

For a nonelectrolyte, $\ln a_1 \to \ln x_1 \to \ln(1 - x_2) \to -x_2$ as $x_2 \to 0$. According to Equation 11.21, $x_2 \to M_1 m/(1000\ \text{g}\cdot\text{kg}^{-1})$ as $m \to 0$, so ϕ defined by $\ln a_1 = -M_1 m\phi/(1000\ \text{g}\cdot\text{kg}^{-1})$ (Equation 11.13) becomes $\phi = (\ln a_1)/x_2 = 1$ as $x_2 \to 0$ or $m \to 0$. For an electrolyte, $x_2 \to \nu M_1 m/(1000\ \text{g}\cdot\text{kg}^{-1})$ as $x_2 \to 0$ or $m \to 0$. Therefore, ϕ defined by $\ln a_1 = -\nu M_1 m/(1000\ \text{g}\cdot\text{kg}^{-1})$ (Equation 11.41) becomes $\phi = (\ln a_1)/x_2 = 1$ as $x_2 \to 0$ or $m \to 0$.

11–35. Use Equation 11.41 and the Gibbs-Duhem equation to derive Equation 11.42.

Consider an aqueous solution consisting of 1000 g of water. The Gibbs-Duhem equation is

$$n_1 d\ln a_1 + n_2 \ln a_2 = 0$$

or

$$(55.506\ \text{mol}\cdot\text{kg}^{-1})d\ln a_1 + md\ln a_2 = 0$$

Use Equation 11.41 to obtain

$$-\nu d(m\phi) + md\ln a_2 = 0$$

Equation 11.37 gives $a_2 = a_\pm^\nu = m_\pm^\nu \gamma_\pm^\nu$, and so we have

$$-\nu d(m\phi) + m\nu d\ln m_\pm\gamma_\pm = 0$$

But generally $m_\pm = cm$, where c is a constant whose value depends upon the type of electrolyte (see Table 11.3), and so

$$\nu d(m\phi) = m\nu d\ln(cm\gamma_\pm)$$
$$= m\nu d\ln(m\gamma_\pm)$$

Thus

$$d(m\phi) = md\ln(m\gamma_\pm)$$

or

$$md\phi + \phi dm = m(d\ln\gamma_\pm + d\ln m)$$

Division by m gives

$$d\phi + \phi\frac{dm}{m} = d\ln\gamma_\pm + \frac{dm}{m}$$

or

$$d\ln\gamma_\pm = d\phi + \frac{\phi - 1}{m}dm$$

Now integrate from $m = 0$ (where $\gamma_\pm = \phi = 1$) to m to obtain Equation 11.42.

11–36. The osmotic coefficient of $CaCl_2(aq)$ solutions can be expressed as

$$\phi = 1.0000 - (1.2083\ \text{kg}^{1/2}\cdot\text{mol}^{-1/2})m^{1/2} + (3.2215\ \text{kg}\cdot\text{mol}^{-1})m$$
$$- (3.6991\ \text{kg}^{3/2}\cdot\text{mol}^{-3/2})m^{3/2} + (2.3355\ \text{kg}^2\cdot\text{mol}^{-2})m^2$$
$$- (0.67218\ \text{kg}^{5/2}\cdot\text{mol}^{-5/2})m^{5/2} + (0.069749\ \text{kg}^3\cdot\text{mol}^{-3})m^3$$
$$0 \le m \le 5.00\ \text{mol}\cdot\text{kg}^{-1}$$

Use this expression to calculate and plot $\ln\gamma_\pm$ as a function of $m^{1/2}$.

Substitute the expression for ϕ given in the problem into Equation 11.42. The result is shown in the following figure.

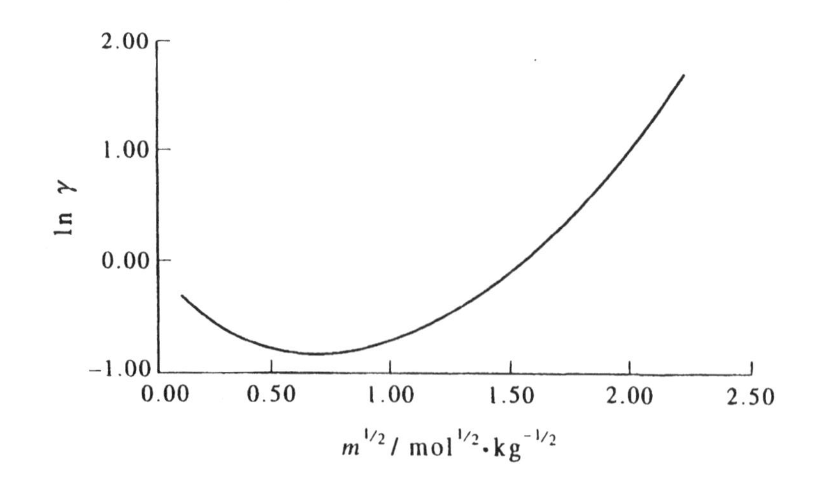

11–37. Use Equation 11.43 to calculate $\ln \gamma_{\pm}$ for NaCl(aq) at 25°C as a function of molality and plot it versus $m^{1/2}$. Compare your results with those in Table 11.4.

Substitute Equation 11.43 into Equation 11.42. The result is shown in the following figure. The calculated and experimental values are indistinguishable on the graph.

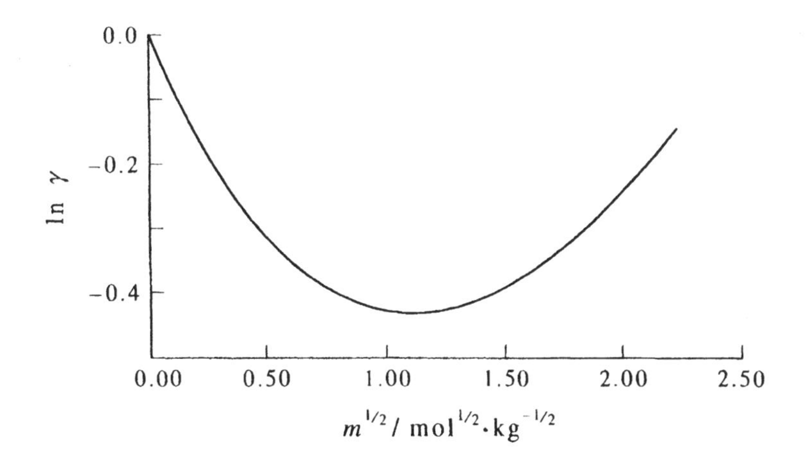

11–38. In Problem 11–19, you determined $\ln \gamma_{2m}$ for sucrose by calculating the area under the curve of $\phi - 1$ versus m. When dealing with solutions of electrolytes, it is better numerically to plot $(\phi - 1)/m^{1/2}$ versus $m^{1/2}$ because of the natural dependence of ϕ on $m^{1/2}$. Show that

$$\ln \gamma_{\pm} = \phi - 1 + 2\int_0^{m^{1/2}} \frac{\phi - 1}{m^{1/2}} dm^{1/2}$$

Start with Equation 11.42, and let $x = m^{1/2}$ and $dx = dm/2m^{1/2} = dm/2x$ to obtain

$$\int_0^m \frac{\phi - 1}{m'} dm' = \int_0^x \frac{\phi - 1}{x'^2} 2x' dx'$$
$$= 2\int_0^{m^{1/2}} \frac{\phi - 1}{m^{1/2}} dm^{1/2}$$

The full expression for $\ln \gamma_{\pm}$ is

$$\ln \gamma_{\pm} = \phi - 1 + 2\int_0^{m^{1/2}} \frac{\phi - 1}{m^{1/2}} dm^{1/2}$$

11–39. Use the data in Table 11.4 to calculate $\ln \gamma_{\pm}$ for NaCl(aq) at 25°C by plotting $(\phi - 1)/m^{1/2}$ against $m^{1/2}$ and determine the area under the curve by numerical integration (Mathchapter A). Compare your values of $\ln \gamma_{\pm}$ with those you obtained in Problem 11–37, where you calculate $\ln \gamma_{\pm}$ from a curve-fit expression of ϕ as a polynomial in $m^{1/2}$.

The plot is essentially identical to the one obtained in Problem 11–37.

11–40. Don Juan Pond in the Wright Valley of Antarctica freezes at −57°C. The major solute in the pond is $CaCl_2$. Estimate the concentration of $CaCl_2$ in the pond water.

We say "estimate" because the concentration will be too large for Equation 11.45 to be quantitative. Nevertheless, we can "estimate" the molality to be

$$m \approx \frac{57\ \text{K}}{(3)(1.84\ \text{K}\cdot\text{kg}\cdot\text{mol}^{-1})} = 10\ \text{mol}\cdot\text{kg}^{-1}$$

where the factor of 3 in the denominator results because $\nu = 3$ for $CaCl_2$.

11–41. A solution of mercury(II) chloride is a poor conductor of electricity. A 40.7-g sample of $HgCl_2$ is dissolved in 100.0 g of water, and the freezing point of the solution is found to be −2.83°C. Explain why $HgCl_2$ in solution is a poor conductor of electricity.

Because 40.7 g $HgCl_2$ corresponds to 0.150 mol $HgCl_2$, the molality of the solution is 1.50 mol·kg^{-1}. Using Equation 11.45, we find ν to be

$$\nu = \frac{\Delta T_{\text{fus}}}{K_f m} = \frac{2.83\ \text{K}}{(1.84\ \text{K}\cdot\text{kg}\cdot\text{mol}^{-1})(1.50\ \text{mol}\cdot\text{kg}^{-1})} = 1.02$$

This result indicates that $HgCl_2$ is not dissociated under these conditions, and so is a poor conductor of electricity.

11–42. The freezing point of a 0.25-molal aqueous solution of Mayer's reagent, K_2HgI_4, is found to be −1.41°C. Suggest a possible dissociation reaction that takes place when K_2HgI_4 is dissolved in water.

Use Equation 11.45 to obtain $\nu = 3$. The equation for the dissociation reaction is

$$K_2KgI_4(aq) \longrightarrow 2\ K^+(aq) + HgI_4^{2-}(aq)$$

11–43. Given the following freezing-point depression data, determine the number of ions produced per formula unit when the indicated substance is dissolved in water to produce a 1.00-molal solution.

Formula	$\Delta T/\text{K}$
$PtCl_2\cdot 4NH_3$	5.58
$PtCl_2\cdot 3NH_3$	3.72
$PtCl_2\cdot 2NH_3$	1.86
$KPtCl_3\cdot NH_3$	3.72
K_2PtCl_4	5.58

Interpret your results.

Use Equation 11.45 to obtain

formula	ν	ions
$PtCl_2 \cdot 4NH_3$	3	$Pt(NH_3)_4^{2+}$ 2 Cl^-
$PtCl_2 \cdot 3NH_3$	2	$Pt(NH_3)_3Cl^+$ Cl^-
$PtCl_2 \cdot 2NH_3$	1	$Pt(NH_3)_2Cl_2$
$KPtCl_3 \cdot NH_3$	2	K^+ $Pt(NH_3)Cl_3^-$
K_2PtCl_4	3	2 K^+ $PtCl_4^{2-}$

11–44. An aqueous solution of NaCl has an ionic strength of 0.315 mol·L^{-1}. At what concentration will an aqueous solution of K_2SO_4 have the same ionic strength?

The ionic strength, I_c, equals c for a 1–1 electrolyte and $3c$ for a 2–1 electrolyte. Therefore, a solution of K_2SO_4(aq) would have an ionic strength of 0.315 mol·L^{-1} when its molarity is 0.105 mol·L^{-1}.

11–45. Derive the "practical" formula for κ^2 given by Equation 11.53.

Start with

$$\kappa^2 = \frac{1}{\varepsilon_0 \varepsilon_r k_B T} \sum_{j=1}^{s} q_j^2 \frac{N_j}{V} = \frac{N_A e^2}{\varepsilon_0 \varepsilon_r k_B T} \sum_{j=1}^{s} z_j^2 \frac{n_j}{V}$$

Now

$$\frac{n_j}{V} = (1000\ \text{L}\cdot\text{m}^{-3})c_j$$

because V, being in SI units, has units of m^3. Therefore,

$$\kappa^2 = \frac{2e^2 N_A (1000\ \text{L}\cdot\text{m}^{-3})}{\varepsilon_0 \varepsilon_r k_B T}(I_c/\text{mol}\cdot\text{L}^{-1})$$

11–46. Some authors define ionic strength in terms of molality rather than molarity, in which case

$$I_m = \frac{1}{2}\sum_{j=1}^{s} z_j^2 m_j$$

Show that this definition modifies Equation 11.53 for dilute solutions to be

$$\kappa^2 = \frac{2e^2 N_A (1000\ \text{L}\cdot\text{m}^{-3})\rho}{\varepsilon_0 \varepsilon_r kT}(I_m/\text{mol}\cdot\text{kg}^{-1})$$

where ρ is the density of the solvent (in g·mL^{-1}).

For dilute solutions, $c = \rho m$ (see Problem 11–5), and so $I_c = \rho I_m$. Therefore,

$$\kappa^2 = \frac{2e^2 N_A (1000\ \text{L}\cdot\text{m}^{-3})\rho}{\varepsilon_0 \varepsilon_r kT}(I_m/\text{mol}\cdot\text{kg}^{-1})$$

11–47. Show that

$$\ln \gamma_{\pm} = -1.171|z_+z_-|(I_m/\text{mol}\cdot\text{kg}^{-1})^{1/2}$$

for an aqueous solution at 25°C, where I_m is the ionic strength expressed in terms of molality. Take ε_r to be 78.54 and the density of water to be 0.99707 g·mL^{-1}.

We use the equation for κ^2 that is derived in Problem 11–46.

$$\kappa^2 = \frac{(2)(1.6022\times10^{-19}\text{ C})^2(6.0221\times10^{23}\text{ mol}^{-1})(1000\text{ L}\cdot\text{m}^{-3})(0.99707\text{ g}\cdot\text{mL}^{-1})}{(8.8542\times10^{-12}\text{ C}^2\cdot\text{s}^2\cdot\text{kg}^{-1}\cdot\text{m}^{-3})(78.54)(1.3806\times10^{-23}\text{ J}\cdot\text{K}^{-1})(298.2\text{ K})}$$
$$\times(I_m/\text{mol}\cdot\text{kg}^{-1})$$
$$= (1.077\times10^{19}\text{ g}\cdot\text{L}\cdot\text{mL}^{-1}\cdot\text{J}^{-1}\cdot\text{s}^{-2})(I_m/\text{mol}\cdot\text{kg}^{-1})$$
$$= (1.077\times10^{19}\text{ g}\cdot\text{L}\cdot\text{mL}^{-1})\frac{1\text{ kg}}{1000\text{ g}}\frac{1000\text{ mL}}{1\text{ L}}(\text{J}^{-1}\cdot\text{s}^{-2})(I_m/\text{mol}\cdot\text{kg}^{-1})$$
$$= (1.077\times10^{19}\text{ m}^{-2})(I_m/\text{mol}\cdot\text{kg}^{-1})$$

The expression for $\ln \gamma_{\pm}$ is

$$\ln \gamma_{\pm} = -|z_+z_-|\frac{e^2\kappa}{8\pi\varepsilon_0\varepsilon_r kT}$$
$$= -|z_+z_-|\frac{(1.6022\times10^{-19}\text{ C})^2(1.077\times10^{19}\text{ m}^{-2})^{1/2}(I_m/\text{mol}\cdot\text{kg}^{-1})^{1/2}}{8\pi(8.8542\times10^{-12}\text{ C}^2\cdot\text{s}^2\cdot\text{kg}^{-1}\cdot\text{m}^{-3})(78.54)(1.3806\times10^{-23}\text{ J}\cdot\text{K}^{-1})(298.2\text{ K})}$$
$$= -1.171|z_+z_-|(I_m/\text{mol}\cdot\text{kg}^{-1})^{1/2}$$

11–48. Use the Debye-Hückel theory to calculate $\ln \gamma_{\pm}$ for a 0.010-molar NaCl(aq) solution at 25°C. Take $\varepsilon_r = 78.54$ for H_2O(l) at 25.0°C. The experimental value of $\gamma_{\pm}$ is 0.902.

We can use Equation 11.56 directly.

$$\ln \gamma_{\pm} = -1.173(0.010)^{1/2} = -0.1173$$

and so $\gamma_{\pm} = 0.889$.

11–49. Derive the general equation

$$\phi = 1 + \frac{1}{m}\int_0^m m'd\ln\gamma_{\pm}$$

(*Hint*: See the derivation in Problem 11–35.) Use this result to show that

$$\phi = 1 + \frac{\ln\gamma_{\pm}}{3}$$

for the Debye-Hückel theory.

Start with (see Problem 11–35)

$$d(m\phi) = md\ln(m\gamma_{\pm}) = m(d\ln m + d\ln\gamma_{\pm})$$
$$= dm + md\ln\gamma_{\pm}$$

and integrate from $m = 0$ to arbitrary m to obtain

$$m\phi = m + \int_0^m m' d \ln \gamma_\pm$$

or

$$\phi = 1 + \frac{1}{m} \int_0^m m' d \ln \gamma_\pm \tag{1}$$

Now use Equation 11.49 to write $\ln \gamma_\pm$ as

$$\ln \gamma_\pm = -|q_+ q_-| \frac{(\kappa/m^{1/2})}{8\pi\varepsilon_0\varepsilon_r k_B T} m^{1/2}$$

where $\kappa/m^{1/2}$ is *independent of m*. Then

$$d \ln \gamma_\pm = -|q_+ q_-| \frac{(\kappa/m^{1/2})}{8\pi\varepsilon_0\varepsilon_r k_B T} \frac{dm}{2m^{1/2}} = \frac{\ln \gamma_\pm}{2m} dm$$

and

$$m d \ln \gamma_\pm = -\frac{1}{2} |q_+ q_-| \frac{(\kappa/m^{1/2})}{8\pi\varepsilon_0\varepsilon_r k_B T} m^{1/2} dm = \frac{\ln \gamma_\pm}{2} dm$$

Substitute these results into Equation 1 to obtain

$$\begin{aligned} \phi - 1 &= \frac{1}{m} \int_0^m m' d \ln \gamma_\pm \\ &= -\frac{1}{2} |q_+ q_-| \frac{(\kappa/m^{1/2})}{8\pi\varepsilon_0\varepsilon_r k_B T} \frac{1}{m} \int_0^m m'^{1/2} dm' \\ &= \frac{\ln \gamma_\pm}{3} \end{aligned}$$

11–50. In the Debye-Hückel theory, the ions are modeled as point ions and the solvent is modeled as a continuous medium (no structure) with a relative permittivity ε_r. Consider an ion of type i (i = a cation or an anion) situated at the origin of a spherical coordinate system. The presence of this ion at the origin will attract ions of opposite charge and repel ions of the same charge. Let $N_{ij}(r)$ be the number of ions of type j (j = a cation or an anion) situated at a distance r from the central ion of type i (a cation or anion). We can use a Boltzmann factor to say that

$$N_{ij}(r) = N_i e^{-w_{ij}(r)/k_B T}$$

where N_j/V is the bulk number density of j ions and $w_{ij}(r)$ is the interaction energy of an i ion with a j ion. This interaction energy will be electrostatic in origin, so let $w_{ij}(r) = q_j \psi_i(r)$, where q_j is the charge on the j ion and $\psi_i(r)$ is the electrostatic potential due to the central i ion.

A fundamental equation from physics that relates a spherically symmetric electrostatic potential $\psi_i(r)$ to a spherically symmetric charge density $\rho_i(r)$ is Poisson's equation

$$\frac{1}{r^2} \frac{d}{dr} \left[r^2 \frac{d\psi_i(r)}{dr} \right] = -\frac{\rho_i(r)}{\varepsilon_0 \varepsilon_r} \tag{1}$$

where ε_r is the relative permittivity of the solvent. In our case, $\rho_i(r)$ is the charge density around the central ion (of type i). First, show that

$$\rho_i(r) = \frac{1}{V} \sum_j q_j N_{ij}(r) = \sum_j q_j C_j e^{-q_j \psi_i(r)/k_B T}$$

where C_j is the bulk number density of species j ($C_j = N_j/V$). Linearize the exponential term and use the condition of electroneutrality to show that

$$\rho_i(r) = -\psi_i(r)\sum_j \frac{q_j^2 C_j}{k_B T} \tag{2}$$

Now substitute $\rho_i(r)$ into Poisson's equation to get

$$\frac{1}{r^2}\frac{d}{dr}\left[r^2\frac{d\psi_i(r)}{dr}\right] = \kappa^2\psi_i(r) \tag{3}$$

where

$$\kappa^2 = \sum_j \frac{q_j^2 C_j}{\varepsilon_0\varepsilon_r k_B T} = \sum_j \frac{q_j^2}{\varepsilon_0\varepsilon_r k_B T}\left(\frac{N_j}{V}\right) \tag{4}$$

Show that Equation 3 can be written as

$$\frac{d^2}{dr^2}[r\psi_i(r)] = \kappa^2[r\psi_i(r)]$$

Now show that the only solution for $\psi_i(r)$ that is finite for large values of r is

$$\psi_i(r) = \frac{Ae^{-\kappa r}}{r} \tag{5}$$

where A is a constant. Use the fact that if the concentration is very small, then $\psi_i(r)$ is just Coulomb's law and so $A = q_i/4\pi\varepsilon_0\varepsilon_r$ and

$$\psi_i(r) = \frac{q_i e^{-\kappa r}}{4\pi\varepsilon_0\varepsilon_r r} \tag{6}$$

Equation 6 is a central result of the Debye-Hückel theory. The factor of $e^{-\kappa r}$ modulates the resulting Coulombic potential, so Equation 6 is called a *screened Coulombic potential.*

The number of ions of type j situated at a distance r from the central ion of type i is given by

$$N_{ij}(r) = N_j e^{-q_j\psi_i(r)/k_B T}$$

The charge about a central ion of type i due to ions of type j is given by $q_j N_{ij}(r)$ and the net charge is given by $\sum_j q_j N_{ij}(r)$. The charge density at a distance r from the central ion of type i is given by

$$\rho_i(r) = \frac{1}{V}\sum_j q_j N_{ij}(r) = \sum_j q_j C_j e^{-q_j\psi_i(r)/k_B T}$$

Now expand the exponential using the expansion $e^{-x} = 1 - x + \cdots$ to obtain

$$\rho_i(r) = \sum_j q_j C_j - \frac{\psi_i(r)}{k_B T}\sum_j q_j^2 C_j + \cdots$$

$$= 0 \text{ (by electroneutrality)} - \frac{\psi_i(r)}{k_B T}\sum_j q_j^2 C_j^2 + \cdots$$

Now substitute $\rho_i(r)$ into Poisson's equation to get

$$\frac{1}{r^2}\frac{d}{dr}\left[r^2\frac{d\psi_i(r)}{dr}\right] = \kappa^2\psi_i(r) \tag{3}$$

where

$$\kappa^2 = \frac{1}{\varepsilon_0 \varepsilon_r k_B T} \sum_j q_j^2 C_j$$

Now

$$\frac{1}{r^2}\frac{d}{dr}\left[r^2 \frac{d\psi_i(r)}{dr}\right] = \frac{d^2\psi_i(r)}{dr^2} + \frac{2}{r}\frac{d\psi_i(r)}{dr}$$

and

$$\frac{d^2[r\psi_i(r)]}{dr^2} = r\frac{d^2\psi_i(r)}{dr^2} + 2\frac{d\psi_i(r)}{dr}$$

so

$$\frac{1}{r^2}\frac{d}{dr}\left[r^2 \frac{d\psi_i(r)}{dr}\right] = \frac{1}{r}\frac{d^2[r\psi_i(r)]}{dr^2}$$

Therefore, Equation 3 can be written as

$$\frac{d^2[r\psi_i(r)]}{dr^2} = \kappa^2[r\psi(r)]$$

This differential equation has the general solution

$$r\psi_i(r) = Be^{\kappa r} + Ae^{-\kappa r}$$

or

$$\psi_i(r) = \frac{B}{r}e^{\kappa r} + \frac{A}{r}e^{-\kappa r}$$

But B must be zero for $\psi_i(r)$ to be finite as $r \to \infty$. Therefore, we have simply

$$\psi_i(r) = \frac{A}{r}e^{-\kappa r}$$

If the concentration is very small, then $\kappa \to 0$ and $\psi_i(r) \to q_i/4\pi\varepsilon_0\varepsilon_r r$. Therefore,

$$\psi_i(r) \longrightarrow \frac{A}{r} = \frac{q_i}{4\pi\varepsilon_0\varepsilon_r r}$$

and we see that $A = q_i/4\pi\varepsilon_0\varepsilon_r$. Finally then, we have

$$\psi_i(r) = \frac{q_i e^{-\kappa r}}{4\pi\varepsilon_0\varepsilon_r r}$$

11–51. Use Equations 2 and 6 of the previous problem to show that the net charge in a spherical shell of radius r surrounding a central ion of type i is

$$p_i(r)dr = \rho_i(r)4\pi r^2 dr = -q_i\kappa^2 re^{-\kappa r}dr$$

as in Equation 11.54. Why is

$$\int_0^\infty p_i(r)dr = -q_i$$

Start with

$$p_i(r)dr = \rho_i(r)4\pi r^2 dr$$

Equations 2 and 4 of Problem 11–50 show that

$$-\frac{\rho_i(r)}{\varepsilon_0\varepsilon_r} = \kappa^2\psi_i(r)$$

so that

$$p_i(r)dr = -\varepsilon_0\varepsilon_r\kappa^2\psi_i(r)4\pi r^2 dr$$

Using Equation 6 of Problem 11–50, we have

$$p_i(r)dr = -q_i\kappa^2 re^{-\kappa r}dr$$

Therefore,

$$\int_0^\infty p_i(r)dr = -q_i\kappa^2\int_0^\infty re^{-\kappa r}dr = -q_i$$

which it must be because of electroneutrality.

11–52. Use the result of the previous problem to show that the most probable value of r is $1/\kappa$.

Problem 11–51 shows that $p_i(r) \approx re^{-\kappa r}$. Therefore, the most probable value of r is given by

$$\frac{dp_i}{dr} \approx e^{-\kappa r} - \kappa re^{-\kappa r} = 0$$

or $r_{\text{mp}} = 1/\kappa$.

11–53. Show that

$$r_{\text{mp}} = \frac{1}{\kappa} = \frac{304 \text{ pm}}{(c/\text{mol}\cdot\text{L}^{-1})^{1/2}}$$

where c is the molarity of an aqueous solution of a 1–1 electrolyte at 25°C. Take $\varepsilon_r = 78.54$ for $H_2O(l)$ at 25°C.

Use Equation 11.53

$$\begin{aligned}\kappa^2 &= \frac{2(1.602\times10^{-19}\text{ C})^2(6.022\times10^{23}\text{ mol}^{-1})(1000\text{ L}\cdot\text{m}^{-3})(I_c/\text{mol}\cdot\text{L}^{-1})}{(8.8542\times10^{-12}\text{ C}\cdot\text{s}^2\cdot\text{kg}^{-1}\cdot\text{m}^{-3})(78.54)(1.3806\times10^{-23}\text{ J}\cdot\text{K}^{-1})(298.15\text{ K})}\\
&= (1.080\times10^{19}\text{ g}\cdot\text{L}\cdot\text{mL}^{-1}\cdot\text{s}^{-2}\cdot\text{J}^{-1})(I_c/\text{mol}\cdot\text{L}^{-1})\\
&= (1.080\times10^{19}\text{ g}\cdot\text{L}\cdot\text{mL}^{-1})\frac{1\text{ kg}}{1000\text{ g}}\cdot\frac{1000\text{ mL}}{1\text{ L}}(\text{s}^{-2}\cdot\text{J}^{-1})(I_c/\text{mol}\cdot\text{L}^{-1})\\
&= (1.080\times10^{19}\text{ m}^{-2})(I_c/\text{mol}\cdot\text{L}^{-1})\end{aligned}$$

$$\kappa = (3.29\times10^9\text{ m}^{-1})(I_c/\text{mol}\cdot\text{L}^{-1})^{1/2}$$

For a 1–1 electrolyte, $I_c = c$, and so

$$\frac{1}{\kappa} = \frac{3.04 \times 10^{-10}\ \text{m}}{(c/\text{mol}\cdot\text{L}^{-1})^{1/2}} = \frac{304\ \text{pm}}{(c/\text{mol}\cdot\text{L}^{-1})^{1/2}}$$

11–54. Show that

$$r_{\text{mp}} = \frac{1}{\kappa} = 430\ \text{pm}$$

for a 0.50-molar aqueous solution of a 1–1 electrolyte at 25°C. Take $\varepsilon_r = 78.54$ for $H_2O(l)$ at 25°C.

Use Equation 11.55 and the result of Problem 11–52:

$$r_{\text{mp}} = \frac{1}{\kappa} = \frac{304\ \text{pm}}{(c/\text{mol}\cdot\text{L}^{-1})^{1/2}} = \frac{304\ \text{pm}}{(0.50)^{1/2}} = 430\ \text{pm}$$

11–55. How does the thickness of the ionic atmosphere compare for a 1–1 electrolyte and a 2–2 electrolyte?

Equation 11.50 shows that $\kappa^2_{2\text{-}2} = 4\kappa^2_{1\text{-}1}$, or that $\kappa_{2\text{-}2} = 2\kappa_{1\text{-}1}$. Because $1/\kappa$ is a measure of the thickness of an ionic atmosphere, we see that the thickness of the ionic atmosphere of a 2–2 electrolyte is one half that of a 1–1 electrolyte.

11–56. In this problem, we will calculate the total electrostatic energy of an electrolyte solution in the Debye-Hückel theory. Use the equations in Problem 11–50 to show that the number of ions of type j in a spherical shell of radii r and $r + dr$ about a central ion of type i is

$$\left(\frac{N_{ij}(r)}{V}\right) 4\pi r^2 dr = C_j e^{-q_j\psi_i(r)/k_BT} 4\pi r^2 dr \approx C_j \left(1 - \frac{q_j\psi_i(r)}{k_BT}\right) 4\pi r^2 dr \tag{1}$$

The total Coulombic interaction between the central ion of type i and the ions of type j in the spherical shell is $N_{ij}(r)u_{ij}(r)4\pi r^2 dr/V$ where $u_{ij}(r) = q_iq_j/4\pi\varepsilon_0\varepsilon_r r$. To determine the electrostatic interaction energy of all the ions in the solution with the central ion (of type i), U_i^{el}, sum $N_{ij}(r)u_{ij}(r)/V$ over all types of ions in a spherical shell and then integrate over all spherical shells to get

$$\begin{aligned} U_i^{\text{el}} &= \int_0^\infty \left(\sum_j \frac{N_{ij}(r)u_{ij}(r)}{V}\right) 4\pi r^2 dr \\ &= \sum_j \frac{C_j q_i q_j}{\varepsilon_0\varepsilon_r} \int_0^\infty \left(1 - \frac{q_j\psi_i(r)}{k_BT}\right) r dr \end{aligned}$$

Use electroneutrality to show that

$$U_i^{\text{el}} = -q_i\kappa^2 \int_0^\infty \psi_i(r) r dr$$

Now, using Equation 6 of Problem 11–50, show that the interaction of all ions with the central ion (of type i) is given by

$$U_i^{\text{el}} = -\frac{q_i^2\kappa^2}{4\pi\varepsilon_0\varepsilon_r}\int_0^\infty e^{-\kappa r} dr = -\frac{q_i^2\kappa}{4\pi\varepsilon_0\varepsilon_r}$$

Now argue that the total electrostatic energy is

$$U^{\text{el}} = \frac{1}{2}\sum_i N_i U_i^{\text{el}} = -\frac{Vk_{\text{B}}T\kappa^3}{8\pi}$$

Why is there a factor of 1/2 in this equation? Wouldn't you be overcounting the energy otherwise?

According to Problem 11–50, the number of ions of type j in a spherical shell of radii r and $r+dr$ about a central ion of type i is given by

$$\frac{N_{ij}(r)4\pi r^2 dr}{V} = C_j e^{-q_j\psi_i(r)/k_{\text{B}}T} 4\pi r^2 dr$$

Linearize the exponential term to obtain

$$\frac{N_{ij}(r)4\pi r^2 dr}{V} = C_j\left(1 - \frac{q_j\psi_i(r)}{k_{\text{B}}T}\right)4\pi r^2 dr$$

The Coulombic interaction between the ions in the spherical shell and the central ion (of type i) is $u_{ij}(r)N_{ij}(r)4\pi r^2 dr/V$, where $u_{ij}(r) = q_i q_j/4\pi\varepsilon_0\varepsilon_r r$. The interaction of all ions with the central ion is given by

$$\begin{aligned}
U_i^{\text{el}} &= \int_0^\infty \sum_j \frac{u_{ij}(r)N_{ij}(r)4\pi r^2 dr}{V} = \sum_j \int_0^\infty \left(\frac{q_i q_j}{4\pi\varepsilon_0\varepsilon_r r}\right) C_j \left(1 - \frac{q_j\psi_i(r)}{k_{\text{B}}T}\right) 4\pi r^2 dr \\
&= \frac{q_i}{4\pi\varepsilon_0\varepsilon_r}\sum_j \int_0^\infty q_j C_j 4\pi r dr - \frac{q_i}{4\pi\varepsilon_0\varepsilon_r k_{\text{B}}T}\sum_j q_j^2 C_j \int_0^\infty \psi_i(r)4\pi r dr \\
&= 0 \text{ (by electroneutrality)} - q_i\left(\sum_j \frac{q_j^2 C_j}{\varepsilon_0\varepsilon_r k_{\text{B}}T}\right)\int_0^\infty \psi_i(r) r dr \\
&= -q_i\kappa^2 \int_0^\infty \psi_i(r) r dr
\end{aligned}$$

Using Equation 6 of Problem 11–50,

$$U_i^{\text{el}} = -\frac{q_i^2\kappa^2}{4\pi\varepsilon_0\varepsilon_r}\int_0^\infty e^{-\kappa r} dr = -\frac{q_i^2\kappa}{4\pi\varepsilon_0\varepsilon_r}$$

The total electrostatic energy is given by

$$U^{\text{el}} = \frac{1}{2}\sum_i N_i U_i^{\text{el}} = -\frac{Vk_{\text{B}}T\kappa}{8\pi}\sum_i \frac{q_i^2 C_i}{\varepsilon_0\varepsilon_r k_{\text{B}}T} = -\frac{Vk_{\text{B}}T\kappa^3}{8\pi}$$

The factor of 1/2 is needed in the second term in the above equation because in the summation over i, each ion occurs both as a central ion and as an ion in the spherical shell.

11–57. We derived an expression for U^{el} in the previous problem. Use the Gibbs-Helmholtz equation for A (Problem 8–23) to show that

$$A^{\text{el}} = -\frac{Vk_{\text{B}}T\kappa^3}{12\pi}$$

Use the Gibbs-Helmholtz equation for A written in the form

$$\left(\frac{\partial\beta A^{\text{el}}}{\partial\beta}\right) = U^{\text{el}}$$

with (see Problem 11–56)

$$U^{\mathrm{el}} = -\frac{Vk_{\mathrm{B}}T\kappa^3}{8\pi} = -\frac{V}{8\pi(k_{\mathrm{B}}T)^{1/2}}\left(\sum_j \frac{q_j^2}{\varepsilon_0\varepsilon_r}C_j\right)^{3/2}$$

$$= -\frac{V\beta^{1/2}}{8\pi}\left(\sum_j \frac{q_j^2 C_j}{\varepsilon_0\varepsilon_r}\right)^{3/2}$$

Substitute this result into the Gibbs-Helmholtz equation and integrate from 0 to β to obtain

$$\beta A^{\mathrm{el}} = -\frac{V\beta^{3/2}}{12\pi}\left(\sum_j \frac{q_j^2 C_j}{\varepsilon_0\varepsilon_r}\right)^{3/2} = -\frac{V\kappa^3}{12\pi}$$

or

$$A^{\mathrm{el}} = -\frac{Vk_{\mathrm{B}}T\kappa^3}{12\pi}$$

11–58. If we assume that the electrostatic interactions are the sole cause of the nonideality of an electrolyte solution, then we can say that

$$\mu_j^{\mathrm{el}} = \left(\frac{\partial A^{\mathrm{el}}}{\partial n_j}\right)_{T,V} = RT\ln\gamma_j^{\mathrm{el}}$$

or that

$$\mu_j^{\mathrm{el}} = \left(\frac{\partial A^{\mathrm{el}}}{\partial N_j}\right)_{T,V} = k_{\mathrm{B}}T\ln\gamma_j^{\mathrm{el}}$$

Use the result you got for A^{el} in the previous problem to show that

$$k_{\mathrm{B}}T\ln\gamma_j^{\mathrm{el}} = -\frac{\kappa q_j^2}{8\pi\varepsilon_0\varepsilon_r}$$

Use the formula

$$\gamma_\pm^{\nu} = \gamma_+^{\nu_+}\gamma_-^{\nu_-}$$

to show that

$$\ln\gamma_\pm = -\left(\frac{\nu_+q_+^2 + \nu_-q_-^2}{\nu_+ + \nu_-}\right)\frac{\kappa}{8\pi\varepsilon_0\varepsilon_r k_{\mathrm{B}}T}$$

Use the electroneutrality condition $\nu_+q_+ + \nu_-q_- = 0$ to rewrite $\ln\gamma_\pm$ as

$$\ln\gamma_\pm = -|q_+q_-|\frac{\kappa}{8\pi\varepsilon_0\varepsilon_r k_{\mathrm{B}}T}$$

in agreement with Equation 11.49.

Using the final result from Problem 11–57,

$$\mu_j^{\text{el}} = \left(\frac{\partial A^{\text{el}}}{\partial N_j}\right)_{\beta,V}$$

$$= -\frac{V\beta^{1/2}}{12\pi}\frac{\partial}{\partial N_j}\left(\sum_j \frac{q_j^2 C_j}{\varepsilon_0\varepsilon_r V}\right)^{3/2} = -\frac{V\beta^{1/2}}{12\pi}\cdot\frac{3}{2}\left(\sum_j \frac{q_j^2 C_j}{\varepsilon_0\varepsilon_r}\right)^{1/2}\frac{q_j^2}{\varepsilon_0\varepsilon_r V}$$

$$= -\frac{\kappa q_j^2}{8\pi\varepsilon_0\varepsilon_r} = k_{\text{B}}T\ln\gamma_j^{\text{el}}$$

Now, take the logarithm of the equation $\gamma_\pm^\nu = \gamma_+^{\nu_+}\gamma_-^{\nu_-}$ and the previous result to obtain

$$\ln\gamma_\pm = \frac{\nu_+\ln\gamma_+ + \nu_-\ln\gamma_-}{\nu_+ + \nu_-} = -\left(\frac{\kappa}{8\pi\varepsilon_0\varepsilon_r k_{\text{B}}T}\right)\left(\frac{\nu_+q_+^2 + \nu_-q_-^2}{\nu_+ + \nu_-}\right)$$

But

$$\nu_+q_+^2 + \nu_-q_-^2 = q_+(\nu_+q_+) + \nu_-q_-^2 = q_+(|\nu_-q_-|) + \nu_-|q_-|^2$$

$$= q_+|q_-|\left(\nu_- + \nu_-\frac{|q_-|}{q_+}\right) = |q_+q_-|(\nu_- + \nu_+)$$

where we have used the electroneutrality condition, $\nu_+q_+ = \nu_-|q_-|$, and so finally

$$\ln\gamma_\pm = -|q_+q_-|\frac{\kappa}{8\pi\varepsilon_0\varepsilon_r k_{\text{B}}T}$$

11–59. Derive Equation 11.56 from Equation 11.49.

See the solution to Problem 11–47, but do not include the factor $\rho = 0.99707\ \text{g}\cdot\text{mL}^{-1}$.

11–60. Show that Equation 11.59 reduces to Equation 11.49 for small concentrations.

We want to show that Equation 11.59 reduces to Equation 11.49 as $\rho \to 0$ or as $\kappa \to 0$. Let's consider $\ln\gamma_\pm^{\text{el}}$ first. Use the fact that $(1+x)^{1/2} = 1 + x/2 - x^2/8 + x^3/16 + \cdots$ to write

$$x(1+2x)^{1/2} - x - x^2 = x\left[1 + x - \frac{(2x)^2}{8} + \frac{(2x)^3}{16} + O(x^4)\right] - x - x^2$$

$$= -\frac{x^3}{2}$$

Using the fact that $x = \kappa d$, Equation 11.60 becomes

$$\ln\gamma_\pm^{\text{el}} = -\frac{\kappa^3}{8\pi\rho}$$

For a 1–1 electrolyte, $\kappa^2 = \rho/\varepsilon_0\varepsilon_r k_{\text{B}}T$, so we have

$$\ln\gamma_\pm^{\text{el}} = -\frac{\kappa}{8\pi\varepsilon_0\varepsilon_r k_{\text{B}}T}$$

in agreement with Equation 11.49 for a 1–1 electrolyte. The $\ln \gamma^{\text{HS}}$ contribution to Equation 11.59 is negligible when $\rho \to 0$ because $y = \pi\rho d^3/6$.

11–61. In this problem, we will investigate the temperature dependence of activities. Starting with the equation $\mu_1 = \mu_1^* + RT \ln a_1$, show that

$$\left(\frac{\partial \ln a_1}{\partial T}\right)_{P,x_1} = \frac{\overline{H}_1^* - \overline{H}_1}{RT^2}$$

where $\overline{H}_1^*$ is the molar enthalpy of the pure solvent (at one bar) and $\overline{H}_1$ is its partial molar enthalpy in the solution. The difference between $\overline{H}_1^*$ and $\overline{H}_1$ is small for dilute solutions, so a_1 is fairly independent of temperature.

Starting with $\mu_1 = \mu_1^* + RT \ln a_1$, differentiate μ_1/T with respect to T to obtain

$$\left(\frac{\partial \mu_1/T}{\partial T}\right)_{P,x_1} - \left(\frac{\partial \mu_1^*/T}{\partial T}\right)_P = R\left(\frac{\partial \ln a_1}{\partial T}\right)_{P,x_1}$$

Now use the equation (see Example 10–1)

$$\left(\frac{\partial \mu_j/T}{\partial T}\right)_{P,x_1} = -\frac{\overline{H}_j}{T^2}$$

to write

$$\left(\frac{\partial \ln a_1}{\partial T}\right)_{P,x_1} = \frac{\overline{H}_1^* - \overline{H}_1}{RT^2}$$

11–62. Henry's law says that the pressure of a gas in equilibrium with a non-electrolyte solution of the gas in a liquid is proportional to the molality of the gas in the solution for sufficiently dilute solutions. What form do you think Henry's law takes on for a gas such as HCl(g) dissolved in water? Use the following data for HCl(g) at 25°C to test your prediction.

$P_{\text{HCl}}/10^{-11}$ atm	$m_{\text{HCl}}/10^{-3}$ mol·kg^{-1}
0.147	1.81
0.238	2.32
0.443	3.19
0.663	3.93
0.851	4.47
1.080	5.06
1.622	6.25
1.929	6.84
2.083	7.12

A plot of pressure against molality is not a straight line, but a plot of pressure against molality squared is almost a straight line. This is due to the fact that HCl(aq) dissociates into H^+(aq) and Cl^-(aq).

11–63. When the pressures in Problem 11–62 are plotted against molality squared, the result is almost a straight line. Curve fit the data to polynomials of the form

$$P = k_{\text{K}} m^2 (1 + c_1 m^{1/2} + c_2 m + c_3 m^{3/2} + \cdots)$$

of increasing degree and evaluate k_{H}.

If the data are fitted to $P = k_{\text{H}} m^2$, k_{H} turns out to be (in units of atm·kg^2·mol^{-2}) 4.15×10^{-7}. The subsequent fits are (suppressing the units)

k_{H}	c_1	c_2	c_3	c_4
4.83×10^{-7}	-1.77			
4.92×10^{-7}	-2.24	3.48		
4.93×10^{-7}	-2.33	4.75	-6.21	
4.93×10^{-7}	-2.34	5.07	-9.45	10.3

Thus we see that $k_{\text{H}} = 4.93 \times 10^{-7}$ atm·kg^2·mol^{-2}.

11–64. When the data in Problem 11–62 are plotted in the form of P/m^2 against $m^{1/2}$, the result is essentially a straight line with a negative slope. Why is this so? Use Debye-Hückel theory to calculate the slope of this line and compare your result with the final value of c_1 in Problem 11–63.

The activity of the HCl(aq) is given by $a_{\text{HCl}} = P/k_{\text{H}}$. Using the fact that $a_{\text{HCl}} = a_{\pm}^2 = m^2\gamma_{\pm}^2$, we have

$$P = k_{\text{H}} m^2 \gamma_{\pm}^2$$

Note that as $m \to 0$, $\gamma_{\pm} \to 1$, and $P \to k_{\text{H}} m^2$, as expected. The Debye-Hückel expression for $\gamma_{\pm}$ in this case is

$$\ln \gamma_{\pm} = -1.171 m^{1/2}$$

Substitute this expression for $\gamma_{\pm}$ into $P = k_{\text{H}} m^2 \gamma_{\pm}^2$ and linearize the exponential according to $e^{-x} = 1 - x + \cdots$ to obtain

$$P = k_{\text{H}} m^2 (1 - 2.342 m^{1/2} + \cdots)$$

Thus, we predict that c_1 in Problem 11–63 is equal to -2.34, in excellent agreement.

CHAPTER 12

Chemical Equilibrium

PROBLEMS AND SOLUTIONS

12–1. Express the concentrations of each species in the following chemical equations in terms of the extent of reaction, ξ. The initial conditions are given under each equation.

a.

	$SO_2Cl_2(g)$	$\rightleftharpoons$	$SO_2(g)$	+	$Cl_2(g)$
(1)	n_0		0		0
(2)	n_0		n_1		0

b.

	$2\,SO_3(g)$	$\rightleftharpoons$	$2\,SO_2(g)$	+	$O_2(g)$
(1)	n_0		0		0
(2)	n_0		0		n_1

c.

	$N_2(g)$	+	$2\,O_2(g)$	$\rightleftharpoons$	$N_2O_4(g)$
(1)	n_0		$2n_0$		0
(2)	n_0		n_0		0

We can use Equation 12.1 in all cases to express the concentrations of each species.

a.

	$SO_2Cl_2(g)$	$\rightleftharpoons$	$SO_2(g)$	+	$Cl_2(g)$
(1)	$n_0 - \xi$		ξ		ξ
(2)	$n_0 - \xi$		$n_1 + \xi$		ξ

b.

	$2\,SO_3(g)$	$\rightleftharpoons$	$2\,SO_2(g)$	+	$O_2(g)$
(1)	$n_0 - 2\xi$		2ξ		ξ
(2)	$n_0 - 2\xi$		2ξ		$n_1 + \xi$

c.

	$N_2(g)$	+	$2\,O_2(g)$	$\rightleftharpoons$	$N_2O_4(g)$
(1)	$n_0 - \xi$		$2n_0 - 2\xi$		ξ
(2)	$n_0 - \xi$		$n_0 - 2\xi$		ξ

12–2. Write out the equilibrium-constant expression for the reaction that is described by the equation

$$2\,SO_2(g) + O_2(g) \rightleftharpoons 2\,SO_3(g)$$

Compare your result to what you get if the reaction is represented by

$$SO_2(g) + \frac{1}{2}O_2(g) \rightleftharpoons SO_3(g)$$

Using Equation 12.12, we write K_P for the first chemical equation as

$$K_P(T) = \frac{P_{SO_3}^2}{P_{O_2}P_{SO_2}^2}$$

For the second chemical equation, we again use Equation 12.12 to find

$$K'_P(T) = \frac{P_{SO_3}}{P_{O_2}^{1/2} P_{SO_2}}$$

which is the square root of K_P.

12–3. Consider the dissociation of $N_2O_4(g)$ into $NO_2(g)$ described by

$$N_2O_4(g) \rightleftharpoons 2\,NO_2(g)$$

Assuming that we start with n_0 moles of $N_2O_4(g)$ and no $NO_2(g)$, show that the extent of reaction, ξ_{eq}, at equilibrium is given by

$$\frac{\xi_{eq}}{n_0} = \left(\frac{K_P}{K_P + 4P}\right)^{1/2}$$

Plot ξ_{eq}/n_0 against P given that $K_P = 6.1$ at 100°C. Is your result in accord with Le Châtelier's principle?

At equilibrium, $n_{N_2O_4} = n_0 - \xi_{eq}$ and $n_{NO_2} = 2\xi_{eq}$. The partial pressures of the species are then

$$P_{N_2O_4} = \frac{n_0 - \xi_{eq}}{n_0 - \xi_{eq} + 2\xi_{eq}} P = \frac{n_0 - \xi_{eq}}{n_0 + \xi_{eq}} P$$

and

$$P_{NO_2} = \frac{2\xi_{eq}}{n_0 + \xi_{eq}} P = \frac{2\xi_{eq}}{n_0 + \xi_{eq}} P$$

Substituting into the expression for K_P, we find

$$K_P = \frac{P_{NO_2}^2}{P_{N_2O_4}} = \frac{4\xi_{eq}^2(n_0 + \xi_{eq})}{(n_0 + \xi_{eq})^2(n_0 - \xi_{eq})} P = \frac{4(\xi_{eq}/n_0)^2}{1 - (\xi_{eq}/n_0)^2} P$$

Solving this expression for ξ_{eq}/n_0, we find that

$$K_P - K_P(\xi_{eq}/n_0)^2 = 4P(\xi_{eq}/n_0)^2$$

$$(4P + K_P)(\xi_{eq}/n_0)^2 = K_P$$

$$\frac{\xi_{eq}}{n_0} = \left(\frac{K_P}{K_P + 4P}\right)^{1/2}$$

A plot of ξ_{eq}/n_0 against P is shown below.

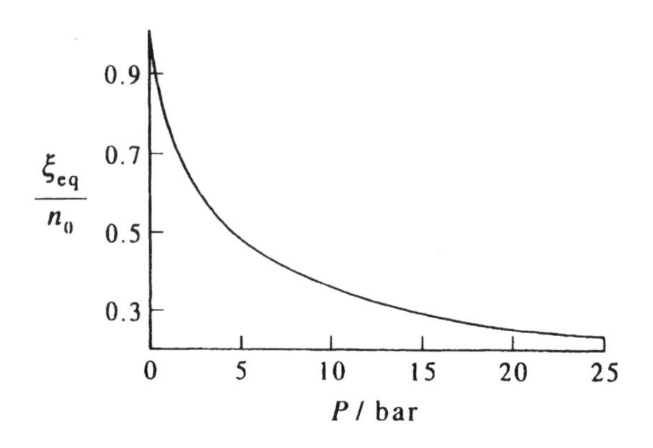

This is in accord with Le Châtelier's principle: as the pressure increases, the reaction occurs to a lesser extent and ξ_{eq} decreases.

12–4. In Problem 12–3 you plotted the extent of reaction at equilibrium against the total pressure for the dissociation of $N_2O_4(g)$ to $NO_2(g)$. You found that ξ_{eq} decreases as P increases, in accord with Le Châtelier's principle. Now let's introduce n_{inert} moles of an inert gas into the system. Assuming that we start with n_0 moles of $N_2O_4(g)$ and no $NO_2(g)$, derive an expression for ξ_{eq}/n_0 in terms of P and the ratio $r = n_{inert}/n_0$. As in Problem 12–3, let $K_P = 6.1$ and plot ξ_{eq}/n_0 versus P for $r = 0$ (Problem 12–3), $r = 0.50$, $r = 1.0$, and $r = 2.0$. Show that introducing an inert gas into the reaction mixture at constant pressure has the same effect as lowering the pressure. What is the effect of introducing an inert gas into a reaction system at constant volume?

At equilibrium, as before, $n_{N_2O_4} = n_0 - \xi_{eq}$ and $n_{NO_2} = 2\xi_{eq}$. However, the total number of moles present has changed to $n_0 + \xi_{eq} + n_{inert}$. The partial pressures of the species are then

$$P_{N_2O_4} = \frac{n_0 - \xi_{eq}}{n_0 + \xi_{eq} + n_{inert}} P \quad \text{and} \quad P_{NO_2} = \frac{2\xi_{eq}}{n_0 + \xi_{eq} + n_{inert}} P$$

Substituting into the expression for K_P, we find

$$K_P = \frac{P_{NO_2}^2}{P_{N_2O_4}} = \frac{4\xi_{eq}^2 P}{(n_0 - \xi_{eq})(n_0 + n_{inert} + \xi_{eq})} = \frac{4(\xi_{eq}/n_0)^2 P}{(1 - \xi_{eq}/n_0)(1 + r + \xi_{eq}/n_0)}$$

Solving for ξ_{eq}/n_0 gives

$$\frac{\xi_{eq}}{n_0} = -\frac{K_P r}{2(K_P + 4P)} \pm \frac{1}{2(K_P + 4P)} \left[K_P^2 r^2 + 4K_P(1 + r)(K_P + 4P)\right]^{1/2}$$

where we have let $r = n_{inert}/n_0$. When $r = 0$, as in Problem 12–3, this expression becomes

$$\frac{\xi_{eq}}{n_0} = \pm\frac{[4K_P(K_P + 4P)]^{1/2}}{2(K_P + 4P)} = \pm\left[\frac{K_P}{(K_P + 4P)}\right]^{1/2}$$

For ξ_{eq} to be positive, we take the positive root. Now we can plot this expression for various values of r:

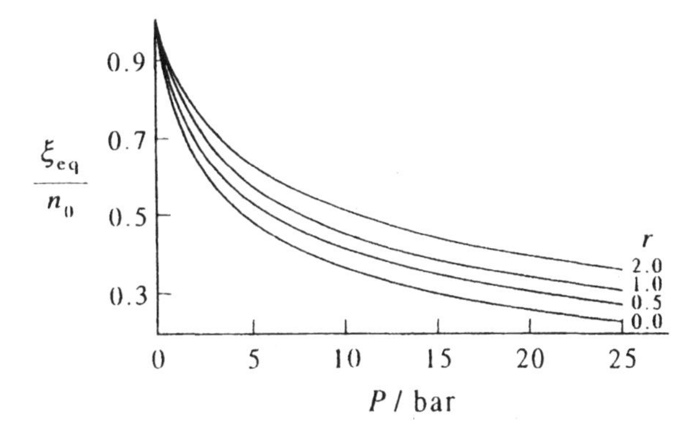

We see that introducing an inert gas into a constant-pressure reaction system increases the value of ξ_{eq} and so has the same effect as lowering the pressure. Introducing an inert gas into a constant-volume reaction system has no effect on the value of ξ_{eq}.

12–5. Re-do Problem 12–3 with n_0 moles of $N_2O_4(g)$ and n_1 moles of $NO_2(g)$ initially. Let $n_1/n_0 = 0.50$ and 2.0.

Now, at equilibrium, $n_{N_2O_4} = n_0 - \xi_{eq}$ and $n_{NO_2} = n_1 + 2\xi_{eq}$. The total number of moles of gas present will be $n_0 + n_1 + \xi_{eq}$. The partial pressures of the species are then, letting $s = n_1/n_0$,

$$P_{N_2O_4} = \frac{n_0 - \xi_{eq}}{n_0 + n_1 + \xi_{eq}} P = \frac{1 - \xi_{eq}/n_0}{1 + s + \xi_{eq}/n_0} P$$

and

$$P_{NO_2} = \frac{n_1 + 2\xi_{eq}}{n_0 + n_1 + \xi_{eq}} P = \frac{s + 2\xi_{eq}/n_0}{1 + s + \xi_{eq}/n_0} P$$

Substituting into the expression for K_P, we find

$$K_P = \frac{P_{NO_2}^2}{P_{N_2O_4}} = \frac{(s + 2\xi_{eq}/n_0)^2}{(1 + s + \xi_{eq}/n_0)(1 - \xi_{eq}/n_0)} P$$

and solving for ξ_{eq}/n_0 gives

$$\frac{\xi_{eq}}{n_0} = -\frac{s}{2} \pm \frac{1}{2}\left[s^2 - 4\left(\frac{Ps^2 - K_P s - K_P}{K_P + 4P}\right)\right]^{1/2}$$

When $s = 0$, as in Problem 12–3, this expression becomes

$$\frac{\xi_{eq}}{n_0} = \pm \frac{K_P^{1/2}}{2(K_P + 4P)^{1/2}} = \pm\left[\frac{K_P}{(K_P + 4P)}\right]^{1/2}$$

For ξ_{eq} to be positive, we take the positive root. Now we can plot this expression for various values of s:

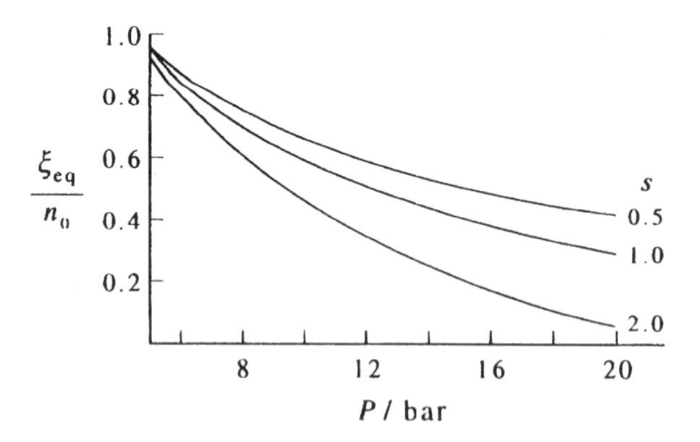

12–6. Consider the ammonia-synthesis reaction, which can be described by

$$N_2(g) + 3H_2(g) \rightleftharpoons 2NH_3(g)$$

Suppose initially there are n_0 moles of $N_2(g)$ and $3n_0$ moles of $H_2(g)$ and no $NH_3(g)$. Derive an expression for $K_P(T)$ in terms of the equilibrium value of the extent of reaction, ξ_{eq}, and the pressure, P. Use this expression to discuss how ξ_{eq}/n_0 varies with P and relate your conclusions to Le Châtelier's principle.

At equilibrium, there will be $n_0 - \xi_{eq}$ moles of $N_2(g)$, $3n_0 - 3\xi_{eq}$ moles of $H_2(g)$, and $2\xi_{eq}$ moles of $NH_3(g)$, yielding a total of $4n_0 - 2\xi_{eq}$ moles of gas. Then

$$P_{N_2} = \frac{n_0 - \xi_{eq}}{4n_0 - 2\xi_{eq}} P \qquad P_{H_2} = \frac{3n_0 - 3\xi_{eq}}{4n_0 - 2\xi_{eq}} P \qquad \text{and} \qquad P_{NH_3} = \frac{2\xi_{eq}}{4n_0 - 2\xi_{eq}} P$$

We then express K_P as

$$\begin{aligned} K_P &= \frac{P_{NH_3}^2}{P_{H_2}^3 P_{N_2}} \\ &= \frac{4\xi_{eq}^2(4n_0 - 2\xi_{eq})^2}{(3n_0 - 3\xi_{eq})^3(n_0 - \xi_{eq})P^2} \\ &= \frac{16(\xi_{eq}/n_0)^2(2 - \xi_{eq}/n_0)^4}{27(2 - \xi_{eq}/n_0)^2(1 - \xi_{eq}/n_0)^4 P^2} \\ &= \frac{16(\xi_{eq}/n_0)^2(2 - \xi_{eq}/n_0)^2}{27(1 - \xi_{eq}/n_0)^4 P^2} \end{aligned}$$

The following plot of ξ_{eq}/n_0 against P shows that ξ_{eq}/n_0 increases as P increases, as Le Châtelier's principle would dictate.

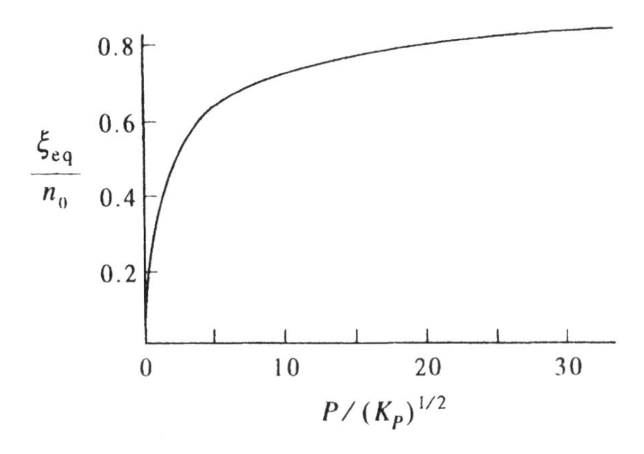

12–7. Nitrosyl chloride, NOCl, decomposes according to

$$2\,NOCl(g) \rightleftharpoons 2\,NO(g) + Cl_2(g)$$

Assuming that we start with n_0 moles of NOCl(g) and no NO(g) or $Cl_2(g)$, derive an expression for K_P in terms of the equilibirum value of the extent of reaction, ξ_{eq}, and the pressure, P. Given that $K_P = 2.00 \times 10^{-4}$, calculate ξ_{eq}/n_0 when $P = 0.080$ bar. What is the new value of ξ_{eq}/n_0 at equilibrium when $P = 0.160$ bar? Is this result in accord with Le Châtelier's principle?

At equilibrium, there will be $n_0 - 2\xi_{eq}$ moles of NOCl(g), $2\xi_{eq}$ moles of NO(g), and ξ_{eq} moles of $Cl_2(g)$, making a total of $n_0 + \xi_{eq}$ moles of gas present. Then

$$P_{NOCl} = \frac{n_0 - 2\xi_{eq}}{n_0 + \xi_{eq}} P \qquad P_{NO} = \frac{2\xi_{eq}}{n_0 + \xi_{eq}} P \qquad \text{and} \qquad P_{Cl_2} = \frac{\xi_{eq}}{n_0 + \xi_{eq}} P$$

We then write K_P as

$$K_P = \frac{P_{Cl_2} P_{NO}^2}{P_{NOCl}^2} = \frac{4\xi_{eq}^3 P}{(n_0 + \xi_{eq})(n_0 - 2\xi_{eq})^2} = \frac{4(\xi_{eq}/n_0)^3 P}{(1 + \xi_{eq}/n_0)(1 - 2\xi_{eq}/n_0)^2}$$

A plot of ξ_{eq}/n_0 against P looks like

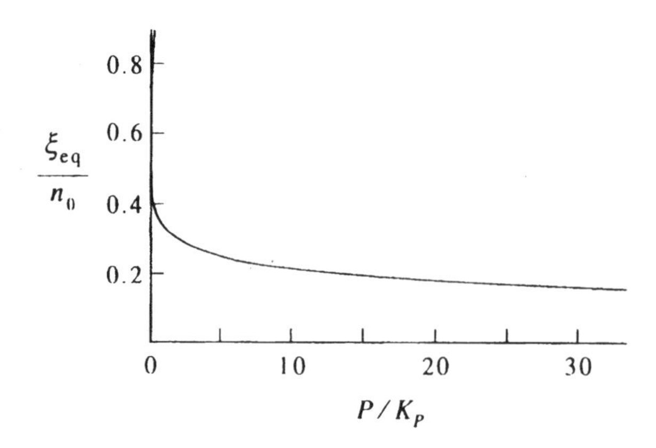

We see that ξ_{eq}/n_0 decreases as P increases, as Le Châtelier's principle would dictate. Letting $K_P = 2.00 \times 10^{-4}$, we find $\xi_{eq}/n_0 = 0.0783$ at $P = 0.080$ bar and $\xi_{eq}/n_0 = 0.0633$ at $P = 0.160$ bar, again in accord with Le Châtelier's principle.

12–8. The value of K_P at 1000°C for the decomposition of carbonyl dichloride (phosgene) according to

$$COCl_2(g) \rightleftharpoons CO(g) + Cl_2(g)$$

is 34.8 if the standard state is taken to be one bar. What would the value of K_P be if for some reason the standard state were taken to be 0.500 bar? What does this result say about the numerical values of equilibrium constants?

Use the definition of K_P to find the value of $K_P(0.500 \text{ bar})$ at the new standard state:

$$K_P(1 \text{ bar}) = \frac{(P_{CO}/1 \text{ bar})(P_{Cl_2}/1 \text{ bar})}{(P_{COCl_2}/1 \text{ bar})} = 34.8$$

$$K_P(0.500 \text{ bar}) = \frac{(P_{CO}/0.500 \text{ bar})(P_{Cl_2}/0.500 \text{ bar})}{(P_{COCl_2}/0.500 \text{ bar})}$$

$$= 0.500 K_P(1 \text{ bar}) = 17.4$$

The numerical values of equilibrium constants are dependent on the standard states chosen.

12–9. Most gas-phase equilibrium constants in the recent chemical literature were calculated assuming a standard state pressure of one atmosphere. Show that the corresponding equilibrium constant for a standard state pressure of one bar is given by

$$K_P(\text{bar}) = K_P(\text{atm})(1.01325)^{\Delta\nu}$$

where $\Delta\nu$ is the sum of the stoichiometric coefficients of the products minus that of the reactants.

Consider the reaction described by

$$\nu_A A(g) + \nu_B B(g) \rightleftharpoons \nu_Y Y(g) + \nu_Z Z(g)$$

(We can extend this case to include as many reactants and products as we desire.) Now write K_P:

$$\begin{aligned}
K_P(\text{bar}) &= \frac{(P_Z/1\ \text{bar})^{\nu_Z}(P_Y/1\ \text{bar})^{\nu_Y}}{(P_B/1\ \text{bar})^{\nu_B}(P_A/1\ \text{bar})^{\nu_A}} \\
&= \frac{P_Z^{\nu_Z}P_Y^{\nu_Y}}{(1\ \text{bar})^{\nu_Z+\nu_Y}}\,\frac{(1\ \text{bar})^{\nu_A+\nu_B}}{P_B^{\nu_B}P_A^{\nu_A}} \\
&= \frac{P_Z^{\nu_Z}P_Y^{\nu_Y}}{P_B^{\nu_B}P_A^{\nu_A}}\left(\frac{1}{1\ \text{bar}}\right)^{\Delta\nu}\left(\frac{1.01325\ \text{bar}}{1\ \text{atm}}\right)^{\Delta\nu} \\
&= \frac{(P_Z/1\ \text{atm})^{\nu_Z}(P_Y/1\ \text{atm})^{\nu_Y}}{(P_B/1\ \text{atm})^{\nu_B}(P_A/1\ \text{atm})^{\nu_A}}(1.01325\ \text{bar})^{\Delta\nu} \\
&= K_P(\text{atm})(1.01325\ \text{bar})^{\Delta\nu}
\end{aligned}$$

12–10. Using the data in Table 12.1, calculate $\Delta_r G^\circ(T)$ and $K_P(T)$ at 25°C for

(a) $N_2O_4(g) \rightleftharpoons 2\,NO_2(g)$
(b) $H_2(g)+I_2(g) \rightleftharpoons 2\,HI(g)$
(c) $3\,H_2(g) + N_2(g) \rightleftharpoons 2\,NH_3(g)$

Use Equations 12.19 and 12.11 to find $\Delta_r G^\circ$ and K_P.

a.
$$\begin{aligned}
\Delta_r G^\circ &= 2(51.258\ \text{kJ}\cdot\text{mol}^{-1}) - 97.787\ \text{kJ}\cdot\text{mol}^{-1} \\
&= 4.729\ \text{kJ}\cdot\text{mol}^{-1} \\
K_P &= e^{-\Delta_r G^\circ/RT} = 0.148
\end{aligned}$$

b.
$$\begin{aligned}
\Delta_r G^\circ &= 2(1.560\ \text{kJ}\cdot\text{mol}^{-1}) - 19.325\ \text{kJ}\cdot\text{mol}^{-1} \\
&= -16.205\ \text{kJ}\cdot\text{mol}^{-1} \\
K_P &= e^{-\Delta_r G^\circ/RT} = 690
\end{aligned}$$

c.
$$\begin{aligned}
\Delta_r G^\circ &= 2(-16.637\ \text{kJ}\cdot\text{mol}^{-1}) = -33.274\ \text{kJ}\cdot\text{mol}^{-1} \\
K_P &= e^{-\Delta_r G^\circ/RT} = 6.80 \times 10^5
\end{aligned}$$

12–11. Calculate the value of $K_c(T)$ based upon a one $\text{mol}\cdot\text{L}^{-1}$ standard state for each of the equations in Problem 12–10.

Use Equation 12.17, recalling that R must be in units of $\text{L}\cdot\text{bar}\cdot\text{K}^{-1}\cdot\text{mol}^{-1}$ because c° is 1 $\text{mol}\cdot\text{L}^{-1}$ and P° is 1 bar:

a. $K_c = K_P(RT)^{-\Delta\nu} = (0.148)[(298.15)(0.083145)]^{-1} = 5.97 \times 10^{-3}$

b. $K_c = K_P(RT)^{-\Delta\nu} = (690)[(298.15)(0.083145)]^{0} = 690$

c. $K_c = K_P(RT)^{-\Delta\nu} = (6.80 \times 10^5)[(298.15)(0.083145)]^{2} = 4.17 \times 10^8$

12–12. Derive a relation between K_P and K_c for the following:

a. $CO(g) + Cl_2(g) \rightleftharpoons COCl_2(g)$

b. $CO(g) + 3\,H_2(g) \rightleftharpoons CH_4(g) + H_2O(g)$

c. $2\,BrCl(g) \rightleftharpoons Br_2(g) + Cl_2(g)$

Again, use Equation 12.17.

a. $K_P = K_c\left(\frac{c^\circ RT}{P^\circ}\right)^{-1}$

b. $K_P = K_c\left(\frac{c^\circ RT}{P^\circ}\right)^{-2}$

c. $K_P = K_c\left(\frac{c^\circ RT}{P^\circ}\right)^{0} = K_c$

12–13. Consider the dissociation reaction of $I_2(g)$ described by

$$I_2(g) \rightleftharpoons 2\,I(g)$$

The total pressure and the partial pressure of $I_2(g)$ at 1400°C have been measured to be 36.0 torr and 28.1 torr, respectively. Use these data to calculate K_P (one bar standard state) and K_c (one $mol{\cdot}L^{-1}$ standard state) at 1400°C.

First we express P_{I_2} and P_I in bars:

$$P_{I_2} = 28.1\text{ torr}\left(\frac{1.01325\text{ bar}}{760\text{ torr}}\right) = 0.0375\text{ bar}$$

$$P_I = P_{tot} - P_{I_2} = 7.9\text{ torr}\left(\frac{1.01325\text{ bar}}{760\text{ torr}}\right) = 0.0105\text{ bar}$$

Now use the definitions of K_P and K_c to write

$$K_P = \frac{P_I^2}{P_{I_2}} = 2.94\times10^{-3}$$

$$\begin{aligned}K_c &= K_P\left(\frac{P^\circ}{c^\circ RT}\right)\\ &= (2.94\times10^{-3})\left[\frac{1\text{ bar}}{(1\text{ mol}{\cdot}\text{L}^{-1})(1673\text{ K})(0.083145\text{ L}{\cdot}\text{bar}{\cdot}\text{mol}^{-1}{\cdot}\text{K}^{-1})}\right]\\ &= 2.11\times10^{-5}\end{aligned}$$

12–14. Show that

$$\frac{d\ln K_c}{dT} = \frac{\Delta_r U^\circ}{RT^2}$$

for a reaction involving ideal gases.

We know that

$$K_P = K_c\left(\frac{c^\circ RT}{P^\circ}\right)^{\Delta\nu} \qquad (12.17)$$

Now begin with Equation 12.29:

$$\begin{aligned}\frac{\Delta_r H^\circ}{RT^2} &= \frac{d\ln K_P}{dT} = \frac{d}{dT}\left[\ln K_c + \Delta\nu\ln\left(\frac{c^\circ RT}{P^\circ}\right)\right]\\ &= \frac{d\ln K_c}{dT} + \frac{\Delta\nu}{T}\end{aligned}$$

or

$$\frac{d\ln K_c}{dT} = \frac{\Delta_r H^\circ - \Delta\nu RT}{RT^2} = \frac{\Delta_r H^\circ - \Delta(PV)}{RT^2} = \frac{\Delta_r U^\circ}{RT^2}$$

because $U = H + PV$.

12–15. Consider the gas-phase reaction for the synthesis of methanol from CO(g) and H_2(g)

$$\mathrm{CO(g)} + 2\,\mathrm{H_2(g)} \rightleftharpoons \mathrm{CH_3OH(g)}$$

The value of the equilibrium constant K_P at 500 K is 6.23×10^{-3}. Initially equimolar amounts of CO(g) and H_2(g) are introduced into the reaction vessel. Determine the value of ξ_{eq}/n_0 at equilibrium at 500 K and 30 bar.

At equilibrium, the number of moles of CO(g) will be $n_0 - \xi_{eq}$, the number of moles of H_2(g) will be $n_0 - 2\xi_{eq}$, and the number of moles of CH_3OH(g) will be ξ_{eq}. The total moles of gas present will therefore be $2n_0 - 2\xi_{eq}$. We can now find the partial pressures of each of the components of the mixture:

$$P_{CO} = \frac{n_0 - \xi_{eq}}{2(n_0 - \xi_{eq})}P = \frac{1}{2}P \qquad P_{H_2} = \frac{n_0 - 2\xi_{eq}}{2(n_0 - \xi_{eq})}P \qquad \text{and} \qquad P_{CH_3OH} = \frac{\xi_{eq}}{2(n_0 - \xi_{eq})}P$$

We then express K_P as

$$K_P = \frac{P_{CH_3OH}}{P_{H_2}^2 P_{CO}} = \frac{4\xi_{eq}(n_0 - \xi_{eq})}{P^2(n_0 - 2\xi_{eq})^2} = \frac{4x(1-x)}{P^2(1-2x)^2}$$

where $x = \xi_{eq}/n_0$. The value of K_P is 6.23×10^{-3}, so

$$\frac{4x(1-x)}{(1-2x)^2} = (30\ \text{bar})^2(6.23 \times 10^{-3})$$

which we can solve numerically (using the Newton-Raphson method) to find

$$x = 0.305 \qquad \text{or} \qquad x = 0.695$$

Since $x < 0.50$ (otherwise the amount of H_2 present will be a negative quantity), $\xi_{eq}/n_0 = 0.31$.

12–16. Consider the two equations

(1) $\mathrm{CO(g)} + \mathrm{H_2O(g)} \rightleftharpoons \mathrm{CO_2(g)} + \mathrm{H_2(g)}$ $\quad K_1$

(2) $\mathrm{CH_4(g)} + \mathrm{H_2O(g)} \rightleftharpoons \mathrm{CO(g)} + 3\,\mathrm{H_2(g)}$ $\quad K_2$

Show that $K_3 = K_1 K_2$ for the sum of these two equations

(3) $\mathrm{CH_4(g)} + 2\,\mathrm{H_2O(g)} \rightleftharpoons \mathrm{CO_2(g)} + 4\,\mathrm{H_2(g)}$ $\quad K_3$

How do you explain the fact that you would add the values of $\Delta_r G^\circ$ but multiply the equilibrium constants when adding Equations 1 and 2 to get Equation 3?

Use Equation 12.12 to express K_1, K_2, and K_3.

$$K_1 = \frac{P_{CO_2}P_{H_2}}{P_{H_2O}P_{CO}} \qquad K_2 = \frac{P_{H_2}^3 P_{CO}}{P_{H_2O}P_{CH_4}}$$

$$K_3 = \frac{P_{H_2}^4 P_{CO_2}}{P_{H_2O}^2 P_{CH_4}} = \left(\frac{P_{CO_2} P_{H_2}}{P_{H_2O} P_{CO}}\right)\left(\frac{P_{H_2}^3 P_{CO}}{P_{H_2O} P_{CH_4}}\right) = K_1 K_2$$

We multiply the equilibrium constants because of their logarithmic relationship with $\Delta_r G^\circ$. Recall that (Equation 12.11) $\Delta_r G^\circ = -RT \ln K_P$. Adding $\Delta_r G_1^\circ$ and $\Delta_r G_2^\circ$ would give

$$-RT \ln K_1 - RT \ln K_2 = -RT \ln K_1 K_2$$

12–17. Given:

$$2\,BrCl(g) \rightleftharpoons Cl_2(g) + Br_2(g) \qquad K_P = 0.169$$
$$2\,IBr(g) \rightleftharpoons Br_2(g) + I_2(g) \qquad K_P = 0.0149$$

Determine K_P for the reaction

$$BrCl(g) + \tfrac{1}{2} I_2(g) \rightleftharpoons IBr(g) + \tfrac{1}{2} Cl_2(g)$$

We number the equations in order of appearance. Equation 3 can be expressed by

$$\text{Equation 3} = \tfrac{1}{2}\text{Equation 1} - \tfrac{1}{2}\text{Equation 2}$$

This means that

$$\Delta_r G_3^\circ = \tfrac{1}{2}\Delta_r G_1^\circ - \tfrac{1}{2}\Delta_r G_2^\circ$$

or

$$K_3 = \frac{K_1^{1/2}}{K_2^{1/2}} = \frac{(0.169)^{1/2}}{(0.0149)^{1/2}} = 3.37$$

12–18. Consider the reaction described by

$$Cl_2(g) + Br_2(g) \rightleftharpoons 2\,BrCl(g)$$

at 500 K and a total pressure of one bar. Suppose that we start with one mole each of $Cl_2(g)$ and $Br_2(g)$ and no $BrCl(g)$. Show that

$$G(\xi) = (1-\xi)G_{Cl_2}^\circ + (1-\xi)G_{Br_2}^\circ + 2\xi G_{BrCl} + 2(1-\xi)RT \ln \frac{1-\xi}{2} + 2\xi RT \ln \xi$$

where ξ is the extent of reaction. Given that $G_{BrCl}^\circ = -3.694\ \text{kJ}\cdot\text{mol}^{-1}$ at 500 K, plot $G(\xi)$ versus ξ. Differentiate $G(\xi)$ with respect to ξ and show that the minimum value of $G(\xi)$ occurs at $\xi_{eq} = 0.549$. Also show that

$$\left(\frac{\partial G}{\partial \xi}\right)_{T,P} = \Delta_r G^\circ + RT \ln \frac{P_{BrCl}^2}{P_{Cl_2} P_{Br_2}}$$

and that $K_P = 4\xi_{eq}^2/(1-\xi_{eq})^2 = 5.9$.

As the reaction progresses, the amount of $Cl_2(g)$ and $Br_2(g)$ can be expressed as $1-\xi$ and the amount of $BrCl(g)$ will be 2ξ. We can then write the Gibbs energy of the reaction mixture as (Equation 12.20)

$$G(\xi) = (1-\xi)\overline{G}_{Cl_2} + (1-\xi)\overline{G}_{Br_2} + 2\xi\overline{G}_{BrCl}$$

Since the reaction is carried out at a total pressure of 1 bar, we can write

$$P_{Cl_2} = P_{Br_2} = \frac{1-\xi}{2} \qquad \text{and} \qquad P_{BrCl} = \frac{2\xi}{2} = \xi$$

We can use these expressions and Equation 8.59 to write $G(\xi)$ as

$$\begin{aligned} G(\xi) &= (1-\xi)\left[G^{\circ}_{Cl_2} + RT\ln P_{Cl_2}\right] + (1-\xi)\left[G^{\circ}_{Br_2} + RT\ln P_{Br_2}\right] \\ &\qquad + 2\xi\left[G^{\circ}_{BrCl} + RT\ln P_{BrCl}\right] \\ &= (1-\xi)G^{\circ}_{Cl_2} + (1-\xi)G^{\circ}_{Br_2} + 2\xi G^{\circ}_{BrCl} \\ &\qquad + 2(1-\xi)RT\ln\frac{1-\xi}{2} + 2\xi RT\ln\xi \end{aligned} \tag{1}$$

Substituting the value given in the problem for G°_{BrCl} and zero for $G^{\circ}_{Cl_2}$ and $G^{\circ}_{Br_2}$ gives

$$\begin{aligned} G(\xi) &= 2\xi(-3694\ \text{J}\cdot\text{mol}^{-1}) + 2(1-\xi)(8.3145\ \text{J}\cdot\text{mol}^{-1}\cdot\text{K}^{-1})(500\ \text{K})\ln\frac{1-\xi}{2} \\ &\qquad + 2\xi(8.3145\ \text{J}\cdot\text{mol}^{-1}\cdot\text{K}^{-1})(500\ \text{K})\ln\xi \\ &= (-7388\ \text{J}\cdot\text{mol}^{-1})\xi + (8314.5\ \text{J}\cdot\text{mol}^{-1})\left[(1-\xi)\ln\frac{1-\xi}{2} + \xi\ln\xi\right] \end{aligned} \tag{2}$$

A plot of $G(\xi)$ against ξ is shown below.

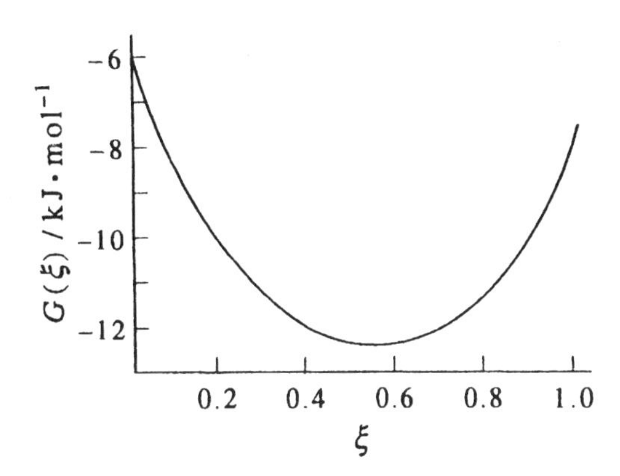

At equilibrium, $0 = (\partial G/\partial\xi)_{T,P}$. We use Equation 2 to express $G(\xi)$ and find that, at equilibrium,

$$0 = -7388\ \text{J}\cdot\text{mol}^{-1} + (8314.5\ \text{J}\cdot\text{mol}^{-1})\left[-\ln\frac{1-\xi_{eq}}{2} - 1 + \ln\xi_{eq} + 1\right]$$

$$0 = -7388\ \text{J}\cdot\text{mol}^{-1} + (8314.5\ \text{J}\cdot\text{mol}^{-1})\ln\frac{2\xi_{eq}}{1-\xi_{eq}}$$

$$\frac{2\xi_{eq}}{1-\xi_{eq}} = 2.432$$

Solving for ξ_{eq} gives $\xi_{eq} = 0.549$.

Differentiating Equation 1 for $G(\xi)$ explicitly, we find that

$$\left(\frac{\partial G}{\partial \xi}\right)_{T,P} = -G^\circ_{Cl_2} - G^\circ_{Br_2} + 2G^\circ_{BrCl} - 2RT\ln\frac{1-\xi}{2}$$

$$+2(1-\xi)RT\left(\frac{2}{1-\xi}\right)\left(-\frac{1}{2}\right) + 2RT\ln\xi + \frac{2\xi RT}{\xi}$$

$$= \Delta_r G^\circ + RT\ln\frac{4}{(1-\xi)^2} - 2RT + RT\ln\xi^2 + 2RT$$

$$= \Delta_r G^\circ + RT\ln\left(\frac{2\xi}{1-\xi}\right)^2 = \Delta_r G^\circ + RT\ln\left(\frac{P^2_{BrCl}}{P_{Cl_2}P_{Br_2}}\right)$$

Note that

$$K_P = \frac{P^2_{BrCl}}{P_{Cl_2}P_{Br_2}} = \frac{4\xi^2_{eq}}{(1-\xi_{eq})^2}$$

so, at equilibrium,

$$K_P = \frac{4(0.549)^2}{(1-0.549)^2} = 5.9$$

12–19. Consider the reaction described by

$$2\,H_2O(g) \rightleftharpoons 2\,H_2(g) + O_2(g)$$

at 4000 K and a total pressure of one bar. Suppose that we start with two moles of $H_2O(g)$ and no $H_2(g)$ or $O_2(g)$. Show that

$$G(\xi) = 2(1-\xi)G^\circ_{H_2O} + 2\xi G^\circ_{H_2} + \xi G^\circ_{O_2} + 2(1-\xi)RT\ln\frac{2(1-\xi)}{2+\xi}$$

$$+2\xi RT\ln\frac{2\xi}{2+\xi} + \xi RT\ln\frac{\xi}{2+\xi}$$

where ξ is the extent of reaction. Given that $\Delta_f G^\circ[H_2O(g)] = -18.334\ \text{kJ·mol}^{-1}$ at 4000 K, plot $G(\xi)$ against ξ. Differentiate $G(\xi)$ with respect to ξ and show that the minimum value of $G(\xi)$ occurs at $\xi_{eq} = 0.553$. Also show that

$$\left(\frac{\partial G}{\partial \xi}\right)_{T,P} = \Delta_r G^\circ + RT\ln\frac{P^2_{H_2}P_{O_2}}{P^2_{H_2O}}$$

and that $K_P = \xi^3_{eq}/(2+\xi_{eq})(1-\xi_{eq})^2 = 0.333$ at one bar.

The amount of H_2O can be expressed by $2-2\xi$, the amount of H_2 as 2ξ, and the amount of O_2 as ξ. We can then write the Gibbs energy of the reaction mixture as (Equation 12.20)

$$G(\xi) = (2-2\xi)\overline{G}_{H_2O} + 2\xi\overline{G}_{H_2} + \xi\overline{G}_{O_2}$$

Since the reaction is carried out at a total pressure of one bar, we can write

$$P_{H_2O} = \frac{2(1-\xi)}{2+\xi} \qquad P_{H_2} = \frac{2\xi}{2+\xi} \qquad \text{and} \qquad P_{O_2} = \frac{\xi}{2+\xi}$$

We can use these expressions and Equation 8.59 to write $G(\xi)$ as

$$G(\xi) = 2(1-\xi)\left[G^{\circ}_{H_2O} + RT\ln P_{H_2O}\right] + 2\xi\left[G^{\circ}_{H_2} + RT\ln P_{H_2}\right] + \xi\left[G^{\circ}_{O_2} + RT\ln P_{O_2}\right]$$

$$= 2(1-\xi)G^{\circ}_{H_2O} + 2\xi G^{\circ}_{H_2} + \xi G^{\circ}_{O_2} + 2(1-\xi)RT\ln\frac{2(1-\xi)}{2+\xi} + 2\xi RT\ln\frac{2\xi}{2+\xi} + \xi RT\ln\frac{\xi}{2+\xi}$$

Substituting the value given in the problem for $G^{\circ}_{H_2O}$ and zero for $G^{\circ}_{O_2}$ and $G^{\circ}_{H_2}$ gives

$$\begin{aligned}G(\xi) &= 2(1-\xi)(-18\,334\ \text{J}\cdot\text{mol}^{-1}) \\ &\quad + 2(1-\xi)(8.3145\ \text{J}\cdot\text{mol}^{-1}\cdot\text{K}^{-1})(4000\ \text{K})\ln\frac{2(1-\xi)}{2+\xi} \\ &\quad + 2\xi(8.3145\ \text{J}\cdot\text{mol}^{-1}\cdot\text{K}^{-1})(4000\ \text{K})\ln\frac{2\xi}{2+\xi} \\ &\quad + \xi(8.3145\ \text{J}\cdot\text{mol}^{-1}\cdot\text{K}^{-1})(4000\ \text{K})\ln\frac{\xi}{2+\xi} \\ &= (-36\,668\ \text{J}\cdot\text{mol}^{-1})(1-\xi) + (66\,516\ \text{J}\cdot\text{mol}^{-1})(1-\xi)\ln\frac{2(1-\xi)}{2+\xi} \\ &\quad + (66\,516\ \text{J}\cdot\text{mol}^{-1})\xi\ln\frac{2\xi}{2+\xi} + (33\,258\ \text{J}\cdot\text{mol}^{-1})\xi\ln\frac{\xi}{2+\xi}\end{aligned}$$

A plot of $G(\xi)$ against ξ is shown below.

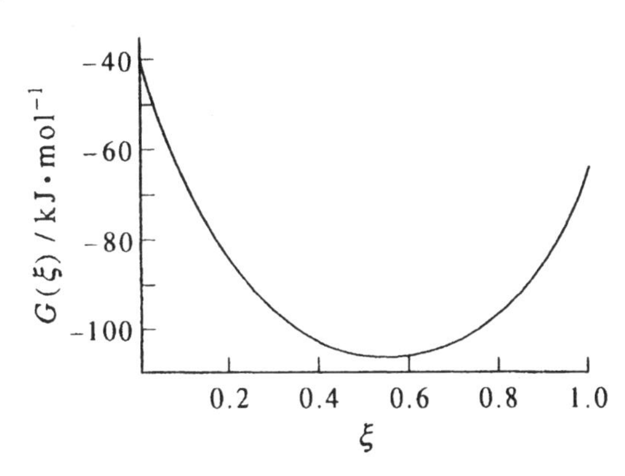

We now differentiate $G(\xi)$ with respect to ξ and find

$$\begin{aligned}\left(\frac{\partial G}{\partial \xi}\right)_{T,P} &= (36\,668\ \text{J}\cdot\text{mol}^{-1}) \\ &\quad -(66\,516\ \text{J}\cdot\text{mol}^{-1})\left\{\ln\frac{2(1-\xi)}{2+\xi} - \frac{(1-\xi)(2+\xi)}{2(1-\xi)}\left[-\frac{2}{2+\xi} - \frac{2(1-\xi)}{(2+\xi)^2}\right]\right\} \\ &\quad +(66\,516\ \text{J}\cdot\text{mol}^{-1})\left\{\ln\frac{2\xi}{2+\xi} + \frac{\xi(2+\xi)}{2\xi}\left[\frac{2}{2+\xi} - \frac{2\xi}{(2+\xi)^2}\right]\right\} \\ &\quad +(33\,258\ \text{J}\cdot\text{mol}^{-1})\left\{\ln\frac{\xi}{2+\xi} + (2+\xi)\left[\frac{1}{2+\xi} - \frac{\xi}{(2+\xi)^2}\right]\right\} \\ &= (36\,668\ \text{J}\cdot\text{mol}^{-1}) - (66\,516\ \text{J}\cdot\text{mol}^{-1})\left[\ln\frac{2(1-\xi)}{2+\xi} + 1 + \frac{1-\xi}{2+\xi}\right] \\ &\quad +(66\,516\ \text{J}\cdot\text{mol}^{-1})\left[\ln\frac{2\xi}{2+\xi} + 1 - \frac{\xi}{2+\xi}\right]\end{aligned}$$

$$+(33\,258\ \text{J}\cdot\text{mol}^{-1})\left[\ln\frac{\xi}{2+\xi}+1-\frac{\xi}{2+\xi}\right]$$

$$=36\,668\ \text{J}\cdot\text{mol}^{-1}+(33\,258\ \text{J}\cdot\text{mol}^{-1})\left[2\ln\frac{2+\xi}{2(1-\xi)}-\frac{2(1-\xi)}{2+\xi}+2\ln\frac{2\xi}{2+\xi}\right.$$

$$\left.-\frac{2\xi}{2+\xi}+\ln\frac{\xi}{2+\xi}+1-\frac{\xi}{2+\xi}\right]$$

$$=36\,668\ \text{J}\cdot\text{mol}^{-1}+(33\,258\ \text{J}\cdot\text{mol}^{-1})\ln\frac{\xi^3}{(1-\xi)^2(2+\xi)} \tag{1}$$

At equilibrium, $(\partial G/\partial\xi)_{T,P}=0$, so

$$\frac{\xi_{\text{eq}}^3}{(1-\xi_{\text{eq}})^2(2+\xi_{\text{eq}})}=\exp\left(\frac{-36\,668\ \text{J}\cdot\text{mol}^{-1}}{33\,258\ \text{J}\cdot\text{mol}^{-1}}\right)=0.332$$

Solving for ξ_{eq} gives $\xi_{\text{eq}}=0.553$.
Note that substituting for $\Delta_f G^\circ[H_2O(g)]$, R, T, P_{H_2}, P_{O_2}, and P_{H_2O} in Equation 1 gives

$$\left(\frac{\partial G}{\partial\xi}\right)_{T,P}=2\Delta_f G^\circ[H_2O(g)]+RT\ln\frac{P_{H_2}^2P_{O_2}}{P_{H_2O}^2}=\Delta_r G^\circ+RT\ln\frac{P_{H_2}^2P_{O_2}}{P_{H_2O}}$$

Now

$$K_P=\frac{P_{H_2}^2P_{O_2}}{P_{H_2O}^2}=\frac{\xi_{\text{eq}}^3}{(1-\xi_{\text{eq}})^2(2+\xi_{\text{eq}})}$$

so

$$K_P=\frac{(0.553)^3}{(1-0.553)^2(2+0.553)}=0.332$$

12–20. Consider the reaction described by

$$3\,H_2(g)+N_2(g)\rightleftharpoons 2\,NH_3(g)$$

at 500 K and a total pressure of one bar. Suppose that we start with three moles of $H_2(g)$, one mole of $N_2(g)$, and no $NH_3(g)$. Show that

$$G(\xi)=(3-3\xi)G_{H_2}^\circ+(1-\xi)G_{N_2}^\circ+2\xi G_{NH_3}^\circ$$
$$+(3-3\xi)RT\ln\frac{3-3\xi}{4-2\xi}+(1-\xi)RT\ln\frac{1-\xi}{4-2\xi}+2\xi RT\ln\frac{2\xi}{4-2\xi}$$

where ξ is the extent of reaction. Given that $G_{NH_3}^\circ=4.800\ \text{kJ}\cdot\text{mol}^{-1}$ at 500 K (see Table 12.4), plot $G(\xi)$ versus ξ. Differentiate $G(\xi)$ with respect to ξ and show that the minimum value of $G(\xi)$ occurs at $\xi_{\text{eq}}=0.158$. Also show that

$$\left(\frac{\partial G}{\partial\xi}\right)_{T,P}=\Delta_r G^\circ+RT\ln\frac{P_{NH_3}^2}{P_{H_2}^3P_{N_2}}$$

and that $K_P=16\xi_{\text{eq}}^2(2-\xi_{\text{eq}})^2/27(1-\xi_{\text{eq}})^4=0.10$.

The amount of H_2 can be expressed by $3 - 3\xi$, the amount of N_2 as $1 - \xi$, and the amount of NH_3 as 2ξ. We can then write the Gibbs energy of the reaction mixture as

$$G(\xi) = (3 - 3\xi)\overline{G}_{H_2} + (1 - \xi)\overline{G}_{N_2} + 2\xi\overline{G}_{NH_3}$$

Since the reaction is carried out at a total pressure of one bar, we can write

$$P_{H_2} = \frac{3(1-\xi)}{2(2-\xi)} \qquad P_{N_2} = \frac{1-\xi}{2(2-\xi)} \qquad \text{and} \qquad P_{NH_3} = \frac{\xi}{(2-\xi)}$$

We can use these expressions and Equation 8.59 to write $G(\xi)$ as (Equation 12.20)

$$\begin{aligned} G(\xi) &= 3(1-\xi)\left[G^\circ_{H_2} + RT\ln P_{H_2}\right] + (1-\xi)\left[G^\circ_{N_2} + RT\ln P_{N_2}\right] \\ &\quad +2\xi\left[G^\circ_{NH_3} + RT\ln P_{NH_3}\right] \\ &= 3(1-\xi)G^\circ_{H_2} + (1-\xi)G^\circ_{N_2} + 2\xi G^\circ_{NH_3} \\ &\quad +3(1-\xi)RT\ln\frac{3(1-\xi)}{2(2-\xi)} + (1-\xi)RT\ln\frac{1-\xi}{2(2-\xi)} \\ &\quad +2\xi RT\ln\frac{\xi}{2-\xi} \end{aligned}$$

Substituting the appropriate values of G° gives

$$\begin{aligned} G(\xi) &= \xi(9600\ \text{J}\cdot\text{mol}^{-1}) + 3(1-\xi)(4157.2\ \text{J}\cdot\text{mol}^{-1})\ln\frac{3(1-\xi)}{2(2-\xi)} \\ &\quad +(4157.2\ \text{J}\cdot\text{mol}^{-1})\left[(1-\xi)\ln\frac{1-\xi}{2(2-\xi)} + 2\xi\ln\frac{\xi}{2-\xi}\right] \end{aligned}$$

A plot of $G(\xi)$ against ξ is shown below.

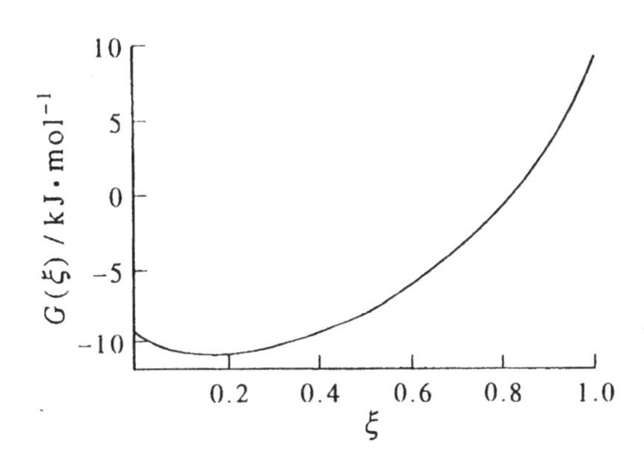

Now we differentiate $G(\xi)$ with respect to ξ:

$$\begin{aligned} \left(\frac{\partial G}{\partial \xi}\right)_{T,P} &= 9600\ \text{J}\cdot\text{mol}^{-1} \\ &\quad +(4157.2\ \text{J}\cdot\text{mol}^{-1})\left\{-3\ln\frac{3(1-\xi)}{2(2-\xi)} + 2(2-\xi)\left[-\frac{3}{2(2-\xi)} + \frac{6(1-\xi)}{4(2-\xi)^2}\right] - \ln\frac{(1-\xi)}{2(2-\xi)}\right. \\ &\quad \left. +2(2-\xi)\left[-\frac{1}{2(2-\xi)} + \frac{2(1-\xi)}{4(2-\xi)^2}\right] + 2\ln\frac{\xi}{2-\xi} + 2(2-\xi)\left[\frac{1}{2-\xi} + \frac{\xi}{(2-\xi)^2}\right]\right\} \\ &= 9600\ \text{J}\cdot\text{mol}^{-1} + (4157.2\ \text{J}\cdot\text{mol}^{-1})\left[\ln\frac{16(2-\xi)^2\xi^2}{27(1-\xi)^4} - 3 + \frac{3-3\xi}{2-\xi} - 1\right. \\ &\quad \left. +\frac{1-\xi}{2-\xi} + 2 + \frac{2\xi}{2-\xi}\right] \\ &= 9600\ \text{J}\cdot\text{mol}^{-1} + (4157.2\ \text{J}\cdot\text{mol}^{-1})\ln\frac{16(2-\xi)^2\xi^2}{27(1-\xi)^4} \end{aligned} \qquad (1)$$

At equilibrium, $(\partial G/\partial \xi)_{T,P} = 0$, so

$$\frac{16(2-\xi_{eq})^2\xi_{eq}^2}{27(1-\xi_{eq})^4} = \exp\left(\frac{-9600\ \mathrm{J\cdot mol^{-1}}}{4157.2\ \mathrm{J\cdot mol^{-1}}}\right) = 0.09934$$

Solving for ξ_{eq} gives $\xi_{eq} = 0.158$.
Note that substituting for $G^\circ_{NH_3}$, R, T, P_{NH_3}, P_{N_2}, and P_{N_2} in Equation 1 gives

$$\left(\frac{\partial G}{\partial \xi}\right)_{T,P} = 2G^\circ_{NH_3} + RT\ln\frac{P^2_{NH_3}}{P^3_{H_2}P_{N_2}} = \Delta_r G^\circ RT\ln\frac{P^2_{NH_3}}{P^3_{H_2}P_{N_2}}$$

Now

$$K_P = \frac{P^2_{NH_3}}{P^3_{H_2}P_{N_2}} = \frac{16\xi^2_{eq}(2-\xi_{eq})^2}{27(1-\xi_{eq})^4}$$

so

$$K_P = \frac{16(0.158)^2(2-0.158)^2}{27(1-0.158)^4} = 0.010$$

12–21. Suppose that we have a mixture of the gases $H_2(g)$, $CO_2(g)$, $CO(g)$, and $H_2O(g)$ at 1260 K, with $P_{H_2} = 0.55$ bar, $P_{CO_2} = 0.20$ bar, $P_{CO} = 1.25$ bar, and $P_{H_2O} = 0.10$ bar. Is the reaction described by the equation

$$H_2(g) + CO_2(g) \rightleftharpoons CO(g) + H_2O(g) \qquad K_P = 1.59$$

at equilibrium under these conditions? If not, in what direction will the reaction proceed to attain equilibrium?

Use Equation 12.25 to find Q_P and Equation 12.26 to find out which way the reaction will proceed.

$$Q_P = \frac{P_{H_2O}P_{CO}}{P_{H_2}P_{CO_2}} = \frac{(0.10)(1.25)}{(0.55)(0.20)} = 1.14$$

$$\Delta_r G = RT\ln\frac{Q_P}{K_P} = RT\ln\frac{1.14}{1.59}$$

$$\Delta_r G < 0$$

The reaction as written will proceed to the right.

12–22. Given that $K_P = 2.21 \times 10^4$ at 25°C for the equation

$$2\,H_2(g) + CO(g) \rightleftharpoons CH_3OH(g)$$

predict the direction in which a reaction mixture for which $P_{CH_3OH} = 10.0$ bar, $P_{H_2} = 0.10$ bar, and $P_{CO} = 0.0050$ bar proceeds to attain equilibrium.

This is done in the same way as the previous problem.

$$Q_P = \frac{P_{CH_3OH}}{P^2_{H_2}P_{CO}} = \frac{10}{(0.10)^2(0.0050)} = 2.00 \times 10^5$$

$$\Delta_r G = RT \ln \frac{Q_P}{K_P} = RT \ln \frac{2.00 \times 10^5}{2.21 \times 10^4}$$

$$\Delta_r G > 0$$

Therefore, the reaction as written will proceed to the left.

12–23. The value of K_P at 500 K for a gas-phase reaction doubles when the temperature is increased from 300 K to 400 K at a fixed pressure. What is the value of $\Delta_r H^\circ$ for this reaction?

Use Equation 12.31, since we assume that $\Delta_r H^\circ$ remains constant over this temperature range.

$$\ln \frac{K_P(T_2)}{K_P(T_1)} = -\frac{\Delta_r H^\circ}{R}\left(\frac{1}{T_2} - \frac{1}{T_1}\right)$$

$$R \ln 2 = -\Delta_r H^\circ \left(\frac{1}{400\text{ K}} - \frac{1}{300\text{ K}}\right)$$

$$\Delta_r H^\circ = 6.91\text{ kJ}\cdot\text{mol}^{-1}$$

12–24. The value of $\Delta_r H^\circ$ is $34.78\text{ kJ}\cdot\text{mol}^{-1}$ at 1000 K for the reaction described by

$$H_2(g) + CO_2(g) \rightleftharpoons CO(g) + H_2O(g)$$

Given that the value of K_P is 0.236 at 800 K, estimate the value of K_P at 1200 K, assuming that $\Delta_r H^\circ$ is independent of temperature.

Again, use Equation 12.31:

$$\ln \frac{K_P(T_2)}{K_P(T_1)} = -\frac{\Delta_r H^\circ}{R}\left(\frac{1}{T_2} - \frac{1}{T_1}\right)$$

$$\ln \frac{K_P(1200\text{ K})}{0.236} = -\frac{34\,780\text{ J}\cdot\text{mol}^{-1}}{8.3145\text{ J}\cdot\text{mol}^{-1}\cdot\text{K}^{-1}}\left(\frac{1}{1200\text{ K}} - \frac{1}{800\text{ K}}\right) = 1.743$$

$$K_P(1200\text{ K}) = 1.35$$

12–25. The value of $\Delta_r H^\circ$ is $-12.93\text{ kJ}\cdot\text{mol}^{-1}$ at 800 K for

$$H_2(g) + I_2(g) \rightleftharpoons 2\,HI(g)$$

Assuming that $\Delta_r H^\circ$ is independent of temperature, calculate K_P at 700 K given that $K_P = 29.1$ at 1000 K.

We do this as we did Problem 12–24, using Equation 12.31.

$$\ln \frac{K_P(T_2)}{K_P(T_1)} = -\frac{\Delta_r H^\circ}{R}\left(\frac{1}{T_2} - \frac{1}{T_1}\right)$$

$$\ln \frac{29.1}{K_P(700\text{ K})} = -\frac{12\,930\text{ J}\cdot\text{mol}^{-1}}{8.3145\text{ J}\cdot\text{mol}^{-1}\cdot\text{K}^{-1}}\left(\frac{1}{1000\text{ K}} - \frac{1}{700\text{ K}}\right) = 0.666$$

$$K_P(700\text{ K}) = 14.9$$

12–26. The equilibrium constant for the reaction described by

$$2\,\mathrm{HBr(g)} \rightleftharpoons \mathrm{H_2(g)} + \mathrm{Br_2(g)}$$

can be expressed by the empirical formula

$$\ln K = -6.375 + 0.6415\ln(T/\mathrm{K}) - \frac{11790\ \mathrm{K}}{T}$$

Use this formula to determine $\Delta_r H^\circ$ as a function of temperature. Calculate $\Delta_r H^\circ$ at 25°C and compare your result to the one you obtain from Table 5.2.

Use Equation 12.29:

$$\frac{d\ln K}{dT} = \frac{\Delta_r H^\circ}{RT^2}$$

$$\frac{0.6415}{T} + \frac{11790\ \mathrm{K}}{T^2} = \frac{\Delta_r H^\circ}{RT^2}$$

$$0.6145RT + (11790\ \mathrm{K})R = \Delta_r H^\circ$$

At 25°C, $\Delta_r H^\circ = 99.6\ \mathrm{kJ\cdot mol^{-1}}$. The value given in Table 5.2 for $\Delta_f H^\circ[\mathrm{HBr(g)}]$ is $-36.3\ \mathrm{kJ\cdot mol^{-1}}$ and that given for $\Delta_f H^\circ[\mathrm{Br_2(g)}]$ is $30.907\ \mathrm{kJ\cdot mol^{-1}}$. We can write $\Delta_r H^\circ$ (using these values) as

$$\Delta_r H^\circ = \Delta_f H^\circ[\mathrm{Br_2(g)}] - 2(\Delta_f H^\circ[\mathrm{HBr(g)}]) = 103.5\ \mathrm{kJ\cdot mol^{-1}}$$

in fairly good agreement with the value of $\Delta_r H^\circ$ found from the equilibrium constant.

12–27. Use the following data for the reaction described by

$$2\,\mathrm{HI(g)} \rightleftharpoons \mathrm{H_2(g)} + \mathrm{I_2(g)}$$

to obtain $\Delta_r H^\circ$ at 400°C.

T/K	500	600	700	800
$K_P/10^{-2}$	0.78	1.24	1.76	2.31

We wish to express K_P in terms of $1/T$ and use Equation 12.29 to find $\Delta_r H^\circ$.

$1000\mathrm{K}/T$	2	1.67	1.43	1.25
$\ln K_P$	−4.85	−4.39	−4.04	−3.77

A linear fit gives

$$\ln K_P = -1.9695 - \frac{1445.73\ \mathrm{K}}{T}$$

$$\frac{d\ln K_P}{dT} = \frac{1445.73\ \mathrm{K}}{T^2}$$

$$\Delta_r H^\circ = R(1445.73\ \mathrm{K}) = 12.02\ \mathrm{kJ\cdot mol^{-1}}$$

and a fit of the form $a + b/T + c\ln T$ gives

$$\ln K_P = -2.33966 - \frac{1020.3\ \mathrm{K}}{T} + 0.6833\ln(T/\mathrm{K})$$

$$\frac{d\ln K_P}{dT} = \frac{1020.3\ \mathrm{K}}{T^2} + \frac{0.6833}{T}$$

$$\Delta_r H^\circ = R(1020.3 + 0.6833T) = 12.31\ \mathrm{kJ\cdot mol^{-1}}$$

12–28. Consider the reaction described by

$$CO_2(g) + H_2(g) \rightleftharpoons CO(g) + H_2O(g)$$

The molar heat capacitites of $CO_2(g)$, $H_2(g)$, $CO(g)$, and $H_2O(g)$ can be expressed by

$$\overline{C}_P[CO_2(g)]/R = 3.127 + (5.231 \times 10^{-3}\ K^{-1})T - (1.784 \times 10^{-6}\ K^{-2})T^2$$

$$\overline{C}_P[H_2(g)]/R = 3.496 - (1.006 \times 10^{-4}\ K^{-1})T + (2.419 \times 10^{-7}\ K^{-2})T^2$$

$$\overline{C}_P[CO(g)]/R = 3.191 + (9.239 \times 10^{-4}\ K^{-1})T - (1.41 \times 10^{-7}\ K^{-2})T^2$$

$$\overline{C}_P[H_2O(g)]/R = 3.651 + (1.156 \times 10^{-3}\ K^{-1})T + (1.424 \times 10^{-7}\ K^{-2})T^2$$

over the temperature range 300 K to 1500 K. Given that

substance	$CO_2(g)$	$H_2(g)$	$CO(g)$	$H_2O(g)$
$\Delta_f H°/kJ\cdot mol^{-1}$	−393.523	0	−110.516	−241.844

at 300 K and that $K_P = 0.695$ at 1000 K, derive a general expression for the variation of $K_P(T)$ with temperature in the form of Equation 12.34.

We first find the values of $\Delta_r C_P^\circ$ and $\Delta_r H^\circ$:

$$\Delta_r C_P^\circ = C_P^\circ[H_2O(g)] + C_P^\circ[CO(g)] - C_P^\circ[H_2(g)] - C_P^\circ[CO_2(g)]$$

$$\Delta_r C_P^\circ/R = 0.219 - (3.051 \times 10^{-3}\ K^{-1})T + (1.544 \times 10^{-6}\ K^{-2})T^2$$

$$\begin{aligned}\Delta_r H^\circ(300\ K) &= \Delta_f H^\circ[H_2O(g)] + \Delta_f H^\circ[CO(g)] - \Delta_f H^\circ[H_2(g)] - \Delta_f H^\circ[CO_2(g)] \\ &= -241.844\ kJ\cdot mol^{-1} - (-110.516\ kJ\cdot mol^{-1}) - (-393.523\ kJ\cdot mol^{-1}) \\ &= 262.195\ kJ\cdot mol^{-1}\end{aligned}$$

Now we use Equation 12.32 to find $\Delta_r H^\circ(T)$:

$$\begin{aligned}\Delta_r H^\circ(T) &= \Delta_r H^\circ(300\ K) + \int_{300}^{T} \Delta_r C_P^\circ(T)dT \\ &= 262.195\ kJ\cdot mol^{-1} + R\int_{300}^{T} [0.219 - (3.051 \times 10^{-3}\ K^{-1})T \\ &\qquad + (1.544 \times 10^{-6}\ K^{-2})T^2]dT \\ &= 262.195\ kJ\cdot mol^{-1} + R[0.219(T - 300) - (1.525 \times 10^{-3}\ K^{-1})(T^2 - 300^2) \\ &\qquad + (5.145 \times 10^{-7}\ K^{-2})(T^3 - 300^3)] \\ &= 262.195\ kJ\cdot mol^{-1} + R[57.681\ K + 0.219T \\ &\qquad - (1.525 \times 10^{-3}\ K^{-1})T^2 + (5.145 \times 10^{-7}\ K^{-2})T^3] \\ &= 262.675\ kJ\cdot mol^{-1} + (1.821 \times 10^{-3}\ kJ\cdot mol^{-1}\cdot K^{-1})T \\ &\qquad - (1.268 \times 10^{-5}\ kJ\cdot mol^{-1}\cdot K^{-2})T^2 + (4.278 \times 10^{-9}\ kJ\cdot mol^{-1}\cdot K^{-3})T^3\end{aligned}$$

This equation is in the form $\alpha + \beta T + \gamma T^2 + \delta T^3$, as was expected (Equation 12.33). Substituting into Equation 12.34, we find that

$$\begin{aligned}\ln K_P(T) = -\frac{31592}{T} + 0.2190\ln(T/K) - (1.525 \times 10^{-3}\ K^{-1})T \\ + (2.573 \times 10^{-7}\ K^{-2})T^2 + A\end{aligned}$$

At $T = 1000$ K, we know that $K_P = 0.695$, so

$$\ln 0.695 = -31.592 + 1.513 - 1.526 + 0.2573 + A$$
$$A = 30.984$$

and so

$$\ln K_P(T) = -\frac{31592}{T} + 0.2190 \ln(T/\mathrm{K}) - (1.525 \times 10^{-3}\,\mathrm{K}^{-1})T + (2.573 \times 10^{-7}\,\mathrm{K}^{-2})T^2 + 30.984$$

12–29. The temperature dependence of the equilibrium constant K_P for the reaction described by

$$2\,C_3H_6(g) \rightleftharpoons C_2H_4(g) + C_4H_8(g)$$

is given by the equation

$$\ln K_P(T) = -2.395 - \frac{2505\ \mathrm{K}}{T} + \frac{3.477 \times 10^6\ \mathrm{K}^2}{T^2} \qquad 300\ \mathrm{K} < T < 600\ \mathrm{K}$$

Calculate the values of $\Delta_r G^\circ$, $\Delta_r H^\circ$, and $\Delta_r S^\circ$ for this reaction at 525 K.

Use Equation 12.11 to find $\Delta_r G^\circ$, Equation 12.29 to find $\Delta_r H^\circ$, and the relation $\Delta_r G^\circ = \Delta_r H^\circ - T\Delta_r S^\circ$ to find $\Delta_r S^\circ$:

$$\Delta_r G^\circ = -RT \ln K_P = -R(525\ \mathrm{K})\left[-2.395 - \frac{2505\ \mathrm{K}}{525\ \mathrm{K}} + \frac{3.477 \times 10^6\ \mathrm{K}^2}{(525\ \mathrm{K})^2}\right]$$
$$= -23.78\ \mathrm{kJ\cdot mol^{-1}}$$
$$\Delta_r H^\circ = RT^2 \frac{d \ln K_P}{dT} = RT^2 \left[\frac{2505\ \mathrm{K}}{T^2} - \frac{6.954 \times 10^6\ \mathrm{K}^2}{T^3}\right]$$
$$= -89.30\ \mathrm{kJ\cdot mol^{-1}}$$
$$\Delta_r S^\circ = \frac{\Delta_r H^\circ - \Delta_r G^\circ}{T} = -124.8\ \mathrm{J\cdot mol^{-1}\cdot K^{-1}}$$

12–30. At 2000 K and one bar, water vapor is 0.53% dissociated. At 2100 K and one bar, it is 0.88% dissociated. Calculate the value of $\Delta_r H^\circ$ for the dissociation of water at one bar, assuming that the enthalpy of reaction is constant over the range from 2000 K to 2100 K.

$$H_2O(g) \rightleftharpoons H_2(g) + \tfrac{1}{2}O_2(g)$$

At 2000 K and one bar, there will be $0.9947n_0$ moles of $H_2O(g)$, $0.0053n_0$ moles of $H_2(g)$, and $0.00265n_0$ moles of $O_2(g)$, for a total of $1.00265n_0$ moles. The partial pressures of the various gases are then

$$P_{H_2O} = \frac{0.9947}{1.00265}P = 0.9921P \qquad P_{H_2} = \frac{0.0053}{1.00265}P = 5.286 \times 10^{-3}P$$

and

$$P_{O_2} = \frac{0.00265}{1.00265}P = 2.643 \times 10^{-3}P$$

and $K_P(2000\text{ K})$ at one bar is

$$K_P(2000\text{ K}) = \frac{(2.643 \times 10^{-3})^{1/2}(5.286 \times 10^{-3})P^{1/2}}{0.9921} = 2.74 \times 10^{-4}$$

Likewise, at 2100 K and one bar, there will be $0.9912n_0$ moles of $H_2O(g)$, $0.0088n_0$ moles of $H_2(g)$, and $0.0044n_0$ moles of $O_2(g)$, for a total of $1.0044n_0$ moles. The partial pressures of the various gases are then

$$P_{H_2O} = \frac{0.9912}{1.0044}P = 0.9868P \qquad P_{H_2} = \frac{0.0088}{1.0044}P = 8.761 \times 10^{-3}P$$

and

$$P_{O_2} = \frac{0.0044}{1.0044}P = 4.381 \times 10^{-3}P$$

and $K_P(2000\text{ K})$ at one bar is

$$K_P(2000\text{ K}) = \frac{(4.381 \times 10^{-3})^{1/2}(8.761 \times 10^{-3})P^{1/2}}{0.9868} = 5.88 \times 10^{-4}$$

Now we can use Equation 12.31 to find $\Delta_r H^\circ$:

$$\ln \frac{K_P(2100\text{ K})}{K_P(2000\text{ K})} = -\frac{\Delta_r H^\circ}{R}\left(\frac{1}{2100\text{ K}} - \frac{1}{2000\text{ K}}\right)$$

$$\Delta_r H^\circ = 266.5\text{ kJ}\cdot\text{mol}^{-1}$$

12–31. The following table gives the standard molar Gibbs energy of formation of Cl(g) at three different temperatures.

T/K	1000	2000	3000
$\Delta_f G^\circ/\text{kJ}\cdot\text{mol}^{-1}$	65.288	5.081	−56.297

Use these data to determine the value of K_P at each temperature for the reaction described by

$$\tfrac{1}{2}Cl_2(g) \rightleftharpoons Cl(g)$$

Assuming that $\Delta_r H^\circ$ is temperature independent, determine the value of $\Delta_r H^\circ$ from these data. Combine your results to determine $\Delta_r S^\circ$ at each temperature. Interpret your results.

Use Equation 12.11 to find K_P at each temperature, then find $\ln K_P$ for use in determining $\Delta_r H^\circ$.

T/K	$\Delta_f G^\circ/\text{kJ}\cdot\text{mol}^{-1}$	K_P	$\ln K_P$
1000	65.288	3.889×10^{-4}	−7.852
2000	5.081	0.7367	−0.3056
3000	−56.297	9.554	2.257

The best-fit line to a plot of $\ln K_P$ vs. $1/T$ is $\ln K_P = 7.290 - 15148/T$. The slope of this line is $-\Delta_r H^\circ/R$, so $\Delta_r H^\circ = 125.9\text{ kJ}\cdot\text{mol}^{-1}$. We can use the expression $\Delta_r G^\circ = \Delta_r H^\circ - T\Delta_r S^\circ$ to find $\Delta_r S^\circ$ at each temperature. These values are tabulated below.

T/K	1000	2000	3000
$\Delta_r S^\circ/\text{J}\cdot\text{mol}^{-1}$	60.61	60.41	60.73

12–32. The following experimental data were determined for the reaction described by

$$SO_3(g) \rightleftharpoons SO_2(g) + \frac{1}{2}O_2(g)$$

T/K	800	825	900	953	1000
$\ln K_P$	−3.263	−3.007	−1.899	−1.173	−0.591

Calculate $\Delta_r G^\circ$, $\Delta_r H^\circ$, and $\Delta_r S^\circ$ for this reaction at 900 K. State any assumptions that you make.

This is done in the same way as the previous problem. We assume that $\Delta_r H^\circ$ does not vary significantly over the temperature range given. A best-fit line to to $\ln K_P$ in $1/T$ is $\ln K_P = 10.216 - (10851 \text{ K})/T$, which gives $\Delta_r H^\circ = 90.2 \text{ kJ}\cdot\text{mol}^{-1}$. Using Equation 12.11 and the relation $\Delta_r G^\circ = \Delta_r H^\circ - T\Delta_r S^\circ$, we find that at 900 K $\Delta_r G^\circ = 14.21 \text{ kJ}\cdot\text{mol}^{-1}$ and $\Delta_r S^\circ = 84.5 \text{ J}\cdot\text{mol}^{-1}\cdot\text{K}^{-1}$.

12–33. Show that

$$\mu = -RT \ln \frac{q(V,T)}{N}$$

if

$$Q(N,V,T) = \frac{[q(V,T)]^N}{N!}$$

Begin with Equation 9.27 and use Stirling's approximation for $N \ln N!$:

$$\begin{aligned}
\mu &= -RT\left(\frac{\partial \ln Q}{\partial N}\right)_{T,V} \\
&= -RT\left[\frac{\partial}{\partial N}(N\ln q - N\ln N!)\right]_{T,V} \\
&= -RT\left[\ln q - \frac{\partial}{\partial N}(-N\ln N + N)\right] \\
&= -RT[\ln q - \ln N - 1 + 1] = -RT\ln\frac{q}{N}
\end{aligned}$$

12–34. Use Equation 12.40 to calculate $K(T)$ at 750 K for the reaction described by $H_2(g) + I_2(g) \rightleftharpoons 2\,HI(g)$. Use the molecular parameters given in Table 4.2. Compare your value to the one given in Table 12.2 and the experimental value shown in Figure 12.5.

We can use Equation 12.40 to calculate $K(T)$, substituting from Equation 4.39 for the partition functions of H_2, I_2, and HI.

$$\begin{aligned}
K &= \frac{q_{HI}^2}{q_{H_2}q_{I_2}} \\
&= \left(\frac{m_{HI}^2}{m_{H_2}m_{I_2}}\right)^{3/2}\left[\frac{4\Theta_{rot}^{H_2}\Theta_{rot}^{I_2}}{(\Theta_{rot}^{HI})^2}\right]\frac{(1-e^{-\Theta_{vib}^{H_2}/T})(1-e^{-\Theta_{vib}^{I_2}/T})}{(1-e^{-\Theta_{vib}^{HI}/T})^2}\exp\frac{2D_0^{HI}-D_0^{H_2}-D_0^{I_2}}{RT}
\end{aligned}$$

$$= \left[\frac{(127.9)^2}{(2.016)(253.8)}\right]^{3/2}\left[\frac{4(85.3\text{ K})(0.0537\text{ K})}{(9.25\text{ K})^2}\right]\frac{(1-e^{-6215/750})(1-e^{-308/750})}{(1-e^{-3266/750})^2}$$

$$\times \exp\frac{2(8.500\text{ kJ}\cdot\text{mol}^{-1})}{(8.3145\times 10^{-3}\text{ kJ}\cdot\text{mol}^{-1}\cdot\text{K}^{-1})(750\text{ K})}$$

$$= 52.29$$

This is in good agreement with the values in the text.

12–35. Use the statistical thermodynamic formulas of Section 12–8 to calculate $K_P(T)$ at 900 K, 1000 K, 1100 K, and 1200 K for the association of Na(g) to form dimers, $Na_2(g)$ according to the equation

$$2\,\text{Na(g)} \rightleftharpoons \text{Na}_2\text{(g)}$$

Use your result at 1000 K to calculate the fraction of sodium atoms that form dimers at a total pressure of one bar. The experimental values of $K_P(T)$ are

T/K	900	1000	1100	1200
K_P	1.32	0.47	0.21	0.10

Plot $\ln K_P$ against $1/T$ to determine the value of $\Delta_r H^\circ$.

We can calculate the partition function of Na using Equation 4.13 (for a monatomic ideal gas) and that of Na_2 using Equation 4.39 (for a diatomic ideal gas):

$$\frac{q_{\text{Na}}}{V} = \left[\frac{2\pi m k_B T}{h^2}\right]^{3/2} q_{\text{elec}} = \left[\frac{2\pi(0.022991\text{ kg}\cdot\text{mol}^{-1})RT}{N_A^2 h^2}\right]^{3/2} \quad (2)$$

$$= 2\left[7.543\times 10^{18}\text{ m}^2\cdot\text{K}^{-1}T\right]^{3/2}$$

$$\frac{q_{\text{Na}_2}}{V} = \left[\frac{2\pi M k_B T}{h^2}\right]^{3/2}\frac{T}{\sigma\Theta_{\text{rot}}}\frac{1}{1-e^{-\Theta_{\text{vib}}/T}}e^{D_0/k_BT}$$

$$= \left[1.508\times 10^{19}\text{ m}^2\cdot\text{K}^{-1}T\right]^{3/2}\frac{T}{0.442\text{ K}}\frac{e^{8707.7\text{ K}/T}}{1-e^{-229\text{ K}/T}}$$

Using Equations 12.39 and 12.17, we have

$$K_P(T) = \frac{(q_{\text{Na}_2}/V)}{(q_{\text{Na}}/V)^2}\left[\frac{(6.022\times 10^{23}\text{ mol}^{-1})(10^5\text{ Pa})}{(1\text{ m}^{-3})(8.3145\text{ J}\cdot\text{K}^{-1}\cdot\text{mol}^{-1})T}\right]$$

We can substitute into the above expressions to find K_P at 900 K, 1000 K, 1100 K, and 1200 K:

T/K	900	1000	1100	1200
K_P	1.47	0.52	0.22	0.11

Given that $K_P = 0.52$ at 1000 K, let us assume that we begin with n_0 moles of Na and no moles of the dimer. Then at equilibrium we will have $n_0 - 2\xi$ moles of Na and ξ moles of the dimer, so that for a total pressure of 1 bar,

$$P_{\text{Na}} = \frac{n_0 - 2\xi}{n_0 - \xi}P = \frac{1-2\xi/n_0}{1-\xi/n_0}P$$

and

$$P_{\text{Na}_2} = \frac{\xi}{n_0 - \xi} P = \frac{\xi/n_0}{1 - \xi/n_0} P$$

Then

$$K_P = \frac{P_{\text{Na}_2}}{P_{\text{Na}}^2} = \frac{\xi/n_0(1 - \xi/n_0)}{(1 - 2\xi/n_0)^2 P}$$

$$0.52(1 - 2x)^2 = x - x^2$$

where $x = \xi/n_0$. Solving for x gives $x = 0.21$ or $x = 0.78$, but $x < 0.50$, so we find that 21% of the sodium atoms will form dimers at 1000 K.

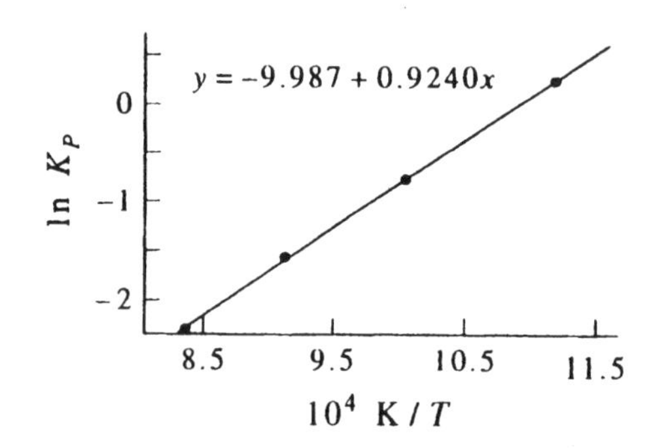

Plotting the experimental values, we find that $\ln K_P = -9.987 + (9240\text{ K})/T$. Therefore, $\Delta_r H^\circ = -R(9240\text{ K}) = -76.8\text{ kJ}\cdot\text{mol}^{-1}$.

12–36. Using the data in Table 4.2, calculate K_P at 2000 K for the reaction described by the equation

$$\text{CO}_2(\text{g}) \rightleftharpoons \text{CO}(\text{g}) + \tfrac{1}{2}\text{O}_2(\text{g})$$

The experimental value is 1.3×10^{-3}.

We can use Equation 4.39 to express the partition functions of O_2 and CO, and Equation 4.57 to express the partition function of CO_2. At 2000 K,

$$\begin{aligned}
\frac{q_{\text{CO}}}{V} &= \left[\frac{2\pi M k_B T}{h^2}\right]^{3/2} \frac{T}{\sigma\Theta_{\text{rot}}} \frac{1}{1 - e^{-\Theta_{\text{vib}}/T}} e^{D_0/k_B T} \\
&= \left[\frac{2\pi(0.02801\text{ kg}\cdot\text{mol}^{-1})(8.3145\text{ J}\cdot\text{mol}^{-1}\cdot\text{K}^{-1})(2000\text{ K})}{(6.022 \times 10^{23}\text{ mol}^{-1})^2(6.626 \times 10^{-34}\text{ J}\cdot\text{s})^2}\right]^{3/2} \\
&\quad \times \frac{2000\text{ K}}{2.77\text{ K}} \frac{1}{1 - e^{-3103/2000}} e^{1072000/(8.3145)(2000)} \\
&= 2.27 \times 10^{64}\text{ m}^{-3} \\
\frac{q_{\text{O}_2}}{V} &= \left[\frac{2\pi M k_B T}{h^2}\right]^{3/2} \frac{T}{\sigma\Theta_{\text{rot}}} \frac{1}{1 - e^{-\Theta_{\text{vib}}/T}} 3e^{D_0/k_B T} \\
&= \left[\frac{2\pi(0.03200\text{ kg}\cdot\text{mol}^{-1})(8.3145\text{ J}\cdot\text{mol}^{-1}\cdot\text{K}^{-1})(2000\text{ K})}{(6.022 \times 10^{23}\text{ mol}^{-1})^2(6.626 \times 10^{-34}\text{ J}\cdot\text{s})^2}\right]^{3/2} \\
&\quad \times \frac{2000\text{ K}}{2(2.07\text{ K})} \frac{1}{1 - e^{-2256/2000}} 3e^{494100/(8.3145)(2000)} \\
&= 5.23 \times 10^{49}\text{ m}^{-3}
\end{aligned}$$

$$\frac{q_{CO_2}}{V} = \left[\frac{2\pi M k_B T}{h^2}\right]^{3/2} \frac{T}{\sigma\Theta_{rot}} \left[\prod_{j=1}^{4}(1 - e^{-\Theta_{vib,j}/T})^{-1}\right] e^{D_0/k_BT}$$

$$= \left[\frac{2\pi(0.04401\ \text{kg}\cdot\text{mol}^{-1})(8.3145\ \text{J}\cdot\text{mol}^{-1}\cdot\text{K}^{-1})(2000\ \text{K})}{(6.022\times10^{23}\ \text{mol}^{-1})^2(6.626\times10^{-34}\ \text{J}\cdot\text{s})^2}\right]^{3/2}$$

$$\times\frac{2000\ \text{K}}{2(0.561\ \text{K})}(1 - e^{-3360/2000})^{-1}(1 - e^{-954/2000})^{-2}(1 - e^{-1890/2000})^{-1}e^{1596\times10^3/(8.3145)(2000)}$$

$$= 5.88\times10^{79}\ \text{m}^{-3}$$

We can now use Equation 12.39 to write (as we did in Section 12–8 for the reaction involving H_2O)

$$K_P(T) = \left[\frac{RT}{N_A(10^5\ \text{Pa})}\right]^{1/2}\frac{(q_{CO}/V)(q_{O_2}/V)^{1/2}}{(q_{CO_2}/V)}$$

$$= \left[\frac{(8.3145\ \text{J}\cdot\text{mol}^{-1}\cdot K^{-1})(2000\ \text{K})}{(6.022\times10^{23}\ \text{mol}^{-1})(10^5\ \text{Pa})}\right]^{1/2}\frac{(2.27\times10^{64}\ \text{m}^{-3})(5.23\times10^{49}\ \text{m}^{-3})^{1/2}}{5.88\times10^{79}\ \text{m}^{-3}}$$

$$= 1.46\times10^{-3}$$

12–37. Using the data in Tables 4.2 and 4.4, calculate the equilibrium constant for the water gas reaction

$$CO_2(g) + H_2(g) \rightleftharpoons CO(g) + H_2O(g)$$

at 900 K and 1200 K. The experimental values at these two temperatures are 0.43 and 1.37, respectively.

We have expressed the partition functions for CO_2 and CO in Problem 12–36, and those for H_2 and H_2O in Section 12–8. At a temperature T, these partition functions are

$$\frac{q_{CO_2}}{V} = \left[\frac{2\pi M k_B T}{h^2}\right]^{3/2} \frac{T}{\sigma\Theta_{rot}} \left[\prod_{j=1}^{4}(1 - e^{-\Theta_{vib,j}/T})^{-1}\right] e^{D_0/k_BT}$$

$$= \left[\frac{2\pi(0.04401\ \text{kg}\cdot\text{mol}^{-1})(8.3145\ \text{J}\cdot\text{mol}^{-1}\cdot\text{K}^{-1})T}{(6.022\times10^{23}\ \text{mol}^{-1})^2(6.626\times10^{-34}\ \text{J}\cdot\text{s})^2}\right]^{3/2}\frac{T}{2(0.561\ \text{K})}(1 - e^{-3360\ \text{K}/T})^{-1}$$

$$\times(1 - e^{-954\ \text{K}/T})^{-2}(1 - e^{-1890\ \text{K}/T})^{-1}e^{1596\times10^3\ \text{K}/8.3145T}$$

$$\frac{q_{H_2}}{V} = \left[\frac{2\pi M k_B T}{h^2}\right]^{3/2}\frac{T}{\sigma\Theta_{rot}}\frac{1}{1 - e^{-\Theta_{vib}/T}}e^{D_0/k_BT}$$

$$= \left[\frac{2\pi(2.016\times10^{-3}\ \text{kg}\cdot\text{mol}^{-1})(8.3145\ \text{J}\cdot\text{mol}^{-1}\cdot\text{K}^{-1})T}{(6.022\times10^{23}\ \text{mol}^{-1})^2(6.626\times10^{-34}\ \text{J}\cdot\text{s})^2}\right]^{3/2}\frac{T}{2(85.3\ \text{K})}\frac{1}{1 - e^{-6215/T}}e^{431800\ \text{K}/8.3145T}$$

$$\frac{q_{CO}}{V} = \left[\frac{2\pi M k_B T}{h^2}\right]^{3/2}\frac{T}{\sigma\Theta_{rot}}\frac{1}{1 - e^{-\Theta_{vib}/T}}e^{D_0/k_BT}$$

$$= \left[\frac{2\pi(0.02801\ \text{kg}\cdot\text{mol}^{-1})(8.3145\ \text{J}\cdot\text{mol}^{-1}\cdot\text{K}^{-1})T}{(6.022\times10^{23}\ \text{mol}^{-1})^2(6.626\times10^{-34}\ \text{J}\cdot\text{s})^2}\right]^{3/2}\frac{T}{2.77\ \text{K}}\frac{1}{1 - e^{-3103\ \text{K}/T}}e^{1072000\ \text{K}/8.3145T}$$

$$\frac{q_{H_2O}}{V} = \left[\frac{2\pi M k_B T}{h^2}\right]^{3/2}\frac{\pi^{1/2}}{\sigma}\left(\frac{T^3}{\Theta_{rot,A}\Theta_{rot,B}\Theta_{rot,C}}\right)^{1/2}\left(\prod_{j=1}^{3}\frac{1}{1 - e^{-\Theta_{vib,j}/T}}\right)e^{D_0/k_BT}$$

$$= \left[\frac{2\pi(0.01801\ \text{kg}\cdot\text{mol}^{-1})(8.3145\ \text{J}\cdot\text{mol}^{-1}\cdot\text{K}^{-1})T}{(6.022\times10^{23}\ \text{mol}^{-1})^2(6.626\times10^{-34}\ \text{J}\cdot\text{s})^2}\right]^{3/2}\frac{\pi^{1/2}}{2}\left[\frac{T^3}{(40.1\ \text{K})(20.9\ \text{K})(13.4\ \text{K})}\right]^{1/2}$$

$$\times(1 - e^{-5360\ \text{K}/T})^{-1}(1 - e^{-5160\ \text{K}/T})^{-1}(1 - e^{-2290\ \text{K}/T})^{-1}e^{917600\ \text{K}/8.3145T}$$

Using Equation 12.39, we can write K_P in terms of the partition functions:

$$K_P = \frac{(q_{H_2O}/V)(q_{CO}/V)}{(q_{CO_2}/V)(q_{H_2}/V)}$$

Below are tabulated values for each partition function and K_P at 900 K and 1200 K.

	900 K	1200 K
$q_{CO_2}/V/\text{m}^{-3}$	1.38×10^{129}	3.22×10^{106}
$q_{H_2}/V/\text{m}^{-3}$	9.17×10^{56}	1.02×10^{51}
$q_{CO}/V/\text{m}^{-3}$	4.15×10^{97}	2.49×10^{82}
$q_{H_2O}/V/\text{m}^{-3}$	1.72×10^{88}	2.18×10^{75}
K_P	0.56	1.66

12–38. Using the data in Tables 4.2 and 4.4, calculate the equilibrium constant for the reaction

$$3\,H_2(g) + N_2(g) \rightleftharpoons 2\,NH_3(g)$$

at 700 K. The accepted value is 8.75×10^{-5} (see Table 12.4).

We have expressed the partition function of H_2 in Problem 12–37, and we can use Equation 4.39 to express the partition function of N_2 and Equation 4.60 to express that of NH_3. At 700 K, these partition functions are

$$\begin{aligned}
\frac{q_{H_2}}{V} &= \left[\frac{2\pi M k_B T}{h^2}\right]^{3/2} \frac{T}{\sigma\Theta_{\text{rot}}} \frac{1}{1 - e^{-\Theta_{\text{vib}}/T}} e^{D_0/k_B T} \\
&= \left[\frac{2\pi(2.016 \times 10^{-3}\ \text{kg}\cdot\text{mol}^{-1})(8.3145\ \text{J}\cdot\text{mol}^{-1}\cdot\text{K}^{-1})(700\ \text{K})}{(6.022 \times 10^{23}\ \text{mol}^{-1})^2(6.626 \times 10^{-34}\ \text{J}\cdot\text{s})^2}\right]^{3/2} \\
&\quad \times \frac{700\ \text{K}}{2(85.3\ \text{K})} \frac{1}{1 - e^{-6215/700}} e^{431800/(8.3145)(700)} \\
&= 7.15 \times 10^{63}\ \text{m}^{-3}
\end{aligned}$$

$$\begin{aligned}
\frac{q_{N_2}}{V} &= \left[\frac{2\pi M k_B T}{h^2}\right]^{3/2} \frac{T}{\sigma\Theta_{\text{rot}}} \frac{1}{1 - e^{-\Theta_{\text{vib}}/T}} e^{D_0/k_B T} \\
&= \left[\frac{2\pi(0.02802\ \text{kg}\cdot\text{mol}^{-1})(8.3145\ \text{J}\cdot\text{mol}^{-1}\cdot\text{K}^{-1})(700\ \text{K})}{(6.022 \times 10^{23}\ \text{mol}^{-1})^2(6.626 \times 10^{-34}\ \text{J}\cdot\text{s})^2}\right]^{3/2} \\
&\quad \times \frac{700\ \text{K}}{2(2.88\ \text{K})} \frac{1}{1 - e^{-3374/700}} e^{941200/(8.3145)(700)} \\
&= 1.08 \times 10^{105}\ \text{m}^{-3}
\end{aligned}$$

$$\begin{aligned}
\frac{q_{NH_3}}{V} &= \left[\frac{2\pi M k_B T}{h^2}\right]^{3/2} \frac{\pi^{1/2}}{\sigma} \left(\frac{T^3}{\Theta_{\text{rot,A}}\Theta_{\text{rot,B}}\Theta_{\text{rot,C}}}\right)^{1/2} \prod_1^6 (1 - e^{-\Theta_{\text{vib},j}/T})^{-1} e^{D_0/k_B T} \\
&= \left[\frac{2\pi(0.03104\ \text{kg}\cdot\text{mol}^{-1})(8.3145\ \text{J}\cdot\text{mol}^{-1}\cdot\text{K}^{-1})(700\ \text{K})}{(6.022 \times 10^{23}\ \text{mol}^{-1})^2(6.626 \times 10^{-34}\ \text{J}\cdot\text{s})^2}\right]^{3/2} \frac{\pi^{1/2}}{3} \left[\frac{(700\ \text{K})^3}{(13.6\ \text{K})^2(8.92\ \text{K})}\right]^{1/2} \\
&\quad \times (1 - e^{-4800/700})^{-1}(1 - e^{-1360/700})^{-1}(1 - e^{-4880/700})^{-2}(1 - e^{-2330/700})^{-2} e^{1158000/(8.3145)(700)} \\
&= 2.13 \times 10^{121}\ \text{m}^{-3}
\end{aligned}$$

Using Equations 12.39 and 12.17, we can express K_P as

$$K_P(T) = \left(\frac{k_B T}{10^5\ \text{Pa}}\right)^2 \frac{(q_{NH_3}/V)^2}{(q_{N_2}/V)(q_{H_2}/V)^3}$$
$$= \left[\frac{k_B(700\ \text{K})}{10^5\ \text{Pa}}\right]^2 \frac{(2.13\times10^{121}\ \text{m}^{-3})^2}{(1.08\times10^{105}\ \text{m}^{-3})(7.15\times10^{63}\ \text{m}^{-3})^3}$$
$$= 1.23\times10^{-4} = 12.3\times10^{-5}$$

The discrepancy between the calculated value and the experimental value (about 40%) is due to the use of the rigid rotator-harmonic oscillator approximation.

12–39. Calculate the equilibrium constant K_P for the reaction

$$I_2(g) \rightleftharpoons 2\,I(g)$$

using the data in Table 4.2 and the fact that the degeneracy of the ground electronic state of an iodine atom is 4 and that the degeneracy of the first excited electronic state is 2 and that its energy is 7580 cm^{-1}. The experimental values of K_P are

T/K	800	900	1000	1100	1200
K_P	3.05×10^{-5}	3.94×10^{-4}	3.08×10^{-3}	1.66×10^{-2}	6.79×10^{-2}

Plot $\ln K_P$ against $1/T$ to determine the value of $\Delta_r H^\circ$. The experimental value is 153.8 $\text{kJ}\cdot\text{mol}^{-1}$.

The degeneracy of the ground electronic state of an iodine atom is 4. The first excited state is 90.677 $\text{kJ}\cdot\text{mol}^{-1}$ above that and its degeneracy is 2, so (using Equations 4.13 and 4.39)

$$\frac{q_I}{V} = \left[\frac{2\pi m k_B T}{h^2}\right]^{3/2}\left(4 + 2e^{-\varepsilon_{e2}/k_BT}\right)$$
$$= \left[\frac{2\pi(0.1269\ \text{kg}\cdot\text{mol}^{-1})(8.3145\ \text{J}\cdot\text{mol}^{-1}\cdot\text{K}^{-1})T}{(6.022\times10^{23}\ \text{mol}^{-1})^2(6.626\times10^{-34}\ \text{J}\cdot\text{s})^2}\right]^{3/2}\left(4 + 2e^{-90677\ \text{K}/8.3145T}\right)$$
$$\frac{q_{I_2}}{V} = \left[\frac{2\pi M k_B T}{h^2}\right]^{3/2}\frac{T}{\sigma\Theta_{rot}}\frac{1}{1-e^{-\Theta_{vib}/T}}e^{D_0/k_BT}$$
$$= \left[\frac{2\pi(0.2538\ \text{kg}\cdot\text{mol}^{-1})(8.3145\ \text{J}\cdot\text{mol}^{-1}\cdot\text{K}^{-1})T}{(6.022\times10^{23}\ \text{mol}^{-1})^2(6.626\times10^{-34}\ \text{J}\cdot\text{s})^2}\right]^{3/2}\frac{T}{2(0.0537\ \text{K})}\frac{1}{1-e^{-308\ \text{K}/T}}e^{148800\ \text{K}/8.3145T}$$

Using Equations 12.39 and 12.17, we can write K_P as

$$K_P = \left(\frac{k_B T}{10^5\ \text{Pa}}\right)\frac{(q_I/V)^2}{(q_{I_2}/V)}$$

The calculated values of K_P using the partition functions above are

T/K	800	900	1000	1100	1200
K_P	3.14×10^{-5}	4.08×10^{-4}	3.19×10^{-3}	1.72×10^{-2}	7.07×10^{-2}

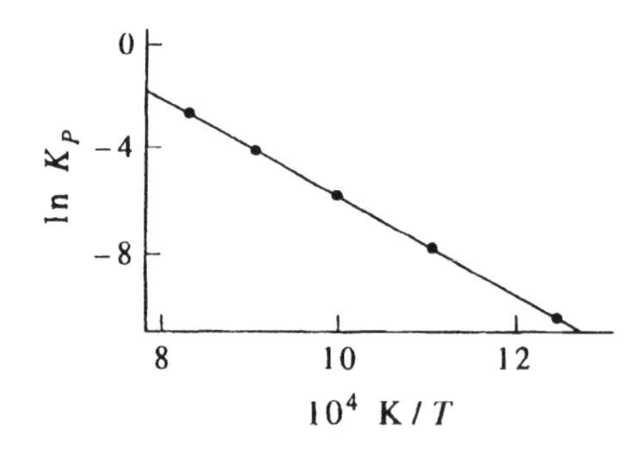

Plotting the calculated values, we find that $\ln K_P = 12.785 - (18498\text{ K})/T$. Therefore, $\Delta_r H^\circ = -R(-18498\text{ K}) = 154.0\text{ kJ}\cdot\text{mol}^{-1}$.

12–40. Consider the reaction given by

$$H_2(g) + D_2(g) \rightleftharpoons 2\,HD(g)$$

Using the Born-Oppenheimer approximation and the molecular parameters in Table 4.2, show that

$$K(T) = 4.24e^{-77.7\text{ K}/T}$$

Compare your predictions using this equation to the data in the JANAF tables.

We have an expression for K_P from Equation 12.39:

$$K_P = \frac{(q_{HD}/V)^2}{(q_{H_2}/V)(q_{D_2}/V)}$$

The relevant partition functions are (Equation 4.39)

$$\frac{q_{H_2}}{V} = \left[\frac{2\pi m_{H_2} k_B T}{h^2}\right]^{3/2} \frac{T}{2\Theta_{rot}^{H_2}} \frac{e^{-\Theta_{vib}^{H_2}/2T}}{1 - e^{-\Theta_{vib}^{H_2}/T}} e^{D_e^{H_2}/RT}$$

$$\frac{q_{D_2}}{V} = \left[\frac{2\pi m_{D_2} k_B T}{h^2}\right]^{3/2} \frac{T}{2\Theta_{rot}^{D_2}} \frac{e^{-\Theta_{vib}^{D_2}/2T}}{1 - e^{-\Theta_{vib}^{D_2}/T}} e^{D_e^{D_2}/RT}$$

$$\frac{q_{HD}}{V} = \left[\frac{2\pi m_{HD} k_B T}{h^2}\right]^{3/2} \frac{T}{\Theta_{rot}^{HD}} \frac{e^{-\Theta_{vib}^{HD}/2T}}{1 - e^{-\Theta_{vib}^{HD}/T}} e^{D_e^{HD}/RT}$$

Substituting into K_P gives

$$K_P = \left(\frac{2\pi m_{HD} k_B T}{h^2}\right)^{3/2} \left(\frac{2\pi m_{H_2} k_B T}{h^2}\right)^{-3/2} \left(\frac{2\pi m_{D_2} k_B T}{h^2}\right)^{-3/2} \left(\frac{T}{\Theta_{rot}^{HD}}\right)^2 \left(\frac{4\Theta_{rot}^{H_2}\Theta_{rot}^{D_2}}{T^2}\right)$$

$$\times \frac{e^{-2\Theta_{vib}^{HD}/2T}}{e^{-\Theta_{vib}^{H_2}/2T} e^{-\Theta_{vib}^{D_2}/2T}} (1 - e^{-\Theta_{vib}^{H_2}/T})(1 - e^{-\Theta_{vib}^{D_2}/T})(1 - e^{-\Theta_{vib}^{HD}/T})^{-2} e^{(2D_e^{HD} - D_e^{D_2} - D_e^{H_2})/RT}$$

$$= \left(\frac{m_{HD}^2}{m_{H_2} m_{D_2}}\right)^{3/2} \left[\frac{4\Theta_{rot}^{H_2}\Theta_{rot}^{D_2}}{(\Theta_{rot}^{HD})^2}\right] \left[\frac{(1 - e^{-\Theta_{vib}^{H_2}/T})(1 - e^{-\Theta_{vib}^{D_2}/T})}{(1 - e^{-\Theta_{vib}^{HD}/T})^2}\right] e^{-(2\Theta_{vib}^{HD} - \Theta_{vib}^{H_2} - \Theta_{vib}^{D_2})/2T}$$

$$\times e^{(2D_e^{HD} - D_e^{D_2} - D_e^{H_2})/RT} \qquad (1)$$

Under the Born-Oppenheimer approximation, $D_e^{\mathrm{HD}} = D_e^{\mathrm{D_2}} = D_e^{\mathrm{H_2}}$, so the last exponential term becomes 1. Also, k and R_e are the same for HD, H_2, and D_2. Then, since $\nu = (k/\mu)^{1/2}/2\pi$ and $I = \mu R_e^2$, we can write Θ_{vib} and Θ_{rot} as

$$\Theta_{\mathrm{vib}} = \frac{h\nu}{k_{\mathrm{B}}} = \frac{hk^{1/2}\pi}{2k_{\mathrm{B}}}\mu^{-1/2} \propto \mu^{-1/2} \qquad \text{and} \qquad \Theta_{\mathrm{rot}} = \frac{\hbar^2}{2Ik_{\mathrm{B}}} = \frac{\hbar^2}{2R_e^2k_{\mathrm{B}}}\mu^{-1} \propto \mu^{-1}$$

Recall that $\mu_{\mathrm{AB}} = (m_{\mathrm{A}}m_{\mathrm{B}})/(m_{\mathrm{A}} + m_{\mathrm{B}})$. Applying this formula, we find that $\mu_{\mathrm{H_2}} = 0.5$ amu, $\mu_{\mathrm{HD}} = 2/3$ amu, and $\mu_{\mathrm{D_2}} = 1$ amu. We can now write

$$\frac{4\Theta_{\mathrm{rot}}^{\mathrm{H_2}}\Theta_{\mathrm{rot}}^{\mathrm{D_2}}}{(\Theta_{\mathrm{rot}}^{\mathrm{HD}})^2} = \frac{4\mu_{\mathrm{H_2}}^{-1}\mu_{\mathrm{D_2}}^{-1}}{\mu_{\mathrm{HD}}^{-2}} = \frac{4(2/3\ \text{amu})^2}{(0.5\ \text{amu})(1\ \text{amu})} = \frac{32}{9}$$

We can also express $\Theta_{\mathrm{vib}}^{\mathrm{HD}}$ and $\Theta_{\mathrm{vib}}^{\mathrm{D_2}}$ in terms of $\Theta_{\mathrm{vib}}^{\mathrm{H_2}}$:

$$\frac{\Theta_{\mathrm{vib}}^{\mathrm{HD}}}{\Theta_{\mathrm{vib}}^{\mathrm{H_2}}} = \left(\frac{\mu_{\mathrm{H_2}}}{\mu_{\mathrm{HD}}}\right)^{1/2} = \left(\frac{1/2}{2/3}\right)^{1/2} = \frac{\sqrt{3}}{2}$$

$$\Theta_{\mathrm{vib}}^{\mathrm{HD}} = \frac{\sqrt{3}}{2}\Theta_{\mathrm{vib}}^{\mathrm{H_2}}$$

$$\frac{\Theta_{\mathrm{vib}}^{\mathrm{D_2}}}{\Theta_{\mathrm{vib}}^{\mathrm{H_2}}} = \left(\frac{\mu_{\mathrm{H_2}}}{\mu_{\mathrm{D_2}}}\right)^{1/2} = \left(\frac{1/2}{1}\right)^{1/2} = \frac{\sqrt{2}}{2}$$

$$\Theta_{\mathrm{vib}}^{\mathrm{D_2}} = \frac{\sqrt{2}}{2}\Theta_{\mathrm{vib}}^{\mathrm{H_2}}$$

Then

$$2\Theta_{\mathrm{vib}}^{\mathrm{HD}} - \Theta_{\mathrm{vib}}^{\mathrm{H_2}} - \Theta_{\mathrm{vib}}^{\mathrm{D_2}} = \left(3^{1/2} - 1 - 2^{-1/2}\right)\Theta_{\mathrm{vib}}^{\mathrm{H_2}} = 155.0\ \mathrm{K}$$

where $\Theta_{\mathrm{vib}}^{\mathrm{H_2}} = 6332$ K. Substituting into Equation 1, we find

$$K = \left(\frac{9}{8}\right)^{3/2}\left(\frac{32}{9}\right)\left[\frac{(1 - e^{-\Theta_{\mathrm{vib}}^{\mathrm{H_2}}/T})(1 - e^{-\Theta_{\mathrm{vib}}^{\mathrm{D_2}}/T})}{(1 - e^{-\Theta_{\mathrm{vib}}^{\mathrm{HD}}/T})^2}\right]e^{-77.7\ \mathrm{K}/T}$$
$$= 4.24e^{-77.7\ \mathrm{K}/T}$$

where we have neglected factors such as $1 - e^{-\Theta_{\mathrm{vib}}/T}$, since they do not contribute significantly to K for $T < 1000$ K. The table below compares calculated values of K with values from the JANAF tables.

T/K	K_p(calc.)	K_p(JANAF)
200	2.88	2.90
400	3.49	3.48
600	3.73	3.72
800	3.85	3.84
1000	3.92	3.91

12–41. Using the harmonic oscillator-rigid rotator approximation, show that

$$K(T) = \left(\frac{m_{\mathrm{H_2}}m_{\mathrm{Br_2}}}{m_{\mathrm{HBr}}^2}\right)^{3/2}\left(\frac{\sigma_{\mathrm{HBr}}^2}{\sigma_{\mathrm{H_2}}\sigma_{\mathrm{Br_2}}}\right)\left(\frac{(\Theta_{\mathrm{rot}}^{\mathrm{HBr}})^2}{\Theta_{\mathrm{rot}}^{\mathrm{H_2}}\Theta_{\mathrm{rot}}^{\mathrm{Br_2}}}\right)\frac{(1 - e^{-\Theta_{\mathrm{vib}}^{\mathrm{HBr}}/T})^2}{(1 - e^{-\Theta_{\mathrm{vib}}^{\mathrm{H_2}}/T})(1 - e^{-\Theta_{\mathrm{vib}}^{\mathrm{Br_2}}/T})}e^{(D_0^{\mathrm{H_2}} + D_0^{\mathrm{Br_2}} - 2D_0^{\mathrm{HBr}})/RT}$$

for the reaction described by

$$2\,\text{HBr(g)} \rightleftharpoons \text{H}_2\text{(g)} + \text{Br}_2\text{(g)}$$

Using the values of Θ_{rot}, Θ_{vib}, and D_0 given in Table 4.2, calculate K at 500 K, 1000 K, 1500 K, and 2000 K. Plot $\ln K$ against $1/T$ and determine the value of $\Delta_r H^\circ$.

We have an expression for K_P from Equation 12.39:

$$K_P = \frac{(q_{\text{H}_2}/V)(q_{\text{Br}_2}/V)}{(q_{\text{HBr}}/V)^2}$$

The relevant partition functions are

$$\frac{q_{\text{H}_2}}{V} = \left[\frac{2\pi m_{\text{H}_2} k_B T}{h^2}\right]^{3/2} \frac{T}{\sigma_{\text{H}_2}\Theta_{\text{rot}}^{\text{H}_2}} \frac{1}{1-e^{-\Theta_{\text{vib}}^{\text{H}_2}/T}} e^{D_0^{\text{H}_2}/RT}$$

$$\frac{q_{\text{Br}_2}}{V} = \left[\frac{2\pi m_{\text{Br}_2} k_B T}{h^2}\right]^{3/2} \frac{T}{\sigma_{\text{Br}_2}\Theta_{\text{rot}}^{\text{Br}_2}} \frac{1}{1-e^{-\Theta_{\text{vib}}^{\text{Br}_2}/T}} e^{D_0^{\text{Br}_2}/RT}$$

$$\frac{q_{\text{HBr}}}{V} = \left[\frac{2\pi m_{\text{HBr}} k_B T}{h^2}\right]^{3/2} \frac{T}{\sigma_{\text{HBr}}\Theta_{\text{rot}}^{\text{HBr}}} \frac{1}{1-e^{-\Theta_{\text{vib}}^{\text{HBr}}/T}} e^{D_0^{\text{HBr}}/RT}$$

so we write K_P as

$$K(T) = \left(\frac{m_{\text{H}_2} m_{\text{Br}_2}}{m_{\text{HBr}}^2}\right)^{3/2} \left(\frac{\sigma_{\text{HBr}}^2}{\sigma_{\text{H}_2}\sigma_{\text{Br}_2}}\right) \left[\frac{(\Theta_{\text{rot}}^{\text{HBr}})^2}{\Theta_{\text{rot}}^{\text{H}_2}\Theta_{\text{rot}}^{\text{Br}_2}}\right] \frac{(1-e^{\Theta_{\text{vib}}^{\text{HBr}}/T})^2}{(1-e^{\Theta_{\text{vib}}^{\text{H}_2}/T})(1-e^{\Theta_{\text{vib}}^{\text{Br}_2}/T})} e^{(D_0^{\text{H}_2}+D_0^{\text{Br}_2}-2D_0^{\text{HBr}})/RT}$$

Using the values from Table 4.2, we find that

T/K	500	1000	1500	2000
K_P	8.96×10^{-12}	1.20×10^{-6}	6.63×10^{-5}	4.97×10^{-4}

We can use these values to create a graph of $\ln K_P$ vs. $1/T$ and curve-fit the points linearly to obtain the equation $\ln K_P = -1.70367 - (11876\text{ K})/T$.

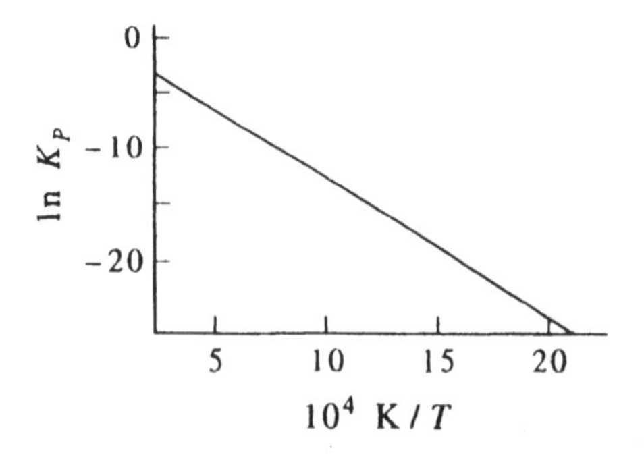

Therefore, $\Delta_r H^\circ = -R(-11876\text{ K}) = 98.8\text{ kJ}\cdot\text{mol}^{-1}$ for the reaction (compared to an experimental value of $106.0\text{ kJ}\cdot\text{mol}^{-1}$).

12–42. Use Equation 12.49*b* to calculate $H^\circ(T) - H_0^\circ$ for NH_3(g) from 300 K to 6000 K and compare your values to those given in Table 12.4 by plotting them on the same graph.

$$H^\circ(T) - H_0^\circ = 4RT + \sum_j \frac{R\Theta_{\text{vib},j}}{e^{-\Theta_{\text{vib},j}/T} - 1} \tag{12.49b}$$

We use Table 4.4 for the appropriate values of $\Theta_{\text{vib},j}$ to produce the graph below. The data points are from the JANAF tables, and the line is the function represented by Equation 12.49b.

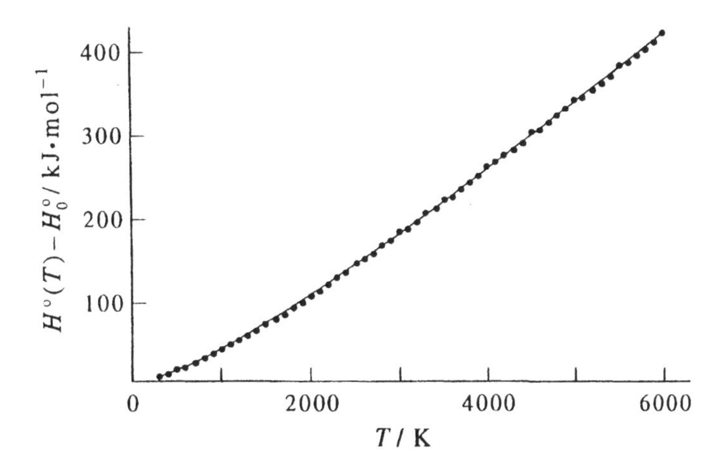

This is a very good fit to the JANAF data.

12–43. Use the JANAF tables to calculate K_P at 1000 K for the reaction described by

$$H_2(g) + I_2(g) \rightleftharpoons 2\,HI(g)$$

Compare your results to the value given in Table 12.2.

The JANAF tables give $\log K_f = 0.732$ for HI(g) at 1000 K. The equation given represents the formation of two moles of HI(g) from its consituent elements, and so $\log K = 2 \log K_f = 1.464$ and $\ln K = 3.37$. The value of $\ln K$ in Table 12.2 is 3.55.

12–44. Use the JANAF tables to plot $\ln K_P$ versus $1/T$ from 900 K to 1200 K for the reaction described by

$$2\,Na(g) \rightleftharpoons Na_2(g)$$

and compare your results to those obtained in Problem 12–35.

We can use Equation 12–11 to find K_P from the values given in the JANAF tables. From the JANAF tables,

T/K	900	1000	1100
$\Delta_f G^\circ[Na_2(g)]$/kJ·mol^{-1}	43.601	34.740	26.068
$\Delta_f G^\circ[Na(g)]$/kJ·mol^{-1}	43.601	34.740	26.068
$\Delta_r G^\circ$/kJ·mol^{-1}	−1.819	6.248	14.338
K_P(JANAF)	1.28	0.472	0.208

compared to $K_P(900\text{ K}) = 1.47$, $K_P(1000\text{ K}) = 0.52$, and $K_P(1100\text{ K}) = 0.22$ from Problem 12–35.

12–45. In Problem 12–36 we calculated K_P for the decomposition of $CO_2(g)$ to CO(g) and $O_2(g)$ at 2000 K. Use the JANAF tables to calculate K_P and compare your result to the one that you obtained in Problem 12–36.

$$CO_2(g) \rightleftharpoons CO(g) + \frac{1}{2}O_2(g)$$

From the JANAF tables,

	CO(g)	$O_2(g)$	$CO_2(g)$
$\Delta_f G^\circ/\text{kJ}\cdot\text{mol}^{-1}$	−286.034	0	−396.333

These values give a $\Delta_r G^\circ = 110.299\ \text{kJ}\cdot\text{mol}^{-1}$, and (Equation 12.11)

$$K_P = e^{-\Delta_r G^\circ/RT} = 1.32 \times 10^{-3}$$

compared to $K_P = 1.46 \times 10^{-3}$ from Problem 12–36.

12–46. You calculated K_P at 700 K for the ammonia synthesis reaction in Problem 12–38. Use the data in Table 12.4 to calculate K_P and compare your result to the one that you obtained in Problem 12–38.

$$3\,H_2(g) + N_2(g) \rightleftharpoons 2\,NH_3(g)$$

From the JANAF tables, we know that at 700 K $\Delta_f G^\circ[NH_3(g)] = 27.190\ \text{kJ}\cdot\text{mol}^{-1}$. Therefore, for the reaction above, $\Delta_r G^\circ = 2(27.190\ \text{kJ}\cdot\text{mol}^{-1}) = 54.380\ \text{kJ}\cdot\text{mol}^{-1}$, and (Equation 12.11)

$$K_P = e^{-\Delta_r G^\circ/RT} = 8.75 \times 10^{-5}$$

compared to $K_P = 12.3 \times 10^{-5}$ from Problem 12–38.

12–47. The JANAF tables give the following data for I(g) at one bar:

T/K	800	900	1000	1100	1200
$\Delta_f G^\circ/\text{kJ}\cdot\text{mol}^{-1}$	34.580	29.039	24.039	18.741	13.428

Calculate K_P for the reaction described by

$$I_2(g) \rightleftharpoons 2\,I(g)$$

and compare your results to the values given in Problem 12–39.

The energy of the reaction above will be twice the energy of formation of iodine, or $\Delta_r G^\circ = 2\Delta_f G^\circ[I(g)]$. Then, using Equation 12.11, we can calculate K_P at each temperature above:

T/K	800	900	1000	1100	1200
K_P(JANAF)	3.05×10^{-5}	4.26×10^{-4}	3.08×10^{-3}	1.66×10^{-2}	6.78×10^{-2}
K_P(calc)	3.14×10^{-5}	4.08×10^{-4}	3.19×10^{-3}	1.72×10^{-2}	7.07×10^{-2}

where we calculated the values of K_P in Problem 12–39.

12–48. Use Equation 4.60 to calculate the value of $q^0(V, T)/V$ given in the text (page 504) for $NH_3(g)$ at 500 K.

We can write Equation 4.60 in terms of D_0 as

$$q(V) = \left(\frac{2\pi M k_B T}{h^2}\right)^{3/2} V \frac{\pi^{1/2}}{\sigma}\left(\frac{T^3}{\Theta_{\text{rot,A}}\Theta_{\text{rot,B}}\Theta_{\text{rot,C}}}\right)^{1/2}\left[\prod_{j=1}^{6}(1 - e^{-\Theta_{\text{vib},j}/T})^{-1}\right] e^{D_0/k_B T}$$

We can ignore the last exponential term when we look at q^0, since q^0 is the energy relative to the ground-state energy. For $NH_3(g)$, this becomes

$$\frac{q}{V} = \left[\frac{2\pi(0.01709\ \text{kg}\cdot\text{mol}^{-1})(8.3145\ \text{J}\cdot\text{mol}^{-1}\cdot\text{K}^{-1})(500\ \text{K})}{(6.022\times10^{23}\ \text{mol}^{-1})^2(6.626\times10^{-34}\ \text{J}\cdot\text{s})^2}\right]^{3/2}\frac{\pi^{1/2}}{3}\left[\frac{(500\ \text{K})^3}{(13.6\ \text{K})^2(8.92\ \text{K})}\right]^{1/2}$$
$$\times(1 - e^{-48/5})^{-1}(1 - e^{-136/50})^{-1}(1 - e^{-488/50})^{-2}(1 - e^{-233/50})^{-2}$$
$$= 2.59\times10^{34}\ \text{m}^{-3}$$

12–49. The JANAF tables give the following data for Ar(g) at 298.15 K and one bar:

$$-\frac{G^\circ - H^\circ(298.15\ \text{K})}{T} = 154.845\ \text{J}\cdot\text{mol}^{-1}\cdot\text{K}^{-1}$$

and

$$H^\circ(0\ \text{K}) - H^\circ(298.15\ \text{K}) = -6.197\ \text{kJ}\cdot\text{mol}^{-1}$$

Use these data to calculate $q^0(V, T)/V$ and compare your result to what you obtain using Equation 4.13.

Use Equation 12.52*b* to find the exponential term in Equation 12.52*a*:

$$-\frac{(G^\circ - H_0^\circ)}{T} = -\frac{[G^\circ - H^\circ(298.15\ \text{K})]}{T} + \frac{[H^\circ - H^\circ(298.15\ \text{K})]}{T}$$
$$= 154.845\ \text{J}\cdot\text{mol}^{-1}\cdot\text{K}^{-1} - \frac{6197\ \text{J}\cdot\text{mol}^{-1}}{298.15\ \text{K}}$$
$$= 134.06\ \text{J}\cdot\text{mol}^{-1}\cdot\text{K}^{-1}$$

Now substitute into Equation 12.52*a*:

$$\frac{q^0(V,T)}{V} = \frac{N_A P^\circ}{RT}e^{-(G^\circ - H_0^\circ)/RT}$$
$$= \frac{(6.022\times10^{23}\ \text{mol}^{-1})(10^5\ \text{Pa})}{(8.3145\ \text{J}\cdot\text{mol}^{-1}\cdot\text{K}^{-1})(298.15\ \text{K})}e^{134.06/8.3145}$$
$$= 2.443\times10^{32}\ \text{m}^{-3}$$

Using Equation 4.13 and looking at q^0 as we did in Section 9–5, we find that

$$\frac{q^0}{V} = \left(\frac{2\pi M k_B T}{h^2}\right)^{3/2}$$
$$= \left[\frac{2\pi(0.039948\ \text{kg}\cdot\text{mol}^{-1})(8.3145\ \text{J}\cdot\text{mol}^{-1}\cdot\text{K}^{-1})(298.15\ \text{K})}{(6.022\times10^{23}\ \text{mol}^{-1})^2(6.626\times10^{-34}\ \text{J}\cdot\text{s})^2}\right]^{3/2}$$
$$= 2.443\times10^{32}\ \text{m}^{-3}$$

12–50. Use the JANAF tables to calculate $q^0(V, T)/V$ for $CO_2(g)$ at 500 K and one bar and compare your result to what you obtain using Equation 4.57 (with the ground state energy taken to be zero).

Using Equation 12.52*b*,

$$-\frac{(G^\circ - H_0^\circ)}{T} = 218.290\ \text{J}\cdot\text{mol}^{-1}\cdot\text{K}^{-1} + \frac{-9364\ \text{J}\cdot\text{mol}^{-1}}{500\ \text{K}}$$
$$= 199.562\ \text{J}\cdot\text{mol}^{-1}\cdot\text{K}^{-1}$$

Now substitute into Equation 12.52*a*:

$$\frac{q^0(V,T)}{V} = \frac{N_A P^\circ}{RT} e^{-(G^\circ - H_0^\circ)/RT}$$
$$= \frac{(6.022 \times 10^{23}\ \text{mol}^{-1})(10^5\ \text{Pa})}{(8.3145\ \text{J}\cdot\text{mol}^{-1}\cdot\text{K}^{-1})(298.15\ \text{K})} e^{199.562/8.3145}$$
$$= 3.84 \times 10^{35}\ \text{m}^{-3}$$

Using Equation 4.57 and looking at q^0 as we did in Section 9–5, we find that

$$\frac{q^0}{V} = \left(\frac{2\pi M k_B T}{h^2}\right)^{3/2} \left(\frac{T}{\sigma \Theta_{\text{rot}}}\right) \prod_{j=1}^{4} (1 - e^{-\Theta_{\text{vib},j}/T})^{-1}$$
$$= \left[\frac{2\pi(0.04400\ \text{kg}\cdot\text{mol}^{-1})(8.3145\ \text{J}\cdot\text{mol}^{-1}\cdot\text{K}^{-1})(500\ \text{K})}{(6.022 \times 10^{23}\ \text{mol}^{-1})^2(6.626 \times 10^{-34}\ \text{J}\cdot\text{s})^2}\right]^{3/2} \left[\frac{500\ \text{K}}{2(0.561\ \text{K})}\right]$$
$$\times (1 - e^{336/500})^{-1}(1 - e^{-954/500})^{-2}(1 - e^{-189/500})^{-1}$$
$$= 3.86 \times 10^{35}\ \text{m}^{-3}$$

12–51. Use the JANAF tables to calculate $q^0(V, T)/V$ for $CH_4(g)$ at 1000 K and one bar and compare your result to what you obtain using Equation 4.60 (with the ground state energy taken to be zero).

Using Equation 12.52*b*,

$$-\frac{(G^\circ - H_0^\circ)}{T} = 209.370\ \text{J}\cdot\text{mol}^{-1}\cdot\text{K}^{-1} + \frac{-10024\ \text{J}\cdot\text{mol}^{-1}}{1000\ \text{K}}$$
$$= 199.35\ \text{J}\cdot\text{mol}^{-1}\cdot\text{K}^{-1}$$

Now substitute into Equation 12.52*a*:

$$\frac{q^0(V,T)}{V} = \frac{N_A P^\circ}{RT} e^{-(G^\circ - H_0^\circ)/RT}$$
$$= \frac{(6.022 \times 10^{23}\ \text{mol}^{-1})(10^5\ \text{Pa})}{(8.3145\ \text{J}\cdot\text{mol}^{-1}\cdot\text{K}^{-1})(298.15\ \text{K})} e^{199.35/8.3145}$$
$$= 1.87 \times 10^{35}\ \text{m}^{-3}$$

Using Equation 4.60 and looking at q^0 as we did in Section 9–5, we find that

$$\frac{q^0}{V} = \left(\frac{2\pi M k_B T}{h^2}\right)^{3/2} \frac{\pi^{1/2}}{\sigma}\left(\frac{T^3}{\Theta_{\text{rot,A}}\Theta_{\text{rot,B}}\Theta_{\text{rot,C}}}\right)^{1/2} \prod_{j=1}^{9}(1 - e^{-\Theta_{\text{vib},j}/T})^{-1}$$

$$= \left[\frac{2\pi(0.01604\ \text{kg}\cdot\text{mol}^{-1})(8.3145\ \text{J}\cdot\text{mol}^{-1}\cdot\text{K}^{-1})(1000\ \text{K})}{(6.022\times10^{23}\ \text{mol}^{-1})(6.626\times10^{-34}\ \text{J}\cdot\text{s})^2}\right]^{3/2}\frac{\pi^{1/2}}{3}$$

$$\times\left[\frac{(1000\ \text{K})^3}{(7.54\ \text{K})^3}\right]^{1/2}(1-e^{-417/100})^{-1}(1-e^{-218/100})^{-2}(1-e^{-432/100})^{-3}(1-e^{-187/100})^{-3}$$

$$= 1.91\times10^{35}\ \text{m}^{-3}$$

12–52. Use the JANAF tables to calculate $q^0(V,T)/V$ for $H_2O(g)$ at 1500 K and one bar and compare your result to what you obtain using Equation 12.45. Why do you think there is some discrepancy?

Using Equation 12.52*b*,

$$-\frac{(G^\circ - H_0^\circ)}{T} = 218.520\ \text{J}\cdot\text{mol}^{-1}\cdot\text{K}^{-1} + \frac{-9904\ \text{J}\cdot\text{mol}^{-1}}{1500\ \text{K}}$$

$$= 211.9\ \text{J}\cdot\text{mol}^{-1}\cdot\text{K}^{-1}$$

Now substitute into Equation 12.52*a*:

$$\frac{q^0(V,T)}{V} = \frac{N_A P^\circ}{RT}e^{-(G^\circ - H_0^\circ)/RT}$$

$$= \frac{(6.022\times10^{23}\ \text{mol}^{-1})(10^5\ \text{Pa})}{(8.3145\ \text{J}\cdot\text{mol}^{-1}\cdot\text{K}^{-1})(298.15\ \text{K})}e^{211.9/8.3145}$$

$$= 5.66\times10^{35}\ \text{m}^{-3}$$

Using Equation 4.60 and looking at q^0 as we did in Section 9–5, we find that

$$\frac{q^0}{V} = \left(\frac{2\pi M k_B T}{h^2}\right)^{3/2} \frac{\pi^{1/2}}{\sigma}\left(\frac{T^3}{\Theta_{\text{rot,A}}\Theta_{\text{rot,B}}\Theta_{\text{rot,C}}}\right)^{1/2} \prod_{j=1}^{3}(1 - e^{-\Theta_{\text{vib},j}/T})^{-1}$$

$$= \left[\frac{2\pi(0.018015\ \text{kg}\cdot\text{mol}^{-1})(8.3145\ \text{J}\cdot\text{mol}^{-1}\cdot\text{K}^{-1})(1500\ \text{K})}{(6.022\times10^{23}\ \text{mol}^{-1})^2(6.626\times10^{-34}\ \text{J}\cdot\text{s})^2}\right]^{3/2}\frac{\pi^{1/2}}{3}$$

$$\times\left[\frac{(1500\ \text{K})^3}{(40.1\ \text{K})(20.9\ \text{K})(13.4\ \text{K})}\right]^{1/2}(1-e^{-536/150})^{-1}(1-e^{-516/150})^{-1}(1-e^{-229/150})^{-1}$$

$$= 5.51\times10^{35}\ \text{m}^{-3}$$

The small discrepancy between these two results is probably due to the use of the harmonic-oscillator approximation in obtaining Equation 4.60.

12–53. The JANAF tables give the following data:

	H(g)	Cl(g)	HCl(g)
$\Delta_f H^\circ(0\ \text{K})/\text{kJ}\cdot\text{mol}^{-1}$	216.035	119.621	−92.127

Use these data to calculate D_0 for HCl(g) and compare your value to the one in Table 4.2.

	$\Delta_r H°/\text{kJ}\cdot\text{mol}^{-1}$	
$\frac{1}{2}H_2(g) \rightleftharpoons H(g)$	216.035	(1)
$\frac{1}{2}Cl_2(g) \rightleftharpoons Cl(g)$	119.621	(2)
$\frac{1}{2}H_2(g) + \frac{1}{2}Cl_2(g) \rightleftharpoons HCl(g)$	−92.127	(3)

We can obtain the reaction $HCl(g) \rightleftharpoons H(g) + Cl(g)$ by subtracting Equation 3 from the sum of Equations 1 and 2, to find

$$D_0 = \Delta_r H° = (216.035 + 119.621 + 92.127)\ \text{kJ}\cdot\text{mol}^{-1} = 427.8\ \text{kJ}\cdot\text{mol}^{-1}$$

compared to a value of $427.8\ \text{kJ}\cdot\text{mol}^{-1}$ in Table 4.2.

12–54. The JANAF tables give the following data:

	C(g)	H(g)	CH_4(g)
$\Delta_f H°(0\ \text{K})/\text{kJ}\cdot\text{mol}^{-1}$	711.19	216.035	−66.911

Use these data to calculate D_0 for CH_4(g) and compare your value to the one in Table 4.4.

	$\Delta_r H°/\text{kJ}\cdot\text{mol}^{-1}$	
$C(s) \rightleftharpoons C(g)$	711.19	(1)
$\frac{1}{2}H_2(g) \rightleftharpoons H(g)$	216.035	(2)
$C(s) + 2H_2(g) \rightleftharpoons CH_4(g)$	−66.911	(3)

We can obtain the reaction $CH_4(g) \rightleftharpoons 4H(g) + C(g)$ by subtracting Equation 3 from the sum of Equations 1 and four times Equation 2, to find

$$D_0 = \Delta_r H° = [66.911 + 711.19 + 4(216.035)]\ \text{kJ}\cdot\text{mol}^{-1} = 1642\ \text{kJ}\cdot\text{mol}^{-1}$$

compared to a value of $1642\ \text{kJ}\cdot\text{mol}^{-1}$ from Table 4.4.

12–55. Use the JANAF tables to calculate D_0 for CO_2(g) and compare your result to the one given in Table 4.4.

	$\Delta_r H°/\text{kJ}\cdot\text{mol}^{-1}$	
$C(s) \rightleftharpoons C(g)$	711.19	(1)
$\frac{1}{2}O_2(g) \rightleftharpoons O(g)$	246.790	(2)
$C(s) + O_2(g) \rightleftharpoons CO_2(g)$	−393.115	(3)

We can obtain the reaction $CO_2(g) \rightleftharpoons 2O(g) + C(g)$ by subtracting Equation 3 from the sum of Equations 1 and two times Equation 2, to find

$$D_0 = \Delta_r H° = [393.115 + 711.19 + 2(246.790)]\ \text{kJ}\cdot\text{mol}^{-1} = 1598\ \text{kJ}\cdot\text{mol}^{-1}$$

compared to a value of $1596\ \text{kJ}\cdot\text{mol}^{-1}$ in Table 4.4.

12–56. A determination of K_γ (see Example 12–11) requires a knowledge of the fugacity of each gas in the equilibrium mixture. These data are not usually available, but a useful approximation is to take the fugacity coefficient of a gaseous constituent of a mixture to be equal to the value for the pure gas at the *total pressure of the mixture*. Using this approximation, we can use Figure 8.11 to determine γ for each gas and then calculate K_γ. In this problem we shall apply this approximation to the data in Table 12.5. First use Figure 8.11 to estimate that $\gamma_{H_2} = 1.05$, $\gamma_{N_2} = 1.05$, and that $\gamma_{NH_3} = 0.95$ at a total pressure of 100 bar and a temperature of 450°C. In this case $K_\gamma = 0.86$, in fairly good agreement with the value given in Example 12–11. Now calculate K_γ at 600 bar and compare your result with the value given in Example 12–11.

First, we must find the reduced temperatures and reduced pressures of each species at a pressure of 100 bar and a temperature of 450°C (we can use Table 2.5 for critical values):

$$P_R(N_2) = \frac{100 \text{ bar}}{34.0 \text{ bar}} = 2.94 \qquad T_R(N_2) = \frac{723 \text{ K}}{126.2 \text{ K}} = 5.73$$

$$P_R(H_2) = \frac{100 \text{ bar}}{12.838 \text{ bar}} = 7.79 \qquad T_R(H_2) = \frac{723 \text{ K}}{32.938 \text{ K}} = 22.0$$

$$P_R(NH_3) = \frac{100 \text{ bar}}{111.30 \text{ bar}} = 0.898 \qquad T_R(NH_3) = \frac{723 \text{ K}}{405.30 \text{ K}} = 1.78$$

Using Figure 8.11, it looks as if $\gamma_{H_2} = 1.05$, $\gamma_{N_2} = 1.05$, and $\gamma_{NH_3} = 0.95$. At 600 bar, $P_R(N_2) = 17.6$, $P_R(H_2) = 46.7$, and $P_R(NH_3) = 5.4$, so $\gamma_{H_2} = 1.3$, $\gamma_{N_2} = 1.3$, and $\gamma_{NH_3} = 0.9$. Then

$$K_\gamma = \frac{\gamma_{NH_3}}{\gamma_{H_2}^{3/2}\gamma_{N_2}^{3/2}} = 0.53$$

as compared to the value in Example 12–11 of 0.496. This is within the margin of error created by estimating the values of γ from Figure 8.11.

12–57. Recall from general chemistry that Le Châtelier's principle says that pressure has no effect on a gaseous equilibrium system such as

$$CO(g) + H_2O(g) \rightleftharpoons H_2(g) + CO_2(g)$$

in which the total number of moles of reactants is equal to the total number of moles of product in the chemical equation. The thermodynamic equilibrium constant in this case is

$$K_f = \frac{f_{CO_2} f_{H_2}}{f_{CO} f_{H_2O}} = \frac{\gamma_{CO_2}\gamma_{H_2}}{\gamma_{CO}\gamma_{H_2O}} \frac{P_{CO_2}P_{H_2}}{P_{CO}P_{H_2O}} = K_\gamma K_P$$

If the four gases behaved ideally, then pressure would have no effect on the position of equilibrium. However, because of deviations from ideal behavior, a shift in the equilibrium composition will occur when the pressure is changed. To see this, use the approximation introduced in Problem 12–56 to estimate K_γ at 900 K and 500 bar. Note that K_γ under these conditions is greater than K_γ at one bar, where $K_\gamma \approx 1$ (ideal behavior). Consequently, argue that an increase in pressure causes the equilibrium to shift to the left in this case.

At 900 K and 500 bar,

$$P_R(\mathrm{CO}) = \frac{500 \text{ bar}}{34.935 \text{ bar}} = 14.3 \qquad T_R(\mathrm{CO}) = \frac{900 \text{ K}}{132.85 \text{ K}} = 6.77$$

$$P_R(\mathrm{H_2O}) = \frac{500 \text{ bar}}{220.55 \text{ bar}} = 2.27 \qquad T_R(\mathrm{H_2O}) = \frac{900 \text{ K}}{647.126 \text{ K}} = 1.39$$

$$P_R(\mathrm{H_2}) = \frac{500 \text{ bar}}{12.838 \text{ bar}} = 38.9 \qquad T_R(\mathrm{H_2}) = \frac{900 \text{ K}}{32.938 \text{ K}} = 27.3$$

$$P_R(\mathrm{CO_2}) = \frac{500 \text{ bar}}{73.843 \text{ bar}} = 6.77 \qquad T_R(\mathrm{CO_2}) = \frac{900 \text{ K}}{304.14 \text{ K}} = 2.96$$

and so $\gamma_{\mathrm{CO}} \approx 1.3$, $\gamma_{\mathrm{H_2O}} \approx 0.8$, $\gamma_{\mathrm{H_2}} \approx 1.15$, and $\gamma_{\mathrm{CO_2}} \approx 1.1$. Then

$$K_\gamma = \frac{\gamma_{\mathrm{CO_2}}\gamma_{\mathrm{H_2}}}{\gamma_{\mathrm{H_2O}}\gamma_{\mathrm{CO}}} = 1.1$$

Since K_f must remain constant, K_P at 500 bar must be smaller than K_P at one bar and so the equilibrium will shift to the left.

12–58. Calculate the activity of $H_2O(l)$ as a function of pressure from one bar to 100 bar at 20.0°C. Take the density of $H_2O(l)$ to be 0.9982 $g\cdot mL^{-1}$ and assume that it is incompressible.

Use Equation 12.69,

$$\begin{aligned}\ln a &= \frac{\overline{V}}{RT}(P-1)\\ &= \left(\frac{1\times10^{-3}\text{ dm}^3}{0.9982\text{ g}}\right)\left(\frac{18.015\text{ g}}{1\text{ mol}}\right)\left[\frac{1}{(0.083145\text{ dm}^3\cdot\text{bar}\cdot\text{mol}^{-1}\cdot\text{K}^{-1})(293.15\text{ K})}\right](P-1)\\ &= (7.40\times10^{-4}\text{ bar}^{-1})(P-1)\end{aligned}$$

Below are some values of $\ln a$ and a at representative temperatures through the range 1 to 100 bar.

P/bar	$\ln a$	a
1	0	1.00
10	6.67×10^{-3}	1.01
50	3.63×10^{-2}	1.04
100	7.33×10^{-2}	1.08

12–59. Consider the dissociation of HgO(s,red) to Hg(g) and $O_2(g)$ according to

$$\mathrm{HgO(s, red)} \rightleftharpoons \mathrm{Hg(g)} + \tfrac{1}{2}\mathrm{O_2(g)}$$

If we start with only HgO(s,red), then assuming ideal behavior, show that

$$K_P = \frac{2}{3^{3/2}}P^{3/2}$$

where P is the total pressure. Given the following "dissociation pressure" of HgO(s,red) at various temperatures, plot $\ln K_P$ versus $1/T$.

$t/°C$	P/atm	$t/°C$	P/atm
360	0.1185	430	0.6550
370	0.1422	440	0.8450
380	0.1858	450	1.067
390	0.2370	460	1.339
400	0.3040	470	1.674
410	0.3990	480	2.081
420	0.5095		

An excellent curve fit to the plot of $\ln K_P$ against $1/T$ is given by

$$\ln K_P = -172.94 + \frac{4.0222 \times 10^5\ \text{K}}{T} - \frac{2.9839 \times 10^8\ \text{K}^2}{T^2} + \frac{7.0527 \times 10^{10}\ \text{K}^3}{T^3}$$

$$630\ \text{K} < T < 750\ \text{K}$$

Use this expression to determine $\Delta_r H°$ as a function of temperature in the interval $630\ \text{K} < T < 750\ \text{K}$. Given that

$$C_P^\circ[\text{O}_2(\text{g})]/R = 4.8919 - \frac{829.931\ \text{K}}{T} - \frac{127962\ \text{K}^2}{T^2}$$

$$C_P^\circ[\text{Hg(g)}]/R = 2.500$$

$$C_P^\circ[\text{HgO(s, red)}]/R = 5.2995$$

in the interval $298\ \text{K} < T < 750\ \text{K}$, calculate $\Delta_r H°$, $\Delta_r S°$, and $\Delta_r G°$ at 298 K.

We can write K_P in terms of the partial pressures of mercury and oxygen (assuming an activity of unity for the solid):

$$K_P = P_{\text{O}_2}^{1/2} P_{\text{Hg}} = \left(\frac{1/2}{3/2}P\right)^{1/2}\left(\frac{P}{3/2}\right) = \frac{2}{3^{3/2}}P^{3/2}$$

Below is a plot of the experimental values of $\ln K_P$ against $1/T$.

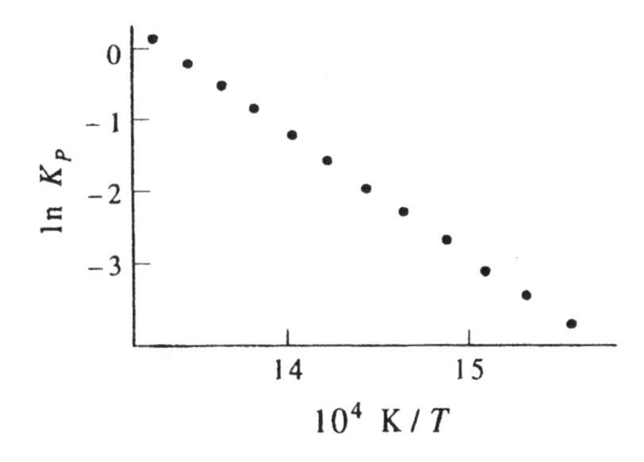

We can now use the equation given in the problem for $\ln K_P$ in Equation 12.29 to find $\Delta_r H°$:

$$\frac{d \ln K_P}{dT} = -\frac{4.0222 \times 10^5\ \text{K}}{T^2} + \frac{5.9678 \times 10^8\ \text{K}^2}{T^3} - \frac{2.1158 \times 10^{11}\ \text{K}^3}{T^4}$$

$$\Delta_r H° = RT^2 \frac{d \ln K_P}{dT}$$

$$\Delta_r H° = R\left(-4.022 \times 10^5\ \text{K} + \frac{5.9678 \times 10^8\ \text{K}^2}{T} - \frac{2.1158 \times 10^9\ \text{K}^3}{T^2}\right)$$

Likewise, we can use Equation 12.11 to find $\Delta_r G^\circ$:

$$\Delta_r G^\circ = -RT \ln K_P$$
$$= R\left(172.94T - 4.0222 \times 10^5\ \text{K} + \frac{2.9839 \times 10^8\ \text{K}^2}{T} - \frac{7.0527 \times 10^{10}\ \text{K}^3}{T^2}\right)$$

We can also find an empirical expression for $\Delta_r S^\circ$ using the equation

$$\Delta_r S^\circ = \frac{\Delta_r H^\circ - \Delta_r G^\circ}{T}$$

The expression for $\ln K_P$ used in the above equalities holds for temperatures ranging from 630 K to 750 K. To find the values of $\Delta_r H^\circ$, $\Delta_r G^\circ$, and $\Delta_r S^\circ$ at 298 K, we can use the equation

$$\Delta_r H^\circ(298\ \text{K}) = \Delta_r H^\circ(700\ \text{K}) - \int_{298\ \text{K}}^{700\ \text{K}} \Delta C_P^\circ(T)dT \tag{5.57}$$

and the similar equation

$$\Delta_r S^\circ(298\ \text{K}) = \Delta_r S^\circ(700\ \text{K}) - \int_{298\ \text{K}}^{700\ \text{K}} \frac{\Delta C_P^\circ(T)}{T} dT$$

Substituting into the high-temperature expressions for $\Delta_r G^\circ$ and $\Delta_r H^\circ$ at 700 K, we find that $\Delta_r G^\circ(700\ \text{K}) = 9.78\ \text{kJ·mol}^{-1}$, $\Delta_r H^\circ(700\ \text{K}) = 154.0\ \text{kJ·mol}^{-1}$, and $\Delta_r S^\circ(700\ \text{K}) = 206.1\ \text{J·mol}^{-1}\text{·K}^{-1}$. Then

$$\begin{aligned}
\Delta_r S^\circ(298\ \text{K}) &= 206.1\ \text{J·mol}^{-1}\text{·K}^{-1} \\
&\quad - R\int_{298\ \text{K}}^{700\ \text{K}} \left[\frac{1}{2}\left(\frac{4.8919}{T} - \frac{829.931\ \text{K}}{T^2} - \frac{127962\ \text{K}^2}{T^3}\right) + \frac{2.500}{T} - \frac{5.2995}{T}\right] dT \\
&= 206.1\ \text{J·mol}^{-1}\text{·K}^{-1} \\
&\quad - R\int_{298\ \text{K}}^{700\ \text{K}} \left(-\frac{0.35355}{T} - \frac{414.966\ \text{K}}{T^2} - \frac{63981\ \text{K}^2}{T^3}\right) dT \\
&= 206.1\ \text{J·mol}^{-1}\text{·K}^{-1} + 11.6\ \text{J·mol}^{-1}\text{·K}^{-1} = 217.7\ \text{J·mol}^{-1}\text{·K}^{-1}
\end{aligned}$$

$$\begin{aligned}
\Delta_r H^\circ(298\ \text{K}) &= 154.0\ \text{kJ·mol}^{-1} \\
&\quad - R\int_{298\ \text{K}}^{700\ \text{K}} \left[\frac{1}{2}\left(4.8919 - \frac{829.931\ \text{K}}{T} - \frac{127962\ \text{K}^2}{T^2}\right) + 2.500 - 5.2995\right] dT \\
&= 154.0\ \text{kJ·mol}^{-1} - R\int_{298\ \text{K}}^{700\ \text{K}} \left(-0.35355 - \frac{414.966\ \text{K}}{T} - \frac{63981\ \text{K}^2}{T^2}\right) dT \\
&= 154.0\ \text{kJ·mol}^{-1} + 5.15\ \text{kJ·mol}^{-1} = 159.2\ \text{kJ·mol}^{-1}
\end{aligned}$$

$$\begin{aligned}
\Delta_r G^\circ(298\ \text{K}) &= \Delta_r H^\circ(298\ \text{K}) - (298\ \text{K})\Delta_r S^\circ(298\ \text{K}) \\
&= 159.2\ \text{kJ·mol}^{-1} - (298\ \text{K})(217.7\ \text{J·mol}^{-1}\text{·K}^{-1}) \\
&= 94.3\ \text{kJ·mol}^{-1}
\end{aligned}$$

12–60. Consider the dissociation of $Ag_2O(s)$ to $Ag(s)$ and $O_2(g)$ according to

$$Ag_2O(s) \rightleftharpoons 2\,Ag(s) + \frac{1}{2}O_2(g)$$

Given the following "dissociation pressure" data:

t/°C	173	178	183	188
P/torr	422	509	605	717

Express K_P in terms of P (in torr) and plot $\ln K_P$ versus $1/T$. An excelllent curve fit to these data is given by

$$\ln K_P = 0.9692 + \frac{5612.7\ \mathrm{K}}{T} - \frac{2.0953 \times 10^6\ \mathrm{K}^2}{T^2}$$

Use this expression to derive an equation for $\Delta_r H^\circ$ from 445 K $< T <$ 460 K. Now use the following heat capacity data:

$$C_P^\circ[\mathrm{O_2(g)}]/R = 3.27 + (5.03 \times 10^{-4}\ \mathrm{K}^{-1})T$$

$$C_P^\circ[\mathrm{Ag(s)}]/R = 2.82 + (7.55 \times 10^{-4}\ \mathrm{K}^{-1})T$$

$$C_P^\circ[\mathrm{Ag_2O(s)}]/R = 6.98 + (4.48 \times 10^{-3}\ \mathrm{K}^{-1})T$$

to calculate $\Delta_r H^\circ$, $\Delta_r S^\circ$, and $\Delta_r G^\circ$ at 298 K.

We can write K_P in terms of the partial pressure of oxygen (assuming an activity of unity for the solids):

$$K_P = P_{\mathrm{O_2}}^{1/2} = P^{1/2}$$

Below is a plot of the experimental $\ln K_P$ versus $1/T$ for this reaction.

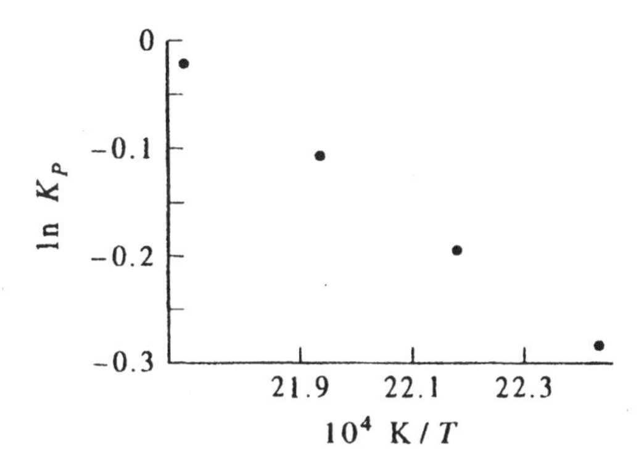

We can now use the equation given in the problem for $\ln K_P$ in Equation 12.29 to find $\Delta_r H^\circ$ and in Equation 12.11 to find $\Delta_r G^\circ$ (as in the previous problem):

$$\frac{d \ln K_P}{dT} = -\frac{5612.7\ \mathrm{K}}{T^2} + \frac{4.1906 \times 10^6\ \mathrm{K}^2}{T^3}$$

$$\Delta_r H^\circ = RT^2 \frac{d \ln K_P}{dT}$$

$$\Delta_r H^\circ = R\left(-5612.7\ \mathrm{K} + \frac{4.1906 \times 10^6\ \mathrm{K}^2}{T}\right)$$

$$\Delta_r G^\circ = -RT \ln K_P$$

$$= R\left(-0.9692T - 5612.7\ \mathrm{K} + \frac{2.0953 \times 10^6\ \mathrm{K}^2}{T}\right)$$

$$\Delta_r S^\circ = \frac{\Delta_r H^\circ - \Delta_r G^\circ}{T}$$

The expression for $\ln K_P$ used in the above equalities holds for temperatures ranging from 445 K to 460 K. To find the values of $\Delta_r H^\circ$, $\Delta_r G^\circ$, and $\Delta_r S^\circ$ at 298 K, we can use the equation

$$\Delta_r H^\circ(298\text{ K}) = \Delta_r H^\circ(450\text{ K}) - \int_{298\text{ K}}^{450\text{ K}} \Delta C_P^\circ(T)dT \tag{5.57}$$

and the similar equation

$$\Delta_r S^\circ(298\text{ K}) = \Delta_r S^\circ(450\text{ K}) - \int_{298\text{ K}}^{450\text{ K}} \frac{\Delta C_P^\circ(T)}{T}dT$$

Substituting into the high-temperature expressions for $\Delta_r S^\circ$ and $\Delta_r H^\circ$ at 450 K, we find that $\Delta_r S^\circ(450\text{ K}) = 94.09\text{ J}\cdot\text{mol}^{-1}\cdot\text{K}^{-1}$ and $\Delta_r H^\circ(450\text{ K}) = 30.76\text{ kJ}\cdot\text{mol}^{-1}$. Then

$$\begin{aligned}
\Delta_r S^\circ(298\text{ K}) &= 94.09\text{ J}\cdot\text{mol}^{-1}\cdot\text{K}^{-1} - R\int_{298\text{ K}}^{450\text{ K}} \left[\frac{1}{2}\left(\frac{3.27}{T} + 5.03\times 10^{-4}\text{ K}^{-1}\right)\right. \\
&\quad \left. +2\left(\frac{2.82}{T} + 7.55\times 10^{-4}\text{ K}^{-1}\right) - \left(\frac{6.98}{T} + 4.48\times 10^{-3}\text{ K}^{-1}\right)\right]dT \\
&= 94.09\text{ J}\cdot\text{mol}^{-1}\cdot\text{K}^{-1} - R\int_{298\text{ K}}^{450\text{ K}} \left(\frac{0.295}{T} + 1.31\times 10^{-3}\text{ K}^{-1}\right)dT \\
&= 94.09\text{ J}\cdot\text{mol}^{-1}\cdot\text{K}^{-1} + 2.42\text{ J}\cdot\text{mol}^{-1}\cdot\text{K}^{-1} = 96.51\text{ J}\cdot\text{mol}^{-1}\cdot\text{K}^{-1} \\
\Delta_r H^\circ(298\text{ K}) &= 30.76\text{ kJ}\cdot\text{mol}^{-1} - R\int_{298\text{ K}}^{450\text{ K}} \left\{\frac{1}{2}\left[3.27 + (5.03\times 10^{-4}\text{ K}^{-1})T\right]\right. \\
&\quad \left. 2\left[2.82 + (7.55\times 10^{-4}\text{ K}^{-1})T\right] - \left[6.98 + (4.48\times 10^{-3}\text{ K}^{-1})T\right]\right\}dT \\
&= 30.76\text{ kJ}\cdot\text{mol}^{-1} - R\int_{298\text{ K}}^{450\text{ K}} \left[0.295 + (1.31\times 10^{-3}\text{ K}^{-1})T\right]dT \\
&= 30.76\text{ kJ}\cdot\text{mol}^{-1} + 0.912\text{ kJ}\cdot\text{mol}^{-1} = 31.67\text{ kJ}\cdot\text{mol}^{-1} \\
\Delta_r G^\circ(298\text{ K}) &= \Delta_r H^\circ(298\text{ K}) - (298\text{ K})\Delta_r S^\circ(298\text{ K}) \\
&= 31.67\text{ kJ}\cdot\text{mol}^{-1} - (298\text{ K})(96.51\text{ J}\cdot\text{mol}^{-1}\cdot\text{K}^{-1}) \\
&= 2.910\text{ kJ}\cdot\text{mol}^{-1}
\end{aligned}$$

12–61. Calcium carbonate occurs as two crystalline forms, calcite and aragonite. The value of $\Delta_r G^\circ$ for the transition

$$CaCO_3(\text{calcite}) \rightleftharpoons CaCO_3(\text{aragonite})$$

is $+1.04\text{ kJ}\cdot\text{mol}^{-1}$ at 25°C. The density of calcite at 25°C is $2.710\text{ g}\cdot\text{cm}^{-3}$ and that of aragonite is $2.930\text{ g}\cdot\text{cm}^{-3}$. At what pressure will these two forms of $CaCO_3$ be at equilbrium at 25°C.

The molar volume of aragonite is

$$\frac{1\times 10^{-3}\text{ dm}^3}{2.930\text{ g}}\left(\frac{100.09\text{ g}}{1\text{ mol}}\right) = 0.0342\text{ dm}^3\cdot\text{mol}^{-1}$$

and the molar volume of calcite is

$$\frac{1\times 10^{-3}\text{ dm}^3}{2.710\text{ g}}\left(\frac{100.09\text{ g}}{1\text{ mol}}\right) = 0.0369\text{ dm}^3\cdot\text{mol}^{-1}$$

We can use Equation 12.65 for $\Delta_r G^\circ$ and Equation 12.69 to express the logarithmic terms. Note that when the forms are in equilibrium, the pressures will be equal.

$$\Delta_r G^\circ = -RT \ln K_a = -RT \ln \frac{a_{\text{aragonite}}}{a_{\text{calcite}}}$$

$$1040\ \text{J}\cdot\text{mol}^{-1} = -RT\left[\frac{(0.0342\ \text{dm}^3\cdot\text{mol}^{-1})(P-1)}{RT} - \frac{(0.0369\ \text{dm}^3\cdot\text{mol}^{-1})(P-1)}{RT}\right]$$

$$= (0.0027\ \text{dm}^3\cdot\text{mol}^{-1})(P-1)$$

Solving this equation for P gives

$$P - 1 = \left(\frac{1040\ \text{J}\cdot\text{mol}^{-1}}{0.0027\ \text{dm}^3\cdot\text{mol}^{-1}}\right)\left(\frac{0.083145\ \text{dm}^3\cdot\text{bar}}{8.3145\ \text{J}}\right)$$

$$P = 3800\ \text{bar}$$

12–62. The decomposition of ammonium carbamate, NH_2COONH_4 takes place according to

$$NH_2COONH_4(s) \rightleftharpoons 2\,NH_3(g) + CO_2(g)$$

Show that if all the $NH_3(g)$ and $CO_2(g)$ result from the decomposition of ammonium carbamate, then $K_P = (4/27)P^3$, where P is the total pressure at equilibrium.

We can assume that the activity of the ammonium carbamate is unity, which means it makes no contribution to the equilibrium constant expression. We can write the number of moles of carbon dioxide present at equilibrium as ξ_{eq} and the number of moles of ammonia present at equilibrium as $2\xi_{eq}$, for a total of $3\xi_{eq}$ moles. (Since the ammonium carbamate is in solid phase and we have assumed that its activity is unity, it does not contribute to the total number of moles of gas present.) Then

$$P_{NH_3} = \frac{2\xi}{3\xi}P = \frac{2}{3}P \quad \text{and} \quad P_{CO_2} = \frac{\xi}{3\xi}P = \frac{1}{3}P$$

Then

$$K_P = P_{NH_3}^2 P_{CO_2} = \frac{4P^2}{9}\left(\frac{P}{3}\right) = \frac{4P^3}{27}$$

at equilibrium.

12–63. Calculate the solubility of LiF(aq) in water at 25°C. Compare your result to the one you obtain by using concentrations instead of activities. Take $K_{sp} = 1.7 \times 10^{-3}$.

The equation for the dissolution of LiF(s) is

$$\text{LiF(s)} \rightleftharpoons \text{Li}^+(\text{aq}) + \text{F}^-(\text{aq})$$

and the equilibrium-constant expression is

$$a_{Li^+}a_{F^-} = c_{Li^+}c_{F^-}\gamma_\pm^2 = K_{sp} = 1.7 \times 10^{-3}$$

or

$$c_{\mathrm{Li}^+}c_{\mathrm{F}^-} = \frac{K_{\mathrm{sp}}}{\gamma_\pm^2}$$

Let the solubility of LiF(s) be s, then $c_{\mathrm{Li}^+} = c_{\mathrm{F}^-} = s$. Therefore, we have

$$s^2 = \frac{K_{\mathrm{sp}}}{\gamma_\pm^2} \tag{1}$$

Set $\gamma_\pm = 1$ and solve Equation 1 for s to obtain $s = (1.7 \times 10^{-3})^{1/2}\mathrm{mol\cdot L^{-1}} = 0.0412\ \mathrm{mol\cdot L^{-1}}$. Now substitute this result into Equation 12.56 (with $I_c = s$) to calculate $\ln \gamma_\pm = -0.198$, or $\gamma_\pm = 0.820$. Substitute this value into Equation 1 to obtain $s = 0.0503\ \mathrm{mol\cdot L^{-1}}$. The next iteration gives $\gamma_\pm = 0.807$, and then $s = 0.0511\ \mathrm{mol\cdot L^{-1}}$. Once more gives $\gamma_\pm = 0.806$, so the final result is then $s = 0.0512\ \mathrm{mol\cdot L^{-1}}$. Thus, $s = 0.051\ \mathrm{mol\cdot L^{-1}}$ to two significant figures.

12–64. Calculate the solubility of CaF_2(aq) in a solution that is 0.0150 molar in $MgSO_4$(aq). Take $K_{\mathrm{sp}} = 3.9 \times 10^{-11}$ for CaF_2(aq).

The equation for the dissolution of CaF_2(s) is

$$\mathrm{CaF_2(s)} \rightleftharpoons \mathrm{Ca^{2+}(aq)} + 2\ \mathrm{F^-(aq)}$$

and the equilibrium-constant expression is

$$c_{\mathrm{Ca}^{2+}}c_{\mathrm{F}^-}^2\gamma_\pm^3 = K_{\mathrm{sp}} = 3.9 \times 10^{-11}$$

Let the solubility of CaF_2(s) be s, then Ca^{2+}(aq) $= s$ and F^-(aq) $= 2s$. Therefore, we have

$$s(2s)^2 = 4s^3 = \frac{K_{\mathrm{sp}}}{\gamma_\pm^3} \tag{1}$$

Set $\gamma_\pm = 1$ to obtain $s = 2.14 \times 10^{-4}\ \mathrm{mol\cdot L^{-1}}$. This value of s gives

$$\begin{aligned} I_c &= \frac{1}{2}\left[4s + 2s + (4)(0.0150\ \mathrm{mol\cdot L^{-1}}) + (4)(0.0150\ \mathrm{mol\cdot L^{-1}})\right] \\ &= 0.000641\ \mathrm{mol\cdot L^{-1}} + 0.0600\ \mathrm{mol\cdot L^{-1}} = 0.0606\ \mathrm{mol\cdot L^{-1}} \end{aligned}$$

Substituting this result into Equation 12.56 gives $\gamma_\pm = 0.629$. Substitute this value into Equation 1 to obtain $s = 3.40 \times 10^{-4}\ \mathrm{mol\cdot L^{-1}}$. Now $I_c = 0.0610\ \mathrm{mol\cdot L^{-1}}$, and $\gamma_\pm = 0.628$. Now use this result in Equation 1 to get $s = 3.40 \times 10^{-4}\ \mathrm{mol\cdot L^{-1}}$. Another iteration gives $\gamma_\pm = 0.628$ and $s = 3.40 \times 10^{-4}\ \mathrm{mol\cdot L^{-1}}$. So, to two significant figures, $s = 3.4 \times 10^{-4}\ \mathrm{mol\cdot L^{-1}}$.

12–65. Calculate the solubility of CaF_2(aq) in a solution that is 0.050 molar in NaF(aq). Compare your result to the one you obtain by using concentrations instead of activities. Take $K_{\mathrm{sp}} = 3.9 \times 10^{-11}$ for CaF_2(aq).

The equation for the dissolution of CaF_2(s) is

$$\mathrm{CaF_2(s)} \rightleftharpoons \mathrm{Ca^{2+}(aq)} + 2\ \mathrm{F^-(aq)}$$

and the equilibrium-constant expression is (Problem 12–64)

$$4s^3 = \frac{K_{sp}}{\gamma_\pm^3} \tag{1}$$

Set $\gamma_\pm = 1$ to obtain $s = 2.14 \times 10^{-4}$ mol·L^{-1}. This value of s gives

$$I_c = \frac{1}{2}\left[4s + 2s + (1)(0.050 \text{ mol·L}^{-1}) + (1)(0.050 \text{ mol·L}^{-1})\right] = 3s + 0.050 \text{ mol·L}^{-1}$$

Because s is much smaller than 0.050 mol·L^{-1}, we initially let $I_c = 0.050$ mol·L^{-1}. Substituting this value into Equation 12.56 gives $\gamma_\pm = 0.651$. Substitute this result into Equation 1 to obtain $s = 3.28 \times 10^{-4}$ mol·L^{-1}. Now $I_c = 0.0510$ mol·L^{-1}, $\gamma_\pm = 0.649$, and $s = 3.29 \times 10^{-4}$ mol·L^{-1}. Thus, $s = 3.3 \times 10^{-4}$ mol·L^{-1} to two significant figures.

CHAPTER 13

Thermodynamics of Electrochemical Cells

PROBLEMS AND SOLUTIONS

13–1. Write the equations for the electrode reactions and the overall cell reaction for the electrochemical cells whose cell diagrams are

a. $Pb(s)|PbI_2(s)|HI(aq)|H_2(g)|Pt(s)$

b. $Cu(s)|Cu(ClO_4)_2(aq)||AgClO_4(aq)|Ag(s)$

c. $In(s)|In(NO_3)_3(aq)||CdCl_2(aq)|Cd(s)$

d. $Sn(s)|SnCl_2(aq)||AgNO_3(aq)|Ag(s)$

---

a. $Pb(s) + 2\ I^-(aq) \longrightarrow PbI_2(s) + 2\ e^-$
$H^+(aq) + e^- \longrightarrow \frac{1}{2}\ H_2(g)$
$Pb(s) + 2\ HI(aq) \longrightarrow PbI_2(s) + H_2(g)$

b. $Cu(s) \longrightarrow Cu^{2+}(aq) + 2\ e^-$
$Ag^+(aq) + e^- \longrightarrow Ag(s)$
$Cu(s) + 2\ Ag^+(aq) \longrightarrow Cu^{2+}(aq) + 2\ Ag(s)$

c. $In(s) \longrightarrow In^{3+}(aq) + 3\ e^-$
$Cd^{2+}(aq) + 2\ e^- \longrightarrow Cd(s)$
$2\ In(s) + 3\ Cd^{2+}(aq) \longrightarrow 2\ In^{3+}(aq) + 3\ Cd(s)$

d. $Sn(s) \longrightarrow Sn^{2+}(aq) + 2\ e^-$
$Ag^+(aq) + e^- \longrightarrow Ag(s)$
$Sn(s) + 2\ Ag^+(aq) \longrightarrow Sn^{2+}(aq) + 2\ Ag(s)$

13–2. Write the equations for the electrode reactions and the overall cell reaction for the electrochemical cells whose cell diagram are

a. $Pb(s)|PbSO_4(s)|K_2SO_4(aq)|Hg_2SO_4(s)|Hg(l)$

b. $Pt(s)|H_2(g)|HCl(aq)|Hg_2Cl_2(s)|Hg(l)$

c. $Zn(s)|ZnO(s)|NaOH(aq)|HgO(s)|Hg(l)$

d. $Cd(s)|CdSO_4(aq)|Hg_2SO_4(s)|Hg(l)$

---

a. $Pb(s) + SO_4^{2-}(aq) \longrightarrow PbSO_4(s) + 2\ e^-$
$Hg_2SO_4(s) + 2\ e^- \longrightarrow 2\ Hg(l) + SO_4^{2-}(aq)$
$Pb(s) + Hg_2SO_4(s) \longrightarrow PbSO_4(s) + 2\ Hg(l)$

b. $H_2(g) \longrightarrow 2\ H^+(aq) + 2\ e^-$
$Hg_2Cl_2(s) + 2\ e^- \longrightarrow 2\ Hg(l) + 2\ Cl^-(aq)$
$H_2(g) + Hg_2Cl_2(s) \longrightarrow 2\ HCl(aq) + 2\ Hg(l)$

c. $Zn(s) + 2\ OH^-(aq) \longrightarrow ZnO(s) + H_2O(l) + 2\ e^-$
$HgO(s) + H_2O(l) + 2\ e^- \longrightarrow Hg(l) + 2\ OH^-(aq)$
$Zn(s) + HgO(s) \longrightarrow ZnO(s) + Hg(l)$

d. $Cd(s) \longrightarrow Cd^{2+}(aq) + 2\ e^-$
$Hg_2SO_4(s) + 2\ e^- \longrightarrow 2\ Hg(l) + SO_4^{2-}(aq)$
$Cd(s) + Hg_2SO_4(s) \longrightarrow CdSO_4(s) + 2\ Hg(l)$

13–3. Consider an electrochemical cell in which the reaction is described by the equation

$$2\ HCl(aq) + Ca(s) \rightleftharpoons CaCl_2(aq) + H_2(g)$$

Predict the effect of the following changes on the cell voltage:

a. decrease in the amount of $Ca(s)$

b. increase in the pressure of $H_2(g)$

c. increase in the concentration of $HCl(aq)$

d. dissolution of $Ca(NO_3)_2(s)$ in the $CaCl_2(aq)$ solution

---

a. no effect

b. decrease

c. increase

d. decrease

13–4. Given the following equation for an electrochemical cell reaction

$$H_2(g) + PbSO_4(s) \rightleftharpoons 2\ H^+(aq) + SO_4^{2-}(aq) + Pb(s)$$

predict the effect of the following changes on the cell voltage:

a. increase in the pressure of $H_2(g)$

b. increase in the size of the lead electrode

c. decrease in the pH of the cell electrolyte

d. dissolution of $Na_2SO_4(s)$ in the cell electrolyte

e. decrease in the amount of $PbSO_4(s)$

f. dissolution of a small amount of $NaOH(s)$ in the cell eletrolyte

---

a. increase

b. no effect (cell voltage is an intensive property)

c. decrease

d. decrease

e. no effect

f. increase

13–5. Determine the value of *n* in the Nernst equation for the following equations:

a. $CH_4(g) + 2\ O_2(g) \longrightarrow CO_2(g) + 2\ H_2O(l)$

b. $2\ Zn(s) + Ag_2O_2(s) + 2\ H_2O(l) + 4\ OH^-(aq) \longrightarrow 2\ Ag(s) + 2\ Zn(OH)_4^{2-}(aq)$

c. $Cd(s) + Hg_2SO_4(s) \longrightarrow 2\,Hg(l) + Cd^{2+}(aq) + SO_4^{2-}(aq)$

d. $C_3H_8(g) + 5\,O_2(g) \longrightarrow 3\,CO_2(g) + 4\,H_2O(l)$

---

a. The oxidation state of the carbon goes from −4 in CH_4 to +4 in CO_2 and the oxidation state of each oxygen goes from 0 in O_2 to −2 in CO_2, and so $n = 8$.

b. The oxidation state of each silver atom goes from +1 in Ag_2O_2 to 0 in Ag(s), and the oxidation state of each oxygen atom goes from −1 in Ag_2O_2 to −2 in OH^-. Therefore, $n = 4$.

c. The oxidation state of the cadmium goes from 0 in Cd(s) to +2 in Cd^{2+}, and the oxidation state of each mercury goes from +1 in Hg_2SO_4 to 0 in Hg(l). Therefore, $n = 2$.

d. The oxidation state of each carbon goes from −8/3 in C_3H_8 to +4 in CO_2 and the oxidation state of each oxygen goes from 0 in O_2 to −2 in CO_2 and H_2O, and so $n = 20$.

13–6. Consider the electrochemical cell whose cell diagram is $Pb(s)|PbBr_2(s)|HBr(aq)|H_2(1\text{ bar})|Pt(s)$. Given emf versus molality of HBr(aq) data for this cell, describe how you would determine the value of E°.

The equations for the two electrode reactions are

$$Pb(s) + 2\,Br^-(aq) \longrightarrow PbBr_2(s) + 2\,e^-$$

and

$$H^+(aq) + e^- \longrightarrow \tfrac{1}{2}\,H_2(g)$$

The equation for the cell reaction is

$$Pb(s) + 2\,HBr(aq) \longrightarrow PbBr_2(s) + H_2(g)$$

The corresponding Nernst equation is

$$E = E^\circ - \frac{RT}{2F}\ln\frac{a_{PbBr_2}a_{H_2}}{a_{Pb}a_{HBr}^2}$$

Setting the activities of $PbBr_2(s)$, Pb(s), and $H_2(g)$ equal to unity gives

$$E = E^\circ + \frac{RT}{F}\ln a_{H^+}a_{Br^-}$$

Use the fact that $a_{H^+}a_{Br^-} = \gamma_\pm^2 m^2$ to get

$$E = E^\circ + \frac{2RT}{F}\ln\gamma_\pm m$$

or

$$E - \frac{2RT}{F}\ln m = E^\circ + \frac{2RT}{F}\ln\gamma_\pm$$

Now use

$$\ln\gamma_\pm = -1.171\frac{m^{1/2}}{1+m^{1/2}} + Cm$$

to write

$$E - \frac{2RT}{F}\ln m + 1.171\frac{2RT}{F}\frac{m^{1/2}}{1+m^{1/2}} = E^\circ + \frac{2RT}{F}Cm$$

Using E versus molality of HBr(aq) data, plot the left side of this equation against m and extrapolate to $m = 0$ to obtain E° (see Figure 13.9b).

13–7. Consider the electrochemical cell whose cell diagram is $Pt(s)|H_2(g)|HCl(aq)|Hg_2Cl_2(s)|Hg(l)$. Given emf versus molality of HCl(aq) data for this cell, describe how you would determine the value of E°.

The equations for the two electrode reactions are

$$H_2(g) \longrightarrow 2\,H^+(aq) + 2\,e^-$$

and

$$Hg_2Cl_2(s) + 2\,e^- \longrightarrow 2\,Hg(l) + 2\,Cl^-(aq)$$

The equation for the cell reaction is

$$H_2(g) + Hg_2Cl_2(s) \longrightarrow 2\,Hg(l) + 2\,HCl(aq)$$

The corresponding Nernst equation is

$$E = E^\circ - \frac{RT}{2F}\ln\frac{a_{\mathrm{Hg}}^2 a_{\mathrm{HCl}}^2}{a_{\mathrm{H_2}} a_{\mathrm{Hg_2Cl_2}}}$$

Setting the activities of Hg(l), Hg_2Cl_2, and $H_2(g)$ equal to unity gives

$$E = E^\circ - \frac{RT}{F}\ln a_{\mathrm{HCl}}$$

Use the fact that $a_{\mathrm{HCl}} = a_{\mathrm{H^+}} a_{\mathrm{Cl^-}} = \gamma_\pm^2 m^2$ to get

$$E = E^\circ - \frac{2RT}{F}\ln \gamma_\pm m$$

Now the procedure is the same as that from Equation 13.12 to 13.16 in the text.

13–8. After having determined the value of E° in Problem 13–6, describe how you would determine the mean ionic activity coefficient of HBr(aq) as a function of molality.

Use the value of E° that you determined, the same data, and the Nernst equation in the form

$$E - E^\circ - \frac{2RT}{F}\ln m = \frac{2RT}{F}\ln \gamma_\pm$$

or

$$\ln \gamma_\pm = \frac{F}{2RT}(E - E^\circ) - \ln m$$

to calculate $\ln \gamma_\pm$ as a function of m.

13–9. After having determined the value of E° in Problem 13–7, describe how you would determine the mean ionic activity coefficient of HCl(aq) as a function of molality.

Use the value of E° that you determined, the same data, and the Nernst equation in the form

$$E - E^\circ + \frac{2RT}{F}\ln m = -\frac{2RT}{F}\ln \gamma_\pm$$

or

$$\ln \gamma_\pm = -\frac{F}{2RT}(E - E^\circ) - \ln m$$

to calculate $\ln \gamma_\pm$ as a function of m.

13–10. Without consulting Table 11.3, derive an expression for the activity of each of the following electrolytes in terms of the mean ionic activity coefficient and the molality:

a. KCl(aq) b. $CaCl_2$(aq)

c. K_2SO_4(aq) d. $ZnSO_4$(aq)

---

a. $a_{KCl} = a_{K^+}a_{Cl^-} = a_\pm^2 = \gamma_\pm^2 m^2$

b. $a_{CaCl_2} = a_{Ca^{2+}}a_{Cl^-}^2 = a_\pm^3 = \gamma_\pm^3(4m^3)$

c. $a_{K_2SO_4} = a_{K^+}^2 a_{SO_4^{2-}} = a_\pm^3 = \gamma_\pm^3(4m^3)$

d. $a_{ZnSO_4} = a_{Zn^{2+}}a_{SO_4^{2-}} = a_\pm^2 = \gamma_\pm^2 m^2$

13–11. Use the following data at 298.15 K for a cell whose cell diagram is Pt(s)|H_2(1 bar)|HCl(m)|AgCl(s)|Ag(s) to determine the value of E°.

$m/\text{mol·kg}^{-1}$	E/V	$m/\text{mol·kg}^{-1}$	E/V
0.003215	0.52042	0.011195	0.45860
0.004488	0.50380	0.013407	0.44974
0.005619	0.49262	0.01710	0.43783
0.007311	0.47957	0.02563	0.41824
0.009138	0.46859	0.05391	0.38222

The equation for the celll reaction is

$$H_2(g) + 2\ AgCl(s) \longrightarrow 2\ Ag(s) + 2\ HCl(aq)$$

The corresponding Nernst equation (with $a_{H_2} = a_{AgCl} = a_{Ag} = 1$) is

$$E = E^\circ - \frac{RT}{2F}\ln a_{HCl}^2 = E^\circ - \frac{RT}{F}\ln a_{H^+}a_{Cl^-} = E^\circ - \frac{2RT}{F}\ln \gamma_\pm m$$

Now use Equation 13.15 for $\ln \gamma_\pm$ to write

$$E + \frac{2RT}{F}\ln m - 1.171\frac{2RT}{F}\frac{m^{1/2}}{1+m^{1/2}} = E^\circ - \frac{2RT}{F}Cm$$

or

$$E + 0.05139\ln m - 0.06017\frac{m^{1/2}}{1+m^{1/2}} = E^\circ - 0.05139Cm$$

(See Equation 13.16.)

A linear curve-fit of the left side of the above equation against m gives $E^\circ = 0.2223$ V.

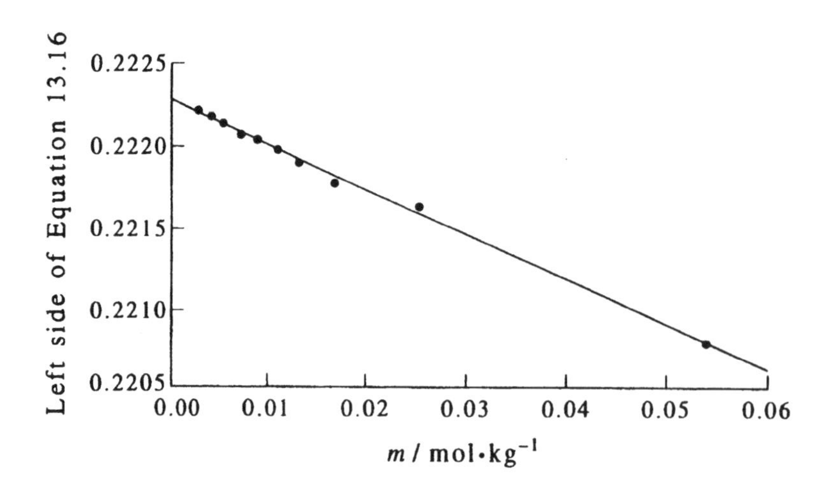

13–12. Use the following data at 298.15 K for a cell whose cell diagram is $Zn(s)|ZnCl_2(aq)|AgCl(s)|Ag(s)$ to determine the value of E°.

$m/\text{mol}\cdot\text{kg}^{-1}$	E/V	$m/\text{mol}\cdot\text{kg}^{-1}$	E/V
0.002941	1.1983	0.04242	1.1090
0.007814	1.1650	0.09048	1.0844
0.01236	1.1495	0.2211	1.0556
0.02144	1.1310	0.4499	1.0328

The equation for the cell reaction is

$$Zn(s) + 2\,AgCl(s) \longrightarrow 2\,Ag(s) + ZnCl_2(aq)$$

and the corresponding Nernst equation (with $a_{Zn} = a_{AgCl} = a_{Ag} = 1$) is

$$E = E^\circ - \frac{RT}{2F}\ln a_{ZnCl_2} = E^\circ - \frac{RT}{2F}\ln a_{Zn^{2+}}a^2_{Cl^-}$$
$$= E^\circ - \frac{3RT}{2F}\ln m - \frac{RT}{2F}\ln 4 - \frac{3RT}{2F}\ln\gamma_\pm \tag{1}$$

Now use Equation 13.14 for the case of a 1–2 electrolyte such as $ZnCl_2$. The factor 1.171 becomes (2)(1.171) because of the the factor $|z_+z_-|$ and the ionic strength is equal to $3m$ for a 1–2 electrolyte. Thus Equation 13.14 becomes

$$\ln\gamma_\pm = -\frac{2.342(3m)^{1/2}}{1+(3m)^{1/2}} + Cm$$

Substitute this equation into Equation 1 to obtain

$$E + \frac{3RT}{2F}\ln m + \frac{RT}{2F}\ln 4 - 2.342\frac{3RT}{2F}\frac{(3m)^{1/2}}{1+(3m)^{1/2}} = E^\circ - \frac{3RT}{2F}Cm$$

or

$$E + 0.03854\ln m + 0.01781 - 0.09026\frac{(3m)^{1/2}}{1+(3m)^{1/2}} = E^\circ - 0.03854Cm$$

A plot of the left side of this equation against m gives $E^\circ = 0.984$ V when extrapolated to $m = 0$.

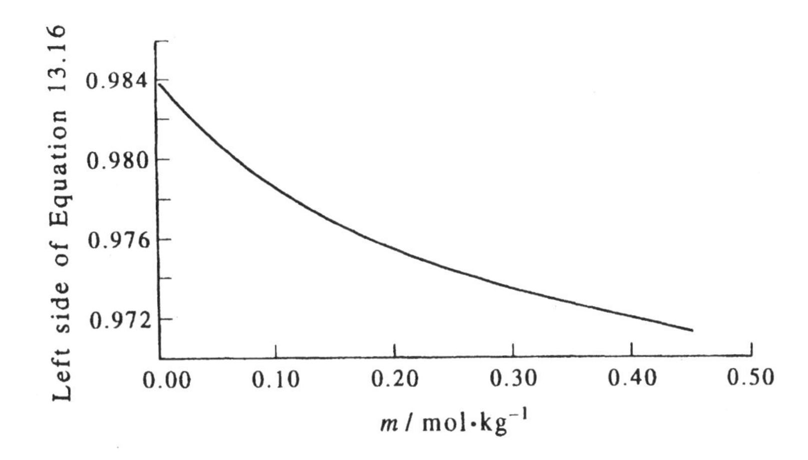

13–13. Using the data in Table 13.2, calculate the emf at 298.15 K of the electrochemical cell whose cell diagram is $Pb(s)|PbI_2(s)|HI(a=0.800)|H_2(1\text{ bar})|Pt(s)$.

The equations for the two electrode reactions are

$$Pb(s) + 2\,I^-(aq) \longrightarrow PbI_2(s) + 2\,e^-$$

and

$$2\,H^+(aq) + 2\,e^- \longrightarrow H_2(g)$$

The equation for the cell reaction is

$$Pb(s) + 2\,HI(aq) \longrightarrow PbI_2(aq) + H_2(g)$$

and so the Nernst equation (with $a_{Pb} = a_{PbI_2} = a_{H_2} = 1$) is

$$E = E^\circ - \frac{RT}{F}\ln a_{HI}$$

Using Table 13.2, $E^\circ = 0\text{ V} - (-0.364)\text{ V} = 0.364$ V, and so

$$\begin{aligned} E &= 0.364\text{ V} + (0.02569\text{ V})\ln 0.800 \\ &= 0.358\text{ V} \end{aligned}$$

13–14. Use the data in Problem 13–12 and the resulting value of E° (0.9845 V) to determine the value of the mean ionic activity coefficient of $ZnCl_2(aq)$ at each given molality.

Solve Equation 1 of Problem 13–12 to obtain

$$\begin{aligned} \ln\gamma_\pm &= \frac{2F}{3RT}(E^\circ - E) - \ln m - \frac{1}{3}\ln 4 \\ &= (25.95\text{ V}^{-1})(0.9845\text{ V} - E) - \ln m - \frac{1}{3}\ln 4 \end{aligned}$$

Using the data in Problem 13–12 gives

$m/\text{mol}\cdot\text{kg}^{-1}$	$\gamma_\pm$
0.002941	0.834
0.007814	0.745
0.01236	0.704
0.02144	0.656
0.04242	0.587
0.09048	0.521
0.2211	0.450
0.4499	0.400

13–15. Use the data in Table 13.1 and the resulting value of E° (0.2224 V) to determine the value of the mean ionic activity coefficient of HCl(aq) at each given molality.

Solve Equation 13.12 for $\ln \gamma_\pm$ to obtain

$$\ln \gamma_\pm = \frac{F}{2RT}(E^\circ - E) - \ln m$$

or

$$\ln \gamma_\pm = (19.46\ \text{V}^{-1})(0.2224\ \text{V} - E) - \ln m$$

Using the data in Table 13.1 gives

$m/\text{mol}\cdot\text{kg}^{-1}$	0.1238	0.05391	0.02563	0.013407	0.009138	0.005619	0.003215
$\gamma_\pm$	0.788	0.827	0.863	0.894	0.909	0.927	0.940

13–16. Use the data in Example 13–4 and the resulting value of E° (0.073 V) to determine the value of the mean ionic activity coefficient of HBr(aq) at each given molality.

Equation 13.12 is applicable to the cell diagrammed in Example 13–4. Solving Equation 13.12 for $\ln \gamma_\pm$ gives

$$\ln \gamma_\pm = \frac{F}{2RT}(E^\circ - E) - \ln m = (19.46\ \text{V}^{-1})(0.073\ \text{V} - E) - \ln m$$

Using the data in Table 13.1 gives

$m/\text{mol}\cdot\text{kg}^{-1}$	0.00100	0.00200	0.00500	0.0100	0.0200	0.0500
$\gamma_\pm$	0.966	0.951	0.930	0.906	0.878	0.838

13–17. Using the data in Table 13.2, calculate the value of the standard emf at 298.15 K of a cell whose cell diagram is $\text{Hg(l)}|\text{Hg}_2\text{Cl}_2\text{(s)}|\text{HCl(aq)}|\text{H}_2\text{(1 bar)}|\text{Pt(s)}$. Which electrode is positive?

The equations for the two electrode reactions are

$$2\,\text{Hg(l)} + 2\,\text{Cl}^-\text{(aq)} \longrightarrow \text{Hg}_2\text{Cl}_2\text{(s)} + 2\,e^- \qquad \text{(left)}$$

$$2\,\text{H}^+\text{(aq)} + 2\,e^- \longrightarrow \text{H}_2\text{(g)} \qquad \text{(right)}$$

Therefore,

$$E^\circ = 0\text{ V} - (+0.268\text{ V}) = -0.268\text{ V}$$

The $H_2(g)$ electrode is positive.

13–18. Using the data in Table 13.2, calculate the value of the standard emf at 298.15 K of a cell whose cell diagram is $Pb(s)|PbSO_4(s)|K_2SO_4(aq)|Hg_2SO_4(s)|Hg(l)$. Which electrode is positive?

The equations for the two electrode reactions are

$$Pb(s) + SO_4^{2-}(aq) \longrightarrow PbSO_4(s) + 2\,e^- \qquad \text{(left)}$$

$$Hg_2SO_4(s) + 2\,e^- \longrightarrow Hg(l) + SO_4^{2-}(aq) \qquad \text{(right)}$$

Therefore,

$$E^\circ = +0.6155\text{ V} - (-0.3583\text{ V}) = +0.9738\text{ V}$$

The Pb(s) electrode is positive.

13–19. Using the data in Table 13.2, calculate the value of the emf at 298.15 K of a cell whose cell diagram is $Ni(s)|Ni^{2+}(a=0.0250)||Zn^{2+}(a=0.300)|Zn(s)$. Which electrode is positive?

The equations for the two electrode reactions are

$$Ni(s) \longrightarrow Ni^{2+}(aq) + 2\,e^-$$

$$Zn^{2+}(aq) + 2\,e^- \longrightarrow Zn(s)$$

Therefore,

$$E^\circ = -0.763\text{ V} - (-0.250\text{ V}) = -0.513\text{ V}$$

The Nernst equation (with $a_{Ni} = a_{Zn} = 1$) is

$$\begin{aligned} E &= E^\circ - \frac{RT}{2F}\ln\frac{a_{Ni^{2+}}a_-^2}{a_{Zn^{2+}}a_-^2} = -0.513\text{ V} - (0.01285\text{ V})\ln\frac{0.0250}{0.300} \\ &= -0.481\text{ V} \end{aligned}$$

The Zn(s) electrode is positive.

13–20. Write the cell diagram for an electrochemical cell whose cell reaction is $H_2(1\text{ bar}) + I_2(s) \longrightarrow 2\,HI(aq)$. Using the data in Table 13.2, calculate the value of E° at 298.15 K for the cell.

The equations for the two electrode reactions are

$$H_2(g) \longrightarrow 2\,H^+(aq) + 2\,e^- \qquad \text{(left)}$$

$$I_2(s) + 2\,e^- \longrightarrow 2\,I^-(aq) \qquad \text{(right)}$$

and the corresponding cell diagram is $H_2(g)|HI(aq)|I_2(s)|Pt(s)$. The value of E° is

$$E^\circ = +0.5345 \text{ V} - (0 \text{ V}) = +0.5345 \text{ V}$$

13–21. Write the cell diagram for an electrochemical cell whose cell reaction is $PbCl_2(s) \longrightarrow Pb^{2+}(aq) + 2\,Cl^-(aq)$. Using the data in Table 13.2, calculate the value of E° at 298.15 K for the cell.

The equations for the two electrode reactions are

$$Pb(s) \longrightarrow Pb^{2+}(aq) + 2\,e^- \qquad \text{(left)}$$

$$PbCl_2(s) + 2\,e^- \longrightarrow Pb(s) + 2\,Cl^-(aq) \qquad \text{(right)}$$

The corresponding cell diagram is $Pb(s)|PbCl_2(aq)|PbCl_2(s)|Pb(s)$. The value of E° is

$$E^\circ = -0.266 \text{ V} - (-0.126 \text{ V}) = -0.140 \text{ V}$$

13–22. Write the cell diagram for an electrochemical cell whose cell reaction is $2\,HCl(aq) + Zn(s) \longrightarrow ZnCl_2(aq) + H_2(g)$. Using the data in Table 13.2, calculate the value of E° at 298.15 K for the cell.

The equations for the two electrode reactions are

$$Zn(s) \longrightarrow Zn^{2+}(aq) + 2\,e^- \qquad \text{(left)}$$

$$2\,H^+(aq) + 2\,e^- \longrightarrow H_2(g) \qquad \text{(right)}$$

The corresponding cell diagram is $Zn(s)|ZnCl_2(aq)||HCl(aq)|H_2(g)$. The value of E° is

$$E^\circ = 0 \text{ V} - (-0.763 \text{ V}) = +0.763 \text{ V}$$

13–23. Calculate the value of the emf at 298.15 K of a cell whose cell diagram is $Pt(s)|Sn^{2+}(a{=}0.200), Sn^{4+}(a{=}0.400)||Fe^{3+}(a{=}0.300)|Fe(s)$.

The equations for the two electrode reactions are

$$Sn^{2+}(aq) \longrightarrow Sn^{4+}(aq) + 2\,e^-$$

$$Fe^{3+}(aq) + 3\,e^- \longrightarrow Fe(s)$$

The value of E° is

$$E^\circ = -0.036 \text{ V} - (+0.15 \text{ V}) = -0.19 \text{ V}$$

The equation for the cell reaction is

$$3\,Sn^{2+}(aq) + 2\,Fe^{3+}(aq) \longrightarrow 3\,Sn^{4+}(aq) + 2\,Fe(s)$$

The Nernst equation (with $a_{Fe} = 1$) is

$$E = E^\circ - \frac{RT}{6F} \ln \frac{a^3_{Sn^{4+}}}{a^3_{Sn^{2+}} a^2_{Fe^{3+}}} = -0.19 \text{ V} - (0.00428 \text{ V}) \ln \frac{(0.400)^3}{(0.200)^3(0.300)^2}$$

$$= -0.21 \text{ V}$$

13–24. Use the data in Table 13.2 to calculate the value of E at 298.15 K for the cell whose cell reaction is

$$Cu(s) + 2\,AgNO_3(0.100 \text{ m}) \rightleftharpoons 2\,Ag(s) + Cu(NO_3)_2(1.00 \text{ m})$$

Take $\gamma_\pm = 0.734$ for $AgNO_3(0.100 \text{ m})$ and 0.338 for $Cu(NO_3)_2(1.00 \text{ m})$. Will the reaction proceed spontaneously?

The equations for the two electrode reactions are

$$Cu(s) \longrightarrow Cu^{2+}(aq) + 2\,e^-$$

and

$$Ag^+(aq) + e^- \longrightarrow Ag(s)$$

The equation for the cell reaction is

$$Cu(s) + 2\,Ag^+(aq) \longrightarrow Cu^{2+}(aq) + 2\,Ag(s)$$

The Nernst equation (with $a_{Cu} = a_{Ag} = 1$) is

$$E = E^\circ - \frac{RT}{2F} \ln \frac{a_{Cu(NO_3)_2}}{a^2_{AgNO_3}}$$

$$= E^\circ - (0.01285 \text{ V}) \ln \frac{4m^3_{Cu(NO_3)_2} \gamma^3_{\pm,Cu(NO_3)_2}}{m^2_{AgNO_3} \gamma^2_{\pm,AgNO_3}}$$

$$= E^\circ - (0.01285 \text{ V}) \ln \frac{(4)(1.00)^3(0.338)^3}{(0.100)^2(0.734)^2}$$

$$= E^\circ - (0.01285 \text{ V}) \ln 28.67$$

$$= E^\circ - 0.0431 \text{ V}$$

But

$$E^\circ = +0.799 \text{ V} - (+0.337 \text{ V}) = +0.462 \text{ V}$$

and so $E = +0.419$ V. The reaction will proceed spontaneously.

13–25. Use the data in Table 13.2 to calculate the value of E at 298.15 K for the cell whose cell reaction is

$$Cd(s) + ZnCl_2(1.00 \text{ m}) \rightleftharpoons Zn(s) + CdCl_2(1.00 \text{ m})$$

Take $\gamma_\pm = 0.341$ for $ZnCl_2(1.00 \text{ m})$ and 0.0669 for $CdCl_2(1.00 \text{ m})$. Will the reaction proceed spontaneously?

The equations for the two electrode reactions are

$$\text{Cd(s)} \longrightarrow \text{Cd}^{2+}\text{(aq)} + 2\,e^-$$

and

$$\text{Zn}^{2+}\text{(aq)} + 2\,e^- \longrightarrow \text{Zn(s)}$$

The equation for the cell reaction is

$$\text{Cd(s)} + \text{Zn}^{2+}\text{(aq)} \longrightarrow \text{Cd}^{2+}\text{(aq)} + \text{Zn(s)}$$

The Nernst equation (with $a_{\text{Cd}} = a_{\text{Zn}} = 1$) is

$$\begin{aligned} E &= E^\circ - \frac{RT}{2F} \ln \frac{a_{\text{CdCl}_2}}{a_{\text{ZnCl}_2}} \\ &= E^\circ - (0.01285\ \text{V}) \ln \frac{4m^3_{\text{CdCl}_2}\gamma^3_{\pm,\text{CdCl}_2}}{4m^3_{\text{ZnCl}_2}\gamma^3_{\pm,\text{ZnCl}_2}} \\ &= E^\circ - (0.01285\ \text{V}) \ln \frac{(0.0669)^3}{(0.341)^3} \\ &= E^\circ + 0.0628\ \text{V} \end{aligned}$$

But

$$E^\circ = -0.763\ \text{V} - (-0.403\ \text{V}) = -0.360\ \text{V}$$

and so $E = -0.297$ V. The reaction will not proceed spontaneously.

13–26. Consider the electrochemical cell whose cell diagram is $\text{Zn(s)}|\text{ZnI}_2(m)|\text{I}_2\text{(s)}|\text{Pt(s)}$. Given that $\gamma_\pm = 0.605$ for 0.500 molal ZnI_2(aq), calculate the value of the emf of the cell at 298.15 K.

The equations for the two electrode reactions are

$$\text{Zn(aq)} \longrightarrow \text{Zn}^{2+}\text{(aq)} + 2\,e^-$$

$$\text{I}_2\text{(s)} + 2\,e^- \longrightarrow 2\,\text{I}^-\text{(aq)}$$

and the equation for the cell reaction is

$$\text{Zn(s)} + \text{I}_2\text{(s)} \longrightarrow \text{ZnI}_2\text{(aq)}$$

The Nernst equation (with $a_{\text{Zn}} = a_{\text{I}_2} = 1$) is

$$\begin{aligned} E &= E^\circ - \frac{RT}{2F} \ln a_{\text{ZnI}_2} = E^\circ - (0.01285\ \text{V}) \ln(4m^3\gamma^3_\pm) \\ &= E^\circ - (0.01285\ \text{V}) \ln 0.1107 \\ &= E^\circ + 0.0283\ \text{V} \end{aligned}$$

But

$$E^\circ = +0.5345\ \text{V} - (-0.763\ \text{V}) = 1.298\ \text{V}$$

and so $E = 1.326$ V.

13–27. Consider the electrochemical cell whose cell diagram is
$Cd(s)|CdCl_2(m)|AgCl(s)|Ag(s)$. Given that $\gamma_\pm = 0.0669$ for 1.00 molal $CdCl_2(aq)$, calculate the value of the emf of the cell at 298.15 K.

The equations for the two electrode reactions are

$$Cd(s) \longrightarrow Cd^{2+}(aq) + 2\,e^-$$
$$AgCl(s) + e^- \longrightarrow Ag(s) + Cl^-(aq)$$

and the equation for the cell reaction is

$$Cd(s) + 2\,AgCl(s) \longrightarrow 2\,Ag(s) + CdCl_2(aq)$$

The Nernst equation (with $a_{Zn} = a_{AgCl} = a_{Ag} = 1$) is

$$\begin{aligned} E = E^\circ - \frac{RT}{2F}\ln a_{CdCl_2} &= E^\circ - (0.01285\text{ V})\ln(4m^3\gamma_\pm^3) \\ &= E^\circ - (0.01285\text{ V})\ln(1.198\times10^{-3}) \\ &= E^\circ + 0.0864\text{ V} \end{aligned}$$

But

$$E^\circ = +0.2224\text{ V} - (-0.403\text{ V}) = 0.6254\text{ V}$$

and so $E = +0.712$ V.

13–28. The standard emf for the reaction described by

$$HClO(aq) + H^+(aq) + 2\,Cr^{2+}(aq) \longrightarrow 2\,Cr^{3+}(aq) + Cl^-(aq) + H_2O(l)$$

is $E^\circ = 1.90$ V. Use the data in Table 13.2 to calculate the value of E°_{red} at 298.15 K for the electrode reaction

$$HClO(aq) + H^+(aq) + 2\,e^- \longrightarrow Cl^-(aq) + H_2O(l)$$

The equations for the two electrode reactions are

$$Cr^{2+}(aq) \longrightarrow Cr^{3+}(aq) + e^-$$
$$HClO(aq) + H^+(aq) + 2\,e^- \longrightarrow Cl^-(aq) + H_2O(l)$$

The value of E° for the cell is given by

$$E^\circ = E^\circ[HClO(aq)/Cl^-(aq)] - E^\circ[Cr^{3+}(aq)/Cr^{2+}(aq)]$$

Using the entry is Table 13.2 for $E^\circ[Cr^{3+}(aq)/Cr^{2+}(aq)]$, we have

$$1.90\text{ V} = E^\circ[HClO(aq)/Cl^-(aq)] - (-0.408\text{ V})$$

or

$$E^\circ[HClO(aq)/Cl^-(aq)] = +1.49\text{ V}$$

13–29. Design an electrochemical cell that can be used to determine the activity coefficients of $H_2SO_4(aq)$ solutions. Write its cell diagram and the corresponding Nernst equation, relating E to $\gamma_\pm$. Determine the value of E° at 298.15 K from Table 13.2.

Consider the two electrode reactions described by

$$PbSO_4(s) + 2\,e^- \longrightarrow Pb(s) + SO_4^{2-}(aq) \qquad \text{(right)}$$

$$H_2(g) \longrightarrow 2\,H^+(aq) + 2\,e^- \qquad \text{(left)}$$

The equation for the cell reaction is

$$PbSO_4(s) + H_2(g) \longrightarrow 2\,Pb(s) + 2\,H^+(aq) + SO_4^{2-}(aq)$$

The corresponding cell diagram is

$$H_2(g)|H_2SO_4(aq)|PbSO_4(s)|Pb(s)$$

and the Nernst equation (with $a_{H_2} = a_{PbSO_4} = a_{Pb} = 1$) is

$$E = E^\circ - \frac{RT}{2F}\ln a_{H_2SO_4} = E^\circ - \frac{RT}{2F}\ln(4m^3\gamma_\pm^3)$$

The value of E° is given by

$$E^\circ = -0.3583\text{ V} - (0\text{ V}) = -0.3583\text{ V}$$

13–30. Data are given below for the emf versus concentration at 298.15 K for the cell whose cell diagram is $Pt(s)|H_2(1\text{ bar})|HBr(aq)|AgBr(s)|Ag(s)$. Determine the value of E° and calculate the value of $\gamma_\pm$ of HBr(aq) for each concentration.

$m/10^{-4}\text{ mol}\cdot\text{kg}^{-1}$	3.198	4.042	8.444	13.55	18.50
E/V	0.48775	0.47604	0.43850	0.41450	0.39891

The equations for the two electrode reactions are

$$H_2(g) \longrightarrow 2\,H^+(aq) + 2\,e^-$$

$$AgBr(s) + e^- \longrightarrow Ag(s) + Br^-(aq)$$

and the equation for the cell reaction is

$$H_2(g) + 2\,AgBr(s) \longrightarrow 2\,Ag(s) + 2\,HBr(aq)$$

We use Equation 13.15 to evaluate E°. A plot of

$$E + 0.05139\ln m - 0.06018\frac{m^{1/2}}{1 + m^{1/2}}$$

against m is shown below. A linear curve fit of the left side of Equation 13.15 against m gives an intercept of 0.0732 V.

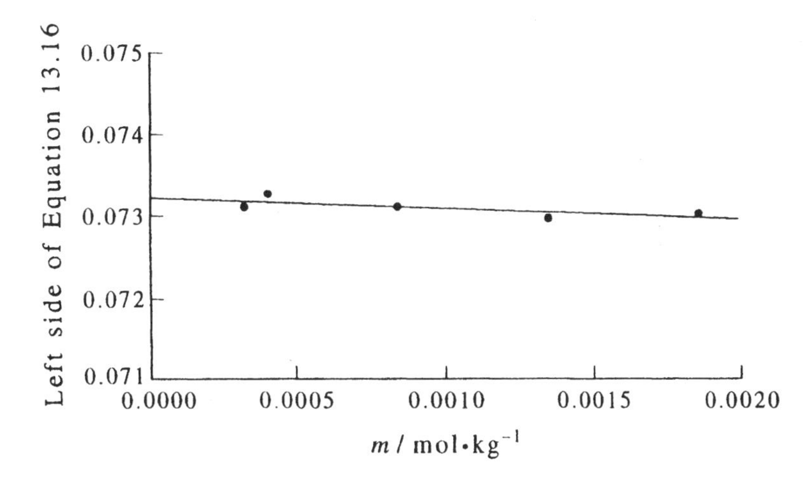

The Nernst equation (with $a_{H_2} = a_{AgBr} = a_{Ag} = 1$) for this system is

$$E = E^\circ - \frac{RT}{2F}\ln a_{HBr}^2 = E^\circ - \frac{RT}{F}\ln a_{H^+}a_{Br^-}$$
$$= E^\circ - \frac{RT}{F}\ln m^2\gamma_\pm^2 = E^\circ - \frac{2RT}{F}\ln m\gamma_\pm$$

Solving for $\ln\gamma_\pm$ gives

$$\ln\gamma_\pm = \frac{F}{2RT}(E^\circ - E) - \ln m$$

Using the data given in the problem, we have

$m/\times 10^{-4}$ mol·kg^{-1}	3.198	4.042	8.444	13.55	18.50
$\gamma_\pm$	0.981	0.975	0.969	0.963	0.955

13–31. The temperature dependence of the standard emf of a cell whose cell diagram is Pt(s)|H_2(1 bar)|HCl(a = 1.00)|AgCl(s)|Ag(s) is given by

$$E^\circ/\text{V} = 0.23659 - (4.8564\times10^{-4}\,^\circ\text{C}^{-1})t - (3.4205\times10^{-6}\,^\circ\text{C}^{-2})t^2$$
$$+ (5.869\times10^{-9}\,^\circ\text{C}^{-3})t^3 \qquad 0 \le t \le 50^\circ\text{C}$$

where the temperature is in degrees Celsius. Write the associated cell reaction, and determine the standard change in Gibbs energy, the enthalpy, and the entropy at 298.15 K.

The equations for the two electrode reactions are

$$H_2(g) \longrightarrow 2\,H^+(aq) + 2\,e^-$$

$$AgCl(s) + e^- \longrightarrow Ag(s) + Cl^-(aq)$$

and the equation for the cell reaction is

$$H_2(g) + 2\,AgCl(s) \longrightarrow 2\,Ag(s) + 2\,HCl(aq)$$

$$\Delta_r G^\circ = -nFE^\circ = -(2)(96\,485\ \text{C}\cdot\text{mol}^{-1})(0.2224\ \text{V})$$
$$= -42.92\ \text{kJ}\cdot\text{mol}^{-1}$$

$$\Delta_r S^\circ = nF\left(\frac{\partial E^\circ}{\partial T}\right)_P = -(2)(96\,485\ \text{C}\cdot\text{mol}^{-1})(-6.457\times10^{-4}\ \text{V}\cdot\text{K}^{-1})$$
$$= -124.6\ \text{J}\cdot\text{K}^{-1}\cdot\text{mol}^{-1}$$

$$\Delta_r H^\circ = \Delta_r G^\circ + T\Delta_r S^\circ = -42.92\ \text{kJ}\cdot\text{mol}^{-1} + (298.15\ \text{K})(-124.6\ \text{J}\cdot\text{K}^{-1}\cdot\text{mol}^{-1})$$
$$= -80.07\ \text{kJ}\cdot\text{mol}^{-1}$$

13–32. The temperature dependence of the standard emf of a cell whose cell diagram is $\text{Pt(s)}|\text{H}_2(1\ \text{bar})|\text{HBr}(a=1.00)|\text{Hg}_2\text{Br}_2\text{(s)}|\text{Hg(l)}$ is given by

$$E^\circ/\text{V} = 0.13970 - (1.54\times10^{-4}\ {}^\circ\text{C}^{-1})(t-25.0^\circ\text{C}) - (3.6\times10^{-6}\ {}^\circ\text{C}^{-2})(t-25.0^\circ\text{C})^2$$

where t is the Celsius temperature. Determine the change in the standard Gibbs energy, the standard enthalpy, and the standard entropy at 298.15 K. Given that $S^\circ[\text{H}_2\text{(g)}] = 130.684\ \text{J}\cdot\text{K}^{-1}\cdot\text{mol}^{-1}$, $S^\circ[\text{Hg}_2\text{Br}_2\text{(s)}] = 218.0\ \text{J}\cdot\text{K}^{-1}\cdot\text{mol}^{-1}$, and $S^\circ[\text{Hg(l)}] = 76.02\ \text{J}\cdot\text{K}^{-1}\cdot\text{mol}^{-1}$ at 298.15 K, calculate the value of S° for $\text{Br}^-\text{(aq)}$. Compare your answer with the value in the *NBS Tables of Chemical Thermodynamic Properties*.

The equation for the cell reaction is

$$\text{H}_2\text{(g)} + \text{Hg}_2\text{Br}_2\text{(s)} \longrightarrow 2\ \text{Hg(l)} + 2\ \text{HBr(aq)}$$

$$\Delta_r G^\circ = -nFE^\circ = -(2)(96\,485\ \text{C}\cdot\text{mol}^{-1})(0.13970\ \text{V})$$
$$= -26.96\ \text{kJ}\cdot\text{mol}^{-1}$$

$$\Delta_r S^\circ = nF\left(\frac{\partial E^\circ}{\partial T}\right)_P = -(2)(96\,485\ \text{C}\cdot\text{mol}^{-1})(-1.54\times10^{-4}\ \text{V}\cdot\text{K}^{-1})$$
$$= -29.7\ \text{J}\cdot\text{K}^{-1}\cdot\text{mol}^{-1}$$

$$\Delta_r H^\circ = \Delta_r G^\circ + T\Delta_r S^\circ = -26.96\ \text{kJ}\cdot\text{mol}^{-1} + (298.15\ \text{K})(-29.7\ \text{J}\cdot\text{K}^{-1}\cdot\text{mol}^{-1})$$
$$= -35.8\ \text{kJ}\cdot\text{mol}^{-1}$$

$$\Delta_r S^\circ = 2S^\circ[\text{Hg(l)}] + 2S^\circ[\text{H}^+\text{(aq)}] + 2S^\circ[\text{Br}^-\text{(aq)}] - S^\circ[\text{H}_2\text{(g)}] - S^\circ[\text{Hg}_2\text{Br}_2\text{(s)}]$$

$$S^\circ[\text{Br}^-\text{(aq)}] = \frac{1}{2}\{\Delta_r S^\circ + S^\circ[\text{H}_2\text{(g)}] + S^\circ[\text{Hg}_2\text{Br}_2\text{(s)}] - 2S^\circ[\text{Hg(l)}] - 2S^\circ[\text{H}^+\text{(aq)}]\}$$
$$= \frac{1}{2}\{-29.7 + 130.684 + 218.0 - (2)(76.02) - 0\}\ \text{J}\cdot\text{K}^{-1}\cdot\text{mol}^{-1}$$
$$= 83.5\ \text{J}\cdot\text{K}^{-1}\cdot\text{mol}^{-1}$$

The value given in the *NBS Tables of Chemical Thermodynamic Properties* is $82.4\ \text{J}\cdot\text{K}^{-1}\cdot\text{mol}^{-1}$.

13–33. Use the data in Table 13.2 to calculate the value of the solubility product of $\text{Hg}_2\text{Cl}_2\text{(s)}$ at 298.15 K.

The equation for the dissolution reaction is

$$Hg_2Cl_2(s) \longrightarrow Hg_2^{2+}(aq) + 2\,Cl^-(aq)$$

A suitable cell is diagrammed by

$$Hg(l)|Hg_2Cl_2(aq)|Hg_2Cl_2(s)|Hg(l)$$

whose two electrode reactions can be described by

$$2\,Hg(l) \longrightarrow Hg_2^{2+}(aq) + 2\,e^-$$

$$Hg_2Cl_2(s) + 2\,e^- \longrightarrow 2\,Hg(l) + 2\,Cl^-(aq)$$

The value of E° from Table 13.2 is

$$E^\circ = +0.268\text{ V} - (-0.796\text{ V}) = -0.528\text{ V}$$

and the value of K_{sp} is given by

$$\begin{aligned} K_{sp} &= e^{nFE^\circ/RT} \\ &= \exp\left[\frac{(2)(96485\text{ C})(-0.528\text{ V})}{(8.314\text{ J}\cdot\text{mol}^{-1}\cdot\text{K}^{-1})(298.15\text{ K})}\right] \\ &= \exp(-41.101) = 1.41 \times 10^{-18} \end{aligned}$$

13–34. Use the data in Table 13.2 to calculate the value of the solubility product of $Ag_2SO_4(s)$ at 298.15 K.

The equation for the dissolution reaction is

$$Ag_2SO_4(s) \longrightarrow 2\,Ag^+(aq) + SO_4^{2-}(aq)$$

A suitable cell is diagrammed by

$$Ag(s)|Ag_2SO_4(aq)|Ag_2SO_4(s)|Ag(s)$$

whose two electrode reactions can be described by

$$Ag(s) \longrightarrow Ag^+(aq) + e^-$$
$$Ag_2SO_4(s) + 2\,e^- \longrightarrow 2\,Ag(s) + SO_4^{2-}(aq)$$

The value of E° from Table 13.2 is

$$E^\circ = +0.653\text{ V} - (0.799\text{ V}) = -0.146\text{ V}$$

and the value of K_{sp} is given by

$$\begin{aligned} K_{sp} &= e^{nFE^\circ/RT} \\ &= \exp\left[\frac{(2)(96485\text{ C})(-0.146\text{ V})}{(8.314\text{ J}\cdot\text{mol}^{-1}\cdot\text{K}^{-1})(298.15\text{ K})}\right] \\ &= \exp(-11.36) = 1.16 \times 10^{-5} \end{aligned}$$

13–35. Use the data in Table 13.2 to calculate the value of the solubility product of $PbSO_4(s)$ at 298.15 K.

The equation for the dissolution reaction is

$$PbSO_4(s) \longrightarrow Pb^{2+}(aq) + SO_4^{2-}(aq)$$

A suitable cell is diagrammed by

$$Pb(s)|PbSO_4(aq)|PbSO_4(s)|Pb(s)$$

whose two electrode reactions can be described by

$$Pb(s) \longrightarrow Pb^{2+}(aq) + 2\,e^-$$

$$PbSO_4(s) + 2\,e^- \longrightarrow Pb(s) + SO_4^{2-}(aq)$$

The value of E° from Table 13.2 is

$$E^\circ = -0.3583\text{ V} - (-0.126\text{ V}) = -0.126\text{ V}$$

and the value of K_{sp} is given by

$$\begin{aligned} K_{sp} &= e^{nFE^\circ/RT} \\ &= \exp\left[\frac{(2)(96485\text{ C})(-0.126\text{ V})}{(8.314\text{ J}\cdot\text{mol}^{-1}\cdot\text{K}^{-1})(298.15\text{ K})}\right] \\ &= \exp(-18.08) = 1.40 \times 10^{-8} \end{aligned}$$

13–36. Given that the standard reduction potential of $2\,D^+(aq) + 2\,e^- \longrightarrow D_2(g)$ is -0.0034 V, calculate the value of the equilibrium constant of

$$2\,H^+(aq) + D_2(g) \longrightarrow 2\,D^+(aq) + H_2(g)$$

at 298.15 K.

A suitable cell is diagrammed by

$$D_2(g)|DCl(aq)||HCl(aq)|H_2(g)$$

whose two electrode reactions can be described by

$$D_2(g) \longrightarrow 2\,D^+(aq) + 2\,e^-$$

$$2\,H^+(aq) + 2\,e^- \longrightarrow H_2(g)$$

The value of E° for the cell is -0.0034 V, and so

$$\begin{aligned} K_{sp} &= e^{nFE^\circ/RT} \\ &= \exp\left[\frac{(2)(96485\text{ C})(-0.0034\text{ V})}{(8.314\text{ J}\cdot\text{mol}^{-1}\cdot\text{K}^{-1})(298.15\text{ K})}\right] \\ &= 1.30 \end{aligned}$$

13–37. Given the reduction electrode-reaction data at 298.15 K

$$HClO(aq) + H^+(aq) + 2\,e^- \longrightarrow Cl^-(aq) + H_2O(l) \qquad E^\circ = 1.49\ V$$

$$ClO^-(aq) + H_2O(l) + 2\,e^- \longrightarrow Cl^-(aq) + 2\,OH^-(aq)\,E^\circ = 0.90\ V$$

calculate the value of the acid-dissociation constant of HClO(aq). *Hint*: You need to use the fact that $K_w = 1.00 \times 10^{-14}$.

Subtract the second equation given in the problem from the first to obtain

$$HClO(aq) + H^+(aq) + 2\,OH^-(aq) \longrightarrow ClO^-(aq) + 2\,H_2O(l)$$

The value of E° associated with this equation is

$$E^\circ = 1.49\ V - 0.90\ V = 0.59\ V$$

and the value of K is given by

$$K = e^{nFE^\circ/RT}$$

$$= \exp\left[\frac{(2)(96485\ C)(0.59\ V)}{(8.314\ J\cdot mol^{-1}\cdot K^{-1})(298.15\ K)}\right]$$

$$= 8.8 \times 10^{19}$$

To obtain the equation for the dissociation of HClO(aq), add $2\,H_2O(l) \rightarrow 2\,H^+(aq) + 2\,OH^-(aq)$ to the first equation above, and so

$$K_a = (8.8 \times 10^{19})K_w^2 = 8.8 \times 10^{-9}$$

13–38. Given the reduction electrode-reaction data at 298.15 K

$$ClO_2(aq) + e^- \longrightarrow ClO_2^-(aq) \qquad E^\circ = 1.15\ V$$

$$ClO_2(aq) + H^+(aq) + e^- \longrightarrow HClO_2(aq) \qquad E^\circ = 1.27\ V$$

calculate the value of the acid-dissociation constant of $HClO_2$(aq).

Subtract the second equation given in the problem from the first to obtain

$$HClO_2(aq) \longrightarrow H^+(aq) + ClO_2^-(aq)$$

The value of E° associated with this equation is

$$E^\circ = 1.15\ V - 1.27\ V = -0.12\ V$$

and the value of K_a is given by

$$K_a = e^{nFE^\circ/RT}$$

$$= \exp\left[\frac{(2)(96485\ C)(-0.12\ V)}{(8.314\ J\cdot mol^{-1}\cdot K^{-1})(298.15\ K)}\right]$$

$$= 9.4 \times 10^{-3}$$

13–39. The value of the acid-dissociation constant of propanoic acid can be determined using the cell

$$\mathrm{Pt(s)|H_2(g)|HP(aq),\ NaP(aq),\ NaCl(aq)|AgCl(s)|Ag(s)}$$

Use the following data at 10°C to determine the value of K_a for propanoic acid at 10°C. Take E° for the cell to be 0.23142 V at 10°C.

$m_{\mathrm{HP}} = m_{\mathrm{NaP}} = m_{\mathrm{NaCl}}$	E/V
0.006442	0.62854
0.007055	0.62622
0.009225	0.61973
0.010812	0.61590
0.01660	0.60544
0.02274	0.59757
0.026833	0.59360
0.03067	0.59037
0.03108	0.59001

We use Equation 13.47 and calculate $-\ln K_a'$ as a function of ionic strength, which in this case, is given by $I_m = 2m_{\mathrm{NaP}} = 2m_{\mathrm{NaCl}}$ [neglecting the ions from HP(aq)]. In this case, we plot $-\ln K_a' = (40.983\ \mathrm{V^{-1}})(E - 0.23142\ \mathrm{V}) + \ln m_{\mathrm{HP}}$ against I_m. The results are

$-\ln K_a'$	$I_m/\mathrm{mol\cdot kg^{-1}}$
11.230	0.012884
11.226	0.014110
11.220	0.018450
11.230	0.021624
11.230	0.033200
11.222	0.045480
11.225	0.053666
11.226	0.061340
11.225	0.062160

A linear curve-fit of the plot of $-\ln K_a'$ against I_m (shown below) gives $-\ln K_a' = 11.23 - 0.0853 I_m$. Therefore, $-\ln K_a' = 11.23$ or $K_a = 1.33 \times 10^{-5}$.

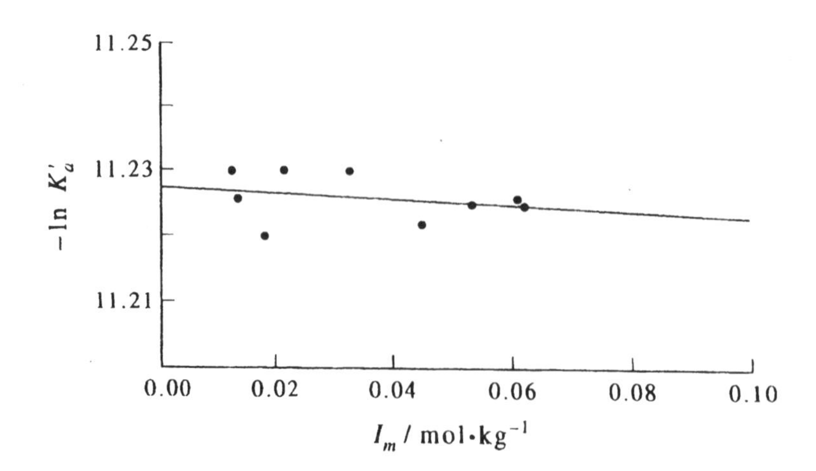

13–40. The value of the acid-dissociation constant of acetic acid can be determined using the cell

$$\text{Pt(s)}|\text{H}_2\text{(g)}|\text{HAc(aq), NaAc(aq), NaCl(aq)}|\text{AgCl(s)}|\text{Ag(s)}$$

Use the following data at 0°C to determine the value of K_a for acetic acid at 0°C. Take $E°$ for the cell to be 0.23655 V at 0°C.

$m_{\text{HAc}}/\text{mol}\cdot\text{kg}^{-1}$	$m_{\text{NaAc}}/\text{mol}\cdot\text{kg}^{-1}$	$m_{\text{NaCl}}/\text{mol}\cdot\text{kg}^{-1}$	E/V
0.0047790	0.0045990	0.0048960	0.61995
0.012035	0.011582	0.012426	0.59826
0.021006	0.020216	0.021516	0.58528
0.049220	0.047370	0.050420	0.56545
0.081010	0.077960	0.082970	0.55388
0.090560	0.087160	0.092760	0.55128

We use Equation 13.47 and calculate $-\ln K_a'$ as a function of ionic strength, which, in this case, is given by $I_m = m_{\text{NaAc}} + m_{\text{NaCl}}$ [neglecting the ions from HAc(aq)]. In this case, we plot

$$-\ln K_a' = (42.484\ \text{V}^{-1})(E - 0.23655\ \text{V}) + \ln \frac{m_{\text{HAc}} m_{\text{NaCl}}}{m_{\text{NaAc}}}$$

against I_m. The results are

$-\ln K_a'$	$I_m/\text{mol}\cdot\text{kg}^{-1}$
11.007	0.009495
11.017	0.024008
11.015	0.041732
11.024	0.097790
11.030	0.16093
11.031	0.17992

A linear curve-fit of the plot of $-\ln K_a'$ against I_m (shown below) gives $-\ln K_a' = 11.01 + 0.125 I_m$. Therefore, $-\ln K_a' = 11.01$, or $K_a = 1.65 \times 10^{-5}$.

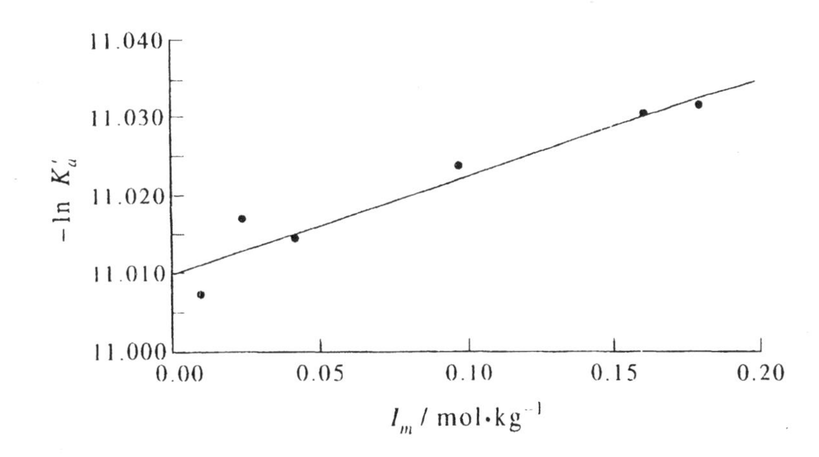

13–41. The value of the acid-dissociation constant of formic acid can be determined using the cell

$$\text{Pt(s)}|\text{H}_2\text{(g)}|\text{HFo(aq), NaFo(aq), NaCl(aq)}|\text{AgCl(s)}|\text{Ag(s)}$$

Determine the equation for the overall reaction of this cell and write the corresponding Nernst equation. Now use the relation

$$K_a = \frac{a_{H^+} a_{Fo^-}}{a_{HFo}}$$

to show that

$$E = E^\circ - \frac{RT}{F} \ln \frac{m_{HFo}\gamma_{HFo} m_{Cl^-}\gamma_{Cl^-}}{m_{Fo^-}\gamma_{Fo^-}} - \frac{RT}{F} \ln K_a$$

Show that this equation can be written as

$$\frac{F}{RT}(E - E^\circ) + \ln \frac{m_{HFo} m_{Cl^-}}{m_{Fo^-}} = -\ln K_a - \ln \frac{\gamma_{HFo}\gamma_{Cl^-}}{\gamma_{Fo^-}}$$
$$= -\ln K_a'$$

Describe how the value of K_a can be determined by plotting

$$\frac{F}{RT}(E - E^\circ) + \ln \frac{m_{HFo} m_{Cl^-}}{m_{Fo^-}} = -\ln K_a'$$

against ionic strength and extrapolating to zero. Unlike the case of propanoic acid presented in Example 13–10, the $H^+(aq)$ and $Fo^-(aq)$ from the dissociation of HFo(aq) cannot be ignored. The way to proceed is as follows. First, calculate the values of $\ln K_a'$ and I_m neglecting the $H^+(aq)$ and $Fo^-(aq)$ from the dissociation of formic acid. Plot $\ln K_a'$ against I_m, and obtain a preliminary value of K_a by extrapolating to zero ionic strength. Now use the preliminary value of K_a to calculate m_{H^+} using

$$K_a = \frac{m_{H^+} m_{Fo^-}}{m_{HFo}} \frac{\gamma_{H^+}\gamma_{Fo^-}}{\gamma_{HFo}} = \frac{m_{H^+}(m_{NaFo} + m_{H^+})}{m_{HFo} - m_{H^+}} \gamma_\pm^2$$

where m_{HFo} and m_{NaFo} are the stoichiometric concentrations of HFo(aq) and NaFo(aq). Realizing that m_{H^+} will be fairly small, neglect m_{H^+} with respect to m_{HFo} and m_{NaFo} and write

$$m_{H^+} = \frac{m_{HFo}}{m_{NaFo}} \frac{K_a}{\gamma_\pm^2}$$

The value of $\gamma_\pm$ can be estimated using Equation 11.56 on a molality scale

$$\ln \gamma_\pm = -\frac{1.171(I_m/\text{mol}\cdot\text{kg}^{-1})^{1/2}}{1 + (I_m/\text{mol}\cdot\text{kg}^{-1})^{1/2}}$$

Using the above procedure and the following data at 25°C, calculate the value of K_a for formic acid at 25°C.

$m_{HFo}/\text{mol}\cdot\text{kg}^{-1}$	$m_{NaFo}/\text{mol}\cdot\text{kg}^{-1}$	$m_{NaCl}/\text{mol}\cdot\text{kg}^{-1}$	E/V
0.0065380	0.0081509	0.0070544	0.57842
0.011760	0.014661	0.012689	0.56294
0.024450	0.030482	0.026381	0.54398
0.035750	0.044570	0.038574	0.53427
0.048630	0.060627	0.052471	0.52647
0.098760	0.12312	0.10656	0.50882

The equations for the two electrode reactions are

$$\tfrac{1}{2}\,H_2(g) \longrightarrow H^+(aq) + e^-$$

$$AgCl(s) + e^- \longrightarrow Ag(s) + Cl^-(aq)$$

and the equation for the cell reaction is

$$\tfrac{1}{2}\,H_2(g) + AgCl(s) \longrightarrow Ag(s) + H^+(aq) + Cl^-(aq)$$

The corresponding Nernst equation is

$$E = E^\circ - \frac{RT}{F}\ln a_{H^+}a_{Cl^-}$$

where $E^\circ = 0.2224$ V. Substitute

$$a_{H^+} = \frac{a_{HFo}K_w}{a_{Fo^-}} = \frac{m_{HFo}\gamma_{HFo}K_w}{m_{Fo^-}\gamma_{Fo^-}}$$

into the Nernst equation to obtain the equation given in the problem. The numerical results obtained by following the procedure given in the problem are

m_{HFo}	0.006538	0.01176	0.02445	0.03575	0.04863	0.09876
m_{Fo^-}	0.008151	0.01466	0.03048	0.04457	0.06063	0.12312
m_{Cl^-}	0.007054	0.01269	0.02638	0.03857	0.05247	0.10656
E/V	0.5784	0.5629	0.5440	0.5343	0.5265	0.5088
$\ln K'$	8.682	8.667	8.661	8.663	8.667	8.688
I_m	0.01520	0.02735	0.05686	0.08314	0.1131	0.2297
m_{H^+}	0.0001789	0.0001929	0.0002172	0.0002337	0.0002495	0.0002955
$\ln K'$	8.633	8.637	8.645	8.651	8.658	8.683
I_m	0.01538	0.02754	0.05708	0.08338	0.1133	0.2300
m_{H^+}	0.0001764	0.0001940	0.0002213	0.0002391	0.0002559	0.0003042
$\ln K'$	8.633	8.637	8.644	8.650	8.657	8.683
I_m	0.01538	0.02754	0.05708	0.08338	0.1133	0.2300

A linear curve-fit of a plot of $-\ln K_a'$ against I_m (shown below) gives $-\ln K_a' = 8.631 + 0.228 I_m$, or $K_a' = 1.78 \times 10^{-4}$.

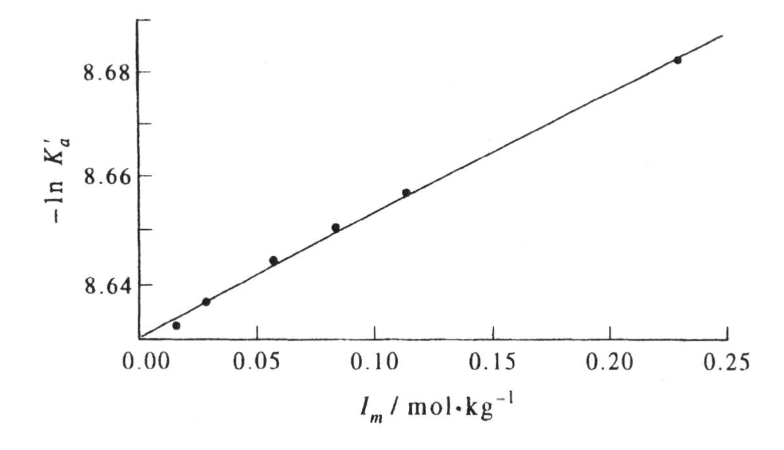

13–42. The value of the dissociation constant of water, K_w, can be determined using the cell

$$\mathrm{Pt(s)|H_2(g)|KOH(aq),\ KCl(aq)|AgCl(s)|Ag(s)} \tag{1}$$

Show that the emf of this cell is given by

$$E = E^\circ - \frac{RT}{F}\ln m_{\mathrm{H}^+} m_{\mathrm{Cl}^-}\gamma_{\mathrm{H}^+}\gamma_{\mathrm{Cl}^-}$$

where E° is the standard emf of the cell

$$\mathrm{Pt(s)|H_2(g)|HCl(aq)|AgCl(s)|Ag(s)}$$

Using the relation

$$K_w = m_{\mathrm{H}^+} m_{\mathrm{OH}^-}\frac{\gamma_{\mathrm{H}^+}\gamma_{\mathrm{OH}^-}}{a_{\mathrm{H_2O}}}$$

show that

$$\frac{F}{RT}(E - E^\circ) + \ln\frac{m_{\mathrm{KCl}}}{m_{\mathrm{KOH}}} = \ln\frac{\gamma_{\mathrm{OH}^-}}{\gamma_{\mathrm{Cl}^-}a_{\mathrm{H_2O}}} - \ln K_w$$
$$= -\ln K_w'$$

Using the following data for the cell given by Equation 1 with $m_{\mathrm{KOH}} = 0.0100\ \mathrm{mol\cdot kg^{-1}}$, determine the value of K_w at each temperature by plotting the left side of the above equation against ionic strength and extrapolating to zero. Take E° to be given by

$$E^\circ/\mathrm{V} = 0.23634 - 0.00047396(t/^\circ\mathrm{C}) - 3.2331\times10^{-6}(t/^\circ\mathrm{C})^2$$
$$- 9.1128\times10^{-9}(t/^\circ\mathrm{C})^3 + 1.477\times10^{-10}(t/^\circ\mathrm{C})^4$$

$m_{\mathrm{KCl}}/\mathrm{mol\cdot kg^{-1}}$	E/V at 0°C	E/V at 10°C	E/V at 20°C	E/V at 30°C	E/V at 40°C	E/V at 50°C	E/V at 60°C
0.0100000	1.0462	1.0478	1.04947	1.0512	1.0530	1.0548	1.0567
0.020000	1.0299	1.0309	1.03203	1.0332	1.0344	1.0356	1.0370
0.030000	1.0204	1.0211	1.02183	1.0226	1.0235	1.0244	1.0253
0.040000	1.0136	1.0141	1.01458	1.0151	1.0158	1.0164	1.0171
0.050000	1.0084	1.0087	1.00903	1.0094	1.0098	1.0102	1.0107
0.070000	1.0005	1.0005	1.00057	1.0007	1.0008	1.0010	1.0012
0.10000	0.99217	0.99192	0.991700	0.99151	0.99136	0.99123	0.99114
0.20000	0.97605	0.97529	0.974550	0.97380	0.97309	0.97239	0.97170
0.30000	0.96680	0.96571	0.964640	0.96357	0.96251	0.96145	0.96041
0.40000	0.96029	0.95901	0.957730	0.95643	0.95513	0.95383	0.95250
0.50000	0.95537	0.95391	0.952440	0.95098	0.94948	0.94798	0.94648
0.60000	0.95138	0.94978	0.948170	0.94655	0.94492	0.94328	0.94164
0.70000	0.94805	0.94634	0.944620	0.94288	0.94113	0.93936	0.93758
0.80000	0.94517	0.94337	0.941540	0.93970	0.93784	0.93596	0.93406
0.90000	0.94271	0.94082	0.938910	0.93462	0.93257	0.93049	0.92839
1.2500	0.93604	0.93389	0.931710	0.92950	0.92726	0.92500	0.92271
1.5000	0.93234	0.93005	0.927740	0.92539	0.92301	0.92060	0.91817
1.7500	0.92955	0.92714	0.924690	0.92204	0.91971	0.91717	0.91460
2.0000	0.92703	0.92454	0.922010	0.91934	0.91435	0.91159	0.90880
2.5000	0.92309	0.92042	0.917720	0.91497	0.91218	0.90935	0.90648
3.0000	0.92004	0.91723	0.914370	0.91146	0.90851	0.90555	0.90249
3.2500	0.91875	0.91587	0.912940	0.90997	0.90675	0.90388	0.90077
3.5000	0.91751	0.91458	0.911600	0.90857	0.90550	0.90237	0.89920

Now plot $\ln K_w$ against $1/T$, curve fit your result to

$$\ln K_w = -\frac{a}{T} - b\ln T + cT + d$$

and use Equation 12.29 to determine the value of $\Delta_r H^\circ$ for the dissociation of water as a quadratic polynomial in T. The experimental value of $\Delta_r H^\circ$ at 25°C is 55.9 kJ·mol^{-1}.

The equations for the two electrode reactions are

$$\tfrac{1}{2}\,H_2(g) \longrightarrow H^+(aq) + e^-$$

$$AgCl(s) + e^- \longrightarrow Ag(s) + Cl^-(aq)$$

and the equation for the cell reaction is

$$\tfrac{1}{2}\,H_2(g) + AgCl(s) \longrightarrow Ag(s) + H^+(aq) + Cl^-(aq)$$

The Nernst equation (with $a_{H_2} = a_{AgCl} = a_{Ag} = 1$) for this sytem is

$$\begin{aligned} E &= E^\circ - \frac{RT}{F}\ln a_{H^+}a_{Cl^-} \\ &= E^\circ - \frac{RT}{F}\ln m_{H^+}m_{Cl^-}\gamma_{H^+}\gamma_{Cl^-} \end{aligned}$$

Solve the relation for K_w given in the problem for $m_{H^+}\gamma_{H^+}$ to obtain

$$m_{H^+}\gamma_{H^+} = \frac{K_w a_{H_2O}}{m_{OH^-}\gamma_{OH^-}}$$

Let $a_{H_2O} = 1$ and subsititute this result into the Nernst equation to obtain

$$E = E^\circ - \frac{RT}{F}\ln\frac{m_{Cl^-}\gamma_{Cl^-}K_w}{m_{OH^-}\gamma_{OH^-}}$$

or

$$\frac{F}{RT}(E - E^\circ) + \ln\frac{m_{KCl}}{m_{KOH}} = \ln\frac{\gamma_{OH^-}}{\gamma_{Cl^-}} - \ln K_w = -\ln K_w'$$

We now plot $-\ln K_w'$ against ionic strength and extrapolate to zero ionic strength to obtain $\ln K_w$. The numerical results and graphs of $-\ln K_w'$ against ionic strength are given below.

	at 0°C		at 10°C	
I_m/mol·kg^{-1}	E/V	$-\ln K_w'$	E/V	$-\ln K_w'$
0.0200	1.0462	34.405	1.0478	33.464
0.0300	1.0299	34.407	1.0309	33.465
0.0400	1.0204	34.408	1.0211	33.468
0.0500	1.0136	34.409	1.0141	33.468
0.0600	1.0084	34.410	1.0087	33.471
0.0800	1.0005	34.412	1.0005	33.473
0.110	0.99217	34.413	0.99192	33.477
0.210	0.97605	34.421	0.97529	33.488
0.310	0.96680	34.434	0.96571	33.501
0.410	0.96029	34.445	0.95901	33.514
0.510	0.95537	34.459	0.95391	33.528
0.610	0.95138	34.472	0.94978	33.541
0.710	0.94805	34.485	0.94634	33.554
0.810	0.94517	34.496	0.94337	33.566
0.910	0.94271	34.509	0.94082	33.580
1.01	0.94065	34.527	0.93886	33.605
1.26	0.93604	34.554	0.93389	33.624
1.51	0.93234	34.579	0.93005	33.649
1.76	0.92955	34.615	0.92714	33.684
2.01	0.92703	34.641	0.92454	33.711
2.26	0.92501	34.673	0.92240	33.741
2.51	0.92309	34.697	0.92042	33.765
3.01	0.92004	34.750	0.91723	33.817
3.26	0.91875	34.775	0.91587	33.841
3.51	0.91751	34.797	0.91458	33.862

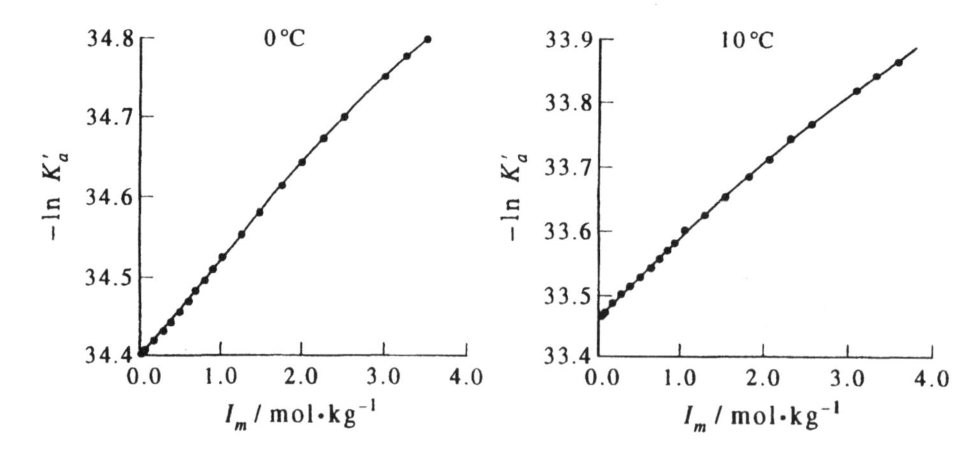

I_m/mol·kg^{-1}	at 20°C E/V	$-\ln K'_w$	at 30°C E/V	$-\ln K'_w$
0.0200	1.04947	32.616	1.0512	31.853
0.0300	1.03203	32.619	1.0332	31.856
0.0400	1.02183	32.621	1.0226	31.858
0.0500	1.01458	32.621	1.0151	31.859
0.0600	1.00903	32.625	1.0094	31.863
0.0800	1.00057	32.626	1.0007	31.864
0.110	0.991700	32.632	0.99151	31.870
0.210	0.974550	32.646	0.97380	31.886
0.310	0.964640	32.659	0.96357	31.900
0.410	0.957730	32.674	0.95643	31.914
0.510	0.952440	32.687	0.95098	31.928
0.610	0.948170	32.701	0.94655	31.941
0.710	0.944620	32.714	0.94288	31.955
0.810	0.941540	32.726	0.93970	31.967
0.910	0.938910	32.740	0.93698	31.980
1.01	0.936650	32.755	0.93462	31.995
1.26	0.931710	32.783	0.92950	32.022
1.51	0.927740	32.808	0.92539	32.047
1.76	0.924690	32.842	0.92204	32.073
2.01	0.922010	32.869	0.91934	32.104
2.26	0.919750	32.897	0.91706	32.134
2.51	0.917720	32.922	0.91497	32.159
3.01	0.914370	32.972	0.91146	32.207
3.26	0.912940	32.996	0.90997	32.230
3.51	0.911600	33.017	0.90857	32.251

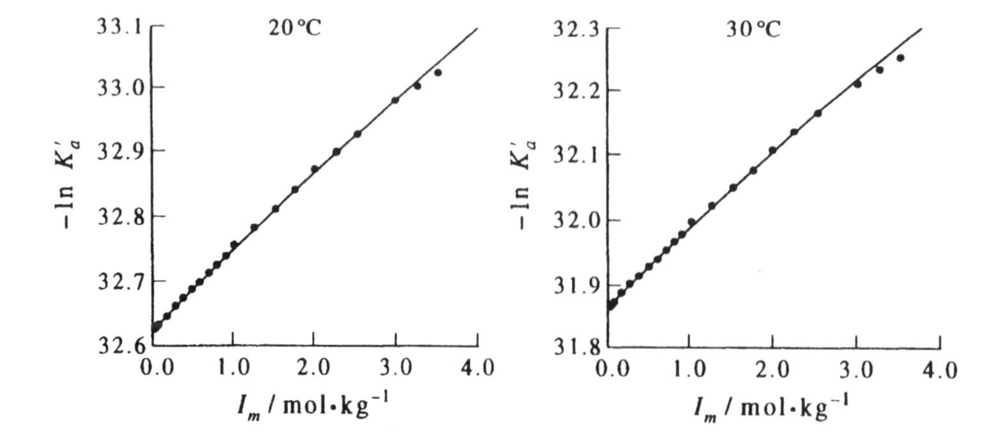

I_m/mol·kg^{-1}	at 40°C E/V	$-\ln K'_w$	at 50°C E/V	$-\ln K'_w$	at 60°C E/V	$-\ln K'_w$
0.0200	1.0530	31.164	1.0548	30.541	1.0567	29.974
0.0300	1.0344	31.168	1.0356	30.543	1.0370	29.981
0.0400	1.0235	31.170	1.0244	30.547	1.0253	29.980
0.0500	1.0158	31.174	1.0102	30.549	1.0107	29.981
0.0800	1.0008	31.177	1.0010	30.554	1.0012	29.985
0.110	0.99136	31.183	0.99123	30.560	0.99114	29.992
0.210	0.97309	31.200	0.97239	30.577	0.97170	30.008
0.310	0.96251	31.213	0.96145	30.589	0.96041	30.020
0.410	0.95513	31.227	0.95383	30.603	0.95250	30.032
0.510	0.94948	31.241	0.94798	30.616	0.94648	30.046
0.610	0.94492	31.268	0.93936	30.643	0.93758	30.072
0.810	0.93784	31.280	0.93596	30.655	0.93406	30.083
0.910	0.93502	31.293	0.93304	30.668	0.93104	30.096
1.01	0.93257	31.307	0.93049	30.681	0.92839	30.109
1.26	0.92726	31.334	0.92500	30.707	0.92271	30.134
1.51	0.92301	31.359	0.92060	30.732	0.91817	30.158
1.76	0.91971	31.390	0.91717	30.763	0.91460	30.188
2.01	0.91673	31.414	0.91408	30.785	0.91139	30.210
2.26	0.91435	31.443	0.91159	30.814	0.90880	30.237
2.51	0.91218	31.468	0.90935	30.839	0.90648	30.262
3.01	0.90851	31.514	0.90555	30.884	0.90249	30.305
3.26	0.90675	31.529	0.90388	30.905	0.90077	30.325
3.51	0.90550	31.557	0.90237	30.924	0.89920	30.345

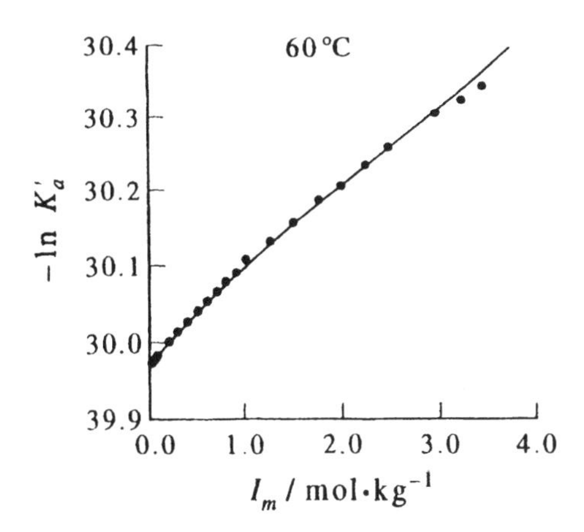

The numerical results and the plot of K_w as a function of temperature are

$t/^\circ C$	$\ln K_w$	T/K	$1/T/K^{-1}$
0.00	−34.402	273.15	0.0036610
10.0	−33.462	283.15	0.0035317
20.0	−32.615	293.15	0.0034112
30.0	−31.865	303.15	0.0032987
40.0	−31.166	313.15	0.0031934
50.0	−30.540	323.15	0.0030945
60.0	−29.975	333.15	0.0030017

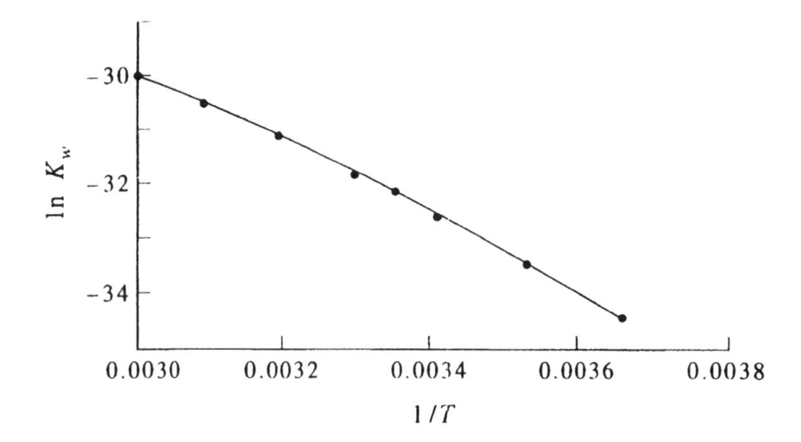

A curve fit of a plot of $\ln K_w$ versus $1/T$ fitted to the expression

$$\ln K_w = -\frac{a}{T} - b \ln T + cT + d$$

gives

$$\ln K_w = -\frac{42007}{T} - 205.61 \ln T + 0.2927T + 1192.9$$

The value of $\Delta_r H^\circ$ is given by (Equation 12.29)

$$\begin{aligned}\Delta_r H^\circ &= RT^2 \left(\frac{d \ln K_w}{dT}\right) \\ &= R(42007 - 205.61T + 0.2927T^2)\end{aligned}$$

The results are

$t/^\circ C$	$\Delta_r H^\circ/\text{kJ}\cdot\text{mol}^{-1}$
0.00	63.88
10.0	60.33
20.0	57.26
30.0	54.67
40.0	52.58
50.0	50.97
60.0	49.84

and a plot of $\Delta_r H^\circ$ against t is

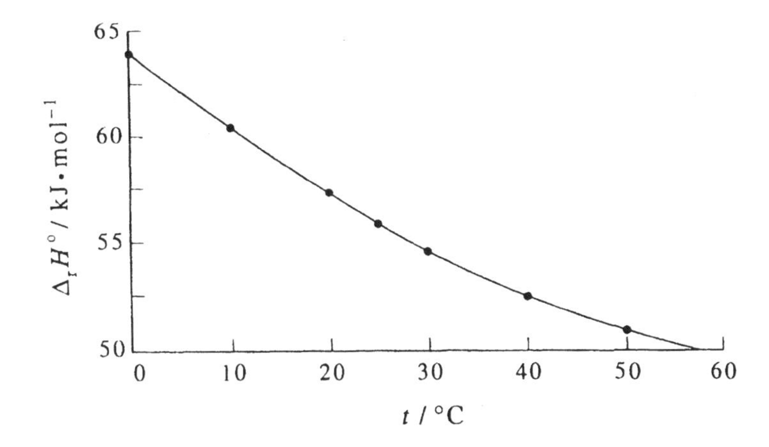

A curve fit of $\Delta_r H^\circ$ to a quadratic polynomial in $(t - 25.0°C)$ gives

$$\Delta_r H^\circ/\text{kJ}\cdot\text{mol}^{-1} = 55.90 - 0.2584(t - 25.0°\text{C}) + 0.002434(t - 25.0°\text{C})^2$$

The value of $\Delta_r H^\circ$ at 25.0°C is 55.90 kJ·mol^{-1}.

13–43. Use the data in Table 13.2 to calculate the value of the solubility product of $PbCl_2(s)$ at 298.15 K, and compare your result with the value that you obtain using the data in the *NBS Tables of Chemical Thermodynamic Properties* [*J. Phys. Chem. Data*, vol. II, suppl. 2 (1982)].

The equation for the dissolution reaction is

$$\text{PbCl}_2(\text{s}) \longrightarrow \text{Pb}^{2+}(\text{aq}) + 2\,\text{Cl}^-(\text{aq})$$

The appropriate cell is diagrammed by

$$\text{Pb(s)}|\text{PbCl}_2(\text{aq})|\text{PbCl}_2(\text{s})|\text{Pb(s)}$$

whose two electrode reactions can be described by

$$\text{Pb(s)} \longrightarrow \text{Pb}^{2+}(\text{aq}) + 2\,e^-$$

$$\text{PbCl}_2(\text{s}) + 2\,e^- \longrightarrow \text{Pb(s)} + 2\,\text{Cl}^-(\text{aq})$$

The value of E° is

$$E^\circ = -0.266\text{ V} - (-0.126\text{ V}) = -0.140\text{ V}$$

and the value of K_{sp} is given by

$$\begin{aligned} K_{sp} &= e^{nFE^\circ/RT} \\ &= \exp\left[\frac{(2)(96485\text{ C})(-0.140\text{ V})}{(8.314\text{ J}\cdot\text{mol}^{-1}\cdot\text{K}^{-1})(298.15\text{ K})}\right] \\ &= \exp(-11.015) = 1.85 \times 10^{-5} \end{aligned}$$

From the *NBS Tables of Chemical Thermodynamic Properties*,

$$\begin{aligned} \Delta_r G^\circ &= \Delta_f G^\circ[\text{Pb}^{2+}(\text{aq})] + 2\Delta_f G^\circ[\text{Cl}^-(\text{aq})] - \Delta_f G^\circ[\text{PbCl}_2(\text{s})] \\ &= -24.43\text{ kJ}\cdot\text{mol}^{-1} + (2)(-131.228\text{ kJ}\cdot\text{mol}^{-1}) - (-314.10\text{ kJ}\cdot\text{mol}^{-1}) \\ &= 27.214\text{ kJ}\cdot\text{mol}^{-1} \end{aligned}$$

and so

$$K_{sp} = e^{-\Delta_r G^\circ/RT} = \exp\left[-\frac{27.214\times 10^3\ \text{J}\cdot\text{mol}^{-1}}{(8.314\ \text{J}\cdot\text{mol}^{-1}\cdot\text{K}^{-1})(298.15\ \text{K})}\right]$$
$$= \exp(-10.978) = 1.71\times 10^{-5}$$

13–44. Use the data in Table 13.2 to calculate the value of the solubility product of $Hg_2SO_4(s)$ at 298.15 K, and compare your result with the value that you obtain using the data in the *NBS Tables of Chemical Thermodynamic Properties* [*J. Phys. Chem. Data*, vol. II, suppl. 2 (1982)].

The equation for the dissolution reaction is

$$Hg_2SO_4(s) \longrightarrow Hg_2^{2+}(aq) + SO_4^{2-}(aq)$$

The appropriate cell is diagrammed by

$$Hg(l)|Hg_2SO_4(aq)|Hg_2SO_4(s)|Hg(l)$$

whose two electrode reactions can be described by

$$2\ Hg(l) \longrightarrow Hg_2^{2+}(aq) + 2\ e^-$$

$$Hg_2SO_4(s) + 2\ e^- \longrightarrow 2\ Hg(l) + SO_4^{2-}(aq)$$

The value of E° is

$$E^\circ = 0.6155\ \text{V} - (0.796\ \text{V}) = -0.1805\ \text{V}$$

and the value of K_{sp} is given by

$$K_{sp} = e^{nFE^\circ/RT}$$
$$= \exp\left[\frac{(2)(96485\ \text{C})(-0.1805\ \text{V})}{(8.314\ \text{J}\cdot\text{mol}^{-1}\cdot\text{K}^{-1})(298.15\ \text{K})}\right]$$
$$= \exp(-14.051) = 7.90\times 10^{-7}$$

From the *NBS Tables of Chemical Thermodynamic Properties*,

$$\Delta_r G^\circ = \Delta_f G^\circ[Hg_2^{2+}(aq)] + \Delta_f G^\circ[SO_4^{2-}(aq)] - \Delta_f G^\circ[Hg_2SO_4(s)]$$
$$= 153.52\ \text{kJ}\cdot\text{mol}^{-1} + (-744.53\ \text{kJ}\cdot\text{mol}^{-1}) - (-625.815\ \text{kJ}\cdot\text{mol}^{-1})$$
$$= 34.805\ \text{kJ}\cdot\text{mol}^{-1}$$

and so

$$K_{sp} = e^{-\Delta_r G^\circ/RT} = \exp\left[-\frac{34.805\times 10^3\ \text{J}\cdot\text{mol}^{-1}}{(8.314\ \text{J}\cdot\text{mol}^{-1}\cdot\text{K}^{-1})(298.15\ \text{K})}\right]$$
$$= 7.99\times 10^{-7}$$

13–45. Given that $\Delta_f G^\circ[Pb^{2+}(aq)] = -24.43\ \text{kJ}\cdot\text{mol}^{-1}$, $\Delta_f G^\circ[SO_4^{2-}(aq)] = -744.53\ \text{kJ}\cdot\text{mol}^{-1}$, and $\Delta_f G^\circ[PbSO_4(s)] = -813.4\ \text{kJ}\cdot\text{mol}^{-1}$ at 298.15 K, calculate the value of the solubility product of $PbSO_4(s)$. Compare your result with the one obtained in Problem 13–35.

The equation for the dissolution of $PbSO_4(s)$ is

$$PbSO_4(s) \longrightarrow Pb^{2+}(aq) + SO_4^{2-}(aq)$$

$$\begin{aligned}\Delta_r G^\circ &= \Delta_f G^\circ[Pb^{2+}(aq)] + \Delta_f G^\circ[SO_4^{2-}(aq)] - \Delta_f G^\circ[PbSO_4(s)] \\ &= -24.43\ \text{kJ}\cdot\text{mol}^{-1} - 744.53\ \text{kJ}\cdot\text{mol}^{-1} - (-813.4\ \text{kJ}\cdot\text{mol}^{-1}) \\ &= 44.4\ \text{kJ}\cdot\text{mol}^{-1}\end{aligned}$$

and so

$$\begin{aligned}K_{sp} &= e^{-\Delta_r G^\circ/RT} = \exp\left[-\frac{44.4\times 10^3\ \text{J}\cdot\text{mol}^{-1}}{(8.314\ \text{J}\cdot\text{mol}^{-1}\cdot\text{K}^{-1})(298.15\ \text{K})}\right] \\ &= 1.64\times 10^{-8}\end{aligned}$$

13–46. Use the data in Figure 13.15 along with $\Delta_f G^\circ[Br^-(aq)] = -103.96\ \text{kJ}\cdot\text{mol}^{-1}$ to calculate the value of the solubility product of thallium(I) bromide at 298.15 K.

The equation for the dissolution of TlBr(s) is

$$TlBr(s) \rightleftharpoons Tl^+(aq) + Br^-(aq)$$

$$\begin{aligned}\Delta_r G^\circ &= \Delta_f G^\circ[Tl^+(aq)] + \Delta_f G^\circ[Br^-(aq)] - \Delta_f G^\circ[TlBr(s)] \\ &= -32.40\ \text{kJ}\cdot\text{mol}^{-1} - 103.96\ \text{kJ}\cdot\text{mol}^{-1} - (-167.36\ \text{kJ}\cdot\text{mol}^{-1}) \\ &= 31.00\ \text{kJ}\cdot\text{mol}^{-1}\end{aligned}$$

and so

$$\begin{aligned}K_{sp} &= e^{-\Delta_r G^\circ/RT} = \exp\left[-\frac{31.00\times 10^3\ \text{J}\cdot\text{mol}^{-1}}{(8.314\ \text{J}\cdot\text{mol}^{-1}\cdot\text{K}^{-1})(298.15\ \text{K})}\right] \\ &= 3.70\times 10^{-6}\end{aligned}$$

13–47. Use the data in *NBS Tables of Chemical Thermodynamic Properties* [*J. Phys. Chem. Data*, vol. II, suppl. 2 (1982)] to calculate the value of the solubility product of barium sulfate at 298.15 K.

The equation for the dissolution of $BaSO_4(s)$ is

$$BaSO_4(s) \rightleftharpoons Ba^{2+}(aq) + SO_4^{2-}(aq)$$

$$\begin{aligned}\Delta_r G^\circ &= \Delta_f G^\circ[Ba^{2+}(aq)] + \Delta_f G^\circ[SO_4^{2-}(aq)] - \Delta_f G^\circ[BaSO_4(s)] \\ &= -560.77\ \text{kJ}\cdot\text{mol}^{-1} - 744.53\ \text{kJ}\cdot\text{mol}^{-1} - (-1362.2\ \text{kJ}\cdot\text{mol}^{-1}) \\ &= 56.9\ \text{kJ}\cdot\text{mol}^{-1}\end{aligned}$$

and so

$$\begin{aligned}K_{sp} &= e^{-\Delta_r G^\circ/RT} = \exp\left[-\frac{56.9\times 10^3\ \text{J}\cdot\text{mol}^{-1}}{(8.314\ \text{J}\cdot\text{mol}^{-1}\cdot\text{K}^{-1})(298.15\ \text{K})}\right] \\ &= 1.08\times 10^{-10}\end{aligned}$$

13–48. Given that $\Delta_f G^\circ[HSO_4^-(aq)] = -755.91\ kJ\cdot mol^{-1}$ and that $\Delta_f G^\circ[SO_4^{2-}(aq)] = -744.53\ kJ\cdot mol^{-1}$, calculate the value of the second acid-dissociation constant of sulfuric acid at 298.15 K.

The equation for the second acid dissociation of $H_2SO_4(aq)$ is

$$HSO_4^-(aq) \longrightarrow H^+(aq) + SO_4^{2-}(aq)$$

$$\begin{aligned}\Delta_r G^\circ &= \Delta_f G^\circ[H^+(aq)] + \Delta_f G^\circ[SO_4^{2-}(aq)] - \Delta_f G^\circ[HSO_4^-(aq)]\\ &= 0\ kJ\cdot mol^{-1} - 744.53\ kJ\cdot mol^{-1} - (-755.91\ kJ\cdot mol^{-1})\\ &= 11.38\ kJ\cdot mol^{-1}\end{aligned}$$

and so

$$\begin{aligned}K_a &= e^{-\Delta_r G^\circ/RT} = \exp\left[-\frac{11.38\times 10^3\ J\cdot mol^{-1}}{(8.314\ J\cdot mol^{-1}\cdot K^{-1})(298.15\ K)}\right]\\ &= 1.01\times 10^{-2}\end{aligned}$$

13–49. Use the data in *NBS Tables of Chemical Thermodynamic Properties* [*J. Phys. Chem. Data*, vol. II, suppl. 2 (1982)] to calculate the value of the first acid-dissociation constant of arsenic acid, $H_3AsO_4(aq)$ at 298.15 K.

The equation for the first acid dissociation of $H_3AsO_4(aq)$ is

$$H_3AsO_4(aq) \longrightarrow H^+(aq) + H_2AsO_4^-(aq)$$

$$\begin{aligned}\Delta_r G^\circ &= \Delta_f G^\circ[H^+(aq)] + \Delta_f G^\circ[H_2AsO_4^-(aq)] - \Delta_f G^\circ[H_3AsO_4(aq)]\\ &= 0\ kJ\cdot mol^{-1} - 753.17\ kJ\cdot mol^{-1} - (-766.0\ kJ\cdot mol^{-1})\\ &= 12.83\ kJ\cdot mol^{-1}\end{aligned}$$

and so

$$\begin{aligned}K_a &= e^{-\Delta_r G^\circ/RT} = \exp\left[-\frac{12.83\times 10^3\ J\cdot mol^{-1}}{(8.314\ J\cdot mol^{-1}\cdot K^{-1})(298.15\ K)}\right]\\ &= 5.65\times 10^{-3}\end{aligned}$$

13–50. Use the data in *NBS Tables of Chemical Thermodynamic Properties* [*J. Phys. Chem. Data*, vol. II, suppl. 2 (1982)] to calculate the value of the first base-protonation constant of hydrazine at 298.15 K.

The equation for the base protonation of hydrazine is

$$N_2H_4(aq) + H_2O(l) \longrightarrow N_2H_5^+(aq) + OH^-(aq)$$

$$\begin{aligned}\Delta_r G^\circ &= \Delta_f G^\circ[N_2H_5^+(aq)] + \Delta_f G^\circ[OH^-(aq)] - \Delta_f G^\circ[N_2H_4(aq)] - \Delta_f G^\circ[H_2O(l)]\\ &= 82.5\ kJ\cdot mol^{-1} - 157.244\ kJ\cdot mol^{-1} - (-128.1\ kJ\cdot mol^{-1}) - (-237.129\ kJ\cdot mol^{-1})\\ &= 34.29\ kJ\cdot mol^{-1}\end{aligned}$$

and so

$$K_b = e^{-\Delta_r G^\circ/RT} = \exp\left[-\frac{34.29\times 10^3\ \mathrm{J\cdot mol^{-1}}}{(8.314\ \mathrm{J\cdot mol^{-1}\cdot K^{-1}})(298.15\ \mathrm{K})}\right]$$
$$= 9.85\times 10^{-7}$$

13–51. Compare the idealized amount of work available when one mole of propane is burned in a fuel cell at 298 K and in a heat engine with $T_h = 900$ K and $T_c = 300$ K. Calculate the value of E° for this fuel cell.

The equation for the combustion of propane is

$$C_3H_8(g) + 5\ O_2(g) \longrightarrow 3\ CO_2(g) + 4\ H_2O(l)$$

and from Table 5.2,

$$\Delta_r H^\circ = 3\Delta_f H^\circ[CO_2(g)] + 4\Delta_f H^\circ[H_2O(l)] - \Delta_f H^\circ[C_3H_8(g)] - 5\Delta_f H^\circ[O_2(g)]$$
$$= 3(-393.509\ \mathrm{kJ\cdot mol^{-1}}) + 4(-285.83\ \mathrm{kJ\cdot mol^{-1}}) - (-103.8\ \mathrm{kJ\cdot mol^{-1}}) - 5(0)$$
$$= -2220.0\ \mathrm{kJ\cdot mol^{-1}}$$

and so the maximum amount of energy available when one mole of propane is burned in a heat engine is

$$\text{maximum energy} = \left(\frac{T_h - T_c}{T_h}\right)|\Delta_r H^\circ| = \left(\frac{900-300}{300}\right)(2200\ \mathrm{kJ}) = 1480\ \mathrm{kJ}$$

Using Table 12.1, we find that the maximum amount of energy available in a fuel cell is

$$-w_{rev} = \Delta_r G^\circ = 3\Delta_f G^\circ[CO_2(g)] + 4\Delta_f G^\circ[H_2O(l)] - \Delta_f G^\circ[C_3H_8(g)] - 5\Delta_f G^\circ[O_2(g)]$$
$$= 3(-394.389\ \mathrm{kJ\cdot mol^{-1}}) + 4(-237.141\ \mathrm{kJ\cdot mol^{-1}}) - (-23.47\ \mathrm{kJ\cdot mol^{-1}}) - 5(0)$$
$$= -2108.26\ \mathrm{kJ\cdot mol^{-1}}$$

The value of n in the above equation for the combustion of one mole of propane is 20, and so

$$E^\circ = \frac{-w_{rev}}{nF} = \frac{2108.26\times 10^3\ \mathrm{J\cdot mol^{-1}}}{(20)(96\,485\ \mathrm{C\cdot mol^{-1}})} = 1.09\ \mathrm{V}$$

13–52. Derive a general relation for the pressure dependence of the emf of an electrochemical cell at constant temperature. Show that for the cell whose diagram is Pt(s)|H_2(g)|HCl(aq)|AgCl(s)|Ag(s) that

$$\left(\frac{\partial E}{\partial P}\right)_T = -\frac{2(\overline{V}_{HCl} + \overline{V}_{Ag}) - (\overline{V}_{H_2} + 2\overline{V}_{AgCl})}{F}$$

where $\overline{V}_{HCl}$ is the partial molar volume of the HCl(aq) at molality m, $\overline{V}_{H_2}$ is the partial molar volume of H_2(g) at pressure P, and $\overline{V}_{Ag}$ and $\overline{V}_{AgCl}$ are the molar volumes of Ag(s) and AgCl(s). Argue that at relatively low pressures

$$E_2 - E_1 \approx \frac{RT}{2F}\ln\frac{P_2}{P_1}$$

where E_j is the emf of the cell at pressure P_j.

Start with the equation

$$\Delta G = -nFE$$

and differentiate with respect to pressure with the temperature held constant

$$\left(\frac{\partial \Delta G}{\partial P}\right)_T = \Delta_r V = -nF\left(\frac{\partial E}{\partial P}\right)_T$$

or

$$\left(\frac{\partial E}{\partial P}\right)_T = -\frac{\Delta_r V}{nF}$$

The equation for the reaction associated with the cell diagram $Pt(s)|H_2(g)|HCl(aq)|AgCl(s)|Ag(s)$ is

$$H_2(g) + 2\,AgCl(s) \longrightarrow 2\,Ag(s) + 2\,HCl(aq)$$

and so

$$\Delta_r V = 2\overline{V}_{Ag} + 2\overline{V}_{HCl} - \overline{V}_{H_2} - 2\overline{V}_{AgCl}$$
$$\approx -\overline{V}_{H_2}$$

Using the fact that $n = 2$ in the above chemical equation, we have

$$\left(\frac{\partial E}{\partial P}\right)_T = \frac{\overline{V}_{H_2}}{2F}$$

and assuming that $H_2(g)$ behaves ideally,

$$\left(\frac{\partial E}{\partial P}\right)_T = \frac{RT}{2F}P$$

Integrating from P_1 to P_2 gives

$$E_2 - E_1 = \frac{RT}{2F}\ln\frac{P_2}{P_1}$$

13–53. Consider the cell whose cell diagram is

$$Pt(s)|H_2(g)|HCl(0.100\text{ m})|Hg_2Cl_2(s)|Hg(l)$$

Show that $\Delta_r V \approx -\overline{V}_{H_2}$. Using a virial equation of state through the second virial coefficient (Equation 2.23), show that (see Problem 13–52)

$$\left(\frac{\partial E}{\partial P}\right)_T = \frac{RT}{2F}\left[\frac{1}{P} + B_{2P}(T) + \cdots\right]$$

Now use the data in Table 2.7 and Figure 2.15 to estimate the value of $B_{2P}(T) = B_{2V}(T)/RT$ for $H_2(g)$ at 25°C. [The value of $B^*_{2V}(T^*)$ from numerical tables is 0.416.] Finally, use the following data at 25°C for the above cell to calculate the value of E as a function of pressure, and compare the result of your calculation with the experimental results graphically.

P/atm	E/V
1.0000	0.39900
37.800	0.44560
51.600	0.44960
110.20	0.45960
204.70	0.46830
439.30	0.48040
568.80	0.48500
701.80	0.48910
754.40	0.49030
893.90	0.49380
1035.2	0.49750

The equation for the cell reaction for the cell diagrammed in the problem is

$$H_2(g) + 2\,Hg_2Cl_2(s) \longrightarrow 2\,Hg(l) + 2\,HCl(aq)$$

Hydrogen is the only gaseous species in this reaction, so the value of $\Delta_r V$ for this reaction is essentially $-\overline{V}_{H_2}$. According to Problem 13–52,

$$\left(\frac{\partial E}{\partial P}\right)_T = -\frac{\Delta_r V}{nF} = \frac{\overline{V}_{H_2}}{2F}$$

and according to Equation 2.23

$$\overline{V} = \frac{RT}{P}[1 + B_{2P}(T)P + \cdots]$$

Substituting this result into the equation for $(\partial E/\partial P)_T$ gives

$$\left(\frac{\partial E}{\partial P}\right)_T = \frac{RT}{2F}\left[\frac{1}{P} + B_{2P}(T) + \cdots\right]$$

At 25°C, $T^* = 298.15\ \text{K}/37.0\ \text{K} = 8.06$ and so $B_{2V}^*(T^*) \approx 0.5$, $B_{2V}(T) \approx (0.5)(31.7\ \text{cm}^3\cdot\text{mol}^{-1}) =$ $15\ \text{cm}^3\cdot\text{mol}^{-1}$, and
$B_{2P}(T) \approx 15\ \text{cm}^3\cdot\text{mol}^{-1}/(82.06\ \text{cm}^3\cdot\text{atm}^{-1}\cdot\text{mol}^{-1}\cdot\text{K}^{-1})(298.15\ \text{K}) = 6.1 \times 10^{-4}\ \text{atm}^{-1}$. If we use the tabulated value of $B^*(T^*) = 0.416$, then $B_{2P}(T) = 5.39 \times 10^{-4}\ \text{atm}^{-1}$. Integrating the expression for $(\partial E/\partial P)_T$ from $P = 1$ to P gives

$$E - 0.39900\ \text{V} = \frac{RT}{2F}[\ln P + (5.39 \times 10^{-4}\ \text{atm}^{-1})(P - 1)]$$

or

$$E = (0.01285\ \text{V})[\ln P + (5.39 \times 10^{-4}\ \text{atm}^{-1})(P - 1)] + 0.39900\ \text{V}$$

The calculated values of E are

P/atm	ln P	E/V	E(calc)/V	E(ideal gas)/V
1.000	0.0000	0.3990	0.3990	0.3990
37.80	3.6323	0.4456	0.4459	0.4457
51.60	3.9435	0.4496	0.4500	0.4497
110.20	4.7023	0.4596	0.4602	0.4594
204.70	5.3215	0.4683	0.4688	0.4674
439.30	6.0852	0.4804	0.4802	0.4772
568.80	6.3435	0.4850	0.4844	0.4805
701.80	6.5536	0.4891	0.4880	0.4832
754.40	6.6259	0.4903	0.4893	0.4841
893.90	6.7956	0.4938	0.4925	0.4863
1035.2	6.9423	0.4975	0.4953	0.4882

and a plot of E(data) (crosses), E(calc) (solid line), and E(ideal gas only) (dotted line) against ln P is shown below.

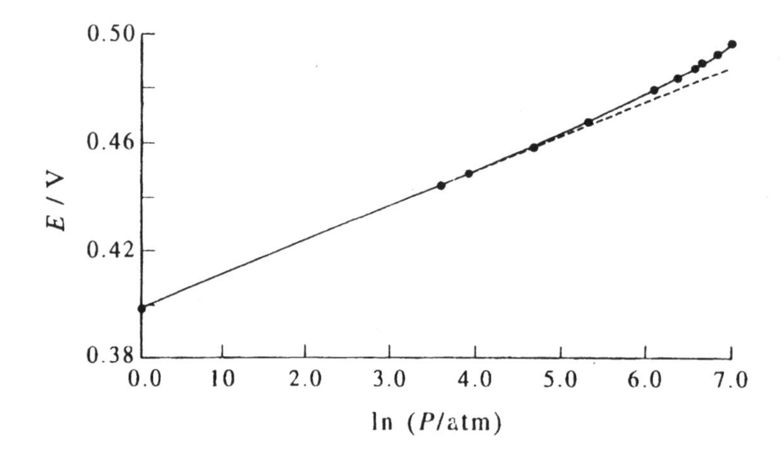

13–54. Sometimes it might be necessary to calculate the value of E° for a reduction electrode reaction that is a combination of other reduction electrode reactions. For example, consider the two reduction electrode reactions

(1) $8\,H^+(aq) + MnO_4^-(aq) + 5\,e^- \longrightarrow 4\,H_2O(l) + Mn^{2+}(aq)$

$$E_1^\circ = 1.491\text{ V}$$

(2) $4\,H^+(aq) + MnO_2(s) + 2\,e^- \longrightarrow 2\,H_2O(l) + Mn^{2+}(aq)$

$$E_2^\circ = 1.208\text{ V}$$

We can use these data to calculate the value of E° for

(3) $4\,H^+(aq) + MnO_4^-(aq) + 3\,e^- \longrightarrow MnO_2(s) + 2\,H_2O(l)$

We first note that Equation 3 results from subtracting Equation 2 from Equation 1. It is important to realize that E_3° is *not* equal to $E_1^\circ - E_2^\circ$, however, because E° is an *intensive* property. The standard Gibbs energy change for Equation 3 *is* given by $\Delta G_3^\circ = \Delta G_1^\circ - \Delta G_2^\circ$, however. Using this fact, show that

$$E_3^\circ = \frac{5E_1^\circ - 2E_2^\circ}{3} = 1.680\text{ V}$$

Using the relation $\Delta G = -nFE$, we have

$$\Delta G_1^\circ = -5FE_1^\circ$$
$$\Delta G_2^\circ = -2FE_2^\circ$$

and

$$\Delta G_3^\circ = -3FE_3^\circ$$

Using the fact that $\Delta G_3^\circ = \Delta G_1^\circ - \Delta G_2^\circ$ gives

$$-3FE_3^\circ = -5FE_1^\circ + 2FE_2^\circ$$

or

$$E_3^\circ = \frac{5E_1^\circ - 2E_2^\circ}{3} = \frac{5(1.491\text{ V}) - 2(1.208\text{ V})}{3} = 1.680\text{ V}$$

13–55. Given that

$$Hg_2^{2+}(aq) + 2\,e^- \longrightarrow 2\,Hg(l) \qquad E^\circ = +0.796\text{ V}$$
$$2\,Hg^{2+}(aq) + 2\,e^- \longrightarrow Hg_2^{2+}(aq)\,E^\circ = +0.907\text{ V}$$

calculate the value of E° for

$$Hg^{2+}(aq) + 2\,e^- \longrightarrow Hg(l)$$

Following the procedure outlined in Problem 13–54, we note that $\Delta G_3^\circ = \frac{1}{2}\Delta G_1^\circ + \frac{1}{2}\Delta G_2^\circ$. BuTque repl

$$\Delta G_1^\circ = -2FE_1^\circ$$
$$\Delta G_2^\circ = -2FE_2^\circ$$
$$\Delta G_3^\circ = -2FE_3^\circ$$

and so

$$-2FE_3^\circ = \frac{1}{2}(-2FE_1^\circ - 2FE_2^\circ)$$

or

$$E_3^\circ = \frac{1}{2}(E_1^\circ + E_2^\circ) = 0.852\text{ V}$$

13–56. Given that

$$Cr^{2+}(aq) + 2\,e^- \longrightarrow Cr(s) \quad E^\circ = -0.91\text{ V}$$
$$Cr^{3+}(aq) + e^- \longrightarrow Cr^{2+}(aq)\,E^\circ = -0.41\text{ V}$$

calculate the value of E° for

$$Cr^{3+}(aq) + 3\,e^- \longrightarrow Cr(s)$$

Following the procedure outlined in Problem 13–54, we note that $\Delta G_3^\circ = \Delta G_1^\circ + \Delta G_2^\circ$. But

$$\Delta G_1^\circ = -2FE_1^\circ$$
$$\Delta G_2^\circ = -FE_2^\circ$$
$$\Delta G_3^\circ = -3FE_3^\circ$$

and so

$$-3FE_3^\circ = -2FE_1^\circ - FE_2^\circ$$

or

$$E_3^\circ = \frac{2E_1^\circ + E_2^\circ}{3} = -0.74\text{ V}$$

13–57. The previous three problems develop the idea that you must use values of ΔG° to calculate values of E° of reduction electrode reactions from the values of E° from other electrode reactions. When the combination is such that the electrons on each side of the equation cancel, however, we do not have to use ΔG° as an intermediate quantity and can write $E^\circ_{\text{rxn}} = E^\circ_{\text{ox}} + E^\circ_{\text{red}}$ directly as we did in Equation 13.22. In this problem, we derive Equation 13.22.

Consider the two reduction electrode reactions

(1) $A + n_1\, e^- \longrightarrow X \qquad E^\circ_{1,\text{red}}$

(2) $B + n_2\, e^- \longrightarrow Y \qquad E^\circ_{2,\text{red}}$

Show that the combination in which the electrons cancel is

(3) $n_2\, A + n_1\, Y \longrightarrow n_1\, B + n_2\, X$

Show that ΔG_3° for Equation 3 is

$$\Delta G_3^\circ = n_2 \Delta G_1^\circ - n_1 \Delta G_2^\circ$$
$$= -n_1 n_2 F E^\circ_{1,\text{red}} + n_1 n_2 F E^\circ_{2,\text{red}}$$

and that

$$E_3^\circ = E^\circ_{1,\text{red}} - E^\circ_{2,\text{red}} = E^\circ_{1,\text{red}} + E^\circ_{2,\text{ox}}$$

which is Equation 13.22

Multiply Equation 1 by n_2 and Equation 1 by n_1, reverse the second result, and add to obtain

$$n_2\, A + n_1\, Y = n_1\, B + n_2\, X \tag{3}$$

Therefore,

$$\Delta G_3^\circ = n_2 \Delta G_1^\circ - n_1 \Delta G_2^\circ$$

Using $\Delta G = -nFE$ gives

$$\Delta G_3^\circ = -n_1 n_2 F E_1^\circ + n_2 n_1 F E_2^\circ$$
$$= n_1 n_2 F(-E_1^\circ + E_2^\circ)$$

But $\Delta G_3^\circ = -n_1 n_2 F E_3^\circ$, so

$$E_3^\circ = E^\circ_{1,\text{red}} - E^\circ_{2,\text{red}} = E^\circ_{1,\text{red}} + E^\circ_{2,\text{ox}}$$

CHAPTER 14

Nonequilibrium Thermodynamics

PROBLEMS AND SOLUTIONS

14–1. Consider a two-compartment system in contact with a heat bath at temperature T, so that $T_1 = T_2 = T$. Let the two-compartment system be surrounded by rigid, impermeable walls, but let the wall separating the two compartments be permeable and flexible. Show that $dA = dn_1(\mu_1 - \mu_2) - dV_1(P_1 - P_2) \leq 0$. Now show that

$$dS_{\text{prod}} = -\frac{dA}{T} = -dn_1(\mu_1 - \mu_2) + dV_1(P_1 - P_2) \geq 0$$

Start with the equations for each compartment

$$dA_1 = -P_1 dV_1 + \mu_1 dn_1 \quad \text{(constant } T\text{)}$$
$$dA_2 = -P_2 dV_2 + \mu_2 dn_2 \quad \text{(constant } T\text{)}$$

Because $A = A_1 + A_2$, we have

$$dA = -P_1 dV_1 - P_2 dV_2 + \mu_1 dn_1 + \mu_2 dn_2 \leq 0$$

The two-compartment system is at constant volume and constant number of particles, so

$$V_1 + V_2 = \text{constant} \quad \text{and} \quad n_1 + n_2 = \text{constant}$$

or

$$dV_1 + dV_1 = 0 \quad \text{and} \quad dn_1 + dn_2 = 0$$

Substituting these results into dA gives

$$dA = dn_1(\mu_1 - \mu_2) - dV_1(P_1 - P_2) \leq 0$$

Because $V_1 + V_2$ is fixed, we can write

$$dU = \delta_e q = TdS_{\text{exch}} = T(dS - dS_{\text{prod}})$$

for the two-compartment system, and so

$$dS_{\text{prod}} = \frac{TdS - dU}{T} = -\frac{dA}{T}$$

and

$$dS_{\text{prod}} = -\frac{dA}{T} = -dn_1(\mu_1 - \mu_2) + dV_1(P_1 - P_2) \geq 0$$

14–2. Discuss the physical meaning of each term on the right side of Equation 14.22.

The first term represents the flow of energy as heat due to a temperature difference and the second term represents a flow of matter due to a difference in μ/T.

14–3. Extend Equation 14.23 to include a flexible wall between the two compartments.

Start with Equation 14.16 and divide by dt to obtain

$$\frac{dS_{\text{prod}}}{dt} = \frac{dU_1}{dt}\left(\frac{1}{T_1} - \frac{1}{T_2}\right) + \frac{dV_1}{dt}\left(\frac{P_1}{T_1} - \frac{P_2}{T_2}\right) + \frac{dn_1}{dt}\left(\frac{\mu_2}{T_2} - \frac{\mu_1}{T_1}\right) \geq 0$$

14–4. Show that the two terms on the right side of $\dot{S}_{\text{prod}} = J_U X_U + J_n X_n$ have the same units as $\dot{S}_{\text{prod}}$.

The units of the two fluxes and forces are

J_U	X_U	J_n	X_n
$\text{J}\cdot\text{s}^{-1}$	K^{-1}	$\text{mol}\cdot\text{s}^{-1}$	$\text{J}\cdot\text{mol}^{-1}\cdot\text{K}^{-1}$

and so we see that the units of $J_U X_U$ and $J_n X_n$ are both $\text{J}\cdot\text{K}^{-1}\cdot\text{s}^{-1}$.

14–5. Show that Fick's law in the form $J_n \propto \mu_2 - \mu_1$ can be written as $J_n \propto c_2 - c_1$, when $\mu_2 - \mu_1$ is small.

Start with $\mu = \mu^\circ + RT \ln c$ and write

$$\mu_2 - \mu_1 = RT \ln \frac{c_2}{c_1}$$

But $c_2 - c_1$ is small, so we write $c_2 = c_1 + \Delta$ where Δ is small,

$$\ln \frac{c_2}{c_1} = \ln \frac{c_1 + \Delta}{c_1} = \ln\left(1 + \frac{\Delta}{c_1}\right)$$

But $\ln(1 + x) \approx x$, so

$$\ln\left(1 + \frac{\Delta}{c_1}\right) \approx \frac{\Delta}{c_1} = \frac{c_2 - c_1}{c_1}$$

Therefore,

$$\mu_2 - \mu_1 = \frac{RT}{c_1}(c_2 - c_1) \propto c_2 - c_1$$

14–6. Extend Equations 14.24 to 14.26 to include a flexible wall between the two compartments.

Start with Equation 14.16 and divide by dt to obtain (see Problem 14–3)

$$\frac{dS_{\text{prod}}}{dt} = \frac{dU_1}{dt}\left(\frac{1}{T_1} - \frac{1}{T_2}\right) + \frac{dV_1}{dt}\left(\frac{P_1}{T_1} - \frac{P_2}{T_2}\right) + dn_1\left(\frac{\mu_2}{T_2} - \frac{\mu_1}{T_1}\right) \geq 0$$

Let

$$J_U = \frac{dU_1}{dt} \qquad X_U = \frac{1}{T_1} - \frac{1}{T_2}$$
$$J_V = \frac{dV_1}{dt} \qquad X_V = \frac{P_1}{T_1} - \frac{P_2}{T_2}$$
$$J_n = \frac{dn_1}{dt} \qquad X_n = \frac{\mu_2}{T_2} - \frac{\mu_1}{T_1}$$

to get

$$\dot{S}_{\text{prod}} = J_U X_U + J_V X_V + J_n X_n$$

14–7. If the system in Figure 14.3 has a flexible wall between the two compartments, then

$$\dot{S}_{\text{prod}} = J_U X_U + J_n X_n + J_V X_V > 0$$

where $J_V = dV_1/dt$ and $X_V = (P_1/T_1) - (P_2/T_2)$. Write out the linear flux-force relations for this system. Show that L_{UV} and L_{VU} have the same units.

The linear flux-force relations for this system are

$$J_U = L_{UU} X_U + L_{Un} X_n + L_{UV} X_V$$
$$J_n = L_{nU} X_U + L_{nn} X_n + L_{nV} X_V$$
$$J_V = L_{VU} X_U + L_{Vn} X_n + L_{VV} X_V$$

To prove that L_{UV} and L_{VU} have the same units, consider

$$J_U \leftarrow \text{units} \rightarrow L_{UV} X_V \quad \text{and} \quad J_V \leftarrow \text{units} \rightarrow L_{VU} X_U$$

and so

$$L_{UV} \leftarrow \text{units} \rightarrow \frac{J_U}{X_V} \quad \text{and} \quad L_{VU} \leftarrow \text{units} \rightarrow \frac{J_V}{X_U}$$

Multiply the first expression by X_U/X_U and the second by X_V/X_V to get

$$L_{UV} \leftarrow \text{units} \rightarrow \frac{J_U X_U}{X_V X_U} \quad \text{and} \quad L_{VU} \leftarrow \text{units} \rightarrow \frac{J_V X_V}{X_U X_V}$$

But $J_U X_U$ and $J_V X_V$ have units of $\dot{S}_{\text{prod}}$, so $L_{UV} \leftarrow \text{units} \rightarrow L_{VU}$.

14–8. Prove that $L_{11} L_{22} > L_{12}^2$.

Start with

$$L_{11} X_1^2 + 2L_{12} X_1 X_2 + L_{22} X_2^2 > 0$$

This expression is valid for any values of X_1 and X_2. Set $X_1 = -L_{12}/L_{11}$ and $X_2 = 1$ to get

$$\frac{L_{12}^2}{L_{11}} - \frac{2L_{12}^2}{L_{11}} + L_{22} > 0$$

or

$$-L_{12}^2 > L_{11}L_{22}$$

or

$$L_{11}L_{22} > L_{12}^2$$

14–9. Prove that $L_{ii} > 0$, $L_{jj} > 0$, and $L_{ii}L_{jj} > L_{ij}^2$ if $\dot{S}_{\text{prod}} = \sum_i \sum_j L_{ij}X_iX_j$.

Start with

$$\dot{S}_{\text{prod}} = \sum_i \sum_j L_{ij}X_iX_j > 0$$

This expression is valid for any values of the Xs. First set all the Xs equal to zero, except two of them, say X_k and X_l. Then

$$\dot{S}_{\text{prod}} = L_{kk}X_k^2 + 2L_{kl}X_kX_l + L_{ll}X_l^2 > 0$$

Now this problem is the same as Problem 14–8.

14–10. Prove that the largest phenomenological coefficient must be one of the diagonal ones.

Assume that the largest phenomenological coefficient is not one of the diagonal ones. Let this phenomenological oefficient be L_{ij} with $i \neq j$. Then $L_{ij} > L_{ii}$ and $L_{ij} > L_{jj}$. Therefore,

$$L_{ij}^2 > L_{ii}L_{jj}$$

But this contradicts the result of Problem 14–9, so the largest phenomenological coefficient must be one of the diagonal ones.

14–11. In this problem, we will show that we can use various linear combinations of fluxes and forces and still preserve the Onsager reciprocal relations. In many applications of nonequilibrium thermodynamics, certain linear flux-force relations are more convenient than others, and the result of this problem says that we can use any convenient linear combinations that we want. We will prove this result for only a special case, but the result is general.

First start with $\dot{S}_{\text{prod}} = J_1X_1 + J_2X_2$ with

$$J_1 = L_{11}X_1 + L_{12}X_2$$
$$J_2 = L_{21}X_1 + L_{22}X_2 \tag{1}$$

To keep the algebra to a minimum, define new fluxes by

$$J_1' = aJ_1 + bJ_2 \qquad J_2' = J_2$$

where a and b are constants. Now solve these two equations for J_1 and J_2 and substitute them into $\dot{S}_{\text{prod}}$ to obtain

$$\dot{S}_{\text{prod}} = J_1'\frac{X_1}{a} + J_2'\left(X_2 - \frac{b}{a}X_1\right) = J_1'X_1' + J_2'X_2'$$

which serves to define X_1' and X_2'. Now define the phenomenological coefficients M_{ij} by

$$J_1' = M_{11}X_1' + M_{12}X_2'$$

$$J_2' = M_{21}X_1' + M_{22}X_2'$$

Convert these equations into the form of Equations 1, and show that $M_{12} = M_{21}$ follows from $L_{12} = L_{21}$.

Start with

$$\dot{S}_{\text{prod}} = J_1X_1 + J_2X_2$$

with

$$J_1 = L_{11}X_1 + L_{12}X_2$$
$$J_2 = L_{21}X_1 + L_{22}X_2$$

Now let

$$J_1' = aJ_1 + bJ_2$$

and

$$J_2' = J_2$$

Solving for J_1 and J_2 gives

$$J_1 = \frac{J_1'}{a} - \frac{b}{a}J_2'$$
$$J_2 = J_2'$$

Substitute this result into $\dot{S}_{\text{prod}}$ above to get

$$\dot{S}_{\text{prod}} = J_1'\frac{X_1}{a} + J_2'\left(X_2 - \frac{b}{a}X_1\right)$$
$$= J_1'X_1' + J_2'X_2'$$

We now write

$$J_1 = \frac{J_1'}{a} - \frac{b}{a}J_2' = \frac{1}{a}(M_{11}X_1' + M_{12}X_2') - \frac{b}{a}(M_{21}X_1' + M_{22}X_2')$$
$$= \left(\frac{M_{11} - bM_{21}}{a}\right)\left(\frac{X_1}{a}\right) + \left(\frac{M_{12} - bM_{22}}{a}\right)\left(X_2 - \frac{b}{a}X_1\right)$$
$$= \left(\frac{M_{11} - bM_{21} - bM_{12} + b^2M_{22}}{a^2}\right)X_1 + \left(\frac{M_{12} - bM_{22}}{a}\right)X_2$$

and

$$J_2 = J_2' = M_{21}X_1' + M_{22}X_2'$$
$$= \frac{M_{21}}{a}X_1 + M_{22}X_2 - \frac{bM_{22}}{a}X_1$$
$$= \left(\frac{M_{21} - bM_{22}}{a}\right)X_1 + M_{22}X_2$$

These equations show that

$$L_{12} = \left(\frac{M_{12} - bM_{22}}{a}\right)$$

and

$$L_{21} = \left(\frac{M_{21} - bM_{22}}{a}\right)$$

But since $L_{12} = L_{21}$, we see that

$$M_{12} - bM_{22} = M_{21} - bM_{22}$$

or

$$M_{12} = M_{21}$$

14–12. The diagonal terms in Equation 14.46 can be directly related to experimentally measurable quantities. Show that the *mechanical conductance*, $(J_V/\Delta P)_{\Delta\psi=0}$ is equal to L_{VV}. Show that the *electrical conductance*, $(I/\Delta\psi)_{\Delta P=0}$ is equal to L_{II}.

Set $\Delta\psi = 0$ in Equation 14.46*a* to obtain $L_{VV} = (J_V/\Delta P)_{\Delta\psi=0}$ and $\Delta P = 0$ in Equation 14.46*a* to obtain $L_{II} = (I/\Delta\psi)_{\Delta P=0}$.

14–13. The *second electroosmotic flow* is defined as $(J_V/\Delta\psi)_{\Delta P=0}$, and the *second streaming current* is defined as $(I/\Delta P)_{\Delta\psi=0}$. Show that these quantities are equal.

Set $\Delta P = 0$ in Equation 14.46*a* to obtain $L_{VI} = (J_V/\Delta\psi)_{\Delta P=0}$ and $\Delta\psi = 0$ in Equation 14.46*b* to obtain $L_{IV} = (I/\Delta P)_{\Delta\psi=0}$. But $L_{VI} = L_{IV}$, and so $(J_V/\Delta\psi)_{\Delta P=0} = (I/\Delta P)_{\Delta\psi=0}$.

14–14. Instead of writing the fluxes as linear combinations of the forces, we can write the forces as linear combinations of the fluxes. For two forces and two fluxes, we have

$$X_1 = R_{11}J_1 + R_{12}J_2$$

$$X_2 = R_{21}J_1 + R_{22}J_2$$

Show that

$$R_{11} = \frac{L_{22}}{\Delta} \qquad R_{12} = -\frac{L_{12}}{\Delta} \qquad R_{21} = -\frac{L_{21}}{\Delta} \qquad R_{22} = \frac{L_{11}}{\Delta}$$

where $\Delta = L_{11}L_{22} - L_{12}L_{21}$. Note that $R_{12} = R_{21}$ as a consequence of the Onsager reciprocal relations, $L_{12} = L_{21}$.

Start with

$$J_1 = L_{11}X_1 + L_{12}X_2$$
$$J_2 = L_{21}X_1 + L_{22}X_2$$

Solve these two equations for X_1 and X_2 to obtain

$$X_1 = \frac{L_{22}J_1 - L_{12}J_2}{\Delta} = R_{11}J_1 + R_{12}J_2$$

and

$$X_2 = \frac{-L_{21}J_1 - L_{22}J_2}{\Delta} = R_{21}J_1 + R_{22}J_2$$

where $\Delta = L_{11}L_{22} - L_{12}L_{21}$. We see that

$$R_{11} = \frac{L_{22}}{\Delta} \qquad R_{12} = -\frac{L_{12}}{\Delta}$$

$$R_{21} = -\frac{L_{21}}{\Delta} \qquad R_{22} = \frac{L_{11}}{\Delta}$$

Note that $L_{12} = L_{21}$ implies that $R_{12} = R_{21}$.

14–15. Show that the two electrokinetic quantities $(\Delta\psi/J_V)_{I=0}$ (*second streaming potential*) and $(\Delta P/I)_{J_V=0}$ (*second elelctroosmotic pressure*) are equal to each other. *Hint*: See the previous problem.

Following Problem 14–14, we write Equations 14.46 as

$$\Delta P = R_{11}J_V + R_{12}I$$

$$\Delta\psi = R_{21}J_V + R_{22}I$$

Note that $(\Delta\psi/J_V)_{I=0} = R_{21}$ and that $(\Delta P/I)_{J_V=0} = R_{12}$. Problem 14–14 shows that $R_{12} = R_{21}$, so we see that $(\Delta\psi/J_V)_{I=0} = (\Delta P/I)_{J_V=0}$.

14–16. Derive Equation 14.51.

Solve Equations 14.50 for [X] and [Y] and substitute into Equation 14.49 to get

$$J = k_{XY}\alpha_X + k_{XY}[X]_{eq} - k_{YX}\alpha_Y - k_{YX}[Y]_{eq}$$

But the condition of equilibrium says that $k_{XY}[X]_{eq} = k_{YX}[Y]_{eq}$, so

$$J = k_{XY}\alpha_X - k_{YX}\alpha_Y$$

Because $[X] + [Y] = [X]_{eq} + [Y]_{eq}$, $\alpha_X + \alpha_Y = 0$, and so

$$J = k_{XY}\alpha_X + k_{YX}\alpha_X$$

in agreement with Equation 14.51.

14–17. At the beginning of Section 14–6, we discussed the chemical equation $X \rightleftharpoons Y$ and took the reaction system to be isolated. Derive the same final result considering the reaction system to be held at a fixed temperature and volume.

Because the volume and temperature are fixed, we start with

$$dA = -SdT - PdV + \mu_X dn_X + \mu_Y dn_Y \le 0$$
$$= \mu_X dn_X + \mu_Y dn_Y \le 0 \qquad \text{(constant } T \text{ and } V\text{)}$$

Using $n_X + n_Y$ = constant gives $dn_X = -dn_Y$, and so

$$dA = dn_X(\mu_X - \mu_Y) \le 0$$

Because $V_1 + V_2$ = constant, we can write

$$dU = \delta_e q = TdS_{\text{exch}} = T(dS - dS_{\text{prod}})$$

$$dS_{\text{prod}} = \frac{TdS - dU}{T} = -\frac{dA}{T}$$

Substituting the above expression for dA into this result gives

$$dS_{\text{prod}} = -dn_X\left(\frac{\mu_X}{T} - \frac{\mu_Y}{T}\right) \ge 0$$

Divide by V and dt to obtain

$$\frac{\dot{S}_{\text{prod}}}{V} = -\frac{d[X]}{dt}\left(\frac{\mu_X}{T} - \frac{\mu_Y}{T}\right) \ge 0$$

in agreement with Equation 14.53.

14–18. Show that $\mathcal{A} = (\mu_X - \mu_Y)/T$ is equal to $R\{(\alpha_X/[X]_{eq}) - (\alpha_Y/[Y]_{eq})\} = R\alpha_X(k_{XY} + k_{YX})/k_{YX}[Y]_{eq}$ where $\alpha_X = [X] - [X]_{eq}$ and when the reaction system is near equilibrium.

Start with

$$\mathcal{A} = \frac{1}{T}(\mu_X - \mu_Y)$$

Substitute $\mu_i = \mu_i^\circ + RT\ln[i]$ into $\mathcal{A}$ to obtain

$$\mathcal{A} = \frac{1}{T}(\mu_X^\circ + RT\ln[X] - \mu_Y^\circ - RT\ln[Y])$$

Now substitute $[X] = [X]_{eq} + \alpha_X$ and $[Y] = [Y]_{eq} + \alpha_Y$ into $\mathcal{A}$ to get

$$\mathcal{A} = \frac{1}{T}\{\mu_X^\circ + RT\ln([X]_{eq} + \alpha_X) - \mu_Y^\circ - RT\ln([Y]_{eq} + \alpha_Y)\}$$
$$= \frac{1}{T}\left\{\mu_X^\circ + RT\ln[X]_{eq} + RT\ln\left(1 + \frac{\alpha_X}{[X]_{eq}}\right)\right.$$
$$\left. - \mu_Y^\circ - RT\ln[Y]_{eq} - RT\ln\left(1 + \frac{\alpha_Y}{[Y]_{eq}}\right)\right\}$$

The equation

$$\mu_X^\circ + RT\ln[X]_{eq} = \mu_Y^\circ + RT\ln[Y]_{eq}$$

is the general condition for chemical equilibrium, so $\mathcal{A}$ becomes

$$\mathcal{A} = R\left\{\ln\left(1 + \frac{\alpha_X}{[X]_{eq}}\right) - \ln\left(1 + \frac{\alpha_X}{[X]_{eq}}\right)\right\}$$

Using the expansion $\ln(1 + x) \approx x$ gives

$$\mathcal{A} = R\left(\frac{\alpha_X}{[X]_{eq}} - \frac{\alpha_Y}{[Y]_{eq}}\right)$$

14–19. Consider the elementary chemical reaction described by

$$\nu_A\,A + \nu_B\,B \rightleftharpoons \nu_Y\,Y + \nu_Z\,Z$$

Show that the affinity in this case is given by $\mathcal{A} = (\nu_A\mu_A + \nu_B\mu_B - \nu_Y\mu_Y - \nu_Z\mu_Z)/T$.

The flow of the reaction from left to right is given by

$$J = -\frac{1}{\nu_A}\frac{d[A]}{dt} = -\frac{1}{\nu_B}\frac{d[B]}{dt} = \frac{1}{\nu_X}\frac{d[X]}{dt} = \frac{1}{\nu_Y}\frac{d[Y]}{dt}$$

Following Example 14–5, we write

$$\begin{aligned}\frac{\dot{S}_{prod}}{V} &= -\frac{\mu_A}{T}\frac{d[A]}{dt} - \frac{\mu_B}{T}\frac{d[B]}{dt} - \frac{\mu_X}{T}\frac{d[X]}{dt} - \frac{\mu_Y}{T}\frac{d[Y]}{dt} \\ &= -\frac{\mu_A}{T}(-\nu_A J) - \frac{\mu_B}{T}(-\nu_B J) - \frac{\mu_X}{T}(\nu_X J) - \frac{\mu_Y}{T}(\nu_Y J) \\ &= J\left(\frac{\nu_A\mu_A + \nu_B\mu_B - \nu_X\mu_X - \nu_Y\mu_Y}{T}\right)\end{aligned}$$

Comparing this result to $\dot{S}_{prod}/V = J\mathcal{A}$ gives

$$\mathcal{A} = \frac{\nu_A\mu_A + \nu_B\mu_B - \nu_X\mu_X - \nu_Y\mu_Y}{T}$$

14–20. Derive Equation 14.67.

First note that J_3 given by the third of Equations 14.59 is

$$\begin{aligned}J_3 &= k_{ZX}[Z] - k_{XZ}[X] = k_{ZX}([Z]_{eq} + \alpha_Z) - k_{XZ}([X]_{eq} + \alpha_X) \\ &= k_{ZX}\alpha_Z - k_{XZ}\alpha_X\end{aligned}$$

because $k_{ZX}[Z]_{eq} = k_{XZ}[X]_{eq}$ according to the condition of detailed balance (Equations 14.64 and 14.65). Now using the second of Equations 14.66 for α_Z and the first for α_X gives

$$\begin{aligned}J_3 &= k_{ZX}\left(\frac{k_{YZ}\alpha_Y - J_2}{k_{ZY}}\right) - k_{XZ}\left(\frac{k_{YZ}\alpha_Y + J_1}{k_{XY}}\right) \\ &= \left(\frac{k_{ZX}k_{YZ}\alpha_Y}{k_{ZY}} - \frac{k_{XZ}k_{YZ}\alpha_Y}{k_{XY}}\right) - \frac{k_{XZ}}{k_{XY}}J_1 - \frac{k_{ZX}}{k_{ZY}}J_2\end{aligned}$$

But the term in parentheses is equal to zero because (Example 14–6)

$$k_{XY}k_{YZ}k_{ZX} = k_{XZ}k_{ZY}k_{YZ}$$

14–21. Show that $\mathcal{A}_1 = RJ_1/k_{XY}[X]_{eq}$ and that $\mathcal{A}_2 = RJ_2/k_{YZ}[Y]_{eq}$ for the triangular reaction scheme discussed in Section 14–6 when it is near equilibrium.

From Equation 14.60,

$$\begin{aligned}\mathcal{A}_1 &= \frac{\mu_X}{T} - \frac{\mu_Y}{T} \\ &= \frac{\mu_X^\circ}{T} + R\ln[X] - \frac{\mu_Y^\circ}{T} - R\ln[Y]\end{aligned}$$

Using $[X] = [X]_{eq} + \alpha_X$ and $[Y] = [Y]_{eq} + \alpha_Y$ gives

$$\begin{aligned}\mathcal{A}_1 &= \frac{\mu_X^\circ}{T} - \frac{\mu_Y^\circ}{T} + R\ln([X]_{eq} + \alpha_X) - R\ln([Y]_{eq} + \alpha_Y) \\ &= \frac{\mu_X^\circ + RT\ln[X]_{eq}}{T} - \frac{\mu_Y^\circ + RT\ln[Y]_{eq}}{T} + R\ln\left(1 + \frac{\alpha_X}{[X]_{eq}}\right) - R\ln\left(1 + \frac{\alpha_Y}{[Y]_{eq}}\right)\end{aligned}$$

The first two terms cancel as the result of the condition for chemcial equilibrium. If we expand the two logarithm terms according to $\ln(1 + x) \approx x$, we get

$$\mathcal{A}_1 = R\left(\frac{\alpha_X}{[X]_{eq}} - \frac{\alpha_Y}{[Y]_{eq}}\right)$$

Now use the detailed balance condition (Equation 14.65), $k_{XY}[X]_{eq} = k_{YX}[Y]_{eq}$ to eliminate $[Y]_{eq}$ from $\mathcal{A}_1$

$$\begin{aligned}\mathcal{A}_1 = R\left(\frac{\alpha_X}{[X]_{eq}} - \frac{k_{YX}\alpha_Y}{k_{XY}[X]_{eq}}\right) &= \frac{R}{k_{XY}[X]_{eq}}(k_{XY}\alpha_X - k_{YX}\alpha_Y) \\ &= \frac{R}{k_{XY}[X]_{eq}}J_1\end{aligned}$$

To show that

$$\mathcal{A}_2 = \frac{R}{k_{YZ}[Y]_{eq}}J_2$$

we start with

$$\mathcal{A}_2 = \frac{\mu_Y}{T} - \frac{\mu_Z}{T}$$

Following the same procedure that we just used for $\mathcal{A}_1$, we get

$$\mathcal{A}_2 = R\left(\frac{\alpha_Y}{[Y]_{eq}} - \frac{\alpha_Z}{[Z]_{eq}}\right)$$

Now use the detailed balance condition (Equation 14.65), $k_{YZ}[Y]_{eq} = k_{ZY}[Z]_{eq}$ to eliminate $[Z]_{eq}$ from $\mathcal{A}_2$

$$\begin{aligned}\mathcal{A}_2 = R\left(\frac{\alpha_Y}{[Y]_{eq}} - \frac{k_{ZY}\alpha_Z}{k_{YZ}[Y]_{eq}}\right) &= \frac{R}{k_{YZ}[Y]_{eq}}(k_{YZ}\alpha_Y - k_{ZY}\alpha_Z) \\ &= \frac{R}{k_{YZ}[Y]_{eq}}J_2\end{aligned}$$

14–22. Discuss why the reaction scheme

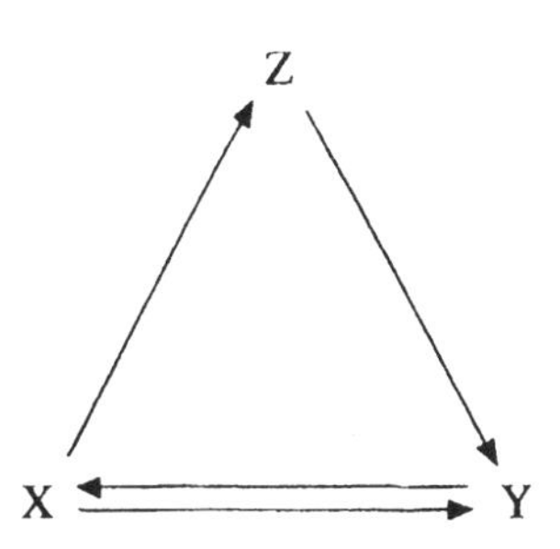

is not allowed.

According to the principle of detailed balance, each pathway must be reversible, which is not the case in the reaction scheme proposed in the problem.

14–23. This problem illustrates an alternate derivation of the Onsager reciprocal relations for the triangular kinetic scheme discussed in Section 14–6. First show that $\mathcal{A}_3$ given by Equation 14.60 can be written as

$$\mathcal{A}_3 = \frac{R}{k_{ZX}[Z]_{eq}} J_3$$

Now use Equations 14.68 and the fact that $\mathcal{A}_3 = -(\mathcal{A}_1 + \mathcal{A}_2)$ to derive

$$J_1 - J_3 = \frac{k_{XY}[X]_{eq} + k_{ZX}[Z]_{eq}}{R}\mathcal{A}_1 + \frac{k_{ZX}[Z]_{eq}}{R}\mathcal{A}_2$$

$$J_2 - J_3 = \frac{k_{ZX}[Z]_{eq}}{R}\mathcal{A}_1 + \frac{k_{YZ}[Y]_{eq} + k_{ZX}[Z]_{eq}}{R}\mathcal{A}_2$$

so that $L_{12} = L_{21}$.

Start with the third of Equations 14.60

$$\mathcal{A}_3 = \frac{\mu_Z}{T} - \frac{\mu_X}{T}$$

Follow the procedure in Problem 14–21 to derive

$$\mathcal{A}_3 = \frac{R}{k_{ZX}[Z]_{eq}} J_3$$

Solve this equation for J_3 and Equations 14.68 for J_1 and J_2 to write

$$J_1 - J_3 = \frac{k_{XY}[X]_{eq}\mathcal{A}_1}{R} - \frac{k_{ZX}[Z]_{eq}\mathcal{A}_3}{R}$$

and

$$J_2 - J_3 = \frac{k_{YZ}[Y]_{eq}\mathcal{A}_2}{R} - \frac{k_{ZX}[Z]_{eq}\mathcal{A}_3}{R}$$

Now use the fact that $\mathcal{A}_3 = -\mathcal{A}_1 - \mathcal{A}_2$ to write

$$J_1 - J_3 = \left(\frac{k_{XY}[X]_{eq} + k_{ZX}[Z]_{eq}}{R}\right)\mathcal{A}_1 + \frac{k_{ZX}[Z]_{eq}}{R}\mathcal{A}_2$$

and

$$J_2 - J_3 = \frac{k_{ZX}[Z]_{eq}}{R} A_1 + \left(\frac{k_{YZ}[Y]_{eq} + k_{ZX}[Z]_{eq}}{R} \right) A_2$$

Note that

$$L_{12} = \frac{k_{ZX}[Z]_{eq}}{R} = L_{21}$$

14–24. Calculate the value of the transmembrane potential at 298.15 K of a membrane that is permeable only to potassium ions if the solution on the two sides of the membrane are 0.200 M and 0.020 M in KCl(aq). Assume that the activity coefficients in the two solutions are the same.

We use Equation 14.78 with $z_i = 1$

$$\Delta\psi = \frac{RT}{F} \ln \frac{0.200}{0.020} = 59 \text{ mV}$$

14–25. Consider a membrane that is permeable to sodium ions. Show that the Gibbs energy required to transport one mole of sodium ions across the membrane is given by

$$\Delta G = RT \ln \frac{a_{\mathrm{Na^+},2}}{a_{\mathrm{Na^+},1}} + F\Delta\psi$$

where $\Delta\psi = \psi_2 - \psi_1$. Take $\Delta\psi$ to be 70 mV, $a_{\mathrm{Na^+},2}/a_{\mathrm{Na^+},1}$ to be 10, and T to be 37°C, and calculate the value of ΔG.

The electrochemical potential on side j is given by

$$\tilde{\mu}_j = \mu_j^{\circ} + RT \ln a_j + F\psi_j$$

The Gibbs energy required to transport one mole of sodium ions across the membrane is

$$\Delta G = \tilde{\mu}_2 - \tilde{\mu}_1 = RT \ln \frac{a_{\mathrm{Na^+},2}}{a_{\mathrm{Na^+},1}} + F(\psi_2 - \psi_1)$$

For $\Delta\psi = 70$ mV, $a_{\mathrm{Na^+},2}/a_{\mathrm{Na^+},1} = 10$, and $T = 37$°C,

$$\Delta G = 12.7 \text{ kJ}\cdot\text{mol}^{-1}$$

14–26. In this problem, we will discuss the *Donnan effect*, which occurs for systems like the one shown in Figure 14.3. Two solutions are separated by a membrane that is permeable to small ions but not to polymers or protein molecules.

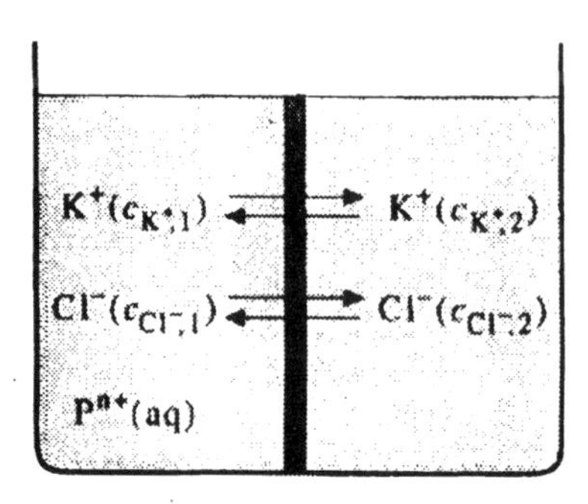

If we ignore any small transport of water molecules, the condition for equilibrium across the membrane is

$$\tilde{\mu}_{K^+,1} = \tilde{\mu}_{K^+,2} \qquad \text{and} \qquad \tilde{\mu}_{Cl^-,1} = \tilde{\mu}_{Cl^-,2}$$

Show that if we replace activities by concentrations, then these two conditions give us

$$\Delta\psi = \psi_2 - \psi_1 = \frac{RT}{F} \ln \frac{c_{K^+,1}}{c_{K^+,2}} = \frac{RT}{F} \ln \frac{c_{Cl^-,2}}{c_{Cl^-,1}}$$

The potential, $\Delta\psi$, is called the *Donnan potential*. Show that $c_{K^+,1}c_{Cl^-,1} = c_{K^+,2}c_{Cl^-,2} = c^2$. Now show that electroneutrality gives

$$c_{K^+,1} + nc_{P^{n+}} = c_{Cl^-,1} \qquad \text{and} \qquad c_{K^+,2} = c_{Cl^-,2} = c$$

and that

$$c_{K^+,2}^2 = c_{K^+,1}(c_{K^+,1} + nc_{P^{n+}})$$

Use these equations to verfiy the entries in the last four columns in the following table (all concentrations are mol·L^{-1} and the temperature is 298.15 K).

$nc_{P^{n+}}$	$c_{K^+,2} = c_{Cl^-,2}$	$c_{K^+,1}$	$c_{Cl^-,1}$	$c_{K^+,1}/c_{K^+,2}$	$-\Delta\psi/\text{mV}$
0.0020	0.0010	0.00041	0.0024	0.41	23
	0.010	0.0091	0.0111	0.91	2.4
	0.100	0.099	0.101	0.99	0.26
0.020	0.0010	0.000050	0.020	0.050	77
	0.010	0.0041	0.024	0.41	23
	0.100	0.091	0.111	0.91	2.4

It is often desirable to suppress the Donnan effect and the above table shows that this can be done by adding relatively high concentrations of salt.

Start with

$$\tilde{\mu}_{K^+,1} = \mu_{K^+}^\circ + RT \ln c_{K^+,1} + F\psi_1$$
$$\tilde{\mu}_{K^+,2} = \mu_{K^+}^\circ + RT \ln c_{K^+,2} + F\psi_2$$

Solving for $\Delta\psi = \psi_2 - \psi_1$ gives

$$\Delta\psi = \psi_2 - \psi_1 = \frac{RT}{F} \ln \frac{c_{K^+,1}}{c_{K^+,2}}$$

Similarly

$$\tilde{\mu}_{Cl^-,1} = \mu_{Cl^-}^\circ + RT \ln c_{Cl^-,1} - F\psi_1$$
$$\tilde{\mu}_{Cl^-,2} = \mu_{Cl^-}^\circ + RT \ln c_{Cl^-,2} - F\psi_2$$

Solving for $\Delta\psi = \psi_2 - \psi_1$ gives

$$\Delta\psi = \psi_2 - \psi_1 = \frac{RT}{F} \ln \frac{c_{Cl^-,2}}{c_{Cl^-,1}}$$

Equate the two expressions for $\Delta\psi$ to get

$$\frac{c_{K^+,1}}{c_{K^+,2}} = \frac{c_{Cl^-,2}}{c_{Cl^-,1}}$$

or

$$c_{K^+,1}c_{Cl^-,1} = c_{K^+,2}c_{Cl^-,2} = c^2$$

The two compartments across the membrane must be electrically neutral, so

$$c_{K^+,1} + nc_{P^{n+}} = c_{Cl^-,1}$$

and

$$c_{K^+,2} = c_{Cl^-,2} = c$$

Multiply the first of these two equations by $c_{K^+,1}$ and use the fact that $c_{K^+,1}c_{Cl^-,1} = c^2 = c^2_{K^+,2}$ to get

$$c^2_{K^+,2} = c_{K^+,1}(c_{K^+,1} + nc_{P^{n+}})$$

One way to solve this equaiton for $c_{K^+,1}$ is by iteration:

$$c_{K^+,1} = \frac{c^2_{K^+,2}}{c_{K^+,1} + nc_{P^{n+}}}$$

Let $nc_{P^{n+}} = 0.0020$, $c_{K^+,2} = 0.0010$, and let $c_{K^+,1} = 0.0010$ initially to get $c_{K^+,1} = 0.000333$, 0.000429, 0.000412, 0.000415, 0.000414, and 0.000414, in agreement with the table in the problem. Furthermore,

$$c_{Cl^-,1} = c_{K^+,1} + nc_{P^{n+}} = 0.000414 + 0.0020 = 0.0024$$

$$\frac{c_{K^+,1}}{c_{K^+,2}} = \frac{0.000414}{0.0010} = 0.414 = \frac{c_{Cl^-,2}}{c_{Cl^-,1}} = \frac{0.0010}{0.0024}$$

and

$$\Delta\psi = \frac{RT}{F}\ln\frac{c_{K^+,1}}{c_{K^+,2}} = (0.02569\text{ V})\ln 0.414 = -0.023\text{ V}$$
$$= -23\text{ mV}$$

The entries in the rest of the table are obtained in the same way.

14–27. Consider the cell whose cell diagram is

$$\text{Pt(s)}|\text{H}_2(\text{g}, P_1)|\text{HCl(aq)}|\text{H}_2(\text{g}, P_2)|\text{Pt(s)}$$

Write the equations for the electrode reactions and the overall reaction of this cell. Would you call this a concentration cell?

The equations for the two electrode reactions are

$$\text{H}_2(P_1) \longrightarrow 2\,\text{H}^+(\text{aq}) + 2\,e^-$$
$$2\,\text{H}^+(\text{aq}) + 2\,e^- \longrightarrow \text{H}_2(P_2)$$

The equation for the overall reaction is

$$\text{H}_2(P_1) \longrightarrow \text{H}_2(P_2)$$

Therefore, the cell diagrammed in the problem is a concentration cell.

14–28. This problem illustrates a method for determining transport numbers experimentally. Consider the schematic diagram below.

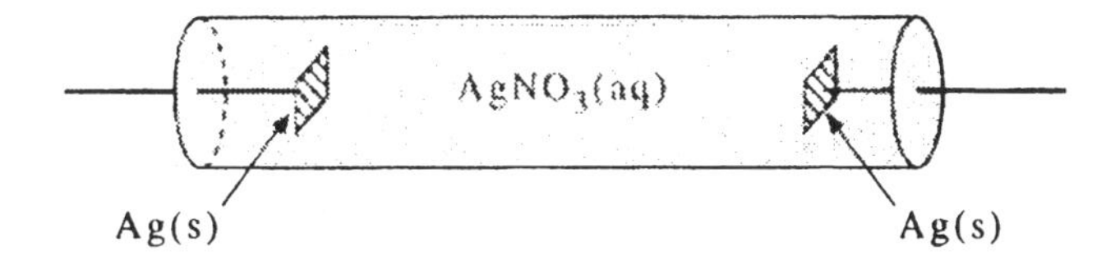

The tube is filled with $AgNO_3$(aq) and there are silver electrodes at each end of the tube. When a current is passed through the cell (electrolysis), the reactions at the two electrodes are described by $Ag(s) \rightarrow Ag^+(aq) + e^-$ and $Ag^+(aq) + e^- \rightarrow Ag(s)$. Thus, if one mole of charge (one faraday) is passed through the solution, then one mole of Ag^+(aq) ions will form at one electrode and one mole will be removed at the other. The current between the electrodes is carried by Ag^+(aq) ions moving in one direction and NO_3^-(aq) ions moving in the other direction. Because the Ag^+(aq) ions do not carry all the current, they do not move away from the electrode at which they are formed as fast as they form. Similarly, they do not arrive at the other electrode as fast as they are removed. Therefore, Ag^+(aq) ions accumulate around the electrode at which they are produced and are depleted around the electrode at which they are removed.

Consider the electrolysis cell pictured above to be divided into two electrodes compartments, and suppose that each electrode compartment initially contains 0.1000 moles of $AgNO_3$(aq). Now suppose that 0.0100 faradays are passed through the $AgNO_3$(aq) solution, and we find that there are 0.1053 moles of $AgNO_3$(aq) in the electrode compartment in which Ag^+(aq) is produced and 0.0947 moles in the compartment in which Ag^+(aq) is removed. If there were no migration of Ag^+(aq) ions from one electrode compartment to the other, the passage of 0.0100 faradays would result in an increase of 0.0100 moles Ag^+(aq). The observed increase, however, is only 0.0053 moles; therefore, (0.0100 – 0.0053) moles = 0.0047 moles of Ag^+(aq) must have migrated out of the electrode compartment. The fraction of the total current flow (0.0100 faradays) carried by the Ag^+(aq) ions is

$$t_+ = \frac{0.0047 \text{ mol}}{0.0100 \text{ mol}} = 0.47$$

Similarly, at the other electrode, the decrease of the amount of Ag^+(aq) is (0.1000 – 0.0947) moles = 0.0053 moles. But in the absence of migration, the passage of 0.0100 faradays would result in a decrease of 0.0100 moles, so 0.0047 moles must have migrated into the electrode compartment. So we see once again that

$$t_+ = \frac{0.0047 \text{ mol}}{0.0100 \text{ mol}} = 0.47$$

Suppose now that 638 C are passed through a $AgNO_3$(aq) solution and that 3.412 g of $AgNO_3$(aq) are found in one electrode compartment and 4.602 g in the other. Calculate the transport number of Ag^+(aq) in $AgNO_3$(aq). Assume that the initial amount of $AgNO_3$(aq) in the two compartment is the same.

Because the total increase in the mass of $AgNO_3$(aq) in one electrode compartment must equal the total decrease in the mass of $AgNO_3$(aq) in the other, the initial mass of $AgNO_3$(aq) in each electrode compartment must be $(3.412 \text{ g} + 4.602 \text{ g})/2 = 4.007$ g. The increase in the number of moles in one electrode compartment is $(4.602 \text{ g} - 4.007 \text{ g})/(169.87 \text{ g}\cdot\text{mol}^{-1}) = 3.503 \times 10^{-3}$ mol. If there were no migration of Ag^+(aq) ions from one electrode compartment to the other, the increase in the number of moles would be $(638 \text{ C})/(96\,485 \text{ C}\cdot\text{mol}^{-1}) = 6.61 \times 10^{-3}$ mol. The difference (6.61×10^{-3} mol $-$ 3.503×10^{-3} mol) $= 3.11 \times 10^{-3}$ mol must be due to the migration of $Ag^{+}(aq)$ ions, and so

$$t_+ = \frac{3.11 \times 10^{-3}\ \text{mol}}{6.61 \times 10^{-3}\ \text{mol}} = 0.470$$

14–29. Show that the Debye-Hückel theory says that $a_{M^+} = a_{A^-}$ for a 1–1 electrolyte. *Hint*: See Equation 11.48.

Equation 11.48 is

$$\ln \gamma_j = -\frac{\kappa q_j^2}{8\pi \varepsilon_0 \varepsilon_r k_B T}$$

Now

$$a_+ = m_+\gamma_+ \qquad \text{and} \qquad m_- = m_-\gamma_-$$

and $m_+ = m_- = m$ for a 1–1 electrolyte. But note that γ_+ and γ_- are the same for a 1–1 electrolyte because $q_+^2 = q_-^2$. Therefore, $a_+ = a_-$.

14–30. Use Equation 14.101 and the following data to calculate the values of the liquid junction potentials at 298.15 K and compare your results with the experimental values given below. Use concentrations in place of activities.

Electrolyte	$c_1/\text{mol·L}^{-1}$	$c_2/\text{mol·L}^{-1}$	t_+	$\Delta\psi/\text{mV}$
NaCl(aq)	0.040	0.020	0.389	−3.70
	0.030	0.020	0.389	−2.17
KCl(aq)	0.040	0.020	0.490	−0.34
	0.030	0.020	0.490	−0.20
HCl(aq)	0.040	0.020	0.827	+11.01

Equation 14.101 is

$$\begin{aligned}\Delta\psi &= (t_- - t_+)\frac{RT}{F}\ln\frac{c_2}{c_1}\\ &= (1 - 2t_+)\frac{RT}{F}\ln\frac{c_2}{c_1}\end{aligned}$$

For the first line of data given in the problem,

$$\begin{aligned}\Delta\psi &= [1 - (2)(0.389)]\frac{(8.314\ \text{J·mol}^{-1}\text{·K}^{-1})(298.15\ \text{K})}{96\,485\ \text{C·mol}^{-1}}\ln\frac{0.020}{0.040}\\ &= -3.95 \times 10^{-3}\ \text{V} = -3.95\ \text{mV}\end{aligned}$$

The results of the remaining calculations are

c_1	c_2	t_+	$\Delta\psi/\text{mV}$
0.030	0.020	0.389	−2.31
0.040	0.020	0.490	−0.356
0.030	0.020	0.490	−0.208
0.040	0.020	827	+11.6

14–31. Show that $E = \Delta\tilde{\mu}_j / z_j F$ for an electrode that reacts reversibly with an ion of valence z_j.

Equation 14.80 is modified according to

$$\mu_{e^-}^{\alpha} - \mu_{e^-}^{\beta} = \tilde{\mu}_j^{\alpha} - \tilde{\mu}_j^{\beta} = \Delta\tilde{\mu}_j$$

and Equation 14.81 becomes

$$\mu_{e^-}^{\alpha} - \mu_{e^-}^{\beta} = z_j F(\phi^{\alpha} - \phi^{\beta}) = z_j FE$$

Equating these two results gives

$$E = \frac{\Delta\tilde{\mu}_j}{z_j F}$$

14–32. Show that the flux of an ion of a salt $M_{\nu+}A_{\nu-}$ is given by $J_+ = \nu_+ J_s$ or $J_- = \nu_- J_s$, where J_s is the flux of neutral salt, as long as neither ion is produced or removed in an electrode reaction.

The number of cations (anions) produced per mole of electrolyte is ν_+ (ν_-), and so if the flux of electrolyte is J_s mol·s^{-1}, then the corresponding flux of cation (anion) is $\nu_+ J_s$ ($\nu_- J_s$) mol·s^{-1}.

14–33. Derive Equation 14.88 from Equation 14.87.

First realize that $J_- \neq \nu_- J_s$ because the electrodes react with the anion in this case. The only difference between Equation 14.88 and 14.87 is the middle term, so let's focus on that.

$$\text{middle term} = (J_- - \nu_- J_s)\Delta\tilde{\mu}_-$$

Use Equation 14.83 to write $\Delta\tilde{\mu}_- = z_- FE$ to write

$$\text{middle term} = (z_- FJ_- - \nu_- z_- FJ_s)E$$

Now use the electroneutrality condition $z_+\nu_+ + z_-\nu_- = 0$ to get

$$\begin{aligned}\text{middle term} &= (z_- FJ_- + z_+ F\nu_+ J_s)E \\ &= (z_- FJ_- + z_+ FJ_+)E = IE\end{aligned}$$

14–34. Show that the volume flow defined by Equation 14.90 has units of L·s^{-1}.

According to Equation 14.90,

$$J_V = J_w \overline{V}_w + J_s \overline{V}_s$$

Both J_w and J_s have units of mol·s^{-1} and both $\overline{V}_w$ and $\overline{V}_s$ have units of L·mol^{-1}, so J_V has units of (mol·s^{-1})(L·mol^{-1}) = L·s^{-1}.

14–35. Derive Equation 14.99.

Start with

$$\Delta\psi = -\frac{t_+}{\nu_+ z_+ F}\Delta\mu_s - \frac{\Delta\mu_-}{z_- F}$$

Substitute $\Delta\mu_s = \nu_+\Delta\tilde{\mu}_+ + \nu_-\Delta\tilde{\mu}_- = \nu_+\Delta\mu_+ + \nu_-\Delta\mu_- + (\nu_+ z_+ + z_-\nu_-)F\Delta\psi$ with $\nu_+ z_+ + z_-\nu_- = 0$ into the above equation to get

$$F\Delta\psi = -\frac{t_+}{z_+}\Delta\mu_+ - \frac{t_+\nu_-}{\nu_+ z_+}\Delta\mu_- - \frac{\Delta\mu_-}{z_-}$$

Now replace $\nu_-/\nu_+ z_+$ by $-1/z_-$

$$F\Delta\psi = -\frac{t_+}{z_+}\Delta\mu_+ + \frac{t_+}{z_-}\Delta\mu_- - \frac{\Delta\mu_-}{z_-}$$

Now use $t_+ + t_- = 1$ to obtain Equation 14.99.

14–36. Show that Equation 14.107 becomes

$$E = 2t_-\frac{RT}{F}\ln\frac{a_{\pm,2}}{a_{\pm,1}}$$

if the electrodes react reversibly with the cation instead of the anion.

Start with the equation just before Equation 14.103

$$-Fd\psi = t_{M^+}d\mu_{M^+} - t_{Cl^-}d\mu_{Cl^-}$$

Substitute $t_{M^+} = 1 - t_{Cl^-}$ and write

$$-Fd\psi = d\mu_{M^+} - t_{Cl^-}(d\mu_{M^+} + d\mu_{Cl^-}) = d\mu_{M^+} - t_{Cl^-}d\mu_{MCl} \qquad (1)$$

According to Equation 14.83, if the electrode reacts reversibly with $M^+(aq)$, then

$$E = \frac{\Delta\tilde{\mu}_{M^+}}{F} = \frac{\Delta\mu_{M^+} + F\psi}{F}$$

or, in differential form,

$$FdE = d\mu_{M^+} + Fd\psi$$

If we solve this equation for $-Fd\psi$ and equate the result to Equation 1, we obtain

$$d\mu_{M^+} - FdE = d\mu_{M^+} - t_{Cl^-}d\mu_{MCl}$$

or

$$FdE = t_{Cl\_}\,RT\ln a_{MCl}$$

If we assume that t_{Cl^-} is constant and integrate from electrode compartment 1 to electrode compartment 2, we obtain

$$E = t_{Cl^-}\frac{RT}{F}\ln\frac{a_{MCl,2}}{a_{MCl,1}} = 2t_{Cl\_}\frac{RT}{F}\ln\frac{a_{\pm,2}}{a_{\pm,1}}$$

14–37. Use Equation 14.107 and the data below to calculate the value of the emf at 298.15 K of each cell given. Compare your results with the experimental values. Assume that the activity coefficients are unity in each case.

Electrolyte	$c_1/\text{mol}\cdot\text{L}^{-1}$	$c_2/\text{mol}\cdot\text{L}^{-1}$	t_+	$E_{\text{exptl}}/\text{mV}$
NaCl(aq)	0.020	0.010	0.391	13.41
	0.040	0.0050	0.391	39.63
KCl(aq)	0.020	0.010	0.490	16.56
	0.040	0.0050	0.490	49.63
HCl(aq)	0.020	0.010	0.826	28.05
	0.040	0.0050	0.826	84.16

The calculation for the first line of data is

$$E = -2(0.391)\frac{(8.314\ \text{J}\cdot\text{mol}^{-1}\cdot\text{K}^{-1})(298.15\ \text{K})}{96\,485\ \text{C}\cdot\text{mol}^{-1}}\ln\frac{0.010}{0.020}$$

$$= 13.93\ \text{mV}$$

The rest of the calculations are

Electrolyte	$c_1/\text{mol}\cdot\text{L}^{-1}$	$c_2/\text{mol}\cdot\text{L}^{-1}$	t_+	$E_{\text{calc}}/\text{mV}$
NaCl(aq)	0.040	0.0050	0.391	41.78
KCl(aq)	0.020	0.010	0.490	17.45
	0.040	0.0050	0.490	52.36
HCl(aq)	0.020	0.010	0.826	29.42
	0.040	0.0050	0.826	88.26

14–38. We claimed in Section 14–9 that an ion in solution in an electric field will quickly come to a constant velocity because of the viscous drag on the ion from the solvent molecules. Let the viscous drag be linearly proportional to the velocity of the ion but in the opposite direction. Then Newton's equation for the motion of the ion is

$$m\frac{dv}{dt} = -fv + zeE$$

where f is the *friction constant*. Show that if the ion is at rest initially, then the solution to this equation is

$$v(t) = \frac{zeE}{f}(1 - e^{-ft/m})$$

Note that when $ft/m \gg 1$, then v has a constant value $v = zeE/f$. We can obtain this result by letting v be a constant in Newton's equation above. The mobility is defined as $u = v/E$, so we see that $u = Ze/f$.

Substitute the proposed solution into the differential equation to obtain

$$m\left(\frac{zeE}{m}e^{-ft/m}\right) \stackrel{?}{=} -zeE(1 - e^{-ft/m}) + zeE$$

$$= zeEe^{-ft/m}$$

Another way to do this problem is to use the general solution to a linear first-order differential equation of the form (*CRC Handbook of Chemistry and Physics*)

$$\frac{dy}{dx} + P(x)y(x) = Q(x)$$

The general solution to this differential equation is

$$y(x) = e^{-\int P(x)dx}\left[\int \left(Q(x)e^{\int P(x')dx'}dx\right) + C\right]$$

where C is a constant. In our case, $v \to y$, $t \to x$, $P(x) = f/m$, and $Q(x) = zeE/m$. therefore, we have

$$v(t) = e^{-ft/m}\left(\frac{zeE}{f}e^{ft/m} + C\right)$$

The particle is at rest initially, so $v(t) = 0$ when $t = 0$. Therefore,

$$0 = \frac{zeE}{f} + C$$

and so $C = -zeE/f$. the solution to the differential equation is

$$v(t) = \frac{zeE}{f}\left(1 - e^{-ft/m}\right)$$

14–39. A famous expression for the friction constant is $f = 6\pi\eta a$, where η is the viscosity of the solvent and a is the radius of the ion. This expression for f is called *Stokes's law*. Given that the viscosity of water is 8.9×10^{-4} kg·m^{-1}·s^{-1} at 25°C, estimate the magnitude of f for an ion such as Na^+(aq). Now estimate f/m and use the result of the previous problem to show that v attains its steady value in about 10^{-12} s to 10^{-13} s.

Take the radius of a hydrated Na^+(aq) ion to be about 100 pm. Then

$$f = 6\pi\eta a = 6\pi(8.9 \times 10^{-4}\ \text{kg}\cdot\text{m}^{-1}\cdot\text{s}^{-1})(100 \times 10^{-12}\ \text{m})$$
$$= 1.7 \times 10^{-12}\ \text{kg}\cdot\text{s}^{-1}$$

In addition, m is about $(100\ \text{amu})(1.67 \times 10^{-27}\ \text{kg}\cdot\text{amu}) = 1.7 \times 10^{-25}$ kg

$$\frac{f}{m} = \frac{1.7 \times 10^{-12}\ \text{kg}\cdot\text{s}^{-1}}{1.7 \times 10^{-25}\ \text{kg}} = 1 \times 10^{13}\ \text{s}^{-1}$$

The steady state is attained when $e^{-ft/m}$ is small compared to 1, or when ft/m is around 3, or when t is around $3m/f$, or 3×10^{-13} s.

14–40. Show that the transport number of ion i in a mixture of ions is given by

$$t_i = \frac{c_i u_i}{\sum_j c_j u_j}$$

In a mixture of 1–1 electrolytes, the generalization of Equation 14.110 for the total ionic current density is

$$j = F\mathcal{E}\sum_k c_k|z_k|u_k$$

The transport number of ion i is

$$t_i = \frac{c_i|z_i|u_i F\mathcal{E}}{\sum_k c_k|z_k|u_k F\mathcal{E}} = \frac{c_i|z_i|u_i}{\sum_k c_k|z_k|u_k}$$

Now all the $|z_k| = 1$ for a 1–1 electrolyte, so

$$t_i = \frac{c_i u_i}{\sum_k c_k u_k} = \frac{c_i u_i}{\sum_j c_j u_j}$$

14–41. Use the Henderson integration scheme to derive Equation 14.105 (with activities replaced by concentrations) for a liquid junction $\text{MCl}(c_1) \vdots \text{MCl}(c_2)$.

Equation 14.102 for this case is

$$\begin{aligned} -F d\psi &= t_+ RT d\ln c_+ - t_- RT d\ln c_- \\ &= \frac{u_+}{u_+ + u_-} RT d\ln c_+ - \frac{u_-}{u_+ + u_-} RT d\ln c_- \end{aligned}$$

Integrating from compartment 1 to compartment 2 (assuming that u_+ and u_- are constant) gives

$$-F\psi = \frac{u_+}{u_+ + u_-} RT \ln \frac{c_{+,2}}{c_{+,1}} - \frac{u_-}{u_+ + u_-} RT \ln \frac{c_{-,2}}{c_{-,1}}$$

Now $c_+ = c_- = c$ for a 1–1 electrolyte, so

$$-F\psi = \frac{u_+ - u_-}{u_+ + u_-} RT \ln \frac{c_2}{c_1}$$

or

$$\begin{aligned} \psi &= \frac{u_+ - u_-}{u_+ + u_-} \frac{RT}{F} \ln \frac{c_2}{c_1} \\ &= (t_- - t_+) \frac{RT}{F} \ln \frac{c_2}{c_1} \end{aligned}$$

which is Equation 14.105.

14–42. Use the Henderson integration scheme to derive Equation 14.115 for a liquid junction $\text{HCl}(c) \vdots \text{KCl}(c)$.

Equation 14.102 for this case is

$$-F d\psi = t_{\text{H}^+} RT d\ln c_{\text{H}^+} + t_{\text{K}^+} RT d\ln c_{\text{K}^+} - t_{\text{Cl}^-} RT d\ln c_{\text{Cl}^-} \quad (1)$$

Using Equations 14.113 as a guide, we write (with $(c_1 = c_2 = c)$

$$\begin{aligned} c_{\text{H}^+} &= c(1-x) \\ c_{\text{K}^+} &= cx \\ c_{\text{Cl}^-} &= c \end{aligned}$$

According to Equation 14.114

$$t_{H^+} = \frac{cu_{H^+}(1-x)}{c(1-x)u_{H^+} + cxu_{K^+} + cu_{Cl^-}} = \frac{u_{H^+}(1-x)}{u_{H^+} + u_{Cl^-} + (u_{K^+} - u_{H^+})x}$$

$$t_{K^+} = \frac{u_{K^+}x}{u_{H^+} + u_{Cl^-} + (u_{K^+} - u_{H^+})x}$$

$$t_{Cl^-} = \frac{u_{Cl^-}}{u_{H^+} + u_{Cl^-} + (u_{K^+} - u_{H^+})x}$$

Let's integrate each term in Equation 1 in turn

$$\text{first term} \longrightarrow RT\int_0^1 \left[\frac{u_{H^+}(1-x)}{u_{H^+} + u_{Cl^-} + (u_{K^+} - u_{H^+})x}\right]\left[-\frac{dx}{(1-x)}\right]$$

$$\text{second term} \longrightarrow RT\int_0^1 \left[\frac{u_{K^+}x}{u_{H^+} + u_{Cl^-} + (u_{K^+} - u_{H^+})x}\right]\left[\frac{dx}{x}\right]$$

$$\text{third term} \longrightarrow 0 \qquad \text{because } d\ln c_{Cl^-} = d\ln c = 0$$

Combining terms gives

$$-F\Delta\psi = RT\int_0^1 \frac{(u_{K^+} - u_{H^+})dx}{u_{H^+} + u_{Cl^-} + (u_{K^+} - u_{H^+})x}$$

The integrand here is of the form dy/y, so

$$-F\Delta\psi = RT\Big|\ln[u_{H^+} + u_{Cl^-} + (u_{K^+} - u_{H^+})x]\Big|_0^1$$

$$= RT\ln\frac{u_{K^+} + u_{Cl^-}}{u_{H^+} + u_{Cl^-}}$$

or

$$\Delta\psi = \frac{RT}{F}\ln\frac{u_{H^+} + u_{Cl^-}}{u_{K^+} + u_{Cl^-}}$$

which is Equation 14.115.

14–43. Use the data in Table 14.1 to calculate the value of $\Delta\psi$ at 25°C for the following liquid junctions. Compare your results with the given experimental values. All concentrations are 0.0100 mol·L^{-1}.

Junction	$\Delta\psi$/mV	Junction	$\Delta\psi$/mV
HCl(c)⋮KCl(c)	25.73	NaCl(c)⋮NH_4Cl(c)	−4.26
HCl(c)⋮NH_4Cl(c)	27.02	NaCl(c)⋮KCl(c)	−5.65
HCl(c)⋮NaCl(c)	31.16	NaCl(c)⋮CsCl(c)	−5.39

We use Equation 14.115

$$\Delta\psi = \frac{RT}{F}\ln\frac{u_{M^+} + u_{Cl^-}}{u_{N^+} + u_{Cl^-}} = \frac{(8.314\ \mathrm{J\cdot mol^{-1}\cdot K^{-1}})(298.15\ \mathrm{K})}{96\,485\ \mathrm{C\cdot mol^{-1}}}\ln\frac{u_{M^+} + u_{Cl^-}}{u_{N^+} + u_{Cl^-}}$$

$$= (25.69\ \mathrm{mV})\ln\frac{u_{M^+} + u_{Cl^-}}{u_{N^+} + u_{Cl^-}}$$

For HCl(c):KCl(c), $\Delta\psi = (25.69\text{ mV})\ln\dfrac{36.3+7.91}{7.62+7.91} = 26.9\text{ mV}$.

For HCl(c):NH_4Cl(c), $\Delta\psi = 26.9$ mV.

For HCl(c):NaCl(c), $\Delta\psi = 31.2$ mV.

For NaCl(c):NH_4Cl(c), $\Delta\psi = -4.37$ mV.

For NaCl(c):KCl(c), $\Delta\psi = -4.37$ mV.

For NaCl(c):CsCl(c), $\Delta\psi = -5.01$ mV.

14–44. Prove to yourself that a material flux, J, is equal to cv, where c is concentration and v is velocity.

A material flux J is the transport of matter across a unit area per unit time. Consider a fluid flowing through an area of 1 cm^2. If the fluid has a velocity, v, then all the fliud within a perpendicular distance of v will flow through the unit area in one second. Therefore, if the density of the fluid is ρ, then ρv will be the mass flowing through the 1 cm^2 area in one second. If we have a solution instead of a pure fluid, then we replace ρ by c, the concentration of solute, to calculate the flow of solute through a 1 cm^2 area.

14–45. Show for a mixture of two 1–1 electrolytes, 2 and 3, that the Debye-Hückel theory gives

$$\begin{aligned}\mu_2 &= \mu_2^\circ + 2RT\ln\gamma_{\pm,2}c_2\\ &= \mu_2^\circ + 2RT\ln c_2 - 1.659RT(c_2+c_3)^{1/2}\end{aligned}$$

Evaluate and compare the magnitude of $\mu_{22} = (\partial\mu_2/\partial c_2)_T$ and $\mu_{23} = (\partial\mu_2/\partial c_3)_T$ at $c_2 = c_3 =$ 0.0010 M, 0.010 M, and 0.10 M.

We start with

$$\begin{aligned}\mu_2 &= \mu_2^\circ + RT\ln a_2 = \mu_2^\circ + RT\ln a_{\pm,2}^2\\ &= \mu_2^\circ + 2RT\ln a_{\pm,2} = \mu_2^\circ + 2RT\ln\gamma_{\pm,2}c_2\end{aligned}$$

According to Equation 11.56

$$\gamma_{\pm,2} = -1.173|z_{+,2}z_{-,2}|(I_c/\text{mol}\cdot\text{L}^{-1})^{1/2}$$

For a mixture of two 1–1 electrolytes,

$$\begin{aligned}I_c &= \frac{1}{2}[(1)^2c_{+,2} + (-1)^2c_{-,2} + (1)^2c_{+,3} + (-1)^2c_{-,3}]\\ &= \frac{1}{2}(2c_2+2c_3) = c_2 + c_3\end{aligned}$$

So

$$\ln\gamma_{\pm,2} = -1.173(c_2+c_3)^{1/2}$$

and

$$\mu_2 = \mu_2^\circ + 2RT\ln c_2 - 2.346RT(c_2+c_3)^{1/2}$$

Now

$$\mu_{22} = \left(\frac{\partial \mu_2}{\partial c_2}\right)_T = \frac{2RT}{c_2} - \frac{1.173RT}{(c_2 + c_3)^{1/2}}$$

and

$$\mu_{23} = \left(\frac{\partial \mu_2}{\partial c_3}\right)_T = -\frac{1.173RT}{(c_2 + c_3)^{1/2}}$$

The ratio of μ_{23} to μ_{22}

$$\frac{\mu_{23}}{\mu_{22}} = \frac{1}{1 - \dfrac{2(c_2 + c_3)^{1/2}}{1.173c_2}}$$

is equal to -0.013 for $c_2 = c_3 = 0.0010$ M, to -0.043 for $c_2 = c_3 = 0.010$ M, and to -0.15 for $c_2 = c_3 = 0.10$ M.

14–46. Under what conditions do $(\partial \mu_2/\partial c_3)_T$ and $(\partial \mu_3/\partial c_2)_T$ equal zero?

The quantity $(\partial \mu_2/\partial c_3)_T$ will equal zero if μ_2 does not depend upon c_3. Such is the case for an ideal solution. A similar argument holds for $(\partial \mu_3/\partial c_2)_T$.

14–47. Use the following experimental data for a NaCl(0.250 M)–KCl(0.500 M) solution at 25°C to verify the Onsager reciprocal relations.

$$\begin{array}{ll} D_{22} = 1.35 \times 10^{-5}\ \text{cm}^2\cdot\text{s}^{-1} & D_{32} = 0.22 \times 10^{-5}\ \text{cm}^2\cdot\text{s}^{-1} \\ D_{23} = 0.013 \times 10^{-5}\ \text{cm}^2\cdot\text{s}^{-1} & D_{33} = 1.86 \times 10^{-5}\ \text{cm}^2\cdot\text{s}^{-1} \\ \mu_{22}/RT = 5.251\ \text{M}^{-1} & \mu_{32}/RT = 1.129\ \text{M}^{-1} \\ \mu_{23}/RT = 1.149\ \text{M}^{-1} & \mu_{33}/RT = 3.105\ \text{M}^{-1} \end{array}$$

This problem is similar to Example 14–13. We use the equations in the Example

$$RT\,L_{23}/\text{M}\cdot\text{cm}^2\cdot\text{s}^{-1} = \frac{D_{23}\mu_{22} - D_{22}\mu_{23}}{\mu_{22}\mu_{33} - \mu_{23}\mu_{32}} = \frac{-1.48 \times 10^{-5}}{15.0} = -9.9 \times 10^{-7}$$

$$RT\,L_{32}/\text{M}\cdot\text{cm}^2\cdot\text{s}^{-1} = \frac{D_{32}\mu_{33} - D_{33}\mu_{32}}{\mu_{22}\mu_{33} - \mu_{23}\mu_{32}} = \frac{-1.42 \times 10^{-5}}{15.0} = -9.5 \times 10^{-7}$$

which agree with each other within experimental error.

14–48. Use the following experimental data for a NaCl(0.500 M)–KCl(0.500 M) solution at 25°C to verify the Onsager reciprocal relations.

$$\begin{array}{ll} D_{22} = 1.40 \times 10^{-5}\ \text{cm}^2\cdot\text{s}^{-1} & D_{32} = 0.17 \times 10^{-5}\ \text{cm}^2\cdot\text{s}^{-1} \\ D_{23} = 0.018 \times 10^{-5}\ \text{cm}^2\cdot\text{s}^{-1} & D_{33} = 1.87 \times 10^{-5}\ \text{cm}^2\cdot\text{s}^{-1} \\ \mu_{22}/RT = 2.989\ \text{M}^{-1} & \mu_{32}/RT = 0.870\ \text{M}^{-1} \\ \mu_{23}/RT = 0.892\ \text{M}^{-1} & \mu_{33}/RT = 2.851\ \text{M}^{-1} \end{array}$$

This problem is similar to Example 14–13. We use the equations in the Example

$$RTL_{23}/\text{M}\cdot\text{cm}^2\cdot\text{s}^{-1} = \frac{D_{23}\mu_{22} - D_{22}\mu_{23}}{\mu_{22}\mu_{33} - \mu_{23}\mu_{32}} = \frac{-1.19 \times 10^{-5}}{7.75} = -1.5 \times 10^{-6}$$

$$RTL_{32}/\text{M}\cdot\text{cm}^2\cdot\text{s}^{-1} = \frac{D_{32}\mu_{33} - D_{33}\mu_{32}}{\mu_{22}\mu_{33} - \mu_{23}\mu_{32}} = \frac{-1.14 \times 10^{-5}}{15.0} = -1.5 \times 10^{-6}$$

14–49. Prove that $\dot{S}_{\text{prod}}$ given by Equation 14.133 is necessarily a minimum at $X_U = X_n = 0$.

We shall prove that $\dot{S}_{\text{prod}}$ given by Equation 14.133 is necessarily a minimum at $X_U = X_n = 0$. We must first show that the two first partial derivatives equal zero at $X_U = X_n = 0$.

$$\left(\frac{\partial \dot{S}_{\text{prod}}}{\partial X_U}\right)_{X_n} = 2L_{UU}X_U + 2L_{nU}X_n = 2J_U = 0$$

$$\left(\frac{\partial \dot{S}_{\text{prod}}}{\partial X_n}\right)_{X_U} = 2L_{nU}X_U + 2L_{nn}X_n = 2J_n = 0$$

where we have used the fact that both fluxes, J_U and J_n, are equal to zero at equilibrium. To show that this is a minimum, and not a maximum or a saddle point, we must show that $(\partial^2 \dot{S}_{\text{prod}}/\partial X_U^2) > 0$ and $(\partial^2 \dot{S}_{\text{prod}}/\partial X_n^2) > 0$ and that

$$\begin{vmatrix} \left(\frac{\partial^2 \dot{S}_{\text{prod}}}{\partial X_U^2}\right)_{X_n} & \left(\frac{\partial^2 \dot{S}_{\text{prod}}}{\partial X_U \partial X_n}\right) \\ \left(\frac{\partial^2 \dot{S}_{\text{prod}}}{\partial X_U \partial X_n}\right) & \left(\frac{\partial^2 \dot{S}_{\text{prod}}}{\partial X_n^2}\right)_{X_U} \end{vmatrix} > 0$$

or that

$$\left(\frac{\partial^2 \dot{S}_{\text{prod}}}{\partial X_U^2}\right)_{X_n}\left(\frac{\partial^2 \dot{S}_{\text{prod}}}{\partial X_n^2}\right)_{X_U} - \left(\frac{\partial^2 \dot{S}_{\text{prod}}}{\partial X_U \partial X_n}\right)\left(\frac{\partial^2 \dot{S}_{\text{prod}}}{\partial X_U \partial X_n}\right) > 0$$

at $X_U = X_n = 0$. Using Equation 14.133, these derivatives come out to be

$$\left(\frac{\partial^2 \dot{S}_{\text{prod}}}{\partial X_U^2}\right) = 2L_{UU} > 0 \left(\frac{\partial^2 \dot{S}_{\text{prod}}}{\partial X_n^2}\right) = 2L_{nn} > 0$$

$$\left(\frac{\partial^2 \dot{S}_{\text{prod}}}{\partial X_U \partial X_n}\right) = 2L_{nU}$$

The above determinant is

$$\Delta = \begin{vmatrix} 2L_{UU} 2L_{nU} \\ 2L_{nU}\ 2L_{nn} \end{vmatrix} > 4(L_{UU}L_{nn} - L_{nU}^2)$$

But Equation 14.32 says that $L_{UU}L_{nn} > L_{nU}^2$, so we see that $\Delta > 0$ and that $\dot{S}_{\text{prod}}$ is a minimum at $X_U = X_n = 0$.

14–50. In this problem, we will prove that a steady state, once established, is stable with respect to small fluctuations in the variable forces. We do this here only for a system with two fluxes and two forces. (The following problem treats the problem more generally.) Let X_1 be fixed and X_2 be variable, so that $J_1 \neq 0$ but $J_2 = 0$. Now let the variable force X_2 vary from its steady-state value by an amount δX_2. Show that $\delta J_2 = L_{22}\delta X_2$. Because $L_{22} > 0$, however, δJ_2 must have the same sign as δX_2. Now argue that if $\delta X_2 > 0$, then $\delta J_2 > 0$ means that X_2 is made smaller, thus causing X_2 to go back toward its steady-state value. Now argue that the same is true if $\delta X_2 < 0$.

Start with

$$J_2 = L_{21}X_1 + L_{22}X_2$$

Because X_1 is fixed,

$$\delta J_2 = L_{22}\delta X_2$$

If $\delta X_2 > 0$, then $\delta J_2 > 0$, which means that the flux J_2 will tend to reduce the perturbation and the system will return to its original steady state. Similarly, if $\delta X_2 < 0$, then $\delta J_2 < 0$, and so the flux J_2 will once again tend to reduce the perturbation.

14–51. In this problem, we will develop a general proof about the stability of the steady state for a system near equilibrium. Let $X_1, X_2, \ldots, X_k$ be fixed forces and $X_{k+1}, \ldots, X_n$ be variable forces, so that $J_{k+1}^{ss} = J_{k+2}^{ss} = \cdots = J_n^{ss} = 0$. Now let one of the variable forces, X_m, change slightly by δX_m, so that $X_m^{ss} \rightarrow X_m^{ss} + \delta X_m$, with $k + 1 \leq m \leq n$. Show that $\delta J_m = L_{mm}\delta X_m$. Argue as we did in the previous problem that the sign of δJ_m will always be such that the flux J_m reduces X_m back to its steady-state value.

Start with

$$J_m^{ss} = \sum_{i=k+1}^{n} L_{mi}X_i^{ss}$$

If one of the X_i^{ss}, say X_m^{ss}, is changed to $X_m^{ss} + \delta X_m^{ss}$, then J_m^{ss} changes to $J_m^{ss} + \delta J_m$, where

$$\delta J_m = L_{mm}\delta X_m$$

But $L_{mm} > 0$, so δJ_m and δX_m must have the same sign. The flow will always tend to reduce the perturbation and the system will return to its original stable state.

14–52. In this problem, we prove that

$$\dot{S}_{\text{prod}} = \sum_{i=1}^{n}\sum_{j=1}^{n} L_{ij}X_iX_j$$

attains a minimum value with respect to each of the nonfixed forces, $X_{k+1}, \ldots, X_n$, in a steady state. The mathematical condition that $\dot{S}_{\text{prod}}$ be a minimum (and not a maximum or an inflection point) is that

$$\frac{\partial \dot{S}_{\text{prod}}}{\partial X_\alpha} = 0 \qquad \frac{\partial \dot{S}_{\text{prod}}}{\partial X_\beta} = 0 \qquad \begin{array}{l} k + 1 \leq \alpha \leq n \\ k + 1 \leq \beta \leq n \end{array}$$

and

$$\left(\frac{\partial^2 \dot{S}_{\text{prod}}}{\partial X_\alpha^2}\right)\left(\frac{\partial^2 \dot{S}_{\text{prod}}}{\partial X_\beta^2}\right) > \left(\frac{\partial^2 \dot{S}_{\text{prod}}}{\partial X_\alpha \partial X_\beta}\right)^2 \qquad \begin{array}{l} k + 1 \leq \alpha \leq n \\ k + 1 \leq \beta \leq n \end{array}$$

Now show that

$$\frac{\partial \dot{S}_{\text{prod}}}{\partial X_\alpha} = \sum_{j=1}^{n} L_{\alpha j} X_j + \sum_{i=1}^{n} L_{i\alpha} X_i = 2J_\alpha = 0 \qquad k+1 \le \alpha \le n$$

and

$$\frac{\partial \dot{S}_{\text{prod}}}{\partial X_\beta} = \sum_{j=1}^{n} L_{\beta j} X_j + \sum_{i=1}^{n} L_{i\beta} X_i = 2J_\beta = 0 \qquad k+1 \le \beta \le n$$

Now show that

$$\frac{\partial^2 \dot{S}_{\text{prod}}}{\partial X_\alpha^2} = 2L_{\alpha\alpha} \qquad \frac{\partial^2 \dot{S}_{\text{prod}}}{\partial X_\beta^2} = 2L_{\beta\beta} \qquad \begin{array}{c} k+1 \le \alpha \le n \\ k+1 \le \beta \le n \end{array}$$

and

$$\frac{\partial^2 \dot{S}_{\text{prod}}}{\partial X_\alpha \partial X_\beta} = \frac{\partial^2 \dot{S}_{\text{prod}}}{\partial X_\beta \partial X_\alpha} = 2L_{\alpha\beta} \qquad \begin{array}{c} k+1 \le \alpha \le n \\ k+1 \le \beta \le n \end{array}$$

The condition that $\dot{S}_{\text{prod}}$ be a minimum is that $L_{\alpha\alpha}L_{\beta\beta} > L_{\alpha\beta}^2$, which is exactly the same as Equation 14.32, which is a consequence of the positive definite nature of $\dot{S}_{\text{prod}}$ (see Problem 14–8).

Consider the double summation

$$\dot{S}_{\text{prod}} = \sum_{i=1}^{n} \sum_{j=1}^{n} L_{ij} X_i X_j \qquad (1)$$

and its differention with respect to X_α. When we differentiate the first factor in the summation over i, we pick out the $i = \alpha$ term to obtain $\sum_{j=1}^{n} L_{\alpha j} X_j$ and when we differntiate the second factor in the summation of j, we pick out the $j = \alpha$ term to obtain $\sum_{i=1}^{n} L_{i\alpha} X_i$. The derivative of $\dot{S}_{\text{prod}}$ with respect to X_α then is

$$\frac{\partial \dot{S}_{\text{prod}}}{\partial X_\alpha} = \sum_{j+1}^{n} L_{\alpha j} X_j + \sum_{i=1}^{n} L_{i\alpha} X_i \qquad (2)$$

But according to Equation 14.29, each of these summations is equal to J_α

$$J_\alpha = \sum_{j+1}^{n} L_{\alpha j} X_j = \sum_{j=1}^{n} L_{j\alpha} X_j$$

(Remember that $L_{\alpha j} = L_{j\alpha}$.) Therefore,

$$\frac{\partial \dot{S}_{\text{prod}}}{\partial X_\alpha} = 2J_\alpha = 0 \qquad k+1 \le \alpha \le n$$

The flux, J_α, is equal to zero because it is the flux associated with a nonfixed force. By analogy,

$$\frac{\partial \dot{S}_{\text{prod}}}{\partial X_\beta} = 2J_\beta = 0 \qquad k+1 \le \beta \le n$$

A more formal way to differentiate $\dot{S}_{\text{prod}}$ is to use the result

$$\frac{\partial X_i}{\partial X_\alpha} = \delta_{i\alpha} \quad \text{and} \quad \frac{\partial X_i}{\partial X_\beta} = \delta_{i\beta}$$

where $\delta_{i\alpha}$ is the so-called Krönecker delta, which has the property

$$\delta_{ij} = \begin{matrix} 0 & \text{if } i \neq j \\ 1 & \text{if } i = j \end{matrix}$$

Let's now formally differentiate $\dot{S}_{\text{prod}}$ given by Equation 1 with respect to X_α to get

$$\begin{aligned} \frac{\partial \dot{S}_{\text{prod}}}{\partial X_\alpha} &= \sum_{i=1}^{n}\sum_{j=1}^{n} L_{ij}\delta_{i\alpha}X_j + \sum_{i=1}^{n}\sum_{j=1}^{n} L_{ij}X_j\delta_{i\alpha} \\ &= \sum_{j=1}^{n} L_{\alpha i}X_j + \sum_{i=1}^{n} L_{i\alpha}X_i \end{aligned}$$

The three second partial derivatives of $\dot{S}_{\text{prod}}$ given by Equation 1 can be obtained from Equation 2 and its analog for $\partial \dot{S}_{\text{prod}}/\partial X_\beta$

$$\frac{\partial \dot{S}_{\text{prod}}}{\partial X_\beta} = \sum_{j=1}^{n} L_{\beta i}X_j + \sum_{i=1}^{n} L_{i\beta}X_i$$

Therefore,

$$\frac{\partial^2 \dot{S}_{\text{prod}}}{\partial X_\alpha^2} = L_{\alpha\alpha} + L_{\alpha\alpha} = 2L_{\alpha\alpha}$$

$$\frac{\partial^2 \dot{S}_{\text{prod}}}{\partial X_\beta^2} = L_{\beta\beta} + L_{\beta\beta} = 2L_{\beta\beta}$$

and

$$\frac{\partial^2 \dot{S}_{\text{prod}}}{\partial X_\alpha \partial X_\beta} = L_{\alpha\beta} + L_{\alpha\beta} = 2L_{\alpha\beta}$$

Using these results, we have

$$\left(\frac{\partial^2 \dot{S}_{\text{prod}}}{\partial X_\alpha^2}\right)\left(\frac{\partial^2 \dot{S}_{\text{prod}}}{\partial X_\beta^2}\right) - \left(\frac{\partial^2 \dot{S}_{\text{prod}}}{\partial X_\alpha \partial X_\beta}\right) = 4(L_{\alpha\alpha}L_{\beta\beta} - L_{\alpha\beta}^2)$$

But this quantity is always greater than zero (Equation 14.33), so the steady-state condition is that $\dot{S}_{\text{prod}}$ be minimum with respect to the nonfixed forces.

14–53. Derive Equation 14.147.

Consider Figure 14.11. The difference between the flux of solute at $x + \Delta x$ and that at x, $J_s^d(x + \Delta x) - J_s^d(x)$, is equal to the rate of change in the concentration of solute in the volume $A\Delta x$. If we let the concentration of solute be c_s, then we have

$$[J_s^d(x + \Delta x) - J_s^d(x)]A = -\frac{\partial c_s}{\partial t} A\Delta x$$

Now divide by Δx and take the limit as $\Delta x \to 0$ to obtain Equation 14.147.

14–54. Use the approach of Problem 8–57 to repartition $n_2 = n_{\text{solute}}$ between two initially identical compartments to show that $(\partial^2 A/\partial n_2^2)_{V,T,n_1} > 0$ and that $(\partial \mu_2/\partial n_2)_{V,T,n_1} = V^{-1}(\partial \mu_2/\partial c_2)_{V,T,n_1} > 0$.

We use the repartition approach developed in Problem 8–57. Consider a two-compartment system at constant V and T. The Helmholtz energy is $A = A(V, T, n_1, n_2)$. Now consider a two-component system initially in equilibrium, and repartition n_2 between the two compartments so that n_2 becomes $n_2 + \Delta n_2$ in one compartment and $n_2 - \Delta n_2$ in the other. Because the Helmholtz energy is a minimum at equilibrium, we have

$$A(V, T, n_1, n_2 + \Delta n_2) + A(V, T, n_1, n_2 - \Delta n_2) > 2A(V, T, n_1, n_2)$$

Using a Taylor expansion in Δn_2 for both terms on the left gives

$$\left(\frac{\partial^2 A}{\partial n_2^2}\right)_{V,T,n_1} > 0$$

But $\mu_2 = (\partial A/\partial n_2)_{V,T,n_1}$, so we have

$$\left(\frac{\partial \mu_2}{\partial n_2}\right)_{V,T,n_1} > 0$$

Because V is held constant, we can write this inequality as

$$\left(\frac{\partial \mu_2}{\partial c_2}\right)_{V,T,n_1} > 0$$

14–55. Prove the Glansdorff-Prigogine inequality for the chemical scheme $X \rightleftharpoons Y$. Do not assume that the system is near equilibrium.

Start with $\partial_X \mathcal{P}/\partial t$, which is given by

$$\frac{\partial_X \mathcal{P}}{\partial t} = \int \sum_{j=1}^{n} J_i \frac{\partial X_i}{\partial t} dV$$

In this case, $\dot{S}_{\text{prod}}/V$ is given by Equation 14.53

$$\frac{\dot{S}_{\text{prod}}}{V} = -\frac{d[X]}{dt}\left(\frac{\mu_X}{T} - \frac{\mu_Y}{T}\right)$$

and so

$$J_i = -\frac{d[X]}{dt} \qquad \text{and} \qquad X_i = \mathcal{A} = \frac{\mu_X - \mu_Y}{T}$$

Therefore,

$$\frac{\partial_X \mathcal{P}}{\partial t} = \int J \frac{\partial \mathcal{A}}{\partial t} dV$$

Let $\mu_X = \mu_X^\circ + RT \ln[X]$ and $\mu_Y = \mu_Y^\circ + RT \ln[Y]$ and differentiate $\mathcal{A}$ with respect to t to get

$$\begin{aligned}\frac{\partial \mathcal{A}}{\partial t} &= R\left(\frac{1}{[X]}\frac{d[X]}{dt} - \frac{1}{[Y]}\frac{d[Y]}{dt}\right) \\ &= R\left(\frac{1}{[X]} + \frac{1}{[Y]}\right)\frac{d[X]}{dt}\end{aligned}$$

Now multiply by $J = -d[X]/dt$ to obtain

$$J\frac{\partial \mathcal{A}}{\partial t} = -R\left(\frac{1}{[X]} + \frac{1}{[Y]}\right)\left(\frac{d[X]}{dt}\right)^2$$

This quantity is always negative, so

$$\frac{\partial_{\mathrm{X}}\mathcal{P}}{\partial t} = \int J\frac{\partial \mathcal{A}}{\partial t}dV \leq 0$$

14–56. Prove the Glansdorff-Prigogine inequality for the chemical scheme given by Equation 14.58. Do not assume that the reaction is near equilibrium.

Start with $\partial_{\mathrm{X}}\mathcal{P}/\partial t$, which is given by

$$\frac{\partial_{\mathrm{X}}\mathcal{P}}{\partial t} = \int \sum_{j=1}^{n} J_i \frac{\partial X_i}{\partial t} dV$$

In this case (see Equation 14.61)

$$\frac{\dot{S}_{\text{prod}}}{V} = (J_1 - J_3)\mathcal{A}_1 + (J_2 - J_3)\mathcal{A}_2$$

where (Equations 14.59)

$$\begin{aligned} J_1 - J_3 &= k_{\mathrm{XY}}[\mathrm{X}] - k_{\mathrm{YX}}[\mathrm{Y}] - k_{\mathrm{ZX}}[\mathrm{Z}] + k_{\mathrm{XZ}}[\mathrm{X}] \\ &= -\frac{d[\mathrm{X}]}{dt} \end{aligned}$$

and

$$\begin{aligned} J_2 - J_3 &= k_{\mathrm{YZ}}[\mathrm{Y}] - k_{\mathrm{ZY}}[\mathrm{Z}] - k_{\mathrm{ZX}}[\mathrm{Z}] + k_{\mathrm{XZ}}[\mathrm{X}] \\ &= \frac{d[\mathrm{Z}]}{dt} \end{aligned}$$

The affinities $\mathcal{A}_1$ and $\mathcal{A}_2$ are given by (Equation 14.60)

$$\mathcal{A}_1 = \frac{\mu_{\mathrm{X}} - \mu_{\mathrm{Y}}}{T} \qquad \mathcal{A}_2 = \frac{\mu_{\mathrm{Y}} - \mu_{\mathrm{Z}}}{T}$$

The time derivatives of $\mathcal{A}_1$ and $\mathcal{A}_2$ are given by (see the solution to Problem 14–55)

$$\frac{\partial \mathcal{A}_1}{\partial t} = R\left(\frac{1}{[\mathrm{X}]}\frac{d[\mathrm{X}]}{dt} - \frac{1}{[\mathrm{Y}]}\frac{d[\mathrm{Y}]}{dt}\right)$$

$$\frac{\partial \mathcal{A}_2}{\partial t} = R\left(\frac{1}{[\mathrm{Y}]}\frac{d[\mathrm{Y}]}{dt} - \frac{1}{[\mathrm{Z}]}\frac{d[\mathrm{Z}]}{dt}\right)$$

Therefore,

$$\begin{aligned} \frac{\partial_{\mathrm{X}}\mathcal{P}}{\partial t} &= R\int \left\{\left(\frac{1}{[\mathrm{X}]}\frac{d[\mathrm{X}]}{dt} - \frac{1}{[\mathrm{Y}]}\frac{d[\mathrm{Y}]}{dt}\right)\left(\frac{-d[\mathrm{X}]}{dt}\right)\right. \\ &= \qquad \left. + \left(\frac{1}{[\mathrm{Y}]}\frac{d[\mathrm{Y}]}{dt} - \frac{1}{[\mathrm{Z}]}\frac{d[\mathrm{Z}]}{dt}\right)\left(\frac{d[\mathrm{Z}]}{dt}\right)\right\} dV \end{aligned}$$

But

$$\frac{d[\mathrm{X}]}{dt} + \frac{d[\mathrm{Y}]}{dt} + \frac{d[\mathrm{Z}]}{dt} = 0$$

because [X] + [Y] + [Z] = constant. Therefore,

$$-\frac{1}{[\mathrm{Y}]}\frac{d[\mathrm{Y}]}{dt}\left(\frac{-d[\mathrm{X}]}{dt}\right)+\frac{1}{[\mathrm{Y}]}\frac{d[\mathrm{Y}]}{dt}\left(\frac{d[\mathrm{Z}]}{dt}\right)=\frac{1}{[\mathrm{Y}]}\frac{d[\mathrm{Y}]}{dt}\left(\frac{-d[\mathrm{Y}]}{dt}\right)$$

and so the integrand of $\partial_{\mathrm{X}}\mathcal{P}/\partial t$ is

$$-\frac{1}{[\mathrm{X}]}\left(\frac{d[\mathrm{X}]}{dt}\right)^2-\frac{1}{[\mathrm{Y}]}\left(\frac{d[\mathrm{Y}]}{dt}\right)^2-\frac{1}{[\mathrm{Z}]}\left(\frac{d[\mathrm{Z}]}{dt}\right)^2$$

This integrand is always negative, so

$$\frac{\partial_{\mathrm{X}}\mathcal{P}}{\partial t}\leq 0$$

14–57. Combine Equations 14.125 and 14.147 to derive the so-called diffusion equation

$$\frac{\partial c}{\partial t}=D\frac{\partial^2 c}{\partial x^2}$$

The solutions to this equation give the concentration of a diffusing substance at any point and time, starting from some intitial distribution of concentration. The solution to the diffusion equation if all the diffusing substance is initially located at the origin is

$$c(x,t)=\frac{c_0}{(4\pi Dt)^{1/2}}e^{-x^2/4Dt}$$

where c_0 is the total concentration, initially located at the origin. First show that $c(x,t)$ does indeed satisfy the diffusion equation. Plot $c(x,t)/c_0$ versus x for increasing values of Dt and interpret the result physically. What do you think the following integral equals?

$$\int_{-\infty}^{\infty}c(x,t)dx=\frac{c_0}{(4\pi Dt)^{1/2}}\int_{-\infty}^{\infty}e^{-x^2/4Dt}dx=?$$

Equation 14.125 can be written as

$$J_n=-D\frac{\partial c}{\partial x}$$

Substitute this into equation 14.147 to get

$$\frac{\partial c}{\partial t}=D\frac{\partial^2 c}{\partial x^2}$$

To show that

$$c(x,t)=\frac{c_0}{(4\pi Dt)^{1/2}}e^{-x^2/4Dt}$$

is a solution to the diffusion equation, find the derivatives of $c(x,t)$

$$\frac{\partial c}{\partial t}=-\frac{1}{2}\left[\frac{c_0}{(4\pi Dt)^{1/2}t}\right]e^{-x^2/4Dt}+\frac{c_0}{(4\pi Dt)^{1/2}}\frac{x^2}{4Dt^2}e^{-x^2/4Dt}$$

$$\frac{\partial c}{\partial x}=-\frac{c_0}{(4\pi Dt)^{1/2}}\frac{x}{2Dt}e^{-x^2/4Dt}$$

$$\frac{\partial^2 c}{\partial x^2}=-\frac{c_0}{(4\pi Dt)^{1/2}}\frac{1}{2Dt}e^{-x^2/4Dt}+\frac{c_0}{(4\pi Dt)^{1/2}}\left(\frac{x}{2Dt}\right)^2e^{-x^2/4Dt}$$

Now note that

$$D\frac{\partial^2 c}{\partial x^2} = \frac{\partial c}{\partial t}$$

Plots of $c(x, t)$ against x for increasing values of Dt are shown below. As time increases, the substance diffuses away from the origin and spreads out.

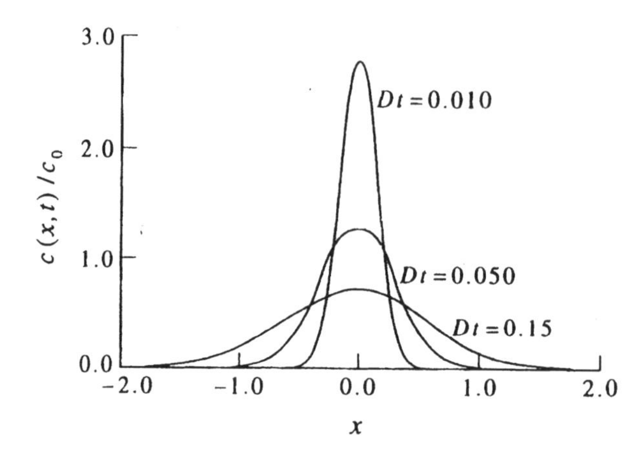

The integral of $c(x, t)$ is the total concentration of the diffusing species, and so it must equal c_0 for all values of t.

14–58. We can interpret $c(x, t)dx/c_0$ in the previous problem as the probability that a diffusing particle is located between x and $x + dx$ at time t (MathChapter B). First show that $c(x, t)/c_0$ is normalized. Now calculate the average distance that the particle will be found from the origin. Interpret your result physically. Now show that

$$\langle x^2 \rangle = \int_{-\infty}^{\infty} \frac{c(x, t)dx}{c_0} = \frac{1}{(4\pi Dt)^{1/2}} \int_{-\infty}^{\infty} e^{-x^2/4Dt} dx = 2Dt$$

The root-mean-square distance traveled by a diffusing particle is

$$x_{\text{rms}} = \langle x^2 \rangle^{1/2} = (2Dt)^{1/2}$$

Given that $D \approx 10^{-5}\ \text{cm}^2\cdot\text{s}^{-1}$ for an ion diffusing in water, calculate the value of x_{rms} at 10^{-12} s, 10^{-9} s, and 10^{-6} s. Discuss your results.

To show that $c(x, t)/c_0$ is normalized, we write

$$\begin{aligned}\int_{-\infty}^{\infty} \frac{c(x, t)}{c_0} dx &= \frac{1}{(4\pi Dt)^{1/2}} \int_{-\infty}^{\infty} e^{-x^2/4Dt} dx \\ &= \frac{1}{(\pi Dt)^{1/2}} \int_{0}^{\infty} e^{-x^2/4Dt} dx = \frac{1}{(\pi Dt)^{1/2}} (\pi Dt)^{1/2} = 1\end{aligned}$$

The average value of x, $\langle x \rangle$,

$$\langle x \rangle = \frac{1}{(4\pi Dt)^{1/2}} \int_{-\infty}^{\infty} x e^{-x^2/4Dt} dx = 0$$

because the particle is equally likely to diffuse to the left or the right.

$$\langle x^2 \rangle = \frac{1}{(4\pi Dt)^{1/2}} \int_{-\infty}^{\infty} x^2 e^{-x^2/4Dt} dx = \frac{1}{(\pi Dt)^{1/2}} \int_0^{\infty} x^2 e^{-x^2/4Dt} dx$$
$$= \frac{1}{(\pi Dt)^{1/2}} (Dt)(4\pi Dt)^{1/2} = 2Dt$$

Using $D = 10^{-5}\ \text{cm}^2\cdot\text{s}^{-1}$, we get

t/s	10^{-12}	10^{-9}	10^{-6}
$\langle x \rangle_{\text{rms}}/\text{cm}$	4×10^{-9}	1×10^{-7}	4×10^{-6}

These values give a good estimate of how far a particle diffuses at the given time intervals.

14–59. Consider a closed circuit consisting of two different metallic conductors where the junctions between the dissimilar metals are maintained at different temperatures, T_1 and T_2. It turns out that a voltage with a resulting electric current will be generated in this circuit. This effect was first observed by T.J.Seebeck in 1822 and is now known as the *Seebeck effect*. The corresponding emf in the circuit (measured with a potentiometer) depends upon the temperatures at the two junctions and upon the two metals. Thus, if one junction is fixed at a known temperature, then the temperature at the other junction can be determined by measuring the emf. Such a device is called a *thermocouple* and is used extensively to measure temperautre.

Let's now consider the inverse effect, where the two junctions are at the same temperature and an electric current is maintained across the junctions between the two metals. In this case, the temperature at one junction will increase and the temperature at the other will decrease, unless heat is supplied or removed at the junctions. The flux of energy as heat that must be supplied to a junction to maintain its initial temperature is directly proportional to the electric current and changes sign when the direction of the current is reversed. The evolution or absorption of energy as heat when an electric current passes across the junction of two dissimilar metals is called the *Peltier effect*, after its discoverer, J.C.A.Peltier. We define the Peltier coefficient π by

$$J_U = \pi I \qquad (\Delta T = 0)$$

where J_U is the flux of energy as heat $(\text{J}\cdot\text{m}^{-2}\cdot\text{s}^{-1})$ and I is the flux of the electric current $(\text{C}\cdot\text{m}^{-2}\cdot\text{s}^{-1})$.

The entropy production equation for a thermocouple can be written as

$$\sigma = J_U \left[\frac{d(1/T)}{dx}\right] + I\left(\frac{1}{T}\frac{d\phi}{dx}\right)$$

where ϕ is the electric potential in the circuit. Show that both sides of this equation have the same units. Write out the two linear flux-force relations and show that

$$\pi = T\frac{dE}{dT}$$

(Recall that the emf, E, is measured under the condition of no current flow.) This equation relating the Seebeck effect and the Peltier effect has been known experimentally since the middle of the 1800s. Not until Onsager's formulation of the reciprocal relations was it derived correctly, however. There were several previous derivations, none of which was correct, and some of which were quite bizarre.

The units of the two terms are

$$(\text{J}\cdot\text{m}^{-2}\cdot\text{s}^{-1})(\text{K}^{-1}\cdot\text{m}^{-1}) \leftarrow \text{units} \rightarrow (\text{C}\cdot\text{m}^{-2}\cdot\text{s}^{-1})(\text{V}\cdot\text{K}^{-1}\cdot\text{m}^{-1})$$

The units are the same because of the relation J = C·V. The two linear flux-force relations are

$$J_U = L_{UU}\left[\frac{d(1/T)}{dx}\right] + L_{UI}\left(\frac{1}{T}\frac{d\phi}{dx}\right)$$

$$I = L_{IU}\left[\frac{d(1/T)}{dx}\right] + L_{II}\left(\frac{1}{T}\frac{d\phi}{dx}\right)$$

The Peltier coefficient can be obtained by setting $d(1/T)/dx$ equal to zero in both equations and then dividing one equation by the other to obtain

$$\Pi = \left(\frac{J_U}{I}\right)_{\Delta T=0} = \frac{L_{UI}}{L_{II}}$$

The temperature dependence of the emf is obtained from the second flux-force relation by setting I equal to zero (in whcih case ϕ becomes E)

$$\frac{L_{IU}}{L_{II}} = -\frac{\dfrac{1}{T}\dfrac{dE}{dx}}{\dfrac{d(1/T)}{dx}} = T\frac{dE}{dT}$$

Using the fact that $L_{UI} = L_{IU}$, we see that

$$\Pi = T\frac{dE}{dT}$$

Some Mathematical Formulas

$\sin(x \pm y) = \sin x \cos y \pm \cos x \sin y$

$\cos(x \pm y) = \cos x \cos y \mp \sin x \sin y$

$\sin x \sin y = \frac{1}{2}\cos(x - y) - \frac{1}{2}\cos(x + y)$

$\cos x \cos y = \frac{1}{2}\cos(x - y) + \frac{1}{2}\cos(x + y)$

$\sin x \cos y = \frac{1}{2}\sin(x + y) + \frac{1}{2}\sin(x - y)$

$\sin^2 x + \cos^2 x = 1$

$$\cosh x = \frac{e^x + e^{-x}}{2} \qquad \sinh x = \frac{e^x - e^{-x}}{2} \qquad \tanh x = \frac{e^x - e^{-x}}{e^x + e^{-x}}$$

$$f(x) = f(a) + f'(a)(x - a) + \frac{1}{2!}f''(a)(x - a)^2 + \frac{1}{3!}f'''(a)(x - a)^3 + \cdots$$

$$e^x = 1 + x + \frac{x^2}{2!} + \frac{x^3}{3!} + \frac{x^4}{4!} + \cdots$$

$$\cos x = 1 - \frac{x^2}{2!} + \frac{x^4}{4!} - \frac{x^6}{6!} + \cdots$$

$$\sin x = x - \frac{x^3}{3!} + \frac{x^5}{5!} - \frac{x^7}{7!} + \cdots$$

$$\ln(1 + x) = x - \frac{x^2}{2} + \frac{x^3}{3} - \frac{x^4}{4} + \cdots \qquad -1 < x \le 1$$

$$\frac{1}{1 - x} = 1 + x + x^2 + x^3 + x^4 + \cdots \qquad x^2 < 1$$

$$(1 \pm x)^n = 1 \pm nx \pm \frac{n(n-1)}{2!}x^2 \pm \frac{n(n-1)(n-2)}{3!}x^3 + \cdots \qquad x^2 < 1$$

$$\int_0^\infty x^n e^{-ax}\,dx = \frac{n!}{a^{n+1}} \qquad (n \text{ positive integer})$$

$$\int_0^\infty e^{-ax^2}\,dx = \left(\frac{\pi}{4a}\right)^{1/2}$$

$$\int_0^\infty x^{2n} e^{-ax^2}\,dx = \frac{1 \cdot 3 \cdot 5 \cdots (2n-1)}{2^{n+1}a^n}\left(\frac{\pi}{a}\right)^{1/2} \qquad (n \text{ positive integer})$$

$$\int_0^\infty x^{2n+1} e^{-ax^2}\,dx = \frac{n!}{2a^{n+1}} \qquad (n \text{ positive integer})$$

$$\int \frac{dx}{ax + b} = \frac{1}{a}\ln(ax + b)$$

$$\int \frac{dx}{ax^2 + bx + c} = \frac{1}{\sqrt{-q}}\ln\frac{2ax + b - \sqrt{-q}}{2ax + b + \sqrt{-q}} \qquad q = 4ac - b^2$$

$$\int \frac{dx}{x(ax + b)} = -\frac{1}{b}\ln\frac{ax + b}{x}$$